# Guide to Wireless Sensor Networks

The **Computer Communications and Networks** series is a range of textbooks, monographs and handbooks. It sets out to provide students, researchers and non-specialists alike with a sure grounding in current knowledge, together with comprehensible access to the latest developments in computer communications and networking.

Emphasis is placed on clear and explanatory styles that support a tutorial approach, so that even the most complex of topics is presented in a lucid and intelligible manner.

For other titles published in this series, go to
www.springer.com/series/4198

Sudip Misra · Isaac Woungang
Subhas Chandra Misra

Editors

# Guide to Wireless
# Sensor Networks

*Editors*
Sudip Misra
School of Information Technology
Indian Institute of Technology
Kharagpur, India

Subhas Chandra Misra
Department of Industrial & Management
  Engineering
Indian Institute of Technology
Kanpur, India

Isaac Woungang
Department of Computer Science
Ryerson University
Toronto, Canada

*Series Editor*
Professor A.J. Sammes, BSc, MPhil, PhD, FBCS, CEng
Centre for Forensic Computing
Cranfield University
DCMT, Shrivenham
Swindon SN6 8LA
UK

ISSN 1617-7975
ISBN 978-1-84996-827-0          e-ISBN 978-1-84882-218-4
DOI 10.1007/978-1-84882-218-4
Springer Dordrecht Heidelberg London New York

British Library Cataloguing in Publication Data
A catalogue record for this book is available from the British Library

Springer is part of Springer Science+Business Media (www.springer.com)

*Dedicated to the newborns: Tultuli (Subhas's daughter) and Babai (Sudip's son)
and Isaac's grand ma: Maria Happi*

# Preface

## Overview and Goals

Wireless communication technologies are undergoing rapid advancements. The last few years have experienced a steep growth in research in the area of wireless sensor networks (WSNs). In WSNs, communication takes place with the help of spatially distributed autonomous sensor nodes equipped to sense specific information. WSNs, especially the ones that have gained much popularity in the recent years, are, typically, ad hoc in nature and they inherit many characteristics/features of wireless ad hoc networks such as the ability for infrastructure-less setup, minimal or no reliance on network planning, and the ability of the nodes to self-organize and self-configure without the involvement of a centralized network manager, router, access point, or a switch. These features help to set up WSNs fast in situations where there is no existing network setup or in times when setting up a fixed infrastructure network is considered infeasible, for example, in times of emergency or during relief operations. WSNs find a variety of applications in both the military and the civilian population worldwide such as in cases of enemy intrusion in the battlefield, object tracking, habitat monitoring, patient monitoring, fire detection, and so on.

Even though sensor networks have emerged to be attractive and they hold great promises for our future, there are several challenges that need to be addressed. Some of the well-known challenges are attributed to issues relating to coverage and deployment, scalability, quality-of-service, size, computational power, energy efficiency, and security.

This handbook attempts to provide a comprehensive guide on fundamental concepts, new ideas, and results in the area of WSNs. This book has been prepared keeping in mind that it needs to prove itself to be a valuable resource dealing with both the important core and the specialized issues in this area. We have attempted to offer a wide coverage of topics. We hope that it will be a valuable reference for students, instructors, researchers, and industry practitioners. We believe that this is particularly an attractive feature of this book, as the very limited selection of books available on WSNs we are aware of are written primarily for academicians/researchers. We have attempted to make this book useful for both the academicians and the practitioners alike.

## Organization and Features

The book is broadly divided into 27 chapters. Chapter 1 is dedicated to the energy efficiency issues in information processing in WSNs, which is probably and definitely arguably one of the challenges of great concern amongst researchers/practitioners working with sensor networks. Chapters 2 and 3 discuss the issues of topology management and coverage in these networks. Chapters 4–7 relate to the issues of routing, data centricity, and cooperation. Chapters 8 and 9 are dedicated to transport control issues including flow-control and congestion-control issues. As sensor network environments are often characterized by noise and error prone-ness, we have included a separate chapter, Chapter 10, relating to the issue of fault tolerance in these networks. Chapter 11 discusses the self-organizing and self-healing behavior/characteristics desirable of sensor networks. Chapter 12 focuses on the challenges concerning offering quality-of-service guarantees in sensor networks. As sensor nodes are operated by specialized operating systems, we have included a separate chapter, Chapter 13, on this topic. Chapters 14–18 relate to discussions about issues concerning medium access control, scheduling, and resource allocation. Chapters 19–21 concern security issues in sensor networks – this is another set of chapters, which would definitely attract many readers, as successfully enabling security in most types of emerging networks and definitely, sensor networks, is considered very challenging. The last few chapters, Chapters 22–27, are relatively specialized and they cover such topics as multimedia sensor networks, middleware for sensor networks, and biologically inspired communication in sensor networks.

We list below some of the important features of this book, which, we believe, would make this book a valuable resource for our readers:

- Most of the chapters of the book are authored by prominent academicians/researchers/practitioners in WSNs who have been working with these topics for several years and have thorough understanding of the concepts.
- The authors of this book are distributed in a large number of countries and most of them are affiliated with institutions of worldwide repute. This gives this book an international flavor. The readers of this book can get absorbed by perspectives, suggestions, experiences, and issues projected forward by authors from different countries.
- Almost all the chapters in this book have a distinct section providing *directions for future research*, which particularly targets researchers working in these areas. We believe that this section in each chapter should provide insight to the researchers about some of the current research issues.

- The authors of each chapter have also attempted to the extent possible to provide a comprehensive bibliography, which should greatly help the researchers and readers interested further to dig into the topic.
- Almost all chapters of this book have a separate section outlining *thoughts for practitioners*. We believe that this section in every chapter will be particularly useful for industry practitioners working directly with the practical aspects behind enabling these technologies in the field.
- Most of the chapters provide a list of important terminologies and their brief definitions.
- Most of the chapters also provide a set of questions at the end that can help in assessing the understanding of the readers.
- In order to make the book useful for pedagogical purposes, almost all chapters of the book also have a corresponding set of presentation slides. The slides can be obtained as a supplementary resource by contacting the publisher, Springer.

We have made attempts, in all possible way we could, to make the different chapters of the book look as much coherent and synchronized as possible. However, it cannot be denied that as the chapters were written by different authors, it was not fully possible to fully achieve this task. We believe that this is a limitation of most edited books of this sort.

## Target Audience

The book is written by primarily targeting the student community. This includes the students of all levels – those getting introduced to these areas, those having an intermediate level of knowledge of the topics, and those who are already knowledgeable about many of the topics. In order to keep up with this goal, we have attempted to design the overall structure and content of the book in such a manner that makes it useful at all learning levels. To aid in the learning process, almost all chapters have a set of questions at the end of the chapter. Also, in order that teachers can use this book for classroom teaching, the book also comes with presentation slides and sample solutions to exercise questions, which are available as supplementary resources.

The secondary audience for this book is the research community, whether they are working in the academia or in the industry. To meet the specific needs to this audience group, most chapters of the book also have a section in which attempts have been made to provide directions for future research.

Finally, we have also taken into consideration the needs to those readers, typically from the industries, who have quest for getting insight into the practical significance of the topics, i.e., how the spectrum of knowledge and ideas are relevant for real-life sensor networks.

## Supplementary Resources

As mentioned earlier, the book comes with the following supplementary resources:

- Solution manual, having sample solutions to most questions provided at the end of the chapters
- Presentation slides, which can be used for classroom instruction by teachers

Teachers can contact the publisher, Springer, in order to get access to these resources.

## Acknowledgments

We are extremely thankful to the roughly 74 authors of the 27 chapters of this book, who have worked very hard to bring this unique resource forward for help of the student, researcher, and practitioner community. The authors were very much inter-active at all stages of preparation of the book from initial development of concept to finalization. We feel it is contextual to mention that as the individual chapters of this book are written by different authors, the responsibility of the contents of each of the chapters lies with the concerned authors.

We are also very much thankful to our colleagues in the Springer publishing and marketing teams, in particular, Mr. Wayne Wheeler and Ms. Catherine Brett, who tirelessly worked with us and guided us in the publication process. Special thanks also go to them for taking special interest in publishing this book, considering the current worldwide market needs for such a book.

Finally, we would like to thank our parents, Prof. J.C. Misra, Mrs. Shorasi Misra, Mr. John Sime, Mrs. Christine Seupa, our wives Satamita, Sulagna, and Clarisse, and our children, Babai, Tultuli, Clyde, Lenny, and Kylian, for the continuous support and encouragement they offered during this project.

Kharagpur, India                                                      *Sudip Misra*
Toronto, Canada                                                       *Isaac Woungang*
Kanpur, India                                                         *Subhas C. Misra*

# Contents

1   Energy Efficient Information Processing in Wireless
    Sensor Networks............................................................... 1
    Bang Wang, Minghui Li, Hock Beng Lim, Di Ma,
    and Cheng Fu

2   Topology Management for Wireless Sensor Networks .................. 27
    Lisa Frye and Liang Cheng

3   Coverage in Wireless Sensor Networks ................................ 47
    Jennifer C. Hou, David K.Y. Yau, Chris Y.T. Ma, Yong Yang,
    Honghai Zhang, I Hong Hou, Nageswara S.V. Rao,
    and Mallikarjun Shankar

4   Routing in Wireless Sensor Networks .................................. 81
    Hannes Frey, Stefan Rührup, and Ivan Stojmenović

5   Geometric Routing in Wireless Sensor Networks......................113
    Jie Gao

6   Cooperative Relaying in Wireless Sensor Networks ...................159
    Robin Doss and Wolfgang Schott

7   Data-Centricity in Wireless Sensor Networks ..........................183
    Abdul-Halim Jallad and Tanya Vladimirova

8   Congestion and Flow Control in Wireless Sensor Networks.............205
    Vikram P. Munishwar, Sameer S. Tilak,
    and Nael B. Abu-Ghazaleh

9   Data Transport Control in Wireless Sensor Networks ...................239
    Hongwei Zhang and Vinayak Naik

10  Fault-Tolerant Algorithms/Protocols in Wireless
    Sensor Networks ................................................................. 261
    Hai Liu, Amiya Nayak, and Ivan Stojmenović

11  Self-Organizing and Self-Healing Schemes in Wireless
    Sensor Networks ................................................................. 293
    Doina Bein

12  Quality of Service in Wireless Sensor Networks ........................... 305
    Can Basaran and Kyoung-Don Kang

13  Embedded Operating Systems in Wireless Sensor Networks ............. 323
    Mohamed Moubarak and Mohamed K. Watfa

14  Adaptive Distributed Resource Allocation for Sensor
    Networks .......................................................................... 347
    Hock Beng Lim, Di Ma, Cheng Fu, Bang Wang,
    and Meng Joo Er

15  Scheduling Activities in Wireless Sensor Networks ....................... 379
    Yu Chen and Eric Fleury

16  Energy-Efficient Medium Access Control in Wireless
    Sensor Networks ................................................................. 419
    Gang Li and Robin Doss

17  Energy-Efficient Resource Management Techniques
    in Wireless Sensor Networks .................................................. 439
    Xiao-Hui Lin, Yu-Kwong Kwok, and Hui Wang

18  Transmission Power Control Techniques
    in Ad Hoc Networks ............................................................ 469
    Luiz Henrique Andrade Correia, Daniel Fernandes Macedo,
    Aldri Luiz dos Santos, and José Marcos Silva Nogueira

19  Security in Wireless Sensor Networks ....................................... 491
    Eric Sabbah and Kyoung-Don Kang

20  Key Management in Wireless Sensor Networks ............................. 513
    Yee Wei Law and Marimuthu Palaniswami

21  Secure Data Aggregation in Wireless Sensor Networks .................. 533
    Yee Wei Law, Marimuthu Palaniswami, and Raphael
    Chung-Wei Phan

**22 Wireless Multimedia Sensor Networks**......................................561
Ivan Lee, William Shaw, and Xiaoming Fan

**23 Middleware for Wireless Sensor Networks:
The Comfortable Way of Application Development**......................583
Kirsten Terfloth, Mesut Güneş, and Jochen H. Schiller

**24 Wireless Mobile Sensor Networks: Protocols and Mobility
Strategies**.................................................................................607
Jung Hyun Jun, Bin Xie, and Dharma P. Agrawal

**25 Analysis Methods for Sensor Networks**...................................635
Peter J. Hawrylak, J.T. Cain, and Marlin H. Mickle

**26 Bio-inspired Communications in Wireless Sensor Networks**............659
Barış Atakan, Özgür B. Akan, and Tuna Tuğcu

**27 Mobile Ad Hoc and Sensor Systems for Global
and Homeland Security Applications**.....................................687
Raffaele Bruno, Marco Conti, and Antonio Pinizzotto

**Biography**...................................................................................709

**Index**......................................................................................713

# Contributors

**Nael B. Abu-Ghazaleh** Department of Computer Science, Watson School of Engineering and Applied Sciences, Binghamton University, T12, Engineering Building, Binghamton, NY 13902, USA

**Dharma P. Agrawal** OBR Center of Distributed and Mobile Computing, Department of Computer Science, University of Cincinnati, Cincinnati, OH 45221-0030, USA

**Özgür B. Akan** Department of Electrical and Electronics Engineering, Middle East Technical University, Ankara 06531, Turkey

**Barış Atakan** Department of Electrical and Electronics Engineering, Middle East Technical University, Ankara 06531, Turkey

**Can Basaran** Department of Computer Science, Thomas J. Watson School of Engineering and Applied Science, State University of New York at Binghamton, P.O. Box 6000, Binghamton, NY 13902-6000, USA

**Doina Bein** Applied Research Laboratory, Information Science and Technology Division, The Pennsylvania State University, University Park, PA 16802, USA

**Raffaele Bruno** Institute for Informatics and Telematics (IIT), Italian National Research Council (CNR), Via G. Moruzzi 1, 56124 Pisa, Italy

**J.T. Cain** Department of Electrical and Computer Engineering, University of Pittsburgh RFID Center of Excellence, 348 Benedum Hall, 3700 O'Hara Street, Pittsburgh, PA 15261, USA

**Yu Chen** ARES/INRIA, Lyon INSA, Villeurbanne 69100, France

**Liang Cheng** Laboratory of Networking Group (LONGLAB), Department of Computer Science and Engineering, Lehigh University, 19 Memorial Drive West, Bethlehem, PA 18015, USA

**Marco Conti** Institute for Informatics and Telematics (IIT), Italian National Research Council (CNR), Via G. Moruzzi 1, 56124 Pisa, Italy

**Luiz Henrique Andrade Correia** Department of Computer Science, Campus Universitário – Caixa Postal 3037, 37200-000 Lavras –, Federal University of Lavras, Brazil

**Robin Doss** School of Information Technology, Deakin University, Burwood, VIC 3125, Australia

**Aldri Luiz dos Santos** Department of Informatics, Federal University of Paraná, 81531-990 Curitiba, PR, Brazil

**Meng Joo Er** School of Electrical and Electronic Engineering, Nanyang Technological University, Block S1-B1C-90, 50 Nanyang Avenue, Singapore 639798, Republic of Singapore

**Xiaoming Fan** Electrical and Computer Engineering, Ryerson University, 350 Victoria Street, Toronto, ON, Canada M5B 2K3
and
Institute of Computer Science, Computer Systems and Telematics (CST), Freie Universität Berlin, Takustr. 9, 14195 Berlin, Germany

**Eric Fleury** Ecole Normale Supérieure de Lyon, 46, allée d'Italie, 69364 Lyon Cedex 07, France

**Hannes Frey** Department of Computer Science, University of Paderborn, Warburger Str. 100, 33098 Paderborn, Germany

**Lisa Frye** Department of Computer Science, Kutztown University, Lytle 267, Kutztown, PA 19530, USA
and
Laboratory of Networking Group (LONGLAB), Department of Computer Science and Engineering, Lehigh University, 19 Memorial Drive West, Bethlehem, PA 18015, USA

**Cheng Fu** Intelligent Systems Centre, Nanyang Technological University, Research Techno Plaza, BorderX Block, Level 7, 50 Nanyang Drive, Singapore 637553, Republic of Singapore

**Jie Gao** 1415 Computer Science Building, Stony Brook University, Stony Brook, NY 11794, USA

**Mesut Güneş** Institute of Computer Science, Computer Systems and Telematics (CST), Distributed embedded Systems (DeS), Freie Universität Berlin, Takustr. 9, 14195 Berlin, Germany

**Peter J. Hawrylak** Department of Electrical and Computer Engineering, University of Pittsburgh RFID Center of Excellence, 3700 O'Hara Street, 348 Benedum Hall, Pittsburgh, PA 15261, USA

**I-Hong Hou** Department of Computer Science, University of Illinois at Urbana-Champaign, 201 N. Goodwin Ave., 3111SC, Urbana, IL 61801, USA

**Jennifer C. Hou** Deceased

**Abdul-Halim Jallad** Surrey Space Centre, Department of Electronic Engineering, University of Surrey, Guildford, Surrey GU2 7XH, UK

**Jung Hyun Jun** OBR Center of Distributed and Mobile Computing, Department of Computer Science, University of Cincinnati, Cincinnati, OH 45221-0030, USA

**Kyoung-Don Kang** Department of Computer Science, Thomas J. Watson School of Engineering and Applied Science, State University of New York at Binghamton, P.O. Box 6000, Binghamton, NY 13902-6000, USA

**Yu-Kwong Kwok** Department of Electrical and Computer Engineering, Colorado State University, Fort Collins, CO 80526-1373, USA

**Yee Wei Law** Department of Electrical and Electronic Engineering, The University of Melbourne, Parkville, VIC 3052, Australia

**Ivan Lee** School of Computer and Information Science, University of South Australia, Mawson Lakes, SA 5095, Australia

**Gang Li** School of Information Technology, Deakin University, Burwood, VIC 3125, Australia

**Minghui Li** Intelligent Systems Centre, Nanyang Technological University, The 7th Storey Research Techno Plaza, 50 Nanyang Drive, Singapore 637553, Republic of Singapore

**Hock Beng Lim** Intelligent Systems Centre, Nanyang Technological University, Research Techno Plaza, BorderX Block, Level 7, 50 Nanyang Drive, Singapore 637553, Republic of Singapore

**Xiao-Hui Lin** Department of Communication Engineering, Shenzhen University, Guangdong, China

**Hai Liu** Department of Computer Science, Hong Kong Baptist University, Kowloon Tong, Kowloon, Hong Kong

**Chris Y.T. Ma** Department of Computer Science, Purdue University, 305 N. University Street, West Lafayette, IN 47906, USA

**Di Ma** Intelligent Systems Centre, Nanyang Technological University, Research Techno Plaza, BorderX Block, Level 7, 50 Nanyang Drive, Singapore 637553, Republic of Singapore

**Daniel Fernandes Macedo** Laboratoire d'Informatique Paris VI, Université Pierre et Marie Curie, 104 Avenue du Président Kennedy, 75016 Paris, France

**Marlin H. Mickle** Department of Electrical and Computer Engineering, University of Pittsburgh RFID Center of Excellence, 3700 O'Hara Street, 348 Benedum Hall, Pittsburgh, PA 15261, USA

**Mohamed Moubarak** Computer Science Department, American University of Beirut, P.O. Box 11-0236, Riad El Solh, Beirut 1107 2020, Lebanon

**Vikram P. Munishwar** Department of Computer Science, Watson School of Engineering and Applied Sciences, Binghamton University, Binghamton, NY 13902, USA

**Vinayak Naik** Department of Computer Science and Automation, Indian Institute of Science, Bangalore 560012, India

**Amiya Nayak** School of Information Technology and Engineering, University of Ottawa, Ottawa, ON, Canada K1N 6N5

**José Marcos Silva Nogueira** Department of Computer Science, Federal University of Minas Gerais, UFMG-ICEX-DCC, Caixa Postal 702, 30123-970 Belo Horizonte, MG, Brazil

**Marimuthu Palaniswami** Department of Electrical and Electronic Engineering, The University of Melbourne, Parkville, VIC 3052, Australia

**Raphael Chung-Wei Phan** Electronic and Electrical Engineering, Loughborough University, Leicestershire LE11 3TU, UK

**Antonio Pinizzotto** Institute for Informatics and Telematics (IIT), Italian National Research Council (CNR), Via G. Moruzzi 1, 56124 Pisa, Italy

**Nageswara S.V. Rao** Oak Ridge National Laboratory, MS 6016, Bldg 5600, Oak Ridge, TN 37831-6016, USA

**Stefan Rührup** Department of Computer Science, University of Freiburg, Georges-Koehler-Allee 51, 79110 Freiburg im Breisgau, Germany

**Eric Sabbah** Department of Computer Science, Thomas J. Watson School of Engineering and Applied Science, State University of New York at Binghamton, P.O. Box 6000, Binghamton, NY 13902-6000, USA

**Jochen H. Schiller** Institute of Computer Science, Computer Systems and Telematics (CST), Freie Universität Berlin, Takustr. 9, 14195 Berlin, Germany

**Wolfgang Schott** Zurich Research Laboratory, IBM Research GmbH, Säumerstrasse 4, CH-8803 Rüschlikon, Switzerland

**Mallikarjun Shankar** Oak Ridge National Laboratory, 1 Bethel Valley Rd., MS 6085, Oak Ridge, TN 37831-6085, USA

**William Shaw** Electrical and Computer Engineering, Ryerson University, 350 Victoria Street, Toronto, ON, Canada M5B 2K3

**Ivan Stojmenović** School of Information Technology and Engineering, University of Ottawa, 800 King Edward, Ottawa, ON, Canada K1N 6N5

**Kirsten Terfloth** Institute of Computer Science, Computer Systems and Telematics (CST), Freie Universität Berlin, Takustr. 9. 14195 Berlin, Germany

**Sameer S. Tilak**  San Diego Supercomputer Center, California Institute
for Telecommunications and Information Technology, University of California
at San Diego, MC 0505, 9500 Gilman Drive, La Jolla, CA 92093-0505, USA

**Tuna Tuğcu**  Department of Computer Engineering, Boğaziçi University,
Bebek-Istanbul 34342, Turkey

**Tanya Vladimirova**  Surrey Space Centre, Department of Electronic Engineering,
University of Surrey, Guildford, Surrey GU2 7XH, UK

**Bang Wang**  Intelligent Systems Centre, Nanyang Technological University,
Research Techno Plaza, BorderX Block, Level 7, 50 Nanyang Drive, Singapore
637553, Republic of Singapore

**Hui Wang**  Department of Communication Engineering, Shenzhen University,
Guangdong, China

**Mohamed K. Watfa**  Computer Science Department, American University
of Beirut, P.O. Box 11-0236, Riad El Solh, Beirut 1107 2020, Lebanon

**Bin Xie**  OBR Center of Distributed and Mobile Computing, Department
of Computer Science, University of Cincinnati, Cincinnati, OH 45221-0030, USA

**Yong Yang**  Department of Computer Science, University of Illinois
at Urbana-Champaign, 201 N. Goodwin Ave., 3111SC, Urbana, IL 61801, USA

**David K.Y. Yau**  Department of Computer Science, Purdue University,
305 N. University Street, West Lafayette, IN 47906, USA

**Honghai Zhang**  143 W Farrell Ave., Apt. A3, Ewing, NJ 08618, USA

**Hongwei Zhang**  Department of Computer Science, Wayne State University,
431 State Hall, 5143 Cass Ave., Detroit, MI 48202, USA

# Chapter 1
# Energy Efficient Information Processing in Wireless Sensor Networks

Bang Wang, Minghui Li, Hock Beng Lim, Di Ma, and Cheng Fu

**Abstract** Wireless sensor networks (WSN), which normally consist of hundreds or thousands of sensor nodes each capable of sensing, processing, and transmitting environmental information, are deployed to monitor certain physical phenomena or to detect and track certain objects in an area of interests. Since the sensor nodes are equipped with battery only with limited energy, energy efficient information processing is of critical importance to operate the deployed networks as long as possible. This chapter presents how some classical information processing problems, mainly focusing on estimation and classification, need to be reexamined in such energy constrained WSNs. We first present the basics of estimation and classification and certain typical solutions. We then introduce the requirements for supporting their counterparts in WSNs. Some recent energy efficient information processing algorithms are then reviewed to illustrate how to enforce energy efficient information processing in WSNs. Examples, questions, and solutions are also provided to help the understanding of the topic in this chapter.

## 1.1 Introduction

Recent advances in micro-electro-mechanical systems, digital electronics, and wireless communications have led to the emergence of *wireless sensor networks* (WSNs), which consist of a large number of sensing devices each capable of sensing, processing, and transmitting environmental information. A single sensor node may only be equipped with limited computation and communication capabilities; However, nodes in a WSN, when properly programmed and networked, can collaboratively perform signal processing tasks to obtain information of a remote and

B. Wang (✉)
Intelligent Systems Center, Nanyang Technological University
e-mail: wangbang@ntu.edu.sg

S. Misra et al. (eds.), *Guide to Wireless Sensor Networks*, Computer Communications and Networks, DOI: 10.1007/978-1-84882-218-4_1,
© Springer-Verlag London Limited 2009

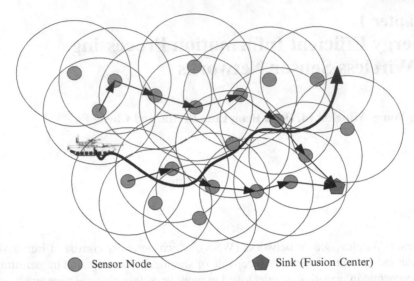

        ⬤ Sensor Node            ⬟ Sink (Fusion Center)

**Fig. 1.1** A target tracking scenario: Sensor nodes send their measurements to the sink (fusion center) via wireless multi-hop communications. The *dashed circle* is the radio range of a sensor node

probably dangerous area in an untended and robust way. Applications of wireless sensors networks include battlefield surveillance, environmental monitoring, biological detection, smart spaces, industrial diagnostics, etc. [1].

Figure 1.1 depicts a canonical application of WSNs: target detection, tracking, and classification [2,3]. In this application scenario, the information processing tasks are to let the sink infer, based on the collected information from the deployed sensor nodes, what type the target is and where the target is. To accomplish these information processing tasks, a naive approach is to let nodes send their measurements (e.g., an acoustic sensor measures the amplitude of the received sound signal) to the sink, possibly via multi-hop communications as shown in Fig. 1.1, and let the sink process the measurements. However, this approach is not energy efficient. It has been widely argued that the transmission and reception energy per bit is much larger than sensing and processing energy per bit [4, 5]. In general, the raw data of a node's measurements is of large volume. Transmitting raw measured data not only consumes large amount of energy but also increases network traffic which poses high bandwidth demand.

Energy efficiency has been deemed as the main challenge in the WSN society. Generally, the power supply of a single sensor node relies on a battery with limited energy (e.g., an AAA battery). Changing or recharging nodes' battery is very difficult, if not impossible, after sensor nodes have been deployed. Therefore, it is desirable to design energy efficient protocols run on individual nodes such that the operation time of the deployed WSN can be maintained as long as possible. Some classical information processing approaches, however, do not consider such energy efficiency issue and need to be reexamined when applied in resource constrained

WSNs. Geographically distributed nodes in a WSN may have different views of the physical phenomenon in the sensor field and their measurements may have some correlations. A well-designed algorithm should also exploit this to accomplish the information processing task via collaboration among nodes.

In this chapter, we review some recent advances on designing energy efficient information processing algorithms in WSNs. One approach for energy efficiency is to transmit less to reduce communication power consumption. This approach can also help to reduce retransmissions and collisions in time-varying and unreliable radio channels. Another approach is to select only the necessary nodes to perform the information processing tasks (such as the nodes close to an event source) and transmit the final results to the sink. This chapter examines how the two approaches are used to design energy efficient estimation and classification in WSNs. We first review some basics of the classical solution approaches to estimation and classification problems in Sect. 1.2. Section 1.3 first presents architectural and energy efficiency considerations for information processing in WSNs and then overviews recent advances on energy efficiency algorithms for both estimation and classification. Section 1.4 introduces how information processing impacts network protocol design in WSNs and some concluding remarks are given in Sect. 1.5.

## 1.2 Backgrounds

In a WSN for monitoring a field or obtaining some information within a field, the geographically distributed sensor nodes can cooperate with each other to improve the performance for the information processing. Most classical information processing algorithms do not consider the resource constraints of individual sensor nodes. In energy-constrained WSN, distributed and energy efficient information processing algorithms are needed to reduce energy consumption of individual nodes and to prolong the network operation time. In this section, two information processing paradigms – estimation and classification, and typical solutions are reviewed in the context of discrete time.

### 1.2.1 Estimation Basics

Estimation theory [6] deals with estimating the values of parameters that are used to describe the physical scenario based on measured data. Estimation theory assumes that some information-bearing quantity is contained in the measured data and thereby assumes that detection-based processing has been performed. For example, the parameter of the propagation direction or amplitude of a reflected signal may disclose the location or the size of a target that reflects the signal. Let us consider a general framework for parameter estimation. Suppose

$K$ independent and geographically distributed sensor nodes are used to estimate $\theta = (\theta_1, \theta_2, \ldots, \theta_p)^T \in \mathbb{R}^p$, a vector of unknown parameters from their noise corrupted observations $x_k$:

$$x_k = \phi_k(\theta) + w_k, \quad k = 1, \ldots, K, \qquad (1.1)$$

where $\phi_k : \mathbb{R}^p \to \mathbb{R}$ is a function describing the parameter propagation characteristics, and the additive noise $w_k$ is assumed as a random variable with zero mean and variance $\sigma_k^2$. The objective of estimation is to find an estimator as a function of $\mathbf{x} = (x_1, \ldots x_K)^T$ to provide the estimate of $\theta$, denoted by $\hat{\theta} = (\hat{\theta}_1, \hat{\theta}_2, \ldots, \hat{\theta}_p)^T$.

An estimator is said to be *unbiased* if the expected value of the estimate equals to the true value of the parameter, i.e.,

$$E[\hat{\theta}_i] = \theta_i, i = 1, \ldots, p. \qquad (1.2)$$

Otherwise, the estimate is said to be *biased*. In searching for optimal estimators, the commonly used optimality criterion is the *mean square error* (MSE)

$$\text{mse}(\hat{\theta}) = E[(\hat{\theta} - \theta)^2], \qquad (1.3)$$

which measures the average mean squared deviation of the estimate from the true value. In case of unbiased estimators, we have $\text{mse}(\hat{\theta}) = \text{var}(\hat{\theta}) + (E[\hat{\theta}] - \theta)^2 = \text{var}(\hat{\theta})$. The performance of different unbiased estimators can be compared by their estimation error variance and an unbiased estimator is optimal in the mean-squared sense if it has the minimum error variance, that is, it is a *minimum variance unbiased* (MVU) estimator.

The *Cramer-Rao Lower Bound* (CRLB) provides a lower bound of the error variance and the estimator achieves this bound for all values of the unknown parameters is the MVU estimator. The statistical information of the measurements $\mathbf{x}$ can be described by the parameterized *probability distribution function* (PDF) $p(\mathbf{x}; \theta)$. Although $p(\mathbf{x}; \theta)$ may be unknown in practice, it provides the CRLB

$$\mathbf{C}_{\hat{\theta}} - \mathbf{I}^{-1}(\theta) \geq 0, \qquad (1.4)$$

i.e., $\mathbf{C}_{\hat{\theta}} - \mathbf{I}^{-1}(\theta)$ is positive semidefinite, where $\mathbf{C}_{\hat{\theta}} = E[(\hat{\theta} - E[\hat{\theta}])^T (\hat{\theta} - E[\hat{\theta}])]$ is the covariance matrix, $\mathbf{I}$ is the **Fisher** matrix:

$$[\mathbf{I}(\theta)]_{ij} = -E\left[\frac{\partial^2 \ln p(\mathbf{x}; \theta)}{\partial \theta_i \partial \theta_j}\right]. \qquad (1.5)$$

Furthermore, an unbiased estimator attains the bound in that $\mathbf{C}_{\hat{\theta}} = \mathbf{I}^{-1}(\theta)$ if and only if

$$\frac{\partial \ln p(\mathbf{x}; \theta)}{\partial \theta} = \mathbf{I}(\theta)(g(\mathbf{x}) - \theta), \qquad (1.6)$$

for some $p$ dimensional function $g$ and some $p \times p$ matrix $\mathbf{I}$. The MVU estimator is $\hat{\theta}_{\text{MVU}} = g(\mathbf{x})$ and the minimum covariance is $\mathbf{I}^{-1}(\theta)$.

*Example 1.* Consider the following *linear model* for the measurements

$$\mathbf{x} = \mathbf{D}\boldsymbol{\theta} + \mathbf{w}, \tag{1.7}$$

where $\mathbf{x}$ is the $K \times 1$ measurements, $\mathbf{D}$ is the $K \times p$ observation matrix (with $K > p$ and invertible), $\boldsymbol{\theta}$ is the $p \times 1$ unknown parameters, and $\mathbf{w}$ is the $K \times 1$ identical independent Gaussian noises each with zero mean and variance $\sigma^2$. The parameterized PDF $p(\mathbf{x}; \boldsymbol{\theta})$ is

$$p(\mathbf{x}; \boldsymbol{\theta}) = \frac{1}{(2\pi\sigma^2)^{\frac{K}{2}}} \exp\left[-\frac{1}{2\sigma^2}(\mathbf{x} - \mathbf{D}\boldsymbol{\theta})^T(\mathbf{x} - \mathbf{D}\boldsymbol{\theta})\right]$$

Taking the first and second derivative:

$$\frac{\partial \ln p(\mathbf{x}; \boldsymbol{\theta})}{\partial \boldsymbol{\theta}} = -\frac{1}{2\sigma^2}\frac{\partial}{\partial \boldsymbol{\theta}}\left[\mathbf{x}^T\mathbf{x} - 2\mathbf{x}^T\mathbf{D}\boldsymbol{\theta} + \boldsymbol{\theta}^T\mathbf{D}^T\mathbf{D}\boldsymbol{\theta}\right]$$

$$= \frac{1}{\sigma^2}\left[\mathbf{D}^T\mathbf{x} - \mathbf{D}^T\mathbf{D}\boldsymbol{\theta}\right]$$

$$= \frac{\mathbf{D}^T\mathbf{D}}{\sigma^2}\left[(\mathbf{D}^T\mathbf{D})^{-1}\mathbf{D}^T\mathbf{x} - \boldsymbol{\theta}\right] \tag{1.8}$$

$$\frac{\partial^2 \ln p(\mathbf{x}; \boldsymbol{\theta})}{\partial \boldsymbol{\theta}^2} = -\frac{\mathbf{D}^T\mathbf{D}}{\sigma^2}$$

In the first derivative, the third equation is due to that $\mathbf{D}$ is invertible and hence $\mathbf{D}^T\mathbf{D}$. For an MVU estimator the lower bound has to apply,

$$\text{var}(\hat{\boldsymbol{\theta}}_{\text{MVU}}) = \frac{1}{-E\left[\frac{\partial^2 \ln p(\mathbf{x};\boldsymbol{\theta})}{\partial \boldsymbol{\theta}^2}\right]} = \sigma^2(\mathbf{D}^T\mathbf{D})^{-1}$$

Hence by considering the first derivative and let $\mathbf{I}(\boldsymbol{\theta}) = \mathbf{D}^T\mathbf{D}/\sigma^2$, the MVU estimator

$$\hat{\boldsymbol{\theta}}_{\text{MVU}} = (\mathbf{D}^T\mathbf{D})^{-1}\mathbf{D}^T\mathbf{x}, \tag{1.9}$$

and the covariance matrix is $\mathbf{C}_{\hat{\boldsymbol{\theta}}} = \mathbf{I}^{-1}(\boldsymbol{\theta}) = \sigma^2(\mathbf{D}^T\mathbf{D})^{-1}$.

In some situations where the MVU estimator does not exist or cannot be found even it does exist, an alternative to the MVU estimator is often the *maximum likelihood estimator* (MLE). The MLE for a vector parameter $\boldsymbol{\theta}$ is defined to be the value that maximizes the likelihood function $\ln p(\mathbf{x}; \boldsymbol{\theta})$ over the allowable domain for $\boldsymbol{\theta}$. Assuming a differentiable likelihood function, the first derivative of the likelihood function is also called *Fisher's score function*

$$\mathbf{s}(\mathbf{x}; \boldsymbol{\theta}) := \frac{\partial \ln p(\mathbf{x}; \boldsymbol{\theta})}{\partial \boldsymbol{\theta}} \tag{1.10}$$

and the MLE is found from

$$\frac{\partial \ln p(\mathbf{x}; \boldsymbol{\theta})}{\partial \boldsymbol{\theta}} = \mathbf{0} \tag{1.11}$$

For example, the MLE for the linear model (1.7) can be derived by letting (1.9) equal to 0 and is the same as the MVU estimator (1.9). In some cases, we even do not have complete knowledge of $p(\mathbf{x}; \boldsymbol{\theta})$ and cannot evaluate (1.11) to find a MLE. In such cases, we can resort to a type of linear estimator, which is a linear combination of the measurements and easy to be implemented. The *best linear unbiased estimator* (BLUE) is such a linear estimator that can be determined with knowledge of only the first and second moments of $p(\mathbf{x}; \boldsymbol{\theta})$. Consider the linear model given in (1.7), where we now assume that $\mathbf{w}$ is a $p \times 1$ noise vector with zero mean and covariance $\mathbf{C}$ (the PDF of $\mathbf{w}$ is otherwise unknown). The BLUE for $\boldsymbol{\theta}$ is

$$\hat{\boldsymbol{\theta}}_{\text{BLUE}} = (\mathbf{D}^T \mathbf{C}^{-1} \mathbf{D})^{-1} \mathbf{D}^T \mathbf{C}^{-1} \mathbf{x}, \tag{1.12}$$

and the covariance matrix of $\hat{\boldsymbol{\theta}}$ is $\mathbf{C}_{\hat{\boldsymbol{\theta}}} = (\mathbf{D}^T \mathbf{C}^{-1} \mathbf{D})^{-1}$. If $\mathbf{w}$ are independent Gaussian noises each with zero mean and $\sigma_k^2$ variance, then $\hat{\boldsymbol{\theta}}_{\text{BLUE}}$ is also a MLE estimator, and furthermore, if $\sigma_k^2 = \sigma^2$ for all $k$, the BLUE (1.12) equals to the MVU estimator (1.9).

### 1.2.2 Classification Basics

Estimation is to estimate the values of physical phenomena, while classification can be regarded as to interpret the phenomena. For example, acoustic sensors measure the amplitude of some acoustic signal by some sampling rate and use classification to infer whether the acoustic signal is from a wheeled vehicle or a tracked vehicle. Other types of sensors, such as seismic sensor and magnetometer, are also often used for classification. To accomplish classification, sensors need first to extract *features* based on its measured data and then labels each feature to a class according to its classification algorithm. Feature extraction/selection is itself an important topic. For acoustic signals, the time, frequency, or the time–frequency domain features are often used for feature extraction. For example, the acoustic signal emitted by a moving vehicle normally comes from two main sources: the engine and the propulsion gear. Different engines may have different spectrum characteristics, which can be used for classification.

Classification can be generally divided into two groups: *supervised* and *unsupervised* classification. For supervised classification, there is a set of *training features* each with known ground truth of its class. The commonly used supervised classification algorithms include k-nearest neighbor, Gaussian mixture model, support vector machines, neural networks, etc. [7]. However, not all of them are applicable for wireless sensor networks. For example, the k-nearest neighbor classifier needs to store a large number of training features to achieve reasonable classification rate, and it also needs high volume computation to conclude the classification result. The

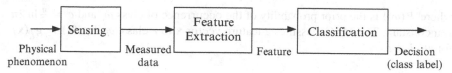

**Fig. 1.2** Classification in a single node

unsupervised classification, on the other hand, does not have a prior training set. So the goal of unsupervised classification is to group features into *clusters* such that features in each cluster share some important properties. Some typical unsupervised classifiers include the k-means clustering and mixture modeling using the expectation maximization (EM) algorithm, etc.

In wireless sensor networks, classification can be made at different levels: node level by individual nodes, group level by a group of nodes close to the physical phenomenon (e.g., a vehicle) and network level by all nodes in the network. For node level classifiers, each node performs classification based on its own extracted features from its measurements, as illustrated by Fig. 1.2. The local classifier at each node can be implemented with the same classification algorithm or different algorithms. Furthermore, individual nodes can be trained either separately or they can be trained together. To conclude the final result, on the one hand, fusion is needed to combine the classification results from different nodes. This is called *decision fusion* where only the decisions other than the measured data or extracted features need to be transmitted to a fusion center. On the other hand, *data fusion* refers that a fusion center combines individual nodes' data or features for its classification. The volume of data or feature is normally much higher than that of the classification decision; hence data fusion is less energy efficient than decision fusion.

*Example 2.* A *maximum likelihood classifier* (MLC) for node-level classification. Suppose that a feature can be classified into one of $M$ classes. Let $\Omega = \{\omega_1, \ldots, \omega_M\}$ denote the class space and $\mathbf{x} = (x^1, x^2, \ldots, x^n)^T$ a $n \times 1$ feature vector to be classified. A classifier can be considered as a function mapping from the feature space to the class space: $c(\mathbf{x}) : \mathbb{R}^n \rightarrow \Omega$. The MLC assumes that the underlying features of each class follow a multivariate Gaussian distribution:

$$p(\mathbf{x}|\omega_i) = \frac{1}{(2\pi)^{\frac{n}{2}} |\sum_i|^{\frac{1}{2}}} \exp\left(-\frac{1}{2}(\mathbf{x} - \mu_i)^T \sum_k^{-1} (\mathbf{x} - \mu_i)\right), \qquad (1.13)$$

where $\mu_i$ and $\sum_i$ are the mean and covariance matrix for class $\omega_i$, respectively. Given $L$ training features for class $\omega_i$, the Maximum Likelihood estimates of $\mu_i$ and $\sum_i$ are given by $\hat{\mu}_i = \frac{1}{L}\sum_{l=1}^{L} \mathbf{x}_l$ and $\hat{\sum}_i = \frac{1}{L}(\mathbf{x}_l - \hat{\mu}_i)(\mathbf{x}_l - \hat{\mu}_i)^T$, respectively. For practical computations, the logarithm form of the discriminant function $g_i()$ is used for classification and given by

$$g_i(\mathbf{x}) = -\frac{1}{2}(\mathbf{x} - \hat{\mu}_i)^T \hat{\sum}_i^{-1} (\mathbf{x} - \hat{\mu}_i) + \ln P(\omega_i) - \frac{1}{2}\ln|\sum_i| - c, \qquad (1.14)$$

where $P(\omega_i)$ is the prior probability of the occurrence of class $\omega_i$ and $c = \frac{n}{2} \ln 2\pi$ is a constant. The MLC classifies a feature vector $\mathbf{x}$ to a class $\omega_i$ if $g_i(\mathbf{x}) > g_j(\mathbf{x})$, for all $j \neq i$.

## 1.3 Thoughts for Practitioners

### 1.3.1 Considerations for Energy Efficient Information Processing

#### 1.3.1.1 Energy Efficiency Considerations

Each sensor node is generally composed of four components: sensing unit, data processing unit, data communication unit and power unit [1]. The power unit supplies power to the other three units. Any activity of the other three units including sensing, data processing, data transmitting, and data receiving will consume the battery energy. Experiments show that wireless communication (data transmitting and receiving) contributes a major part to energy consumption rather than sensing and data processing [8, 9]. For example, it takes on the order of $1 \, \mu J$ of energy to transmit a single bit and on the order of $0.5 \, \mu J$ of energy to receive a bit. During this time, the processor can execute 208 cycles (roughly 100 instructions) and can consume up to $0.8 \, \mu J$ in [8]. The power usage for WINS Rockwell seismic sensor for transmit receive sensing operational modes is 0.38–0.7:0.36:0.02 [9]. Therefore, reducing the energy consumption of wireless radios is the key to conserve battery energy and prolong network operation time in WSNs.

*Example 3.* A widely accepted first order energy consumption model for radios of wireless sensors is shown in Fig. 1.3 [4]. The energy consumed for receiving each bit of data is assumed as a constant $e_r$. The energy consumed for transmitting each

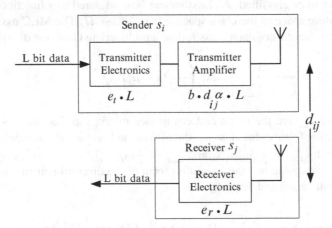

**Fig. 1.3** The first order energy consumption model [4]

bit of data depends on the distance between the transmitter and the receiver. Let $e^t_{ij}$ denote the energy consumed by sender $s_i$ for transmitting a bit to receiver $s_j$ : $e^t_{ij} = e_t + b \cdot d^\alpha_{ij}$. The part of $e_t$ is the energy consumed by transmitter electronics. The part of $b \cdot d^\alpha_{ij}$ is the energy consumed by transmitter amplifier, where $b$ is a constant, $\alpha$ is the path loss factor and $d_{ij}$ is the Euclidean distance between node $s_i$ and $s_j$. In [4], the value of different parameters are set as $e_t = e_r = 50nJ/bit$, $\alpha = 2$ and $b = 100pJ/bit/m^2$.

### 1.3.1.2   Architectural Considerations

Depending on application scenarios, the sensor nodes that are required to be involved in the information processing tasks can be either all the nodes in the sensor field or only the nodes closest to the event source. In what follows, we use $K$ to denote the number of sensor nodes involved in the information processing task and these $K$ nodes can be either a fraction of or all of the deployed nodes in a WSN. A parallel architecture or a tandem architecture can be applied to these $K$ nodes.

Figure 1.4 illustrates a parallel architecture for estimation: Each sensor node first processes its measurements $x_k$ to generate an intermediate result as a message $m_k$ that is sent over radio and may be corrupted. A fusion center performs the estimation based on the received messages $m'_k$ to conclude the final estimation result $\hat{\theta}$. The fusion center can be located at the sink or is simply a sensor node with advanced computation capability. In the parallel architecture, individual nodes may not require sophisticated processing capabilities. For example, the processing part may only need to compare the measurements to a threshold and outputs binary messages. Figure 1.5 illustrates a tandem architecture for estimation: A sensor node performs estimation based on its own measurements and the results from previous nodes. In

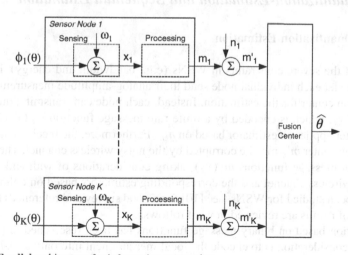

**Fig. 1.4** Parallel architecture for information processing

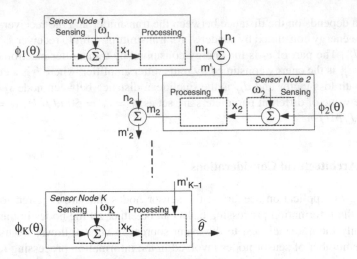

**Fig. 1.5** Tandem architecture for information processing

this case, each sensor node may require more sophisticated computation capability. If only a fraction of nodes close to the event source is selected for estimation, the tandem architecture enables messages exchange locally and only the last estimation result is needed to be transmitted to the sink. The parallel architecture and the tandem architecture are also applicable to classification and other information processing algorithms.

## 1.3.2 Quantization-Estimation and Sequential Estimation

### 1.3.2.1 Quantization-Estimation

Because of the severe constraints in WSNs (e.g., bandwidth and energy), it is not favorable to let each individual node send their analog-amplitude measurements $x_k$ to the fusion center for the estimation. Instead, each node can transmit a quantized version of $x_k$, which is encoded by a finite rate message function $m_k(x_k)$ and the fusion center applies an estimator based on $m_k$. Furthermore, the received messages at the fusion center $m'_k$ may be corrupted by the noisy wireless channel. The design of the local message functions $m_k(x_k)$ taking considerations of with and without the noise wireless channel and the corresponding estimators based on either $m_k$ or $m'_k$ have been studied for WSNs (see [10] for a good survey and references therein) and some of results are reviewed in what follows.

Estimation based on binary message functions has been researched in [11–15]. The basic consideration is to encode the local measurement into only an 1-bit message to reduce bandwidth and energy consumption. Consider a simple estimation model in a WSN, where an unknown parameter $\theta$ is to be estimated and the measurement at each sensor node is

$$x_k = \theta + w_k, \quad k = 1, 2, \ldots, K. \tag{1.15}$$

If the noise PDF $p_k(w)$ is known, the binary message function is designed as an indicator function [11]:

$$m_k(x_k) = \begin{cases} 1 & x_k \in (\tau_k, \infty); \\ 0 & otherwise. \end{cases} \tag{1.16}$$

That is, if the measurement $x_k$ is larger than the threshold $\tau_k$, then it is encoded as 1 and otherwise, 0. The message $m_k$ is a Bernoulli random variable with parameter

$$q_k(\theta) := \Pr\{x_k \in (\tau_k, \infty)\} = F_k(\tau_k - \theta), \tag{1.17}$$

where $F_k(x) := 1/(\sqrt{2\pi}\sigma_k) \int_x^{+\infty} \exp(-u^2/2\sigma_k^2) du$ is the complementary cumulative distribution function of $w_k$.

*Example 4 [11].* Assume that all the noises $w_k$ are independent and identical Gaussian with PDF $p(w) = \frac{1}{\sqrt{2\pi}\sigma} \exp\left(-\frac{w^2}{2\sigma^2}\right)$ and further assume that all sensors use the same message function $m(x_k) = 1\{x_k \in (\tau_c, \infty)\}, k = 1, \ldots, K$. Hence $q_k(\theta) = q(\theta) := F(\tau_c - \theta)$ and $\frac{\partial q(\theta)}{\partial \theta} = p(\tau_c - \theta)$. The PDF of $\mathbf{m} := (m_1, \ldots, m_K)$ with respect to $\theta$ is

$$p(\mathbf{m}; \theta) = \prod_{k=1}^{K} [q(\theta)]^{m(x_k)} [1 - q(\theta)]^{(1-m(x_k))} \tag{1.18}$$

and the first derivative of the likelihood function with respect to $\theta$ is

$$\frac{\partial \ln p(\mathbf{m}; \theta)}{\partial \theta} = \frac{\partial}{\partial \theta} \sum_{k=1}^{K} [m(x_k) \ln(q(\theta)) + (1 - m(x_k)) \ln(1 - q(\theta))]$$

$$= \sum_{k=1}^{K} \left[ \frac{m(x_k) p(\tau_c - \theta)}{q(\theta)} - \frac{(1 - m(x_k)) p(\tau_c - \theta)}{1 - q(\theta)} \right]$$

Let $\frac{\partial \ln p(\mathbf{m}, \theta)}{\partial \theta} = 0$, we have the MLE as:

$$\hat{\theta}_{\text{MLE}} = \tau_c - F^{-1}\left( \frac{1}{K} \sum_{k=1}^{K} m(x_k) \right). \tag{1.19}$$

The second derivative is

$$
\frac{\partial^2 \ln p(\mathbf{m}; \theta)}{\partial \theta^2} = \sum_{k=1}^{K} m(x_k) \left[ -\frac{p^2(\tau_c - \theta)}{q^2(\theta)} + \frac{\partial p(\tau_c - \theta)/\partial \theta}{q(\theta)} \right]
$$

$$
+ \sum_{k=1}^{K} [1 - m(x_k)] \left[ -\frac{p^2(\tau_c - \theta)}{[1 - q(\theta)]^2} - \frac{\partial p(\tau_c - \theta)/\partial \theta}{1 - q(\theta)} \right].
$$

Since for a Bernoulli variable $E[m(x_k)] = q(\theta)$, the CRLB is found as

$$
\mathrm{var}(\hat{\theta}_{\mathrm{MLE}}) = -\frac{1}{E\left[ \frac{\partial^2 \ln p(\mathbf{m};\theta)}{\partial \theta^2} \right]} = \frac{1}{K} \left[ \frac{p^2(\tau_c - \theta)}{F(\tau_c - \theta)[1 - F(\tau_c - \theta)]} \right]^{-1} := B(\theta)
$$

The minimum of $B(\theta)$ is achieved when $\tau_c = \theta$ and $B(\theta)_{\min} = \frac{\pi}{2} \frac{\sigma^2}{K}$. Note that $\frac{\sigma^2}{K}$ is the minimum of the error variance of a MLE without using the one-bit message function. Therefore, if $\tau_c$ is chosen optimally, the variance increases only by a factor of $\pi/2$ compared with a MLE using uncompressed measurements.

The aforementioned MLE requires the knowledge of the noise PDF, which may not be available in practice. The *decentralized estimation scheme* (DES) proposed in [13], however, does not require the knowledge of the noise PDF. Assume that $\theta \in [-V, V]$ and $w_k \in [-U, U]$ for $V, U > 0$ as constants. Hence $x_k \in [-V - U, V + U]$. The basic idea of the (DES) [13] is to divide the $K$ nodes into different groups $S_i$ of different sizes and each node in the $i$th group encodes its observation to the $i$th *most significant bit* (MSB). Specifically, it allocates 1/2 nodes into the 1 st group, 1/4 of the nodes into the 2 nd group and so on. For a nonnegative real number $u \in [0, 2(U + V)]$, it can be decomposed into such binary expansions: $u = \sum_{i=1}^{\infty} u_i 2^{i_0 - i}$ with $u_i \in \{0, 1\}$ and $i_0 = \lceil \log_2(2(U + V)) \rceil$. Hence the $i$th MSB bit of $u$ is $u_i$. The group $S_i$ is defined as the subset of $[0, 2(U + V)]$ in which the $i$th MSB is 1, i.e., $S_i = \{u \in [0, 2(U + V)] : u_i = 1\}$. Note that these sets $S_i, i = 1, 2, \ldots$, do overlap. The local message function $m_k^i$ for a sensor $k$ in the $S_i$ group is given as

$$
m_k^i(x_k) = \begin{cases} 1, & x_k + U + V \in S_i \\ 0, & x_k + U + V \in S_i^c, \end{cases} \tag{1.20}
$$

where $S_k^c$ is the complement of $S_k$ in $\mathbb{R}$. Let $|S_i|$ denote the number of nodes in the group $S_i$ and $I_S$ denote the number of such groups. The fusion center then uses a linear estimator to average $m_k$ s and to conclude the final estimate

$$
\hat{\theta}_{\mathrm{DES}} = -(U + V) + 2(U + V) \sum_{i=1}^{I_S} 2^{-i} \frac{\sum_{m_k^i \in S_i} m_k^i}{|S_i|}, \tag{1.21}
$$

where $m_k^i$ is the binary output of the message function from the sensor $k$ who belongs to the group $S_i$. We use a numerical example to illustrate how DES works.

*Example 5.* Suppose we have 15 sensors and $V = 5$ and $U = 10$. Hence we can divide sensors to 4 groups $S_1, S_2, S_3, S_4$ containing 8, 4, 2, and 1 sensors respectively. Furthermore, $S_1 = [16, 30]$, $S_2 = [24, 30] \cup [8, 16]$ and so on for $S_3$ and $S_4$. For $x_k = 21$, the message function output will be $m_k^1(21) = 1$ (or $m_k^2 = 0, m_k^3 = 1$, $m_k^4 = 0$) if the sensor $k$ is in group $S_1$ (or group $S_2, S_3, S_4$, respectively). Now suppose for sensors in the $S_1, S_2, S_3$ and $S_4$, the message outputs are $\{1, 0, 1, 1, 1, 1, 0, 1\}, \{1, 1, 0, 1\}, \{1, 1\}$ and $\{1\}$, respectively. Then from Eq. (1.21), the final estimate of the fusion center is

$$\hat{\theta} = -15 + 30 \times \left( 2^{-1} \times \frac{6}{8} + 2^{-2} \times \frac{3}{4} + 2^{-3} \times \frac{2}{2} + 2^{-4} \times \frac{1}{1} \right)$$

$$= -15 + 30 \times 0.75 = 7.5.$$

The DES is an unbiased estimator (this is required to be proven as the question) and the upper and lower bound for MSE are derived as [13, 14]:

$$\frac{U^2}{4K} \le E\left[ (\hat{\theta}_{\text{DES}} - \theta)^2 \right] \le \frac{(U + V)^2}{K}.$$

Furthermore, it is independent of the noise or parameter distributions and hence is called as *universal* estimation. However, the aforementioned DES needs to specify which sensor belongs to which group and requires a fusion center to conclude the final estimate. The *isotropic* DES scheme [15] solves this problem by using a probabilistic approach for a sensor to quantize the measurement. For each new measurement $x_k$, the sensor $k$ flips a coin and, with probability 1/2, quantizes $x_k$ to the 1st MSB, with probability 1/4, quantizes $x_k$ to the 2nd MSB, and so on. With this coin flipping strategy for each sensor, there will be roughly 1/2 of the sensors quantizing measurements to the 1st MSB, 1/4 to the 2nd MSB, and so on. This probabilistic DES [15] is hence isotropic in that all sensors operate identically and independently of network topology.

In an inhomogeneous sensing environment, the assumption that the sensors' noises have the same distribution (or the same mean and variance) may be invalid and different sensors may have different quality of observations. For example, the parameter $\theta$ to be estimated is the amplitude of an acoustic signal, which attenuates with distances. Hence a sensor close to the signal source may have a larger local *Signal to Noise Ratio* (SNR) than the one father way. In such cases, it is not necessary to require all sensors to encode the measurements with the same amount of bits and a sensor has higher local SNR can use more bits to encode its messages. In [16], the length of a local message $L_k$ is designed to be proportional to the logarithm of its local SNR and the first $L_k$ bits of the binary expansion of the normalized measurements are used as the message.

The messages are sent over radio channel, which is time-varying and unreliable. To further reduce energy consumption, the design of the message function may also take the quality of the radio channel into consideration. For example, if the radio channel of a node is very poor, it may choose not to send its messages or not to compress its measurements at all, even if its quality of observation is high. In [17], the wireless link between a sensor and the fusion center is modeled as an additive white Gaussian noise with known path gain. With quadrature amplitude modulation and no channel coding, the message length is designed to be proportional to the local observation SNR scaled by the channel path gain.

### 1.3.2.2 Sequential Estimation

Instead of sending all (quantized) measurements to a central estimator located at the fusion center, estimation can also be done sequentially. In sequential estimation, a node not only makes its measurements but also works as an estimator to output an estimate based on its own measurements and the estimation results from other nodes. Although sequential estimation requires high computation capability and consumes more energy for individual nodes, it may help to reduce total energy consumption for the whole network. If the parameter to be estimated is from a point source (and is attenuated with distance), such as the sound energy level emitted by a moving vehicle, the estimation can be done by those nodes close to the event source. These nearby nodes use local message exchange in sequential estimation and only send the final estimation result to the sink. If the sink is far away from these nodes and multi-hop communications have to be used, sequential estimation may produce less data traffic in the network compared with that all nodes send their measurements to a fusion center for estimation and hence consumes less energy network wide.

Sequential estimation can be considered as a kind of decentralized incremental optimization where the estimate is circulated and improved incrementally by the iterating nodes. Rabbat and Nowak [18] apply the theory of incremental subgradient optimization to sequential estimation and analyze the energy-accuracy tradeoffs. Blatt and Hero [19] propose a sequential ML estimation based on the Fisher scoring method and analyze the asymptotic performance if the individual nodes only produce suboptimal estimates. The iterative version for the maximum likelihood estimator can be obtained by the Fisher's scoring method as follows

$$\hat{\boldsymbol{\theta}}(k) = \hat{\boldsymbol{\theta}}(k-1) + \mathbf{I}^{-1}(\hat{\boldsymbol{\theta}}(k-1))\mathbf{s}(\mathbf{x}(k); \hat{\boldsymbol{\theta}}(k-1)), \qquad (1.22)$$

where $\hat{\boldsymbol{\theta}}(k)$ is the estimate after the $k$ th iteration, $\hat{\boldsymbol{\theta}}(0)$ the randomly chosen initial estimate, $\mathbf{x}(k) := (x_1, \ldots, x_k)^T$ the measurements vector, $\mathbf{I}$ is the Fisher information matrix defined in (1.5) and $\mathbf{s}$ the Fisher's score function defined in (1.10). Obviously, to apply (1.22) for sequential estimation, we need to know the parameterized PDF $p(\mathbf{x}; \boldsymbol{\theta})$. In the case of independent sensor nodes and measurements, the Fisher's score function takes a sum form,

$$\mathbf{s}(\mathbf{x}(k); \hat{\boldsymbol{\theta}}(k-1)) = \sum_{i=1}^{k} \frac{\partial \ln p(x_i; \boldsymbol{\theta})}{\partial \boldsymbol{\theta}}\Big|_{\boldsymbol{\theta} = \hat{\boldsymbol{\theta}}(i-1)}$$

and is updated by previous estimate and current measurement. The sequential estimation can be ended if the difference between two successive estimates is small than
a predefined threshold. In cases of unknown $p(\mathbf{x}; \boldsymbol{\theta})$, the linear estimator BLUE can
also be used and its sequential version can be derived for given linear data models.

*Example 6   [20].* Consider the following linear data model

$$\mathbf{x} = \mathbf{D}\boldsymbol{\theta} + \mathbf{w},$$

where $x = (x_1, \ldots, x_K)^T, \mathbf{D} = (d_1^{-\alpha}, \ldots, d_K^{-\alpha})$ and $\mathbf{w} = (w_1, \ldots, w_K)^T$. This
model can be used in applications where sensor nodes are deployed to estimate the
temperature of a point heat source and $d_k$ is the distance between the $k$ th node and
the event source. The additive noises are assumed to be spatially uncorrelated white
noise with zero mean and $\sigma_k^2$ variance, but otherwise unknown. The BLUE for this
model is given by (1.12). The $k$ th sensor can make an estimate $\hat{\theta}(k)$ by using the
following recursive structure

$$\hat{\theta}(k) = \hat{\theta}(k-1) + \frac{B(k)}{d_k^\alpha \sigma_k^2} \left( x_k - \frac{\hat{\theta}(k-1)}{d_k^\alpha} \right), \quad k = 1, 2, \ldots, K, \qquad (1.23)$$

where

$$B(k) = \left( \frac{1}{B(k-1)} + \frac{1}{d_k^\alpha \sigma_k^2} \right)^{-1}. \qquad (1.24)$$

These equations are initialized by $\hat{\theta}(0) = 0$ and $B(0)$ equal to a very large number.

Zhao and Nehorai [21] develop a sequential Bayesian estimation method for applications of localizing diffusive sources. For the Bayesian estimation, they use a
Gaussian density approximation and a linear combination of polynomial Gaussian
density functions to represent the state belief. In their sequential Bayesian estimation, the state belief is exchanged and updated using the measurements from a newly
selected node. Four information utility measures are used in the node selection:
mutual information, posterior Cramér-Rao bound, Mahalanobis distance, and the
covariance-based measure. In fact, the node selection in sequential estimation is
also very important since it determines the convergence rate of the estimation and
the total energy consumption for the estimation. Wang et al. [20] provide examples
where a sequence of nodes with fastest convergence rate is not necessarily the one
with the minimum energy consumption if adjustable transmission power is used.
Quan et al. [22] propose a greedy heuristic to select a new node if the inclusion
of its measurements for the reduction of the estimation error is the greatest among
other possible neighbors.

### 1.3.3 Classification Fusion

#### 1.3.3.1 Fusion for Independent Local Classifiers

Since the volume of measurements or extracted features are much higher than local classification results, data fusion is not suitable for WSNs. For example, consider the following product rule for data fusion. According to the Bayesian theory, classification fusion of $K$ features from $K$ sensor nodes is to assign the class with the maximum a posteriori probability

$$\arg\max_{\omega_j} P(\omega_j | \mathbf{x}_1, \ldots, \mathbf{x}_K). \tag{1.25}$$

Obviously, this fusion rule requires that individual sensor nodes transmit their features to the fusion center. If we assume that all $K$ features from $K$ distinct nodes are statistically independent, then the product rule of fusion is expressed as

$$\arg\max_{\omega_j} P(\omega_j) \prod_{k=1}^{K} p(\mathbf{x}_k | \omega_j), \tag{1.26}$$

where $P(\omega_j)$ is the a priori probability of occurrence of class $\omega_j$ and $p(\mathbf{x}_k | \omega_j)$ is the PDF of $\mathbf{x}$ given class $\omega_j$. The equivalent sum rule is to select a class with the maximum likelihood:

$$\arg\max_{\omega_j} [l_{\omega_j}(\mathbf{x}_1, \ldots, \mathbf{x}_k) := \ln P(\omega_j) + \sum_{k=1}^{K} \ln p(\mathbf{x}_k | \omega_j)]. \tag{1.27}$$

The above fusion rule only require individual nodes to transmit their intermediate fusion results (i.e., $\ln p(\mathbf{x}_k | \omega_j)$). When the feature dimension is larger than the class space, the above sum rule helps to save energy. Some other fusion rules based on the a posterior probability include the max-rule, the min-rule, the median rule and the majority voting [23]. The majority voting is to let each node use a binary function to represent its classification result

$$\delta_{ik} = \begin{cases} 1 & \text{if } P(\omega_i | \mathbf{x}_k) = max_{\omega_j} P(\omega_j | \mathbf{x}_k) \\ 0 & \text{otherwise} \end{cases} \tag{1.28}$$

and the fusion center selects the class with the largest votes

$$\arg\max_{\omega_i} \sum_{k=1}^{K} \delta_{ik}, k = 1, \ldots, K. \tag{1.29}$$

The assumption that all features from different nodes are independent may be too strong in real situations. For example, the nodes close to the signal source may produce very correlated measured data while nodes farther away have uncorrelated measurements. To solve this problem, references [24–26], propose to divide the whole sensor field into small subregions, called *spatial coherence regions* (SCRs). In each SCR, the extracted features (by distinct nodes within the SCR from their respective measurements) are considered to be correlated while the features in distinct SCRs are considered to be independent. A two-step fusion method is then applied to obtain the final result: The features in each SCR are first averaged and the averaged features are then used to perform region-level classification; the intermediate results (e.g., $p(\mathbf{x}_k|\omega_j)$) are then transmitted to the fusion center and the fusion center uses the fusion rule (1.27) to obtain the final classification result. The performance of the proposed fusion method have been analyzed and evaluated in [26,27] for single target or multitarget classification in the presence of ideal or noisy radio channels. It has been shown that the probability of error decreases exponentially with the number of independent node measurements, as long as each node communicates with a nonvanishing power.

In real situations, there may have some other factors that can be exploited in target classification fusion, for example, the distance to the target and the SNR of the extracted feature. Intuitively, the measured data from a node close to the target have high SNR and may well reflect the characteristics of the target. To this end, the majority voting (1.29) can be modified to a weighted version

$$\arg\max_{\omega_i} \sum_{k=1}^{K} w_k \delta_{ik}, \quad k = 1, \ldots, K, \tag{1.30}$$

and the weights $w_k$ is set to

$$w_k = P(\mathbf{x}_k|d_k, \gamma_k) P(d_k, \gamma_k), \tag{1.31}$$

where $d_k$ is the distance between the target and the $k$ th sensor node and $\gamma_k$ is the feature SNR. The conditional probability $P(\mathbf{x}_k|d_k, \gamma_k)$ and the a priori probability $P(d_k, \gamma_k)$ can be obtained from the training set or from empirical data. For the linearly weighted majority voting, the weights can also take other forms and annotations [28]. For example, it can only let nodes whose distances to the target are less than a threshold and assign them the same weight value. That is, $w_k = 1$, if $d_k \leq d_{\text{thres}}$; and $w_k = 0$, otherwise. However, from the results reported in [28] for real-life vehicle classification, no particular weighting method can produce consistent improvement over others. In some cases, the classification rate after fusion is even smaller than the one without using classification fusion.

The aforementioned fusion rules use (weighted) linear combination of the node-level classification results to conclude the final decision. Another approach is to treat the WSNs as an expert system with each node as an expert, and uses the Dempster-Shafer theory of evidence (or D-S theory) [29] to combine the node-level classification results. In the framework of D-S theory, a finite set containing all possible answers to a certain question is called the *frame of discernment*, usually denoted by $\Theta$, which is the class space $\{\omega_1, \ldots, \omega_M\}$ in our context. Instead of assigning probability to each element in $\Theta$, it uses *basic probability assignment* (BPA) function $b$ to map the power set of $\Theta(2^\Theta)$ to the closed interval $[0,1]$, i.e., $b : 2^\Theta \to [0, 1]$ and satisfies

$$b(\emptyset) = 0, \sum_{A \subseteq \Theta} b(A) = 1, \tag{1.32}$$

where $\emptyset$ is the empty set. For any $A \subseteq \Theta, b(A)$ represents the *belief* that one is willing to commit exactly to $A$ (but not to any subset of $A$) given a certain piece of evidence. The Dempster's rule of combination provides a method to combine evidences from different nodes to produce a new BPAs. Let $b_1$ and $b_2$ on $\Theta$ denote two BPAs used by two nodes, the orthogonal sum of $b_1$ and $b_2$ defines a new BPA $b = b_1 \oplus b_2$ by

$$b(C) = b_1 \oplus b_2 = \frac{\sum_{A \cap B = C} b_1(A)b_2(B)}{\sum_{A \cap B \neq \emptyset} b_1(A)b_2(B)}, C \neq \emptyset. \tag{1.33}$$

The combination rule is communicative and associative. The combination of multiple evidences is

$$b = b_1 \oplus b_2 \oplus \cdots \oplus b_K. \tag{1.34}$$

and the fusion center uses the combined BPA $b$ to select the class $\omega_i$ with the highest basic probabilities

$$\arg\max_{\omega_i} b(\omega_i). \tag{1.35}$$

Reference [30] applies the above D-S theory based fusion rule for the classification fusion, where individual node uses *k-nearest neighbor* (k-NN) classifiers and the weights from the output of the k-NN classifier are used to assign BPA to each node. Reference [30] uses the same data sets in [28], but uses a wavelet-based feature extraction method other than the FFT-based feature extraction used in [28]. Their results show that the D-S fusion has generally higher classification rate than that of the majority voting in all of their testing scenarios.

The aforementioned fusion rules, though in different forms, are based on the results of separately trained local classifiers. In other words, the decision rule derived for one local classifier is independent of other local classifiers; and so does for its local decision result. In many cases, local classifiers can also be trained by the same training set. In the next subsection, we review a recently proposed fusion rule, which is based on the results of local classifiers whose decision rules are dependent with each other.

### 1.3.3.2 Fusion for Dependent Local Classifiers

As discussed in previous subsection, classification based on decision fusion helps to conserve energy consumption since the data volume to be transmitted over radio channel for the result of a local classifier is much less than that for the features extracted by a node. For example, if there are $M$ types of targets to be classified, then for the majority voting based decision fusion (1.29), each local classifier needs to transmit $\log_2 M$ bits to represent the decision results. Wang et al. [31] proposes to further reduce the information bits to represent local decision result by only binary local decision.

The fusion scheme proposed in [31], called DCFECC (fault-tolerant distributed multi-class classification fusion approach using error correcting codes), is based on the use of error correcting codes as the fusion rule to achieve fault tolerance capability. Suppose we use $K$ nodes as local classifiers to classify $M$ distinct targets and $K > M$. Let $\mathbf{T}$ denote a $M \times K$ code matrix with elements $t_{ij} \in \{0,1\}, i = 1, \ldots, M, j = 1, \ldots, K$ and $\mathbf{t}_i := (t_{i1}, t_{i2}, \ldots, t_{iK})$ the $i$ th row of $\mathbf{T}$. The fusion center uses the following fusion rule to classify a $1 \times K$ binary vector $\mathbf{u}$ to class $\omega_i$: Assign class $\omega_i$ to $\mathbf{u}$ if the Hamming distance between $\mathbf{u}$ and $\mathbf{t}_i$ is the minimum, i.e.,

$$\arg \min_{1 \leq i \leq M} d_H(\mathbf{u}, \mathbf{t}_i), \tag{1.36}$$

where $d_H(\mathbf{u}, \mathbf{t}_i)$ is the Hamming distance. Tie is broken randomly if more than one row produces the same minimum Hamming distance. For example, if the fusion center uses the following code matrix

$$\mathbf{T} = \begin{pmatrix} 0 & 1 & 0 & 1 & 0 & 1 & 0 & 1 \\ 1 & 0 & 1 & 1 & 0 & 0 & 1 & 1 \\ 0 & 0 & 1 & 0 & 1 & 1 & 0 & 0 \end{pmatrix}$$

and receives $\mathbf{u} = (1, 0, 0, 1, 0, 0, 0, 1)$, then the Hamming distances are $d_H(\mathbf{u}, \mathbf{t}_1) = 3, d_H(\mathbf{u}, \mathbf{t}_2) = 2$, and $d_H(\mathbf{u}, \mathbf{t}_3) = 6$, and hence the fusion center concludes the class $\omega_2$ as the final result.

The design objective of a code matrix and local classifiers is to minimize the final classification error with the consideration of transmission error. Let $C(\mathbf{u}, \omega_i)$ denote the cost when the fusion center receives $\mathbf{u}$ and the ground truth is $\omega_i$. Let $\mathbf{u}_k^0$ and $\mathbf{u}_k^1$ denote the received vector with the $k$ th element being 0 and 1, respectively. Let $u_k^*$ denote the decision of the $k$ th local classifier and $u_k$ the received version of $u_k^*$ at the fusion center. Assuming that the conditional PDFs $p(\mathbf{x}_1|\omega_i), \ldots, p(\mathbf{x}_K|\omega_i), i = 1, \ldots, M$ are independent, the $k$ th local classifier uses the following binary classification rule

$$\sum_i p(x_k|\omega_i)\mathbf{K}_{ki} \underset{u_k^*=0}{\overset{u_k^*=1}{\gtrless}} 0 \tag{1.37}$$

and

$$\mathbf{K}_{ki} = \sum_{j_1,\dots,j_{k-1},j_{k+1},\dots,j_K} P(\omega_i)p(u_1 = j_1|\omega_i)$$

$$\times \dots \times p(u_{k-1} = j_{k-1}|\omega_i)p(u_{k+1} = j_{k+1}|\omega_i)$$

$$\times \dots \times p(u_K = j_K|\omega_i) \times [C(\mathbf{u}_k^0, \omega_i) - C(\mathbf{u}_k^1, \omega_i)] \times (1 - p_{1k} - p_{0k}),$$

where $j_1, \dots, j_K \in \{0, 1\}$ and $p_{1k} = P(u_k = 1|u_k^* = 0)$ and $p_{0k} = P(u_k = 0|u_k^* = 1)$ denote the two transmission error probabilities. Since $\mathbf{K}_{ki}$ depends on the decision rules of other nodes, the classification rule at node $k$ is hence also dependent on other classifiers'. Furthermore, whenever the code matrix $\mathbf{T}$ at the fusion center is changed, the corresponding local decision rules also need to be modified.

The design of the code matrix is coupled with the local decision rules and hence is very complicated. Wang et al. [31] proposes two heuristic algorithms for the code matrix design. One is based a cyclic column replacement approach, which is usually fast but may converge to a local optimum. Another is to use simulated annealing, which is time-consuming but is robust and has better performance.

## 1.4 Directions for Future Research

Traditional network protocol design applies a layered architecture to separate the functionalities of different layers. But for WSNs, since the ultimate objective is to let resource-limited sensor nodes collaboratively accomplish some specific tasks, the application specific and energy efficient protocol design may be more appropriate and preferred. In some cases, cross-layer protocol design is required for WSNs and information processing may also play a role in the design of network architecture and protocols.

Figure 1.6 illustrates a schematic diagram for application based cross-layer protocol design. For example, the media access for parameter estimation can be based on the types of the measurements other than the raw measurements and such a data-centric MAC allocates network resources to data types other than individual nodes [32]. As most applications in WSNs involve with information processing, how to design network protocols to facilitate information processing and to prolong network operation time will be one of directions for future research. In what follows, we provide some examples that design network protocols based on application specificity and information processing in WSNs.

### 1.4.1 Information Processing and Sensor Node Management

Sensor activity management is one of middleware services for WSNs to coordinate sensor sensing states. If the number of deployed sensor nodes is more than the

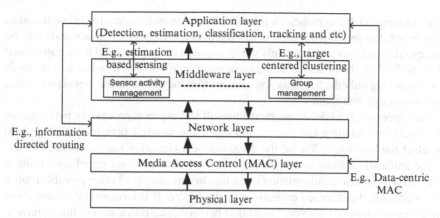

**Fig. 1.6** Application based cross-layer protocol design

optimum, sensor activity management is used to choose only a fraction of the deployed nodes to perform sensing tasks while the application requirement such as the estimation accuracy can still be guaranteed. Wang et al. [33,34] propose the concept of information coverage based on estimation and design sensor activity management to schedule sensor nodes' sensing state to prolong network lifetime [35]. Generally speaking, if the parameter of a target can be estimated by $K$ sensors with required estimation accuracy, then this target is said to be information covered by these $K$ sensor nodes. By exploiting the correlation among nodes' measurements, a target can be information covered by $K$ nodes even if it cannot be information covered by any of them. The sensing activity management for targets' information coverage is to partition sensor nodes into different sensor covers each providing information coverage for all targets.

The design of dynamic group management protocol for mobile target tracking is also much related to information processing. When a target traverses a sensor field, a group of sensor nodes needs to be formed and dynamically reconfigured to track the location of the moving target. Using a group of nodes for collaborative information processing helps to improve the localization performance. Group leader initialization can be based on the detection time or detection reliability and group member recruitment can be based on the quality of the tracking application. The group leader needs to handover its leadership if the tracked target is about to move farther away from it. The group leader can handover its leadership to a node who has the largest mutual information [36] (or the maximum utility [37]). That is, the leader selects the node whose measurement would provide the greatest amount of information about the target location.

## 1.4.2  Information Processing and Routing Strategy

Routing strategy in WSNs may also differ much from the ones in traditional wireless networks in that in-network data processing (or data aggregation) is often needed in

WSNs to exploit the correlations in between the measurements of different nodes. In-network data processing refers to that the data generated by a sensor node may be processed by another node and only the aggregated data are needed to be transmitted to the sink. With the in-network data processing in mind, how to select a route needs to consider not only the transmission energy efficiency but also the signal processing performance of that route.

One approach is to decomposite the overall information processing performance of a route into additive link metric and apply the shortest path methodology [38] to select the best route. Taking the detection of a Gaussian random field as an example application, Sung et al. [39, 40] introduce Chernoff routing where a route is selected if its Chernoff information is the maximum among all other possible routes. Unfortunately, the standard expression of the Chernoff information does not allow the decomposition of the overall detection performance into a sum of the incremental performance gains at each link. Sung et al. show that the Chernoff information of a route with a Gaussian signal is approximately equal to the sum of the logarithm of the innovations variance at each link. Hence, the innovations representation of the log-likelihood function can be used as an additive link metric and the classic Bellman-Ford algorithm can be used to implement a distributed version of the Chernoff routing.

Liu et al. [41] consider the query routing problem where a query node enquires the sensor network to collect information about a phenomenon of interest. However, the query node may not know *a priori* where such information is located. Liu et al. formulate the query routing problem as to find a route with lowest communication cost and largest information gain. A straightforward approach is to use a greedy node selection strategy, which always selects the next node with the minimum cost among other possible one-hop neighboring nodes. However, this approach may introduce routing hole problem as the information gain of individual nodes may be time-varying. To solve this hole problem, they propose to use an information-directed multiple step look-ahead approach where more than one-hop away nodes are searched in each selection step.

## 1.5 Concluding Remarks

Wireless sensor networks, which can help people to monitor and supervise an area of interests from distant locations, are expected to support a broad range of applications, including battle field monitoring, environmental science, health care, smart home, etc. Perhaps most of WSNs' applications are related with information processing to detect, estimate, classify, track, or collect physical phenomena. Because of the resource constraints of individual nodes and the network, energy efficient information process techniques are needed for WSNs. This chapter reviewed recent advances on energy efficient estimation and classification algorithms for WSNs. The focuses were put on two approaches. One is how to reduce the data volume of each individual node to be transmitted over the network while preserving the network information processing performance. Another is to let only nodes close to event source perform processing while only the final results are sent to the sink. Not only information processing algorithms but also other network protocols are needed to be energy efficient for WSNs. For application specific WSNs, information processing also plays an important role in such energy efficient network protocols such as node management and routing protocols. It is envisioned that information networking which integrates information processing into network design will be one of future research directions.

## Terminologies

*Wireless sensor node.* A wireless sensor node consists of, at least, a sensing unit, a power unit, a radio communication unit and a computation and storage unit, and is used to sense and process physical phenomena, and transmit the processed data via wireless.

*Wireless sensor network.* A wireless sensor network consists of, normally a large amount of, spatially distributed sensor nodes to cooperatively monitor and report physical phenomena of an area, such as temperature, pressure of a room, or target moving in an open space.

*Information processing.* Information processing describes the change of information, from one representation to some another representation, or to extract some useful data/information via the process of the original data set. For example, estimate a parameter from timer series and classify an object type from the feature data.

*Energy efficient information processing.* Information processing consumes energy. Energy efficient information processing tries to conserve energy consumption while maintaining reasonable performance loss.

*Estimation.* Estimation deals with estimating the values of parameters (or signals) that are used to described the physical scenario based on measured data. Estimation theory assumes that some information-bearing quantity is contained in the measured data.

*Unbiased estimation.* Unbiased estimation means that the expected value of the estimate equals to the true value of the parameter.

*Minimum variance unbiased estimator.* The performance of an estimator is often measured by the minimum square error of the estimate and the true value. A minimum variance unbiased estimator has the minimum error variance and is unbiased.

*Cramer-Rao lower bound.* The Cramer-Rao lower bound provides a lower bound of the error variance for an estimator. An estimator achieves this bound for all values of the unknown parameters is a minimum variance unbiased estimator.

*Classification.* Classification aims at classifying data set to some class (i.e., put a label) based on either a priori knowledge or on statistical information of each class.

*Classification fusion.* Classification fusion is to combine the classification results
(or intermediate results) of individual classifiers to conclude a final classification
result.

*Weighted majority voting.* Weighted majority voting is one of the classification fusion methods. The outputs of individual classifier are weighted and linearly
summed up and the class label with the largest weight is chosen as the final
classification fusion result.

# Questions

1. List some applications of wireless sensor networks.
2. Why energy efficiency is important to WSNs?
3. What are the approaches to design energy efficient information processing
   algorithms in WSNs?
4. Compare the advantages and disadvantages of the parallel architecture and the
   tandem architecture for information processing in WSN.
5. What is an unbiased estimator and what is the commonly used optimality criterion for comparing estimators?
6. Show that DES is an unbiased estimator for the data model given in (1.15).
   Suppose that $\theta \in [-U, U]$, the additive noises $w_i$s are identical and independent
   random variables with zero mean and support in $[-U, U]$.
7. Derive the sequential BLUE given by (1.23) and (1.24).
8. Derive the product rule of fusion (1.26) for statistically independent features.
9. An example of using weighted majority voting for classification fusion. The
   fusion results from 5 sensors for 2 target types (A and B): $\delta_1 = (A, B) =
   (1, 0)$, $\delta_2 = (1, 0)$, $\delta_3 = (0, 1)$, $\delta_4 = (0, 1)$, $and$ $\delta_5 = (1, 0)$. What is the final
   fusion result if $\mathbf{w} = (w_1, w_2, w_3, w_4, w_5) = (1, 1, 1, 1, 1), = (0, 1, 0, 0, 1), =
   (0.2, 0.4, 0.3, 0.5, 0.1)$?
10. Give examples to show that a cross-layer networking protocol design takes the
    application requirement and information processing into consideration.

# References

1. I. Akyildiz, W. Su, Y. Sankarasubramaniam and E. Cayirci, "Wireless sensor networks: A survey," Computer Networks, Elsevier Publishers, vol. 39, no. 4, pp. 393–422, 2002.
2. D. Li, K. D. Wong, Y. H. Hu and A. M. Sayeed, "Detection, classification, and tracking of
   targets," IEEE Signal Processing Magazine, vol. 19, no. 2, pp. 17–29, 2002.
3. F. Zhao and L. Guibas, Wireless Sensor Networks: An Information Processing Approach,
   Elsevier Inc., New York, USA, 2004.
4. W. R. Heinzelman, A. Chandrakasan and H. Balakrishnan, "Energy-efficient communication
   protocol for wireless microsensor networks," in IEEE Proceedings of Hawaii International
   Conference on System Sciences, 2000, pp. 1–10.

5. Q. Wang, M. Hempstead and W. Yang, "A realistic power consumption model for wireless sensor network devices," in IEEE 3rd Annual Communications Society on Sensor and Ad Hoc Communications and Networks (SECON), 2006, pp. 286–295.
6. S. M. Kay, Fundamentals of Statistical Signal Processing: Estimation Theory, Prentice Hall Inc., New Jersey, USA, 1993.
7. S. Theodoridis and K. Koutroumbas, Pattern Recognition (2nd Edition), Academic Press, San Diego, USA, 2003.
8. J. Hill, R. Szewczyk, A. Woo, S. Hollar, D. Culler and K. Pister, "System architecture directions for networked sensors," in the 9th International Conference on Architectural Support for Programming Languages and Operating Systems, 2000.
9. V. Raghunathan, C. Schurgers, S. Park and M. B. Srivastava, "Energy-aware wireless microsensor networks," IEEE Signal Processing Magazine, no. 19, pp. 45–50, 2002.
10. J.-J. Xiao, A. Ribeiro, Z.-Q. Luo and G. B. Giannakis, "Distributed compression-estimation using wireless sensor networks," IEEE Signal Processing Magazine, vol. 23, no. 4, pp. 27–41, 2006.
11. A. Ribeiro and G. B. Giannakis, "Bandwidth-constrained distributed estimation for wireless sensor networks–part i: Gaussian case," IEEE Transactions on Signal Processing, vol. 54, no. 3, pp. 1131–1143, 2006.
12. A. Ribeiro and G. B. Giannakis, "Bandwidth-constrained distributed estimation for wireless sensor networks–part ii: Unknown probability density function," IEEE Transactions on Signal Processing, vol. 54, no. 7, pp. 2784 – 2796, 2006.
13. Z.-Q. Luo, "Universal decentralized estimation in a bandwidth constrained sensor network," IEEE Trans. on Information Theory, vol. 51, no. 6, pp. 2210–2219, 2005.
14. J.-J. Xiao, Z.-Q. Luo and G. B. Giannakis, "Performance bounds for the rate-constrained universal decentralized estimators," IEEE Signal Processing Letters, vol. 14, no. 1, pp. 47–50, 2007.
15. Z.-Q. Luo, "An isotropic universal decentralized estimation scheme for a bandwidth constrained ad hoc sensor network," IEEE Journal on Selected Areas in Communications, vol. 23, no. 4, pp. 735–744, 2005.
16. J.-J. Xiao and Z.-Q. Luo, "Decentralized estimation in an inhomogeneous environment," IEEE Transactions on Information Theory, vol. 51, no. 10, pp. 3564–3575, 2005.
17. J.-J. Xiao, S. Cui, Z.-Q. Luo and A. J. Goldsmith, "Power scheduling of universal decentralized estimation in sensor networks," IEEE Transactions on Signal Processing, vol. 54, no. 2, pp. 413–422, 2006.
18. M. Rabbat and R. Nowak, "Distributed optimization in sensor networks," in The 3rd International Symposium on Information processing in sensor networks (IPSN), 2004, pp. 20–27.
19. D. Blatt and A. Hero, "Distributed maximum likelihood estimation for sensor networks," in IEEE International Conference on Acoustic, Speech, and Signal Processing (ICASSP), 2004.
20. B. Wang, K. C. Chua and V. Srinivasan, "Localized recursive estimation in energy constrained wireless sensor networks," Journal of Networks, vol. 1, no. 2, pp. 18–26, 2006.
21. T. Zhao and A. Nehorai, "Distributed sequential bayesian estimation of a diffusive source in wireless sensor networks," IEEE Transactions on Signal Processing, vol. 55, no. 4, pp. 1511–1524, 2007.
22. Z. Quan, W. J. Kaiser and A. H. Sayed, "A spatial sampling scheme based on innovations diffusion in sensor networks," in The 6th international conference on Information processing in sensor networks (IPSN), 2007, pp. 323–330.
23. J. Kittler, M. Hatef, R. P. Duin and J. Matas, "On combining classifers," IEEE Transactions on Pattern Analysis and Machine Intelligence, vol. 20, no. 3, pp. 226–239, 1998.
24. A. M. D'Costa and A. M. Sayeed, "Data versus decision fusion for distributed classification in sensor networks," in IEEE Military Communications Conference (Milcom), 2003, pp. 585–590.
25. A. D'Costa and A. M. Sayeed, "Collaborative signal processing for distributed classification in sensor networks," in International Conference on Information Processing in Sensor Networks (IPSN), 2003, pp. 193–208.

26. A. D'Costa, V. Ramachandran and A. M. Sayeed, "Distributed classification of gaussian space-time sources in wireless sensor networks," IEEE Journal on Selected Areas in Communications, vol. 22, no. 6, pp. 1026–1036, 2004.

27. J. H. Kotecha, V. Ramachandran and A. M. Sayeed, "Distributed multitarget classification in wireless sensor networks," IEEE Journal on Selected Areas in Communications, vol. 23, no. 4, pp. 703–713, 2005.

28. M. F. Duarte and Y. H. Hu, "Vehicle classification in distributed sensor networks," Journal of Parallel and Distributed Computing, vol. 64, no. 7, pp. 826–838, 2004.

29. G. Shafer, A Mathematical Theory of Evidence, Princeton University Press, Princeton, New Jersey, USA, 1976.

30. C.-T. Liu, H. Huo, T. Fang, D.-R. Li and X. Shen, "Classification fusion in wireless sensor networks," ACTC Automatica Sinica, vol. 32, no. 6, pp. 947–955, 2006.

31. T.-Y. Wang, Y. S. Han, P. K. Varshney and P.-N. Chen, "Distributed fault-tolerant classification in wireless sensor networks," IEEE Journal on Selected Areas in Communications, vol. 23, no. 4, pp. 724–734, 2005.

32. G. Mergen and L. Tong, "Type based estimation over multiaccess channels," IEEE Transactions on Signal Processing, vol. 54, no. 2, pp. 613–626, 2006.

33. B. Wang, W. Wang, V. Srinivasan and K. C. Chua, "Information coverage for wireless sensor networks," IEEE Communications Letters, vol. 9, no. 11, pp. 967–969, 2005.

34. B. Wang, K. C. Chua, V. Srinivasan and W. Wang, "Information coverage in randomly deployed wireless sensor networks," IEEE Transactions on Wireless Communications, vol. 6, no. 8, pp. 2994–3004, 2007.

35. B. Wang, K. C. Chua, V. Srinivasan and W. Wang, "Scheduling sensor activity for point information coverage in wireless sensor networks," in International Symposium on Modelling and Optimization in Mobile, Ad Hoc, and Wireless Networks (WiOpt), 2006.

36. J. Liu, J. Reich and F. Zhao, "Collaborative in-network processing for target tracking," EURASIPJornal on Applied Signal Processing, vol. 2003, no. 4, pp. 378–391, 2003.

37. F. Zhao, J. Liu, J. Liu, L. Guibas and J. Reich, "Collaborative signal and information processing: an information-directed approach," Proceedings of the IEEE, vol. 91, no. 8, pp. 1199–1209, 2003.

38. D. P. Bertsekas and R. G. Gallager, Data Networks (2nd Edition), Prentice-Hall, Englewood Cliffs, New Jersey, USA, 1992.

39. Y. Sung, L. Tong and A. Ephremides, "A new metric for routing in multi-hop wireless sensor networks for detection of correlated random fields," in IEEE Military Cmmmunications Conference (Milcom), 2005, pp. 2327–2332.

40. Y. Sung, S. Misra, L. Tong and A. Ephremides, "Signal processing for application-specific ad hoc networks–the role of signal processing in protocol design," IEEE Signal Processing Magazine, vol. 23, no. 5, pp. 74–83, 2006.

41. J. Liu, F. Zhao and D. Petrovic, "Information-directed routing in ad hoc sensor networks," IEEE Journal on Selected Areas in Communications, vol. 23, no. 4, pp. 851–861, 2005.

# Chapter 2
# Topology Management for Wireless Sensor Networks

Lisa Frye and Liang Cheng

**Abstract** Topology management is a key component of network management of wireless sensor networks. The primary goal of topology management is to conserve energy while maintaining network connectivity. Topology management consists of knowing the physical connections and logical relationships among the sensors. Another important concept of topology management is to have only a subset of nodes actively participating in the network, thus creating less communication and conserving energy in nodes. This chapter provides a detailed survey of existing topology management algorithms proposed for wireless sensor networks in three categories: topology discovery (learning the layout of the nodes), sleep cycle management (allowing some nodes to sleep to conserve energy), and clustering (grouping nodes to conserve energy).

## 2.1 Introduction

Networks require constant monitoring in order to ensure consistent and efficient operations. This is also true of wireless sensor networks (WSNs). The International Standards Organization (ISO) developed a network model consisting of five functional areas: fault management, configuration management, security management, performance management, and accounting management. Configuration management entails initial set-up of the network devices and continuous monitoring and controlling of these devices. One key aspect of configuration management for WSNs is topology management, which considers how the nodes are arranged.

The primary goal of topology management is to maintain network connectivity in an energy-efficient manner. Nodes in a WSN have minimal resources, including

L. Frye (✉)
Department of Computer Science, Kutztown University, Lytle 267, Kutztown, PA 19530
e-mail: frye@kutztown.edu

S. Misra et al. (eds.), *Guide to Wireless Sensor Networks*, Computer Communications
and Networks, DOI: 10.1007/978-1-84882-218-4_2,
© Springer-Verlag London Limited 2009

processing, memory, and energy. Typical sensor nodes are powered by batteries and are deployed in networks that receive little direct human interaction. To allow for lengthy deployments of WSNs, the battery life of the nodes must be extended. This can be accomplished by minimizing the amount of energy consumed by the nodes. One way to minimize energy consumption is to implement a topology management algorithm in the WSN. This algorithm can be in one of three categories of topology management: (1) topology discovery, (2) sleep cycle management, and (3) clustering.

## 2.2 Background

Topology discovery involves a network management station, or a base station, determining the organization or topology of the nodes in the sensor network. The physical connectivity and/or the logical relationship of nodes in the network is reported to the management station, which maintains a topology map of the WSN. The base station, or network management station, will send a topology discovery request to the network. Each node in the network will respond with its information. There are three basic approaches taken for topology discovery. The first one is a direct approach. In this approach, a node will immediately send a response back upon receiving a topology request. The node's response will contain information only about that particular node. The second approach is an aggregated approach in which a node will forward the request but will not send an immediate response. Instead, the node will wait until it gets responses to the request from its children. The node will then aggregate all the data received from its children, include its own information, and then send the response back to its parent or the initiating station. The third approach is a clustered approach, which forms groups or clusters from the nodes. One node in each cluster is selected as the leader. Only the leader will reply to the topology request. The leader's reply will include the topology information about all the nodes in its cluster. No matter which approach is used, knowing the topology of the WSN is important for effective and efficient network management of the WSN. Although topology discovery does not directly conserve energy, knowing the topology of the WSN can be useful for other algorithms that may aim to conserve energy.

One way for topology management algorithms to conserve energy in a node is to only have it powered on when necessary; the node would be powered off or put to sleep all other times. WSNs are typically very densely-deployed networks, meaning there are many nodes in each area of the network. For example, an area that has ten nodes deployed may only require three nodes to get complete coverage of the area. This means that seven of the nodes are sending data that is duplicated or redundant. These additional nodes are considered redundant nodes and may not always be necessary. When these redundant nodes are not required they will be put to sleep. Determining which nodes are redundant, putting them to sleep, and waking them up again are tasks of sleep cycle management algorithms. The sleep-wake cycle must be properly managed so that the nodes that stay awake, and participate in data reporting and network functionality, are rotating. By doing this all the nodes will last about the same amount of time; this is important in WSN to properly maintain network connectivity. Care must be taken when implementing sleep cycle management algorithms as many sacrifice energy conservation for latency.

The primary use of energy in WSNs is the transmission of data. Another way to conserve energy is to have fewer nodes transmit data to the base station, which is the device collecting the application data. Clustering algorithms are used to decrease the number of nodes that transmit data to the base station (BS). These algorithms arrange the nodes deployed in the WSN into groups or clusters. One node in each cluster is identified as the leader of the cluster or the cluster head (CH). The nodes that are in a cluster, but are not a cluster head, become member nodes of that cluster. The member nodes will transmit their data to their cluster head, which is typically within only a short distance thus consuming less energy. The cluster head will then forward the data received from each of its member nodes to the base station. Only the cluster heads will transmit data to the base station. Many clustering algorithms will also aggregate or fuse the data received from the member nodes at the cluster head resulting in less data being transmitted from each cluster head to the base station. As less data is transmitted, less energy is used.

Another advantage of clustering is that it allows for the spatial reuse of resources. MAC (Media Access Control) protocols are involved in the process of data communication. A property of wireless communication is that there may be a collision if two nodes within each other's broadcast interference range try to transmit at the same time on the same frequency. To avoid collisions, MAC protocols utilize various methods to prevent multiple nodes from transmitting at the same time. If two nodes do not have overlapping radio ranges then they may be able to transmit at the same time and not have a collision. If two nodes exist in different nonneighboring clusters (so they could transmit at the same time and not result in a collision), it may be possible for the two nodes to share the same frequency or time slot (for Time Division Multiplexing protocols). For example, if clusters A and B are nonneighboring clusters, then a node in cluster A can be transmitting at the same time as a node in cluster B. This can save transmission time as multiple nodes can be transmitting at one time. Clustering algorithms can help determine which nodes can transmit simultaneously.

This chapter will review several topology management algorithms in each of the three categories. Figure 2.1 lists the algorithms that will be discussed.

## 2.3 Thoughts for Practitioners

The application domain of the specific deployment must be considered when selecting appropriate algorithms. Energy conservation is critical in all WSNs, but may be more critical in a long-term deployment. If the intended lifetime of the WSN is relatively short, then an algorithm that conserves less energy but sacrifices less latency may be appropriate. If application data is important, then it may be necessary to

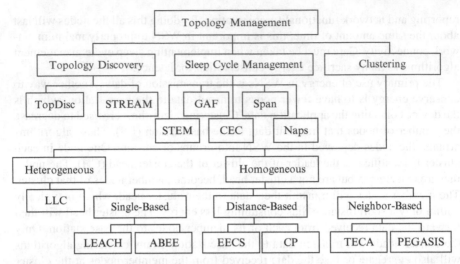

**Fig. 2.1** Taxonomy of topology management algorithms

have sufficient density in every area and have fewer nodes go to sleep in each area. These are just a few of the considerations when selecting the appropriate topology management algorithms for the deployed WSN.

The nature of sensors makes them prone to failures. Many sensors fail simply because they run out of energy, which is limited. Other failures include hardware failures, the destruction of a sensor (being stepped on, dropped, etc.), or link failures. Node failures must be a consideration when designing WSN deployments or new algorithms. For instance, a fault may have an impact on the topology management algorithm. If a member node fails, there should be no impact; however, if the cluster head fails, all member nodes of that cluster would no longer be able to communicate with the base station, at least for the current round. In a sleep-cycle management algorithm, the failure of a nonredundant node may result in the base station obtaining no data for that particular area. These scenarios may result in lost application data. Cluster head and nonredundant node failures should be an important consideration for topology management algorithms.

Many node failures occur quickly and silently, meaning that the node just stops working. However, it is also possible for nodes to fail in other ways. For instance, if a node begins to fail due to the radio malfunctioning or something similar, it may broadcast a signal erratically. This may prevent other nodes from being able to communicate, which could interfere with the application. It may also interfere with topology management algorithms. Another consequence of this type of node failure is that adding new nodes to the network in this area may not help; it may require a different type of solution, such as a new route to avoid this area of the network. Consideration must also be given to the application data in this area.

Another way for nodes to fail is a node failing slowly, meaning that it may still work but not properly, such as the battery or signal weakening. This may prevent

other nodes from being able to reach the weak node, but the nodes may still try
to reach it. The other nodes in the network may retransmit data trying to reach the
failing node. Other nodes may also increase their signal strength to try to reach
the failing node. Both of these may lead to more energy consumption. It may be
necessary for topology management algorithms to mark these types of failing nodes
and not have them participate in the topology management algorithm, such as not
allowing them to be a cluster head or mark them as redundant and have them sleep.
This may actually lead to more energy conservation by other nodes as they will not
retransmit data or strengthen their signal to reach the failing node.

There are many different facets that must be considered when dealing with
topology management algorithms. One important aspect is the application and deployment scenario. This may have a large impact on the most effective topology
management algorithm. When choosing and also developing a topology management algorithm, node failures must be considered as they may play an important
role in the operation of the network and the topology management algorithm.

## 2.4   Topology Discovery Algorithms

### 2.4.1   TopDisc Algorithm

One topology discovery algorithm is the Topology Discovery algorithm or TopDisc
[3]. TopDisc uses the clustered approach to topology discovery and first creates clusters, similar to clustering algorithms, and identifies a cluster head for each cluster.
In TopDisc, the responsibility of the cluster head is to report the network topology
to the monitoring node or base station. The clusters in TopDisc are created by finding the set coverage with a greedy approximation algorithm. The algorithm begins
by the monitoring node sending a broadcast packet containing a topology request
(a packet asking each node for its topology information). This topology request is
propagated throughout the WSN. One of two different coloring algorithms is used
to find the cluster heads while the topology request is propagating throughout the
network.

The first coloring algorithm of TopDisc uses a three color approach. A white node
indicates an undiscovered node, a black node is a cluster head, and a grey node has
a neighbor node that is a black node. The algorithm begins with all nodes colored
white. If a white node receives the topology request packet from a black node, then
the white node becomes grey. If a white node receives the topology request packet
from a grey node, then it waits a random amount of time. If this node receives the
topology request packet from a black node before the time expires, it becomes grey;
otherwise, it becomes black. After a node turns grey or black it will ignore all future
topology request packets it receives. All black nodes become cluster heads and will
report its neighborhood set back to the monitoring node.

The second version of TopDisc uses a four color scheme. White, black, and grey are the same as in the first version. This version adds dark grey, which is a discovered node that is not covered by a black node (it is at least two hops away from a black node). A dark grey node has received a request from a grey node and started a timer to see if it should become a grey node (receives a request from a black node before the timer expires) or a black node (does not receive a request from a black node). The addition of a fourth color allows the black nodes to cover the maximum number of other nodes; thus, black nodes are two hops away from other black nodes. This allows the clusters formed to have less overlap than clusters formed using the three-color version.

TopDisc is a distributed algorithm that is scalable and uses only local information. This means that there is not a lot of information being exchanged, which means less data transmitted and thus less energy consumed. One problem with TopDisc is that there is nothing that guarantees a certain distance between black nodes. Therefore, some black nodes are close to each other and do not cover an optimum number of grey nodes.

## 2.4.2 Sensor Topology Retrieval at Multiple Resolutions

Another topology discovery algorithm is Sensor Topology Retrieval at Multiple Resolutions or STREAM [4]. This algorithm can return the network topology at different degrees, since different applications may require different topology resolutions. Some applications, such as a routing algorithm, may simply need to know about one node in each area showing that the network is connected and there is complete coverage. Other applications may need to know the topology for all nodes in the network. STREAM creates an approximate topology by getting neighborhood lists from a subset of nodes. The minimal set of nodes required to determine sufficient topology is defined as the Minimum Virtual Dominating Set or MVDS. The MVDS is created using a message complexity of $N$ (the number of nodes in the WSN), which does not increase as network density increases. The MVDS tree created is optimal in that the topology responses will travel the minimum number of hops to reach the monitoring node. It does require global knowledge of the neighborhood sets to choose nodes in the MVDS that cover the most undiscovered nodes.

STREAM uses a coloring scheme, similar to TopDisc, to find the network topology. The coloring scheme in STREAM uses four colors: white, black, red, and blue. A white node is an undiscovered node, a black node is a node in the MVDS, a red node is within the virtual range of a black node, and a blue node is a node that is within the communication range of a black node but outside its virtual range. The virtual range is a parameter to the STREAM algorithm that controls the resolution of the topology returned.

The first step of STREAM is for all nodes to be colored white and the monitoring node or base station to send a topology request. If a white node receives the request from a black node and the white node is within the virtual range of the black node, then it will become a red node and forward the request. If the white node is not within the virtual range of the black node it will become a blue node, start a timer, and forward the request. A blue node within the virtual range of a black node that receives the request from that black node will stop its timer and become a red node. It will only forward the request if it has not previously forwarded it. A white node that receives the request from a red or blue node will become a blue node, start a timer, and forward the request. If the timer of a blue node expires then that node will become a black node. Any node that is black or red will not forward additional topology requests it receives.

## 2.5  Sleep-Cycle Management Algorithms

### 2.5.1  Sparse Topology and Energy Management (STEM)

Sparse Topology and Energy Management or STEM [11] manages the sleep cycle by adding a second radio to the sensor nodes. The primary radio, the data radio, is used to transmit application data, routing data, and the majority of other transmitted data. The second radio, the wakeup radio, is only used to transmit data for managing the sleep cycle. This radio is a low duty cycle radio, which uses less energy than typical radios. The data radio is turned off unless it needs to receive or forward data. The wakeup radio will operate at a different frequency and will also follow a sleep-listen cycle with the sleep time being shorter than the sleep time of the data radio.

There are two versions of STEM: STEM-B and STEM-T. In STEM-B, the wakeup radio will send a beacon when it must transmit data. The beacon will contain the MAC address of the target node. If a sensor hears the beacon on its wakeup radio, and it is the target node, then it will power on the data radio and receive the data. If it is not the target node the wakeup radio will go to its sleep state. Because of the possibility of a collision during the transmission of the beacon, a node that detects a collision will also turn on its data radio. This will allow the target radio to be turned on to receive the data. In STEM-T, the wakeup radio will simply send a tone. Any node detecting a tone (detect signal energy on the frequency) will turn on their data radio. In both cases, the node turns off the data radio after it receives and transmits the data or after a specified time expires indicating that it was not the target node.

As the network density increases or as the network spends more time in monitoring state, STEM shows energy savings. However, this energy savings is the result of a sacrifice in latency. If a node must communicate with another node, and the wakeup radio of the other node is sleeping, then the node must wait before sending its data. For this reason, if information is time-sensitive and latency is not acceptable, STEM should not be used.

## 2.5.2 Geographic Adaptive Fidelity (GAF)

Geographic Adaptive Fidelity or GAF [12, 13] is a localized, distributed sleep cycle management algorithm. GAF uses location information, typically from a Global Positioning System (GPS) device, to organize redundant nodes into groups or virtual grids. A virtual grid is "defined such that, for two adjacent grids A and B, all nodes in A can communicate with all nodes in B and vice versa" [12]. Since all nodes in adjacent grids can communicate with each other, the nodes in these two grids are equivalent for the routing protocol.

All nodes begin in a discovery state where it will send out a discovery message and receive replies back to determine the nodes in its same grid. The node will then enter an active state where it stays for a specified period of time before returning to the discovery state. If a node determines that it is a redundant node for the routing protocol, it will enter a sleeping state for a specified period of time. The nodes use a ranking procedure to determine which node will stay awake and handle routing for the grid. The node with the highest priority is the one with the longest lifetime with ties broken by node ID. When a node's timer expires it will transition back to the discovery state.

GAF is a distributed algorithm but it requires location information. GAF guesses at connectivity instead of directly measuring it thus requiring more nodes to remain awake than may be necessary. This algorithm is completely independent of the routing algorithm used. This means that GAF may allow a node to sleep even if that node is actively participating in routing. This may cause interruptions in communication and increase routing latency. Therefore, GAF should only be utilized if latency is acceptable and location information is available.

## 2.5.3 Cluster-Based Energy Conservation (CEC)

Cluster-based Energy Conservation or CEC [13] is an algorithm based on GAF but it directly measures the network connectivity, thus not requiring location information and finding redundancy in the network more accurately. CEC conserves more energy than GAF. However, it does not perform well if the network topology changes frequently.

CEC organizes nodes into overlapping clusters, which are nodes that are at most two hops from each other. The node in the cluster that is one hop from all other nodes and has the most residual energy selects itself to be the cluster head. Since the clusters overlap, some nodes will be members of multiple clusters. These nodes are gateway nodes and will connect the clusters preserving network connectivity. After all cluster heads and gateway nodes are determined, the rest of the nodes in the clusters are considered redundant and are put to sleep. After a specified amount of time, these nodes wake up and cluster heads will once again be selected. The role of cluster head is rotated among all nodes in the cluster so all nodes have a chance to sleep and conserve energy.

CEC is an acceptable sleep cycle management algorithm for applications where there is no location information available. It would be a good sleep cycle management algorithm solution for network deployments that have variable radio ranges since it directly measures the network connectivity.

## 2.5.4  Span

As with other sleep cycle management algorithms, Span [2] is based on the fact that a WSN can be connected with only a subset of the deployed nodes being active with a sufficient density of nodes. The nodes that remain active in Span are called coordinators and are used by the routing protocol. The rest of the nodes are put in power-saving node, which means they are sleeping (their radios are turned off). Coordinators are elected based on local information only. All nodes maintain a neighbor list, which includes a list of coordinators. This neighbor list is maintained by each node periodically broadcasting a HELLO message, or a message simply stating the node is alive. This message contains information about the node, such as if it is a coordinator or not, its current coordinator list, and current neighbor list. A node will become a coordinator if two of its neighbors cannot reach each other. Nodes can reach each other if they can communicate directly or by using one or two coordinators.

Coordinators will periodically demote themselves in order to allow themselves to go into power-saving mode. A demoted coordinator will no longer be designated a coordinator but will continue to participate in routing until another node replaces it as coordinator. This means that there will not be an increase in latency. Demoted coordinators are replaced by noncoordinator nodes becoming a coordinator. This is accomplished by noncoordinator nodes periodically waking up and determining if it needs to become a coordinator (two of its neighbors cannot reach each other as previously described). Changing the role of coordinator among all the nodes helps to prolong the network lifetime. By allowing nodes to reach each other via one or two coordinators, fewer coordinators can be used, which will also increase the network lifetime since more nodes can be put in power-saving mode. Span uses only local information (no centralized decisions); however, it requires nodes to send HELLO messages, which uses bandwidth and energy.

## 2.5.5  Naps

Another sleep cycle algorithm is Naps [6], which is based on message broadcasts. Each node will cycle through a time period. At the beginning of the time period, a node will broadcast a HELLO message and set a counter to zero. The node will then go into a listening state where it listens for HELLO messages from other nodes. Each time it receives a HELLO message from another node it will increment its counter. When the counter reaches a threshold value, which is a parameter to the algorithm, the node will go to sleep. When the time period for the node expires, the node will wake up and start the time period again. If the node's counter does not reach the specified threshold then the node will remain awake for that time period.

Naps is a simple algorithm that does not require any location information, such as neighbor lists or node positions. Naps does introduce extra traffic into the network with its use of HELLO messages. It does not minimize the number of awake nodes since it assumes reliable transmission of the HELLO messages, which is not always the case due to collisions. The role of which nodes are put to sleep is not rotated as the nodes that are put to sleep are just the ones that happened to receive the threshold number of HELLO messages from neighbors. Because of the proximity of nodes to each other, it may always be the same nodes that are allowed to be put to sleep. As network density increases, more nodes will be put to sleep; therefore, low-density networks may not benefit from Naps since few nodes will be allowed to sleep meaning little energy savings.

## 2.6 Clustering Algorithms

There are numerous clustering algorithms that exist today and more being developed. These algorithms can be classified in a variety of ways, such as requiring location information or not, being distributed or centralized, cluster head selection, and cluster formation. Here the algorithms are classified first according to the type of WSN deployed. A WSN may be heterogeneous or homogeneous. A heterogeneous WSN is one in which the network has varying types of nodes in terms of resources. Typically some nodes will have more resources available, such as processing power and energy, than the rest of the nodes. In a homogeneous WSN, all the nodes are the same in terms of available resources.

### 2.6.1 Heterogeneous Clustering Algorithms

LLC [9] or low-energy localized clustering is a clustering algorithm for heterogeneous networks. The network using LLC is a two-tiered network consisting of one tier that is cluster heads and a lower tier consisting of nodes used for sensing only. In this algorithm the cluster heads are determined prior to network deployment and are deployed at random. Cluster heads are devices that have more processing power and more initial energy. When deployed the sink node will know the location of all deployed cluster heads.

This algorithm consists of two phases, an initial phase and the cluster radius control phase. During the initial phase, the cluster heads create a triangle that is used to determine a cluster radius decision point or CRDP. The triangle that is created is an equilateral triangle consisting of three cluster heads, one at each point of the triangle. Using equilateral triangles and the CRDP helps to load-balance the energy consumption of the cluster heads. The radius of the cluster is estimated to be the distance between the CRDP and each of the three cluster heads in the triangle. Calculating an optional cluster radius leads to minimal energy consumption by minimizing distances between the nodes and the cluster heads.

The cluster radius control phase will adjust the cluster radius to minimize the radius and conserve energy in the cluster heads. LLC considers two different types of cluster radius control algorithms. These are the Non-Linear Programming or NLP-based approach and the Vector Computation (VC)-based approach. Both approaches utilize an objective function. The primary difference between the two approaches is that the NLP approach considers the energy of each cluster head to help minimize cluster coverage. The NLP approach also uses an iteration policy to recompute suitable CRDP values, which incurs additional overhead. This computation overhead is reduced in the VC approach by finding the optimal solution for CRDP based on vector computation.

The LLC algorithm requires the location information of all nodes. It also does not rotate the role of the cluster head since cluster heads are predetermined and are more powerful devices. By finding the optimal CRDP, LLC minimizes overlapping clusters, which leads to energy conservation.

## 2.6.2  Homogeneous Clustering Algorithms

A homogeneous WSN includes identical nodes in terms of resources. The cluster algorithms utilize various methods to find appropriate nodes to serve as cluster heads and assign member nodes to the optimal cluster. The algorithms are further categorized according to how they determine cluster heads and form the clusters. Signal-based clustering algorithms form clusters and determine cluster heads based on signal strength. Distance-based clustering algorithms base their decisions on distance metrics. Basing cluster formation decisions on each node's neighbor list is the method used by neighbor-based clustering algorithms.

### 2.6.2.1  Signal-Based Algorithms

Low-Energy Adaptive Clustering Hierarchy (LEACH)

One of the most studied clustering algorithms is LEACH [7]. LEACH manages a WSN so that nodes die at about the same time by rotating the role of the cluster head and basing cluster head selection partly on remaining energy. This extends the network lifetime and leaves little energy in nodes when the network dies.

The operation of LEACH is broken up into rounds. Each round consists of a setup phase and a steady-state phase. The setup phase is when the nodes organize themselves into clusters. A node decides to be a cluster head for that round independent of all other nodes. The node will select a random number and if that number is less than the threshold value then the node will become a cluster head. The threshold value is based on the suggested percentage of cluster heads for that round (determined a priori), the number of times the node has already been a cluster head and the amount of residual energy in the node. The cluster head will broadcast an advertisement message indicating that it is a cluster head. A noncluster head node will join the cluster of which it received the strongest advertised signal from the cluster head. Each node will send a message to its new cluster head informing the cluster head that it is joining its cluster.

After the clusters are formed, the cluster heads create a transmission schedule for its member nodes based on TDMA (Time Division Multiple Access). This allows member nodes to further conserve energy by turning off their radio except during their scheduled transmission time. Another feature of LEACH that helps conserve energy is that after all member nodes transmit their data to the cluster head, the cluster head will fuse these data into a single packet, thus transmitting less data. After a certain time (determined a priori), this round ends and the next round begins, which allows the role of cluster head to rotate among all nodes.

There are several disadvantages to LEACH. For example, there is a large cluster formation overhead. All cluster heads must broadcast advertisement messages to all nodes in their communication radii. Another downfall is that all cluster heads must transmit data to the base station, which is a single hop but may be a long distance, requiring more energy.

### Access-Based Energy Efficient Cluster Algorithm (ABEE)

Access-Based Energy Efficient (ABEE) cluster algorithm [8] is a request-response algorithm that uses a first-come-first-serve method for cluster formation. When a node is deployed it determines its current position (typically from a GPS device) and begins in the idle state. The node will broadcast a request message and begin a timer. If the node receives a response from a cluster head, it will join that cluster by sending a message back to the cluster head. This new member node will ignore any other responses it receives from cluster heads. If the node does not receive any responses from a cluster head, it will broadcast a message that it is a cluster head.

Each member node will periodically send a message to its cluster head informing the cluster head of its current position. The member nodes also maintain information about its cluster head based on the same periodic messages the cluster heads broadcast. This information allows the member node to calculate the distance to its cluster head. Since all cluster heads periodically broadcast messages about its current position, member nodes will also hear broadcast messages from the other cluster heads that are within its radio range. If a member node hears a broadcast from another cluster head that is closer to it than its current cluster head, it will send a message to its current cluster head saying it is leaving the cluster. It will then send a message to the closer cluster head to join that cluster. This allows the member nodes to minimize the distance to their cluster head and conserve energy by transmitting their data to a closer node. Another way to conserve energy is to minimize the number of clusters. If a cluster head receives a message from another cluster head and the distance between the two cluster heads is below a threshold, which is an algorithm parameter, then the two clusters will merge, minimizing the number of clusters.

ABEE tries to balance the residual energy in all the nodes by periodically rotating the role of the cluster head. The new cluster head is selected by treating the "whole cluster as an entity and each node stands for particles with equal mass to form the entity" [8]. This is done by the cluster head collecting all location information from member nodes, accomplished by each member node periodically sending its location information to the cluster head. The cluster head then uses the location information to select the node that is the closest to the mass center of the cluster to be the next cluster head.

Another benefit of ABEE is that the cluster head fuses the member nodes' data before sending it to the base station. There are two major drawbacks to ABEE. First, ABEE requires that all nodes know their location information, usually requiring each node to have a GPS device. The other problem is that the residual energy of the nodes is not a consideration when selecting the cluster heads. This means that the same node may always be selected as a cluster head meaning it will run out of energy before most of the other nodes in its area. This may lead to a shorter network lifetime.

### 2.6.2.2 Distance-Based Algorithms

Energy Efficient Clustering Scheme (EECS)

A clustering algorithm that bases some of the decisions on distance is the Energy Efficient Clustering Scheme or EECS [14]. This algorithm is based on LEACH but tries to achieve a better load balance among the clusters. When the network is deployed the base station will broadcast a message to all nodes in the network. This message will be sent by the base station at a specified power level. This will allow all deployed nodes to determine their appropriate distance from the base station.

There are three phases in EECS: cluster head election phase, cluster formation phase, and data transmission phase. Each node can be either a candidate node, a head node, or a plain node. All nodes start in the plain state indicating they are ordinary nodes. Each round begins in the cluster head election phase where cluster heads are selected. Each node will select a random number, which is the probability of that node becoming a cluster head. If the node's probability is less than a specified threshold then the node will become a candidate node. Candidate nodes will broadcast a message and start a timer. If the candidate node receives a broadcast message from another candidate node and that node has more residual energy, then the candidate node will join that cluster and stop its timer. A candidate node will become a head node or cluster head if it does not hear a broadcast from another candidate node with more residual energy before its timer expires.

Plain nodes join a cluster and load balancing of the clusters occurs during the cluster formation phase. Each cluster head will send a broadcast message. A plain node will join a cluster based on several distances. A plain node will compute the distance to each possible cluster head (all cluster heads within its radio range) and the distance between each possible cluster head and the base station. A plain node will join the cluster that minimizes these two distances and a weight factor. Minimizing the distance between the node and its cluster head helps to conserve energy in the plain nodes. Energy consumption is minimized in the cluster head by transmitting less data or transmitting data a shorter distance. Consequently, cluster heads that are further from the base station should have less plain nodes assigned to its cluster (cluster heads are one hop from the base station so cluster heads that are further from the base station will consume more energy when transmitting the data to the base station). There must be a tradeoff between the two distance metrics used (node to cluster head and cluster head to base station), which is the role of the weight factor. The weight factor is a parameter to EECS and the optimal value depends on the specific network deployment. By using the two distance metrics and the weight factor, the load (energy consumed) can be balanced among all the clusters formed.

The final phase is the data transmission phase. During this phase, plain nodes will transmit their data to their cluster head. Each cluster head will fuse all the data it receives from its plain nodes and then transmit one message to the base station.

By using the distance metrics and a candidate broadcast message, EECS can ensure that there is only one cluster head within a certain range with high probability. The number of clusters in EECS is constant. The energy consumed by nodes is balanced by rotating the role of the cluster head each round. This helps to prolong the network lifetime by having all nodes run out of energy at about the same time. This is also aided by considering the residual energy when selecting the cluster head for each round.

### The Clustering Protocol (CP)

The Clustering Protocol (CP) [5] is based on the covering problem using hexagons. Nodes can be in one of three states: unclustered, clustered, or cluster head. All nodes begin in the unclustered state. The base station becomes the initial cluster head and forms a hexagon around itself with itself as the center of the hexagon. The base station sends a message saying it is a cluster head. This message will be broadcast to all nodes within two hops of the new cluster head (the base station in this case). An unclustered node receiving a broadcast directly from a cluster head will join that cluster and change its state to clustered. Nodes in the clustered state will ignore cluster head broadcasts. An unclustered node receiving a cluster head broadcast message indirectly will compute that cluster head's position and orientation, calculates its distance to the center of that cluster head's hexagon, and starts a timer. If this node receives a message from another node in that cluster then it will join that cluster and change its state to clustered. If it does not receive another broadcast message and the timer expires, then it will become a cluster head itself. In order for all nodes in the cluster to be able to communicate with the cluster head in one hop, the hexagon's arm length is equal to the nodes' radio range.

There is little overhead in CP since there are not a lot of extra messages transmitted to maintain neighbor lists; the only messages transmitted are the cluster head broadcast messages. This allows CP to scale with network density. However, all nodes must know their location, typically from a GPS device. Also, residual energy is not a consideration in selecting the cluster head.

### 2.6.2.3 Neighbor-Based Algorithms

Topology and Energy Control Algorithm (TECA)

Topology and Energy Control Algorithm or TECA [1] will cluster nodes by using one-hop neighbor information. TECA utilizes five states for nodes: initial, sleeping, passive, bridge, and cluster head. All nodes begin in the initial state and start a timer. Nodes in the initial state will listen for data transmissions by other nodes. Based on overhearing these transmissions, a node will build a neighborhood table. A node will measure the link quality of each node in its neighborhood table based on signal strength of the overheard messages. When the initial timer expires the node will enter the passive state. Nodes in the passive state will continue to overhear messages and maintain their neighborhood table. Beacons are also used to help maintain each node's neighborhood table. Beacons are messages that are periodically transmitted by each node containing the node's ID, current state, residual energy, a timeout value, and its one-hop neighborhood information.

Any node not already assigned to a cluster will become a cluster head candidate. These candidate nodes will broadcast that it is a cluster head candidate. The node within the one-hop neighborhood with the most residual energy will become the cluster head. All other nodes in the cluster head's one-hop neighborhood will join that cluster. Each cluster head will start a timer and will remain a cluster head for that time period. When this timer expires then that node will try to find a cluster to join so it does not have to be a cluster head again and can conserve energy. Any node that is not a cluster head remains a passive node.

To maintain network connectivity, TECA also makes use of bridge nodes. Passive nodes start a timer and continue to listen to their neighbors. If a passive node hears a broadcast from at least two cluster heads, then it will become a bridge candidate. Bridge selection from all bridge candidate nodes is a complex process that aims to minimize packet loss between clusters and the number of bridge nodes selected. This process is based on graphs created between each node's two-hop neighborhoods. Each graph has different link costs and the minimum spanning tree is computed. If the passive node is not selected to be a bridge node, then when its timer expires it will go to sleep and will sleep for another specified time period. During this time, only cluster heads and bridge nodes will remain active.

By using one-hop neighborhood information, adjacent cluster heads are at most three hops apart but are never geographically close to each other. The role of the cluster head and bridge node is rotated and residual energy is considered in cluster head selected. These attributes of TECA help prolong the network lifetime. An advantage of TECA over many other clustering algorithms is the use of bridge nodes to maintain network connectivity. However, the use of bridge nodes also has a drawback in it requiring more nodes to remain active, which will use more energy and shorten the lifetime of the network.

Power-Efficient Gathering in Sensor Information Systems (PEGASIS)

Power-efficient gathering in sensor information systems or PEGASIS [10] was developed to improve the efficiency of cluster formation in LEACH. The key idea of PEGASIS is to consume less energy by having each node communicate with only one close neighbor node. This is done by forming a chain of all the nodes in the network. The chain is formed either by the base station computing the chain or by using the greedy algorithm. After the chain is formed, each node will transmit its data to the next node in the chain. The next node will fuse the data it received with its own data and forward this packet to the next node in the chain. The leader of the chain for that round will transmit the final fused data to the base station. When the round ends, a new leader is selected and the new round begins.

To help distribute the load, the role of the leader will be rotated among all the nodes. Only nodes that are a long distance from the base station will not serve as the leader. This is because transmitting data over a long range consumes a lot of energy and would counter-act the energy conservation of utilizing a clustering algorithm. Energy is conserved by having each node transmit and receive data from only one node in each round, and by having the node it receives from and transmits to be next to it in the chain making the transmissions a short distance. Only one node will transmit data to the base station in each round. This also conserves energy. A major drawback of PEGASIS is that it does not consider residual energy when selecting the leader for each round. Also, to form the best chain, it is necessary to have global knowledge of the network, such as the number of nodes and the position of each node.

## 2.7   Directions for Future Research

Research is expected to continue in all three areas of topology management. One area of possible research is to create an optimal clustering algorithm. This can be accomplished by combining the strengths of existing algorithms, eliminating the disadvantages, and relaxing the assumptions. Accomplishing all of this may prove to be a daunting task.

Any clustering algorithm developed should be distributed and should rotate the cluster head nodes so that all nodes will die at approximately the same rate. A distributed algorithm may utilize less energy than a centralized one since there will be less data transmitted (there will be no need to transmit control data, necessary for cluster organization and maintenance, between the base station and the nodes). During cluster head selection, residual energy should be one of the considerations. The cluster head should fuse the data received by the member nodes before sending it to the base station to help conserve energy. Maintaining a relatively short distance from each cluster head to the base station is also an important factor to consider. This will lessen the energy consumption required to transmit the data from the cluster head to the base station and prolong the life of each of those nodes. One possible way to shorten the distance from cluster head to base station is to have each cluster head transmit to the nearest base station if multiple base stations exist.

Noncluster head nodes in the network should select the best cluster to join based on a variety of information. One of these pieces of information should not be location information as that adds complexity and requires more energy consumption by the nodes. The nodes should join the cluster that allows them to consume the least amount of energy while achieving the goal of the application. The clusters should also be load balanced, again so that all nodes die at approximately the same time.

Most of the algorithms developed have assumptions. If some of these assumptions can be addressed, the new algorithm would be more robust. One of the most profound assumptions made by some algorithms that should be addressed is the transmission of data. An algorithm that considers errors in transmission and collisions would be an improvement over some existing algorithms. A new algorithm should also consider node failures in the network, including the failure of a cluster head.

An algorithm that can be developed with these characteristics would be a possible clustering algorithm that could be used in most WSNs. Research on such an algorithm persists.

## 2.8  Conclusions

Topology management is an essential part of Wireless Sensor Networks (WSNs), most often used to conserve energy while maintaining network connectivity. The three categories of topology management, topology discovery, sleep cycle management, and clustering, all have benefits to a WSN. Topology discovery allows the network manager to see various maps of the network, such as physical deployment or logical groupings. To conserve energy, redundant nodes can be put to sleep for a time period. Determining which nodes to put to sleep and for how long is the job of sleep cycle management algorithms. Most deployed WSNs will utilize some type of clustering algorithm, which will conserve energy by grouping the nodes and having a subset of the deployed nodes send application data to the base station.

Most of the topology management algorithms have been tested in simulation environments. This gives researchers insight into how the algorithm may behave when implemented. However, to fully test the functionalities of the algorithms discussed here, they should be implemented in testbeds and evaluated in WSN deployments.

## Terminologies

*Base station.* The node, typically a laptop or desktop that collects network management data and/or application data.

*Clustering.* Organizing the nodes of a network into groups or clusters.

*Covering problem.* A common research area in computer science that determines if a certain area is covered by various shapes, such as hexagons.

*Latency.* A delay; in networking it is a delay involved with transmitting a data packet between two nodes, which may include propagation delay, transmission delay, queuing delay, and processing delay components.

*Minimum virtual dominating set (MVDS).* The smallest number of nodes that must be deployed or awake in order to cover the application area.

*Neighbor list.* The nodes in the network that can be reached from one node in one hop.

*Network lifetime.* The amount of time the network will remain active, either containing at least one node or containing the minimum number of nodes necessary to cover the application area.

*Sleep cycle.* The time period when a node alternates between being awake and being asleep.

*Spatial reuse.* Allowing nodes that are not within radio range of each other to transmit at the same time.

*Topology.* The logical or physical organization of the nodes in the network.

## Questions

1. What is the advantage of a four color coloring scheme over a three color scheme?
2. What is a disadvantage of a distance-based clustering algorithm?
3. What is the primary goal of any topology management algorithm?
4. Why does maintaining a neighborhood list consume additional energy?
5. Which type of approach to topology discovery is the best? Explain.
6. Which clustering algorithm is the best approach? Explain.
7. Why not always deploy a heterogeneous network to take advantage of more robust nodes for cluster heads?
8. How can additional energy be conserved by any of the topology management algorithms discussed?

9. Why is requiring location information for topology management a disadvantage?
10. Which sleep cycle management algorithm results in the lowest latency?

# References

1. M. Busse, T. Haenselmann, and W. Effelsberg (2006) TECA: A topology and energy control algorithm for wireless sensor networks. International Symposium on Modeling, Analysis and Simulation of Wireless and Mobile Systems (MSWiM '06), Torremolinos, Malaga, Spain, ACM, October 2–6, 2006.
2. B. Chen, K. Jamieson, H. Balakrishnan, and R. Morris (2001) Span: An energy-efficient coordination algorithm for topology maintenance in ad hoc wireless networks. MobiCom 2001, Rome, Italy, pp. 70–84, July 2001.
3. B. Deb, S. Bhatnagar, and B. Nath (2001) A topology discovery algorithm for sensor networks with applications to network management. Technical Report dcs-tr-441, Rutgers University, May 2001.
4. B. Deb, S. Bhatnagar, and B. Nath (2003) Multi-resolution state retrieval in sensor networks. Proceedings of the First IEEE. 2003 IEEE International Workshop on Sensor Network Protocols and Applications, 2003, 11 May 2003, pp. 19–29.
5. A. Durresia, V. Paruchuri, and L. Barolli (2006) Clustering protocol for sensor networks. 20th International Conference on Advanced Information Networking and Applications, 2006 (AINA 2006), Volume 2. pp. 18–20, April 2006.
6. B. P. Godfrey and D. Ratajczak (2004) Naps: Scalable, robust topology management in wireless ad hoc networks. ISPN '04, Berkeley, CA, ACM, April 26–27, 2004.
7. W. R. Heinzelman, A. Chandrakasan, and H. Balakrishnan (2000) Energy-efficient communication protocol for wireless microsensor networks. Proceedings of the 33rd Annual Hawaii International Conference on System Sciences, Jan 4–7, 2000.
8. X. Hong and Q. Liang (2004) An access-based energy efficient clustering protocol for ad hoc wireless sensor network. 15th IEEE International Symposium on Personal, Indoor and Mobile Radio Communications, 2004, (PIMRC 2004). Sept. 5–8, 2004, Volume 2, pp. 1022–1026.
9. J. Kim, S. Kim, D. Kim, and W. Lee (2005) Low-energy localized clustering: an adaptive cluster radius configuration scheme for topology control in wireless sensor networks. IEEE 61st Vehicular Technology Conference, 2005. (VTC 2005). 30 May-1 June 2005. Volume 4. pp. 2546–2550.
10. S. Lindsey and C. S. Raghavendra (2002) PEGASIS: Power-efficient gathering in sensor information systems. IEEE Aerospace Conference Proceedings, 2002.
11. C. Schurgers, V. Tsiatsis, S. Ganeriwal, and M. Srivastava (2002) Topology management for sensor networks: Exploiting latency and density. MOBIHOC '02, Lausanne, Switzerland, ACM, June 9–11, 2002.
12. Y. Xu, S. Bien, Y. Mori, J. Heidemann, and D. Estrin (2003) Topology control protocols to conserve energy in wireless ad hoc networks. Technical Report 6, University of California, Los Angeles, Center for Embedded Networked Computing, January 2003.
13. Y. Xu, J. Heidemann, and D. Estrin (2001) Geography-informed energy conservation for ad hoc routing. Proceedings of the 7th annual International Conference on Mobile Computing and Networking, Rome, Italy, June 16–21, 2001, pp. 70–84.
14. M. Ye, C. Li, G. Chen, and J. Wu (2005) EECS: An energy efficient clustering scheme in wireless sensor networks. 24th IEEE International Performance, Computing, and Communications Conference, 2005, (IPCCC 2005), April 7–9, 2005, pp. 535–554.

# Chapter 3
# Coverage in Wireless Sensor Networks

Jennifer C. Hou, David K.Y. Yau, Chris Y.T. Ma, Yong Yang, Honghai Zhang, I-Hong Hou, Nageswara S.V. Rao, and Mallikarjun Shankar

**Abstract** Ad-hoc networks of devices and sensors with (limited) sensing and wireless communication capabilities are becoming increasingly available for commercial and military applications. The first step in deploying these wireless sensor networks is to determine, with respect to application-specific performance criteria, (i) in the case that the sensors are static, where to deploy or activate them; and (ii) in the case that (a subset of) the sensors are mobile, how to plan the trajectory of the mobile sensors. These two cases are collectively termed as the coverage problem in wireless sensor networks. In this chapter, we give a comprehensive treatment of the *coverage problem*. Specifically, we first introduce several fundamental properties of coverage that have been derived in the literature and the corresponding algorithms that will realize these properties. While giving insights on how optimal operations can be devised, most of the properties are derived (and hence their corresponding algorithms are constructed) under the *perfect disk* assumption. Hence, we consider in the second part of the chapter coverage in a more realistic setting, and allow (i) the sensing area of a sensor to be anisotropic and of arbitrary shape, depending on the terrain and the meteorological conditions, and (ii) the utilities of coverage in different parts of the monitoring area to be nonuniform, to account for the impact of a threat on the population, or the likelihood of a threat taking place at certain locations. Finally, in the third part of the chapter, we consider mobile sensor coverage, and study how mobile sensors may navigate in a deployment area to maximize threat-based coverage.

## 3.1 Introduction

Recent technological advances have led to the emergence of pervasive networks of small, low-power devices that integrate sensors and actuators with limited on-board processing and wireless communication capabilities. These sensor networks

D.K.Y. Yau (✉)
Purdue University, 305 N. University Street, West Lafayette, IN 47907
e-mail: yau@cs.purdue.edu

S. Misra et al. (eds.), *Guide to Wireless Sensor Networks*, Computer Communications and Networks, DOI: 10.1007/978-1-84882-218-4_3,
© Springer-Verlag London Limited 2009

open new vistas for many potential applications, such as environmental monitoring (e.g., traffic, habitat, and security), industrial sensing and diagnostics (e.g., factory, appliances), critical infrastructure protection (e.g., power grids, water distribution, waste disposal), and situational awareness for battlefield applications [1–4]. For these algorithms, the sensor nodes are deployed to cover the monitoring area. They collaborate with each other in sensing, monitoring, and tracking events of interests and in transporting acquired data, usually stamped with the time and position information, to one or more sink nodes.

There are usually two deployment modes in wireless sensor networks. On the one hand, if the cost of the sensors is high and deployment with a large number of sensors is not feasible, a small number of sensors are deployed in several preselected locations in the area. In this case, the most important issue is *sensor placement* – where to place the sensors in order to fulfill certain performance criteria. On the other hand, if inexpensive sensors with a limited battery life are available, they are usually deployed with high density (up to $20\,\text{nodes}/\text{m}^3$ [5]). The most important issue in this case is *density control* – how to control the density and relative locations of active sensors at any time so that they properly cover the monitoring area. (Another relevant issue is how to rotate the role of active sensors among all the sensors so as to prolong the network lifetime [6].) Although at first glance, sensor placement and density control are two different issues, they both boil down to the issue of determining a set of locations either to place the sensors or to activate sensors in the vicinity, with the objective of fulfilling the following two requirements: (i) *coverage*: a predetermined percentage of the monitored area is covered; and (ii) *connectivity*: the sensor network remains connected so that the information collected by sensor nodes can be relayed back to data sinks or controllers.

In this chapter, we consider the coverage issue in wireless sensor networks. We consider two operational modes.

**Case I: All sensor nodes are statically deployed.** We consider the issue of determining the minimum set of sensors required to cover the predetermined percentage of the area, assuming that each sensor node can monitor a certain area (e.g., a disk centered at the sensor and with the radius being the *sensing range* of the sensor) on a two-dimensional surface. As indicated in [7], if the radio range is at least twice the sensing range, complete coverage of a convex area implies connectivity among the set of active nodes. This condition actually holds for a wide spectrum of sensor devices that recently emerge [7], and as a result considering only the coverage issue is sufficient.

We approach the coverage issue along two research thrusts. We first introduce several fundamental properties that have been derived in the literature [7–10] and their corresponding algorithms that realize the properties [6, 11–17] (Sect. 3.2.1). Most of the efforts introduced in this thrust focus on minimizing the number of sensors, subject to the requirement of ($k$-)covering the entire monitoring area. While shedding insights on how optimal operations can be devised, most of the algorithms/analysis are derived under the *perfect disk* assumption. As revealed in several deployment efforts [18], the sensing range is in reality highly irregular because of variations in terrain/meteorological conditions. Moreover, while maximizing the geometric coverage is important, it makes more sense to quantify the utility of sensor coverage by considering its ability to manage potential *threats*. For example, a densely populated and poorly ventilated area should be classified as high risk under a chemical plume attack, and therefore receive priority attention in the sensor placement.

We consider in the second research thrust coverage in a more realistic setting (Sect. 3.4). In particular, the sensing area of a sensor can be anisotropic and of arbitrary shape, depending on the material released, its dosage fields and release patterns, the wind speed and direction, and the dispersion model. The expected risks of insufficient coverage (or utilities of coverage) in different parts of the monitoring area can also be nonuniform to account for the impact of a threat on the population, or the likelihood of a threat taking place at certain locations. Under this more general setting, we consider the issue of determining the minimum set of sensors required to minimize the threat.

**Case II: Sensor nodes are mobile.** In the case that some of the sensor nodes are mobile, we add another dimension of coverage – mobile sensor coverage (Sect. 3.5). Once sensors have been deployed in the area according to a sensor placement/density control algorithm, operating conditions may change to render the original results suboptimal or invalid. For example, sensors may fail or their sensing range may weaken, and obstacles may appear that affect a sensor's ability to cover its local area. Mitigating the effects of these unexpected situations could be solved by tasking a mobile sensor that navigates along a trajectory to minimize the detection time of events of interest. Specifically, the monitoring area is divided into a two-dimensional grid of cells. For each cell, the risk is defined as the steady-state presence probability of the event of interest (e.g., a chemical attack) in that cell. The distribution of threat in the area is characterized by a *threat profile*, which considers the impact of a realized risk on the area's population. We introduce a stochastic movement algorithm for sensors to achieve threat-based coverage, so that the cells are covered in proportion to their threat levels.

## 3.2 Background

### 3.2.1 Fundamental Properties of Coverage

Several researchers have carried out studies on the fundamental properties of coverage and sensor placement. We first summarize three representative results in this section. Then, to introduce the methodology taken by this line of research, we give a more detailed account of the first result in Sect. 3.3.

Zhang and Hou [7] focus on the sensor coverage problem of finding the *minimum number* of sensors that maintain full coverage. They prove that, given a region $R$ containing sensors, if each crossing point in $R$ is covered by at least one other sensor in $R$, then $R$ is completely covered. By a crossing point, they mean an intersection point of the two sensing disks of two neighboring sensors, or that of a sensing disk and the boundary of region $R$. They also derive optimal conditions between neighboring sensors for minimizing the number of sensors needed. On the basis of the optimal conditions, they then propose a fully decentralized and localized algorithm, called *Optimal Geographical Density Control* (OGDC), in large scale wireless sensor networks.

Wang et al. [10] take one step further and prove that if all the crossing points in the region $R$ are $k$-covered, then $R$ is $k$-covered. They then propose the *Coverage and Configuration Protocol* (CCP) in which each node collects neighboring information and then uses this information as an eligibility rule to decide if a node can sleep – if all the crossing points inside the sensing range are at least $k$-covered, then a node can be inactive. Huang and Tseng [19] consider the problem from another angle (crossing points vs. perimeter) and prove that an area is $k$-covered if each sensor in the network is $k$-perimeter-covered, where a $k$-perimeter-covered sensor has each point on the perimeter of its sensing disk covered by at least $k$ other sensors. They then devise an algorithm for determining the perimeter $k$-coverage (and hence the $k$-coverage of a region) and use it to determine the redundant sensors and schedule their inactive periods. However, to determine its redundancy, a sensor $s$ has to ask all the sensors within twice its sensing range to reevaluate the coverage of their perimeter without sensor $s$, making the complexity of the algorithm high.

### 3.2.2 Coverage Problem Formulations

Sensor coverage can be formulated as an optimization problem: Given the sensing range $R$ of sensors, how to place the sensors so that the number of sensors $N$ needed to cover the monitoring area is minimized?

We can also formulate the problem as: Given the number of available sensors $N$, how to place the sensors so that the sensing range $R$ needed to cover the monitoring area is minimized? This formulation is used when we are more concerned about the detection time or the energy consumption of sensing, which are highly related to the sensing range $R$. This is actually the well-known $N$-centering problem. A greedy algorithm can be used to solve the problem with an approximation ratio of 2 under the assumption of the triangle inequality [17]. Essentially the algorithm iteratively places a new sensor at a cell that is furthest away from the current set of sensors.

It can be proved that the above two optimization problems are equivalent in the sense that if there exists a solution algorithm to one problem, the other problems can be solved by invoking the solution algorithm a polynomial number of times, subject to the change of the approximation ratio.

## 3.3   Optimal Geographical Density Control (OGDC) and its Fundamental Base

Now to highlight the general methodology for the problem of finding the *minimum* number of sensor locations that maintain full coverage, we discuss in detail the Optimal Geographical Density Control (OGDC) method [7]. Implied in the coverage objective are two requirements. First, the set of sensors deployed or activated should completely cover the region $R$. To derive a sufficient condition for ensuring full coverage, we define a crossing as an intersection point of two circles (boundaries of disks) or that of a circle and the boundary of region $R$. A *crossing* is said to be *covered* if it is an *interior point* of a third disk. The following lemma from [20] pages 59 and 181 provides a sufficient condition for complete coverage. This condition is also necessary if we assume that the circle boundaries of any three disks do not intersect at a point. The assumption is reasonable as the probability of the circle boundaries of three disks intersecting at a point is zero, if all the sensors are randomly placed in a region with uniform distribution. Lemma 1 serves as an important theoretical basis for OGDC.

**Lemma 1.** *Suppose the size of a disk is sufficiently smaller than that of a convex region $R$. If one or more disks are placed within the region $R$, and at least one of those disks intersect another disk, and all the crossings in the region $R$ are covered, then $R$ is completely covered.*

The second requirement is that the set of sensors deployed or activated for coverage should be minimal. To derive conditions under which the second requirement is fulfilled, we first define the *overlap* at a point $x$ as the number of sensors whose sensing ranges cover the point minus $I_R(x)$, where

$$I_R(x) = \begin{cases} 1 & x \in R \\ 0 & \text{otherwise} \end{cases} \tag{3.1}$$

The overlap of sensing areas of all the sensors is then the integral of overlaps of the points over the area covered by all the sensors. Now we show in Lemma 2 that minimizing the number of active sensors is equivalent to minimizing the overlap of sensing areas of all the active sensor nodes.

**Lemma 2.** *If all sensor nodes (i) completely cover a region $R$ and (ii) have the same sensing range, then minimizing the number of working nodes is equivalent to minimizing the overlap of sensing areas of all the active nodes.*

*Proof.* See Appendix 1.                                                              □

Lemma 2 is important as it relates the total number of active sensor nodes to the overlapping areas between the active nodes. Since the latter can be readily measured from a local point of view, this greatly simplifies the task of designing a decentralized and localized sensor placement or density control algorithm.

### 3.3.1 Properties Under the Ideal Case

With Lemmas 1–2, we are now in a position to discuss how to minimize the overlap of sensing areas of all the sensor nodes. Our discussion is built upon the assumption that the region $R$ is large enough compared with the sensing range of each sensor node so that the boundary effects can be ignored. By Lemma 1, in order to totally cover the region $R$, some sensors must be placed inside region R and their coverage areas may intersect one another. If two disks $A$ and $B$ intersect, at least one more disk is needed to cover their crossing points. Consider, for example, in Fig. 3.1, disk $C$ is used to cover the crossing point $O$ of disks $A$ and $B$. To minimize the overlap while covering the crossing point $O$ (and its vicinity not covered by disks $A$ and $B$), disk $C$ should intersect disks $A$ and $B$ at the point $O$; otherwise, one can always move disk $C$ away from disks $A$ and $B$ to reduce the overlap.

Given that two disks $A$ and $B$ intersect, we now investigate the number of disks needed, and their relative locations, to cover a crossing point $O$ of disks $A$ and $B$ and at the same time minimize the overlap. Take the case of three disks (Fig. 3.1) as an example. Let $\angle PAO = \angle PBO \equiv \alpha_1$, $\angle OBQ = \angle OCQ = \alpha_2$, and $\angle OCR = \angle OAR = \alpha_3$. We consider two cases: (i) $\alpha_1$, $\alpha_2$, $\alpha_3$ are all variables; and (ii) $\alpha_1$ is a constant but $\alpha_2$ and $\alpha_3$ are variables. Case (i) corresponds to the case where we can choose all the node locations, while case (ii) corresponds to the case where two nodes (A and B) are already fixed and we need to choose the position of a third node C to minimize the overlap. Both of the above two cases can be extended to the general situation in which $k$-2 additional disks are placed to cover one crossing point of the first two disks (that are placed on a two-dimensional plane), and $\alpha_i$, $1 \le i \le k$, can be defined accordingly. Again, the boundaries of all disks should intersect at point O to reduce the overlap. In the following discussion, we assume for simplicity that the sensing range $r = 1$. Note, however, that the results still hold when $r \ne 1$.

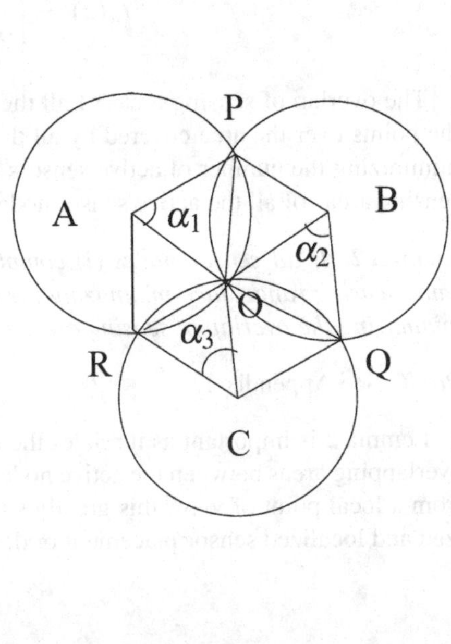

Fig. 3.1 An example that demonstrates how to minimize the overlap while covering the crossing point $O$

*Case 1:* $\alpha_i, 1 \le i \le k$, are all variables.

We first prove the following Lemma.

**Lemma 3.**

$$\sum_{i=1}^{k} \alpha_i = (k-2)\pi, \tag{3.2}$$

*Proof.* See Appendix 2. □

Now the overlap between the ith and *(i mod k)+1*th disks (which are called adjacent disks) is $(\alpha_i - \sin\alpha_i), 1 \le i \le k$. If we ignore the overlap caused by non-adjacent disks, then the total overlap is $L = \sum_{i=1}^{k} \alpha_i - \sin\alpha_i$. The coverage problem can be formulated as

**Problem 1.**

$$\min \sum_{i=1}^{k} (\alpha_i - \sin\alpha_i)$$

$$s.t. \sum_{i=1}^{k} \alpha_i = (k-2)\pi. \tag{3.3}$$

The Lagrangian multiplier method can be used to solve the above optimization problem. The solution is $\alpha_i = (k-2)\pi/k$, $i = 1, 2, \ldots, k$ and the resulting minimum overlap using $k$ disks to cover the crossing point $O$ is

$$L(k) = (k-2)\pi - k\sin\left(\frac{(k-2)\pi}{k}\right) = (k-2)\pi - k\sin\left(\frac{2\pi}{k}\right).$$

Note that the overlap per disk

$$\frac{L(k)}{k} = \pi - \frac{2\pi}{k} - \sin\left(\frac{2\pi}{k}\right) \tag{3.4}$$

monotonically increases with $k$ when $k \ge 3$. Moreover when $k = 3$ (which means that we use one disk to cover the crossing point), the optimal solution is $\alpha_i = \pi/3$ and there is no overlap between *nonadjacent* disks. When $k > 3$, the overlap per disk is always higher than that in the case of $k = 3$, even if we ignore the overlaps between *nonadjacent* disks. This implies that using one disk to cover the crossing point and its vicinity is optimal in the sense of minimizing the overlap. Moreover, the centers of the three disks should form an equilateral triangle with edge $\sqrt{3}$. We state the above result in the following theorem.

**Theorem 1.** *To cover one crossing point of two disks with the minimum overlap, only one disk should be used and the centers of the three disks should form an equilateral triangle of side length $\sqrt{3}r$, where r is the radius of the disks.*

*Case 2*: $\alpha_1$ is a constant, while $\alpha_i, 1 \leq i \leq k$, are variables.

In this case the problem can still be formulated as in Problem 1, except that $\alpha_1$ is fixed. The Lagrangian multiplier method can again be used to solve the problem, and the optimal solution is $\alpha_i = ((k-2)\pi - \alpha_1)/(k-1), 2 \leq i \leq k$. Again a similar conclusion can be drawn that using one disk to cover the crossing point gives the minimum overlap. We state the result in the following theorem.

**Theorem 2.** *To cover one crossing point of two disks whose locations are fixed (i.e., $\alpha_1$ is fixed in Fig. 3.1), only one disk should be used and $\alpha_2 = \alpha_3 = (\pi - \alpha_1)/2$.*

In summary, to cover a large region $R$ with the minimum overlap, one should ensure that (i) at least one pair of disks intersect; (ii) the crossing points of any pair of disks are covered by a third disk; (iii) if the locations of any three sensor nodes are adjustable, then as stated in Theorem 1 the three nodes should form an equilateral triangle of side length $\sqrt{3}\,r$. If the locations of two sensor nodes $A$ and $B$ are already fixed, then as stated in Theorem 2, the third sensor node should be placed on the line that is perpendicular to the line connecting nodes $A$ and $B$ and has a distance $r$ to the intersection of the two circles (i.e., the optimal point in Fig. 3.2 is $C$). These conditions are optimal for the coverage problem in the ideal case.

The notion of overlap can be extended to the heterogeneous case in which sensors have different sensing ranges. Specifically, Theorem 1 and 2 can be generalized to the heterogeneous case. The interested reader is referred to [7] for a detailed account.

### 3.3.2 Optimal Geographical Density Control Algorithm

Now we introduce a completely localized density control algorithm, called OGDC, that makes use of the optimal conditions derived above. Conceptually, OGDC attempts to select as active nodes the sensor nodes that are closest to the optimal locations.

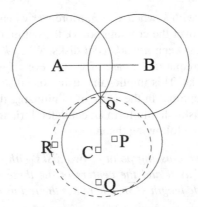

**Fig. 3.2** Although $C$ is the optimal place to cover the crossing $O$ of $A$, $B$, there is no sensor node there. The node closest to $C$, $P$, is selected to cover the crossing $O$

For clarity of presentation, we assume that (i) each node is aware of its own position and (ii) all sensor nodes are time synchronized. The first assumption is not impractical, as many research efforts have been made to address the localization problem [21–23]. The second assumption is made to facilitate the description of the algorithm. A more general algorithm that operates without the assumption can be found in [7].

At any time, a node is in one of the three states: "UNDECIDED," "ON," and "OFF." Time is divided into rounds. Each round has two phases: the *node selection phase* and the *steady-state phase*. At the beginning of the node selection phase, all the nodes wake up, set their states to "UNDECIDED," and carry out the operation of selecting working nodes. By the end of this phase, all the nodes change their states to either "ON" or "OFF". In the steady-state phase, all nodes keep their states fixed until the beginning of the next round. The length of each round is so chosen that it is much larger than that of the node selection phase but much smaller than the average sensor lifetime. As shown in [7], the time it takes to execute the node selection operation for networks of size up to 1,000 nodes in an area of $50 \times 50\,\mathrm{m}^2$ (with timer values appropriately set) is usually well below 1 s and most nodes can decide their states (either "ON" or "OFF") in less than 0.2 s, from the time instant when at least one node volunteers to be a starting node. The interval for each round is usually set to approximately hundreds of seconds, and the overhead of density control is small ($<1\%$).

The node selection phase in each round commences when one or more sensor nodes volunteer to be starting nodes. For example, suppose node $A$ volunteers to be a starting node in Fig. 3.2. Then one of its neighbors with an (approximate) distance of $\sqrt{3}\,r$, say node $B$, will be "selected" to be an active node. To cover the crossing point of disks $A$ and $B$, the node whose position is closest to the optimal position $C$ (e.g., node $P$ in Fig. 3.2) will then be selected, in compliance with Theorem 2, to become an active node. The process continues until all the nodes change their states to either "ON" or "OFF," and the set of nodes with state "ON" form the working set. As a node probabilistically volunteers itself to be a starting node (with a probability that is related to its remaining power) in each round, the set of working sensor nodes is not likely to be the same in each round, thus ensuring uniform (and minimum) power consumption across the network, as well as complete coverage and connectivity. The interested reader is referred to [7] for a detailed description of OGDC.

### 3.3.3 Performance of OGDC

To validate and evaluate the proposed design of OGDC, a simulation study has been conducted in [7] (with ns-2 with the CMU wireless extension) in a $50 \times 50\,\mathrm{m}^2$ region where up to 1,000 sensors are uniformly randomly distributed.

In addition to evaluating OGDC, the study also evaluates the performance of the PEAS algorithm proposed in [6], the CCP algorithm in [10], and a hexagon-based GAF-like algorithm. The hexagon based GAF-like algorithm is built upon GAF [24] and operates as follows. The entire region is divided into square grids and one node is selected to be awake in each grid. To maintain coverage, the grid size must be less than or equal to $r_s/\sqrt{2}$. Thus, for a large area with size $l \times l$, it requires $2l^2/r_s^2$ nodes to operate in the active mode to ensure complete coverage. To maintain coverage in hexagonal grids, the side length of each hexagon is at most $r_s/\sqrt{2}$, and it requires $8l^2/(3\sqrt{3}r_s^2) \approx 1.54l^2/r_s^2$ working nodes to completely cover a large area with size $l \times l$.

In the simulation study, the energy model in [6] is used, where the power consumption ratio for transmitting, receiving (idling), and sleeping is 20:4:0.01. One unit of energy (power) is defined as that required for a node to remain idle for 1 s. Each node has a sensing range of $r_s = 10$ m, and a lifetime of 5,000 s if it is idle all the time. The tunable parameters in OGDC are set as follows: the round time is set to 1,000 s, the power threshold $Pt$ is set to the level that allows a node to be idle for 900 s, the timer values are set to, respectively, $T_d = 10$ ms, $T_s = 1$ s, and $T_e = T_s/5 = 200$ ms, $t_0$ is set to the time it takes to send a power-on packet, 6.8 ms (the wireless communication capacity is 40 Kbps, the packet size is 34 bytes).

The coverage is measured as follows: the area is divided into $50 \times 50$ square grids, and a grid is considered covered if the center of the grid is covered. Coverage is then defined as the ratio of the number of grids that are covered by at least one sensor to the total number of grids.

Figure 3.3 shows the number of working nodes and coverage versus the number of sensor nodes deployed in the network. Both metrics are measured after the density control process is completed. Under most cases, OGDC takes less than 1 s to perform density control in each round, while PEAS [6] and CCP [10] may take up to 100 s. As shown in Fig. 3.3, OGDC needs only half as many nodes to operate in the active mode when compared with the hexagon-based GAF-like algorithm, but achieves almost the same coverage (in most cases OGDC achieves more than 99.5% coverage). Moreover, the number of working nodes required under OGDC modestly increases with the number of sensor nodes deployed, while both PEAS and CCP incur a 50% increase in the number of working nodes, when the number of sensor nodes deployed in the network increases from 100 to 1,000. Another observation is that when the number of working nodes becomes very large, the coverage ratio of CCP actually decreases. This is because a large number of message exchanges are required in CCP to maintain neighborhood information. When the network density is high, packets incur collision more often and the neighborhood information may be inaccurate. In contrast, in OGDC each working node sends out at most one power-on message in each round, and as a result the packet collision problem is not so serious.

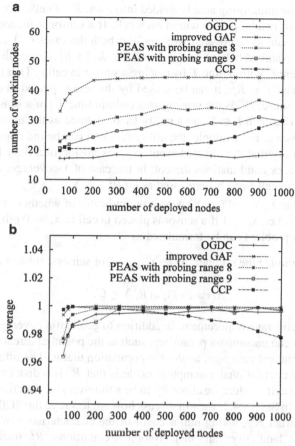

**Fig. 3.3** # of working nodes and coverage versus # of sensor nodes in a $50 \times 50\,\text{m}^2$ area. **a** # of working nodes vs. # of deployed nodes. **b** Coverage vs. # of deployed nodes

## 3.4  Sensor Placement in Realistic Environments

While all the above studies give insightful properties of ($k$-)coverage and shed light on designing coverage algorithms for full ($k$-)coverage, they all make the *perfect disk* assumption. As a result, it is not clear whether or not these results can be readily applied to the case of highly irregular sensing disks. In this section, we consider sensor placement in a more realistic setting – where to place sensors in order to fulfill certain performance criteria, *subject to the number of sensors to be deployed, the distribution of threats, the terrain, land cover and meteorological conditions, and the population distribution.* The performance criteria are either to minimize the maximal detection time (e.g., the time interval from the instant when a dirty bomb explodes to the instant the explosion is detected) or to maximize the population evacuation time (e.g., the time interval between the detection time to the time instant the plume reaches a populated area).

Specifically, the monitoring area is divided into a set, $X$, of cells. We assume that at most one sensor can be placed within each cell. If a sensor is placed in the cell, the whole cell is said to be covered. We consider both the cases of 1-coverage and $k$-coverage (to be defined below). For each cell $i \in X$, let $R_i^T$ denote the set of cells that can be "covered" within time $T$ by placing a sensor in cell $i$. That is if an event occurs in some cell $j \in R_i^T$, it can be sensed by the sensor placed in cell $i$ within time $T$. In some sense, $R_i^T$ is the sensing area (within time $T$) of a sensor placed in cell $i$. Also, for each cell $i \in X$, let a utility $U_i$ be defined as the utility gained by having cell $i$ covered. For example, the utility function can be the population in an area, the probability that the targeted event (e.g., explosion of a dirty bomb) takes place in this area, or combinations thereof. In the case of 1-coverage, the utility of placing a sensor in cell $i$ can be expressed as $U(R_i^T) = \sum_{j \in R_i^T} U_j$.

Let the variable $x_i (\forall$ cell $i \in X)$ denote the indicator of whether or not a sensor is placed in cell $i$, i.e., $x_i = 1$ if a sensor is placed in cell $i$; $x_i = 0$ otherwise. Now the optimization problem can be formulated as

**Problem 2.** Minimize the number, $N = \sum_{i \in X} x_i$, of sensors, subject to

$$U(\cup_{i \in X \wedge x_i = 1} R_i^T) \geq C, \tag{3.5}$$

where $C$ is the coverage requirement. In addition to geometric coverage, the coverage requirement can encompass parameters such as the potential threats that arise in the case of insufficient coverage, and/or the population that will be affected.

Note that the conventional assumption made is that $R_i^0$ is a disk centered at the sensor (placed in cell $i$). Here we allow $R_i^T$ to be a time-varying function of $T$ and of arbitrary shape. In Sect. 3.4.2, we will discuss how we leverage the SCIPUFF model [25, 26] to construct $R_i^T$, taking into account of the characteristics of the released material, terrain, land cover, and meteorological conditions. $R_i^T$ thus constructed will then be fed into the solution algorithm as input. Note also that conventionally $U(\cdot)$ is a uniform function and the utility reduces to geometric sensor coverage. As mentioned in Sect. 3.1, the utility function can be so defined that it quantifies the potential threats reduced or the potential benefits gained by having cell $i$ covered. In Sect. 3.4.2, we will use the real-life population distribution as $U(\cdot)$ to evaluate our proposed algorithms.

In the case of $k$-coverage, the utility of a cell takes effect only if the cell is covered by at least $k$ sensors. In other words, a cell $i$ is considered to be covered only when $\sum_{j:i \in R_j^T} x_j \geq k$. Note that $k$-coverage is required in the case of *inverse/forward prediction* in which the origin of the event (e.g., a plume) is inferred as well as the future regions to be affected by the dispersion is predicted. In this case, the information multiple sensors have gathered will be correlated and fed into certain inverse/forward algorithms. The optimization problem can then be formulated as

**Problem 3.** Minimize the number, $N = \sum_{i \in X} x_i$, of sensors, subject to

$$\sum_{i \in X} U_i \cdot I \left\{ \sum_{j:i \in R_j^T} x_j \geq k \right\} \geq C, \tag{3.6}$$

where $I(\cdot)$ is the indicator function. Note that the constraint in Problem 2 is not a linear expression. In Sect. 3.4.1, we will discuss methods to transform $I(\cdot)$ into a set of linear constraints.

## 3.4.1   Solutions to Problems 1 and 2

In this section, we discuss solution algorithms to Problems 1 and 2. In the case of 1-coverage, the problem (Problem 1) reduces to the *weighted partial set cover* problem, and we introduce a *logC* approximation algorithm. In the case of $k$-coverage, we first discuss a special case in which the coverage requirement is stringent and full $k$-coverage is required. In this case, we can further simplify the formulation of Problem 2 to a linear program. In the more general case, the formulation of Problem 2 can only be reduced to an integer program. We present one algorithm that is built upon the algorithm for partial 1-coverage to solve the problem.

### 3.4.1.1   Solution Algorithm for Problem 1

Algorithm 1 gives the *LogC-Partial-1* algorithm. The algorithm finds the cell $i^*$ with the highest utility $U_{i^*}$, and marks $x_{i^*} = 1$ to denote that a sensor will be placed in the cell. Then the cell $i^*$ is removed from $X$, the coverage of each cell $i \in X$ is updated as $R_i^T = R_i^T \setminus R_{i^*}{}^T$, and the coverage requirement is updated as $C = C_i - U(R_{i^*}{}^T)$. The process repeats until either the coverage requirement is satisfied ($C \leq 0$) or all the cells have been placed with sensors ($X = \emptyset$).

---

**Algorithm 1** LogC-Partial-1 $(X, \{R_i^T\}, U, w, C)$

---

1: $Y = \emptyset$
2: **while** $C > 0$ AND $X \neq \emptyset$ **do**
3:     $i^* = argmax_{i \in X} U(R_i^T)$
4:     $Y = Y \cup \{i^*\}$
5:     $X = X \setminus \{i^*\}$
6:     $C = C - U(R_{i^*}^T)$
7:     for each $i \in X$, $R_i^T = R_i^T \setminus R_{i^*}^T$
8: **end while**
9: return $Y$

---

**Theorem 3.** *The algorithm LogC-Partial-1 has an approximation factor of logC, given the range of $U(\cdot)$ is integer.*

*Proof.* Suppose the optimal placement requires $N_{OPT}$ sensors. Let $i$ be the ith sensors placed by *LogC-Partial-1*. Let $A_i$ be the cells that are covered by the ith sensor, and have not been covered by any $j < i$ sensor. Basically, $A_i$ is a set of new cells covered in the ith iteration. Let $C_i$ be the coverage requirement left to be met by the ith iteration, and $C_0 = C$. Then among cells that are not covered yet, one of those $N_{OPT}$ sensors in the optimal placement at least can cover $C_{i-1}/N_{OPT}$ amount of utility. *LogC-Partial-1* picks the sensor that has the largest utility coverage, and thus $U(A_i)$ is at least $C_{i-1}/N_{OPT}$. Therefore,

$$\sum_{i=1}^{N_{OPT}} U(A_i) \geq \sum_{i=1}^{N_{OPT}} \frac{C_{i-1}}{N_{OPT}} \geq \sum_{i=1}^{N_{OPT}} \frac{C_{OPT}}{N_{OPT}} = C_{OPT} = C - \sum_{i=1}^{N_{OPT}} U(A_i). \qquad (3.7)$$

Thus we have $\sum_{i=1}^{N_{OPT}} U(A_i) \geq C/2$, which means *LogC-Partial-1* can use $N_{OPT}$ sensors to meet at least half of the coverage requirement. Therefore, *LogC-Partial-1* totally needs at most $N_{OPT} \cdot \log C$ sensors to meet the coverage requirement $C$.

### 3.4.1.2  Solution Algorithm for the Full $k$-Coverage Problem

Recall that the constraint in Problem 2 is not a linear expression, because of the indicator function. When the coverage requirement is stringent, i.e., $C = \sum_{i \in X} U_i$, the entire monitoring area has to be $k$-covered and the indicator function can be readily removed. That is, (3.6) can be reduced to $\sum_{j:i \in R_j^T} x_j \geq k$, and Problem 2 reduces to

$$\min \sum_{i \in X} x_i, \qquad (3.8)$$

$$s.t. \sum_{j:i \in R_j^T} x_j \geq k \quad \forall i \in X. \qquad (3.9)$$
$$x_i \in \{0, 1\}$$

Because in general integer programs are NP-hard, we relax the above integer program into a linear program by replacing the last constraint with

$$0 \leq x_i \leq 1 \qquad (3.10)$$

and solve the linear program (named as *Full-k-LP*) in polynomial time. Now the remaining issues are how to construct a feasible solution for the integer program from that of the linear program, and how good the constructed solution to the integer program is. We answer both issues below:

**(1) Constructing a feasible solution for the integer program**
To construct a feasible solution $\{x_i\}$ for the original integer program based the solution $\{\hat{x}_i\}$ returned by the linear program, we define the maximum number of sensing areas by which a cell can be covered as $F = \max_{i \in X} |\{j : i \in R_j^T\}|$, where $| \cdot |$ is the cardinality function. Note that only when $k \leq F$, the $k$-coverage problem has a solution. We assign $x_i = 1$ if $\hat{x}_i \geq 1/(F - k + 1)$; and $x_i = 0$ otherwise.

**Theorem 4.** *The solution $\{x_i\}$ constructed from the solution $f\{\hat{x}_i\}$ obtained from the linear program $(x_i = 1$ if $\hat{x}_i \geq 1/(F - k + 1)$; and $x_i = 0$ otherwise) is a feasible solution to the original integer program (3.9).*

*Proof.* To prove that $\{x_i\}$ is a feasible solution to the original integer program, one needs to show $\sum_{j:i\in R_j^T} x_j \geq k$. This can be proved by contradiction. For some $i \in X$, assume that in $\{\hat{x}_j : i \in R_j^T\}$, $P_i$ elements are no less than $1/(F-k+1)$. Let $O_i \equiv |\{\hat{x}_j : i \in R_j^T\}|$. Then $(O_i - P_i)$ elements in $\{\hat{x}_j : i \in R_j^T\}$ is less than $1/(F-k+1)$. If $P_i \leq k-1$, the following inequality holds

$$\sum_{j:i\in R_j^T} \hat{x}_j < \{O_i - P_i\}\frac{1}{F-k+1} + P_i = O_i\frac{1}{F-k+1} + P_i\frac{F-k}{F-k+1}$$

$$\leq F\frac{1}{F-k+1} + (k-1)\frac{F-k}{F-k+1} = k,$$

which contradicts $\sum_{j:i\in R_j^T} \hat{x}_j \geq k$ (recall that $\{\hat{x}_i\}$ is a feasible solution for the relaxed linear program). Hence $P_i > k-1$ and hence $\sum_{j:i\in R_j^T} x_j \geq k$. $\qquad\square$

**(2) Deriving the approximation ratio of the constructed feasible solution**
Now we discuss the approximation factor of the constructed feasible solution.

**Theorem 5.** $\sum_{i\in X} x_i \leq (F-k+1)\sum_{i\in X} x_i^*$, *where* $\{x_i^*\}$ *is the optimal solution for the integer program and* $\{x_i\}$ *is the solution constructed from that of the linear program.*

*Proof.* First, $\sum_{i\in X} x_i \leq \sum_{i\in X} x_i^*$ because the solution space of the integer program is a subset of the solution space of the relaxed linear program. Thus, the following inequality holds:

$$\sum_{i\in X} x_i \leq \sum_{i\in X}((F-k+1)\cdot\hat{x}_i) = (F-k+1)\sum_{i\in X}\hat{x}_i$$

$$\leq (F-k+1)\sum_{i\in X} x_i^*, \tag{3.11}$$

where the first inequality follows from the construction rule of the feasible solution, i.e., $x_i = 1$ if $\hat{x}_i \geq 1/(F-k+1)$; and $x_i = 0$ otherwise. An example can be carefully constructed to show that $\sum_{i\in X} x_i = (F-k+1)\sum_{i\in X} x_i^*$, i.e., $(F-k+1)$ is a tight approximation ratio. $\qquad\square$

### 3.4.1.3 Solution Algorithm for Problem 2

In the general case, the indicator function in Problem 2 can be "removed" by utilizing the property $I\{x \geq k\} = \max\{0, \min\{1, x-k+1\}\}$. Furthermore, $y = \max\{x_i, x_j\}$ can be replaced by the following constraints:

$$y \geq x_i, \quad y \geq x_j,$$
$$y - x_i \leq c_i M, \quad y - x_j \leq (1-c_i)M,$$
$$c_i \in \{0, 1\},$$

where $M$ is a sufficiently large positive constant. The first pair of constraints ensures that $y$ is no less than either $x_i$ or $x_j$. The second pair of constraints ensures that either $y = x_i$ or $y = x_j$, depending on whether the variable $c_i$ is 0 or 1.

Similarly, $y = \min\{x_i, x_j\}$ can be replaced by the following constrains:

$$y \le x_i, \qquad y \le x_j,$$
$$x_i - y \le c_i M, \qquad x_j - y \le (1 - c_i)M,$$
$$c_i \in \{0, 1\},$$

Therefore Problem 2 can be reduced to the following integer program (named as *Partial-k-IP*):

$$Min \sum_{i \in X} x_i$$

$$s.t. \sum_{i \in X} U_i \cdot y_i \ge C,$$
$$y_i \ge 0, \qquad y_i \ge z_i,$$
$$y_i \le c_i F, \qquad y_i - z_i \le (1 - c_i)F, \qquad (3.12)$$
$$z_i \le 1, \qquad z_i \le \sum_{j:i \in R_j^T} x_j - k + 1,$$
$$1 - z_i \le d_i F,$$
$$\sum_{j:i \in R_j^T} x_j - k + 1 - z_i \le (1 - d_i)F,$$
$$x_i, c_i, d_i \in \{0, 1\}.$$

Unfortunately converting the above integer problem into the linear program by enforcing $0 \le x_i \le 1$, $0 \le c_i \le 1$, and $0 \le d_i \le 1$ and constructing the corresponding solution for the original integer program does not always yield a feasible solution. Actually, allowing $0 \le c_i \le 1$ and $0 \le d_i \le 1$ results in the optimal $\sum_{i \in X} x_i$ equal to zero. Hence, in what follows we discuss a heuristic algorithm based on the algorithm proposed above for partial 1-coverage.

One-Incremental Algorithm for Partial $k$-Cover

One straightforward solution for partial $k$-coverage is to perform the 1-coverage algorithm $k$ times. The pseudo code of the one-incremental algorithm is given in Algorithm 2. By the end of the $(r-1)$th invocation of the 1-coverage algorithm, it is possible that some cells have already been at least $r$-covered, denoted as $X' = \{i \in X : I(\sum_{j:i \in R_j^T} x_j \ge r) = 1\}$. Therefore, in the rth invocation of the 1-coverage algorithm, the utility coverage requirement can be reduced by $\sum_{i \in X'} U_i$. Also, the coverage utility gain for placing a sensor in cell $i$ in the $k$-coverage case is

$$U(R_i^T, k, Y) = \sum_{j \in R_i^T} U_j \cdot I(|\{h : h \in Y \wedge j \in R_h^T\}| = k - 1), \qquad (3.13)$$

which is the total utility of cells that are in $R_i^T$ and have already been exactly $(k - 1)$ covered by the placement $Y$. Hence, if one sensor is placed at cell $i$, $U(R_i^T, k, Y)$ would be the utility gain with respect to $k$-coverage.

### Redundancy Removal

The above heuristic algorithm acts in a greedy manner by choosing the cells that contribute the most utility. It is possible that in the resulting placement, some sensors are redundant, in the sense that their removal will not result in the failure to fulfill the utility coverage requirement. These are the sensors we should remove after invoking *One-Incremental*. The pseudo code of the redundancy removal procedure is given in Algorithm 3. It operates in a greedy manner. Let $Y$ denote the set of cells returned by either *One-Incremental* or *Partial-1+Full-k*. For $\forall\ i \in Y$, the utility loss after the removal of the sensor in cell $i$ is exactly $U(R_i^T, k, Y)$, which equals the total utility of cells in $R_i^T$ that are exactly $k$-covered by $Y$. Thus, removing $i$ from $Y$ results in that amount of utility loss. Iteratively, one searches for the cell $i \in Y$ with the smallest $U(R_i^T, k, Y)$. If $Y \setminus \{I\}$ can still satisfy the requirement $C$, cell $i$ is removed from $Y$.

---

**Algorithm 2** One-Incremental $(X, \{R_i^T\}, U, C)$

---

1: $Y = \emptyset$
2: **for** $r = 1; r \le k; r + +$ **do**
3:    $X' = \{i \in X \cup Y : i$ is at least $r$-covered by $Y\}$
4:    $C' = C - \sum_{i \in X'} U_i$
5:    **while** $C' > 0$ AND $X \ne \emptyset$ **do**
6:       $i^* = argmax_{i \in X} U(R_i^T, r, Y)$
7:       $Y = Y \cup \{i^*\}$
8:       $X = X \setminus \{i^*\}$
9:       $C' = C' - U(R_{i^*}^T, r, Y)$
10:    **end while**
11: **end for**
12: **return** $Y$

---

---

**Algorithm 3** Redundancy Removal $(X, Y, \{R_i^T\}, U, C)$

---

1: **while** $Y \ne \emptyset$ **do**
2:    $i^* = argmin_{i \in Y} U(R_i^T, k + 1, Y)$
3:    **if** $\sum_{i \in X} U_i \cdot I\{|\{h : h \in Y \setminus \{i^*\} \wedge i \in R_h^T\}| \ge k\} \ge C$ **then**
4:       $Y = Y \setminus \{i^*\}$
5:    **else**
6:       break;
7:    **end if**
8: **end while**
9: **return** $Y$

---

### 3.4.2  Gathering and Computing the Input Data

Data gathering and computation to prepare input for the sensor placement algorithms comprise a major part of the sensor placement process. Recall that the most important input to Problems 1 and 2 that characterizes the physical phenomena is the set, $R_i^T$, of cells that can be covered within time $T$ by placing a sensor in cell $i$. In this section, we discuss how one can leverage the SCIPUFF model to calculate $R_i^T$, taking into account of the characteristics of the released material, terrain, land cover, and meteorological conditions.

Calculation of $R_i^T$ is affected by the following parameters: released material (characteristics of the material such as the decay rate and deposition velocity, the release function), terrain, and meteorological conditions. The former can be obtained from the *GLOBE* database [27] by National Geophysical Data Center. GLOBE contains elevation data for the whole world at a latitude-longitude grid spacing of 30 arc-seconds. On the other hand, a useful representation of the meteorological conditions at a location is a *wind rose*. A wind rose gives an information-laden view of how wind *speed* and *direction* are typically distributed at a particular location. Specifically, it specifies wind direction and speed pairs and their percentage of occurrence. The wind rose that is most used is produced by the National Resources Conservation Service (NRCS). NRCS uses data from the Solar and Meteorological Surface Observation Network (SAMSON) that consists of hourly observations from 1961 through 1990 at 237 National Weather Service stations in the United States, Guam, and Puerto Rico.

Given a detection time $T$, the dispersion in a cell that results from a release is computed with the use of SCIPUFF [25, 26]. The dosage field resulting from a dispersion computation is contoured by the exposure levels. After the dispersion contours that result from the release in every cell is obtained, one can compute $R_i^T$ as follows: Let the threshold of the dosage level required for a sensor to detect a plume activity is *Th*. A cell $j$ is added into $R_i{}^T$ if cell $i$ is contained in the contour of the dosage level $\geq Th$ *and* the dispersion contours result from a release in cell $j$ within time $T$.

### 3.4.3  Performance Evaluation

The sensor placement algorithm One-Incremental has been evaluated in the real setting of Port of Memphis, with the objective of protecting people in Memphis and its vicinity against chemical plume attacks. Also, both random placement and grid placement are used as baseline algorithms and their performance (with the use of the same number of sensors as *One-Incremental*) are compared against *One-Incremental*. The coordinate (Longitude, Latitude) for the lower left corner of the monitoring area is (−90.25E, 34.85N), while that for the upper right corner is (−89.75E, 35.35N). The width and length of the monitoring area are both 0.5 arc-degree, which is about $45 \times 55 \, \text{km}^2$. Figure 3.4a is the satellite picture of the area

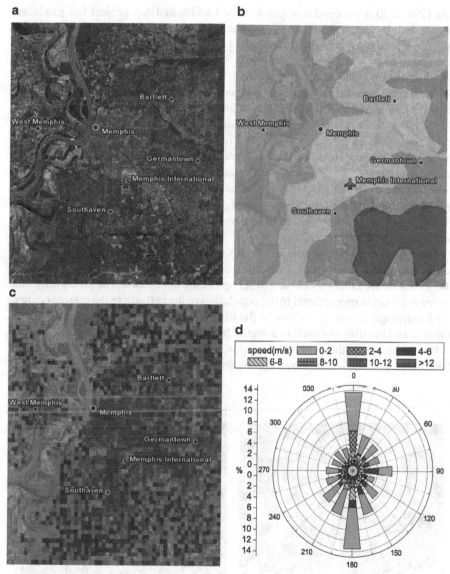

**Fig. 3.4** The terrain, population, and meteorological conditions in Port of Memphis and its vicinity. (**a**) Satellite picture (**b**) Terrain (**c**) Population Distribution and (**d**) Wind rose

provided by Google Earth [28]. The area is divided into $60 \times 60$ cells, and each cell is $0.5' \times 0.5'$ in arc-minute, and $750 \times 917\,m^2$. Figure 3.4c shows the terrain of the monitoring area.

Because the objective is to protect people in Memphis and its vicinity against chemical plume attacks, the utility function is defined to be the population distribution. To obtain the population in each cell, one can leverage the LandScan 2005 data [29] at 30 arc-second resolution. (The LandScan USA project has produced day- and night-time high resolution population distributions at 3 arc-second resolution for some cities, including Memphis, but these data have yet to be vetted and released by the Department of Homeland Security.) Figure 3.4b shows the population distribution in the monitoring area.

The threats are modeled as instantaneous releases of specific materials at specific release rates using the Hazard Prediction and Assessment Capability (HPAC) [30]. The SAMSON data at the Memphis International Airport (which is close to the center of the monitoring area as shown in Fig. 3.4a) is used. As shown in Fig. 3.4d, the wind with speed between 0 and 2 m/s in direction 0° (blowing from North) and in direction 180° (blowing from South) are the most common cases. Thus, the experiment focuses on these two meteorological conditions. Given the terrain and meteorological conditions, the sensing area of a sensor is a function of time. Figure 3.5 gives the sensing areas of sensors in different locations and with different detection times.

The expected detection time has been used as the performance metric. Given a sensor placement, the expected detection time is calculated as follows. A total of 100 locations are randomly chosen to set up chemical releases. The probability that a release occurs is proportional to the population of the cell where the release occurs. For $k$-coverage, the detection time is the time between the time instant the threat is released and the time instant it is detected by at least $k$ sensors.

Figure 3.6 shows the expected detection time for $T = 30, 60$, and $90$ min under the case of $k = 3$. Several observations are in order. First, the expected detection

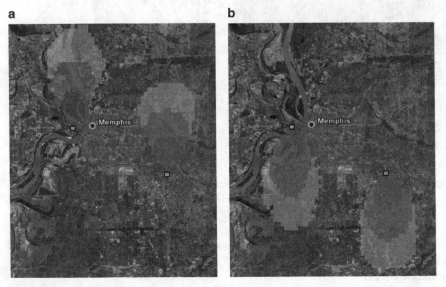

**Fig. 3.5** Sensing areas within detection time $T = 30, 60$, and $90$ min, under different meteorological conditions. Note the sensing areas are prolonged along the opposite direction of the wind. (**a**) Wind speed $= 1$ m/s from North (**b**) Wind speed $= 1$ m/s from South

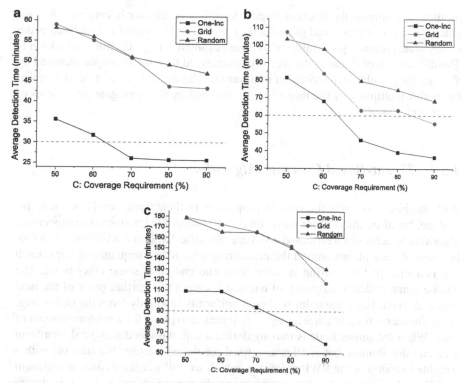

**Fig. 3.6** Average detection time given different maximum allowable detection time $T$. $k = 3$ for these experiments. (a) $T = 30$, (b) $T = 60$ (c) $T = 90$

time of the placement by *One-Incremental* decreases as $C$ increases, and eventually becomes smaller than the maximum allowable detection time $T$. Second, *One-Incremental* incurs 30% ∼ 50% smaller detection time than random or grid placement. Third, the expected detection time of the placement by *One-Incremental* appears to converge to a value that is less than $T$ when $C$ becomes large. This implies that partial coverage with a reasonable high coverage requirement has comparable performance to full coverage.

## 3.5  Coverage with the Use of Mobile Sensors

Once sensors have been deployed in the area according to certain sensor placement/density control algorithm, operating conditions may change to render the original results suboptimal or invalid. Either additional sensors have to be statically deployed or mobile sensors can be dispatched to monitor the area and/or detect events of interest. In the case that (some of) the sensors are mobile, coverage reduces to the problem of laying out sensor movement trajectories, subject to threat profiles, to minimize the effect of threats. Specifically, the monitoring area $R$ is divided into a two-dimensional grid of cells. For each cell $i$, the *risk* $r_i$ is defined as the steady-state presence probability of the event of interest (e.g., chemical attack) in $i$. The distribution of threat in the area is characterized by a *threat profile*, denoted by $\Phi$ (e.g., the population distribution of the area). The threat level of a cell $i$ is given by its risk multiplied by the threat in $i$, normalized by the aggregate threat level of the coverage area.

### 3.5.1 Threat-Based Coverage Algorithm

With the intent to cover the cells in proportion to their threat levels, we now introduce, based on the random waypoint (RWP) model [31], a stochastic movement algorithm to achieve threat-based coverage. Specifically, in the RWP model, a mobile sensor node moves around the monitoring area $R$ in a sequence of trips. Each trip is a straight line starting at some point and ending at some other point. The ending point, called a *waypoint*, of one trip becomes the starting point of the next trip, and so on. Each waypoint is chosen uniformly randomly from the entire area. Once the sensor node reaches a waypoint, it may also pause for a random amount of time. When the sensor node is moving during a trip, the speed may be drawn from a certain distribution, but is otherwise fixed for the whole trip. We start off with a weighted version of the RWP algorithm, which we call *weighted random waypoint* (WRW) [32]. Suppose that the sensor is currently at cell $i$. A cell $j$, $j \neq i$, is chosen to be the next waypoint with probability $\Phi(j)$. The choice of the waypoint is made according to the threat profile, instead of the uniform random distribution.

The basic WRW algorithm is simple, but its coverage profile fails to accurately match the threat profile, because it fails to consider the *intermediate* cells covered between the source and destination. For example, consider a coverage area with a few high threat hotspots. In moving between the hotspots to give them adequate coverage, the sensor will also visit frequently all the cells between the hotspots, thus overcovering the intermediate cells. To solve the problem, the basic algorithm is augmented with the following features:

- **Maximum trip length:** The distance of one trip is not allowed to exceed a parameter $L$ (in distance units). Hence, when one chooses the next waypoint, we restrict the candidate cells to be within the disc of radius $L$ and centered at the current cell. Limiting the trip length forces the algorithm to consider more possible routes to go between any two hotspots, thus reducing the possibility of "warming up" the intermediate cells.
- **Adaptivity to prior coverage:** Because of the probabilistic nature of the algorithm, the correlations between the cells visited, and the finite speed of the sensor, the algorithm's actual coverage at any time may deviate from the given threat profile. To correct the deviation, the notion of *undercoverage* is introduced and

computed for each cell $i$ as $\bar{C}_t(i) = \max\{0, \Phi(i) - \Pi_t(i)\}$, where $\Pi_t(i)$ is the fraction of time that cell $i$ was visited by the sensor up until the end of the $t$th trip. Then, the probability that a candidate cell, say $i$, is chosen as the next waypoint is proportional to $\bar{C}_t(i)$. Hence, an undercovered cell is more likely to be chosen as the next waypoint than a cell that has received too much coverage.

- **Random pause time:** If the sensor is at an undercovered cell, one way to correct the undercoverage is for the sensor to stay in the cell for some pause time $p$. The time $p$ is drawn randomly from a distribution determined by a pause time parameter denoted by $P$ (in time units). Specifically, at the end of the $t$th trip at destination cell $i$, $p \sim Uniform(0, \Omega_t(i))$, where $\Omega_t(i) = \frac{P \times \bar{C}_t(i)}{\sum_{j \in \ell} \bar{C}_t(j)}$ and $\ell$ is the set of cells that are candidates as the next waypoint. The range of the pause time is controlled by $P$. In general, the pause time is expected to be larger when the undercoverage is higher. After the pause, the selection of the next waypoint that defines the next trip occurs as before. The pause time attempts to correct the undercoverage in an extremely efficient way – with zero movement overhead and no possibility of inadvertently changing the coverage of other cells.

Notice that the set of features augmenting the WRW algorithm can be picked *à la carte*. For the convenience of notation, we denote a particular augmented algorithm by WRW-*feat*, where *feat* is a list of letters enumerating the augmentations in alphabetical order, and the letters L, a, and P, are for the "maximum trip length", "adaptivity to prior coverage", and "random pause time" features, respectively. For example, WRW-L denotes the WRW algorithm with the maximum trip length constraint, and WRW-aLP denotes the algorithm with all the three features enabled.

### 3.5.1.1  Matching Performance

A simulation study has been carried out in [32] to illustrate the performance of the algorithms introduced above. The coverage of a number of metropolitan cities, including San Francisco, Los Angeles (LA), Atlanta, Paris, London, and Tokyo, is considered. Figure 3.7a gives the threat profile of Atlanta. Figures 3.7b–e show the achieved steady-state coverage profiles of the WRW, WRW-a, WRW-aL, and WRW-aLP algorithms, respectively, also for Atlanta. Visually, the matching with the threat profile improves as we progress from Fig. 3.7b to Fig. 3.7e. The visual observation can be quantitatively confirmed by computing the root mean square error (RMSE) of the matching. The RMSE achieved by each algorithm is shown in Fig. 3.8, normalized to the RMSE of the WRW algorithm. For the five cities shown, including Atlanta, the normalized RMSE consistently decreases from left to right. Hence, the progression of features, namely, a, aL, and aLP, each contributes to increased matching accuracy, and WRW-aLP is the most powerful algorithm in the matching respect.

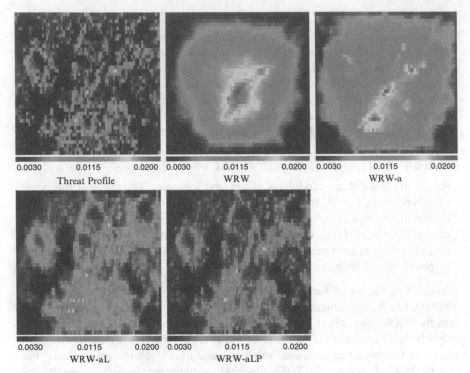

**Fig. 3.7** Threat profiles and steady-study coverage profiles of mobility algorithms for Atlanta

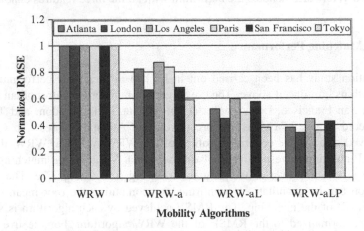

**Fig. 3.8** Normalized RMSE of mobility algorithms for six different cities

## 3.5.2 Temporal Dimension of Uncertainty Reduction

In the previous discussion, we assume that a source is detected whenever it falls within the range of a sensor. In real life, the sensing process is unreliable, and the sensing environment is noisy. A single sensor reading, obtained at one point in time, generally does not give all the useful information about the environment.

Specifically, consider the detection of a point radiation source of strength $A$, in counts per minute (CPM), such that an ideal detector without background radiation located at a distance $d$ from the source will register a count of $c$ in a one second time interval. By radiation physics [33], $c$ is Poisson distributed with parameter $A/d^2$. However, because of background radiation, a detector may register a radiation count even when there is no identifiable source present. Moreover, these counts are random. Hence, a method is needed to ensure that a sensor count is due to a radiation source, and not due to random fluctuations of the background radiation, which can be modeled as a point source of strength $B$.

A reliable detection method can be derived based on the Neyman-Pearson test [34]. The method will allow us to conclude that a sensor reading is from a radiation source with false alarm rate $\alpha$. Consider a sensor, say $i$, that registers a radiation count of $c_i$ in a unit time interval. The source detection problem is formulated as the following hypothesis testing [35]:

- $H_0$ : $c_i$ is Poisson distributed with parameter $B$
- $H_1$ : $c_i$ is Poisson distributed with parameter $B + A/d^2$

We can then formulate the Neyman-Pearson test with a false alarm probability of $\alpha$ by computing a threshold $\tau$ such that if $\Pr(c_i|H_1)\,/\,\Pr(c_i|H_0) > \tau$, then $H_1$ is chosen and otherwise, $H_0$ is chosen. The value of $t$ that yields the desired a can be computed using the Lagrangian method.

The implication of hypothesis testing, as it occurs in the Neyman-Pearson method, is that the confidence, or the *utility*, of a detection result is increased if the time interval of the sensing is increased. The utility function of the sensing against the sensing time is of the form in Fig. 3.9, which illustrates an interesting *temporal*

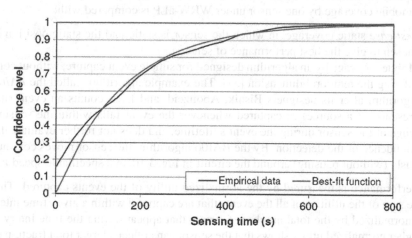

**Fig. 3.9** Confidence (i.e., utility) of measurement as a function of the sensing time in radiation detection: empirical characterization and least-square fit function

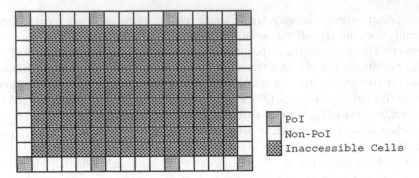

**Fig. 3.10** The ring topology

*dimension* of the sensing problem. The utility function shown is concave, which is representative of many real life sensing activities.

To investigate whether or not the temporal dimension matters for sensor mobile coverage, the following experiment has been carried out in the ring topology in Fig. 3.10 [36]. The area is divided into a circular sequence of 50 cells. It has ten points of interest (PoIs), at which a dynamic radiation source to be detected may appear. The PoIs are uniformly placed on the ring, and each adjacent pair of PoIs is the same distance apart. A source is dynamic because its appearance is a *transient* event – i.e., the source is alternately present and absent – at a PoI, as controlled by given Poisson processes.

The WRW-aLP algorithm discussed above considers a PoI as high threat, and therefore will target the PoIs specifically for coverage. Moreover, since a sensor under WRW-aLP may pause for significant time intervals at the PoIs, the algorithm has the ability to increase the utility of detection results having a temporal dimension. The mobile coverage by one sensor under WRW-aLP is compared with:

- *Best-case* static coverage, in which the sensor is static and the static position is chosen to give the best performance of sensing.
- Mobile coverage by an algorithm designed for simple event capture, without considering the temporal dimension [37]. The example algorithm, called the *BAI06* algorithm after its designers Bisnik, Abouzeid, and Isler, counts an event (the presence of a source) as captured whenever the event falls within the sensing range of the sensor during the event's lifetime, and does not further evaluate the confidence of the detection. By the BAI06 algorithm, the sensor moves continuously (without pausing) around the circuit in Fig. 3.10, at a specifiable speed $v$.

Performance is measured by the normalized utility of the events captured. This is the sum of the utilities of all the events that are captured within a given time interval, normalized by the total number of events that appear during the time interval. A higher normalized utility shows that the sensor can collect a larger total fraction of the interesting information. Figure 3.11 plots the normalized utility achieved by the

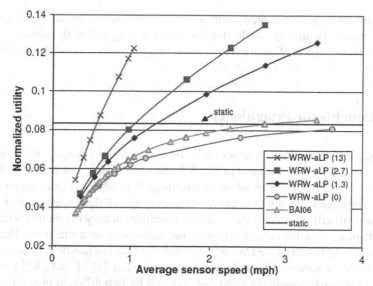

**Fig. 3.11** Normalized utility as a function of the average sensor speed

different algorithms as a function of the sensor's average speed. Two observations are in order:

- Notice that WRW-aLP(0) has a similar performance as BAI06. This is because $P = 0$ ensures that the sensor will continuously move between the PoIs, similar to the BAI06 algorithm. When $P$ increases to 2.7 time units, however, WRW-aLP(2.7) can perform significantly better than BAI06. For example, when the average speed is about 2.2 mph, WRW-aLP(2.7) achieves a 50% higher normalized utility of 0.12, compared with 0.08 for BAI06. The results show that pausing at PoIs can improve the quality of the sensing by allowing the events to be measured for longer and therefore with higher confidence.
- Static coverage is extremely efficient. Hence, while it is inherently unfair (i.e., it completely ignores some of the PoIs), it might perform the best purely from a utility standpoint. Figure 3.11, however, shows that WRW-aLP always outperforms static coverage when the average speed exceeds a modest value. This is partly due to the concavity of the utility function. When the utility function is concave, much of the utility is obtained during the initial period of observing a new event. This encourages the sensor to occasionally move from one PoI to another to catch more new events, as long as the moving speed is not too low to make the travel overhead too high.

Interestingly, the above results show how the temporal dimension can fundamentally impact the performance of mobile coverage. To illustrate, it is known that for the BAI06 algorithm, a faster sensor always gives better performance in the sense of more events captured. The results here show that, while increasing the speed of the sensor can increase the fraction of the events captured [37, 38], the sensing *uncertainty* about each captured event also increases, when the temporal dimension is present. Intuitively, while moving quickly may allow the sensor to see more events, the vision of each event becomes increasingly blurred due to the fast movement.

## 3.6  Thoughts for Practitioners

In reality, the sensing area of sensors may be dynamically changing with various physical conditions. As illustrated in Fig. 3.5, the sensing areas (or equivalently the calculation of $R_i^T$) highly depend on meteorological conditions. One cannot use a fixed meteorological condition to generate the contours of the dispersion and place sensors accordingly; otherwise the coverage requirement may not be met when the meteorological conditions change through the seasons or over the years. Here we propose one way to extend the *One-Incremental* algorithm to handle various sensing areas induced by various meteorological conditions. Let $\{R_i'^T\}$ and $\{R_i''^T\}$ denote two sets of sensing areas in the same *FoI*, induced by two different meteorological conditions. Note that in general $R_i'^T \neq R_i''^T$. The two sets of sensing areas can be merged to define a new merged sensing area of a sensor (placed in a cell $i$) as $R_i^T = R_i'^T \cup R_i''^T$. Moreover, the utility of a cell under a certain meteorological condition is defined as its original utility multiplied by the probability that this meteorological condition occurs. In this way, the sensor placement problem that accommodates various meteorological conditions is essentially the same as that for one specific meteorological condition (fixed wind speed and direction).

We have provided threat-based mobile coverage in which the coverage time is accurately apportioned by the threat profile. Ideally, one would like the sharing to be realized over extremely fine time scales, so that the sensor will return to every PoI quickly and detect any interesting event with small delay. In practice, such fine time-scale sharing is limited by the speed of the sensor and the time/energy overheads of travel between the PoIs. The travel overhead is more generally a primary issue in mobile coverage, namely its advantages must be properly balanced against the costs of supporting the mobility.

## 3.7  Directions for Future Research

In the chapter, the sensing models considered is deterministic, which assumes that events within the sensing area are always detected and there is no false positive either. Another type of sensing models is probability-based, which specifies the confidence interval of detections. So one future work is to study how to place sensors under a probability-based model such that the overall false alarm rate and target-missing rate are minimized.

Second, power consumption and network lifetime are important performance criteria for sensor networks.

In OGDC, each node probabilistically volunteers itself to be a starting node in each round. To ensure uniform power consumption across the network, a node chooses this probability based on its remaining power. One future work is to extend OGDC (or other density control algorithms) to achieve the maximal network lifetime, while still satisfying the coverage requirement.

For mobile sensor coverage, we have considered differential coverage based on the importance levels of different sub-areas. Another important consideration is the type of event being covered and the event dynamics. Optimal mobile coverage algorithms can be designed to maximize the amount of the information captured, based on such information. Also, if we have multiple sensors, the coverage of the sensors may overlap. If the number of sensors is large relative to the coverage area, and these sensors independently try to cover the whole area, the redundancy of the coverage may be significant, resulting in inefficient resource use. In this case, a coordination protocol that will enable the sensors to work well together as a group becomes important.

## 3.8 Conclusion

In this chapter, we first introduce several fundamental properties of coverage and show the formulations of the coverage problem in different ways. Then we discuss in detail a decentralized and localized density control algorithm, OGDC. A simulation study shows that the number of working nodes required under OGDC modestly increases with the number of sensor nodes deployed, while both PEAS and CCP incur a 50% increase in the number of working nodes.

Next, we consider the problem of sensor placement in a more realistic setting: we acknowledge nonnegligible detection time; we allow the sensing area of a sensor (at certain time instant) to be anisotropic and of arbitrary shape, and we define the utility function $U(\cdot)$ to model the expected utilities of coverage (or risks of insufficient coverage) in different parts of the area. The proposed sensor placement algorithm *One-Incremental* is evaluated in the realistic setting of Port of Memphis. The results show that *One-Incremental* incurs 30% ~ 50% smaller detection time than random or grid placement.

Finally, we consider threat-based mobile coverage, and evaluate how the temporal dimension of real-life sensing tasks will impact the performance of the mobile coverage.

## Terminologies

*Coverage problem.* How to deploy sensor nodes to cover a monitoring area in order to fulfill certain performance criteria, such as minimizing the number of sensors, or minimizing the sensing range.

*Sensing Range.* The sensing range of a sensor is the range within which events of interest can be detected by the sensor.

*Detection time T.* The time interval between the instant when the event of interest happens and the instant when the event is detected by *any* of the sensor nodes.

*Coverage requirement C.* In the case of partial coverage, Coverage Requirement $C$ is used to lowerbound the coverage performance, such as the area or the population covered.

*Neyman-Pearson method.* A hypothesis testing method to increase the reliability of detecting a point radiation source in the presence of background radiation.

*OGDC.* Optimal Geographical Density Control [7].

*SCIPUFF.* Second-order Closure Integrated Gaussian Puff (SCIPUFF) is a dispersion model [25,26], which can be to calculate the dispersed material in space and time, subject to terrain, land cover, and meteorological conditions.

*Temporal dimension of sensing.* The effects of the sensing time on the reduction of the sensing uncertainty, by producing a sequence of measurements enabling the removal of noise and statistical outliers.

*Threat-based mobile coverage.* Mobile coverage by a sensor, with the goal of matching the coverage profile with a given threat profile.

*Wind rose.* A wind rose gives an information-laden view of how wind *speed* and *direction* are typically distributed at a particular location. Specifically, it specifies wind direction/speed pairs and their percentage of occurrence.

## Questions

1. What is sensor placement problem and what is sensor density control problem. And explain how these two are related.
2. Why we need $k$-coverage ($k > 1$)?
3. Give one sufficient condition for complete coverage by using sensors with *convex* sensing area. Does your condition hold for sensors with arbitrary sensing area?
4. Sketch a proof that complete coverage of a convex area implies connectivity among the sensors, given that the radio range $r_c$ is at least as twice the sensing range $r_s$.
5. For simplicity of algorithm discussion in OGDC, we have assumed that all nodes are time synchronized. Find a way to relax this by only requiring relative time synchronization.
6. Prove that the two problem formulations in Sect. 3.2.2 are equivalent.
7. Propose another heuristic algorithm, other than *One-Incremental*, for the partial $k$-coverage problem.
8. How does the instantaneous coverage of a mobile sensor differ from a static sensor? In what situations might mobile sensors be desired?
9. If a mobile sensor uses the random waypoint algorithm in an open rectangular area, what do you think would be the coverage profile? Explain why.

10. Suppose three cells, 1, 2, and 3 form a horizontal grid in that order, i.e., Cell 2 is in the middle between 1 and 3. Suppose the threat profile is (0.5, 0, 0.5). What pause time values would be effective for accurate matching? What is the problem in practice of using such a pause time value?

11. For the dynamic events described in Sect. 3.5.2, why is it in general advantageous for the sensor to move as quickly as possible between the PoIs in order to detect as many events as possible?

## Appendix 1. Proof of Lemma 2

The Lemma is proved by showing that given the conditions stated in the lemma, the number of working sensor nodes and the overlap have a linear relationship with a positive slope.

Let the indicator function of a working node $i$, $I_i(x)$, be defined as

$$I_R(x) = \begin{cases} 1 & \text{if} \quad x \in R, \\ 0 & \text{otherwise.} \end{cases}$$

Let $R'$ be a region that contains $R$ and the coverage areas of all sensor nodes. Then the coverage area of a sensor node $i$ is a disk with the size $\int_{R'} I_i(x)\,dx \triangleq |S_i|$, where $|S_i|$ denotes the size of the area $S_i$ covered by sensor node $i$. By condition (ii), $|S_i| = |S|$ for all $i$. With the definition of $I_i(x)$, the overlap at point $x$ can be written as

$$L(x) = \sum_{i=1}^{N} I_i(x) - I_R(x), \qquad (3.14)$$

where $N$ is the number of working nodes, and the overlap of sensing areas of all the sensor nodes, $L$, can be written as

$$\begin{aligned} L &= \int_{R'} L(x) \\ &= \int_{R'} \left( \sum_{i=1}^{N} I_i(x) - I_R(x) \right) dx \qquad (3.15) \\ &= \sum_{i=1}^{N} \int_{R'} I_i(x)\,dx - |R| \\ &= N|S| - |R|, \end{aligned}$$

where condition (i) is implied in the first equality and condition (ii) is implied in the fourth equality. Equation (3.15) states that minimizing the number of working nodes $N$ is equivalent to minimizing the overlap of sensing areas of all the sensor nodes $L$.                                                                     □

# Appendix 2. Proof of Lemma 3

There are multiple coverage areas centered at $C_i$'s and they all intersect at point $O$. The centers of these coverage areas are labeled as $C_i$, with the index $i$ increasing clockwise. (Figure 3.1 gives the case of $k = 3$, where $C_1 = A$, $C_2 = B$, and $C_3 = C$.) Now $\sum_{i=1}^{k} \angle C_i O C_{(i \bmod k)+1} = 2\pi$ and $\angle C_i O C_{(i \bmod k)+1} + \alpha_i = \pi$. From the above equations, one can derive that $\sum_{i=1}^{k} \alpha_i = (k-2)\pi$.                □

# References

1. D. Estrin, R. Govindan, J. S. Heidemann, and S. Kumar. Next century challenges: Scalable coordination in sensor networks. In *Proc. of ACM MobiCom'99*, Washington, August 1999.
2. J. M. Kahn, R. H. Katz, and K. S. J. Pister. Next century challenges: Mobile networking for "smart dust". In *Proc. of ACM MobiCom'99*, August 1999.
3. I. F. Akyildiz, W. Su, Y. Sankarasubramaniam, and E. Cayirci. *Wireless Sensor Networks: A Survey, Computer Networks*. March 2002.
4. A. Mainwaring, J. Polastre, R. Szewczyk, and D. Culler. Wireless sensor networks for habitat monitoring. In *First ACM International Workshop on Wireless Workshop in Wireless Sensor Networks and Applications (WSNA 2002)*, August 2002.
5. E. Shih, S. Cho, N. Ickes, R. Min, A. Sinha, A. Wang, and A. Chandrakasan. Physical layer driven protocol and algorithm design for energy-efficient wireless sensor networks. In *Proc. of ACM MobiCom'01*, Rome, Italy, July 2001.
6. F. Ye, G. Zhong, S. Lu, and L. Zhang. PEAS: A robust energy conserving protocol for long-lived sensor networks. In *The 23nd International Conference on Distributed Computing Systems (ICDCS)*, 2003.
7. H. Zhang and J. C. Hou, "Maintaining sensing coverage and connectivity in large sensor networks," *Wireless Ad Hoc and Sensor Networks: An International Journal*, Vol. 1, No. 1–2, pp. 89–123, January 2005.
8. S. Slijepcevic and M. Potkonjak. Power efficient organization of wireless sensor networks. In *Proc. of ICC*, Helsinki, Finland, June 2001.
9. H. Gupta, S. Das, and Q. Gu. Connected sensor cover: Self-organization of sensor networks for efficient query execution. In *Proc. of ACM MOBIHOC*, 2003.
10. X. Wang, G. Xing, Y. Zhang, C. Lu, R. Pless, and C. Gill. Integrated coverage and connectivity configuration in wireless sensor networks. In *Proc. of SENSYS*, 2003.
11. A. Cerpa and D. Estrin. Ascent: Adaptive self-configuring sensor networks topologies. In *Proc. of IEEE INFOCOM*, March 2002.
12. D. Tian and N. D. Georganas. A coverage-preserving node scheduling scheme for large wireless sensor networks. In *First ACM International Workshop on Wireless Sensor Networks and Applications*, Georgia, GA, 2002.
13. F. Ye, H. Zhang, S. Lu, L. Zhang, and J. C. Hou. A randomized energy-conservation protocol for resilient sensor networks. *ACM Wireless Network (WINET)*, 12(5):637–652, Oct. 2006.
14. K. Chakrabarty, S. Iyengar, H. Qi, and E. Cho. Grid coverage for surveillance and target location in distributed sensor networks. *IEEE Trans. on Computers*, 51(12), 2002.
15. Z. Zhou, S. Das, and H. Gupta. Connected k-coverage problem in sensor networks. In *Proc. of International Conference on Computer Communication and Networks (ICCCN'04)*, Chicago, IL, October 2004.
16. S. Yang, F. Dai, M. Cardei, and J. Wu. On multiple point coverage in wireless sensor networks. In *Proc. of MASS*, Washington, DC, November 2005.
17. T. Feder and D. Greene. Optimal algorithms for approximate clustering. In *Proc. of the 20th Annual ACM Symposium on Theory of Computing (STOC'88)*, New York, NY, 1988.

18. R. W. Lee and J. J. Kulesz. A risk-based sensor deployment methodology. Technical report, Oak Ridge National Laboratory, 2006.
19. C.-F. Huang and Y.-C. Tseng. The coverage problem in a wireless sensor network. In *Proc. of 2nd ACM International Conf. on Wireless Sensor Networks and Applications (WSNA'03)*, pages 115–121, 2003.
20. P. Hall. *Introduction to the Theory of Coverage Processes*. 1988.
21. A. Savvides, C. Han, and M. Strivastava. Dynamic fine-grained localization in ad-hoc networks of sensors. In *Proc. of ACM MOBICOM'01*, pages 166–179. ACM Press, 2001.
22. S. Meguerdichian, F. Koushanfar, M. Potkonjak, and M. B. Srivastava. Coverage problems in wireless ad-hoc sensor networks. In *INFOCOM*, pages 1380–1387, 2001.
23. L. Doherty, L. El Ghaoui, and K. S. J. Pister. Convex position estimation in wireless sensor networks. In *Proc. of IEEE Infocom 2001*, Anchorage, AK, April 2001.
24. Y. Xu, J. Heidemann, and D. Estrin. Geography-informed energy conservation for ad hoc routing. In *Proc. of ACM MOBICOM'01*, Rome, Italy, July 2001.
25. R. I. Sykes, C. P. Cerasoli, and D. S. Henn. The representation of dynamic flow effects in a lagrangian puff dispersion model. *J. Haz. Mat.*, 64:223–247, 1999.
26. R. I. Sykes and R. S. Gabruk. A second-order closure model for the effect of averaging time on turbulent plume dispersion. *J. Appl. Met.*, 36:165–184, 1997.
27. National Geophysical Data Center. Global Land One-km Base Elevation Database. http://www.ngdc.noaa.gov/mgg/topo/globe.html, 2007.
28. Google Eearth. http://earth.google.com/, 2007.
29. Oak Ridge National Laboratory. Landscan main page. http://www.ornl.gov/sci/gist/landscan, 2005.
30. Defense Threat Reduction Agency. Hazard prediction and assessment capability (hpac). http://www.dtra.mil/Toolbox/Directorates/td/programs/acec/hpac.cfm.
31. J. Broch, D. A. Maltz, D. B. Johnson, Y. C. Hu, and J. Jetcheva. A performance comparison of multi-hop wireless ad hoc network routing protocols. In *Proc. of MobiCom*, Dallas, Texas, USA, October 1998.
32. C. Y. T. Ma, J. C. Chin, D. K. Y. Yau, N. S. V. Rao, and M. Shankar. Matching and fairness in threat-based mobile sensor coverage. Technical report, Department of Computer Science, Purdue University, March 2007.
33. R. E. Lapp and H. L. Andrews. *Nuclear Radiation Physics*. Prentice-Hall, 1948.
34. A. Sundaresan, P. K. Varshney, and N. S. V. Rao. Distributed detection of a nuclear radiaoactive source using fusion of correlated decisions. In *Proc. of International Conference on Information Fusion*, Quebec, Canada, July 2007.
35. C. Y. T. Ma, D. K. Y. Yau, J.-C. Chin, N. S. V. Rao, and M. Shankar. *Distributed Detection and Data Fusion*. Springer-Verlag, 1997.
36. C. Y. T. Ma, D. K. Y. Yau, J.-C. Chin, N. S. V. Rao, and M. Shankar. Resource-constrained coverage of radiation threats using limited mobility. Technical report, Department of Computer Science, Purdue University, June 2007.
37. N. Bisnik, A. Abouzeid, and V. Isler. Stochastic event capture using mobile sensors subject to a quality metric. In *Proc. of MobiCom*, Los Angeles, California, USA, September 2006.
38. B. Liu, P. Brass, O. Dousse, P. Nain, and D. Towsley. Mobility improves coverage of sensor networks. In *Proc. of MobiHoc*, Urbana-Champaign, IL, USA, May 2005.

# Chapter 4
# Routing in Wireless Sensor Networks

**Hannes Frey, Stefan Rührup, and Ivan Stojmenović**

**Abstract** Wireless sensor networks are formed by small sensor nodes communicating over wireless links without using a fixed network infrastructure. Sensor nodes have a limited transmission range, and their processing and storage capabilities as well as their energy resources are also limited. Routing protocols for wireless sensor networks have to ensure reliable multi-hop communication under these conditions. We describe design challenges for routing protocols in sensor networks and illustrate the key techniques to achieve desired characteristics, such as energy efficiency and delivery guarantees. We give a survey of state-of-the-art routing techniques with a focus on geographic routing, a paradigm that enables a reactive message-efficient routing without prior route discovery or knowledge of the network topology. Different geographic routing strategies are described as well as beaconless routing techniques. We also show the physical layer impact on routing and outline further research directions.

## 4.1 Introduction

Wireless sensor networks are formed by small devices communicating over wireless links without using a fixed networked infrastructure. Because of limited transmission range, communication between any two devices requires collaborating intermediate forwarding network nodes, i.e. devices act as routers and end systems at the same time. Communication between any two nodes may be trivially based on simply flooding the entire network. However, more elaborate routing algorithms are essential for the applicability of such wireless networks, since energy has to be conserved in low powered devices and wireless communication always leads to increased energy consumption.

H. Frey (✉)
Department of Computer Science, University of Paderborn, Warburger Str. 100,
33098 Paderborn, Germany
e-mail: hannes.frey@uni-paderborn.de

S. Misra et al. (eds.), *Guide to Wireless Sensor Networks*, Computer Communications
and Networks, DOI: 10.1007/978-1-84882-218-4_4,
© Springer-Verlag London Limited 2009

The first routing algorithms for wireless networks followed the traditional approach of *topology-based* routing, i.e. forwarding decisions are based on information about currently available links between network nodes [1–5]. Early proposals are based on *proactive* routing strategies maintaining routing information about all available paths even when these paths are never used. Proactive routing does not scale well in dynamically changing network topologies, thus, *reactive* methods maintaining only these routes which are currently in use have been investigated further on.

Even reactive routing methods may still generate a significant amount of traffic when network topology changes frequently due to device mobility or alternating energy conserving sleep cycles. In recent years, *location awareness* (i.e. nodes know their physical location) has been investigated as a possible solution to the inherent limitations of topology-based methods. Several novel *geographic* (also termed *position-based*) routing algorithms have been proposed, which allow routers to be nearly stateless since packet forwarding is achieved by using information about the position of candidate nodes in the vicinity and the position of the destination node only. Information of physical location might be determined by means of a global positioning technique like GPS, or relative positioning based on distance estimation on incoming signal strengths [6, 7]. Geographic routing requires that location information about the destination node has to be known or acquired in advance. In wireless sensor networks, data are typically collected by a designated node, called *sink*, and the location of the sink can be hard-coded. In general, however, inquiry of destination position is done by using an additional *location service* [8] producing additional network load, which has to be considered if performance of position-based methods are compared with topology-based ones.

## 4.2  Background

Routing in wireless sensor networks differs from conventional routing in fixed networks in various ways: There is no infrastructure, wireless links are unreliable, sensor nodes may fail, and routing protocols have to meet strict energy saving requirements. Many routing algorithms were developed for wireless networks in general. Routing algorithms that perform an end-to-end message delivery with host-based addressing can be classified as *topology-based*, if the destination is given by an ID, or as *position-based*, if the destination is a geographic location. The latter are also called geographic routing algorithms. Both topology-based and geographic routing algorithms are *address-centric*, and besides these types, the *data-centric* routing paradigm has become popular in the area of sensor networks. Data-centric routing is based on queries that are issued by the sink to request data. These requests are not addressed to specific sensor nodes. Instead, the sensor nodes that can deliver the requested data will answer the query. An overview of data-centric routing algorithms is given in [9].

A further classification criterion is the usage of messages: A routing method is called a *single-path* strategy, if there is only one instance of the message in the network at any time. Other forwarding strategies can be classified as *partial flooding* and *multi-path* routing, depending on messages being forwarded to some neighbors in each routing step or when routing is performed along a few recognizable paths, respectively. Single path strategies are more resource saving, since they keep the number of message transmission to a minimum opposed to multi-path or flooding based approaches. However, flooding achieves better results in highly dynamic network scenarios [10], while multipath routing forms a compromise between both extremes. *Guaranteed delivery* serves as further classification of single-path, multi-path, and flooding-based routing strategies reviewed in this chapter. Assuming an ideal, collision free access scheme, these algorithms guarantee delivery of a message, when source and destination are located in the same network partition. Routing algorithms can additionally be classified if they require nodes to maintain state information about ongoing routing tasks, which is termed *memorization* in literature. It is preferable to avoid memorization of past traffic on any node, but however, as long memorization does not lead to an increased message complexity, this is not the crucial part to build resource saving protocols. Memory even on small devices is expected to increase exponentially in future, while communication resources will remain the limiting factor.

## 4.3  Greedy Packet Forwarding

Greedy routing algorithms limit forwarding decisions to information about the location of the current forwarding node, its neighbors, and the message destination. Each intermediate node applies this greedy principle until the destination, if possible, is eventually reached. The characteristics of greedy routing algorithms differ with the optimization criterion applied in each forwarding step.

The distance between a node S and the projection A' of a neighbor node A onto the line connecting S and destination D is defined as *progress* (see Fig. 4.1).

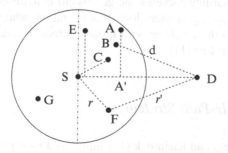

**Fig. 4.1**  Several greedy routing strategies can be defined by the notion of progress, distance and direction. For example, node A is the neighbor of S with most forward progress, B has least distance to D, and C is closest in direction to D

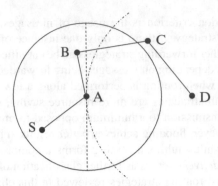

**Fig. 4.2** A packet addressed to node D will be dropped at node A, since each neighbor of A (i.e. S and B) is in backward direction. This routing failure can appear in both distance-based and progress-based greedy forwarding strategies

Neighbors with positive progress are denoted to be in *forward direction*. For example in Fig. 4.1 the neighbors A, B, C, E, and F are in forward direction. The remaining node G is termed to be in *backward direction*. Alternatively, greedy routing can be based on *distance*, considering the Euclidean distance between all neighbors of sender S and the destination D. Finally, *direction*-based methods consider the deviation (angle between next hop, current, and destination node) from the line connecting current sender and destination.

In general, greedy forwarding based on progress or distance considers nodes in forward direction resp. closer to destination only, since choosing a node in backward direction might lead to a routing loop. Consequently, greedy routing cannot guarantee packet delivery even if there exists a path from source to destination. For example in Fig. 4.2 there exists a path from source S to destination D, however, a packet addressed to node D is dropped at node A, since each node within its transmission range is in backward direction. Such a situation is called *local minimum*, and the node where greedy forwarding is stopped is termed a *concave node*.

Distance-based greedy routing methods are inherently loop-free, since the distance from destination is reduced in each forwarding step. In general, each greedy routing algorithm forwarding packets to neighbors closer to the destination or within the most forward progress guarantees loop-free operation, while greedy algorithms forwarding packets to the neighbor with closest direction (and possibly to other neighbors) are not loop-free [11].

### 4.3.1 Basic Single-Path Strategies

In the mid 1980s Takagi and Kleinrock [12] introduced *most forward within radius (MFR)*, the first position-based routing algorithm at all. A packet with destination D is forwarded to the next neighbor in forward direction maximizing the progress towards D (e.g. node A in Fig. 4.1). The widely used greedy forwarding strategy proposed by Finn [13] applies the same principle but considers distance instead of progress, i.e. a node forwards a packet to the neighbor with the smallest distance d to the destination (e.g. node B in Fig. 4.1).

If signal strength can not be adjusted, it is a good choice to maximize the advance in each routing step, since it attempts to minimize the number of hops a packet has to travel. Nevertheless, even if signal strength is a fixed parameter, sending a packet to a distant neighbor in the border area of transmission range results in a higher probability of packet loss due to signal attenuation and node mobility. Hou and Li [14] observed that by adjusting signal strength to nearest neighbors, the probability of message loss due to collision can be reduced significantly. They proposed *nearest with forward progress (NFP)*, where each node sends the packet to the nearest neighbor with forward progress (e.g. node E in Fig. 4.1). Stojmenovic and Lin defined *nearest closer (NC)* [15], which is a modification of NFP considering distance instead of progress, i.e. packets are forwarded to the nearest neighbor among all neighbors closer to the destination (e.g. node C in Fig. 4.1). To overcome the trade-off between progress and transmission success the *random progress method (RPM)* [16] by Nelson and Kleinrock selects randomly (uniformly distributed) one of all neighbors with forward progress.

Kranakis et al. defined *compass routing (DIR)* [17], where source or intermediate node forwards a packet to the neighbor node lying in closest direction compared to the line connecting sender and destination. For example in Fig. 4.1 node C is in closest direction regarding the line connecting node S and destination D. By applying this scheme in each routing step, compass routing attempts to minimize the Euclidean path length a packet has to travel.

## 4.3.2  Improved Single-Path Strategies

Greedy routing based on progress and distance can be improved by allowing a message to travel in backward direction for one hop, i.e. a message is dropped only if it has to be sent back to the node of the previous forwarding step [11]. In combination with this scheme, greedy routing overcomes concave nodes having a 2-hop neighbor closer to the destination. For example in Fig. 4.2 a message addressed to node D overcomes the concave node A, since A forwards it to node B and B has neighbor C which is closer to D than A. Stojmenovic and Lin [11] proposed *geographical distance routing (GEDIR)*, which is an improvement of distance-based routing applying that backward rule.

Delivery rate of existing greedy routing algorithms can furthermore be improved if nodes exchange information about their neighbors and thus each node is aware of its 2-hop neighbors (termed 2-MFR, 2-GEDIR or 2-DIR for instance) [11]. The next forwarding node is selected among all one- and two-hop neighbors. To reach a selected two-hop neighbor C, the forwarding criterion is applied again on all one-hop neighbors connected to the selected node C.

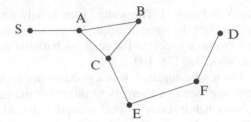

**Fig. 4.3** The paths selected for a message from source S to destination D are SABACBCEFD for alternate-GEDIR and SABCEFD for disjoint-GEDIR, respectively

Memorizing past traffic may also be used to reduce failure rate of existing greedy routing algorithms. Lin and Stojmenovic [18] defined *alternate* and *disjoint* routing schemes, which are based on routing algorithms GEDIR, MFR, and DIR, but allow selection of forwarding nodes in backward direction for more than one hop. Both schemes maintain state information to avoid message loops, but differ in the way concave nodes select next hop nodes. In greedy routing improved by alternate scheme each intermediate node forwards $i$-th received message to the $i$-th best neighbor, according to the forwarding criterion applied. If there is no remaining neighbor the message is dropped by the forwarding node, thus, possible loops produced by this scheme are only temporary and limited by the maximum node degree. In greedy routing improved by the disjoint scheme, each node forwarding a message is memorized by all its neighbors and further eliminated from the set of possible next hop candidate nodes for that message. A node with an empty set of possible candidate nodes will drop the message. Disjoint scheme is inherently loop free, since each node will receive a message at most once. Figure 4.3 gives an example where GEDIR extended by alternate and disjoint schemes is successful, while GEDIR applied on its own will drop a message at concave node B.

### 4.3.3 *Multipath and Flooding-Based Strategies*

Routing strategies where each intermediate node is forwarding messages to possibly more than one neighbor in forward direction are termed *restricted directional flooding* strategies. The rationale behind redundant message transmission is to increase success rate of existing forwarding strategies. A message dropped at a concave node may travel an alternative path leading to its destination. However, even restricted directional flooding does not guarantee delivery.

Basagni et al. [19] proposed the *distance routing effect algorithm for mobility* (DREAM), a strategy based on restricted directional flooding, which requires memorization to avoid loops due to forwarding the same message more than once. Source and any intermediate node will forward a message once to all its one-hop neighbors

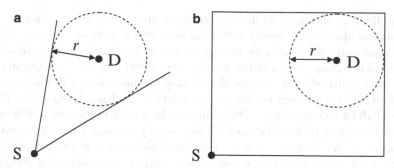

**Fig. 4.4** The destination node D is expected somewhere inside the depicted circle (the expected region). (**a**) DREAM will forward the message to each node lying inside the angular range defined by the tangents passing S and the expected region. (**b**) In LAR 1 scheme each node within the rectangular region is a possible candidate node to forward the message

lying in a certain angular range toward destination D. As depicted in Fig. 4.4a, this range is calculated from the tangents leading from forwarding node to the expected area of destination D, which is a circle centered at D with radius reflecting the maximum possible movement of D since the last location update (i.e. node D can be expected somewhere inside that circle). Independently, similar strategies based on directional flooding and memorization were also described by Ko an Vaidya [20]. They proposed *location aided routing* (LAR), which was originally intended to support topology-based reactive routing protocols in finding routes in an efficient way. Reactive routing protocols frequently use flooding to determine new routes between source and destination. LAR uses location information to restrict flooding to a certain area, termed *request zone*. Only nodes within request zone are allowed to forward route discovery packets. Figure 4.4b depicts the request zone used by LAR 1 scheme, i.e. route discovery is restricted to the rectangular region containing the expected zone of destination D and source node S. The LAR 2 scheme restricts route discovery to nodes with distance to destination D at most some d greater than the distance between previous forwarding node and D.

   Stojmenovic et al. generalized the concept of restricted directional flooding for distance-, progress-, and direction-based greedy routing and defined *V-GEDIR*, *CH-MFR*, and *R-DIR*, respectively [21]. The basic idea of these algorithms is to determine all possible "best" next hop nodes when forwarding criterion was applied for each possible destination position within its expected area. The methods differ how these nodes are determined efficiently. R-DIR determines all neighbors restricted to the angular range and possibly two additional neighbors lying closest in direction to one of the two tangents (see Fig. 4.4a). In V-GEDIR, next hop nodes are determined by intersecting the *Voronoi diagram* of neighbors with the expected area of destination, while CH-MFR calculates the *convex hull* of neighboring nodes (for details see [21]).

In addition to alternate and disjoint greedy routing, Lin and Stojmenovic proposed a class of multipath greedy routing methods based on these concepts [18]. A source node initially forwards the message to $c$ best neighbors according to selection criterion. Several copies of the message may travel along a few recognizable paths, while paths selected are depending if original, alternate, or disjoint routing is used. In *original c-greedy* method (greedy is one of the base algorithms GEDIR, MFR or DIR) a message received by an intermediate node is forwarded only once to its best neighbor (if available) and all successively received copies are ignored. Intermediate nodes in the *alternate c-greedy* and *disjoint c-greedy* method apply the original alternate and disjoint criterion, respectively, i.e., multiple initial copies of the message are treated like the one message in the original algorithms.

### 4.3.4 Energy-Aware Routing

Computational power (even in mobile devices) is increasing rapidly, while battery lifetime is not expected to increase significantly in the future. If signal strength can be adjusted, localized routing algorithms could attempt to reduce energy consumption by choosing forwarding nodes within optimal transmission range. Stojmenovic and Lin [15] proposed a general power metric combining signal attenuation of various exponents, energy loss due to start up, collisions, retransmissions, and acknowledgments in one expression depending on the distance between sender and receiver. Assuming that additional nodes can be set in arbitrary positions between source and destination, there is an optimal number of equally spaced intermediate forwarding nodes producing minimal power consumption [15]. The optimal number of intermediate nodes is calculated from the distance between source and destination and the general power metric parameters.

This result is used to define the *power-routing* algorithm. Nodes cannot be placed arbitrarily, but assuming that power consumption for the rest of the path is equal to the optimal one, each intermediate node S selects a neighbor F closer to D minimizing the sum of power needed to transmit the packet from S to F and the optimal power consumption needed to forward the packet from F to the destination D over the remaining distance $r'$.

Power-routing tries to minimize energy consumption, but single nodes might be selected by many routing tasks, which will result in their premature failure. To maximize the number of successful routing tasks, a cost metric is used in [15] to define the *cost-routing* algorithm. This metric is a function proportional to the inverse of remaining battery power, expressing the reluctance of a node to forward a packet. Each forwarding node chooses a neighbor minimizing the sum of cost metric and an estimated cost for the remaining path. Finally, *power-cost-routing* is investigated there, which tries to minimize a combination of power and cost metric.

Kuruvila et al. [22] proposed localized power and cost aware routing schemes based on the notion of proportional progress. Referring to Fig. 4.1, let the node currently holding the packet be S, let F be one candidate neighbor of S, and let D be the destination. Let $|SF| = r$, $|SD| = m$, and $|FD| = r'$, with $r' < m$. Let us measure the proportional progress as the power used to make a portion of the progress. The power needed to send from S to F is $r^\alpha + c$, where $\alpha$ is the signal attenuation exponent (a value between 2 and 5), and $c$ is a constant that accounts for minimal reception power and computation power. The portion of progress made with it is $m - r'$. With similar advance continuing, there would be $m/(m - r')$ such steps, and the total cost would be $(r^\alpha + c)m/(m - r')$. Therefore the neighbor that minimizes $(r^\alpha + c)/(m - r')$ will be selected for forwarding the message. This rule therefore selects a neighbor that minimizes the power spent per unit of progress made, in terms of getting closer to destination. Power metrics can be similarly replaced by a cost or power-cost metric to define cost or power-cost per unit of progress made. This leads to the algorithms that select forwarding neighbors that minimize $f(F)/(m - r')$, and minimize $f(F)(r^\alpha + c)/(m - r')$, respectively, where $f(F)$ is a measure of cost for using node F to forward (it may be the inverse of its remaining power, for instance).

## 4.4 Planar Graph Routing

The problem of greedy forwarding is that messages are dropped by concave nodes, which have no neighbors closer to the target. In such local minimum situation, a recovery strategy is needed to guarantee delivery.

Bose et al. described *FACE*, the first memoryless single-path recovery mechanism with guaranteed delivery [23] (The integration of this algorithm with IEEE 802.11 was later implemented in the *greedy perimeter stateless routing* (GPSR) protocol by Karp and Kung [24]). The FACE algorithm is an improvement of the planar graph routing algorithm due to Kranakis et al. [17]. FACE routing provides guaranteed delivery if it is applied on planar connected geometric graphs. A geometric graph is termed to be planar if there is no intersection between any two edges of the graph (see the graph depicted in Fig. 4.5 for example).

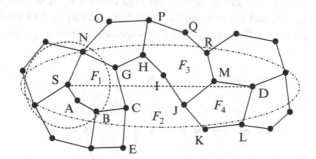

**Fig. 4.5** Illustration of the face routing algorithm applied on a planar subgraph. For instance, traversal of face $F_4$ by applying the *right-hand rule* leads to the path JKLDM when starting at node J. Face routing of a packet sent from source S to destination D leads to the path ABCE...LKJKLD if the right-hand rule is applied on each face intersected by the straight line connecting S and D. Note, the outer face $F_2$ is traversed completely except of nodes G, H, and I

Wireless networks can be modeled as geometric graphs, where the geographical position of each mobile device defines a point in the plane. Different graph types can be defined on that point set. A *unit disk graph* (UDG) reflects a wireless network, where each node has the same transmission radius $R$, i.e. an edge exists between any two nodes $u$ and $v$, if the Euclidean distance between $u$ and $v$ is less than a fixed unit $R$. Unit disk graphs are most commonly used in literature. Variations are *minpower graphs* with links between any two points (and possibly different transmission radii) if bidirectional communication between them is possible, and subsets of unit disk graphs modeling disconnection due to *obstacles* between sender and receiver.

### 4.4.1 Localized Planar Subgraph Construction

In general, the geometric graph reflecting a wireless network is not planar. Thus, before the FACE recovery procedure can be performed, a planar subgraph has to be extracted from the complete network graph. In the description of FACE, Bose et al. [23] propose a distributed algorithm for extracting a planar subgraph from a unit disk graph, which is based on the *Gabriel graph (GG)* [25], a well-known geometric planar graph construction. A Gabriel graph for a finite point set $S$ is constructed by connecting any two nodes $v$ and $w$ of $S$ if and only if the circle with diameter $(v, w)$ contains no other node of $S$ (see Fig. 4.6a). It is proved in [23] that each node checking this condition for its neighbors only is sufficient to locally construct a connected planar subgraph of the unit disk graph.

Alternatively, a *relative neighborhood graph (RNG)* [26] can be constructed by checking emptiness of the intersection between the circles centered at one node and passing the other one, i.e if there is no other node whose distance is less or equal to the distance between node $v$ and $w$ (see Fig. 4.6b). GG and RNG are so-called proximity graphs and both belong to the general class of β-skeletons [27]. Among the β-skeletons, GG and RNG are the extreme cases of localized planar graph construction, since using a greater area than the area defined by RNG might result in a disconnected graph, and using a smaller area than defined by GG might result in intersecting graph edges. Bose et al. [27] investigated the spanning ratio of GG and

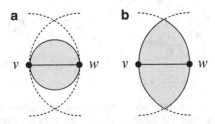

**Fig. 4.6** Planar graph construction based on β-skeletons. (**a**) An edge is preserved in the Gabriel graph construction if the circle with diameter $(v, w)$ contains no other neighbor than $v$ and $w$ respectively. (**b**) A relative neighborhood graph contains an edge, if the intersection of the circles centered at $v$ and $w$ is empty

RNG. The spanning ratio of a graph $G$ is defined by the maximum ratio of the Euclidean length of the shortest path connecting two arbitrary nodes $X$ and $Y$ in $G$ and their direct Euclidean distance. Since GG construction will preserve each edge of RNG construction, the spanning ratio of GG is less than spanning ratio of RNG. It is shown in [27], that spanning ratio of GG is $\Theta(n^{1/2})$ in the worst case, while RNG has a spanning ratio of $\Theta(n)$. The result shows that planar graph construction based on $\beta$-skeletons will produce graphs with spanning ratio depending on the number of network nodes.

The Delaunay triangulation is a good spanner for geometric graphs, since the spanning ratio is known to be constant [28]. Thus, Delaunay triangulation has recently been investigated to serve as an alternative planar graph construction. However, a Delaunay triangulation cannot be constructed locally, since it may contain arbitrary long edges. In recent publications, localized construction of planar graphs using local Delaunay triangulations on neighbor sets has been proposed [29–31]. These algorithms produce a planar graph with a constant spanning ratio but have an increased communication cost opposed to GG and RNG construction.

Li, Stojmenovic, and Wang [32] described *PDT (partial Delaunay triangulation)* as portion of Delaunay triangulation whose edges can be locally confirmed, based on 1-hop or 2-hop knowledge. The structure, like GG, does not require any message exchange between neighbor, and GG is subgraph of it. It is somewhat denser than GG, resulting in somewhat shorter hop counts for face routing, described in the next section.

### 4.4.2   The Face Routing Principle

A geometric planar graph partitions the plane into faces bounded by the polygons made up of the edges of the graph. The main idea of the FACE algorithm is to route a packet along the interiors of the faces intersected by the straight line connecting source node S and destination D (see Fig. 4.5). Each face interior is traversed by applying the well-known *right-hand rule (left-hand rule)*, i.e., a packet is forwarded along the next edge clockwise (counterclockwise) from the edge where it arrived. When the packet arrives at an edge intersecting the line connecting S and D, by skipping that edge the next face intersected by this line is handled in the same way. For example in Fig. 4.5 a packet routed from source S to destination D visits the faces $F_1 \ldots F_4$. The algorithm proceeds until the destination node is eventually reached or if the first edge of current face traversal is traversed twice in the same direction. In the latter case, the destination node is not reachable. Face routing is proved to be loop free and to guarantee delivery in static connected planar geometric graphs [33].

The relation of path length produced by face routing opposed to the length of the shortest path increases as the average degree of the network is increased. This surprising property of face routing can be explained by the fact that the subgraph used by face routing has at most an average degree of 6, while the complete graph used for shortest path finding has an increasing average degree [23]. The path produced by successful greedy routing is comparable to the one produced by Dijkstra's single source shortest path algorithm, thus, Bose et al. [23] proposed a combination of FACE algorithm with distance-based greedy routing, termed as *GFG*. A packet arriving at a concave node is switched into recovery mode and routed along faces until reaching a node closer to the destination than the position of the concave node where recovery mode was entered. At this node routing is performed in greedy mode again. Karp and Kung [24] implemented the GFG algorithm, added medium access layer, renamed the protocol as GPSR, and conducted experiments with moving nodes.

A slight improvement of GFG can be obtained by the *sooner-back* method [34], which additionally considers each neighbor of the current forwarding node during face traversal. If there is a neighbor closer to the destination, face routing is canceled and the packet is sent to that node again in greedy mode.

### 4.4.3  Internal Nodes and Shortcuts

Face routing has an increased hop count opposed to Dijkstra's single source shortest path algorithm, since planar graph construction based on Gabriel graphs favors short edges over long ones. Datta et al. [34] improved the performance of GFG by the concept of *internal nodes* and *shortcut-based routing*. The improvements are termed as *GFG-I*, *GFG-S*, or *GFG-I-S* if both concepts are applied, respectively.

A subset $S$ of all network nodes $G$ is termed as *dominating set*, if each node of $G$ is either element of $S$ or has at least one neighbor in $S$. Nodes that belong to the dominating set are called internal nodes. If the dominating set is connected and nontrivial, GFG constrained on dominating sets will produce shorter paths since the search space of planar graph routing is reduced to a subset of all nodes. More precisely, face routing is performed on edges resulting from Gabriel graph construction on internal nodes only. If a concave node is no internal node it forwards the message to one of its adjacent internal nodes. From there on the message is forwarded along internal nodes only until the local minimum is handled or the destination node is eventually reached.

To construct a dominating set locally, Datta et al. adopted the distributed algorithm proposed by Wu and Li [35], which is further improved in [36]. The algorithm in [36] does not require any communication between neighboring nodes to decide dominating status (other than messages needed to learn the position of neighboring nodes). The algorithm is based on the concept of *intermediate* nodes used to define a dominating set, and two additional rules (based on *inter-gateway* and *gateway* nodes) used to reduce the number of internal nodes while preserving network connectivity. A node is an *intermediate* node if it has two unconnected neighbors. A node $A$ is covered by neighboring node $B$ if each neighbor of $A$ is also neighbor of $B$, and $key(A) < key(B)$, while key is a record $(d, x, y)$ consisting of node degree $d$ (number of neighbors) and node position $(x, y)$. Nodes not covered by any neighbor are *inter-gateway* nodes. A node $A$ is covered by two connected neighboring nodes $B$ and $C$ if each neighbor of $A$ is also neighbor of either $B$ or $C$ (or both), *key*(*A*) < *key*(*B*), and *key*(*A*) < *key*(*C*). An intermediate node not covered by any neighbor becomes an *inter-gateway* node. An inter-gateway node not covered by any pair of connected neighboring nodes becomes a *gateway* node.

In addition to the next forwarding node, there might be more neighbor nodes on the same path produced by FACE routing. For example in Fig. 4.5, the nodes A and B on the path produced by a traversal of face $F_1$ are within the transmission range of node S (the circle around S). When information about 2-hop neighbors is available, the concept of shortcut-based routing can be applied at each node. A forwarding node locally constructs the part of the planar graph seen by all its neighbors. On the basis of this information, a node can make a shortcut by sending the message to the last known hop directly instead of forwarding it to the next hop along the path. For example in Fig. 4.5, node S could send the packet to node B directly.

### 4.4.4 Energy Aware Routing with Guaranteed Delivery

Energy consumption in localized routing algorithms was considered by the greedy routing methods power-routing, cost-routing, and power-cost-routing. However, these methods do not guarantee delivery in connected unit disk graphs. Stojmenovic and Datta investigated *power-face-power (PFP)*, *cost-face-cost (CFC)*, and *power/cost-face-power/cost (PcFPc)* routing [37], which combine energy aware greedy routing schemes with face routing in the same way as applied in GFG.

In [37] also the concept of internal nodes and shortcuts applied to PFP, CFC, and PcFPc are investigated. According to the terminology used for GFG, the algorithms are termed as PFP-I-S, CFC-I-S, and PcFPc-I-S, respectively. Experimental results showed a notable improvement when the recovery procedure FACE is performed on internal nodes only, since the algorithm is applied on a subset of all nodes, thus, producing shorter paths while traversing the faces. Furthermore, an improvement can be observed when the principle of shortcuts are used during recovery mode to choose the best neighbor with respect to the considered energy metric. For example in Fig. 4.5 node S might select node B as power optimal next hop node by applying the same minimization criterion on possible next hop forwarding nodes A and B as it is used on all nodes in forwarding direction when power-routing is used.

Dominating set construction leads to increased energy consumption at internal nodes, since face routing on dominating sets considers internal nodes only. Consequently, a static selection of internal nodes results in a shorter lifetime of these nodes, which finally leads to a shorter lifetime of the whole network. Thus, with the same argument applied to cost-routing, a cost metric might be applied to dominating set construction, considering the nodes remaining battery power. This kind of energy-aware dominating set construction has been proposed by Wu et al. [38]. Roughly, the algorithm is an extension of the basic distributed dominating set construction from [35] with an additional rule for removing redundant nodes having low remaining battery power.

### 4.4.5   Restricting the Searchable Area

Efficient operation of face routing depends on the decision in the starting node if a face is being traversed in clockwise or counterclockwise direction. For example in Fig. 4.5 applying the right-hand rule to traverse the outer face $F_2$ leads to the path CE . . . LKJ until arriving at the edge (J, K) intersecting the line connecting source S and destination D. In contrast, if face traversal was started in the opposite direction the packet is forwarded along the significantly shorter path CGHI before switching to face $F_3$.

To cope with that suboptimality, Kuhn et al. proposed an extension of GFG algorithm limiting the searchable area during face traversal [39]. If the optimal path length $l$ between source S and destination D is known in advance, it suffices to limit exploration of faces to an ellipse with foci S and D (see Fig. 4.5) containing all points with sum of their distances from S and D less than $l$ (i.e., the optimal path is completely covered by the ellipse). When the forwarding algorithm hits the ellipse, it has to turn back and route the packet in the opposite direction. For example in Fig. 4.5, the edge (C, E) hits the ellipse, thus, packet forwarding along face $F_2$ applying the right-hand rule is interrupted at node C and performed in the opposite direction afterwards, which altogether leads to the path SABCGHIJMD.

In general, the length of the optimal path is not predictable. However, the principle of restricting face traversal to a bounding ellipse can be performed by adaptively increasing the size of the ellipse, i.e., if the ellipse is hit by the face traversal algorithm, its size is doubled and traversal is performed in the opposite direction. By using this adaptive mechanism Kuhn et al. defined the *greedy other adaptive face routing (GOAFR)* algorithm.

With the same argument used for GFG, for practical purposes adaptive face routing should fall back to greedy mode as soon as possible. In their publication [40] Kuhn et al. proposed a further improvement *GOAFR+* which, in contrast to GOAFR, uses a circle centered at the destination node D to restrict itself to a searchable area. During the algorithm execution, the radius of that circle is adapted in predefined steps according to the current distance from D. This circle is used to apply an elaborate "early fallback" technique to return to greedy routing as soon as possible (for a detailed description of GOAFR+ see [40]).

## 4.5   Beaconless Routing

Traditional greedy forwarding mechanisms need periodic hello messages (beaconing) transmitted with maximum signal strength by each node to provide current position information about all one-hop neighbors. This proactive component of greedy routing leads to additional energy consumption, which occurs independently of current data traffic.

Heissenbüttel and Braun [41] proposed the *beacon-less routing (BLR)* algorithm. The *contention-based forwarding (CBF)* by Füßler et al. [42] and *implicit geographic forwarding (IGF)* by Blum et al. [43] are implementing the same idea focusing on the integration of beaconless routing with the IEEE 802.11 MAC layer. Since no beacons are transmitted, a node is generally not aware of any of its neighboring nodes and just broadcasts a data packet. The main idea of beacon-less routing is that a neighboring node receiving the packet, calculates a small transmission timeout before forwarding the packet depending on its position relative to the last node and destination. The node located at the "best" position uses the smallest delay and retransmits the packet at first. The remaining nodes cancel their scheduled transmissions.

To ensure that all potential forwarding nodes detect the retransmission, only nodes within a certain *forwarding area* are allowed as candidate nodes for the next forwarding step. The forwarding area has the property that each node is able to overhear the transmission of every other node within that area. If nodes outside the forwarding area participate in the contention process it can happen that messages are duplicated. Suppose all nodes closer to the destination are eligible candidates (see Fig. 4.7), and suppose that the forwarding delay depends on the distance to the destination. Then $C_2$ retransmits the message first, $C_1$ notices the transmission and remains silent. $C_3$ is not able to overhear this transmission and retransmits the message a second time, which would lead to a packet duplication. Thus, only nodes in the forwarding area ($C_1$ and $C_2$) are proper candidates.

Another technique to prevent duplication is the *active selection* [42] of the candidate by the forwarder: Instead of broadcasting the full message, the forwarder broadcasts a control packet including the destination location to request the message transmission ("request to send", RTS). This request is answered by one or probably more than one candidate ("clear to send", CTS). Then the forwarder selects the most suitable candidate and forwards the packet (unicast). The advantage of this technique is the larger set of potential candidates, the disadvantage is the overhead due to additional control messages (RTS/CTS).

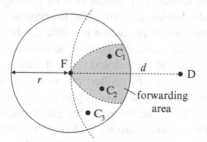

**Fig. 4.7** The beaconless routing principle: Forwarder $F$ broadcasts a packet, candidates $C_1$ and $C_2$ contend for the packet. $C_2$ is closer to the destination ($D$) and re-transmits the packet. $C_3$ is not in the forwarding area

The described algorithms are greedy strategies and decisions are locally optimal, just like in conventional greedy forwarding algorithms. If no candidates are available in the forwarding area (i.e. in a local minimum situation), these algorithms have to be assisted by a recovery strategy to achieve guaranteed delivery.

### 4.5.1  Beaconless Routing with Guaranteed Delivery

The preferred recovery method for conventional geographic routing is the face traversal on a planar subgraph, which is constructed from neighborhood information. But beaconless routing algorithm have no a priori knowledge of the neighborhood. Instead, part of this knowledge has to be gained by exchanging messages, if it is not implicitly given by the location of the nodes.

The BLR protocol uses a simple recovery mechanism, which is called *Request-response* approach [41]: The forwarder broadcasts a request and *all* neighboring nodes answer including their respective position in the response message. If no node is closer to the destination, the forwarder constructs a local planar subgraph (GG) from the position information of the neighbors and forwards the packet according to the right-hand rule. The position when entering backup mode is stored in the packet. Greedy forwarding is resumed as soon as a node is closer to the destination. Request-Response can be regarded as reactive beaconing, because all neighbors are involved in exchanging position information.

This message overhead can be avoided by using a *Select-and-Protest* approach [44]. The forwarder triggers a contention process, where only possible neighbors of a planar subgraph may answer. Afterwards, protest messages are used to correct wrong decisions. There are two possibilities how to solve the beaconless recovery problem by using the Select-and-Protest approach.

The first one is called beaconless forwarder planarization (BFP) and constructs a local planar subgraph, which can be used afterwards by a face routing algorithm (see Fig. 4.8). In BFP, the forwarder $F$ sends an RTS message, then the contention among the candidate nodes begins. In contrast to beaconless greedy forwarding, all the nodes within the transmission range are possible candidates and their timeout is based on the distance to the forwarder (not to the destination). A candidate $C$ is suppressed, i.e. it has to cancel the scheduled reply, if another candidate $C'$ located within the Gabriel circle over $(C, F)$ replied previously. Unfortunately, after this not only the Gabriel edges remain, because suppressed nodes could be the witness against other candidates. Therefore, the resulting graph might contain more edges than the Gabriel subgraph and wrong decisions have to be corrected by protest messages. These protest messages are necessary, even if another subgraph construction, e.g. the relative neighborhood graph, is used. It is shown in [44] that no undirected, planar, and connected proximity graph can be constructed without protests.

The second solution to the beaconless recovery problem is Angular Relaying [44]. It also begins with an RTS message broadcast by the forwarder $F$. This message contains the position of forwarder, the position of the previous hop and the

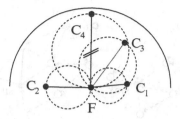

**Fig. 4.8** Beaconless Forwarder Planarization: Candidates answer in the order $C_1C_2C_3C_4$ according to their distance to the forwarder $F$. $C_3$ is suppressed; $C_4$ protests against $C_4$ because $(F, C_4)$ is not a Gabriel edge

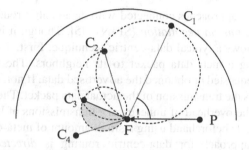

**Fig. 4.9** Angular Relaying: Candidates answer in the order $C_1C_2C_3C_4$ according to the angle $\alpha$. $C_1$ is selected first, then $C_2$ protest against $C_1$, $C_3$ protests against $C_2$. Finally, $C_3$ is selected without further protests

recovery direction (left-hand or right-hand). Now, the candidates answer according to a delay function that is based on the angle between previous hop, forwarder and candidate. This way, the first node $C$ in (counter-)clockwise order replies, but also in this case this is not the final candidate. Other candidates with larger delays may be located within the Gabriel circle over $(F, C)$. Such a candidate may send a protest against the first decision and becomes automatically the selected candidate. This decision can again be corrected by a further protest message, until no protest is issued any more. Then the last selected candidate becomes the next hop and gets the message from the forwarder (see Fig. 4.9).

## 4.6  Data-Centric Routing

Data-centric routing differs from topology-based and geographic routing in that sense, that messages are not forwarded to a specific host, which is determined by a network address or a geographic location. In data-centric routing, the sink issues a request for or interest in sensor data, and the respective sensors will answer this query. As an example, the sink may request to be alerted, if a sensor measures a temperature increase by more than 10 degrees. This request is propagated throughout the network and answered by sensor nodes once the event occurs.

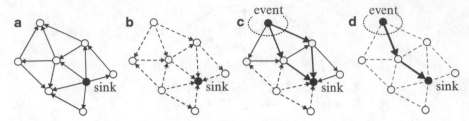

**Fig. 4.10** Directed diffusion [46]: (**a**) Interest propagation (*diffusion*), (**b**) gradient setup, (**c**) data routing along initial paths, (**d**) after reinforcement

One of the earliest approaches connected with data-centric routing is the *sensor protocol for information via negotiation* (SPIN) [45]. Though it is an application-level approach, it shows a typical data-centric technique: First, a sensor advertises new data by sending a meta-data packet to its neighbors. The neighbor checks whether it already requested or obtained the advertised data. If not, it sends a request message that triggers the transmission of the actual data packet. The idea behind this protocol is to save the overhead of unnecessary transmissions of long data packets by negotiating requests beforehand using a small amount of meta-data.

An often-cited approach for data-centric routing is *directed diffusion* [46] (Fig. 4.10). Directed diffusion is based on naming the data by attribute-value pairs. This way nodes can specify their interests for specific data. The message exchange is performed in the following way: First, an interest message is diffused, i.e., prop-agated by flooding algorithm through the network. This sets up so-called gradients that point backwards along the path of the interest propagation. The gradients deter-mine the path back to the originator of the interest. Sensors that hold data matching the interest send the requested data along the gradients. Gradients are maintained for each interest and used for path establishment. This path-memorization technique is combined with a reinforcement mechanism that gradually improves the paths from the initial gradients.

While directed diffusion performs a flooding of the network for initializing the gradients, the ACQUIRE protocol [47] tries to reduce this overhead: The sink issues a query that is forwarded in the network on a random or predetermined path. This query can be complex and consist of multiple interests. Once a sensor node receives this query, it tries to partially resolve the query by using information from a $d$-hop neighborhood. This requires a local information exchange within this neighborhood, which can be performed *on demand* by restricted flooding if the information has be-come obsolete. Once the query is completely resolved, the answer is sent back to the sink. The protocol is designed to answer so-called *one-shot queries* efficiently. The parameter $d$ for the look-ahead of information from the neighborhood repre-sents a tradeoff between latency and energy-efficiency: The larger the look-ahead the faster the query completion, but the higher the overhead for message exchange in the $d$-hop neighborhood, which increases the energy-consumption.

## 4.7  Discussion of the Presented Algorithms

The described routing algorithms follow different ideas and no algorithm can be the best in all disciplines. Thus, the question about efficiency is manifold: How reliable is an algorithm? How efficient does it use resources? And how fast can a message be delivered? This can be quantified by the performance metrics that are described in the following.

The *delivery rate* of a routing algorithm is defined as the fraction of successful delivered messages over the total number of messages created at the source node. This is a reasonable quantity for discussing the performance of greedy routing algorithms, which might fail in connected wireless networks. Additionally, *flooding rate* is used as a measure of communication overhead of multipath and flooding-based strategies. Flooding rate is defined as the ratio of the number of message transmissions needed by the algorithm and the shortest possible path between source and destination node. Algorithms with guaranteed delivery have always a delivery rate of 1. Thus, *dilation* is often used in literature to express the performance of these algorithms. This quantity is defined as the ratio of hop count for the given method and the hop count produced by the shortest path algorithm. The following two subsections discuss both greedy forwarding and combined greedy face routing, respectively. The third subsection discusses physical layer impact on nearly all existing geographic routing algorithms.

### 4.7.1  Characteristics of Greedy Forwarding

The delivery rates for DIR, GEDIR, and MFR are comparable and greatly depend on the network degree [11]. In sparse networks with average degree 4 the delivery rate is only about 50%. Running in dense networks the methods achieve delivery rates over 90%. The 2-hop variants of GEDIR, MFR, and DIR provide a minor improvement below 10% for sparse networks. It can also be observed that GEDIR and MFR methods select the same paths in most cases and when successful are competitive with Dijkstra's single source shortest path algorithm. Paths selected by DIR tend to be slightly longer [11].

NC was introduced as an alternative for NFP, which was experimentally observed to have low success rates due to greedy routing failures [15]. For small network areas, maximizing advance consumes less power than choosing nearest neighbors or neighbors closest in direction to $D$. However, for larger network sizes choosing nearest neighbors performs better than direction-based routing or maximizing advance [15]. It is shown by simulation that delivery rate of power-routing is competitive with MFR, DIR, and GEDIR, while it outperforms all known greedy routing methods regarding minimal power consumption. Additionally, it is shown in [15] that power-, cost-, and power-cost-routing is competitive with Dijkstra's single source shortest path algorithm using the general power, cost, and power-cost metric, respectively.

The forwarding area of known beacon-less routing mechanisms is covering at most 0.25 of the total transmission range. Thus, in sparse networks beacon-less routing will sooner lead to greedy routing failures opposed to traditional greedy routing mechanisms using at most 0.5 of the total transmission range (see transmission range in forward direction in Fig. 4.1). Performance evaluation in dense networks with no or low mobility show that beacon-less routing is comparable to conventional position-based routing mechanisms. However, under high mobility beacon-less routing strongly outperforms conventional position-based routing mechanisms suffering from outdated neighbor information [41].

Simulation results show superiority of V-GEDIR, CH-MFR, and R-DIR over DREAM and LAR. The proposed algorithms have higher delivery rates while flooding rates are reduced. The latter is explained by the fact, that in contrast to DREAM and LAR not all nodes inside the angular range will forward the message in the next routing step [21].

In general, alternate and disjoint scheme does not guarantee delivery, but higher success rates can be observed for disjoint scheme compared with alternate one [18]. Delivery rates of the proposed multipath strategies (in particular disjoint c-greedy) are comparable to the best existing restricted directional flooding algorithms, while the linear communication overhead is reduced to $O(n^{1/2})$. It was observed experimentally that $c < 4$ are reasonable choices for $c$, while the additional success rate for $c > 3$ does not compensate for additional flooding rate [18].

Assuming a network with $n$ nodes and a uniform two-dimensional node distribution, in average each presented single-path greedy algorithm creates $O(n^{1/2})$ packets to deliver a message between two arbitrary selected nodes. Nevertheless, the methods differ in the amount of traffic produced and memory needed to keep neighbor information up to date. Total communication complexity caused by beaconing and state volume per device depends on the *locality* of the presented method, i.e., if 1-hop, 2-hop or no neighbor information is needed at all.

### 4.7.2 Characteristics of Planar Graph Routing

The performance evaluation in [23] shows that average dilation of GFG is depending on both, the number of nodes and the average degree of the network. In sparse networks with a high number of nodes, average dilation is increased significantly. For instance, for 100 nodes and average degree of 4 a path produced by GFG is more than three times larger the shortest path in average. For dense networks, average dilation tends to the optimal value 1.0 and the number of nodes has almost no impact. This property of GFG can be explained by the fact that face routing is performed as a recovery mechanism only and, consequently, in dense networks routing is done almost only by the greedy routing part which has comparable performance to the shortest path algorithm.

Used in combination, the concept of shortcut routing and internal nodes drastically reduces the average path length opposed to FACE and even GFG. In particular, in networks with low node degree the excess of additional path length compared with Dijkstra's single source shortest path is reduced about half of that of GFG algorithm [34]. It can be observed that for sparse networks performance improvement results mainly from the internal node concept. This is due to the dominating set construction used by GFG-I-S. It holds that Dijkstra's single source shortest path algorithm applied only on the internal nodes produced by the distributed dominating set construction (without the second rule) always creates the shortest possible path between any two nodes [35]. Consequently, face routing will produce paths closer to the optimal one, since only a subset of edges lying outside the optimal paths is removed from the graph.

Simulation experiments show that average power consumption is significantly reduced by PFP-I-S, while the advantage of using shortcut procedure is notable, giving more benefits opposed to the concept of internal nodes [37]. For a network with a low degree of 4, the measured excess of power compared with Dijkstra's shortest weighted path algorithm is about 31%, while for a dense network with degree of 10 the excess reduces to 15%. The effect of cost- and power-cost-routing is measured in terms of number of successful routing tasks until the first forwarding node fails. It is observed that a combination of power and cost is better than power or cost alone. Additionally, power-, cost-, and power-cost-aware routing algorithms were superior with respect to all nonpower and noncost aware algorithms. The best performing localized algorithm PcFPc-I-S achieved a network life of about 83%–92% compared with Dijkstra's single source shortest weighted path algorithm.

In the unit disk graph model and under the assumption that the distance between two nodes is always greater than a possibly small but fixed constant, it is shown that the cost produced by GOAFR is bounded by the square of the cost (e.g. hop count) needed to route a packet along the optimal path. This is asymptotical optimal, i.e. asymptotically no local position-based routing algorithm performs better than GOAFR [40]. When GOAFR+ is applied on a precomputed dominating set with bounded average degree, asymptotical optimality of GOAFR+ can even be proved for arbitrary unit disk graphs, but in contrast the cost function considered has to be bound by a linear function from below. Unlike for linearly bounded cost functions, the cost of a local position-based routing algorithm cannot be bounded by the cost of an optimal path for *super-linear* cost functions (i.e., a function which is not linearly bounded). Linearly bounded cost functions are expected to be the relevant ones from a practical point of view [40]. For instance, the cost function for power-adaptive transmission sometimes expressed as a superlinear function in literature will never drop below a certain threshold regarding the energy required for the transmission of a message and thus is also linearly bounded in practice. In addition to the worst case analysis, simulation results in [40] show that even in average GOAFR+ outperforms GFG and all known variants of adaptive face routing. In particular, the most significant performance improvement can be observed for networks with a critical network density around 4.5 nodes per unit disk.

### 4.7.3 Characteristics of Data-Centric Routing

Data-centric routing protocols can be used in sensor networks where no position information is available. Their efficiency depends on the type of requests used for querying sensor data. Repeated queries and frequent sensor readings justify a larger overhead for the setup of an efficient path system, while one-shot queries should be answered with as little communication as possible.

Directed diffusion is a flooding-based approach, which has an inherently high message complexity. SPIN follows also the flooding principle, but it is triggered by events in contrast to directed diffusion, where the sink floods the request message. Directed diffusion is well-suited for repeated querying, as the gradient setup is already done and the reinforcement can gradually increase the route quality. Unlike the flooding-based protocols, ACQUIRE follows the principle of rumor routing [50], where a query is sent on a random walk through the network. This protocol is well-suited for one-shot queries. Flooding can be avoided or restricted to a small neighborhood, but then answering the query takes more time.

## 4.8 Thoughts for Practitioners

Most routing algorithm were designed with respect to certain assumptions, which are not valid in reality. Most prominent example is the unit disk graph assumption, which does not reflect real signal propagation and can only be regarded as a coarse approximation. This has to be considered when routing algorithms are implemented. Geographic routing algorithms usually rely on the unit disk graph assumption and require precise localization. But that does not imply that they do not work when these assumptions are not fulfilled. Certain adaptions have to be made, e.g., if the transmission range is unstable, if one has to cope with the implications of a realistic physical layer. However, delivery guarantees cannot be given for all cases.

### 4.8.1 Unstable Transmission Ranges

The original FACE algorithm guarantees delivery and proper operation on unit disk graphs. In a real life scenario, obstacles between mobile hosts, weather conditions, or unrelated radio transmissions (just to mention a few effects) might lead to some instability in transmission range. Even when only bidirectional links are considered (e.g. a communication link from $A$ to $B$ is valid only if it has been acknowledged by $B$) the algorithm might fail due to slight variations of the transmission ranges [48]. Barierre et al. proposed a routing scheme [48] (termed *robust-GFG*) based on the idea of face routing but ensuring delivery even if the variation in transmission range is at most $\sqrt{2}$, i.e. a transmission range as shown in Fig. 4.11 has to be inside the area defined by a minimum and maximum transmission range $r$ and $R$, respectively, while the ratio of $R$ and $r$ has to be at most $\sqrt{2}$.

**Fig. 4.11** The transmission range of mobile host $S$ varies between minimum and maximum transmission range $r$ and $R$

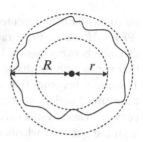

Restricting the variation to $\sqrt{2}$ guarantees that for any two adjacent nodes X and Y each node inside the circle with diameter passing X and Y will be seen at least by one of them [48]. This property is used in the *completion phase* of the algorithm to locally construct a virtual supergraph $G$ of the underlying physical network graph. More precisely, each node X checks for each adjacent edge (X, Y) if there are neighbor nodes possibly not seen by node Y and causing the edge (X, Y) being removed from the local Gabriel graph. Node X announces these nodes to Y, since they are exactly the nodes that might lead to an inconsistent view of the resulting planar graph, if the Gabriel graph construction was applied on the original network graph only.

The resulting supergraph is sufficient to apply the Gabriel graph method, to construct a planar graph containing edges from the original network graph and virtual edges resulting from the completion phase. Finally, routing is performed on the extracted planar graph by using the face routing algorithm extended by a *virtual routing* component to handle the virtual edges (for a detailed description see [48]). The construction of robust planar graph is further improved by Li et al. [51]. The paper describes a fuzzy unit disk graph, which improves the method from [48] in terms of communication overhead and virtual edges added.

The graph model investigated in [48] was also studied by Kuhn et al. [52]. They investigated properties and routing algorithms for this model and proposed an expanded ring flooding protocol, which follows a greedy algorithm as long as possible, and switches to a flooding protocol at failure points. A failure point initiates flooding at increasing (doubling radius each time) hop distances looking for a node that is closer to destination than the failing node. The failing node then asks that node to continue with greedy mode. The approach has some similarity with recovery the phase proposed by Finn [13], which uses fixed arbitrary distance (thus not being able to guarantee delivery) instead of iterating and doubling it.

## 4.8.2 Physical Layer Impact on Routing

Besides unstable transmission regions, other side-effects of the physical layer can be observed in practice, e.g., fluctuations of the received signal strength, which affect a successful reception. In some articles, the probability of a failed or successful transmission is represented by the bit error rate (BER) or the packet reception rate (PRR). However, these rates are often assumed to be constant, independent of the distance between nodes. In practice, one can observe that the probability of reception varies with the distance. There is a region around the sender where it is almost 1, while it becomes 0 at a larger distance. The interesting cases are located in the transitional region, i.e. between the region of full reception and the region of no reception. This is especially important for routing decisions, because it is connected with the question whether it is better to choose close node that does not enable a large progress, but a successful transmission or a distant node, where the probability of reception is low and one possibly has to take re-transmissions into account.

These considerations have led to new aspects under which routing protocols have to be developed [53]. One important aspect is the routing metric, i.e. the decision criterion for choosing a neighbor when forwarding a packet. Routing protocols based on a hop count metric usually favor long links, which are probably lossy and require multiple transmission attempts, i.e., a hop count metric that does not take retransmissions into account fails in minimizing the total number of transmissions.

An alternative metric has been proposed by De Couto et al. [54]. It is based on the packet loss rate for transmissions and acknowledgments and represents the expected transmission count (ETX) instead of the hop count. Experiments with the reactive routing protocol DSR showed a performance improvement over the pure hop-count based metric. Using this metric requires knowledge of the packet loss on the sender and the receiver side, which is done by sending probing messages and evaluating the acknowledgements.

Seada et al. [55] propose different techniques to tackle physical layer problems in geographic routing algorithms. First, the routing metric is not only based on distance, but on the product of distance and packet reception rate (PRR). Then, the nodes with a low PRR or a high distance are blacklisted (based on a threshold or a relative proportion of all neighbors) and become unavailable for forwarding. Thus, the worst candidates are sorted out to avoid many unsuccessful transmission attempts. Using this routing metric requires the PRR to be known to the sender. If this cannot be estimated accurately from channel quality indicators, probing messages for PRR measurement are necessary.

In geographic routing algorithms, the progress or distance to the target can be measured in each step, and each step is connected with a certain transmission cost. Therefore, Kuruvila et al. [56, 57] propose routing metrics based on a cost-over-progress ratio. The cost depends on the expected hop count (EHC), which includes retransmissions and lost acknowledgements.

### 4.8.3  The Effect of Localization Errors

Theoretical models for geographic routing assume exact information about the nodes. However, in practice there is always some estimation error on information about physical device location, which is depending on the environment and the

**Fig. 4.12** Disconnection due to incorrect edge removal. Because of incorrect estimated location information $A'$ about node $A$ the edge $(S, B)$ violates the Gabriel graph condition and is wrongly removed (example from [49])

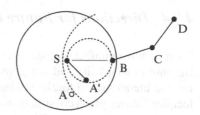

localization system being used. The effect of such localization errors on face routing has been studied by Seada et al. [49] for the first time.

It can be observed that location inaccuracies might lead to errors during the localized planar graph construction, which finally result in pathological forwarding decisions when face routing is applied. For instance, suppose that in Fig. 4.12 the estimated location $A'$ of node $A$ is inside the circle with diameter $(S, B)$, while the remaining nodes $S, B, C$, and $D$ have an exact location estimate. Localized planar graph construction applied in node $S$ would lead to removal of edge $(S, B)$, while on the other hand, node $A$ does not preserve a link to node $B$ since node $A$ and $B$ are not connected. Hence, the resulting planar graph is disconnected and face routing does thus not guarantee message delivery any more even if there is a path from source to destination. In addition to the example of disconnection caused by incorrect edge removal, similar error scenarios are constructed in [49] illustrating delivery failures due to cross links, and destination location inaccuracies.

In general, the forwarding errors result from removal of essential edges leading to network partitions and insufficient removal of edges resulting in forwarding loops. On the basis of an informal analysis which is also supported by simulation results, Seada et al. conjecture that the disconnection problem caused by edge removal seems to have the highest probability, and solving this might give the most gains of performance [49]. They propose a fix to the face algorithm, which requires a modification to the localized planar graph construction (here termed request-GFG). When a node A is about to remove an edge (A, B) due to a node C inside the circle with diameter (A, B), it has to send an inquiry to B whether it sees node C. Node A must not remove the edge until and unless it gets a reply from B, indicating that B indeed sees node C [49]. Note, in contrast to robust-GFG this fix is done by a simple request response protocol between any two neighbor nodes, since the unit disk graph model is still assumed.

Excessive message exchange may improve the accuracy of localized position estimates. For instance, a geographic position system based on iterative relaxation is proposed in [58]. The authors [58] report thousands of iterations in their distributed protocol, which means that each node should send and receive thousands of messages before the position becomes reasonably accurate. The obvious problem with the protocol is its extreme communication cost to derive position information. The routing then follows GFG [23] or its variants.

## 4.8.4   Directions for Future Research

There is still a number of open research issues needed to be addressed in the future. Because of simplicity and low communication cost, Gabriel graphs are preferably used in literature as a localized planar graph construction method. However, other localized planar graph constructions are known, and generally, it remains an open issue to investigate the tradeoff between path lengths produced by face routing on such graphs and the communication cost needed to extract them locally. Furthermore, robustness remains still an open issue. Robust face routing is currently investigated on wireless networks with a limited variation in minimum and maximum sending radius for each device. The applicability of face routing in practice would benefit from techniques for locally extracting FACE routing supporting subgraphs from arbitrary subsets of minpower graphs. In addition, the related problem of location inaccuracies has to be investigated further on and it remains open to find memoryless solutions, which guarantee delivery even in this case. Mobility-caused loops might lead to disconnection even if there is a path from source to destination. Further investigation of techniques coping with dynamic network topologies are of interest. Finally, considering QoS aspects, congestion in neighbor nodes and end-to-end delay are interesting future improvements of existing geographic routing algorithms.

## 4.9   Conclusion

Geographic routing is an appropriate routing paradigm for sensor networks, because it enables the design of efficient and scalable protocols that use only local decisions and need minimum storage capabilities. Location awareness as a main prerequisite is needed anyway in many sensor applications, since knowing the location of a sensed event is crucial for its evaluation. The global objective of a routing task can be achieved by basic greedy forwarding algorithms based on local decisions and without memorizing past traffic. The characteristics of such algorithms depend on the optimization criterion applied in each forwarding step. In general, basic greedy routing has a performance close to the shortest weighted path algorithm in dense networks, while delivery rate decreases significantly for sparse networks. Delivery rate has been improved in several directions, like providing 2-hop neighbor information, message forwarding in backward direction and multipath or flooding-based message forwarding. Memorization can be used to provide guaranteed delivery based on flooding but also for single-path strategies. However, these strategies may require an increased communication overhead and abandon the stateless property of single-path greedy routing.

Face routing is loop-free and guarantees delivery for static wireless networks, while preserving the memoryless property and locality of greedy routing techniques. Face routing applied on its own tends to increased communication overhead opposed to the number of messages needed when greedy routing was successful. Thus, face routing is mainly used as a recovery mechanism to overcome local minima where greedy routing fails. Some improvements of face routing presented here are focused on reducing hop count by extracting *better* planar graphs or enabling nodes to choose more nodes than the next hop node along current face traversal. Additionally, energy considerations are included in planar graph construction as well. A further improvement copes with the suboptimality resulting from the fact that path length depends on the first decision about the direction of face traversal. Finally, face routing was originally defined for an idealized wireless network with uniform transmission ranges and stationary located network nodes. Recent research deals with the applicability of face routing in a practical network by making it robust against unstable transmission ranges and dynamically changing network topologies.

**Acknowledgments** This work is supported by NSERC Strategic Grant on sensor and actuator networks, and the UK Royal Society Wolfson Research Merit Award.

## Terminologies

*Beaconless routing.* Routing scheme based on position-information, where the nodes are not aware of their neighbors, because they do not exchange beacons (hello messages).

*Concave node.* Node that has no neighbor closer to the destination.

*Data-centric routing.* A routing scheme, where certain nodes announce an interest in getting some data and other nodes forward data messages according to the interest.

*Delivery rate.* Fraction of successfully delivered messages; measure for the reliability of a routing algorithm.

*Dilation.* Ratio of the number of hops used by an algorithm and the hop count of the shortest path.

*Dominating set.* A subset of nodes in a network that covers all nodes, i.e., each non-dominating set node has at least one neighbor in the dominating set.

*Face routing.* Routing along the faces of the communication graph, usually used as a recovery mechanism in situations in which greedy routing fails.

*Flooding.* Disseminating a message throughout the whole network.

*Flooding rate.* Ratio of the number of message transmissions and the hop-length of the shortest path; a measure for the message-efficiency of a routing algorithm.

*Greedy forwarding.* Position-based routing paradigm, where a node always forwards a message to one of its neighbors that minimizes the distance to the destination (or fulfills a different local optimization criterion).

*Multipath strategy.* A routing strategy, where a message is sent to the destination by forwarding multiple copies on different paths to the destination.

*Planarization.* Construction of a planar subgraph of the communication graph.

*Right-hand rule.* Rule for forwarding packets along the faces of the communication graph: a packet is forwarded along the next edge in clockwise direction from the edge where it arrived.

*Routing metric.* A function that assigns a weight or cost to a pair of nodes; it is a measure for comparing alternative routes.

*Spanning ratio.* Maximum ratio of shortest path between two nodes and their Euclidean distance.

*Unit disk graph (UDG).* Abstract model for wireless networks with fixed transmission radii: the UDG of a node set contains an edge between two nodes only if the distance between them is covered by the transmission range.

## Questions

1. Which greedy forwarding strategies are loop-free and which are not?
2. Can greedy forwarding find an energy-optimal path by applying a power metric and selecting a neighbor such that the power per unit of progress is minimized?
3. Does face routing work on arbitrary unit disk graphs?
4. Does planarization require global knowledge of the network?
5. Face routing guarantees delivery in unit disk graphs, can it also guarantee short routes?
6. Consider the network in Fig. 4.5. Which route is constructed by a variant of the GFG algorithm that starts distance-based greedy forwarding, switches to face routing (right-hand rule) when arriving at a concave node, and resumes greedy forwarding once a node is found that is closer to the destination than the concave node?
7. Greedy forwarding is more efficient than face routing on a planar subgraph. Should one therefore modify the GFG algorithm and stop face routing/resume greedy forwarding as early as possible, i.e., whenever a node closer to the destination is available?
8. What is the dilation and the flooding rate of face routing in Fig. 4.5?
9. Beaconless routing algorithms are able to perform position-based routing without knowing the neighbors' positions in advance. What is then the benefit of beaconing?
10. Can geographic routing be applied to non-unit disk graphs?

## References

1. S. Ramanathan and Martha Steenstrup. A survey of routing techniques for mobile communications networks. *Mobile Networks and Applications*, 1(2):89–104, 1996.
2. Elizabeth M. Royer and Chai-Koeng Toh. A review of current routing protocols for ad hoc mobile wireless networks. *IEEE Personal Communications*, 6(2):46–55, April 1999.
3. Josh Broch, David A. Maltz, David B. Johnson, Yih-Chun Hu, and Jorjeta Jetcheva. A performance comparison of multi-hop wireless ad hoc network routing protocols. In *Proceedings of the 4th ACM/IEEE International Conference on Mobile Computing and Networking (Mobi-Com'98)*, pages 85–97, 1998.
4. Kemal Akkaya and Mohamed Younis. A survey on routing protocols for wireless sensor networks. *Ad hoc Networks*, 3(3):325–349, May 2005.

5. I.F. Akyildiz, W. Su, Y. Sankarasubramaniam, and E. Cayirci. Wireless sensor networks: a survey. *Computer Networks*, 38:393–422, 2002.
6. Jeffrey Hightower and Gaetano Borriella. Location systems for ubiquitous computing. *IEEE Computer*, 34(8):57–66, 2001.
7. Srdjan Capkun, Maher Hamdi, and Jean-Pierre Hubaux. GPS-free positioning in mobile ad-hoc networks. In *Proceedings of the Hawaii International Conference on System Sciences (HICSS'01)*, 2001.
8. Ivan Stojmenovic. Location updates for efficient routing in ad hoc networks. In Ivan Stojmenovic, editor, *Handbook of Wireless Networks and Mobile Computing*, Chapter 21, pages 451–471. Wiley, 2002.
9. Ivan Stojmenovic and Stephan Olariu. Data-centric protocols for wireless sensor networks. In *Handbook of Sensor Networks*, Chapter 13, pages 417–456. Wiley, 2005.
10. Christopher Ho, Katia Obraczka, Gene Tsudik, and Kumar Viswanath. Flooding for reliable multicast in multi-hop ad hoc networks. In *Proceedings of the 3rd International Workshop on Discrete Algorithms and Methods for Mobile Computing and Communications (DIAL-M'99)*, pages 64–71, 1999.
11. Ivan Stojmenovic and Xu Lin. Loop-free hybrid single-path/flooding routing algorithms with guaranteed delivery for wireless networks. *IEEE Transactions on Parallel and Distributed Systems*, 12(10):1023–1032, October 2001.
12. Hideaki Takagi and Leonard Kleinrock. Optimal transmission ranges for randomly distributed packet radio terminals. *IEEE Transactions on Communications*, 32(3):246–257, March 1984.
13. Gregory G. Finn. Routing and addressing problems in large metropolitan-scale internetworks. Technical Report ISI/RR-87-180, Information Sciences Institute (ISI), March 1987.
14. Ting-Chao Hou and Victor O.K. Li. Transmission range control in multihop packet radio networks. *IEEE Transactions on Communications*, 34(1):38–44, January 1986.
15. Ivan Stojmenovic and Xu Lin. Power-aware localized routing in wireless networks. *IEEE Transactions on Parallel and Distributed Systems*, 12(11):1122–1133, November 2001.
16. Randolph Nelson and Leonard Kleinrock. The spatial capacity of a slotted aloha multihop packet radio network with capture. *IEEE Transactions on Communications*, 32(6):684–694, June 1984.
17. Evangelos Kranakis, Harvinder Singh, and Jorge Urrutia. Compass routing on geometric networks. In *Proceedings of the 11th Canadian Conference on Computational Geometry (CCCG'99)*, pages 51–54, August 1999.
18. Xu Lin and Ivan Stojmenovic. Location-based localized alternate, disjoint and multi-path routing algorithms for wireless networks. *Journal of Parallel and Distributed Computing*, 63:22–32, 2003.
19. Stefano Basagni, Imrich Chlamtac, Violet R. Syrotiuk, and Barry A. Woodward. A distance routing effect algorithm for mobility (DREAM). In *Proceedings of the 4th Annual ACM/IEEE International Conference on Mobile Computing and Networking (MOBICOM-98)*, pages 76–84, October 1998.
20. Young-Bae Ko and Nitin H. Vaidya. Location-aided routing (LAR) in mobile ad hoc networks. In *Proceedings of the 4th Annual ACM/IEEE International Conference on Mobile Computing and Networking (MOBICOM-98)*, pages 66–75, October 1998.
21. Ivan Stojmenovic, Anand Prakash Ruhil, and D. K. Lobiyal. Voronoi diagram and convex hull based geocasting and routing in wireless networks. In *Proceedings of the 8th IEEE Symposium on Computers and Communications ISCC*, pages 51–56, July 2003.
22. Johnson Kuruvila, Amiya Nayak, and Ivan Stojmenovic. Progress and location based localized power aware routing for ad hoc and sensor wireless networks. *International Journal of Distributed Sensor Networks*, 2(2):147–159, July 2006.
23. Prosenjit Bose, Pat Morin, Ivan Stojmenovic, and Jorge Urrutia. Routing with guaranteed delivery in ad hoc wireless networks. In *Proceedings of the 3rd ACM International Workshop on discrete Algorithms and Methods for Mobile Computing and Communications (DIAL-M'99)*, pages 48–55, August 1999.

24. Brad Karp and H. T. Kung. GPSR: Greedy perimeter stateless routing for wireless networks. In *Proceedings of the 6th ACM/IEEE Annual International Conference on Mobile Computing and Networking (MobiCom'00)*, pages 243–254, August 2000.

25. K.R. Gabriel and R.R. Sokal. A new statistical approach to geographic variation analysis. *Applied Zoology*, 18:259–278, 1969.

26. G.T. Touissaint. The relative neighborhood graph of a finite planar set. *Pattern Recognition*, 12:261–268, 1980.

27. Prosenjit Bose, Luc Devroye, William Evans, and David Kirkpatrick. On the spanning ratio of gabriel graphs and beta-skeletons. In *Proceedings of the Latin American Theoretical Informatics (LATIN'02)*, April 2002.

28. J.M. Keil and C.A. Gutwin. Classes of graphs which approximate the complete euclidean graph. *Discrete and Computational Geometry*, 7:13–28, 1992.

29. Jie Gao, Leonidas J. Guibas, John Hershberger, Li Zhang, and An Zhu. Geometric spanner for routing in mobile networks. In *Proceedings of the second ACM International Symposium on Mobile Ad Hoc Networking and Computing (MobiHoc'01)*, pages 45–55, October 2001.

30. Xiang-Yang Li, Gruia Calinescu, and Peng-Jun Wan. Distributed construction of a planar spanner and routing for ad hoc wireless networks. In *Proceedings of the 21st Annual Joint Conference of the IEEE Computer and Communications Society (INFOCOM)*, pages 1268–1277, June 2002.

31. Xiang-Yang Li and Yu Wang. Quality guaranteed localized routing for wireless ad hoc networks. In *IEEE ICDCS 2003 (MWN workshop)*, 2003.

32. Xiang-Yang Li, Ivan Stojmenovic, and Yu Wang. Partial delaunay triangulation and degree limited localized bluetooth scatternet formation. *IEEE Transactions on Parallel and Distributed Systems*, 15(4):350–361, 2004.

33. H. Frey and I. Stojmenovic. On delivery guarantees of face and combined greedy face routing in ad hoc and sensor networks. In *Proceedings of the ACM Annual International Conference on Mobile Computing and Networking (Mobicom'06)*, Los Angeles, USA, 2006

34. Susanta Datta, Ivan Stojmenovic, and Jie Wu. Internal node and shortcut based routing with guaranteed delivery in wireless networks. In *Proceedings of the IEEE International Conference on Distributed Computing and Systems (Wireless Networks and Mobile Computing Workshop WNMC)*, pages 461–466, April 2001.

35. Jie Wu and Hailan Li. On calculating connected dominating set for efficient routing in ad hoc wireless networks. In *Proceedings of the 3rd International Workshop on Discrete Algorithms and Methods for Mobile Computing and Communications (DIAL M '99)*, pages 7–14, August 1999.

36. Ivan Stojmenovic, Mahtab Seddigh, and Jovisa Zunic. Dominating sets and neighbor elimination-based broadcasting algorithms in wireless networks. *IEEE Transactions on Parallel and Distributed Systems*, 13(1):14–25, January 2002.

37. Ivan Stojmenovic and Susanta Datta. Power and cost aware localized routing with guaranteed delivery in wireless networks. In *Proceedings of the Seventh International Symposium on Computers and Communications (ISCC'02)*, pages 31–36, July 2002.

38. Jie Wu, Fei Dai, Ming Gao, and Ivan Stojmenovic. On calculating power-aware connected dominating sets for efficient routing in ad hoc wireless networks. *Journal of Communications and Networks*, 4(1), March 2002.

39. Fabian Kuhn, Roger Wattenhofer, and Aaron Zollinger. Worst-case optimal and average-case efficient geometric ad-hoc routing. In *Proceedings of the 4th ACM International Symposium on Mobile Computing and Networking (MobiHoc 2003)*, pages 267–278, 2003.

40. Fabian Kuhn, Roger Wattenhofer, Yan Zhang, and Aaron Zollinger. Geometric ad-hoc routing: Of theory and practice. In *Proceedings of the 22nd ACM International Symposium on the Principles of Distributed Computing (PODC)*, Boston, Massachusetts, USA, pages 63–72, July 2003.

41. Marc Heissenbüttel and Torsten Braun. BLR: Beacon-less routing algorithm for mobile ad-hoc networks. *Elsevier's Computer Communications Journal*, pages 1076–1086, 2003.

42. Holger Füßler, Jörg Widmer, Michael Käsemann, Martin Mauve, and Hannes Hartenstein. Contention-based forwarding for mobile ad-hoc networks. *Ad Hoc Networks*, 1(4):351–369, November 2003.

43. Brian M. Blum, Tian He, Sang Son, and John A. Stankovic. IGF: A state-free robust communication protocol for wireless sensor networks. Technical Report CS-2003-11, Department of Computer Science, University of Virginia, April 21 2003.

44. H. Kalosha, A. Nayak, S. Rührup, and I. Stojmenovic. Select-and-protest-based beaconless georouting with guaranteed delivery in wireless sensor networks. In *Proceedings of the 27th IEEE International Conference on Computer Communications (INFOCOM)*, April 2008.

45. Wendi Rabiner Heinzelman, Joanna Kulik, and Hari Balakrishnan. Adaptive protocols for information dissemination in wireless sensor networks. In *Proceedings of the 5th annual ACM/IEEE international conference on Mobile computing and networking (MobiCom'99)*, pages 174–185, 1999.

46. Chalermek Intanagonwiwat, Ramesh Govindan, Deborah Estrin, John Heidemann, and Fabio Silva. Directed diffusion for wireless sensor networking. *IEEE/ACM Transactions on Networking*, 11(1):2–16, 2003.

47. Narayanan Sadagopan, Bhaskar Krishnamachari, and Ahmed Helmy. Active query forwarding in sensor networks. *Ad Hoc Networks*, 3(1):91–113, January 2005.

48. Lali Barriere, Pierre Fraigniaud, Lata Narajanan, and Jaroslav Opatrny. Robust position-based routing in wireless ad hoc networks with unstable transmission ranges. In *Proceedings of the fifth ACM International Workshop on Discrete Algorithms and Methods for Mobile Computing and Communications (DIAL-M'01)*, pages 19–27, 2001.

49. Karim Seada, Ahmed Helmy, and Ramesh Govindan. On the effect of localization errors on geographic face routing in sensor networks. Technical Report 03-797, University of Southern California USC, 2003.

50. David Braginsky and Deborah Estrin. Rumor routing algorthim for sensor networks. In *Proceedings of the 1st ACM international workshop on Wireless sensor networks and applications (WSNA'02)*, pages 22–31, 2002.

51. X.Y. Li, K. Moaveninejad, and W.Z. Song. Robust position-based routing for wireless ad hoc networks. *Ad Hoc Networks*, 3(5):546–559, September 2005.

52. Fabian Kuhn, Roger Wattenhofer, and Aaron Zollinger. Ad-hoc networks beyond unit disk graphs. In *Proceedings of the 2003 Joint Workshop on Foundations of Mobile Computing (DIALM-POMC)*, pages 69–78, September 2003.

53. Ivan Stojmenovic, Amiya Nayak, and Johnson Kuruvila. Design guidelines for routing protocols in ad hoc and sensor networks with a realistic physical layer. *IEEE Communications Magazine*, 43(3):101–106, 2005.

54. Douglas S. J. De Couto, Daniel Aguayo, John Bicket, and Robert Morris. A high-throughput path metric for multi-hop wireless routing. *Wireless Networks*, 11(4):419–434, 2005.

55. Karim Seada, Marco Zuniga, Ahmed Helmy, and Bhaskar Krishnamachari. Energy-efficient forwarding strategies for geographic routing in lossy wireless sensor networks. In *Proceedings of the 2nd International Conference on Embedded Networked Sensor Systems (SenSys'04)*, pages 108–121, 2004.

56. Johnson Kuruvila, Amiya Nayak, and Ivan Stojmenovic. Hop count optimal position-based packet routing algorithms for ad hoc wireless networks with a realistic physical layer. *IEEE Journal on Selected Areas in Communications*, 23(6):1267–1275, 2005.

57. Johnson Kuruvila, Amiya Nayak, and Ivan Stojmenovic. Greedy localized routing for maximizing probability of delivery in wireless ad hoc networks with a realistic physical layer. *Journal of Parallel and Distributed Computing*, 66:499–506, 2006.

58. A. Rao, S. Rathasamy, C. Papadimitriou, S. Shenker, and I. Stoica. Geographic routing without location information. In *Proceedings of the 9th Annual International Conference on Mobile Computing and Networking (MobiCom'03)*, pages 96–108, 2003.

# Chapter 5
# Geometric Routing in Wireless Sensor Networks

Jie Gao

**Abstract** This chapter surveys routing algorithms for wireless sensor networks that use geometric ideas and abstractions. Wireless sensor networks have a unique geometric character as the sensor nodes are embedded in, and designed to monitor, the physical space. Thus the geometric embedding of the network can be exploited for scalable and efficient routing algorithm design. This chapter starts with geographical routing that use nodes' geographical locations to guide the choice of the next hop node on the routing path. The scalability of geographical routing motivates more work on the design of virtual coordinates with which greedy routing algorithms are developed and applied to route messages in the network. The last section is concerned about data-centric routing, in which a query is routed to reach the sensor node holding data of interest. Thus the challenge is to discover the "source node" that possess the data as well as route the message there.

## 5.1 Introduction

The eternal goal and the most fundamental problem for any types of networks are to enable efficient information delivery between the peers. A routing protocol, which establishes routes between a pair of nodes efficiently and correctly so that messages can be delivered in a timely manner, is an important component of the network architecture.

Traditional Internet routing achieves scalability by address aggregation, in which the routes to multiple destinations are summarized by a single routing table entry. In multi-hop sensor networks, this is not applicable as the node IDs of nearby sensor nodes are by no means close to each other. However, wireless sensor networks have a unique geometric character as sensor nodes are embedded in, and designed to

J. Gao
Department of Computer Science, Stony Brook University, Stony Brook, NY 11794, USA
e-mail: jgao@cs.sunysb.edu

S. Misra et al. (eds.), *Guide to Wireless Sensor Networks*, Computer Communications and Networks, DOI: 10.1007/978-1-84882-218-4_5,
© Springer-Verlag London Limited 2009

monitor, the physical environment. Thus the physical locations of sensor nodes have provided a lot of opportunities to be exploited for efficient and scalable routing mechanisms in sensor networks.

This chapter surveys two topics in which geometric ideas are used extensively for routing in wireless sensor networks: *point-to-point routing* and *data-centric routing*.

*Point-to-point routing* is to find routes between source and destination efficiently. Because of the limited resources available at sensor nodes (communication, power, and memory constraints), it is important to have a light-weight routing protocol with low maintenance effort and small state information, in order to be scalable. The set of routing protocols covered in this survey achieves scalability by using the geographical locations of the sensor nodes, or virtual coordinates (aka. names) assigned to the nodes, with which routing is often conducted in a local and greedy manner.

The second category of geometric routing focuses on *data-centric routing*, i.e., routing to find the data of interest. This is motivated by the application-specific characteristics of sensor networks and that the node itself is not as interesting as the data it senses and delivers. In data-centric routing, a query carries a description of the data types it is looking for and the routing protocol routes the query to meet the data.

At last we will cover *location service and hierarchical routing* in sensor networks. Location service is an important supporting infrastructure component for point-to-point routing with geographical coordinates or virtual coordinates. Each sensor node often comes with an ID, assigned during manufacture, which uniquely identifies the node but is typically not used to aid routing. A point-to-point routing scheme often assigns a *name* or *routing address* to a node to which a routing protocol is applied. In geographical routing, for example, the geographical location of a node is its routing name. A source routes to a destination with the name of itself and the name of the destination. Location service is to maintain the mapping of IDs and routing names and support queries for node names. Location service can be considered as a special case of data-centric routing, in which the "data of interest" is the name of the destination given its ID.

This chapter focuses on the design principle and the main idea in each algorithm. Please refer to the original papers for implementation details and experimental evaluations. The typical setting is a large homogeneous sensor network with nearby nodes being able to communicate directly and far away nodes using multi-hop routing to communicate with each other. Since many of the routing protocols use the sensor locations, we first survey background knowledge on localization protocols to find the locations of the sensor nodes.

## 5.2  Background

Sensor location information is indispensable for both sensor data integrity and network organization. Traditional approaches to obtain location information include global positioning systems (GPS) [49]. But GPS is not appropriate for large-scale sensor network localization because of its high cost, large form factor, and outdoor constraints. There has been a lot of study on localization algorithms that derive the locations of sensor nodes from local measurements including distance and angle estimations between neighboring nodes [7, 23, 44, 71, 75–77, 86, 87, 89, 91]. The distance between two communicating nodes can be estimated by received signal strength indicator (RSSI) or time of arrival (ToA) techniques. Angles between adjacent edges can be measured by multiple ultrasound receivers [80], or directional antennas and laser transmitters/receivers.

Generally speaking, localization algorithms can be classified as anchor-based and anchor-free methods. Anchor-based methods assume that a (sometimes large) number of anchor nodes know their positions already [23,75–77,86,87,89]. The rest of the sensor nodes derive their positions by using distance or angle measurements to anchor nodes. In anchor-free methods, no node knows their absolute coordinates. The output is the relative positioning of the sensors, subject to a global translation and rotation.

A number of anchor-based localization algorithms use iterative trilateration or its variants. Anchor nodes obtain their locations by GPS or as prespecified. Then the other nodes find their locations by trilateration progressively [75, 86, 87]. In the basic trilateration scheme, a node $p$ can determine its location with distance measurements to three anchor nodes that are not colinear. Then node $p$ becomes a new anchor node. See Fig 5.1. Repeat this process until all the nodes are localized.

Similar incremental localization methods can also be done by using angles [76,77]. For these incremental methods, two issues need to be addressed. One is to deal with cascading error accumulation in large networks. One can adopt optimization techniques such as mass-spring relaxation to smooth out the error distribution, or, use robust statistics to handle outliers in input measurements [66]. The other issue is to handle insufficient number of initial anchor nodes. If there are a few anchors or the anchors are not well distributed, some nodes may not be able to find three neighboring anchor nodes to localize themselves. In this case, one can use distance estimations to anchor nodes via multi-hop paths [75] or adopt collaborative multi-lateration by solving a larger optimization problem [86,87].

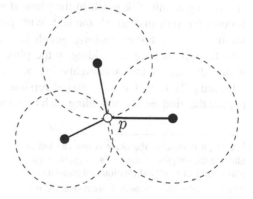

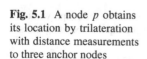

**Fig. 5.1** A node $p$ obtains its location by trilateration with distance measurements to three anchor nodes

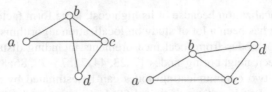

**Fig. 5.2** A connectivity graph with two distinct embeddings having the same set of edge lengths

Existing anchor-free algorithms take either the connectivity graph [89] or the distances between neighboring sensor nodes [7, 42, 71, 91] as input. One major challenge with this approach is localization ambiguity – when the localization solution is not unique, localization algorithms may come up with a different embedding that satisfies all the distance constraints but deviates far from the ground truth. See Fig. 5.2 for an example. Indeed, with range information local optimization such as mass-spring relaxation techniques may get stuck at local minima with part of the network flip over the rest and generate a network layout far away from the ground truth [24,71].

To deal with localization ambiguity, Moore et al. [71] proposed to use robust quadrilaterals as the basic iterative operation. A quadrilateral on four nodes with all pairs of edges is a globally rigid[1] component with a unique realization. The global layout is obtained by gluing locally identified robust quadrilaterals. Similarly, with ideas from rigidity theory, Goldenberg et al. proposed to record, propagate, and verify multiple possible locations of sensors to discover the truth network layout [42].

Global optimization techniques for sensor network localization include multi-dimensional scaling (MDS) [9, 89] and semidefinite programming [7, 91]. They formulate the problem as solving a global optimization problem for the sensor locations such that the distance constraints are satisfied. The results are typically better than local optimization algorithms. Similarly, with angle measurements one can formulate a linear program that solves for the node locations [13]. But these algorithms are centralized solutions. The performances of different localization algorithms have been evaluated in a Mote-based testbed [104, 105].

From a theoretical point of view, one can formulate the localization problem as embedding a unit disk graph in the plane if we assume the communication graph follows the unit disk graph model.[2] With purely connectivity information, determining whether a combinatorial graph is a unit-disk graph is NP-complete, and thus finding such an embedding in the plane (with neighboring nodes embedded within distance 1 and nonneighboring nodes more than distance 1 away) is also NP-hard [12]. In fact, even a relaxed version of the problem is still hard. Kuhn et al. proved that finding an embedding such that nonneighboring pairs are at least 1 away

---

[1] A graph is *rigid* in the plane is one can not continuous deform the shape of the graph without altering the lengths of the edges. A graph is *globally rigid* if it admits a unique embedding in the plane, subject to global rotation and translations.

[2] Two nodes within distance 1 are connected by an edge in a unit disk graph.

and neighboring pairs are within $\sqrt{3/2}$ is NP-hard [60]. Even with distance measurements of neighboring nodes or angle measurements of neighboring edges alone the problem is still NP-hard [3, 5, 13]. Not much is known on approximation algorithms for unit disk graph embedding. So far the only known theoretical result is an algorithm with an upper bound $O(\log^{2.5} n \sqrt{\log\log n})$ on the ratio of the longest distance between neighboring pairs to the shortest distance between nonneighboring pairs [60]. However, this algorithm uses some heavy graph embedding machinery and is not practical.

In practice, there have been a number of systems designed for indoor and outdoor localization. Most of them assume some anchors or beacon nodes with known locations, such as WiFi access points. One prevailing strategy is to use RF fingerprinting a priori to collect signal strength values from different access points tagged with location information. Then a node can be localized by matching the current signal strength values it receives from different access points with the RF fingerprints. A few representative systems include RADAR [6, 58, 63, 107], Intel's Place lab system [20]. Another category of indoor localization uses directional antennas and angle of arrival (AoA) to localize a node. In VORBA [78], the anchor nodes are equipped with rotating directional antennas to estimate AoA information from packets transmitted from a node to be localized. A simple triangulation approach is used to find its position. Approaches for indoor localization using other mediums such as ultra sounds (Active Bat [48]), infrared (Active Badge [102]), optical waves [72] have also been proposed. In this category, a widely known system is the Cricket [80] system from MIT. It uses ultrasound combined with RF. It uses several beacons that transmit ultrasound waves deployed on the ceiling of each room in the building. The nodes receiving these waves infer the range and localize themselves.

## 5.3   Geographical Routing

Geographical routing was originally proposed by Bose et al. [10], and independently by Karp and Kung [53] for routing in resource-constrained and dynamic networks such as ad hoc mobile networks and sensor networks. Further improvement on worst-case efficiency was proposed in the GOAFR+ family protocols [61, 62]. In geographical routing, the physical locations of the sensor nodes are used to guide information routing. Geographical routing has two modes: greedy mode and recovery mode.

### 5.3.1   Greedy Mode

In this mode, the node currently holding the packet "advances" it toward the destination, based only on the location of itself, its immediate neighbors, and the destination. The advance to the destination may be defined in many ways. Examples are, closest to destination [10, 30, 53, 96], most forward within radius (MFR) [96],

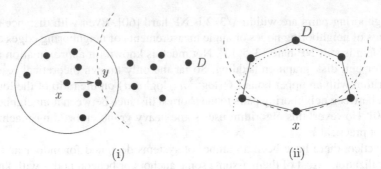

**Fig. 5.3** (i) In the greedy forwarding mode, $x$ sends the packets to $y$, the neighbor closest to the destination $D$; (ii) $x$ is a local minimum whose neighbors are all further away from the destination $D$ than $x$ itself

nearest forward progress (NFP), nearest closer (NC) [95], geographic distance routing (GEDIR) [94], and compass routing [57]. The most popular way of defining the advance is to examine the Euclidean distance to the destination and select the next hop as the node closest to the destination, as shown in Fig. 5.3. The greedy routing often suffices to deliver a packet in a dense network, but may fail in a sparse network. For example, the packet may reach a node called a *local minimum* whose neighbors are all further away from the destination than itself, as shown in Fig. 5.3.

### 5.3.2 Recovery Mode

When the greedy mode fails to advance a packet at some node, the routing process is converted to the recovery mode. The recovery mode defines how to forward the packet at a local minimum, to guarantee delivery of the packets. Some examples of the methods to get out of the local minimum are simple flooding [94], terminode routing [8], bread-first search or depth-first search [51], face routing algorithm [10], and perimeter routing [53].

Here we use a specific routing protocol, the greedy perimeter stateless routing (GPSR) protocol [53], to explain the two modes in detail. In the greedy mode, a node forwards the packet to its neighbor who is closest to the destination. The packet may reach a *local minimum* whose neighbors are all further away from the destination than itself. To get out of the local minimum, the protocol maintains a planar and connected subgraph, e.g., *Gabriel graph* (GG) or *relative neighborhood graph* (RNG) (see Fig. 5.4).

When a packet gets stuck, the recovery mode applies routing along the faces of the subgraph that intersect the imaginary line between the source and the destination (Fig 5.5). When the message reaches a node closer to the destination than the node where it enters the recovery mode, greedy routing is adopted again. Greedy routing combined with the perimeter routing guarantees the delivery of a packet to the destination if there is a path. This recovery mode has been independently discovered by Bose as the face routing algorithm [10]. The construction of the planar

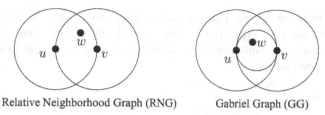

Relative Neighborhood Graph (RNG)          Gabriel Graph (GG)

**Fig. 5.4** In RNG, the edge *uv* is included unless there is a node *w* in the intersection of the two disks centered at *u*, *v* with radii equivalent to the Euclidean distance between *u* and *v*. In GG, the edge *uv* is included unless there is a node *w* in the disk with *u*, *v* as diameter. Both RNG and GG are planar subgraphs of the communication graph, modeled as a unit disk graph

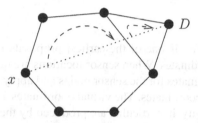

**Fig. 5.5** Perimeter routing along the faces of a planar graph from *x* to the destination

graphs as well as the routing decisions requires localized decisions with only the information of neighboring nodes. This makes geographical routing very attractive for large-scale networks as the state information maintained in each node and in the packet is nearly minimal.

The recovery scheme with face routing or perimeter routing works nicely in theory but encounters a few problems in practice. The correct construction of a planar subgraph assumes accurate location information that is hard to obtain, and that the communication graph is modeled by a unit disk graph, which is not true in practice. Experimental evaluations of the communication model on sensor nodes reveal various spatial and temporal radio irregularities [38, 110]. Close-by nodes may not be able to communicate directly while some long and high-quality links may exist. Over time link quality also varies. In practice, the relative neighborhood graph or Gabriel graph extracted may become disconnected or still contain crossing edges, as shown in [55, 88]. This subsequently causes the delivery rate drop to about 68% on a real testbed [55]. Later, a number of repair mechanisms propose to remove crossing links by probing techniques [45, 54, 55]. These mechanisms improve the delivery rate with extra processing.

Another problem of face routing, especially in a sensor field with big holes, is that the routes are likely to be "hugging" the hole boundaries. Thus the nodes on the hole boundaries experience higher traffic than the average. The load imbalance may cause the boundary nodes run out of battery earlier than the others, enlarging the size of the holes and/or disconnecting the network.

To summarize, the challenge of localization and extracting a planar subgraph and the problem of traffic accumulation on the hole boundaries are the two major motivations to move beyond geographical routing and design virtual coordinates that still retain the good properties of geographical routing, including the localized routing algorithm, near minimum state information, and robustness to topological changes and node failures.

## 5.4 Routing with Virtual Coordinates

### 5.4.1 Virtual Coordinates

#### 5.4.1.1 NoGeo

NoGeo [81], by Rao et al. is one of the earliest proposals for efficient geometric routing with virtual coordinates. When sensor locations are not available, the idea is to generate virtual coordinates for the sensor nodes and apply standard geographical routing on these virtual coordinates. The virtual coordinates are defined with respect to the network connectivity, in particular, are produced by the rubberband representation of a graph [67]. To get a rubberband representation, some nodes (preferably on the network perimeter) are given fixed locations, each node (that is not fixed) runs an iterative algorithm to put itself at the center of mass of the neighbors' current locations. At each iteration, a node's location is updated as the average of the neighbors' current locations:

$$x_i = \frac{\sum\limits_{j \in N(i)} x_j}{n_i}; \quad y_i = \frac{\sum\limits_{j \in N(i)} y_j}{n_i},$$

where $(x_i, y_i)$ is the current location of node $i$, $N(i)$ is the set of neighbors of node $i$ and $n_i = |N_i|$. This algorithm is known to reduce the total energy of the system, if all the edges are considered as rubberbands, and converge to a unique drawing what is called the rubberband representation.

When the nodes on the perimeter are identified but their locations are unavailable, a bootstrap stage is introduced to first embed the nodes on the perimeter. Each node on the perimeter floods the network and every other node on the perimeter records the minimum hop count distance to all other nodes on the perimeter. Now each node on the perimeter runs the following optimization to find an embedding that minimizes the sum of squared difference between the embedded pairwise distances and the hop count measurements:

$$\min \sum_{i,j \in P} [h(i, j) - d(i, j)]^2,$$

where $P$ is the set of nodes on the perimeter, $h(i, j)$ is the hop count between two perimeter nodes $i, j$ and $d(i, j)$ is the distance of $i, j$ in the embedding to be found.

This optimization can be solved by running a standard multidimensional scaling algorithm [9] at each perimeter node.

Last, if the nodes on the perimeter are not known, we can identify them in the following manner. A couple of nodes, chosen arbitrarily, are designated as the bootstrapping beacons. They flood the network and the nodes with maximum hop count values within 2-hop neighborhood are identified as perimeter nodes.

As the virtual coordinates have no reason to be close to the real locations, only the greedy algorithm is applied for routing on these virtual coordinates. When a packet gets stuck at a node $u$, *scoped flooding* is adopted to deliver the packet. In particular, $u$ performs a restricted flooding scheme and floods the network with TTL initially set to 1. Only the nodes within TTL-hops from $u$ receive the message. TTL is then doubled each time until the destination is reached.

The performance of greedy routing on the constructed virtual coordinates is shown in [81] on a number of topologies. Surprisingly, the greedy forwarding scheme is quite robust and achieves a high delivery rate given sufficient node density. When the network has big holes due to obstacles, the performance of greedy routing on virtual coordinates can be even better than the performance of routing on true locations. Intuitively, this is because the virtual coordinates, derived from the network connectivity, reflect the network connectivity better than the true geographical locations.

### 5.4.1.2  Greedy Embeddings

A few theoretical work motivated by NoGeo asks whether one can find an embedding of a given graph in the plane such that greedy routing is guaranteed to work [19, 79]. Such an embedding is called a *greedy embedding*. In other words, for every pair of nodes $p$, $q$, there is a neighbor of $p$ whose distance to $q$ is smaller than the distance from $p$ to $q$. It is known that not every graph admits a greedy embedding, such as a star with 7 leaf nodes [79]. Some graphs are known to have a greedy embedding, for example, graphs with a Hamiltonian path, any complete graphs, any 4-connected planar graphs (as they have Hamiltonian paths [97]), and any Delaunay triangulations. It still remains open to fully characterize the class of graphs that admit a greedy embedding.

Papadimitriou and Ratajczak [79] made the following conjecture that any planar 3-connected graph[3] has a greedy embedding in the plane. They show that a planar 3-connected graph has an embedding in 3D such that with a special distance function greedy routing always works. This is due to a famous graph theory result that a 3-connected planar graph is actually the edge graph of a 3D convex polytope [111]. Thus one can find a distance function to route along the surface of this convex polytope that guarantees delivery. Raghavan Dhandapani discovered that any planar triangulation admits a greedy embedding in the plane for which greedy forwarding always succeeds [22].

---

[3] A graph is 3-connected if it remains connected after the removal of any 2 nodes.

Recently the 3-connected graph conjecture was proved to be true by Leighton and Moitra [64], and independently by Angelini et al. [2]. Later the algorithm in [64] was improved such that the coordinates use $O(\log n)$ bits for a graph with $n$ vertices [43].

A recent observation by Kleinberg [56] shows that if we use hyperbolic space then greedy routing becomes easy. He showed that any connected graph has an embedding in the hyperbolic space such that by using the hyperbolic distance greedy routing from any node to any node always succeeds. The intuition is to embed a tree in a hyperbolic space such that greedy routing works on the tree. Since any connected graph has a spanning tree, greedy routing works for all connected graphs. This observation also leads to a distributed algorithm to assign virtual coordinates to the sensor nodes in a hyperbolic space.

### 5.4.1.3   Virtual Polar Coordinate Routing (GEM)

Virtual polar-coordinate routing (GEM), proposed by Newsome and Song [73], constructs a *polar* coordinate system as the virtual coordinates for the sensor nodes. Specifically, take a spanning tree on the sensor nodes and give each node a polar coordinate with respect to its position in the tree. A node's virtual coordinate includes two components: its *level*, i.e., minimum hop count to the root of the tree, and its *angle range*. The root of the tree is given the largest angle range, e.g., from 0 to $2^{16} - 1$. The child of a node $p$ is given a subset of the range of $p$, proportional to the size of its subtree. The angle ranges of two children do not overlap (Fig. 5.6).

With this polar coordinate system the following greedy routing scheme, named the virtual polar coordinate routing (VPCR), is used to deliver messages from one node to another. In VPCR, a node checks its 1-hop neighbors to look for any node that is "closer" to the destination. The "closeness" or distance of two angle ranges is defined as the minimum distance between two angles in the respective ranges. If a node does not find any neighbor closer to the destination, then the packet is routed to the parent. Notice that when the packet reaches the root of the tree, the packet is delivered down the tree structure to the child whose subtree contains the destination. Therefore, the algorithm eventually delivers the message. Similar with NoGeo [81],

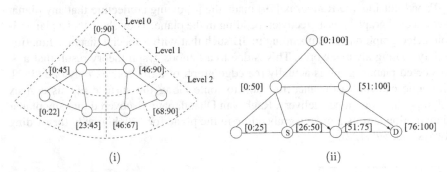

**Fig. 5.6** (i) The polar coordinate assignment; (ii) Virtual polar coordinate routing

GEM also maintains the constant storage requirement and $O(1)$ state information in the packets, as well as the localized routing steps.

One issue with the polar coordinates is that traffic tends to accumulate around the root of the tree and its neighborhood. To alleviate this, multiple trees can be constructed and a packet is routed randomly on one tree and the set of polar coordinates associated with it.

### 5.4.1.4   Face Tracing

Motivated by the challenges of extracting a planar subgraph from the communication network for face routing with geographical locations, Zhang et al. proposed to define a "face" in a generic way and conduct face routing on an arbitrary graph [108]. At each node $v$, a fixed and arbitrary cyclic ordering of the adjacent edges is defined, and named as a "rotation". Each edge of $v$ is given an index as an integer from 0 to $d - 1$, where $d$ is the degree of $v$. The edge $vw$ with index $i + 1$ is said to follow the edge $vu$ with index $i$. The edge with index 0 follows the edge with index $d - 1$. An edge $uv$ has two directions, one is from $u$ to $v$ and one is from $v$ to $u$. With a fixed rotation scheme on the vertices, one defines a "face" by traversing the edges following the rotation scheme. For example, start with a directed edge $uv$, the adjacent edge on the face is determined as the next edge at $v$ following the rotation scheme at $v$. See Fig. 5.7. The face tracing algorithm visits $v$ from the directed edge $uv$ and takes the next edge $vw$ by the cyclic index ordering at $v$.

The face tracing algorithm traces a sequence of directed edges by following the orderings of the rotation schemes at the vertices involved, and stops when the first directed edge is encountered again. Note that the first directed edge that is encountered the second time by the face tracing algorithm must be the first edge on the sequence, i.e., no directed edge can appear twice on a face boundary – if a different directed edge say $vw$ is encountered the second time during face tracing, then the preceding edge $uv$ must have been encountered twice as well, as the only way to take the edge $vw$ is to come from the edge $uv$ unless $vw$ is the first edge on the sequence. Thus each face traced by the algorithm is a closed cycle of directed edges.

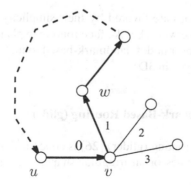

**Fig. 5.7**  Following the rotation scheme at each vertex, traversing a directed edge finds a face

Notice that the face tracing algorithm requires nothing beyond a fixed but arbitrary ordering of the edges at each vertex. Thus it works fine on a general graph and does not assume unit disk graph model. The face tracing algorithm is performed as a preprocessing phase to discover all the faces adjacent to each node. Each face is given a face ID. Now, routing with geographical coordinates consists of the standard greedy routing mode and the *face tracing* mode. When a message gets stuck at a node $u$, unable to make progress towards the destination, it enters the face tracing mode and traverses the graph by following a face adjacent to $u$. If the message discovers a node closer to the destination than $u$, it leaves the face tracing mode and returns to greedy routing mode. Otherwise, the message is routed to an adjacent face, that is, a face that shares an edge with the current face. The message saves the IDs of the faces it has visited, and hops from face to face until either it reaches the destination or traverses the whole graph and declares there is no path to the destination.

Different from the approach of using probing to remove crossing edges in an unsuccessful planar graph subtraction [45, 54, 55], the face tracing technique does not remove any links in face routing mode.

## 5.4.2 Landmark Routing

A number of virtual coordinate routing systems are landmark-based [16, 26, 32, 74]. The idea of using landmarks for generating node names or addresses and routing in large networks appeared two decades ago, such as the landmark hierarchy by Tsuchiya [100]. We will discuss hierarchical routing in Sect. 5.6. All the landmark-based routing schemes described below have a two-level hierarchy.

In sensor network setting, a small subset of the nodes is selected as *landmarks*. The landmarks flood the network and every node records its hop count distances to these landmarks. The landmark distances are then used to generate virtual coordinates for routing between the nodes. Routing is typically guided in a greedy way by a potential function on the landmark-based distances. The major differences among the following schemes are mostly in the design of the potential functions, as will be explained later.

Landmark-based schemes are favored by their simplicity and independence on the dimensionality of the network. Unlike face routing in planar graphs that requires a planar deployment of sensor nodes, landmark-based routing schemes can be easily extended to sensors deployed in 3D.

### 5.4.2.1 Gradient Landmark-Based Routing (glider)

Gradient landmark-based routing (glider) [26] is concerned with routing in a sensor network with a few big holes or an irregular shape. It separates global and local routing by encoding the global topology of the network with a compact combinatorial graph, which provides a rough global routing guidance. The actual route with the global guidance will be realized with a greedy routing rule based on the landmark distances.

In glider, a number of landmarks are selected and the network is partitioned into *Voronoi tiles* such that all the nodes inside the same tile are closest (in terms of hop count distance) to the same landmark. Two Voronoi tiles are adjacent if there are two neighboring nodes in different Voronoi tiles. The tile adjacency graph, denoted by the *combinatorial Delaunay graph*, is abstracted and disseminated to all the nodes, for global routing across tiles. Figure 5.8 shows the landmark Voronoi diagram and the dual combinatorial Delaunay graph.

Each node $p$ is given a virtual coordinate for routing. Once the Voronoi tiles are partitioned, each node belongs to a Voronoi cell. We call that cell the *resident tile* of the node, and we call its landmark the *home landmark*. Nodes with equal distance to multiple landmarks will choose the landmark with smallest ID as the home landmark. The neighbors of $p$'s home landmark on the combinatorial Delaunay graph are called the *reference landmarks*. The *name* of a node includes the ID of its home landmark, and the list of hop count distances to its reference landmarks. Notice that the name of a node is locally defined by the nearby subset of landmarks.

To route to a node in a different tile, a node first consults with the combinatorial Delaunay graph to find a sequence of tiles to visit. The actual routing across tiles is called *inter-tile routing* and can be implemented with the virtual coordinates of a node itself and its neighbors. Recall that every node has the minimum hop count

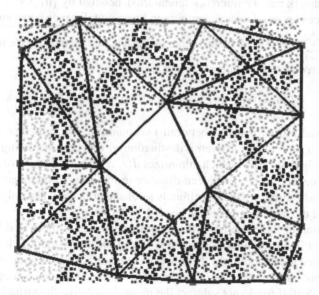

**Fig. 5.8** Landmarks are shown by *large squares*. Sensor nodes are shown as *small dots*. The nodes are divided into tiles. All the nodes in the same tile are drawn in the same color. The dark nodes are the boundaries of the tiles

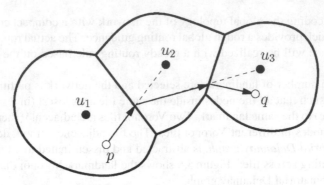

**Fig. 5.9** Routing across tiles: to route toward the tile of $u_2$ the packet is delivered in a greedy fashion towards the adjacent landmark $u_2$ until it enters the tile

information to its reference landmarks. Thus a packet destined to an adjacent tile is routed toward the landmark in that tile, in particular, forwarded to the neighbor whose hop count distance to the landmark of that tile is reduced. See Fig. 5.9. When a packet enters the next tile, it checks whether this tile contains the destination. If not, inter-tile routing is used again to route to the next transit tile. We remark that such transition happens when the packet first enters a tile, and thus the packet does not necessarily go through the landmarks in the transit tiles.

To route to a node in the same tile, the packet is guided by greedy descending on a potential function constructed with distances to a set of local landmarks (including the home landmark and the reference landmarks), denoted by $\{u_1, u_2, \ldots, u_k\}$. For any node $p$, let $\tau(p, u_i)$ denote the minimum hop count distance between $p$ and $u_i$. Let $\overline{\tau}(p) = \sum_{i=1}^{k} \tau(p, u_i)^2 / k$ as the mean of the squared distances. We then assign to $p$ the *centered virtual coordinate vector*

$$C(p) = (\tau(p, u_1)^2 - \overline{\tau}(p), \ldots, \tau(p, u_k)^2 - \overline{\tau}(p)).$$

The centered virtual distance between two points $p$ and $q$ is then defined as $d(p, q) = |C(p) - C(q)|^2$. Given a destination $q$, our greedy routing algorithm chooses the neighbor $r$ of $p$, which minimizes $d(r, q)$. That is, we move packets by greedy minimization of the Euclidean distance to the destination, measured in the virtual coordinate system. This algorithm is local and efficient since only the virtual coordinates of the neighbor nodes are needed.

The selection of the centered virtual coordinate vector is motivated by the fact that the variant in the continuous domain is free of local minimum as long as there are at least three local landmarks that are not colinear. The subtraction of the mean from the squared distance vector in the centered virtual coordinate is important. As shown in Fig. 5.10 if we do not subtract the mean (i.e., define the virtual coordinate vector as the squared distance vector) routing may get stuck at local minimum.

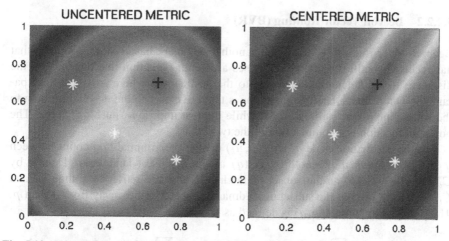

**Fig. 5.10** The distance function for landmark-based greedy routing. There are three landmarks marked by *snowflakes*. The destination is marked by a + sign. The color of a point represents its distance to the destination with respect to uncentered coordinates and centered coordinates, respectively. Note the local minimum in the uncentered case

In a discrete network inter-tile routing is always successful but intra-tile routing with the centered virtual coordinate vector may still get stuck, if the sensors are distributed sparsely. Experiments show that when the randomly distributed nodes have on average about 6 or more neighbors the delivery rate is close to 100%. In the worst case when a packet arrives at a local minimum, the node can simply flood the tile containing the destination to deliver the message.

To summarize, glider separates the global routing (how to get around holes) and local routing (to deliver a packet to the destination) by the two-level hierarchy. In many of the real-world situations where sensor networks may be deployed, the topological features of the layout (e.g., holes) will be only a few and will mostly reflect the underlying structure of the environment (e.g., obstacles). Moreover, this relatively simple global topology is likely to remain stable: nodes may come and go, but such spontaneous changes are unlikely to destroy or create large-scale topological features. It follows, since the global topology is stable, that we can afford to carry out proactive routing at the abstract combinatorial Delaunay graph level and realize the high-level routes as actual paths in the network by using reactive protocols by greedy routing with the virtual landmark-based coordinates.

The design principle of glider hinted that the landmarks should be selected to capture the global topology of the sensor field. Thus the number of landmarks is likely to be dependent on the topological complexity, such as the number of holes in the sensor field, rather than the number of sensors. A very recent work [40] investigated this problem in a continuous domain and established criteria on how the landmarks are placed to guarantee that the combinatorial Delaunay graph (or its continuous variant, the geodesic Delaunay triangulation) represents the exact topology of the underlying (unknown) domain. In companion, a distributed algorithm on how the landmarks are selected in a sensor network setting is also proposed in [40].

### 5.4.2.2 Beacon Vector Routing (BVR)

Beacon vector routing (BVR) [32] is another landmark-based routing scheme that has received significant attention. It uses a potential function that depends on the distances to the $k$ landmarks closest to the destination, where $k$ is a system parameter. Out of the $k$ landmarks, the ones that are closer to the destination than to the source impose a "pulling" force while the rest impose a "pushing" force. The potential function is a combination of the two.

For the destination node $q$, let $\tau(p, u_i)$ denote the minimal hop count between $q$ and a landmark $u_i$. Denote by $C_k(q)$ as $q$'s $k$ closest landmarks. Denote by $C_k^+(p, q) \subseteq C_k(q)$ as the subset of landmarks that are closer to $q$ than $p$; and $C_k^-(p, q) \subseteq C_k(q)$ as the subset of landmarks that are further away from $q$ than $p$. The potential function $\delta(p, q)$ is defined as

$$\delta_k(p, q) = A \sum_{i \in C_k^+(p,q)} [\tau(p, u_i) - \tau(q, u_i)] + \sum_{i \in C_k^-(p,q)} [\tau(q, u_i) - \tau(p, u_i)],$$

where $A$ is a parameter taken as 10. The greedy routing chooses a neighbor that minimizes the above potential function to the destination. When the greedy routing gets stuck, the message is delivered to the landmark closest to the destination, from where a restricted flooding with increasing scope is performed to deliver the packet.

Although the potential function design is a heuristic, the algorithm works well as evaluated in simulations. BVR has been selected in a number of cases as a comparison benchmark [70, 74].

### 5.4.2.3 Greedy Landmark Descent Routing (GLDR)

Greedy landmark descent routing is recently proposed by Nguyen et al. [74], which, combined with a landmark selection strategy, guarantees packet delivery with bounded stretch in the continuous domain.

The landmarks are selected to be an $r$-sampling: any node is within hop count $r$ from a landmark. These landmarks are chosen by a distributed algorithm running in parallel on all the nodes. Each node waits for a random period of time and declares itself to be a landmark, if it has not been suppressed by other landmarks. Once it becomes a landmark it notifies its $r$-hop neighbors and suppresses all nonlandmark nodes. This generates an $r$-sampling and depending on the parameters of the random waiting time the number of landmarks selected can be controlled.

Similar to BVR, a node's virtual coordinate is defined as the distance vector to a small set of nearest landmarks called the *addressing landmarks*. To route a packet to its destination, the source node chooses, among the destination's addressing landmarks, the one that *maximizes* the ratio of the distances to source and destination, and moves toward this landmark until it reaches a node with the same hop count to this landmark as the destination. At this point, the routing procedure is repeated and a different landmark is chosen until the destination is reached. An example is shown

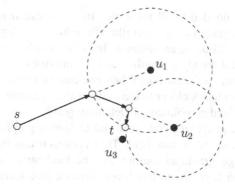

**Fig. 5.11** Greedy landmark descent routing. The source $s$ chooses the landmark $u_1$ and routes toward it until the distance to $u_1$ equals the distance from the destination $t$ to $u_1$. The same procedure is repeated until the destination $t$ is reached

in Fig. 5.11. This scheme is known to always work in the continuous domain and generates a path with constant bounded stretch. In a discrete network it may happen that the packet gets in a loop, which can be detected by checking the last few landmarks toward which the packet moves. When this happens the packet is greedily forwarded to the destination using $L_1$ and $L_\infty$ norms on the distance vectors of the addressing landmarks. If the destination is still not reached, a scoped flooding is used to deliver the packet.

The landmark selection scheme by using $r$ samplings is of independent interest and can be useful for other landmark-based routing schemes. For example, when the landmarks are selected as an $r$-sampling rather than a random sampling, BVR has improved delivery rate [74].

### 5.4.2.4   Macroscopic Geographic Greedy Routing

Funke and Milosavljevic proposed Macroscopic Geographic Greedy Routing that combines the idea of glider and geographical routing to account for inaccurate location information and small network holes and irregularities [36, 37]. In glider, a sparse set of landmarks is selected to capture the big topological features of the sensor field; the combinatorial Delaunay graph on the landmarks is distributed to every node for global routing guidance. In Macroscopic Geographic Greedy Routing, a dense set of landmarks is selected and the combinatorial Delaunay graph is not distributed to all the nodes but rather used for geographical routing.

The idea is to extract a planar subgraph of the combinatorial Delaunay graph and embed it in the plane. Each node $v$ is given a name consisting of the home landmark, i.e., the landmark whose Voronoi tile contains $v$, and its unique ID. The intratile and intertile routing are implemented in the following way. For intratile routing, since the landmarks are dense and each tile has only a constant number of nodes, the packet is simply flooded to all the nodes in the destination tile. For intertile routing, the packet is guided to the next tile whose home landmark is closer to the destination in terms of Euclidean distance. In other words, the next landmark is chosen by geographical greedy forwarding and the intertile routing in glider is used to route a packet to that adjacent tile. When the current home landmark does not have a neighboring landmark closer to the destination, perimeter routing is adopted to route on the planarized combinatorial Delaunay graph.

To summarize, the planarized combinatorial Delaunay graph and its embedding are used to navigate in the sensor field. This approach can be considered as in between glider and geographical routing. As the landmarks are reasonably separated, the navigation is robust to low-level network link variations and location inaccuracies.

### 5.4.2.5 Small State Small Stretch Routing

Mao et al. [70] adopted the idea of compact routing schemes (by Thorup and Zwick [98, 99]) for routing in sensor networks. The basic idea is to select about $O(\sqrt{n})$ landmarks. These landmarks flood the network and every node records the hop count distances to these landmarks. Denote by $u(p)$ the landmark that is closest to $p$ and $r(p)$ the distance from $p$ to $u(p)$. Each node $p$ identifies its *cluster* $C(p)$, consisting of the nodes $q$ whose distance to $p$ is within distance $r(q)$, i.e., all the nodes that are closer to $p$ than their closest landmarks. For every node $p$, it maintains a routing table with hop count values to all the landmarks and the nodes in its cluster $C(p)$.

The routing algorithm is a greedy algorithm with the following rules:

- If the destination $t$ is inside the cluster $C(s)$ of the source $s$, then $s$ routes to $t$ directly with its routing table.
- If the destination $t$ is outside the cluster $C(s)$ of the source $s$, then $s$ first routes the packet toward the landmark closest to $t$. When it encounters a node with routing table entry to $t$, the packet is delivered to $t$.

One can see that this routing algorithm guarantees delivery – $t$ is within the cluster of its closest landmark $u(t)$, thus the packet will eventually get to a node who has a routing table entry for $t$, from where the packet is guided by the local routing tables to the destination. In addition, the length of the routing path is within a factor of 3 of the minimum hop count distance between the source and the destination. Denote by $\tau(p, q)$ the hop count distance between $p$ and $q$. If the destination is within $C(s)$, the packet is delivered through the shortest path with the help of the routing tables. If the destination is outside $C(s)$, as shown in Fig. 5.12, $\tau(s, t) \geq \tau(t, u(t))$. Thus the packet is routed toward $u(t)$. Suppose on the way to $u(t)$ the packet arrives at a node $p$ whose cluster contains the destination $t$. We have $\tau(s, p) \leq \tau(s, u(t))$, $\tau(p, t) \leq \tau(u(t), t)$. Now the length of the routing path is

**Fig. 5.12** An example of compact routing

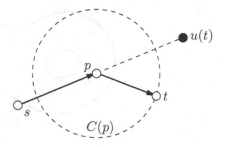

$$\tau(s,p) + \tau(p,t) \qquad \leq \tau(s,u(t)) + \tau(u(t),t)$$
$$\leq [\tau(s,t) + \tau(t,u(t))] + t(u(t),t)$$
$$\leq \tau(s,t) + 2\tau(u(t),t)$$
$$\leq 3\tau(s,t).$$

When the sensor nodes are uniformly deployed, one can place about $O(\sqrt{n})$ landmarks uniformly such that the size of the cluster for each node is roughly $O(\sqrt{n})$ as well. Thus the size of the routing table is in the order of $O(\sqrt{n})$, with which a worst-case stretch 3 path can be established for any pair of nodes.

To conclude this subsection, we surveyed a set of representative ideas in using landmarks to define virtual coordinates for greedy routing in a sensor network. A few other protocols with similar ideas to some of these were not covered in details here. For example, Caruso et al. [16] used three landmarks and a heuristic potential function for routing in a sensor network. Wattenhofer et al. [103] studied the placement of landmarks for routing in special graphs such as rings, grids, trees, butterfly graphs, and hypercubes. At last we remark that these landmarks are mainly used as points of reference, and do not serve the functionalities of gateway nodes (which tends to attract network traffic). After the establishment of the virtual coordinates the landmarks typically behave in the same way as the other nodes.

### 5.4.3 Medial Axis-Based Routing

The medial axis-based routing, proposed by Bruck, Gao, and Jiang [14], considers a setting similar to the setting in glider, in which the sensors are uniformly deployed in a geometric region with possible obstacles or irregular shape. Again the global topology of the sensor field is captured for global routing guidance, the difference is in the use of the medial axis of the underlying sensor field for this purpose.

The *medial axis* of a sensor network is identified as the set of nodes with at least two closest nodes on the network boundary. A point on the medial axis is called a medial vertex if it has three or more closest nodes on the network boundary. See Fig. 5.13 for an example. The medial axis is a "skeleton" of a region that captures all the topological features, such as how many holes there are and how they are

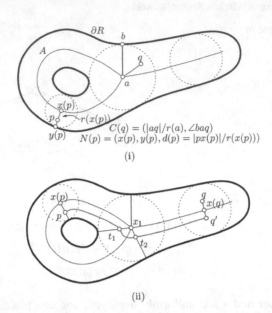

**Fig. 5.13** An example of the medial axis of the boundary of a closed region $R$ (only the part of the medial axis in the interior of $R$ is shown). $\partial R$ is shown by *thick curves*. The medial axis $A$ has a cycle, which means that the region $R$ has a punched hole. (i) The naming scheme. (ii) Routing from $p$ to $q$ by the road system

connected. For the purpose of routing, the medial axis is represented by a medial axis graph (MAG), which is a combinatorial graph connecting the medial vertices, and has a size proportional to the number of large geometric features. This medial axis graph is very compact and is known to every sensor. For example, the medial axis graph of Fig. 5.13 has two medial vertices, one edge and one self-loop. The construction of the medial axis requires the detection of nodes on the network boundaries (or hole boundaries), for which there exist distributed and efficient algorithms [25,28,29,33,35,41,59,101]. The nodes on the medial axis can be identified through local flooding. Specifically, each boundary node initiates the flooding of a message, which contains its ID and a counter that records how many hops the message has traveled. If a node receives a packet from a boundary node that is further away from its current nearest boundary node(s), it stops forwarding this packet. If the boundary nodes initiate their flooding at approximately the same time, and each packet travels at approximately the same speed, then a packet is dropped once it reaches the medial axis. This approach cuts down the total number of packets delivered and keeps the total communication cost low. As a result, each node learns its nearest boundary node(s) and can determine whether it itself is on the medial axis. After the construction of the medial axis, we let a node flood the network, pull the information about the medial axis, and construct the abstracted medial axis graph. This graph is then broadcast to every sensor. In addition, each node on a medial edge remembers which medial edge it stays on, its neighboring nodes on that medial edge, and how many hops it is from each endpoint of the medial edge.

With the medial axis constructed, each node is given a name with respect to the medial axis graph, defined by the closest node on the medial axis and the hop count distance from it. For each sensor $w$ on the medial axis, we define a chord as the shortest path (tree) from $w$ to its closest sensor nodes on the boundary. A sensor's name includes the chord on which it stays, and a normalized distance to its corresponding medial axis node. See point $p$ in Fig. 5.13. Such a naming scheme partitions the sensor field into canonical regions inside each of which a local Cartesian coordinate system is defined with one axis as an edge of the medial axis graph and the other axis as a chord of a vertex on that medial axis edge. The local Cartesian coordinate systems are glued together in exactly the same way as indicated by the edge adjacency of the medial axis graph, and provide a smooth and natural road system for efficient point-to-point routing. To help a packet route from one canonical piece to an adjacent one, we also establish a Polar coordinate system around each medial vertex, see node $q$ in Fig. 5.13.

For a node to find a route to the destination, it only needs to know the medial axis graph and the name of the destination. Routing is first planned on the abstract medial axis graph, which is usually of a small size, and then realized in each canonical region by reactive local gradient descent routing. By using the medial axis graph in a global planning step, a source can find the *reference path*, defined as the shortest path in the medial axis graph, from the node in the medial axis corresponding to the source to the one corresponding to the destination. The actual routing rule is of Manhattan-type, i.e., first trying to match the medial axis point with that of the destination and routing in parallel with the reference path, and then trying to match the distance to the medial axis point with that of the destination and routing along chords. Both routing in parallel with the medial axis and along chords can be realized by efficient local gradient descending in the local coordinate systems of the canonical regions. See Fig. 5.13.

We show in Fig. 5.14 an example of a sensor network with a few large holes, the medial axis and the medial axis graph. Figure 5.15 shows the path generated by the medial axis-based routing and the path generated by GPSR. The medial-axis based routing scheme has accumulatively more balanced traffic load on the sensor nodes as routing paths are not converging to hole boundaries in such a complex network.

### 5.4.4  Ring-Based Routing Scheme

The geometric schemes we discussed so far use the fact that the sensors are typically deployed densely in a 2D region, to provide sufficient coverage and sensing resolution. A ring-based routing scheme, recently proposed by Caesar et al. [15], reduces the problem to routing in a 1D ring.

In virtual ring routing (VRR), the nodes are given arbitrary identifiers and organize themselves into a virtual ring in the cyclic increasing order of their IDs. Each node maintains $r$ adjacent neighbors on the ring ($r/2$ neighbors clockwise and $r/2$ neighbors counter-clockwise). These neighbors are called *virtual neighbors* of a node, to be differentiated from its *physical neighbors* that are directly connected by

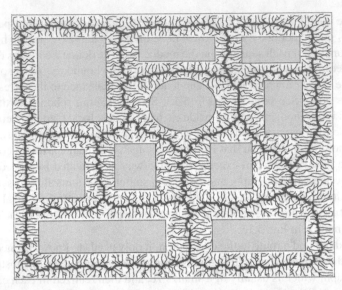

**Fig. 5.14** The medial axis and the chords (shortest path tree rooted at nodes on the medial axis). The obstacles (holes) are colored in *yellow*

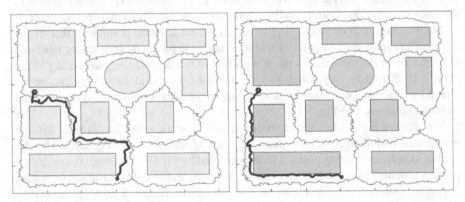

**Fig. 5.15** The path generated by the medial axis based routing (*left*), in comparison with that of geographical routing (*right*)

wireless links. Notice that the virtual neighbors are by no means physically close. In virtual ring routing, the routes from a node to each of its virtual neighbors are established and stored in the routing tables of the nodes on the routes. See Fig. 5.16. These paths are called *vset-paths* and will be maintained when the network topology changes. Specifically, each node $p$ maintains a routing table with the following entries ($S_i$, $T_i$, next hop toward $S_i$, next hop toward $T_i$), for each pair of virtual neighbors ($S_i$, $T_i$) whose path goes through $p$.

When a packet is routed to the destination, if the source node has a routing table entry for the destination then the packet will be delivered accordingly. Otherwise,

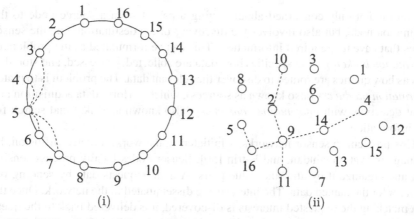

**Fig. 5.16** (i) Virtual ring and the virtual neighbors of node 5. (ii) The actual paths between node 5 and its virtual neighbor 4, and node 10 and its virtual neighbor 11 in the real network. All the nodes on the paths maintain routing table entries for these paths

the source $s$ checks its routing table and picks the path endpoint $q$ closest to the destination ID on the virtual ring. The packet is forwarded to the next hop toward $q$. At the next node this operation is repeated. One can see that with the virtual paths stored in the network one can route from any node to any node – in the worst case the packet can follow the virtual ring to reach the destination. The actual performance can be better as each node also maintains routing information for the virtual paths that go through it. Intuitively, a node is routed toward the node closest to the destination in the ID space (on the virtual ring), although that node might be in a completely different direction from the destination. But the nodes close to the destination on the virtual ring have routing table entries to the destination. Thus the algorithm routes a packet to some nodes with more knowledge of the destination. In fact, it often happens that a packet does not have to visit the virtual neighbors of the destination before it finds routing information to the destination. Simulation results show that the routing path actually has on average a small stretch compared with the shortest paths.

During the network initialization, the virtual paths can be constructed incrementally without flooding the network. The control messages to build a new virtual path are routed with the current virtual paths. Any path failures caused by node mobility or topological changes are repaired in a similar way by using the remaining virtual paths, to adapt to network dynamics.

## 5.5 Information Discovery and Data-Centric Routing

In data-centric routing, a node poses a query for data of certain types, and the routing algorithm routes the query to retrieve the relevant data in the sensor network. Data-centric routing is fundamentally different from traditional routing paradigms as the problem is not only concerned about getting a packet from a source node to the destination node, but also involves the discovery of the destination, i.e., the sensor nodes that have the required information. This can be formulated as the problem of *information brokerage* that specifies how data are collected, processed, and stored as well as how queries are routed to discover the relevant data. The problem is to match *information producers* (also known as sources), which perform data acquisition and event detection, with *information consumers* (also known as sinks) that search for this information.

This problem in sensor networks is initiated in the work, directed diffusion, by Intanagonwiwat, Govindan, and Estrin [50]. Sensor nodes locally process sensing data and organize it by attribute-value pairs. A node requests data by sending out *interests* for the named data. The interests are disseminated in the network. Once the data matching the requested interests is discovered, it is delivered back to the query node. The intermediate nodes may also cache interests and reinforce certain routes from data sources to the query node. A similar routing paradigm is also adopted in TinyDB [69] to support query of aggregated information of the distributed data. A SQL-style query is disseminated to all the nodes in the network during which a tree rooted at the query node is established. As the data are delivered back to the root, it is aggregated at the internal nodes. In this approach, little collaborative pre-processing is performed. Thus the discovery of the desired information usually relies on flooding the network. This approach targets at infrequent queries for streaming data type so that the cost of flooding can be justified and amortized by the following long-term data delivery.

### 5.5.1 Geographical Hash Tables

A parallel approach to data-centric routing is to adopt data-centric storage that targets at large-scale networks with many simultaneously detected events that are not necessarily desirable for all users [83, 90]. The idea is similar with Distributed Hash Tables (DHTs) on the Internet [82, 84, 92, 109]. A producer leaves data on rendezvous nodes for consumers to retrieve. Thus data across space and time can be aggregated at rendezvous nodes.

In geographical hash tables (GHTs) [83], data are hashed by data type to geographical locations. The nodes close to the hashed location serve as rendezvous nodes. The consumer applies the same hash function and retrieves data from the same set of rendezvous nodes. Data and query delivery to the rendezvous nodes is implemented by geographical routing such as GPSR [53]. Specifically, each piece of event, based on its data type, is hashed with a content-based hash function to a geographical location. This hashed location may not have a sensor node right there. The node *geographically closest* to the hashed location is considered as the *home node*. Then the event is routed with GPSR toward the hashed location. This packet will eventually reach the home node, by the property of geographical routing. At the home node, GPSR will not be able to find a neighbor closer to the hashed location and thus enter the perimeter mode to tour the face on which the home node stays. This face is named the *home perimeter*. The nodes on the perimeter, other than the home node, are named the *replica nodes*. After the packet returns to the home node, the data caching component is accomplished. All the nodes on the home perimeter cached the data. To account for possible changes to the network topology, the home node periodically sends refresh packets to the hashed location to repair the home node and the home perimeter.

GHTs have greatly reduced the communication cost and energy consumption by avoiding network-wide flooding for information discovery. Its simplicity is also very appealing. This idea is further developed in [46] to build a hierarchical storage structure that is aware of data correlations, i.e., similar data are hashed close by.

## 5.5.2 Double Rulings

The idea of double rulings is an extension and improvement of GHTs, in terms of *distance sensitivity*. In particular, if the producer is actually close to the data producer (although neither has the knowledge of each other), we would like the data consumer to discover the data producer quickly. This is an attractive feature in many applications, as information will be most useful, thus queried more frequently, in the spatiotemporal locale where it was collected. In GHTs, the hashed location, typically generated by a random hash function uniformly on a sensor field, may be far away from both the producer and the consumer.

In double rulings, the rendezvous nodes are chosen along a continuous curve. The motivation is twofold. Data delivery from data source to a rendezvous node is implemented by multihop routing. Thus it is natural to leave information hints along the trail that the data travels on, at no extra communication cost. Furthermore, data hint replication on multiple nodes provides more flexibility for a consumer to discover relevant data – it is easier to encounter a 1D curve than a 0D node.

### 5.5.2.1  Rectilinear Double Rulings

A basic *double-ruling* scheme, developed in different variations for information discovery and routing [68, 93, 106], works as follows: data or data pointers are stored at nodes that follow a *replication curve* while a data request travels along a *retrieval curve*. Any retrieval curve intersects the replication curve for the desired data. Thus successful retrieval can be guaranteed. For an easy familiar case, assume the network is a two-dimensional grid embedded in the plane with nodes located at all the lattice points (see Fig. 5.17). The information storage curves follow the horizontal lines. The information retrieval curves follow the vertical lines. To be differentiated from the other double rulings schemes, we name this scheme the *rectilinear double rulings*. Notice that the data retrieval curves are independent of the locations of the data sources. In fact, a consumer traveling along the vertical line through itself

**Fig. 5.17** The rectilinear
double ruling scheme on a
grid

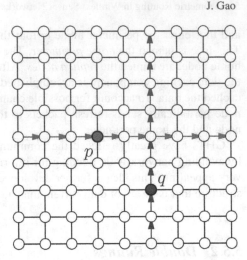

is guaranteed to hit *all* horizontal storage lines, and thus is able to find *all* the data
stored in the network. This double-ruling scheme is also distance sensitive – if the
producer and consumer are actually near each other, they must also be near each
other along the path connecting them using the horizontal and vertical lines.

### 5.5.2.2 Spherical Double Rulings

The spherical double rulings scheme, proposed by Sarkar, Zhu, and Gao [85], has
both GHTs and the rectilinear double rulings as subcases. Same as in GHTs, a data
item is hashed by its data type to a geographical location. However, instead of traveling along the geographical greedy path to the rendezvous node, the producer travels
along a circle that goes through itself and the rendezvous node and replicates data
or data pointers on the way.

For an easy explanation, we will use the stereographic projection to map sensor
nodes onto a sphere [21]. We put a sphere with radius $r$ tangent to the plane at the
origin. Denote this tangent point as the south pole and its antipodal point as the
north pole. A point $h^*$ on the plane is mapped to the intersection of the line through
$h^*$ and the north pole with the sphere. See Fig. 5.18. This provides a one-to-one
mapping of the plane to the sphere, in addition, with the north pole mapped to the
point of infinity. Stereographic projection preserves circularity. Any circle on the
sphere, including great circles, is mapped to a circle in the plane. With this mapping
specified, the replication and retrieval schemes are described on a sphere.

Each data type is hashed to a geographical location $h^*$ as in GHTs. When a
producer routes toward the hashed location, instead of following the geographical
greedy route as in GHTs, it follows the great circle defined by its own location $p$
and the hashed location $h$, denoted by $C(p, h)$. Data from different producers with
the same data type will be routed to the same hashed location where information

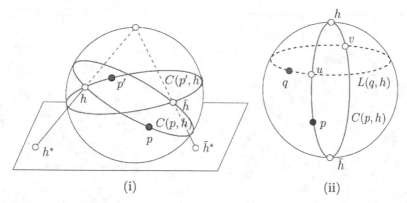

**Fig. 5.18** (i) A point in the plane $h^*$ is projected to a point $h$ on the sphere. The great circles for two producers $p$, $p'$ are drawn in *red*. (ii) The consumer follows the circle with fixed distance (*dashed circle*) to the hashed location $h$ to retrieve all the data with the same data type

aggregation can be performed. All the great circles with type $C(*, h)$ pass through the hashed location $h$, as well as the antipodal point $\bar{h}$. Thus there are actually two rendezvous nodes, $h$ and $\bar{h}$, located far away in the network that both have all the information of the same data type. Notice that the hashed location $h$ depends only on the data type. Thus the location $\bar{h}$ can be derived by a simple geometric computation. See Fig. 5.18 for an example.

Now, this new data replication strategy from producers to hashed locations enables more flexible retrieval scheme for the consumer.

1. *GHTs retrieval rule*: the same as in GHTs, the consumer can route to the hashed location $h^*$ or $\bar{h}^*$, whichever is closer, to retrieve all the data of the same type.
2. *Distance-sensitive retrieval rule:* If the consumer is of distance $d$ from the producer, the consumer can discover the data with a cost of $O(d)$, although neither has the knowledge of each other's location or the bound on $d$. Specifically, if we rotate the sphere so that the hashed location $h$ is at the north pole, then the replication curve is exactly a longitude curve. The distance-sensitive retrieval scheme follows the latitude curve searching for a replication curve. See Fig. 5.18 for an example. The retrieval curve (i.e., the latitude curve) has two intersections with the replication curve, one of which is within distance $O(d)$ from the consumer. Thus the consumer travels along the circle on the sphere with equal distance to the hashed location $h$, and uses a doubling trick[4] to discover the closer intersection with the replication curve. An example is shown in Fig. 5.19.

---

[4] The problem is to find a destination on a circle without knowing whether it is shorter to go clockwise or counterclockwise. With a doubling trick, one can first take 1 step clockwise from the starting point. If the destination is not reached, turn back and go counterclockwise from the starting point for 2 steps. If the destination is still not reached, turn back and go clockwise from the starting point for 4 steps and so on. Within $O(k)$ steps in total, one can find the destination if the destination is $k$ steps away.

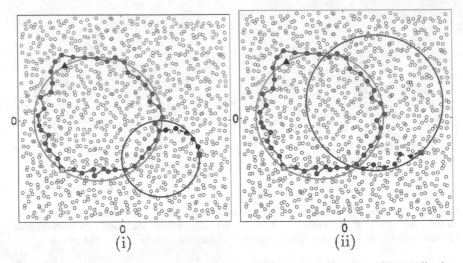

**Fig. 5.19** *Dark triangle* denotes the hashed location; the *red paths* denote producer replication curves; *dashed blue paths* denote retrieval curves; *yellow square* denotes one producer and magenta square denotes one consumer. (i) Consumer latitude curve. (ii) Consumer great circle curve

3. *Aggregated data retrieval rule:* Since *any* closed curve that separates the hashed location $h$ from its antipodal point $\overline{h}$ will intersect all the replication curves with the same data type, a consumer now has great flexibility to choose a retrieval curve for a set of data types $\{T_i\}$, $i = 1, \ldots, m$. In particular, a consumer can follow a data retrieval curve that, for each data type $T_i$,

   - Either goes through the hashed location $h = h(T_i)$ or $\overline{h}$, where the aggregates are computed and stored
   - Or is a closed curve that separates $h$ from $\overline{h}$, collects all the relevant data and computes the aggregates

   The above retrieval rule does not specify a unique retrieval curve but allow infinitely many possibilities. Thus multiple consumers searching for the same data type may choose, by their own decisions, different routes. This flexibility of data retrieval rule enables load balanced traffic patterns and routing robustness.

4. *Full power data retrieval rule:* The consumer travels along any great circle and is able to retrieve *all* the data stored in the network, because any two great circles intersect. An example is shown in Fig. 5.19.

The spherical double rulings scheme, with modestly increased replication, is more robust to node failures. With the flexibility in retrieval curves, the rendezvous node is no longer a bottleneck since retrieval curves may not necessarily visit it. In addition, the data storage scheme admits a local recovery scheme upon node failures. If the sensors in a certain region are destroyed, then all the relevant data are stored on the boundary and thus can be locally recovered.

### 5.5.2.3 Rumor Routing

Rumor routing, proposed by Braginsky and Estrin [11], can be considered as a prob-abilistic double rulings scheme. Information producer takes a walk and leaves data pointers on the trail. A consumer travels along another walk hoping to encounter one of the data pointers. The retrieval curve has a probability to intersect with the producer curve. Thus the consumer sends out enough retrieval walks to have a sufficiently high probability to meet with one of the event curves.

If the paths taken by the producer and consumers are along straight lines, the probability that two random lines intersect inside the sensor field has a constant probability. If the paths taken follow a random walk, the probability that two random walks intersect is lower. The algorithm is randomized with a small probability to fail to discover the producer curve. The attractive property is its robustness to node failures by the built-in randomization in the algorithm design.

### 5.5.2.4 Double Rulings with Virtual Coordinates

Double rulings schemes can also be defined in the virtual coordinate space. In fact, many virtual coordinates naturally admit two sets of rulings orthogonal to each other, e.g., the horizontal lines and vertical lines in a Cartesian coordinate space; the rays emitting from the origin and the cocentric circles in a polar coordinate space. Fang et al. [27] used double rulings scheme for data-centric routing in a general sensor field with nontrivial topology. The idea is to combine double rulings with GHTs on the two-level glider [26]. Recall that in glider, a set of landmarks are selected and the sensors are partitioned into Voronoi tiles, with each tile containing the nodes closest to the same landmark. The top level of the hierarchy is the combinatorial Delaunay graph that describes the adjacency information of the Voronoi tiles. Conceptually, the landmark-based information storage and retrieval scheme has two levels a distributed hash table for information storage at the tile adjacency graph level and a double-ruling scheme at the lower level (inside each tile), which ensures information retrieval within each tile.

In particular, on the top level a data type is hashed to a tile instead of a single node. On the basis of the tile adjacency graph, the shortest path tree rooted at the hashed tile can be computed at each node. All producers and consumers of the same content proceed to the hashed tile following this common shortest path tree. For a producer, data are replicated inside all the tiles (not all the nodes) along the way from where the producer resides to the hashed tile (which we call the *replication path*). The information consumer proceeds toward the hashed tile and checks each tile on its way for the desired data (which we call the *retrieval path*), returns when the retrieval path meets the replication path. At the bottom level, i.e., inside each tile, data storage and retrieval are implemented by a landmark-based double rulings scheme. In the worst case, the consumer visits the final hashed tile to retrieve the desired data. Often on the way to the final tile, the consumer may hit a tile that already has the desired information and stops moving further.

## 5.6 Location Services and Hierarchical Routing Schemes

In geographical routing or routing with virtual coordinates, the source routes a message to a destination with the *name* (the geographical location or the virtual coordinates) of the destination. In certain applications such as routing to a geographical location or region, the source knows the destination's location already. For other applications, such as routing to a node with a particular ID, the source needs to obtain the location or the name of the destination, via a *location service*. A location service supports queries from any node for the name of any other node based on its ID. For the Internet, the mapping between host names and IP addresses is implemented by centralized servers called the DNS. In sensor networks, location service is preferred to be implemented in a distributed fashion, to avoid overloading the sensors identified as location servers.

Location service can be viewed as a special case of data-centric routing. In particular, the data of interest is the name of a node, the producer of this data is the node itself. Data-centric routing or data-centric storage is used to disseminate and store the names of nodes in a distributed way such that any other node can query for this information.

In this section, we cover a few hierarchical routing schemes. The routing schemes described in Sect. 5.5 are more or less "flat", exploiting the two-dimensional embedding of sensor nodes in the physical space. Alternatively hierarchies can be built on the nodes for both point-to-point routing and data-centric storage and information retrieval. Typically when we think of a hierarchical data structure, we are often concerned about load balancing issue and the single point of failure at the root of the hierarchy. The following schemes exploit a few novel ideas to disseminate work load and avoid overloading high-level nodes on the hierarchy.

### 5.6.1 Landmark Hierarchy

#### 5.6.1.1 Landmark Hierarchy

Probably as the first hierarchical routing scheme, Tsuchiya [100] proposed a landmark hierarchy for node naming and routing. Every node is a landmark and the nodes are organized into landmarks of different levels. A landmark at a level $i$ has a radius $r_i$. Nodes within hop count distance $r_i$ from a landmark at a level $i$ have routing table entries to this landmark. The radii $r_i$s are increasing as the level $i$ is increasing. At the highest level $h$, there are a small number of "global" landmarks such that every node has a routing table entry for them. In other words, every node knows how to route to these global landmarks, and they know how to route to nearby "local" landmarks at different resolution.

To construct the landmark hierarchy, every node is a landmark at level 0. Some landmarks of level $i - 1$ are selected to be at level $i$ such that every landmark at level $i - 1$ has at least one level-$i$ landmark within distance $r_{i-1}$. A node may be a landmark at multiple levels. Each node $p$ maintains routing table entries for all the landmarks $\ell$ such that $p$ is within distance $r_i$ of $\ell$ if $\ell$ is a level-$i$ landmark.

The address of a node $p$ is taken as a chain of landmarks $l_0, l_1, \ldots l_h$ such that the landmark $\ell_{i+1}$ is within distance $r_i$ from $\ell_i$, and $\ell_0 = p$. In other words, the landmark $\ell_i$ has a routing table entry to $\ell_{i-1}$. Routing is based on the landmark address in a greedy fashion. To route to a node $p$, a node checks its routing table entry for $p$'s addressing landmarks and takes the lowest level landmark $\ell_i$ (in the worst case the global landmark $\ell_h$). Then the packet is delivered to the next hop toward $\ell_i$. This is repeated until the message is delivered to the destination. When the packet is far away from the destination, it is guided by the high-level addressing landmark of the destination and the routing information is refined as the packet gets close to the destination. A few remarks are in place.

- The addressing landmarks are not necessarily unique. A node may have multiple valid addresses according to the naming scheme.
- The routing scheme guarantees message delivery but the path may not be the shortest path from the source to destination.
- The paths to a destination $p$ do not necessarily go through $p$'s addressing landmarks.

The landmark routing makes use of a landmark hierarchy but routing takes advantages of the cross-branch links in the hierarchy, thus having better robustness to topological changes, compared with routing on a tree. This idea is further developed later in two aspects: refine the construction of the landmark hierarchy to obtain bounded routing path stretch (the ratio of the path length vs. the shortest path length); and use the landmark hierarchy for location service and address lookup [18, 34].

### 5.6.1.2 Discrete Center Hierarchy

Funke et al. [34] refined the hierarchy by taking a discrete center hierarchy [39] defined as follows. Similar to the landmark hierarchy above, each node is a level-0 center. The level-$i$ centers are selected as a subset of the level $i - 1$ centers that satisfies both the *covering* and *packing* properties:

- Every level $i - 1$ center is within distance $2^i$ from at least one level $i$ center
- Two level $i$ centers are at least distance $2^i$ away from each other

The highest level is at most $\log_2 n$, as the covering radius doubles at every level. Each level $i - 1$ center picks one of the level $i$ center within distance $2^i$ and denotes it as its *parent*. Thus the address of a node $p$ is determined by its ancestors in the hierarchy, $l_0, l_1, \ldots l_h$, in which $\ell_0 = P$ and $\ell_i$ is the parent of $\ell_{i-1}$.

With the discrete center hierarchy, cross-branch links are also included to help routing. In particular, a center $v$ at level $i$ has its *cluster* $C(v)$ as the set of descendant nodes. A node $u$ has a cluster $C(v)$ at level $i$ as a *neighboring cluster* if there is a node $q$ in $C(v)$ within distance $\alpha \cdot 2^{i+1}$ for a constant $\alpha > 0$. Each node maintains a routing table entry for each neighboring cluster at each level. Again a node has routing information to nearby neighboring clusters at different resolutions. The following properties can be said to the neighboring clusters.

- The distance from a node to its level $i$ ancestor is bounded by $2^{i+1}$. In particular, we follow the parent link and sum up its maximum possible length, we get $1 + 2 + 2^2 + \ldots + 2^i \leq 2^{i+1}$.
- If we take $\alpha > 3$, a node in cluster $C(v)$ at level $i$ has $C(w)$ as a neighboring cluster, where $w$ is a child of $v$, $w$ has level $i - 1$.

For a node $u$ to route a packet to a destination node $v$, $u$ checks its routing table for the lowest level cluster $C(w)$ that contains the destination. Then $u$ routes the packet toward the cluster $C(w)$. When the packet arrives at the cluster $C(w)$, the same procedure is repeated at the next node. The new hierarchy guarantees that the route taken is within four times the shortest path length $d_{uv}$ from source $u$ to destination $v$. Indeed, say the cluster $C(w)$ at lowest level found in $u$'s routing table entry is at level $k$, then $d_{uv}$ must be greater than $\alpha \cdot 2^k$ (otherwise $u$ has a neighboring cluster at level $k - 1$ that contains the destination). The path taken from $u$ to $C(w)$ is at most $\alpha \cdot 2^{k+1} \leq 2d_{uv}$. Once the packet arrives at $C(w)$ it is able to find a neighboring cluster at level strictly lower than $k$. Thus the path taken to arrive at the next lower level cluster is at most half the length of the first segment on the path, i.e., the path taken so far from $u$ to $C(w)$. Thus the total path length is at most double the length of the first segment and is less than $4d_{uv}$.

In a sensor network setting, when the sensors are reasonably uniformly distributed, the communication graph typically has constant doubling dimension [47], i.e., a ball of radius $R$ can be covered by a constant number of balls of radius $R/2$. In this case, the hierarchy has depth $O(\log n)$ and each node has a constant number of neighboring clusters. Therefore, the number of entries in the routing table for each node is $O(\log n)$ in the worst case and on average a constant number.

The landmark hierarchy can also support data-centric routing and distributed location services. In particular, a data item is hashed and stored at certain other nodes called the *data servers*. Specifically, a producer $u$ has its ancestors at each level, denoted as $l_0 = u, l_1, \ldots l_h$. For each level $i$, the producer sends the data to each of $u$'s neighboring clusters at that level. For one of such cluster $C(p)$, the data are hashed by its data type to one node $q$ inside this cluster. Again the number of copies a data item is stored in the network is at most $O(\log n)$.

Storing data in the landmark hierarchy enables the consumer to find the data item in a distance sensitive fashion, i.e., a producer with distance $d$ from the consumer is able to retrieve the data within cost $4d$. In particular, when a consumer $v$ wants to find a data item, it starts to visit the data server in the clusters of its ancestors at level $i$, with $i$ increasing from 0, until the data item is found or the highest level is reached. Intuitively, the data storage scheme places more data servers in nearby clusters and fewer servers in far away but large clusters. The retrieval scheme starts to search in local neighborhood until it discovers a nearby data server.

At last we remark that similar idea of using a landmark hierarchy for routing and location service was also explored in [4, 17, 18]. These constructions use different ways to construct the hierarchy and admit different quality bounds on the routing performance.

### 5.6.2 Routing and Location Services in Mobile Networks

In a mobile network, the landmark hierarchies described in the previous subsection are not appropriate as the structure is dependent on the network distance which changes frequently as the nodes move around. Geographical routing, on the other hand, is robust in a mobile network given that the source knows the location of the destination. As the nodes constantly move around, we will need a location service for the mobile nodes such that any node can query for the current location of any other node. Notice that in implementing this location service we will need some routing protocol to deliver the query messages. Thus routing and location service appear to be a chicken and egg problem that will need to be solved all at once in a collaborative framework. The grid location service (GLS), proposed by Li et al. [65], represents an example of such a protocol.

GLS implements a distributed location service in which a node serves as a *location server* and stores location information for other nodes. A node's location is stored in multiple other location servers. As a node moves around and changes its location, it will update its location servers of its new location. Any other node looking for the location of a node with only the ID of the node is accomplished with pretty much the same protocol as the node uses to select its location server in the first place. GLS uses consistent hashing [52], originally proposed for hierarchical hashing of web pages in the Internet, for location server assignment.

We assume the nodes know their geographical locations via GPS devices for example. The domain in which the nodes move around is partitioned recursively as a quadtree of depth $k$ such that the smallest square contains only one node. The smallest square is called an order-1 square. The bounding box is called an order-$k$ square and partitioned into 4 equal-size order $k - 1$ squares. In fact, other type of balanced partitioning can also be used here. The selection of location servers follows the same philosophy as the previous section. A node $u$ stays in a unique square of order $i$, for each $i$ from 1 to $k$. $u$ selects a location server in each of the three other sibling order-$i$ squares of the one containing $u$. In particular, inside a square $S$, the location server is chosen as the node *closest* to the ID of $u$, defined as the one with *least ID greater than* $u$. The ID space is circular. For example, ID 2 is closer to 17 than 7 is.

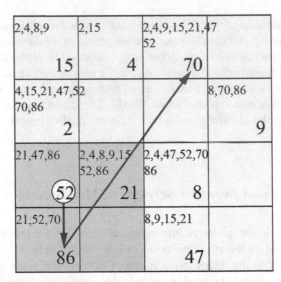

**Fig. 5.20** An example of GLS. In each small square the number at lower right is the ID of the node inside this square. The numbers on the *upper left* are the IDs of other nodes for which this node is the location server. The *arrows* show that a query at 52 looking for the location of a node with ID 70

To perform a query for the location of $v$, $u$ knows only the ID of $v$ and does not know the location servers of $v$. In fact $v$ does not know its location servers either. Now recall that $u$ is also the location server of some other nodes. $u$ will simply send a request by geographical routing to the node closest to $v$ for which $u$ has location information. Say this node is $w$. This request can be delivered by geographical routing as $u$ has the location information of $w$. Now $w$ does the same thing until the message reaches a location server of $v$, from where the message can then be delivered to $v$ via geographical routing. An example is shown in Fig. 5.20.

The correctness of the scheme follows the following claim that the query visits the node closest to $v$ in $u$'s order-$i$ square $S_i(u)$, with increasing $i$. Thus if $v$ is inside node $u$'s order $k$ square, then the closest node to $v$ is $v$ itself. Thus the message will be delivered to $v$. To see that the claim is true, we show by induction. At level $i = 1$, it is trivially true as $u$ is the only node in its order-1 square. Now assume that the message arrives at a node $w$ which is the closest node to $v$ in $u$'s order $i$ square $S_i(u)$. This says that in $S_i(u)$, there is no other nodes in between the ID of $v$ and $w$. Now suppose that the node closest to $v$ in $S_{i+1}(u)$ is $x$. If $x$ is $w$, then the claim is true trivially. If $x$ is not $w$, $x$ stays in a sibling square of $S_i(u)$. $x$ will choose its location server in $S_i(u)$ as the node closest to $x$, which must be $w$. Thus $w$ is the location server of the node closest to $v$ in $S_{i+1}(u)$. The routing protocol will forward the message to $x$. See Fig. 5.21 for an illustration.

When a node moves around, it will update its location at its location servers. This update operation is essentially the same as the query algorithm. When a node

**Fig. 5.21** The node closest to $v$ in $S_{i+1}(u)$, denoted by $x$, if not equal to $w$, must choose $w$ as its location server in $S_i(u)$. Thus $w$ has $x$'s location and the message is delivered to $x$

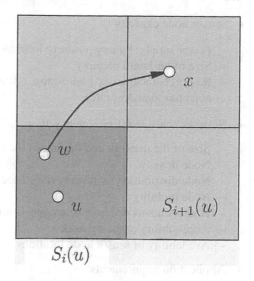

$u$ wants to update the location server at a square $S$, it sends a message to that square with geographical routing. Once the message arrives at $S$, the first node will start with a query for the node $u$ inside $S$. As shown by the query algorithm, this message is eventually forwarded to the node closest to $u$ inside $S$ before it leaves this square, which is precisely the location server for $u$. Thus location information is updated at this location server.

To initialize the system, the location servers are updated from bottom up. The nodes first locate their location servers in order 1 sibling squares. After this is done, the nodes use the location update scheme to locate the location servers in order 2 sibling squares, and so on. Thus no flooding is needed to set up the system.

The combination of location service with geographical routing, with the help of consistent hashing is a very neat idea that requires no global coordination. The scheme is later improved to have worst-case bounded performance in terms of location update and distance-sensitive query (such as LLS by Abraham et al. [1] and MLS by Flury and Wattenhofer [31]). The idea used in LLS is to have a node publish its location along a spiral that exponentially increases in distance. Similarly, the lookup operation also takes a spiral from the query's location. When the lookup spiral visits a square in which the location update has visited, the query will return the location information. The intuition is that the query spiral will intersect the publishing spiral within a traveling distance $O(d)$ if the node is of distance $d$ away.

## 5.7  Thoughts for Practitioners

For practitioners, the selection of routing algorithms may depend on the following considerations.

- Sensor node capacity

  - Power supply: battery power or long-lasting power supplies
  - Size of on-board memory
  - Radio characteristics (radio range, directional antennas or omni-directional antennas, data lost rate)

- Network scale and deployment setting

  - Size of the network and the size of the deployment region
  - Node density
  - Node distribution uniformity (existence of routing voids or not)
  - Node mobility
  - Homogeneous nodes or heterogeneous nodes
  - Accessibility of the network
  - Availability of sensor node locations

- Application requirements

  - Traffic pattern (many-to-one data collection, one-to-many data dissemination, point-to-point routing, and data-centric routing)
  - Traffic load (frequency of message delivery)
  - Expected network operational time
  - Data packet size and format
  - Quality-of-service requirement on latency and throughput

## 5.8  Directions for Future Research

The work surveyed in this chapter revealed the power of using geometric abstractions in sensor network routing. There are still directions for future research on many fundamental problems.

- Communication models: modeling wireless communication channels has a fundamental influence in routing algorithm design Simplified models such as unit disk graphs have been adopted widely in the algorithm and theory community as it is simple yet embraces a rich set of geometric structures. However, the deviation from realistic scenarios makes some algorithms with unit disk graph model fail in practice. More realistic models have been proposed and adopted. But it still remains a challenge to develop different communication models that are more elaborative and realistic, yet still reveals key insights and admit efficient routing algorithm design.
- Supporting infrastructure support: applying the geometric ideas in routing algorithm design often requires some supporting architecture components such as localization, synchronization, and topology understanding.

- Mobility: most algorithms covered in this chapter assume static sensor networks. With node mobility and/or heterogeneity, new ideas are needed on what geometric structure can we abstract from the node mobility and how does that help with routing information in the network efficiently.
- Geometry of the sensor data: when a sensor network monitors a physical signal landscape, there is natural correlation between data from sensor nodes, both in the spatial and temporal spectrum. Abstracting and utilizing the geometry of sensor data can help with data compression, aggregation, and validation in routing algorithms.
- Theoretical foundation: a number of the geometric algorithms discussed in this chapter use ideas from the continuous geometric domain and apply an analog of the ideas in the discrete network. Investigation of rigorous definition of the geometric concepts in the discrete network remains an open problem.

## 5.9   Conclusions

In the past few years, geometric ideas have been used extensively in the design of wireless sensor networks. This chapter covers routing algorithms that make use of various aspects of "sensor network geometry": the nodes' geographical locations, such as geographical routing; as well as the global shape and topology of a sensor field.

## Terminologies

*Network localization.* In a wireless network, network localization finds the geographical location of all the nodes. A network location algorithm can be anchor-based or anchor-free. Anchor-based algorithms assume known location information for some of the nodes (called anchors) in the network. An anchor-free algorithm does not assume the knowledge of any location information and returns the relative positions of the network.

*Geographical routing.* In a wireless ad hoc network, geographical routing is a routing algorithm that uses the nodes' geographical locations to choose a neighbor as the next hop.

*Virtual coordinates.* Each node in a wireless sensor network is assigned a virtual coordinate as its identifier in the routing algorithm. A virtual coordinate routing algorithm makes routing decisions with the virtual coordinates of the source, its neighbors, and the destination. The size of the virtual coordinate is typically a constant or a small polylogarithmic function of the network size.

*Greedy embedding.* Given a graph, embed it in a d-dimensional Euclidean space such that for each pair of node p, q, there is a neighbor of p whose distance to the destination is smaller than the Euclidean distance between p and q. Such an embedding is called a greedy embedding.

*Landmark routing.* A subset of nodes in a wireless sensor network are selected as landmarks such that the hop count distances to these landmarks are used to identify other nodes as well as route messages between them.

*Data centric routing.* The destination of the message is not specified as a fixed destination node, but rather defined as the sensor node(s) that has the information as requested in the message. A data-centric routing algorithm needs to discover the nodes holding data of interest as well as route the message to these nodes. The nodes with data are often called sources or information producers. The nodes who pose queries for the data are called sinks or information consumers.

*Geographical hash tables.* It is a data-centric routing scheme. A source node will take a hash function on the data type and deliver the data to a geographical hash location. The nodes close to the hashed location serve as rendervous and store the data. A sink node searching for this type of data will route the query to the same geographical location and retrieve the data from the rendervous nodes.

*Double rulings.* It is a data-centric routing scheme. The source node will hash data along a data storage curve (possibly dependent on the source location and the data type) in the network. The sink node will hash data along a data retrieval curve (possibly dependent on the sink location and the data type) in the network. If each data storage curve intersects with each data retrieval curve, the data retrieval mechanism guarantees successful data discovery.

*Location service.* For geographical routing, a source node needs to know the destination's location to send a message to it. Location service provides a map between each node and its geographical location. Location service can be implemented at a central location server or at distributed location servers.

*Landmark hierarchy.* All nodes are landmarks at different levels of the landmark hierarchy. A node is given a name according to the network distances to some other landmarks in the landmark hierarchy. Routing decisions are made with the name of the source and the destination. The landmark hierarchy can be considered as multilevel landmark routing.

## Questions

1. In network localization, suppose that $k$ nodes out of a total number of $n$ nodes are anchor nodes. There are $m$ edges in the graph. How large does m have to be such that there is no degree of freedom in the system?
2. Prove the following two claims that will show both relative neighborhood graph and Gabriel graph on a set of points P in the plane are planar, connected subgraphs of unit disk graphs.

   (a) Prove that the relative neighborhood graph is a subgraph of the Gabriel Graph.

(b) Prove that the Gabriel graph is a subgraph of the Delaunay triangulation on the same set of nodes.

For a set of point P in the plane, the Voronoi diagram partitions the plane into connected cells such that all the points in the same cell has the same closest point in P. The Delaunay triangulation is the dual graph of the Voronoi diagram such that there is an edge uv, if the cells of u, v are adjacent. The Delaunay triangulation has what is called "empty-circle property": for any Delaunay edge, one can find a circle that goes through the two endpoints of the edge and has no other points inside. The converse is also true: if all edges in a graph has this empty circle property, the graph is a Delaunay triangulation.

The Delaunay triangulation is known to be a planar graph. Thus there are no crossing edges in either relative neighborhood graph or Gabriel graph.

(c) Prove that the relative neighborhood graph contains the minimum spanning tree. Thus both the relative neighborhood graph and the Gabriel graph are connected.

3. Prove that a Delaunay triangulation of a set of point in the plane admits a greedy embedding.
4. Show that a star with 7 leaf nodes does not have a greedy embedding in the plane.
5. Prove that for any graph with a subset of vertices pinned down, the rubberband representation (defined in Sect 5.4.1.1) is uniquely determined. (Hint: show that the rubberband embedding minimizes a convex energy function.)
6. Prove for the case of Euclidean plane that greedy routing algorithm used in intratile routing of GLIDER by minimizing the centered virtual distance to the destination is free of local minimum, when the landmarks are not collinear.
7. Elaborate how to incorporate node mobility in geographical routing.
8. Prove that in landmark-based routing, if we choose at least three landmarks that are not collinear and define the virtual coordinate of any other point in the Euclidean plane as the Euclidean distances to these landmarks, the virtual coordinate for each point is unique. Also give an example that this may not be true in a discrete wireless network (modeled by unit disk graph).
9. Compare the pros and cons of geographical hash tables and double rulings scheme for data-centric routing, in terms of storage requirement, query latency, robustness to node failures, etc.
10. Elaborate how to handle node insertion in the Grid location service (GLS).

# References

1. I. Abraham, D. Dolev, and D. Malkhi. LLS: A locality aware location service for mobile ad hoc networks. In *DIALM-POMC '04: Proceedings of the 2004 joint workshop on Foundations of mobile computing*, 2004. ACM Press, New York, pages 75–84.

2. P. Angelini, F. Frati, and L. Grilli. An algorithm to construct greedy drawings of triangulations. In *Proceedings of the 16th International Symposium on Graph Drawing*, pages 26–37, 2008.
3. J. Aspnes, D. Goldenberg, and Y. R. Yang. On the computational complexity of sensor network localization. In *The First International Workshop on Algorithmic Aspects of Wireless Sensor Networks (ALGOSENSORS)*, pages 32–44, 2004.
4. B. Awerbuch and D. Peleg. Concurrent online tracking of mobile users. In *SIGCOMM '91: Proceedings of the conference on Communications architecture & protocols*, 1991. ACM Press, New York, pages 221–233.
5. M. Badoiu, E. D. Demaine, M. T. Hajiaghayi, and P. Indyk. Low-dimensional embedding with extra information. In *SCG '04: Proceedings of the twentieth annual symposium on Computational geometry*, 2004. ACM Press, New York, pages 320–329.
6. P. Bahl and V. N. Padmanabhan. RADAR: An in-building RF-based user location and tracking system. In *IEEE INFOCOM*, Volume 2, pages 775–784, 2000.
7. P. Biswas and Y. Ye. Semidefinite programming for ad hoc wireless sensor network localization. In *Proceedings of the 3rd International Symposium on Information Processing in Sensor Networks*, pages 46–54, 2004.
8. L. Blazević, L. Buttyán, S. Capkun, S. Giordano, H. Hubaux, and J. L. Boudec. Self-organization in mobile ad hoc networks: The approach of terminodes. *IEEE Communications Magazine*, pages 166–175, 2001.
9. I. Borg and P. Groenen. *Modern Multidimensional Scaling: Theory and Applications*. Springer-Verlag, Berlin 1997.
10. P. Bose, P. Morin, I. Stojmenovic, and J. Urrutia. Routing with guaranteed delivery in ad hoc wireless networks. *Wireless Networks*, 7(6):609–616, 2001.
11. D. Braginsky and D. Estrin. Rumor routing algorithm for sensor networks. In *Proc. of the 1st ACM Int'l Workshop on Wireless Sensor Networks and Applications (WSNA)*, pages 22–31, September 2002.
12. H. Breu and D. G. Kirkpatrick. Unit disk graph recognition is NP-hard. *Computational Geometry: Theory and Applications*, 9(1–2):3–24, 1998.
13. J. Bruck, J. Gao, and A. Jiang. Localization and routing in sensor networks by local angle information. In *Proceedings of the Sixth ACM International Symposium on Mobile Ad Hoc Networking and Computing (MobiHoc'05)*, pages 181–192, May 2005.
14. J. Bruck, J. Gao, and A. Jiang. MAP: Medial axis based geometric routing in sensor networks. *Wireless Networks*, 13(6):835–853, 2007.
15. M. Caesar, M. Castro, E. B. Nightingale, G. O'Shea, and A. Rowstron. Virtual ring routing: Network routing inspired by dhts. In *SIGCOMM '06: Proceedings of the 2006 conference on Applications, technologies, architectures, and protocols for computer communications*, 2006. ACM Press, New York, pages 351–362.
16. A. Caruso, A. Urpi, S. Chessa, and S. De. GPS free coordinate assignment and routing in wireless sensor networks. In *Proceedings of the 24th Conference of the IEEE Communication Society (INFOCOM)*, Volume 1, pages 150–160, March 2005.
17. H. T.-H. Chan, A. Gupta, B. M. Maggs, and S. Zhou. On hierarchical routing in doubling metrics. In *SODA '05: Proceedings of the sixteenth annual ACM-SIAM symposium on Discrete algorithms*, 2005. Society for Industrial and Applied Mathematics, Pennsylvania, pages 762–771.
18. B. Chen and R. Morris. L+: Scalable landmark routing and address lookup for multi-hop wireless networks. Technical Report MIT-LCS-TR-837, Massachusett Institute of Technology, 2002.
19. M. B. Chen, C. Gotsman, and C. Wormser. Distributed computation of virtual coordinates. In *SCG '07: Proceedings of the twenty-third annual symposium on Computational geometry*, 2007. ACM Press, New York, pages 210–219.
20. Y.-C. Cheng, Y. Chawathe, A. LaMarca, and J. Krumm. Accuracy characterization for metropolitan-scale wi-fi localization. In *MobiSys '05: Proceedings of the 3rd international conference on Mobile systems, applications, and services*, 2005. ACM Press, New York, pages 233–245.

21. H. S. M. Coxeter. *Introduction to Geometry.* Wiley, New York, 2nd edition, 1969.
22. R. Dhandapani. Greedy drawings of triangulations. In *SODA '08: Proceedings of the nineteenth annual ACM-SIAM symposium on Discrete algorithms*, 2008.
23. L. Doherty, L. E. Ghaoui, and S. J. Pister. Convex position estimation in wireless sensor networks. In *IEEE Infocom*, Volume 3, pages 1655–1663, April 2001.
24. A. Efrat, C. Erten, D. Forrester, A. Iyer, and S. G. Kobourov. Force-directed approaches to sensor localization. In *Proceedings of the 8th Workshop on Algorithm Engineering and Experiments (ALENEX)*, pages 108–118, 2006.
25. Q. Fang, J. Gao, and L. Guibas. Locating and bypassing routing holes in sensor networks. In *Mobile Networks and Applications*, Volume 11, pages 187–200, 2006.
26. Q. Fang, J. Gao, L. Guibas, V. de Silva, and L. Zhang. GLIDER: Gradient landmark-based distributed routing for sensor networks. In *Proceedings of the 24th Conference of the IEEE Communication Society (INFOCOM)* Volume 1, pages 339–350, March 2005.
27. Q. Fang, J. Gao, and L. J. Guibas. Landmark-based information storage and retrieval in sensor networks. In *The 25th Conference of the IEEE Communication Society (INFOCOM'06)*, pages 1–12, April 2006.
28. S. P. Fekete, M. Kaufmann, A. Kröller, and N. Lehmann. A new approach for boundary recognition in geometric sensor networks. In *Proceedings 17th Canadian Conference on Computational Geometry*, pages 82–85, 2005.
29. S. P. Fekete, A. Kröller, D. Pfisterer, S. Fischer, and C. Buschmann. Neighborhood-based topology recognition in sensor networks. In *ALGOSENSORS*, Volume 3121 of *Lecture Notes in Computer Science*, 2004 Springer, Berlin, pages 123–136.
30. G. G. Finn. Routing and addressing problems in large metropolitan-scale internetworks. Technical Report ISU/RR-87-180, ISI, March 1987.
31. R. Flury and R. Wattenhofer. MLS: An efficient location service for mobile ad hoc networks. In *MobiHoc '06: Proceedings of the seventh ACM international symposium on Mobile ad hoc networking and computing*, 2006. ACM press, New York, pages 226–237.
32. R. Fonseca, S. Ratnasamy, J. Zhao, C. T. Ee, D. Culler, S. Shenker, and I. Stoica. Beacon vector routing: Scalable point-to-point routing in wireless sensornets. In *Proceedings of the 2nd Symposium on Networked Systems Design and Implementation (NSDI)*, pages 329–342, May 2005.
33. S. Funke. Topological hole detection in wireless sensor networks and its applications. In *DIALM-POMC '05: Proceedings of the 2005 Joint Workshop on Foundations of Mobile Computing*, 2005. ACM Press, New York, pages 44–53.
34. S. Funke, L. J. Guibas, A. Nguyen, and Y. Wang. Distance-sensitive routing and information brokerage in sensor networks. In *IEEE International Conference on Distributed Computing in Sensor System (DCOSS'06)*, pages 234–251, June 2006.
35. S. Funke and C. Klein. Hole detection or: "How much geometry hides in connectivity?". In *SCG '06: Proceedings of the twenty-second annual symposium on Computational geometry*, pages 377–385, 2006.
36. S. Funke and N. Milosavljević. Guaranteed-delivery geographic routing under uncertain node locations. In *Proceedings of the 26th Conference of the IEEE Communications Society (INFOCOM'07)*, pages 1244–1252, May 2007.
37. S. Funke and N. Milosavljević. Network sketching or: "how much geometry hides in connectivity? - part II". In *SODA '07: Proceedings of the eighteenth annual ACM-SIAM symposium on Discrete algorithms*, 2007. Society for Industrial and Applied Mathematics, Pennsylvania, pages 958–967.
38. D. Ganesan, B. Krishnamachari, A. Woo, D. Culler, D. Estrin, and S. Wicker. Complex behavior at scale: An experimental study of low-power wireless sensor networks. Technical Report UCLA/CSD-TR 02-0013, UCLA, 2002.
39. J. Gao, L. Guibas, and A. Nguyen. Deformable spanners and their applications. *Computational Geometry: Theory and Applications*, 35(1–2):2–19, 2006.
40. J. Gao, L. J. Guibas, S. Y. Oudot, and Y. Wang. Geodesic Delaunay triangulations and witness complexes in the plane. In *Proceedings of the ACM–SIAM Symposium on Discrete Algorithms (SODA'08)*, 571–580, January 2008.

41. R. Ghrist and A. Muhammad. Coverage and hole-detection in sensor networks via homology. In *Proceedings the 4th International Symposium on Information Processing in Sensor Networks (IPSN'05)*, pages 254–260, 2005.

42. D. K. Goldenberg, P. Bihler, Y. R. Yang, M. Cao, J. Fang, A. S. Morse, and B. D. O. Anderson. Localization in sparse networks using sweeps. In *MobiCom '06: Proceedings of the 12th annual international conference on Mobile computing and networking*, 2006, ACM Press, New York, pages 110–121.

43. M. T. Goodrich and D. Strash. Succinct greedy geometric routing in r2. Technical report on arXiv:0812.3893, 2008.

44. C. Gotsman and Y. Koren. Distributed graph layout for sensor networks. In *Proceedings of the International Symposium on Graph Drawing*, pages 273–284, September 2004.

45. Y.-J. Kim, R. Govindan, B. Karp, and S. Shenker. Lazy cross-link removal for geographic routing. In *SenSys '06: Proceedings of the 4th international conference on Embedded networked sensor systems*, 2006. ACM Press, New York, pages 112–124.

46. B. Greenstein, D. Estrin, R. Govindan, S. Ratnasamy, and S. Shenker. DIFS: A distributed index for features in sensor networks. In *Proceedings of First IEEE International Workshop on Sensor Network Protocols and Applications*, pages 163–173, Anchorage, Alaska, May 2003.

47. A. Gupta, R. Krauthgamer, and J. R. Lee. Bounded geometries, fractals, and low-distortion embeddings. In *FOCS '03: Proceedings of the 44th Annual IEEE Symposium on Foundations of Computer Science*, 2003. IEEE Computer Society, Washington, page 534–543.

48. A. Harter, A. Hopper, P. Steggles, A. Ward, and P. Webster. The anatomy of a context-aware application. In *MobiCom '99: Proceedings of the 5th annual ACM/IEEE international conference on Mobile computing and networking*, 1999, ACM Press, New York, pages 59–68.

49. B. Hofmann-Wellenhof, H. Lichtenegger, and J. Collins. *Global Positioning Systems: Theory and Practice*. 5th edition, Springer, Berlin, 2001.

50. C. Intanagonwiwat, R. Govindan, and D. Estrin. Directed diffusion: A scalable and robust communication paradigm for sensor networks. In *ACM Conference on Mobile Computing and Networking (MobiCom)*, pages 56–67, 2000.

51. R. Jain, A. Puri, and R. Sengupta. Geographical routing using partial information for wireless ad hoc networks. *IEEE Personal Communications*, 8(1):48–57, Feb. 2001.

52. D. Karger, E. Lehman, T. Leighton, R. Panigrahy, M. Levine, and D. Lewin. Consistent hashing and random trees: Distributed caching protocols for relieving hot spots on the world wide web. In *STOC '97: Proceedings of the twenty-ninth annual ACM symposium on Theory of computing*, 1997. ACM Press, New york, pages 654–663.

53. B. Karp and H. Kung. GPSR: Greedy perimeter stateless routing for wireless networks. In *Proceedings of the ACM/IEEE International Conference on Mobile Computing and Networking (MobiCom)*, pages 243–254, 2000.

54. Y.-J. Kim, R. Govindan, B. Karp, and S. Shenker. Geographic routing made practical. In *Proceedings of the Second USENIX/ACM Symposium on Networked System Design and Implementation (NSDI 2005)*, May 2005.

55. Y.-J. Kim, R. Govindan, B. Karp, and S. Shenker. On the pitfalls of geographic face routing. In *DIALM-POMC '05: Proceedings of the 2005 joint workshop on Foundations of mobile computing*, 2005. ACM Press, New York, pages 34–43.

56. R. Kleinberg. Geographic routing using hyperbolic space. In *Proceedings of the 26th Conference of the IEEE Communications Society (INFOCOM'07)*, pages 1902–1909, 2007.

57. E. Kranakis, H. Singh, and J. Urrutia. Compass routing on geometric networks. In *Proceedings 11th Canadian Conference on Computational Geometry*, pages 51–54, 1999.

58. P. Krishnan, A. S. Krishnakumar, W.-H. Ju, C. Mallows, and S. Ganu. A system for LEASE: System for location estimation assisted by stationary emitters for indoor RF wireless networks. In *IEEE Infocom*, Volume 2, pages 1001–1011, Hongkong, March 2004.

59. A. Kröller, S. P. Fekete, D. Pfisterer, and S. Fischer. Deterministic boundary recognition and topology extraction for large sensor networks. In *Proceedings of the Seventeenth Annual ACM-SIAM Symposium on Discrete Algorithms*, pages 1000–1009, 2006.

60. F. Kuhn, T. Moscibroda, and R. Wattenhofer. Unit disk graph approximation. In *Proceedings of the 2004 Joint Workshop on Foundations of Mobile Computing*, pages 17–23, 2004.
61. F. Kuhn, R. Wattenhofer, Y. Zhang, and A. Zollinger. Geometric ad-hoc routing: Of theory and practice. In *Proceedings 22nd ACM International Symposium on the Principles of Distributed Computing (PODC)*, pages 63–72, 2003.
62. F. Kuhn, R. Wattenhofer, and A. Zollinger. Asymptotically optimal geometric mobile ad-hoc routing. In *Proceedings of the 6th International Workshop on Discrete Algorithms and Methods for Mobile Computing and Communications*, pages 24–33, 2002.
63. A. M. Ladd, K. E. Bekris, A. Rudys, L. E. Kavraki, D. S. Wallach, and G. Marceau. Robotics-based location sensing using wireless ethernet. In *MobiCom '02: Proceedings of the 8th annual international conference on Mobile computing and networking*, 2002. ACM Press, New York, pages 227–238.
64. T. Leighton and A. Moitra. Some results on greedy embeddings in metric spaces. In *Proceeding of the 49th IEEE Annual Symposium on Foundations of Computer Science*, pages 337–346, October 2008.
65. J. Li, J. Jannotti, D. Decouto, D. Karger, and R. Morris. A scalable location service for geographic ad-hoc routing. In *Proceedings of 6th ACM/IEEE International Conference on Mobile Computing and Networking*, pages 120–130, 2000.
66. Z. Li, W. Trappe, Y. Zhang, and B. Nath. Robust statistical methods for securing wireless localization in sensor networks. In *IPSN '05: Proceedings of the 4th international symposium on Information processing in sensor networks*, Piscataway, NJ, USA, 2005. IEEE Press, New York, pages 91–98.
67. N. Linial, L. Lovász, and A. Wigderson. Rubber bands, convex embeddings and graph connectivity. *Combinatorica*, 8(1):91–102, 1988.
68. X. Liu, Q. Huang, and Y. Zhang. Combs, needles, haystacks: Balancing push and pull for discovery in large-scale sensor networks. In *SenSys '04: Proceedings of the 2nd international conference on Embedded networked sensor systems*, 2004, ACM Press, New York pages 122–133.
69. S. Madden, M. J. Franklin, J. M. Hellerstein, and W. Hong. TAG: A tiny aggregation service for ad-hoc sensor networks. *SIGOPS Operating Systems Review*, 36(SI):131–146, 2002.
70. Y. Mao, F. Wang, L. Qiu, S. S. Lam, and J. M. Smith. S4: Small state and small stretch routing protocol for large wireless sensor networks. In *Proceedings of the 4th USENIX Symposium on Networked System Design and Implementation (NSDI 2007)*, April 2007.
71. D. Moore, J. Leonard, D. Rus, and S. Teller. Robust distributed network localization with noisy range measurements. In *SenSys '04: Proceedings of the 2nd international conference on Embedded networked sensor systems*, 2004. ACM Press, New York, pages 50–61.
72. A. Nasipuri and R. E. Najjar. Experimental Evaluation of an Angle Based Indoor Localization System. In *Proceedings of the 4th International Symposium on Modeling and Optimization in Mobile, Ad Hoc and Wireless Networks*, pages 1–9, Boston, MA, April 2006.
73. J. Newsome and D. Song. GEM: Graph embedding for routing and data-centric storage in sensor networks without geographic information. In *SenSys '03: Proceedings of the 1st international conference on Embedded networked sensor systems*, 2003. ACM Press, New York, pages 76–88.
74. A. Nguyen, N. Milosavljevic, Q. Fang, J. Gao, and L. J. Guibas. Landmark selection and greedy landmark-descent routing for sensor networks. In *Proceedings of the 26th Conference of the IEEE Communications Society (INFOCOM'07)*, pages 661–669, May 2007.
75. D. Niculescu and B. Nath. Ad hoc positioning system (APS). In *IEEE GLOBECOM*, pages 2926–2931, 2001.
76. D. Niculescu and B. Nath. *Ad hoc* positioning system (APS) using AOA. In *IEEE INFOCOM*, Volume 22, pages 1734–1743, March 2003.
77. D. Niculescu and B. Nath. Error characteristics of ad hoc positioning systems (APS). In *MobiHoc '04: Proceedings of the 5th ACM International Symposium on Mobile Ad Hoc Networking and Computing*, pages 20–30, 2004.

78. D. Niculescu and B. Nath. VOR base stations for indoor 802.11 positioning. In *MobiCom '04: Proceedings of the 10th annual international conference on Mobile computing and networking*, 2004. ACM Press, New York, pages 58–69.
79. C. H. Papadimitriou and D. Ratajczak. On a conjecture related to geometric routing. *Theoretical Computer Science*, 344(1):3–14, 2005.
80. N. B. Priyantha, A. Chakraborty, and H. Balakrishnan. The cricket location-support system. In *MobiCom '00: Proceedings of the 6th ACM Annual International Conference on Mobile Computing and Networking*, pages 32–43, 2000.
81. A. Rao, C. Papadimitriou, S. Shenker, and I. Stoica. Geographic routing without location information. In *Proceedings of the 9th annual international conference on Mobile computing and networking*, 2003. ACM Press, New York, pages 96–108.
82. S. Ratnasamy, P. Francis, M. Handley, R. Karp, and S. Schenker. A scalable content-addressable network. In *SIGCOMM '01: Proceedings of the 2001 conference on Applications, technologies, architectures, and protocols for computer communications*, 2001. ACM Press, New York, pages 161–172.
83. S. Ratnasamy, B. Karp, L. Yin, F. Yu, D. Estrin, R. Govindan, and S. Shenker. GHT: A geographic hash table for data-centric storage in sensornets. In *Proceedings 1st ACM Workshop on Wireless Sensor Networks and Applications*, pages 78–87, 2002.
84. A. I. T. Rowstron and P. Druschel. Pastry: Scalable, decentralized object location, and routing for large-scale peer-to-peer systems. In *Middleware '01: Proceedings of the IFIP/ACM International Conference on Distributed Systems Platforms Heidelberg*, London, UK, 2001. Springer-Verlag, Berlin, pages 329–350.
85. R. Sarkar, X. Zhu, and J. Gao. Double rulings for information brokerage in sensor networks. In *Proceedings of the ACM/IEEE International Conference on Mobile Computing and Networking (MobiCom)*, pages 286–297, September 2006.
86. A. Savvides, C.-C. Han, and M. B. Strivastava. Dynamic fine-grained localization in *ad-hoc* networks of sensors. In *Proceedings 7th Annual International Conference on Mobile Computing and Networking (MobiCom 2001)*, Rome, Italy, July 2001. ACM Press, New York, pages 166–179.
87. A. Savvides, H. Park, and M. B. Strivastava. The *n*-hop multilateration primitive for node localization problems. *Mobile Networks and Applications*, 8(4):443–451, 2003.
88. K. Seada, A. Helmy, and R. Govindan. On the effect of localization errors on geographic face routing in sensor networks. In *IPSN '04: Proceedings of the third international symposium on Information processing in sensor networks*, 2004. ACM Press, New York, pages 71–80.
89. Y. Shang, W. Ruml, Y. Zhang, and M. P. J. Fromherz. Localization from mere connectivity. In *MobiHoc '03: Proceedings of the 4th ACM International Symposium on Mobile Ad Hoc Networking and Computing*, pages 201–212, 2003.
90. S. Shenker, S. Ratnasamy, B. Karp, R. Govindan, and D. Estrin. Data-centric storage in sensornets. *SIGCOMM Computer Communication Review*, 33(1):137–142, 2003.
91. A. M.-C. So and Y. Ye. Theory of semidefinite programming for sensor network localization. In *SODA '05: Proceedings of the sixteenth annual ACM-SIAM symposium on Discrete algorithms*, 2005. Society for Industrial and Applied Mathematics, Pennsylvania, pages 405–414.
92. I. Stoica, R. Morris, D. Karger, M. F. Kaashoek, and H. Balakrishnan. Chord: A scalable peer-to-peer lookup service for internet applications. In *SIGCOMM '01: Proceedings of the 2001 conference on Applications, technologies, architectures, and protocols for computer communications*, 2001. ACM Press, New York, pages 149–160.
93. I. Stojmenovic. A routing strategy and quorum based location update scheme for ad hoc wireless networks. Technical Report TR-99-09, SITE, University of Ottawa, September, 1999.
94. I. Stojmenovic and X. Lin. Loop-free hybrid single-path/flooding routing algorithms with guaranteed delivery for wireless networks. *IEEE Transactions on Parallel and Distributed Systems*, 12(10):1023–1032, 2001.
95. I. Stojmenovic and X. Lin. Power-aware localized routing in wireless networks. *IEEE Transactions on Parallel and Distributed Systems*, 12(11):1122–1133, 2001.
96. H. Takagi and L. Kleinrock. Optimal transmission ranges for randomly distributed packet radio terminals. *IEEE Transactions on Communications*, 32(3):246–257, 1984.

97. R. Thomas and X. Yu. 4-connected projective-planar graphs are hamiltonian. *Journal of Combinational Theory Series B*, 62(1):114–132, 1994.

98. M. Thorup and U. Zwick. Approximate distance oracles. In *Proceedings ACM Symposium on Theory of Computing*, pages 183–192, 2001.

99. M. Thorup and U. Zwick. Compact routing schemes. In *SPAA '01: Proceedings of the thirteenth annual ACM symposium on Parallel algorithms and architectures*, 2001. ACM Press, New York, pages 1–10.

100. P. F. Tsuchiya. The landmark hierarchy: A new hierarchy for routing in very large networks. In *SIGCOMM '88: Symposium proceedings on Communications architectures and protocols*, 1988. ACM Press, New York, pages 35–42.

101. Y. Wang, J. Gao, and J. S. B. Mitchell. Boundary recognition in sensor networks by topological methods. In *Proceedings of the ACM/IEEE International Conference on Mobile Computing and Networking (MobiCom)*, pages 122–133, September 2006.

102. R. Want, A. Hopper, V. Falcao, and J. Gibbons. The active badge location system. *ACM Transactions on Information Systems*, 10:91–102, January 1992.

103. M. Wattenhofer, R. Wattenhofer, and P. Widmayer. Geometric Routing without Geometry. In *12th Colloquium on Structural Information and Communication Complexity (SIROCCO), Le Mont Saint-Michel, France*, May 2005.

104. K. Whitehouse and D. Culler. A robustness analysis of multi-hop ranging-based localization approximations. In *IPSN '06: Proceedings of the fifth international conference on Information processing in sensor networks*, 2006. ACM Press, New York, pages 317–325.

105. K. Whitehouse, C. Karlof, A. Woo, F. Jiang, and D. Culler. The effects of ranging noise on multihop localization: an empirical study. In *IPSN '05: Proceedings of the 4th international symposium on Information processing in sensor networks*, Piscataway, NJ, USA, 2005. IEEE Press, New York, pages 73–80.

106. F. Ye, H. Luo, J. Cheng, S. Lu, and L. Zhang. A two-tier data dissemination model for large-scale wireless sensor networks. In *MobiCom '02: Proceedings of the 8th annual international conference on Mobile computing and networking*, 2002. ACM Press, New York, pages 148–159.

107. M. Youssef, A. Agrawala, and U. Shankar. WLAN location determination via clustering and probability distributions. Technical report, University of Maryland, College Park, MD, March 2003.

108. F. Zhang, H. Li, A. A. Jiang, J. Chen, and P. Luo. Face tracing based geographic routing in nonplanar wireless networks. In *Proceedings of the 26th Conference of the IEEE Communications Society (INFOCOM'07)*, pages 2243–2251, May 2007.

109. B. Y. Zhao, J. D. Kubiatowicz, and A. D. Joseph. Tapestry: An infrastructure for fault-tolerant wide-area location and. Technical report, Berkeley, CA, USA, 2001.

110. G. Zhou, T. He, S. Krishnamurthy, and J. A. Stankovic. Impact of radio irregularity on wireless sensor networks. In *MobiSys '04: Proceedings of the 2nd international conference on Mobile systems, applications, and services*, 2004. ACM Press, New York, pages 125–138.

111. G. Ziegler. *Lectures on Polytopes*. Springer-Verlag, Berlin 1995.

# Chapter 6
# Cooperative Relaying in Wireless Sensor Networks

Robin Doss and Wolfgang Schott

**Abstract** Cooperative relaying has been shown to be an effective method to significantly improve the error-rate performance in wireless networks. This technique combats fading by exploiting the spatial diversity made available through cooperating nodes that relay signals for each other. In the context of wireless sensor networks, cooperative relaying can be applied to reduce the energy consumption in sensor nodes and thus extend the network lifetime. Realizing this benefit, however, requires a careful incorporation of this technique into the routing process to exploit diversity gains. In this chapter, we introduce the basic concepts required to understand cooperative relaying and review current state of the art energy-efficient routing protocols that realize cooperative relaying.

## 6.1 Introduction

Wireless communication is made possible through the transmission and reception of electromagnetic waves over a wireless channel. A *channel* is defined by Shannon [1] as a medium that is used for data transmission from a sender to a receiver. In the case of wireless communication, the medium is the "radio channel" together with an associated range of radio frequencies over which data communication is attempted. An underlying principle in both wired and wireless communication is that the quality of the channel influences the amount of information that can be successfully communicated from the sender to the receiver. A "stronger" channel thus enables communication at a much higher rate[1] than a "weaker" channel and hence will be preferred.

R. Doss (✉)
Deakin University, 221 Burwood Hwy, Burwood, Victoria 3125, Australia
e-mail: robin.doss@deakin.edu.au

[1] Rate refers to the spectral efficiency measured in bits/sec/Hz.

S. Misra et al. (eds.), *Guide to Wireless Sensor Networks*, Computer Communications and Networks, DOI: 10.1007/978-1-84882-218-4_6,
© Springer-Verlag London Limited 2009

The challenges in wireless communication arise from the nature of the medium. First, the wireless medium is open to influence from atmospheric factors, which can negatively impact on the strength and integrity of the transmitted signal. Second, due to multipath propagation, multiple copies of the attenuated signal arrive at the receiver out-of-phase to combine either constructively or destructively. The resulting variation in received signal strength, referred to as *fading*, together with signal attenuation influence the maximum rate of data transfer over the channel. Third, the broadcast nature of the wireless medium implies that it is open to interference from other signals that are being simultaneously transmitted over the same channel. Some of these factors that impact negatively on radio transmission such as atmospheric effects and signal attenuation cannot be prevented by engineering techniques. However, the negative effects of other factors such as fading and channel contention can be decreased if not prevented by innovative engineering design and protocol formulation. One such innovation is the use of *cooperative relaying* to combat the negative effects of fading in wireless networks.

## 6.2   Background: Relaying in Wireless Networks

It is well known that one of the impediments for wireless communication stems from the decay in the strength of the transmitted signal with an increase in the distance traveled. Consequently, an antenna that is farther away from the transmitter will receive a signal with reduced power, which can prevent the successful reception of the transmitted signal. The *free space loss* for an isotropic antenna is given by the following relation

$$\frac{P_{tx}}{P_{rx}} = \frac{(4\pi d)^2}{\lambda^2}. \tag{6.1}$$

Here, $P_{tx}$ is the transmitted signal power; $P_{rx}$ is the received signal power; $\lambda$ is the carrier wavelength; and $d$ is the distance of separation between the transmitter and receiver. For other types of antenna the individual gains of the transmitter and receiver antennas should also be considered.

From (6.1), the received power is inversely proportional to the distance traveled by the signal, i.e., $P_{rx} \sim P_{tx} d^{-\alpha}$. $\alpha$ is referred to as the path loss exponent and takes on a value between 2 (for outdoor environments) and 4 (for indoor environments) depending on the nature of the wireless environment. Importantly, the practical impact of the path loss is in the limitation of the transmission range of a wireless source.

To increase the transmission range relaying can be employed. Relaying is a technique used to increase the transmission range of wireless nodes and in its simplest form, classical relaying involves the use of an intermediate node to improve the signal received from the original source. This improved signal is then forwarded toward the intended destination. In Fig. 6.1, node R acts as a relay node for the transmission between node S and node D. Node R receives the original transmission from node S,

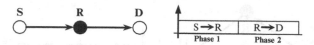

**Fig. 6.1** Classical Relaying: Node R acts as a relay node between node S and node D. In phase 1, transmission is from node S to node R followed in phase 2 with transmission from node R to node D

improves the signal and forwards it to node D. One of the main benefits of classical relaying is an increase in the coverage region of the wireless network.

A more recent form of relaying that is increasing in its popularity is cooperative relaying. Classical relaying aims to increase the transmission range of a wireless source, while cooperative relaying aims at combating fading effects in the radio channel to reduce the number of retransmissions required to successfully decode a particular message at the receiver. The focus of cooperative relaying is therefore on decreasing the error probability of a wireless channel that experiences fading.

### 6.2.1  Cooperative Relaying

Cooperative relaying [2–5] aims to exploit the inherent diversity that is present in wireless networks, and has at its heart the use of independent fading paths and diversity combining techniques. It is based on the observation that channels that are separated in frequency, time, or space usually experience fading effects independently. If the information data is received through multiple independent paths, the probability is high that at least one of those paths will not be in a deep fade. Hence, by optimally combining the signals received from different paths, the receiver is likely to decode transmissions that it otherwise would not have been able to.

The first step to achieving cooperative relaying is the realization of multiple independently fading paths. In wireless systems, we observe three main types of diversity that can be exploited for this purpose. These are:

1. *Temporal diversity*: This is observed as channel conditions vary over time. The level of fading experienced over a channel changes between coherence intervals[2] giving rise to independently fading paths that are separated in time, which can be exploited through interleaving and forward error correction (FEC) codes to achieve diversity gains.
2. *Frequency diversity*: This is observed when the available bandwidth is larger than the coherence bandwidth.[3] Each frequency range corresponding to the coherence

---

[2] The *coherence interval* is defined as the period of time over which the condition of the channel remains approximately the same.

[3] The *coherence bandwidth* is the frequency range over which the channel response is roughly the same.

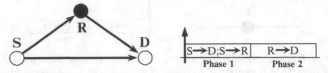

**Fig. 6.2** Cooperative Relaying: Node R acts as a relay node between node S and node D. In phase 1, transmission is from node S to D and from S to node R; in phase 2, transmission is from node R to node D; following phase 2, node D optimally combines the transmission from S in phase 1 with the transmission from R in phase 2 to exploit spatial diversity

   bandwidth represents an independently fading channel separated in frequency, which can be exploited using spread spectrum techniques or interleaving and FEC to achieve diversity gains.
3. *Spatial diversity*: This is provided by the use of multiple antennas at the transmitter and/or receiver ends. Assuming that the antennas are sufficiently separated in space (more than half a wavelength), they give rise to channels whose fading characteristics are independent and can therefore be used to achieve diversity gains. Receive diversity is realized by coherently combining the signals received on different antennas [6]. In transmit diversity [7–9], the transmitter uses channel state information (CSI) to process the signals launched from different antennas so that they arrive in phase at the receiver. If CSI is not available, transmit diversity can still be realized by means of coding over space (i.e., antennas) and time, a technique known as space-time coding.

   Of the three forms of diversity, spatial diversity incurs additional hardware costs as it requires additional antennas together with the corresponding RF circuitry. In systems with stringent complexity constraints such as wireless sensor networks, the use of antenna arrays to achieve spatial diversity may not be possible. However, a *virtual antenna array* that offers diversity gains can be constructed by the cooperation of multiple sensor nodes. Individual antennae from each cooperating sensor node can be combined to virtually constitute such an array and can be used to realize spatial diversity. An example of such a cooperative relaying scheme involving three sensor nodes is presented in Fig. 6.2 where the antennas at the source and relay nodes constitute a virtual array.

## 6.2.2  Relaying Strategies

Relaying strategies can be categorized into three main categories based on the forwarding strategy adopted by the relay node. Depending on the delay constraints of the application and the computing power of the relay node, the relay node can be used to amplify-and-forward (AF), decode-and-forward (DF), or decode-and-reencode (DR) the incoming signal. Each of these methods has advantages and disadvantages. We describe the methods in more detail later.

1. *Amplify-and-forward*: In amplify-and-forward schemes, the relay node receives
   the signal from the source in phase 1 (Fig. 6.2). During phase 2, the relay transmits the signal to the destination node after amplifying the strength of the
   attenuated signal. Before decoding, the destination node combines the relayed
   signal with the original transmission that it has received over the direct channel
   (during phase 1) from the source. One of the main advantages of this scheme
   is that the relay node uses very little computing power in comparison to DF
   schemes as the relay node is not required to decode the signal. Further, the delay introduced as a result of relaying is least with AF. In [4], it has been shown
   that second order diversity gains are possible using AF. It is seen that the outage
   probability[4] is given by $P_{\text{out}} \approx \left[ \frac{2^{2R}-1}{SNR} \right]^2$ for large *SNR*, where $R$ is the spectral
   efficiency of the channel (in bits/sec/Hz). Hence for a 10 dB rise in the *SNR* the
   outage probability will decrease by a factor of 100. However, one of the main
   drawbacks of the scheme is that any noise that is part of the original signal will
   be amplified by the relay and forwarded to the destination.
2. *Decode-and-forward*: In DF schemes, the relay node on receiving the signal in
   phase 1 attempts to decode the signal. It forwards to the destination node in
   phase 2 only if it was successful in phase 1. The destination node will optimally
   combine the relayed transmission with the original signal it has stored from
   phase 1. One of the main advantages of this method is that noise introduced on
   the relay channel is removed during decoding, which results in the transmission
   of an exact copy of the original signal in phase 2. However, in the absence of
   error correcting codes, any errors that have been introduced into the signal will
   be propagated to the destination. The use of error detection techniques (such as a
   CRC checksum) can ensure that erroneous messages are not relayed to the destination node. Depending on the relaying strategy employed the relay node can
   choose to drop any message that arrives at the relay corrupted. The advantages
   of DF schemes come with the added cost of requiring the relay node to decode
   each message that it receives. This added computation at the relay can also
   introduce significant delays into the forwarding process and is not well-suited
   to delay-sensitive applications. In [4], it has been shown that DF does not offer
   diversity gain at high SNR. More specifically, the outage probability satisfies
   $P_{\text{out}} \approx \left[ \frac{2^{2R}-1}{SNR} \right]$ for large *SNR*, where $R$ is the spectral efficiency. Note that requiring the relay to successfully decode every message before relaying limits the
   performance of DF schemes to the performance of the source to relay channel.
3. *Decode-and-reencode*: DR is a variation of DF that follows similar principles. It
   differs from DF in that the codewords used to encode the signal at the relay node
   are different to those used for the source to destination transmission. Therefore,
   additional redundancy is available at the receiver since it receives two copies of
   the same message encoded with different codewords. However, the problems of
   errors propagating from the relay channel to the destination node still persist.

---

[4] The outage probability is the probability that the instantaneous capacity of the relaying system is
insufficient to support the desired rate R.

### 6.2.3  Combining Strategies

The effectiveness of cooperative relaying is dependent on the ability of the receiver to coherently and optimally combine the incoming signals. Various combining strategies are possible. A combining strategy defines how the receiver deals with multiple signals that are assumed to arrive through independently fading paths. It can either define the proportion in which each of the individual signals contribute to the receiver output or how an individual signal is selected as the input to the receiver decoder. If the former of the two strategies is chosen then prior to the combining stage, it is important that the phase associated with the incoming signals are removed. This is referred to as *cophasing*. Since the incoming signals arrive over independent paths, each of them will have a phase $\theta_i$ and arrive at the receiver out-of-phase with each other. The phase difference between the incoming signals needs to be compensated to coherently combine the incoming signals. A failure in cophasing the incoming signals will result into an output signal that still exhibits significant fading due to the constructive and destructive addition of signals via the different fading paths. The phase of an incoming signal is removed by multiplying the signal with its associated phasor, $e^{-j\theta_i}$, where $\theta_i$ is the phase shift of the channel prior to combining [10]. Most combining strategies can be implemented as a linear combiner with different weights associated to the different incoming signals. The system model for diversity combining at the receiver can be represented as in Fig. 6.3.

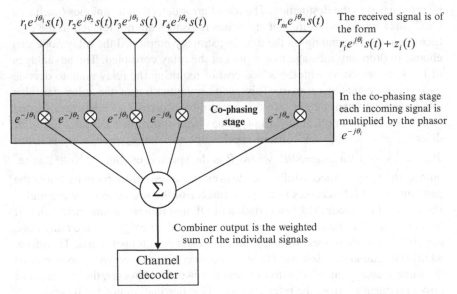

**Fig. 6.3** Combining strategies implemented as a linear combiner with weights assigned to the various branches. The phase associated with the individual branches is removed in the co-phasing stage [10]

We can categorize combining strategies based on the weights that are associated to the individual branches. These are:

1. *Equal-ratio combining (ERC)*: The easiest method of combining the individual incoming signals is by assigning equal weights to each branch. In ERC, after cophasing, the signals are linearly combined with each individual branch contributing equally to the output signal. Hence, the output signal $y_d[t]$ of the linear combiner will be

$$y_d[t] = \sum_{i=1}^{m} e^{-j\theta_i} y_{i,d}[t]. \qquad (6.2)$$

Since the weights do not take into account channel quality, this method is particularly useful if channel state information such as the time varying *SNR* is not available for the individual paths. In the case of the 3-node example in Fig. 6.2, the output signal will be the linear combination (after co-phasing) of the direct signal $y_{s,d}[t]$ from the source to the destination and the relayed signal $y_{r,d}[t]$.

2. *Fixed-ratio combining (FRC)*: In FRC, each individual signal is assigned a fixed weight that does not change for the entire communication. The weight associated with an individual path is an estimate of the perceived average channel quality. Since the weight is fixed for the entire duration of the communication, it is not based on time-varying channel characteristics such as received signal strength (RSS) but rather on factors such as the fixed distance between the relays and the destination. Hence, the weights are a perceived estimate of the channel quality and not an accurate estimate. Since multiple signals are combined, FRC requires cophasing to be completed before the signals are combined. Using FRC, the output of the linear combiner is

$$y_d[t] = \sum_{i=1}^{m} w_{i,d} e^{-j\theta_i} y_{i,d}[t], \qquad (6.3)$$

where $w_{i,d}$ is the weight for the individual signal path.
Applying this method to the 3-node example in Fig. 6.2, the output of the linear combiner will be,

$$y_d[t] = w_{s,d} e^{-j\theta_s} y_{s,d}[t] + w_{r,d} e^{-j\theta_r} y_{r,d}[t]. \qquad (6.4)$$

However, since the relay channel is a multihop path, we need to also take into account the quality of the forward channel from the source to the relay. Hence the linear combiner will need to consider the entire path from source to relay to destination to compute the weight $w_{r,d}$.

3. *Selection combining (SC)*: In selection combining the individual path that has the highest average SNR is chosen. The combiner therefore is not used to coherently combine the individual signals. Athough it has been argued that cophasing between individual signals is not required with SC since only one signal is chosen [10], cophasing is implicit in computing the $SNR_i$ corresponding to $y_{i,d}$.

The selection rule can be represented as,

$$y_d[t] = y_{i^*,d}[t],$$ (6.5)

where $i^* = \arg\max_i \text{SNR}_i$.

We can also combine ERC and SC in such a way that SC is applied for relay selection and ERC is applied to linearly combine the relayed signal with the direct transmission from source to destination. When multiple relays are present and applying this to the 3-node example in Fig. 6.2, we have

$$y_d[t] = e^{-j\theta_s} y_{s,d}[t] + e^{-j\theta_r^*} y_{r^*,d}[t],$$ (6.6)

where $r^* = \arg\max_r \text{SNR}_r$.

Clearly, to achieve this, cophasing of the signals will be need to be completed before the input to the linear combiner.

4. *Threshold combining (TC)*: Threshold combining is a simpler form of selection combining to implement receiver diversity. The individual branches are scanned in sequential order until the first branch that has a SNR value above a predefined threshold $\tau_{\text{SNR}}$ is found. This branch is treated as the preferred branch until the value falls below $\tau_{\text{SNR}}$ when the search is restarted. Since this method outputs only one received signal it does not require cophasing of the individual signals. This can be represented as,

$$y_d[t] = y_{i^*,d}[t],$$ (6.7)

where each branch $i^*$ satisfies $\text{SNR}_i > \tau_{\text{SNR}}$.

If this method is applied to the three node example in Fig. 6.2, it will result in either the direct signal or the relayed signal being chosen as the output of the combiner. If multiple relays are used then there is the possibility to combine the direct signal with a relayed signal. TC can be used for relay selection in combination with ERC or FRC to combine the relayed signal with the direct signal prior to decoding. This can be represented as,

$$y_d[t] = e^{-j\theta_s} y_{s,d}[t] + e^{-j\theta_{i^*}} y_{i^*,d}[t],$$ (6.8)

where each branch $i^*$ satisfies $\text{SNR}_i > \tau_{\text{SNR}}$.

5. *Maximum-ratio combining*: In this, the signals received from all of the individual branches are optimally combined (optimally in the sense that the output SNR is maximized). As noted earlier, in the co-phasing stage multiplication with the phasor, $e^{-j\theta_i}$, removes the phase associated with the individual channels. In addition for the calculation of the optimal weights for each of the individual channels the attenuation factor $(r_i)$, for each channel is to be known as well. MRC is based on the assumption that the receiver knows the channel gain $h_i$ corresponding to the individual branches. We can represent the output of the linear combiner as

$$y_d[t] = \sum_{i=1}^{m} w_{i,d} \cdot y_{i,d}[t], \tag{6.9}$$

where $w_{i,d} = h_i^*$, i.e., the weight equals the complex conjugate of the channel fading gain $h_i$.

In the case of the 3-node example in Fig. 6.2, the combiner output is the optimal combination of the direct signal with the relayed signal and can be represented as,

$$y_d[t] = h_s^* \cdot y_{s,d}[t] + h_r^* \cdot y_{r,d}[t], \tag{6.10}$$

where $h_i^* = r_i e^{-j\theta_i}$.

## 6.3  Proof of Concept for Cooperative Relaying in Sensor Networks

We consider a simple 2-hop network as illustrated in Fig. 6.4 to provide a proof of concept for the benefits of cooperative relaying in wireless networks. We consider two network models – asymmetric and symmetric. In the asymmetric model, the position of the relay node $r$ is such that it is equidistant from the source $s$ and destination $d$ (i.e., normalized distance $d_{s,d} = 1$ and $d_{s,r} = d_{r,d} = 0.5$). In the symmetric model, we position the relay such that the source, destination and the relay are all equidistant from each other (i.e., normalized distance $d_{s,d} = d_{s,r} = d_{r,d} = 1$).

Assuming a Rayleigh flat-fading channel in which the power of the radio signals decays with distance as $d^{-\alpha}$, we calculate the outage probability experienced by the different relaying schemes (classical, AF, and DF). Further, the combining strategy at the receiver is assumed to be maximum ratio combining. The transmission between the source and the destination node can be modeled by the input-output relation

$$y[t] = d^{-\alpha/2} h[t] x[t] + z[t], \tag{6.11}$$

where $h[t]$ is the channel gain describing the fading, $x[t]$ and $y[t]$ are the complex baseband input and output signals, $z[t]$ is a zero-mean white complex Gaussian

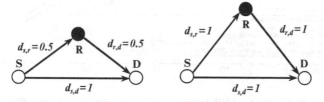

**Fig. 6.4**  Network topology: asymmetric and symmetric network topologies

process with spectral power density $N_0$, and $d$ and $\alpha$ are the distance between the nodes and the path loss exponent, respectively.

For a given channel realization $h[t]$, the maximum rate at which reliable communication is possible at time $t$ is given by the mutual information $I[t]$, which is represented as,

$$I[t] = \log\left(1 + \frac{P_x\,|h[t]|^2}{d^\alpha N_0}\right), \tag{6.12}$$

where a Gaussian codebook with an average transmit power $P_x$ is used. In a fading environment, the channel is random implying that the maximum rate $I[t]$ for reliable communication is random as well. Assuming the coding length is sufficiently large, a lower bound to the packet error rate can be obtained by the outage probability, i.e., the probability that the mutual information falls below the normalized packet data rate $R$. This is given as,

$$\Pr(I[t] < R) = \Pr\left(|h[t]|^2 < \frac{(2^R - 1)N_0}{P_x d^{-\alpha}}\right). \tag{6.13}$$

To identify the best-suited relaying strategy, we compare the performance of the above-mentioned strategies by simulating the outage probability [11]. This performance figure indicates how likely it is that the instantaneous capacity provided by the investigated relaying scheme is less than a given rate or spectral efficiency $R$.

For the asymmetric network topology, Fig. 6.5 shows the outage probability as a function of the SNR for the direct transmission scheme, classical relaying, and

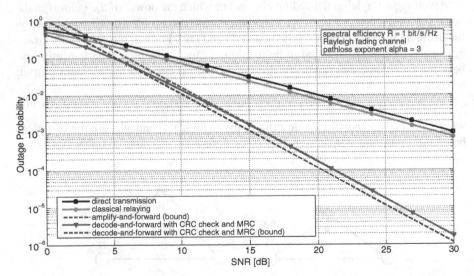

**Fig. 6.5** Outage Probability vs. SNR for the asymmetric network topology. SNR gains of up to 8 dB can be observed which can translate into significant energy savings in wireless sensor networks

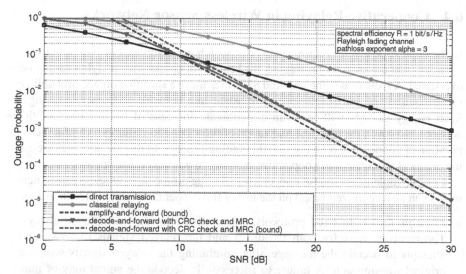

**Fig. 6.6** Outage Probability vs. SNR for the symmetric network topology. SNR gains of up to 4 dB can be observed, which can translate into significant energy savings in wireless sensor networks

cooperative relaying using either an amplify-and-forward or a decode-and-forward relaying strategy. The results indicate the probability that the investigated techniques cannot support a rate of 1 bit/s/Hz for a path loss exponent set to 3. For direct transmission, a SNR of 20 dB is required to guarantee no errors in the data transmission with a probability of 99%. If conventional relaying is used, the same outage probability can be obtained with a slightly smaller SNR. However, significant performance gains are obtained by using cooperative relaying. At an outage probability of $10^{-2}$, the cooperative relaying protocols offer an SNR gain of 8 dB compared to conventional relaying.

Similar trends are also observed in Fig. 6.6 which shows the outage probability as a function of SNR for the symmetric topology with one significant difference. It can be observed that classical relay does not offer any performance gains in such a network. In fact its performance is worse than that of direct transmission requiring a higher SNR to achieve the same outage probability. At an outage probability of $10^{-2}$, the cooperative relaying protocols offer an SNR gain of 4 dB compared with conventional relaying. These are significant SNR gains.

The SNR gains can be exploited in one of two ways. It can be either used to enhance the transmission quality between source and destination or to decrease the transmit-power level in the source and relay node without degrading the error rate performance. From the results, we can also observe that amplify-and-forward outperforms the decode-and-forward scheme.

## 6.4   Cooperative Relaying in Wireless Sensor Networks

In the context of wireless sensor networks, which are composed of individual sensor nodes that are energy-constrained, the benefits of cooperative relaying cannot be overstated. A reduction in the required SNR to achieve a specified outage probability for a required spectral efficiency can translate into network wide energy savings. Such energy savings will result in a significant increase of the lifetime[5] of the network. However, the practical realization of these benefits requires a careful incorporation of cooperative relaying into the routing process. In this section, we discuss the practical application of cooperative relaying in wireless sensor networks.

Current implementations of cooperative relaying protocols in wireless sensor networks can be categorized based on the use of the cooperative relaying technique as,

1. *Fixed protocols*: In fixed protocols, cooperative relaying is employed always. The receiver waits for both the direct and the relayed signals to arrive and attempts to decode the message after combining the relayed signals with the original transmission. A failure to successfully decode the signal may or may not result in a retransmission request to the receiver. A majority of methods that have been proposed in literature are in this category.
2. *Adaptive protocols*: In adaptive protocols cooperative relaying is employed only if the channel between the relay(s) and the destination can be guaranteed to have a low error rate. This requires the relay node to estimate the relay-destination channel and relaying is performed only if the channel quality is above a certain threshold. If relay channels that satisfy this requirement cannot be found retransmission will be attempted by the source. Consequently, adaptive protocols have the drawback that retransmissions with repetition coding over the same channel (source to destination) does not provide any diversity benefits.
3. *On-demand protocols*: In on-demand protocols cooperative relaying is performed only when specifically requested by the destination node. On-demand protocols rely on feedback from the receiver to indicate if the original transmission from the source was unsuccessful. If the original transmission was unsuccessful, then the relay nodes cooperatively relay the data to the destination node.
4. *Opportunistic protocols*: Opportunistic schemes are an extension of on-demand cooperative relaying protocols. They exploit the property that in a multihop communication successful decoding of the data packet at intermediate nodes in the network is not a requirement for successful end-to-end communication. They rely on circumventing weak channels to ensure successful end-to-end communication.

In the following sections, we provide an overview of the current cooperative relaying protocols from a protocol design perspective. We refer the reader to the respective references for a more rigorous treatment of each of the methods.

---

[5] A conservative estimate of *network lifetime* is calculated to be the time at which the first wireless sensor exhausts its energy (i.e., dies).

### 6.4.1  The Detached Cooperative Diversity (DCD) Protocol

The detached cooperative diversity protocol [12] is a fixed cooperative relaying protocol that employs decode and forward as the forwarding strategy and maximum ratio combining as the combining strategy. DCD is designed to exploit the nonlinearity of path loss over wireless channels in asymmetric network topologies. As we alluded to in Sect. 6.3 in an asymmetric topology the proximity of the relay node to the source implies that the average path loss experienced over the source to relay channel will be less than the path loss of the source to destination channel. This translates into a higher receive signal strength at the relay than at the destination node for the same broadcast transmission from the source. The DCD protocol seeks to exploit this property. In addition, to fully exploit the nonlinearity in the path loss DCD employs adaptive power control at the relay nodes. Once a relay node has successfully decoded a signal, it adjusts the transmit power in such a way that the relayed signal will be received by the destination with the same SNR as the direct transmission from the source. The receiver combines the two copies of the received signal to exploit gains due to spatial diversity.

The "detached" nature of the protocol is attributed to the fixed relaying property of the protocol. Once a relay node has successfully decoded the packet, it reencodes the packet and forwards to the destination. It is, therefore, not dependent on the source or destination to trigger its forwarding i.e., receiver feedback is not required.

The outage probability $P_{\text{out}}$ for DCD has been shown to be (we refer the reader to [12] for the details):

$$
P_{\text{out}} = \frac{2d_{s,r}^{\alpha} + 1}{2} \left( \frac{d_{r,d}^{\alpha} + 1}{2} \right)^2 \left( \frac{2^{2R} - 1}{\text{SNR}} \right)^2 , \tag{6.14}
$$

where $\alpha$ is the path loss exponent, $d_{i,j}$ is the distance terms relating to the position of the relay, and $R$ is the desired spectral efficiency.

Since DCD is a fixed protocol there is the need for the protocol to achieve double the desired spectral efficiency to meet normalization requirements. Hence the outage probability in terms of the average mutual information is defined as $P_{\text{out}} = \Pr[I < 2R]$. This definition introduces the $2R$ term in (6.16). From (6.16) we can observe that to achieve a given outage probability while maintaining a desired spectral efficiency the required SNR is dependent on the relay position. It has been found that for optimal performance (i.e., maximum SNR gains in comparison with direct transmission), the position of the relay should be such that $r_{s,r} = 1 - r_{r,d} = 0.5$ [12].

DCD is simple in its design and offers diversity gains but it is not without its drawbacks. The requirement of the relay nodes to be positioned exactly mid-way between the source and destination does not allow for random sensor network topologies. Further, the heavy dependence of DCD on the nonlinear property of the path loss does not allow it to be extended to symmetric networks where the nonlinearity does not hold. Furthermore, the "detached" nature of the protocol prevents a retransmit mechanism either from the source or from the relay node when cooperative relaying fails. This can lead to low packet delivery ratios in fading environments. In addition, DCD proposes adaptive power control on the relays for energy-efficiency. However, adaptive power control on the sensor nodes is challenging (almost impossible) and it is further exacerbated by the requirement of destination SNR values, which introduces added complexity and control overhead into the network.

### 6.4.2 Simple Cooperative Diversity (SCD) Protocol

The simple cooperative diversity protocol [13] is a fixed cooperative relaying protocol that can use either amplify-and-forward or decode-and-forward as the relaying mechanism with maximum ratio combining as the combining strategy. SCD is designed to select the "best" relay node using a distributed mechanism that allows potential relay nodes to contend with each other. Assuming a slow fading channel, SCD uses instantaneous channel state information to select the relay node that offers the best end-to-end path from the source to the destination. This is shown in Fig. 6.7. The nodes $r_1$ to $r_5$ are potential relays as they can communicate with both the source and the destination. Relay $r_3$ is chosen for data forwarding as it offers the best end-to-end path based on channel quality.

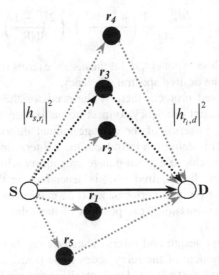

**Fig. 6.7** Relay selection in SCD: On the basis of instantaneous channel state information, the relay $r_3$ is chosen from a set of potential relays because it offers the best end-to-end path from source to destination

Relay selection in SCD is integrated into the MAC protocol, which is assumed to be a CSMA protocol that incorporates the exchange of RTS/CTS packets. The relay selection is done proactively and proceeds as follows:

1. The source S transmits an RTS packet that is received by all of the potential relay nodes and the destination D. Each of the potential relays that receive the RTS use this transmission to estimate the source to relay channel quality i.e., $|h_{s,r}|^2$.
2. The destination node on receiving the RTS packet responds with a CTS packet. All nodes that have previously received the RTS and that receive the CTS from the destination are potential relays. The reception of the RTS and the CTS indicate that these relays share a channel with both the source and the destination. Importantly, the CTS response from D allows a potential relay to estimate relay to destination channel quality i.e., $|h_{r,d}|^2$.
3. Assuming channel reciprocity[6] and the coherence time of the channel to be sufficiently large each relay is able to estimate the quality $q_i$ of the end-to-end path from the source to the destination. In [13], two policies for the estimation of this quality parameter are proposed. These are:

   • *Policy 1: Minimum channel quality*
     In policy 1, the quality parameter is chosen as the minimum of the two channel quality estimates, e.g.,

     $$q_i = \min\left[ |h_{s,r_l}|^2, |h_{r_l,d}|^2 \right]. \qquad (6.15)$$

   • *Policy 2: Harmonic mean of channel quality*
     In policy 2, the quality parameter is chosen as the harmonic mean the two channel quality estimates, e.g.,

     $$q_i = \left[ \frac{2|h_{s,r_i}|^2 |h_{r_i,d}|^2}{|h_{s,r_i}|^2 + |h_{r_i,d}|^2} \right]. \qquad (6.16)$$

4. Clearly, the node with largest $q_i$ should be chosen to forward the packet. To implement choosing of the "best" relay in a distributed fashion, each relay starts a timer with a duration of $T_i$ that is inversely proportional to the value of $q_i$, e.g.,

   $$T_i = \frac{C}{q_i}, \text{ where } C \text{ is a constant}, \qquad (6.17)$$

and starts listing for incoming packets. Equation (6.17) implies that the timer of the relay whose value of $q_i$ is the maximum will have the minimum value of $T_i$ i.e., its timer will expire first. This ensures that the relay with the best end-to-end path between the source and destination will be chosen as the relay for the forwarding process.

---

[6] *Channel reciprocity* is the property that the forward and backward channels exhibit similar characteristics and are of equivalent quality.

5. When the timer of the best relay expires, it broadcasts a short packet to inform other potential relays about its presence. As soon as the other relays receive this packet, they backoff their timer and terminate the contention process.

### 6.4.3  Hybrid ARQ-Based Cooperative Relaying (HACR) Protocol

HACR [14] is an on-demand cooperative relaying protocol that uses decode and forward as the relaying mechanism with maximum ratio combining as the combining strategy. It is referred to as hybrid as it employs both *forward error correction (FEC)* and *automatic repeat request (ARQ)* to recover from decoding errors at the receiver. HACR employs block coding and exploits the broadcast nature of the wireless channel to satisfy ARQs from relay nodes thus achieving spatial diversity. The underlying MAC protocol is assumed to be based on time division multiplexing (TDM). Hence each node transmits or receives in a particular time slot. Further each node is assumed to be location aware. The functioning of the protocol is as follows:

1. The source S prepares the transmit message by encoding the message of $b$ bits into a codeword of $n$ symbols. The codeword is broken into $M$ blocks and if repetition coding is assumed all $M$ blocks will be identical. A node that receives two or more blocks will attempt decoding by diversity combining.
2. The source broadcasts the first block during the first time slot. The destination attempts to decode the message and is successful if $I_{s,d}[1] > R$ (i.e., the required spectral efficiency for successful decoding of the first block is satisfied). The destination responds with an acknowledgment (ACK) if successful. If it is not then retransmission is required. Retransmission will be triggered from a relay node that is chosen following a contention process.
3. Following the original transmission from the source all nodes that received the transmission during slot 1 will attempt to decode it. The nodes that are successful in decoding (i.e., $I_{s,r}[1] > R$) are included in a set $D(S)$, the decode set. The nodes in $D(S)$ contend to serve as the relay to the destination and the node that "wins" the contention process sends an acknowledgment (ACK) to the source. Of course the contention process is entered into by the intermediate nodes only if the destination was not successful in decoding the initial transmission.

   The purpose of the contention process is to ensure that the forwarding of the packet is done by the relay that is the closest to the destination. To achieve this each relay in the network is assigned a window $w_i$ for its ACK. This assignment is achieved by dividing the contention interval into $i$ subintervals one for each relay $r_i (i = 1, 2 \ldots m)$. During the first interval, the relay $r_m$ that is the closest to the destination responds with an ACK packet if it was successful in decoding. It remains silent otherwise. Similarly in the second interval relay $r_{m-1}$ responds if it was successful such that in general in an interval $w_i$ relay $r_{m-i+1}$ responds if and only if it is in the decode set $D(S)$.

4. The first relay that responds with an ACK then sends the second block with an identical contention process following the transmission of that block. This process continues until the destination is able to successfully decode the transmission. In the event that no relays respond after the transmission of a block, following the contention period the source will retransmit.

### 6.4.4  Geographic Routing with Cooperative Relaying and Leapfrogging

In [15], an opportunistic cooperative relaying protocol that circumvents links with weak radio channels is presented. Decode-and-forward is employed as the relaying mechanism with maximum-ratio combining used as the combining technique. It is a location-aware protocol that builds on other geographic routing protocols such as GeRaF [16], is completely distributed, and is energy-efficient.

In multihop wireless sensor networks, data packets are not directly transmitted from the source S to the destination D, but they are forwarded by sensor nodes, which are scattered between the end nodes. Geographic routing is based on the basic assumption that each node knows its own geographical location and that of the destination D. In addition, apart from the source and the destination, the sensor nodes can take on different roles to cooperatively transfer data packets from the source to the destination. Each node can act as a next-hop forwarding node X, as a relay node R, or as a leapfrog node LPF. A next-hop node X forwards data packets hop by hop similar to conventional schemes. However, if the link quality between two forwarding nodes is poor, a relay node R can help transfer the packet to the next hop node X. If a relay transmission can be overheard and successfully decoded by a neighbour node that is geographically more advanced towards the destination than the next-hop node X, this LPF node can take over the task of forwarding the packet.

To ensure that the best-suited set of nodes is selected for the data packet transfer, the nodes compete with their peers to act either as a next-hop, a relay, or a leapfrog node. The different types of nodes are represented in Fig. 6.8.

One of the key elements of the protocol is the optimal selection of forwarding nodes. In networks comprising numerous sensor nodes, each source node has multiple neighbours, who can serve as the next hop node. To choose the most suitable next

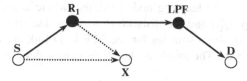

**Fig. 6.8**  For the communication from node S to node D, node X is the next hop node, node $R_1$ is a potential relay node and node LPF is a potential leapfrog node. The dotted lines indicate a diversity combining point

hop node, the neighbours of a source node are forced to contend with each other. This principle of contention also applies to relay selection and leapfrog node selection and the contention process is similar for all three types of nodes. It is achieved by calculating a metric for controlling the transmission of command frames in an RTS/CTS-based CSMA/CA protocol. The metric depends on the role of the node to be assigned, and it always takes into account the energy available in the battery of the sensor. Depending on the role of the node, the metric may also incorporate the distance to the final destination, the distance to the previous or next hop node, and the channel quality of the corresponding radio links. Link qualities are only taken into account if they are readily available. The selection rule between the nodes is implemented in a distributed fashion by calculating the metric in each node and delaying a response command frame for a duration that is inversely proportional to the calculated metric (similar to SCD [13]).

For the selection of the next hop node X out of the set of potential forwarding nodes $x$, the metric takes into account three cost criteria, which are residual node energy $\xi_x[t]$, the distance $d_{x,D}$ to the final destination $D$, and the link quality of the backward link from the source S to $x$. The link quality is characterised by the fading gain $|h_{S,x}[t]|^2$ and the distance $d_{S,x}$. The metric is defined as

$$m_x[t] = \frac{\xi_x[t]}{|h_{S,x}[t]|^{-2} d_{S,x}^\alpha + d_{x,D}^\alpha}, \qquad (6.18)$$

and the best next hop node X is the node that satisfies

$$X[t] = \arg\max_x m_x[t]. \qquad (6.19)$$

For the selection of the relay node R, the metric takes into account the residual node energy $\xi_r[t]$, the link quality of the backward link from the source $S$ to the potential relay node $r$, and the quality of the forward channel from the relay $r$ to X. The metric is therefore defined as

$$m_r[t] = \frac{\xi_r[t]}{|h_{S,r}[t]|^{-2} d_{S,r}^\alpha + |h_{r,x}[t]|^{-2} d_{r,X}^\alpha}, \qquad (6.20)$$

and the best relay node R is the node that satisfies

$$R[t] = \arg\max_r m_r[t]. \qquad (6.21)$$

For the selection of the leapfrog node LPF, the metric takes into account the residual node energy $\xi_l[t]$ and the distance $d_{l,D}$ of the leapfrog node to the final destination D. Link quality estimates for the LPF to D link are not available and hence are not considered. The metric is, therefore, simply defined as

$$m_l[t] = \frac{\xi_l[t]}{d_{l,D}^\alpha}, \qquad (6.22)$$

and the best LPF node is the node that satisfies

$$LPF[t] = \arg\max_l m_l[t]. \tag{6.23}$$

Since the relaying mechanism is decode-and-forward, the set of potential relay and leapfrog nodes is restricted to nodes that have successfully decoded the data packet. Experimental results prove that cooperative relaying with leapfrogging provides us with significant energy gains compared with direct transmission and proactive cooperative relaying.

To illustrate the working of the protocol, we provide in Fig. 6.8 a representative scenario, where communication from S to D is achieved by the use of cooperative relaying with leapfrogging. We refer the reader to [15] for a more detailed protocol description and scenario analysis.

We consider a simple 2-hop scenario, where a source S seeks to communicate with a destination D. Node X is chosen as the best next hop forwarding node based on the distributed contention mechanism as described earlier. S sends the data packet to X. Since this is a broadcast transmission all nodes that are within the communication range of S receive and attempt to decode the packet. All nodes other than X that are successful in decoding the packet are potential relays. If node X is unsuccessful in decoding the initial transmission from S, it will request a retransmission. All potential relays that receive this retransmission request contend based on the metric given in (6.22). The relay whose contention timer expires first is chosen as the relay and will relay the data packet to X. Node X attempts to decode the data packet after diversity combining it with the original transmission. If X is successful, the node will take on the role of a source and forwarding to the next hop will proceed along similar lines. If decoding fails at X, it is possible to circumvent X, if a node that is geographically more advanced than X was able to successfully decode the retransmission from $R_1$ intended for X. This mechanism is referred to as *leapfrogging* and ensures that weak channels in a part of the network do not prevent end-to-end communication (Fig. 6.9).

## 6.5 Thoughts for Practitioners

One of the important observations that can be made is that the benefits of relaying (both classical and cooperative) are dependent on the position of the relay node. From a protocol design point of view, this is an important consideration that needs to be taken into account. When there are multiple potential relay(s), only those relay(s) that can provide diversity gains should be chosen. This relay selection problem is a critical aspect of cooperative relaying protocol design. It can be seen from the results presented that the impact of the position of the relay node is more significant in the case of classical relaying than on cooperative relaying strategies.

The importance of taking into account residual node energy cannot be overstated. For instance, the SCD protocol is a proactive protocol but can be modified to function as a reactive protocol with minimum effort. It has significant advantages over other schemes such as GeRaF [16] that depend on, location information. By taking

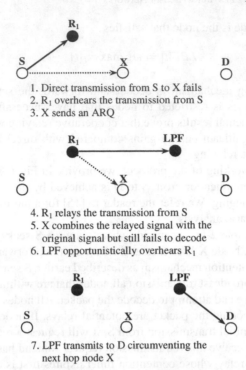

1. Direct transmission from S to X fails
2. $R_1$ overhears the transmission from S
3. X sends an ARQ

4. $R_1$ relays the transmission from S
5. X combines the relayed signal with the
   original signal but still fails to decode
6. LPF opportunistically overhears $R_1$

7. LPF transmits to D circumventing the
   next hop node X

**Fig. 6.9** Cooperative relaying with leapfrogging: Communication from S to D is first attempted through cooperative relaying from $R_1$. When cooperative relaying fails, opportunistic leapfrogging is attempted to successfully reach the destination D. The *dotted lines* indicate failed transmissions

into account the channel state information, SCD implicitly takes into account the position of the relays without requiring accurate location information. Further, the distributed relay selection method makes it a practical scheme for random network topologies. However, the protocol does not take into account the residual energy levels in relay nodes, which can result in early node exhaustion in slow fading environments. In addition, distance-based relay selection other strategies based on instantaneous SNR of the relay to destination channel (instantaneous relaying) and transmit probabilities (random relaying) are also possible [12]. However, a common drawback of current approaches to relay selection/contention is the lack of consideration of residual energy levels in relay nodes.

## 6.6 Directions for Future Research

The realisation of the benefits of cooperative diversity in wireless sensor networks is still in its early stages and presents some potential areas for further research.

It is clear from our discussion that cooperative relaying is dependent on the optimal choice of relay nodes that provide independently fading paths. To achieve this in sensor networks, researchers have mainly focussed on spatial diversity. Spatial diversity is attractive as it allows us to model the sensor network as a virtual distributed multiple antenna system thus enabling useful comparisons to MIMO systems. Further the benefits of spatial diversity are dependent on a sufficient spacing between the sensor nodes. This can be restrictive in dense sensor networks.

Research is also required into combinatorial diversity methods that seek to dynamically exploit temporal, spectral, and spatial diversity. The design of such schemes is yet to be attempted. The benefits and careful integration of such a technique into the cooperative relaying process is yet to be studied.

Current research in cooperative diversity has been focussed on static single-hop communication and/or in multihop communication between a single source and destination. Hence topological constraints do not need to be considered. However in practical implementations, differing communication patterns are possible. In reality, in any sensor network deployment, there will be multiple senders with one or more destinations and sensors capable of mobility. More research is required to support multihop cooperative relaying in mobile sensor network deployments, which require the consideration of topological constraints.

The use of optimal encoding and decoding schemes play a pivotal role in the cooperative relaying process. Cooperative diversity was initially based on space-time codes [17]. The development of practical space-time codes is an active area of research particularly in MIMO systems.

Cooperative diversity can also be cast in the framework of network coding. In addition to efficient channel encoding, we believe an attractive area of research is in network coding. Network coding can improve the benefits of cooperative relaying particularly in deployments that have multiple sources. Initial studies show that there is potential in this area of research [18].

It is also imperative that the current benefits of cooperative relaying that have been identified in literature is verified through experimental test bed evaluations to verify the validity of the assumptions that current solutions have been based on.

## 6.7   Conclusions

Cooperative relaying is a promising strategy that has been proven to offer significant benefits in terms of combating fading in wireless sensor networks. In this chapter, we have provided and discussed the conceptual foundation of cooperative relaying, the realisation of independently fading paths and diversity combining techniques. Through a proof of concept, we have shown that cooperative relaying offers significant savings in the required *SNR* to achieve a specified outage probability while maintaining a desired spectral efficiency. This is of particular significance to energy-constrained wireless sensor networks as by lowering the transmit power of sensor node we can increase the network lifetime significantly. An overview of cooperative relaying protocols that exploit distributed antenna diversity in a practical sense concludes the chapter.

# Terminologies

*Coherence interval.* The coherence interval is defined as the period of time over
  which the condition of the channel remains approximately the same.
*Coherence bandwidth.* The coherence bandwidth is the frequency range over which
  the channel response is roughly the same.
*Outage probability.* The outage probability is the probability that the instantaneous
  capacity of the relaying system is insufficient to support the desired rate R.
*Channel reciprocity.* Channel reciprocity is the property that the forward and backward channels exhibit similar characteristics and are of equivalent quality.
*Cooperative relaying.* Cooperative relaying has been shown to be an effective
  method to significantly improve the error-rate performance in wireless networks.
  This technique combats fading by exploiting the spatial diversity made available
  through cooperating nodes that relay signals for each other.
*Temporal diversity.* Temporal diversity is observed as channel conditions vary over
  time. The level of fading experienced over a channel changes between coherence intervals giving rise to independently fading paths that are separated in time
  which can be exploited through interleaving and forward error correction (FEC)
  codes to achieve diversity gains.
*Frequency Diversity.* Frequency diversity is observed when the available bandwidth
  is larger than the coherence bandwidth. Each frequency range corresponding to
  the coherence bandwidth represents an independently fading channel separated
  in frequency which can be exploited using spread spectrum techniques or interleaving and FEC to achieve diversity gains.
*Spatial Diversity.* Spatial diversity is provided by the use of multiple antennas at the
  transmitter and/or receiver ends. Assuming that the antennas are sufficiently separated in space (more than half a wavelength), they give rise to channels whose
  fading characteristics are independent and can therefore be used to achieve diversity gains.
*Equal-ratio combining (ERC).* The easiest method of combining the individual incoming signals is by assigning equal weights to each branch. In ERC, after
  co-phasing, the signals are linearly combined with each individual branch contributing equally to the output signal.
*Maximum ratio combining (MRC).* In maximum-ratio combining, the signals received from all of the individual branches are optimally combined (optimally
  in the sense that the output SNR is maximized). MRC is based on the assumption that the receiver knows the channel gain $h_i$ corresponding to the individual
  branches.

# Questions

1. Describe the benefits of cooperative relaying.
2. List the different types of diversity observed in wireless networks.
3. Describe the "amplify and forward" relaying strategy.

4. Describe the "decode and forward" relaying strategy.
5. Describe the "decode and reencode" relaying strategy.
6. List the four different categories of cooperative relaying protocols used in wireless sensor networks.
7. Describe the working of fixed cooperative relaying protocols.
8. Describe the working of adaptive cooperative relaying protocols.
9. Describe the working of on-demand cooperative relaying protocols.
10. Describe the working of opportunistic cooperative relaying protocols.

# References

1. C. E. Shannon, "A mathematical theory of communication", *SIGMOBILE Mobile Comp. Commun. Rev.* 5, 1, Jan. 2001, pp. 3–55.
2. A. Sendonaris, E. Erkip, and B. Aazhang, "User cooperation diversity – Part I: System description," *IEEE Trans. Comm.*, vol. 51, no. 11, Nov. 2003, pp. 1927–1938.
3. A. Sendonaris, E. Erkip, and B. Aazhang, "User cooperation diversity – Part II: Implementation aspects and performance analysis," *IEEE Trans. Comm.*, vol. 51, no. 11, Nov. 2003, pp. 1939–1948.
4. J. N. Laneman, "Cooperative Diversity in Wireless Networks: Algorithms and Architectures", PhD Thesis, Massachusetts Institute of Technology, Sep. 2002.
5. R. U. Nabar, H. Boelcskei, and F. W. Kneubuehler, "Fading relay channels: Performance limits and space-time signal design," *IEEE J. Sel. Areas Commun.*, Vol. 22, No. 6, Aug. 2004, pp. 1099.
6. W. C. Jakes, "Microwave mobile communications,"New York: Wiley, 1974.
7 V Tarokh, N. Seshadri and A. R. Calderbank, "Space time codes for high data rate wireless communication: Performance criterion and code construction," *IEEE Trans. Inf. Theory*, vol. 44, March 1998, pp. 744–765.
8. N. Seshadri and J. Winters, "Two signaling schemes for improving the error performance of frequency-division-duplex (FDD) transmission systems using transmitter antenna diversity," *Int. J. Wireless Inform. Netw.*, vol. 1, no. 1, 1994, pp. 49–60.
9. J. Guey, M. Fitz, M. Bell, and W. Kuo, "Signal design for transmitter diversity wireless communication systems over Rayleigh fading channels," in Proc. IEEE VTC, 1996, pp.136–140.
10. A. Goldsmith, "Wireless Communications", Cambridge University Press, 2005.
11. EU project e-SENSE, deliverable D3.2.1, "Efficient Protocol Elements for Light Weight Wireless Sensor Communication Systems," Nov. 2006.
12. E. Zimmermann, P. Herhold, and G. Fettweis, "A Novel Protocol for Cooperative Diversity in Wireless Networks" Proc. of the 5th European Wireless Conference - Mobile and Wireless Systems beyond 3G, Feb. 2004.
13. D. R. A. Beltsas, A. Khisti, and A. Lippman, "A simple cooperative diversity method based on network path selection", *IEEE Trans. on Sel. Areas Commun.*, vol. 24, no. 3, March 2006.
14. B. Zhao and M. C. Valenti, "Practical Relay Networks: A Generalisation of Hybrid-ARQ", *IEEE J. Sel. Areas Commun.*, vol. 23, no.1, Jan. 2005.
15. P. Coronel, R. Doss, and W. Schott, "Geographic Routing with Cooperative Relaying and Leapfrogging", Proc. of IEEE Global Telecommunications conference, Nov. 2007.
16. M. Zorzi and R. Rao, "Geographic Random Forwarding (GeRaF) for Ad hoc and Sensor Networks: Energy and Latency Performance," *IEEE Trans. Mobile Comp.*, vol. 2, no. 4, Oct. 2003, pp. 349–365.
17. J. N. Laneman and G.W. Wornell, "Distributed space-time coded protocols for exploiting cooperative diversity in wireless networks", *IEEE Trans. Inf. Theory*, vol. 51, no. 12, Nov. 2003, pp. 1126–1131.
18. Y. Chen, S. Kishore, and J. Li, "Wireless Diversity through Network Coding", Proc. of IEEE Wireless Communications and Networking Conference (WCNC), March 2006.

# Chapter 7
# Data-Centricity in Wireless Sensor Networks

**Abdul-Halim Jallad and Tanya Vladimirova**

**Abstract** Data-centricity is a feature of wireless sensor networks that distinguishes them from other wireless data networks. Establishing the data as the centre of operation in sensor networks provides better usage of the limited resources available in such networks. In addition, data-centricity matches well the nature of wireless sensor networks. This chapter reviews a number of emerging topics collectively constituting a data-centric view of wireless sensor networks. These topics include data-centric routing, data aggregation, and data-centric storage.

## 7.1 Introduction

In the past two decades, the continuous advancements in processor technology and computer networking fuelled an increasing interest in the field of distributed computing. A distributed computing system consists of several computing entities which are inter-connected via a network and are capable of collaborative processing. The distributed nature of such systems necessitates elaborate system design techniques to tackle their inherent complexity. With the advent of the wireless technology, wireless computer networks became a prominent branch of distributed systems. An *infrastructure network* is a wireless network that connects through an access point to a conventional Local Access Network (LAN) [1]. An alternative configuration for wireless networking is known as *ad hoc networking*. The primary characteristic of ad hoc networks is that they do not include an access point or a base station. They are formed as a result of the mutual detection of two or more mobile devices with wireless interfaces located in the same vicinity.

A.-H. Jallad (✉)
Surrey Space Centre, Department of Electronic Engineering, University of Surrey,
Guildford, Surrey, GU2 7XH, UK
e-mail: A.jallad@surrey.ac.uk

S. Misra et al. (eds.), *Guide to Wireless Sensor Networks*, Computer Communications
and Networks, DOI: 10.1007/978-1-84882-218-4_7,
© Springer-Verlag London Limited 2009

A relatively recent type of networks that have emerged as a special case of ad hoc networks is what is now commonly referred to in the literature as wireless sensor networks (WSNs) [2–4]. Wireless sensor networks are large-scale distributed systems formed of nodes possessing a limited amount of resources that are tightly coupled to the environment in which they operate via sensors and actuators. WSN systems are distributed in nature but must achieve a single goal cooperatively. The node itself may be composed of heterogeneous elements such as various multimode sensing hardware, an embedded processor, memory, a power supply, communications and location determination capabilities. WSNs can, therefore, be characterized as networked embedded systems as follows:

- *Networked* – networking is necessary to coordinate and perform higher-level tasks underpinning collaborative sensing and actuation schemes.
- *Embedded* – numerous embedded distributed devices are used to monitor and interact with the physical world forming small untethered autonomous systems.
- *Systems* – sensing and actuation are tightly coupled to the physical environment.

The traditional communication paradigm focuses on the relationship between communicating peers, i.e., the sender and the receiver of the data. In WSNs, the application is not interested in the identity of the nodes, but rather in the information that the nodes posses about the physical environment that the WSN operates in. The main concept behind *data-centric networking* is that the focal point of the network is the data being communicated and not the identity of the nodes. The consequence of this is that an application makes requests to the network using data (and not nodes) as addresses [5].

A simple illustration of the data-centricity concept is shown in Fig. 7.1. In the address-centric approach, the data source (node 1) sends a data packet preceded by the address of the sink node (node 4) as illustrated by the left-hand diagram in Fig. 7.1. In the data-centric approach illustrated by the right-hand diagram in Fig. 7.1, the source sends the data packet preceded by an identification tag ("A" in this case). Only the nodes waiting for "A" will receive the data packet.

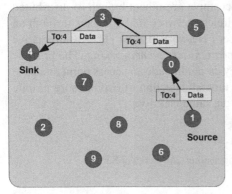 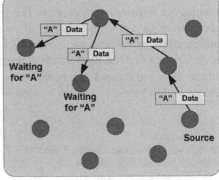

**Fig. 7.1** Data-centric and address-centric approaches to distributed system design

The data-centric approach allows development of very different networking architectures compared with traditional, address-centric networks. Data-centricity in WSNs enables advantageous network properties such as the following:

- In-network aggregation – reduces the amount of traffic flowing in the network [6].
- Data-centric addressing – enables simple expressions of communication relationships [7].
- Decoupling in time – data requests do not specify any timing details for the response, a property that is useful for event-detection sensing applications [7].
- Fault-tolerance – as the nodes are no more the focus of the network, the failure of a node has a limited effect on the network.
- Scalability – the fact that the addressing mechanism does not depend on the number of nodes in the network and the possibility of using localized algorithms, such as clustering, enhances the scalability of the system.

All of the advantages listed above make data-centric design attractive to WSNs where the aim is to achieve optimum usage of the limited resources available. In addition, WSN applications are naturally data-centric [3, 8], with data being continuously collected and integrated from a large number of physically dispersed sensor nodes.

The aim of this chapter is to provide an insight into the nature of data-centric networks, introduce their advantages, and show how establishing the data as the focus of the network design changes considerably the conventional design paradigms. The chapter starts with a discussion on data-centric abstractions in Sect. 7.2. Then it is chapter described how adopting a data-centric approach reflects on to the underlying system architecture. Section 7.3 presents an overview of data-centric routing giving some illustrative examples of protocols that have been reported in the literature. The real power of data-centricity lies in providing the ability to operate on the data itself while flowing in the network. In-network processing techniques, namely data aggregation, are described in detail in Sect. 7.4. Data-centric storage is then explained using examples from the literature in Sect. 7.5. Section 7.6 summarizes the chapter and presents conclusions.

## 7.2 Implementations of Data-Centric Abstractions

There are several approaches to implementation of data-centric networks. Each approach implies a certain set of interfaces that would be useable by an application. This section describes two of the most important schemes: publish/subscribe and databases.

## 7.2.1 Publish/Subscribe Scheme

The Publish/Subscribe paradigm is a communications infrastructure connecting independent nodes in a distributed system. The conceptual idea (illustrated in Fig. 7.2) is simple. All nodes are connected to a "software bus." Nodes make their data available publicly on the software bus via a "publish" action and announce their interest in a particular type of data via a "subscribe" action. Nodes that have previously "subscribed" to some kind of data that have been published are then notified that the data are available on the bus.

The relationship between publishers and subscribers of data imposed by the interaction pattern of the publish/subscribe paradigm is characterised by three features [9, 10]:

- *Decoupling in space* – there is no need for publishers and subscribers to be aware of each other.
- *Decoupling in time* – there is no dependency between the events of publishing and notification of data, with the "software bus" providing intermediate storage.
- *Decoupling in flows* – asynchronous interactions with the software bus can take place without any blocking.

There are several flavours of publish–subscribe systems, based on the methodology used for addressing the data. These flavours are briefly described below [9]:

- *Group-based addressing.* This is the earliest flavour of publish/subscribe systems. In such systems each node (whether publisher or subscriber) participates as a member in one or more predetermined groups. Subscriptions by subscribers (and similarly publications by publishers) are limited to the groups that the subscribers (or publishers) are members of. The disadvantage of this approach is

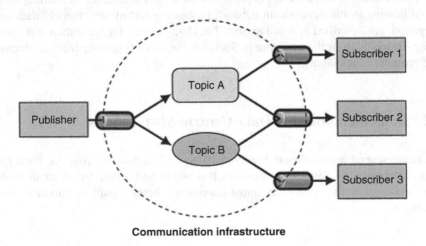

**Communication infrastructure**

**Fig. 7.2** Publish-subscribe system

that it leads to restricted access in the system with the subscribers not being able to receive publications from some publishers. In addition, this approach does not support data-centric communication of messages advocated by the publish/subscribe paradigm.

- *Subject-based addressing.* In subject-based (also called topic-based) publish/subscribe systems, the data publications/subscriptions have a subject. The subject belongs to a predefined namespace of subjects. Subscribers subscribe to the software bus identifying the subjects that they are interested in, and publishers publish messages with subjects associated to them. A typical example of topics are names of stocks traded at a stock exchange; when the price of a given stock changes, a notification for the corresponding topic is generated. Although the subject-based abstraction is simple, it lacks flexibility.
- *Content-based* addressing. In content-based publish/subscribe systems, the subscription matching criteria are extracted from the message content itself. The primary advantage of the content-based approach is that it provides maximum flexibility in giving the subscription criteria. It is this advantage of flexibility in giving the subscription criteria that makes the content-based approach of considerable importance to WSNs. An example for such a predicate from the domain of space-based WSNs would be "Is the amount of fuel available on board the satellite sensor node lower than the threshold value?" Primitive predicates can be combined into more complex ones using standard logical operators (and, or, not) with the usual semantics [11].

### *7.2.2  Databases*

A different view for implementing data-centric abstractions in wireless sensor networks is to consider them as dynamic databases, which is a completely different approach to that used by publish/subscribe systems. The database conceptual view matches quite well with the data-centric approach to design of wireless sensor networks. This is because "being interested" in a certain aspect of the physical environment that is surveyed by a WSN can be viewed as being equivalent to formulating queries in a database.

Two of the most representative sensor database systems are TinyDB [12] and Cougar [13]. In TinyDB, users specify a set of declarative queries that define the information to be gathered from the wireless sensor network. Queries indicate the type of readings to be obtained, including the subset of nodes the user is interested in, and any simple transformations to be performed over the collected data. They are specified using a language such as the structured query language (SQL). A sample query could be expressed as follows:

```
SELECT AVG(temp)
FROM sensors
WHERE location in (0,0,100,100) AND light . 1000 lux
SAMPLE_PERIOD 10 seconds
```

The TinyDB queries are specified on a personal computer (PC) and then the task of distributing them to the WSN is left to the query executor. The query is then disseminated and the results are returned to the query dissemination point in an energy-efficient manner using several techniques, including in-network processing and cross-layer optimizations.

The queries in TinyDB are disseminated through the entire network and collected via a routing tree. The root node of the routing tree is the end point of the query, which is generally where the user that issued the query is located. Nodes within the routing tree maintain a parent–child relationship for the purpose of properly propagating results to the root. The design of an appropriate acquisitional query processor for data collection in WSNs is an active research topic with in the area of query processing techniques development. Information such as where, when, and how often data are physically collected and delivered can be leveraged to significantly reduce the overall power consumption in the sensor network.

## 7.3   Data-Centric Routing

In multihop networks, such as WSNs, packets have to be relayed from the source node to the destination node through intermediate nodes. It is the task of an intermediate node to determine to which neighbouring nodes to forward an incoming packet that is not destined for it. This is usually done using routing tables that lists destination nodes against each of the most appropriate neighbouring nodes. The routing tables are constructed and maintained by a set of rules that form the routing protocol.

Several routing mechanisms have been proposed specifically for WSNs putting into consideration the unique characteristics of such networks [14]. Almost all of these routing protocols can be classified as data-centric, hierarchical, or location-based, although there are some distinct ones based on the network flow or quality of service (QoS) awareness. As this chapter focuses on data-centricity, it is only concerned with data-centric routing protocols. In data centric routing, the routing decision is based on the name(s) associated with the data. A number of data-centric routing protocols exist. In this section, we present a few representative data-centric routing algorithms to assist in understanding the concept.

### 7.3.1   Flooding and Gossiping

Flooding and Gossiping [15] are two classical mechanisms for propagating data in WSNs without the use of any routing algorithms or topology maintenance. The flooding mechanism is based on the following operational concept: each node broadcasts a received data packet to all of its neighbours until the packet reaches its final destination. Gossiping is an enhanced version of flooding in which the receiving node randomly chooses a single neighbour that it will forward the packet to, which will in turn forward the packet to a single randomly chosen neighbour, and so on.

The main advantage of flooding is that it is simple to implement; however, this simplicity comes at the expense of performance. Flooding has several deficiencies [16] as follows:

- Implosion – duplicated messages are sent to the same node.
- Overlap – two nodes sensing the same stimuli at the same time.
- Resource-Blindness – not taking into account the availability of resources.

### 7.3.2 Sensor Protocols for Information via Negotiation

One of the early data-centric protocols, developed particularly targeting WSNs which aimed to overcome the deficiencies in the classic flooding approach were the sensor protocols for information via negotiation (SPIN) [16]. The basic idea behind this set of protocols is that it is more efficient to send information about the data rather than sending the data itself.

The data to be communicated in a SPIN protocol are named using high-level descriptors or metadata. The semantics of the metadata format is application-specific and is not specified in SPIN. The core of the operation of a SPIN protocol is dependant on three types of messages: ADV, REQ, and DATA. The ADV messages are used by the nodes for advertising a particular metadata, the REQ message – to request specific data, and the DATA messages – to carry the actual data. The operation of the SPIN protocol is illustrated in the following example.

*Example 1.* Consider the 6-step sensor network example illustrated in Fig. 7.3. Node "A" initiates the protocol by advertising its data to node "B" using a message of type "ADV." If node "B" is interested in the data, it responds by sending

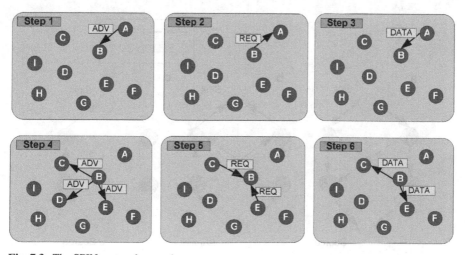

**Fig. 7.3** The SPIN protocol example

back a message of type "REQ." These data are then sent to "B" using a message of the type "DATA" as shown in Step 3 in Fig. 7.3. Node "B" then initiates another cycle of the protocol by advertising the data to its neighbours, "C," "D," and "E." As only nodes "C" and "E" are interested in the data, they request the data from "B", which then sends it to them in the next step using a message of the type "DATA." As this process repeats in the network, the sensor nodes that are interested in the data will get a copy.

SPIN reduces the amount of energy dissipation in the sensor network by a factor of around 3.5 compared with flooding and metadata negotiation almost halves the amount of redundant data. However, a downfall of SPIN is that its advertisement mechanism cannot guarantee the delivery of the data. Consider, for example, the case where the sink (a node the data is to be communicated to) is far away from the source. If the intermediate nodes between the sink and the source are not interested in the data; then the data will not reach the sink.

### 7.3.3 Directed Diffusion

Directed diffusion is one of the pioneering data-centric communication paradigms developed specifically for WSNs [17, 18]. Diffusion is based on a publish/subscribe application programming interface (API) where the details of how published data are delivered to subscribers are hidden from the data producers (sources) and publishers (sinks).

Several protocol variants of directed diffusion exist, each optimised for a different situation. This makes directed diffusion more of a design philosophy rather than a concrete protocol [17]. The basic variant of directed diffusion is the "two-phase pull diffusion," which consists of three phases as shown in Fig. 7.4: interest

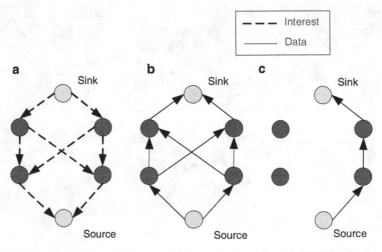

**Fig. 7.4** Three steps in directed diffusion (**a**) Interest propagation; (**b**) Gradient setup; (**c**) Data delivery along reinforced path

propagation, data propagation, and reinforcement. The first phase involves nodes broadcasting their *interests* in certain kinds of named *data* expressed in the form of a set of attribute-value pairs as shown in the following space-based WSN example:

> *// detect reconfiguration of the satellite formation*
> *type = position*
> *// send back results every 20ms*
> *interval = 20s*
> *// for the next 5 minutes*
> *duration = 15 minutes*
> *// from satellites with available power greater than 1.5 Watts*
> *power = >1.5 W*

Interest messages are distributed through the network either using flooding or some other more sophisticated technique. When a node receives the interest packet, it checks whether the packet is new to the node by retrieving the internal cache of the node. If the packet is new to the node, the packet is cached and rebroadcast to the neighbouring nodes. The node also remembers which neighbouring nodes it received the interest packet from such that, later on, once the data have been published, the actual data could be forwarded to all those nodes. This is called the setting up of a gradient toward the sender of an interest. A gradient cache is maintained at each node to store a separate set of gradients for each type of data received in an interest.

The second step of the directed diffusion process involves the propagation of the data packets through the network, which is initiated once the gradients are set up. A node that possesses the actual data required by the sink becomes a source and starts to send data packets. Each node that receives a data packet performs a matching operation according to a list of attributes and their corresponding values. If a match is established, the node packet is passed on to the application module of the node, else the node is considered as an intermediate node.

In its simplest form, intermediate nodes would broadcast all incoming data packets to all their outgoing gradients, while possibly suppressing some of the data messages to adapt to the data rate of each gradient. A problem with this simple scheme is that it results in unnecessary overhead in the network as the data packets are needlessly repeated due to the presence of loops in the gradient graph. Simply checking the source of these data messages is not feasible because of the lack of globally unique identifiers. The problem could be solved by introducing a data cache at each node in the network. The data cache at each node stores the recently received data messages for each known interest. If the sink has multiple neighbours it *reinforces* one of its neighbours (for example, the one that delivered the first copy of the data message). To do this, the sink reinforces the preferred neighbour, which in turn, reinforces its preferred upstream neighbour, and so on.

The fact that the variant of directed diffusion explained above involves two phases (first flooding the interest message through the network and then delivering the data along a reinforced path), in addition to the fact that it is the sink that initiates the "pulling" of data, is the reason behind calling it "two-phase pull" directed diffusion. Other variants of the original form of directed diffusion have been developed [18]. One such alternative is *push diffusion,* which is intended for networks with many receivers and only a few senders. A typical example is an application where sensor nodes need to subscribe to each other frequently to be aware of local events, but where the amount of actual events is quite low.

One-Phase Pull diffusion is another variant of directed diffusion [19]. This one is geared toward networks with many senders and a small number of receivers. As the name indicates, one-phase pull eliminates the second of the flooding phases of the two-phase pull, which constitutes its major overhead. The network is still flooded with interest messages during the first phase of the procedure. However, the interest messages set up direct parent–child relationships between a node and the node from which it first receives an interest message, forming a routing tree in the network.

## 7.4 Data Aggregation

The real power of data-centricity lies in the ability to operate on the data while it is transported in the network [6]. The simplest example of such in-network processing is aggregation. Data aggregation can be perceived as a set of automated methods of combining the data that comes from many sensor nodes into a set of meaningful information. In this context, data aggregation can also be referred to as data fusion. The following example illustrates the advantages of incorporating data aggregation in a data-centric system.

*Example 2.* Figure 7.5 shows an example of data-centric routing where the average temperature is reported to the data sink. The aggregation function is AVG. The

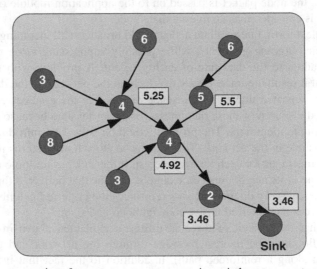

**Fig. 7.5** Data aggregation of temperature measurements in a wireless sensor network

label $x(y)$ at each node represents the local temperature measurement is $x$ while the aggregated (average) value so far is $y$. For example, at node 4(4), the average temperature is $(4 + 6)/2 = 5$.

This section provides an overview of the existing research activities on data aggregation. First, different categories of aggregation functions are described. Second, the importance of appropriate aggregation system design is discussed. This is followed by an outline of the system trade-offs that may be involved in an aggregation system.

### 7.4.1 Aggregation Functions

Depending on the application requirements, three types of data aggregation functions can be used: basic aggregation functions, redundancy suppression, and estimation of a system parameter.

(a) *Basic Operations.* The most basic operations for data aggregation include: COUNT, MIN, MAX, SUM, and AVERAGE [20]. Although the basic functions share the same aggregation structure, the characteristics of different functions differ in three aspects [20]:

1. Duplicate sensitive: A duplicate sensitive function is one which is affected by a duplicate reading from a single sensor. Examples of duplicate-sensitive functions are SUM and AVERAGE.
2. Exemplary or Summary: Exemplary functions, on the one hand, depend on any one value from the set of all sensor readings. Summary functions, on the other hand, depend on the entire set of values, and typically, do not strongly depend on individual values.
3. Monotonic: A function is said to be monotonic if it only increases the magnitude of the partial states it operates on (or, equivalently, decreases). Formally, a function $f$ is monotonic if and only if for any two partial states, $s_1$ and $s_2$, and their aggregate state $s'$ each with a magnitude noted as $m(s)$, either $m(s') \geq \max(m(s_1), m(s_2))$, or $m(s') \leq \min(m(s_1), m(s_2))$.

(b) *Redundancy Suppression.* Data redundancy is an important aggregation operation because sensor data are correlated both in the spatial and temporal domains. Data aggregation in this case is equivalent to data compression [21,22]. A good and simple example of redundancy suppression is the one proposed by Petovic et al. [23] in the context of data funneling. Data funnelling applies the concept of source coding based on the ordering to compress the data of concatenated readings. For instance, suppose there are four nodes with ID's 1, 2, 3, and 4 in a region. The readings of the four sensors are integers within the range of 0–5. Node 4 that receives the concatenation of readings from the other three sensors can send only three readings by implicitly encoding the fourth reading by the order of the three readings. This is possible because there are six (3!) combinations of the ordering relationship. More specifically, the ordering of IDs (1, 2, 3) in the data packet represents that the fourth reading of data are 0, the ordering (1, 3, 2) represents the value of 1, and so on. In this way, compression is achieved by sending out three readings for each four.

(c) *Estimation of a System Parameter.* On the basis of the observations from several pieces of sensor data, the data-fusion function aims to solve an optimization problem minimizing the estimation error of a system parameter [24]. Sensors cooperate to disseminate necessary information to certain nodes, which then proceed to estimate the parameter of interest. An example of such an optimization problem is the averaging of all the temperature readings from sensors within a room to estimate its temperature. The estimation is optimal with respect to the minimum square error (MSE) criterion. In statistics, minimum mean square error (MMSE) describes the statistical estimator with the least possible mean squared error.

### 7.4.2  System Architecture

An aggregation system consists of three primary components: source, sink, and aggregator. There exists a wide variety of ways to determine the location of the data aggregator optimizing the aggregation process. This section describes some approaches that have been used in the literature.

The Data Funnelling routing protocol integrates aggregation and compression techniques. In Data Funnelling, the nodes in any particular region within the WSN, after receiving the interest message, would report their data to a predesignated *border node*. A common schedule according to which the border node acts as the aggregator node during each round of reporting is shared by all the nodes within the region. After the designated aggregator node receives all the reports from its region it compresses the data before sending it to the sink node (Fig. 7.6).

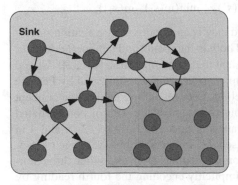 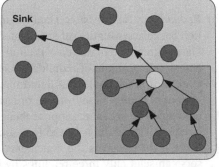

**Fig. 7.6** Data Funnelling: (*left*) directed flooding phase and (*right*) data communication phase

DFuse [25] is a distributed data-fusion framework especially designed for video streaming applications. The first step involves a "naive" assignment of aggregation roles to the nodes within the network. Each node then locally decides whether it wants to transfer the role assigned to it in the previous step to a neighbouring node. This decision is made using a cost function determining the appropriateness of hosting a particular role. The cost function includes *minimize transmission cost (MT)*, *minimize power variance (MPV)*, and *minimize ratio of transmission cost to power (MTP)*. For example, in the case of an aggregation function $f$ with $m$ input data sources (fan-in) and $n$ output data consumers (fan-out), the transmission cost for placing $f$ on node $k$ is formulated as:

$$CMT(k, f) = \sum^{m} t(source_i) \; {}^*hopCount(input_i, k) + \sum^{n} t(f) \; {}^*hopCount(k, output_j),$$

where $t(x)$ represents the transmission rate of the data source $x$, and hopCount$(i, k)$ is the distance (in number of hops) between nodes $i$ and $k$.

*Example 3.* Figure 7.7 illustrates a case where the allocated aggregation point moves in a system running the DFuse data fusion algorithm. If all the inputs to an aggregation node are coming via a relay node (Step 1 in Fig. 7.7), and there is a contraction at the fusion point, then the relay node will become the new aggregation node, and the old aggregation node will transfer its responsibility to the new one (Step 2 in Fig. 7.7). In this case, the aggregation point is moving away from the sink, and is coming closer to the data source points. Similarly, if the output of the aggregation node is going to a relay node, and there is data expansion, then again the relay node will act as the new aggregation node. In this case, the fusion point is coming closer to the sink and is moving away from the data source points.

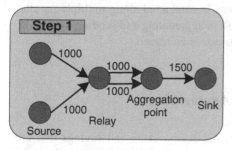 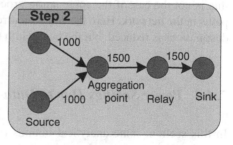

**Fig. 7.7** Aggregation in DFuse

### 7.4.3   Trading-off Resources

Depending on the resource constraints in a sensor network, there exist the following trade-offs: energy vs. estimation accuracy [24, 26], energy vs. aggregation latency [27, 28], and bandwidth vs. aggregation latency [21].

#### 7.4.3.1   Trade-off Between Energy and Accuracy

Energy is proportional to the number of messages flowing through the network. Therefore, reducing the number of messages is an ultimate aim of the aggregation in WSNs to reduce the energy consumption. However, accuracy is also known (in many cases) to be proportional to the number of messages received by the sink. This requires a trade-off to be made between accuracy and energy within the aggregation system.

#### 7.4.3.2   Trade-off Between Energy and Latency

The worse-case latency due to aggregation will be proportional to the number of hops between the sink and the farthest source. When no aggregation is employed, the delay between the time when the various sources transmit data and the sink receives the first packet is proportional to the number of hops between the sink and the nearest source. Hence one way to quantify the effect of the aggregation delay is to examine the difference between these two distances.

#### 7.4.3.3   Trade-off Between Bandwidth and Latency

It is obvious that the bandwidth is more efficiently utilized when data aggregation is used along a path due to the reduction in the amount of data transmitted. However, this is not the case if the aggregation process is conducted along multiple parallel paths in the network. Here the delay incurred in aggregating and sending data to the destinations is reduced, but the bandwidth usage is increased.

### 7.4.4   Power Savings Due to Data Aggregation

In this section, some analytical bounds on the energy costs and savings that are brought about by using data aggregation is presented. Calculations based on the analysis in [5] also formally prove that the efficiency of using a data-centric approach for networking in WSNs is higher compared with the address-centric approach.

The total number of data transmissions, $d_i$, required for the optimal address-centric approach is:

$$N_A = d_1 + d_2 + \cdots d_k = \text{sum}(d_i) \tag{7.1}$$

**Proposition 1.** *The optimal data-centric protocol performance, $N_D$, in terms of data transmissions is lower than $N_A$:*

$$N_D \leq N_A$$

*Proof*: Doing data aggregation optimally can only decrease the minimum number of edges in the aggregation tree that are needed compared with the situation when the sources send information to the sink along the shortest paths.

**Definition 1.** *The diameter X (in number of hops) of a set of nodes S is the maximum of the pair wise shortest paths between these nodes,*

$$X = \max_{i.j \in S} SP(i, j),$$

where $SP(i, j)$ is the shortest number of hops needed to move from node $i$ to $j$.

**Proposition 2.** *If the source nodes $S_1$, $S_2$, ... $S_k$ have a diameter $X \geq 1$, the total number of transmissions $(N_D)$ required for the optimal data-centric protocol satisfies the following bounds:*

$$N_D \leq (k - 1)X + \min(d_i), \tag{7.2}$$
$$N_D \geq \min(d_i) + (k - 1). \tag{7.3}$$

*Proof.* Equation (7.2) can be obtained by constructing the data aggregation tree, which consists of $(k - 1)$ sources sending their packets to the remaining nearest to the sink source. Equation (7.3) is obtained by considering the case when $X = 1$.

**Proposition 3.** *If the diameter $X < \min(d_i)$, then $N_D \leq N_A$, from which it can be concluded that the optimum data-centric protocol will perform better than the address-centric protocol.*
*Proof*:

$$N_D \leq (k - 1)X + \min(d_i) < (k) \min(d_i)$$
$$\Rightarrow N_D < \text{sum}(d_i) = N_A \tag{7.4}$$

**Definition 2.** *The fractional savings obtained using a data-centric protocol as opposed to a address-centric protocol, FS, can be quantified as follows:*

$$FS = (N_A - N_D)/(N_A). \tag{7.5}$$

**Proposition 4.** *The fractional savings FS lies with in the following bounds:*

$$FS \geq 1 - ((k - 1)X + \min(d_i))/\text{sum}(d_i), \tag{7.6}$$
$$FS \leq 1 - (\min(d_i) + k - 1)/\text{sum}(d_i). \tag{7.7}$$

*Assuming that all the sources are located at the same shortest-path distance from the sink, i.e.,* $\min(d_i) = \max(d_i)$, *then FS are bounded as below:*

$$1 - \frac{((k-1)X + d)}{kd} \leq FS \leq 1 - \frac{(d+k-1)}{kd}. \tag{7.8}$$

**Proposition 5.** *Assuming that X and k are fixed, then as d tends to infinity (i.e., as the sink is farther and farther away from the sources) the following is true:*

$$\lim_{d\to\infty} FS = 1 - 1/k. \tag{7.9}$$

*Proof.* In the limit, $X \ll d$, and $k \ll d$. It suffices to show that both the lower and the upper bounds in (7.8) converge to the same right hand side value:

$$\lim_{d\to\infty} \left(1 - \frac{(k-1)X + d}{kd}\right)$$
$$= \lim_{d\to\infty} \left(1 - \frac{(k-1)X}{kd} - \frac{d}{kd}\right)$$
$$= 1 - 1/k$$

and,

$$\lim_{d\to\infty} \left(1 - \frac{(d+k-1)}{kd}\right)$$
$$= \lim_{d\to\infty} \left(1 - \frac{d}{kd} - \frac{k-1}{kd}\right)$$
$$= 1 - 1/k$$

Equation (7.9) shows that if the distance between the sink and the sources is large compared with the distance between the sources, then the optimal data-centric protocol gives $k$-fold savings over the address-centric protocol. When there are four sources that are close together and far away from the sink, then the address-centric protocol will require about four times as many transmissions, i.e., there are roughly 75% fewer transmissions with data aggregation. When there are 10 such sources, the gains are nearly 90% and so on.

## 7.5   Data-Centric Storage Paradigms

Nodes in a WSN produce data related to the event information, which they sense from the environment. These data need to be stored at some location either in situ or externally. Three possible data-storage paradigms are employable for wireless sensor networks as illustrated in Fig. 7.8 [29]:

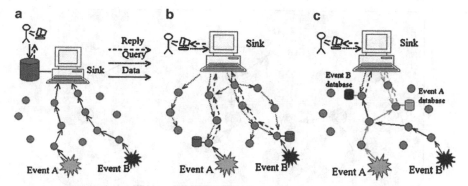

**Fig. 7.8** Three types of storage scenarios [29]. (**a**) External storage; (**b**) Local storage; (**c**) Data-centric storage

- *External storage* – all event data are stored at an external storage point for processing.
- *Local storage* – all the collected information collected is stored locally at the detecting node.
- *Data-centric storage* – the event data are stored at predesignated nodes, associated with a particular event type each.

The third of these paradigms, namely data-centric storage, is a companion method to the data-centric routing explained in Section 7.3. Data-centric storage requires the availability of specific nodes storing all the data with the same general name collected by any node in the network. Any queries that are made to the network for data with a particular name are directly sent to the node that is assigned the task of storing that particular named data. The advantage of this approach is that it avoids the flooding that is required in some data-centric routing protocols. The approach is particularly advantageous for WSNs in which queries are frequently initiated from within the network. Data-centric storage systems may exist along with data-centric routing, whereby the appropriate mode of data access is selected by the applications according to their specific needs.

*Example 4.* Consider the network shown in Fig. 7.9, the nodes of which contain temperature sensors. Assume that the data from the temperature sensors of nodes "A" and "B" indicate a rise in temperature exceeding 70° degrees in their respective areas. As node "C" is already mapped to be associated with the name "temperature >70°" the events produced by nodes "A" and "B" are routed to node "C." Knowing the name, another node "D" can retrieve the events by directly routing a message to "C." Obviously, this kind of system avoids the need of flooding messages through the network.

Two representative approaches to implementing data-centric storage are: geographic hash table (GHT) and distributed index for multidimensional data (DIM). Both of these approaches rely on geographic information. In GHT [29], hash functions are used to transform an attribute or a specific type of event into

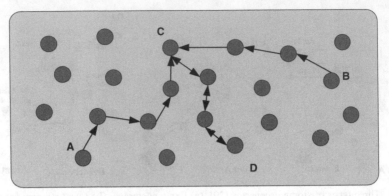

**Fig. 7.9** Data-centric storage concept

a point in the two-dimensional space. If the WSN does not have any node located at the coordinates specified by the result of the hash function, the data are stored at the node that is closest to the hash result.

DIM [30] is designed especially for multidimensional range queries. DIM maps a vector of readings with multiple attributes to a two-dimensional geographic zone. Two assumptions are made in DIM: first, sensors are aware of their own locations and field boundaries, and second, all the sensors are static.

## 7.6 Thoughts for Practitioners

Data-centricity is not a valid option for applications that require a minimal infrastructure, as the infrastructure would require specific addressing of the nodes. Data-centricity is optimal for usage in larger networks where scalability is essential. For smaller networks consisting of a few nodes, the performance of the data-centric paradigm may be found to be poorer than (or equivalent to) that of the address-centric paradigm.

The implementation of data-centricity in wireless sensor networks may be incorporated in the middleware layer, the operating system, or both. Middleware usually implements the higher abstraction layers and provides a suitable API to the programmer. The implementation of certain protocols such as aggregation usually takes place within the middleware layer. The lower layers, such as data-centric routing and data-centric MAC protocols may be implemented as part of the operating system.

## 7.7 Future Research Directions

Data-centric networking for wireless sensor networks is in its early stages of development and poses a number of research challenges. Efficient data-centric protocols are yet to be developed for mobile sensor networks [31] and heterogeneous sensor networks [32]. Data-centric protocols are generally designed for "traditional" sensor networks. Further research is needed to cover wireless sensor networks that have unique attributes, for example WSNs that operate in extreme environments [33]. An example of a promising new direction of research is the design of a data-centric sensor network for spacecraft formation flying missions [34].

## 7.8 Conclusions

This chapter reviews several topics related to data-centric algorithms and mechanisms including data-centric routing, data aggregation, and data-centric storage. It is shown that the nature of data-centric networks differs significantly from traditional networks that rely on address-centric approaches. Using the data itself as the backbone of the network operation considerably changes the existing design paradigms. There are several approaches to the implementation of data-centric networks depending on the programming abstractions that are used. The publish/subscribe abstraction is by far the most commonly used due to its simplicity. Adopting any of the possible data-centric abstractions mainly affects the interaction between the user of the network and the network itself. However it also affects the choice of protocols that are employed by the network. For example, directed diffusion is a routing protocol that is specifically designed for publish/subscribe networks. In data-centric protocol design for wireless sensor networks efficiency is the primary goal given the limited computing resources of such networks.

## Terminology

*Data-centricity.* The usage of data as the means for node addressing in a network.

*Aggregation.* The in-network processing of data while transporting them within the network.

*Publish-subscribe system.* A software communications paradigm that is based on the publication of data to a software bus, where the data is available for access by subscribing nodes in the network.

*Data-centric routing.* A data-centric paradigm in which data is made to flow from a source that provides the data to a sink that requires the data based on the name(s) associates with the data itself.

*Data-centric storage.* A data-centric paradigm where data associated with an event are stored at predesignated nodes, each associated with a particular event type.

*Aggregation point.* The node in a sensor network that performs the process of aggregation on the incoming data.

*Data fusion.* The combination of raw data from more than one source into a meaningful data format.

*Data funnelling.* The process in which data from nodes in a specific region are aggregated at one node before sending the aggregated result to the sink node across
the network.

*Flooding.* The process by which each node broadcasts its data to all of its neighbouring nodes.

## Questions

1. State two possible options for the implementation of data-centric abstractions
   in sensor networks.
2. Briefly list the advantages of adopting data-centricity in wireless sensor networks.
3. What are publish/subscribe systems?
4. Differentiate between push diffusion and pull diffusion.
5. What is the difference between data-centric routing and data-centric storage?
6. Consider the following example. A group of 33 satellite sensor nodes in space
   flying freely at an orbit of 600 km altitude. All the data are relayed from a single
   node that has the ability of communicating with the ground station on Earth.
   Calculate the power savings brought about from using data-centricity compared
   with address-centricity in this network.
7. How does Geographic hash tables (GHT) relate to data-centric storage?
8. Classify the following aggregation cases:

   (a) A sensor network that involves identifying the average number of animals
       at any one time in a $100 \, \text{Km}^2$ area of a forest.
   (b) A sensor network is required to estimate the pollution level in a rural area
       within a range of 10% error.
   (c) A sensor network that needs to find out the average illumination level in an
       agricultural area.

9. Explain the idea behind the SPIN protocol.
10. Describe the relation between databases and data-centricity.

## References

1. George Coulouris, Jean Dollimore, and Tim Kindberg. "Mobile and Ubiquitous Computing." In
   *Distributed Systems: Concepts and Design book edited by George Coulouris, Jean Dollimore,
   and Tim Kindberg, Chapter 16, 2005.*
2. Ian F. Akyildiz, Weilian Su, Yogesh Sankarasubramaniam, and Erdal Cayirci. "A survey on
   Sensor Networks." *IEEE Communications Magazine, 40*(8):102–114, 2002.
3. Deborah Estrin, Ramesh Govindan, John Heidemann, and Satish Kumar. "Next Century
   Challenges: Scalable Coordination in Sensor Networks." In *Proceedings of the ACM/IEEE*

*International Conference on Mobile Computing and Networking*, Seattle, Washington, USA, ACM. August, 1999, pages 263–270.

4. G. J. Pottie and W. J. Kaiser. "Wireless Integrated Network Sensors," *Communications of the ACM, 43*(5):52–58, May 2000.

5. Bhaskar Krishnamachari, Deborah Estrin, and Stephen Wicker. "Modelling Data-Centric Routing in Wireless Sensor Networks." *USC Computer Engineering Technical Report CENG 02–14*, 2002.

6. B. Krishnamachari, D. Estrin, and S. Wicker. "The Impact of Data Aggregation in Wireless Sensor Networks." *International Workshop on Distributed Event-Based Systems, (DEBS '02)*, Vienna, Austria, July 2002.

7. J. Schlesselman, G. Pardo-Castellote, and B. Farabaugh. "OMG data-distribution service (DDS): architectural update." *IEEE Military Communications Conference*, MILCOM 2004.

8. J. A. Stankovic, T. E. Abdelzaher, C. Lu, L. Sha, and J. C. Hou. "Real-time communication and coordination in embedded sensor networks." *Proc. IEEE, 91*(7):1002–1022, 2003.

9. P. Eugster, P. Felber, R. Guerraoui, and A. Kermarrec. "The Many Faces of Publish/Subscribe." ACM *Computing Surveys, 35*(2), 2003.

10. V. Narayanmurthy. "A Publish/Subscribe scheme for networked embedded systems." *Masters Thesis, Dept. of Computer Science, University of Iowa.* August 2002.

11. G. Banavar, T. Chandra, B. Mukherjee, J. Nagarajarao, R. Strom, and D. Sturman. "An Efficient Multicast Protocol for Content-Based Publish-Subscribe Systems." In *Proceedings of the 19th International Conference on Distributed Computing Systems*, pages 262–272, 1999.

12. S. Madden, M. J. Franklin, J. M. Hellerstein, and W. Hong. "The design of an acquisitional query processor for sensor networks." In *Proceedings of the 2003 ACM SIGMOD International Conference on Management of Data*, San Diego, California, June 2003, pages 491–502.

13. W. F. Fung, D. Sun, and J. Gehrke. "COUGAR: The Network is the Database." In *Proceedings of ACM SIGMOD International Conference on Management of Data*, ACM Press, NY, 2002, pages 621–621.

14. Jamal N. Al-Karaki and Ahmed E. Kamal. "Routing Techniques in Sensor Networks: A survey." *IEEE communications, 11*(6):6–28, December 2004.

15. S. Hedetniemi and A. Liestman. "A survey of gossiping and broadcasting in communication networks," *Networks, 18*(4):319–349.

16. W. R. Heinzelman, J. Kulik, and H. Balakrishnan. "Adaptive Protocols for Information Dissemination in Wireless Sensor Networks." *Proc. ACM MobiCom '99*, Seattle, WA, 1999, pages 174–185.

17. C. Intanagonwiwat, R. Govindan, and D. Estrin. "Directed diffusion: A scalable and robust communication paradigm for sensor networks." In *Proceedings of the 6th Annual International Conference on Mobile Computing and Networking*, ACM Press, NY, 2000, pages 56–67.

18. C. Intanagonwiwat, R. Govindan, D. Estrin, J. Heidemann, and F. Silva. "Directed diffusion for wireless sensor networking." *IEEE/ACM Transactions on Networking, 11*(1):2–16, 2003.

19. J. Heidemann, F. Silva, and D. Estrin. "Matching data dissemination algorithms to application requirements." In *Proceedings of the 1st international Conference on Embedded Networked Sensor Systems* (Los Angeles, California, USA, November 05 - 07, 2003). SenSys '03. ACM, New York, NY, pages 218–229.

20. S.R. Madden, M.J. Franklin, J.M. Hellerstein, and W. Hong. "TAG: a Tiny Agegation Service for Ad. Hoc Sensor Networks." In *Proceedings of OSDI*, Boston, MA, December 2002.

21. A. Scaglione and S. D. Servetto. "On the interdependence of routing and data compression in multi-hop sensor networks." In *Proceedings of the 8th Annual ACM/IEEE International Conference on Mobile Computing and Networking (MobiCom '02)*, Atlanta, Georgia, 2002.

22. J. Chou, D. Petrovic, and K. Ramchandran. "A distributed and adaptive signal processing approach to reducing energy consumption in sensor networks." In *Proceedings of INFOCOM 2003*, San Francisco, April 2003.

23. D. Petrovic, R. C. Shah, K. Ramchandran, and J. Rabaey. "Data funneling: Routing with aggregation and compression for wireless sensor networks." *In proceedings of the 1st IEEE International workshop on Sensor Network Protocols and Applications (SNPA)*, Alaska, May 2003.

24. M. Rabbat and R. Nowak. "Distributed optimization in sensor networks." In *Proceedings of the 3rd International Symposium on Information Processing in Sensor Networks (IPSN)*, Berkeley, California, April 2004.
25. R. Kumar, M. Wolenetz, B. Agarwalla, J. Shin, P. Hutto, A. Paul, and U. Ramachandran. "DFuse: A framework for distributed data fusion." In *Proceedings of the 1st ACM Conference on Embedded Networked Sensor Systems (Sensys '03)*, Los Angeles, California, November 2003.
26. A. Boulis, S. Ganeriwal, and M. B. Srivastava. "Aggregation in sensor networks: An energy-accuracy trade-off." In *Proceedings of the 1st IEEE International Workshop on Sensor Network Protocols and Applications (SNPA 2003)*, Anchorage, Alaska, May 2003.
27. C. Schurgers, V. Tsiatsis, S. Ganeriwal, and M. Srivastava. "Optimizing sensor networks in the energy-latency-density design space." *IEEE Transactions on Mobile Computing*, 1(1):70–80, January 2002.
28. Y. Yu, B. Krishnamachari, and V. K. Prasanna. "Energy-latency tradeoffs for data-gathering in wireless sensor networks." *In Proceedings of INFOCOM 2004*, Hong Kong, March 2004.
29. S. Ratnasamy, B. Karp, S. Shenker, D. Estrin, R. Govindan, L. Yin, and F. Yu. "Datacentric storage in sensornets with GHT, a geographic hash table." *Mobile Networks and Applications*, 8(4):427–442, August 2003.
30. X. Li, Y. J. Kim, R. Govindan, and W. Hong. "Multi-dimensional range queries in sensor networks." In *Proceedings of the 1st ACM Conference on Embedded Networked Sensor Systems (Sensys '03)*, Los Angeles, California, November 2003.
31. Y. C. Tseng, S. L. Wu, W. H. Liao, and C. M. Chao. "Location Awareness in Ad Hoc Wireless Mobile Networks." *IEEE Computer*, 34(6):46–52, June 2001.
32. M. Yarvis, N. Kashalnagar, H. Singh, A. Rangarajan, Y. Liu, and S. Singh. "Exploiting Heterogeneity is sensor networks." *IEEE proceedings of INFOCOM 2005. 24th Annual Joint Conference of the IEEE Computer and Communications Societies*, 2005.
33. T. Vladimirova , C. P. Bridges, G. Prassinos, X. Wu, K. Sidibeh, D. J. Barnhart, A. Jallad, J. R. Paul, A. Lappas, A. Baker, K. Maynard, and R. Magness. "Characterising Wireless Sensor Motes for Space Applications," in 2nd NASA/ESA Conference on Adaptive Hardware and Systems (AHS), 2007.
34. A.-H. Jallad, "Space-Based Wireless Sensor Networks", PhD Thesis, Surrey Space Centre, Department of Electronic Engineering, University of Surrey, October 2008.

# Chapter 8
# Congestion and Flow Control in Wireless Sensor Networks

**Vikram P. Munishwar, Sameer S. Tilak, and Nael B. Abu-Ghazaleh**

**Abstract** Wireless sensor networks (WSNs) present a range of unique challenges to protocol designers due to their communication pattern, poor and unpredictable performance of their low-power wireless radios, wireless interference, and resource constrains of individual sensor nodes. One of the challenges is how to address congestion control and reliable data delivery in such environments: the nature of WSN applications (data centric, prone to redundancy due to multiple sensors reporting a single event) and infrastructure (sensor capabilities, and deployment density and strategy) invite significantly different solutions from those present in conventional networks. In this chapter, we present a survey of existing congestion control approaches and classify them based on various parameters such as mechanisms used for congestion detection and control, support for application specific design, target data delivery model, and support for fairness and reliability. Since, WSN applications exhibit a wide variety of communication patterns, existing literature has focused on three types of applications with regards to communication among sensors: one-to-one, one-to-many, and many-to-one. Reliability and congestion management approaches in the case of one-to-one (unicast) and one-to-many (multicast or broadcast) communication have been studied extensively in wired as well as wireless ad hoc networks, providing significant experience to draw on. However, many-to-one communication pattern involves various opportunities (e.g., loss-tolerance) as well as challenges (e.g., congestion management), thereby gaining major attention from the research community. Thus, the main focus of this chapter is on congestion and flow control approaches for many-to-one traffic pattern in WSNs.

V.P. Munishwar (✉)
Department of Computer Science, Watson School of Engineering and Applied Sciences,
State University of New York, Binghamton, Binghamton, NY 13902, USA
e-mail: vmunish1@cs.binghamton.edu

S. Misra et al. (eds.), *Guide to Wireless Sensor Networks*, Computer Communications
and Networks, DOI: 10.1007/978-1-84882-218-4_8,
© Springer-Verlag London Limited 2009

## 8.1  Introduction

Wireless sensor nodes are generally characterized by their limited resources, including processing, storage, communication, and energy resources. The self-organizing and cooperative nature of sensor nodes makes it possible to deploy them in ad hoc fashion in an inaccessible location to obtain required information about the target area. A wireless sensor network can be composed of multiple sensor nodes that self-organize to gather information, and relay this information to observers. An observer can either be a central node (*base station*) or a mobile node that moves around the network and collects the data. In either case, observers are assumed to be resource rich nodes having capabilities to store, process, and analyze the data collected from sources and make the data available over a network, if desired. This makes sensor networks an attractive choice for a wide range of applications in areas such as military, civil, mining, health, and monitoring for scientific as well as commercial purposes.

WSNs have a number of unique characteristics that require protocols and system support that significantly differs from those of traditional networks. Specifically, WSNs differ in the following ways:

1. They are *data-centric*: The observers are interested in timely and accurate gathering of the data. In contrast, traditional networks typically focus on communication between different end-points on the network. There is redundancy among the information reported by physically colocated sensors. The performance of the network is best measured in application specific terms, rather than in networking terms such as throughput.
2. Unique traffic patterns: In WSNs, often data are funneled from a subset of sensors toward an observer interested in collecting the data. Since, such communication involves multiple senders and a single receiver, we term it as many-to-one traffic pattern.

These factors, combined with the nature of the wireless channel, and the emphasis on energy efficiency, require new protocols and algorithms for WSNs that are sensitive to their unique characteristics.

In this chapter, we overview and classify protocols for *congestion management* in WSNs [1]. Congestion is a classical problem in packet switched networks where the sources collectively exceed the network capacity at one or more intermediate nodes (routers), leading to packet drops [2]. If the offered load to the network is not controlled (via a congestion management protocol that typically limits the source rates), congestive collapse can arise. Because of the properties of WSNs overviewed in the previous paragraph, the impact of congestion is best measured in application-specific terms [1]. As such, new congestion management protocols that are aware of the nature of the application can lead to more effective solutions than classical approaches that rely only on networking measures.

Congestion control is essentially a resource allocation problem. A basic congestion control simply ensures that the source rates are regulated to avoid or mitigate congestion. Better approaches are also able to allocate the resources fairly – attempting to ensure that the competing sensor nodes at a congested node get equal shares of the bandwidth [3]. Similarly, handling cases, where sender(s) are sending data faster than a receiver can consume them, is called flow control. Finally, one of the important goals of congestion and flow control algorithms in WSNs is to provide desired level of reliability at the sink node, in an energy efficient manner, where reliability is nothing but a guarantee of successful delivery of data from source to sink. Again, the notion of reliability is best measured here in application-specific terms (e.g., data fidelity, or reliability in detecting an event).

In this chapter, we begin with presenting various congestion and flow control algorithms for WSNs, specifically designed for commonly observed data collection-based application, in which data are funneled from a subset of sensors toward an observer interested in collecting the data. Most of these algorithms focus on avoiding or mitigating congestion in WSNs; however, they use simple ARQ-based protocols to ensure overall reliability of the communication. Since an important goal of providing congestion and flow control in sensor networks is to minimize the overall energy consumption in the network, it is equally important to focus on energy-efficient reliability guaranteeing protocols for WSNs. Therefore, we also describe some efforts specifically targeted toward providing energy-efficient reliable data delivery in WSNs. Finally, we overview protocols that support congestion control for applications involving different (less common) data delivery models for WSNs, and a MAC protocol that implicitly supports rate control in WSNs.

The remainder of the chapter is organized as follows: Sect. 8.2 explains congestion and flow control in detail followed by the description of why existing techniques for congestion management are not directly applicable in case of WSNs. In Sect. 8.3, we describe general challenges and design issues associated with developing energy efficient congestion and flow control mechanisms in WSNs. Section 8.4 gives the classification approach that we have used to differentiate the congestion and flow control protocols discussed in this chapter. Although, congestion control and reliability are two important features of transport protocols, we prefer to discuss them in separate sections because majority of the existing protocols in WSNs focus mainly on an individual feature, with little or no attention given to the other. In Sect. 8.5 we focus on typical data collection-based application, and discuss some important contributions on congestion control, flow control, and fairness guarantee in sensor networks. In Sect. 8.6, we discuss some of the existing approaches that focus mainly on providing reliability guarantees in WSNs. In Sect. 8.7, we describe other related works that either implicitly or explicitly contribute toward congestion and flow control in WSNs. Finally, in Sect. 8.8, we compare some of the approaches discussed in the chapter, based on the classification scheme presented in Sect. 8.4, and give concluding remarks.

## 8.2  Background

In this section, we overview the problems of general congestion management and then those of flow control in traditional networks. We then explain why congestion and flow control requirements are unique for WSNs, motivating the need for alternative protocols and algorithms that are tailored for them.

## 8.2.1 Congestion Management

Congestion is caused by sources exceeding the link or buffer capacity at intermediate nodes. In wired networks, the resource exceeded is the capacity of a given link (which is fixed for point to point links). In wireless networks, the capacity of a link is determined by the number of active sources in interference range with each other. The MAC protocol is responsible for arbitrating the medium among the contending sources. However, as a result, the effective capacity of a link can vary over time even if the traffic going through a congested node does not vary over time.

Congestion control is carried out by either detecting early signs of oncoming congestion and recovering before the onset of congestion (known as *congestion avoidance*) or detecting congestion and then recovering from it (known as *congestion mitigation*). Congestion management algorithms differ in how they accomplish congestion detection and control (or avoidance). Detection may be carried out at the sources or at the intermediate nodes. In either case, control is carried out by reducing the sending rate of the sources. We elaborate on these approaches in the following.

### 8.2.1.1 Congestion Detection

Congestion detection is a primitive and important step in any congestion control algorithm. In general, congestion can be detected either at an intermediate node or at the sink (end-to-end). At an intermediate node, congestion or the possibility of congestion can be inferred based on various factors such as current queue occupancy, packet service time, or a combination of the present and the past channel loading conditions. For instance, congestion detection based on queue occupancy is used in DECbit [4] and Random Early Detection (RED) [5] mechanisms. DECbit gives explicit notification of congestion, while RED gives implicit congestion notification, either by piggybacking congestion state on a data packet to be transmitted or by dropping it. End-to-end congestion detection can consider factors such as timeouts, duplicate ACKs, inter-packet delay, etc. Transmission control protocol (TCP) [6] detects congestion based on duplicate ACKs or timeouts. Therefore, upon receiving three duplicate ACKs, the fast recovery module of TCP considers that the congestion has occurred.

### 8.2.1.2 Congestion Control

Once congestion (or early sign of congestion) is detected, congestion control mechanism needs to be initiated. Congestion control can be performed either in proactive way, by avoiding the congestion collapse to occur, or in reactive way, by mitigating the congestion already occurred in the network. We briefly explain the two mechanisms now.

- *Congestion avoidance*: Congestion avoidance strategy involves detecting early signs of congestion in the network and initiating preventive measures to avoid congestion collapse. For instance, TCP Vegas [7] tries to detect congestion when it is in incipient stage by comparing the measured throughput rate with expected throughput rate at the sender. Depending on this difference, the source can determine whether the sending rate needs to be adjusted to make sure that the *right* amount of extra data is present in the network. Other congestion avoidance mechanisms include DECbit and RED, which detect and inform the early signs of congestion to the end node by monitoring the queue occupancy at an intermediate node.
- *Congestion mitigation*: Once congestion arises and is detected, it is mitigated (recovered from) by having the sources reduce the overall offered load. In case of wired networks and wireless ad hoc networks, data flow is end-to-end and not collaborative (as in case of WSNs). Thus, for such networks, the assumption is that congestion occurs among competing sources, and therefore, ensuring fairness among the sources is an issue. This is especially true since congestion control is carried out in a distributed way by different sources; that is, the source rate control is not coordinated. In WSNs, the competition for the resources is often among multiple sources that respond to the same query; that is, all packets are in service of the same flow and the notion of fairness must be replaced by application specific measures of data quality.

## 8.2.2 Flow Control

Flow control is a mechanism for ensuring that there is no rate-mismatch between one or more transmitter(s) and their receiver. In other words, flow control ensures that the overall transmission rate of the sender(s) never exceeds the reception rate of the receiver. There are two types of flow control mechanisms:

- *Closed-loop flow control*: In this technique, the receiver gives feedbacks directly to the source. Depending on the feedback, source can adjust its transmission rate to handle the rate-mismatch.
- *Open-loop flow control*: In this technique, there is no feedback given by the receiver to the source. On the other hand, it involves hop-by-hop feedbacks.

## 8.2.3 The Need for Congestion and Flow Control in WSNs

Before describing the congestion control problem in WSNs, we define the keywords that will be used throughout the chapter. Since, multiple sensors reporting their readings to the base station is a commonly observed scenario in WSNs, data follows a tree-based routing topology with the base station (sink) as a root of the tree, as shown in Fig. 8.1. Here, node $S$ represents the sink, while the other nodes

**Fig. 8.1** Tree-based routing
topology

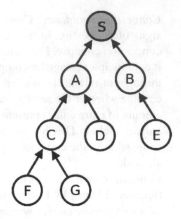

represent data generators and/or forwarders. In this chapter, we term the nodes that
are near the sources as *upstream* nodes while the nodes near the sink as *downstream*
nodes. Thus, if nodes $F$ and $G$ are the sources, then they represent a set of upstream
nodes for node $C$, whereas node $A$ acts as a downstream node for node $C$. In addition, we term the one-hop neighbor that is on the data forwarding path toward the
sink as a *parent* node. Thus, for this topology, node $A$ acts as a parent of nodes $C$
and $D$. Similarly, we term *subtree* of node $A$ as a tree formed by all of its upstream
nodes $(C, D, F, G)$, with node $A$ acting as a root of the tree.

The impact of congestion and the need for congestion avoidance mechanisms in
WSN was identified by Tilak et al. [1]. Specifically, they study the effect of increasing the data reporting rate and the sensor density on the quality of the information
as viewed by the observer in a data funneling application. Although, higher density of sensor nodes presents an opportunity for more accurate sensing, it becomes
harmful to the performance of the overall network if congestion is not controlled;
as sensor density increases, a higher load is placed on shared wireless channel and
the network is saturated faster. In addition to the need to make sure that the overall reporting rates of sensors do not exceed the network capacity, it is important to
make sure that minimum application-specific accuracy requirements are met, and at
the same time the network is not expending energy on achieving excessive accuracy
than the desired accuracy level. To support application's requirement, following inequality should be satisfied.

$$C_{\text{application}} \leq \sum_{i=1}^{M} b(S_i) \leq \alpha C_{\text{total}}, \tag{8.1}$$

where, $C_{\text{application}}$ is the required channel capacity to satisfy application's need,
$C_{\text{total}}$ is the total channel capacity, $\alpha$ is the fraction of the capacity dictated by the
self interference that arises in multihop connections ($\alpha$ is typically around 0.25 [8]).
$\sum_{i=1}^{M} b(S_i)$ is the total data in transit from $M$ event-detecting sensors, where each
sensor, $S_i$ is transmitting at the bit rate of $b(S_i)$, from time $T$ to $T + \delta$, where $\delta$ is

**Fig. 8.2** Funneling effect of traffic in WSNs

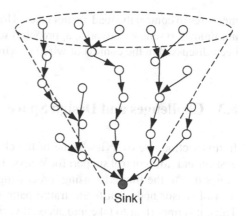

the average latency. The desired accuracy level required by the observer can be different in case of different types of applications with different traffic flow paradigms. For instance, traffic generated by multiple nodes in response to a single event is generally redundant and thus loss-tolerant. This decreases the level of accuracy required at the base station.

In a funneling WSN application, as shown in Fig. 8.2, it is helpful to differentiate between congestion that arises near the sink, vs. that arises near the sources.

- *Congestion Near the Event-Sources*: Detection of an event creates sudden burst of traffic from sensor nodes lying in the event region, leading to collisions and significant packet loss near the sources. This problem is especially prominent if the density of detecting nodes as well as the data generation rate is high, in which case the probability of having persistent hotspots near the sources increases dramatically [9]. To handle this problem, hop-by-hop signaling indicating the need for reducing the reporting rate (backpressure) has been proposed [3, 9].
- *Congestion Near the Sink*: Traffic generated at multiple source nodes travels in multihop fashion toward the sink (base station). Especially in case of detection of an event, data are generated at the source nodes almost at the same time. Such traffic, along the way to the base station, increases traffic-load in the region near to the base station due to the funnel-like communication pattern [10]. This increased traffic load can result in node-level as well as link-level congestion near the sink node. Specifically, in a sparsely deployed network, sources generating data at higher rate can create transient hotspots near the sink [9].

Other forms of traffic such as one-to-one and one-to-many communication may routinely arise in WSNs. These communication patterns also pose congestion management related challenges. For example, one-to-one communication may observe packet loss due to the hidden-terminal problem or interference due to concurrent-flows in the network, whereas one-to-many traffic flows (in case of flooding) exhibit congestion-related issues away from the source node, since downstream nodes forwarding the data simultaneously can result in packet losses due to collisions, if any measures to control the load are not taken. However, since such routing scenarios are not unique to sensor networks, approaches to handle these challenges have already been discussed in the context of wired or wireless ad hoc networks.

## 8.3  Challenges and Design Space

In this section, we overview some of the challenges associated with building congestion and flow control support for WSNs. WSNs present some unique design level issues due to the limited sensing, processing, storing and communicating capabilities of sensor nodes, different traffic patterns, and different deployment patterns. Thus, it is important to take into account various challenges associated with WSNs, before developing efficient congestion and flow control mechanisms. We now briefly describe challenges and design level issues in WSNs.

### 8.3.1  Resource Constraints

One of the major challenges presented by a wireless sensor network is the limited battery power, processing power, memory and storage capacity of sensor nodes. Radio communication is a costly operation in terms of energy consumption, which places an emphasis on in-network processing to reduce the amount of data being communicated, as well as effective networking protocols that reduce unnecessary packet losses.

### 8.3.2  Traffic Patterns

Another difference between traditional networks and WSNs is the unique patterns of communication that can occur in WSNs [11]. In most traditional networks, point-to-point, or unicast, communication dominates. However, in WSNs, many-to-one communication is most common, as data are sent from multiple sensors to an observer continuously or in response to a query or an event. Furthermore, one-to-many communication via optimized broadcast or geocast algorithms can occur as a query is disseminated in the network, or information about an important event (or event summaries) are disseminated proactively in the network [12]. Periods of bursty activity can occur as an important event occurs, while at other times, the network can exhibit low activity. In applications that use storage, other complex patterns of communication can arise. These traffic patterns play an important role in how congestion is manifested, and the solutions for detecting and controlling it.

### 8.3.3 Network Architecture

Sensor network architecture can be broadly classified into flat and multi-tiered architectures. In a *flat architecture*, all sensor nodes are generally of the same type and have similar responsibilities. Although, in such a topology, there is no overhead of topology construction, scalability is a major issue because even a small change needs to be propagated to the whole network. In case of multi-tired topology, resource constrained sensor nodes (motes) are present in the lower tire(s), whereas relatively resource rich nodes (Stargate-NetBridge) are present in the higher tire(s) [13]. Such a mote and Stargate-NetBridge based architecture provides increased network capacity, due to the increased ability to perform computationally intensive processing within the network itself. In addition, it also provides increased manageability and scalability.

### 8.3.4 Alternative Performance Metrics

Since WSNs are application driven networks, whose value is mainly in terms of the quality of sensing provided to the application, ideally congestion and flow control mechanisms should target maintaining application level metrics of data quality. In a general scenario, sources react to the congestion in the network by limiting the data generation rates, by scheduling packets or by dropping packets, thereby resulting in reduced data quality measured at the sink. Metrics such as coverage, data freshness, fidelity, and event detection reliability replace conventional network-centric metrics of throughput, delay, and overhead. Multiple events detected in a sensor network could be of different importance levels. Therefore, it is possible to provide mechanisms that give priority to the data with higher importance, when congestion arises.

### 8.3.5 Data Redundancy

Related to application level metrics, but also of broader applicability, is the observation that typically an observation of interest will generally be detected by multiple sensor nodes that are in within sensing range of it. It is important to attempt to reduce this redundancy via data aggregation, or source rate control to reduce the amount of communication in the network [1]. However, should some redundancy remain, some loss of information may be acceptable if it does not compromise the data quality.

## 8.4 Classification of Congestion and Flow Control Approaches

We differentiate the congestion control protocols along several axes that are described in this section.

### 8.4.1 Congestion Detection Mechanism

Congestion detection mechanisms can be classified into *local* congestion detection and *global* congestion detection mechanisms. Local congestion detection typically takes place at intermediate nodes, where congestion detection can be carried out by monitoring local indicators of congestion such as queue occupancy or channel state. On the other hand, global congestion detection is carried out at the sink (base station) where end-to-end attributes such as inter-packet delays and the frequency or distribution of losses can be used to infer congestion, especially how these attributes vary when the source rates are controlled.

### 8.4.2 Congestion Control Goals

Transport protocols employ a closed loop control process where they attempt to control source transmissions to achieve an effective operating point. Some WSN protocols that are based on traditional congestion control approaches consider network centric performance metrics such as end-to-end throughput or reliability. We term such approaches as *network-centric* congestion control approaches. Alternatively, the application-driven nature of WSNs invites other approaches, where the protocol attempts to achieve application-specific metrics of performance, as explained in Sect. 8.2.3. We term such approaches as *application-specific* approaches to congestion control.

### 8.4.3 Rate Control Mechanisms

Rate control mechanisms in WSNs can be broadly classified into *centralized, source-control* mechanism, and distributed, *hop-by-hop backpressure* mechanism. Source-control mechanism is carried out at the data collecting node (sink). Essentially, when congestion (or early sign of congestion) is detected, the sink instructs the sources to adjust their reporting rates. Whereas, hop-by-hop backpressure mechanism is carried out at intermediate nodes, in which the intermediate node instructs its upstream nodes to adjust their reporting rates based on its local congestion state.

### 8.4.4 Fairness and/or QoS

Congestion control represents the base requirement in terms of resource allocation in the presence of congestion. Specifically, congestion control is simply tasked with reducing the sending rates to avoid congestion. More refined resource control, where additional requirements about the resource allocation are maintained, is possible. This includes, for example, approaches that attempt to maintain fairness among the contending flows when congestion arises. In the same vein, QoS approaches attempt to allocate the resources according to the flow importance or reservation levels. We identify protocols that provide some support for these capabilities.

### 8.4.5  Target Application Model

Most protocols focus on the many-to-one model of communication with data being funneled to the base-station. However, some protocols differ on their *assumptions* within this model, such as assuming high-rate flows, or multiple queries to multiple sinks, and others target *different data delivery models*, such as many-to-many communication of events, or one-to-many communication of queries.

### 8.4.6  Other Metrics

We also differentiate between protocols in terms of additional metrics. For example, some protocols require additional support (specialized MAC, or additional network capacity). Moreover, some protocols pay special attention to energy efficiency.

## 8.5  Congestion and Flow Control for Many-to-One Traffic in WSNs

In this section, we overview WSN transport protocols that feature congestion control and flow control support. Unlike traditional congestion management algorithms from wired or wireless ad hoc networks, most of the existing protocols in WSNs focus on applications involving many-to-one communication pattern with data being funneled from sensors toward a base station or a query generator. Thus, the approaches discussed in this section mainly concentrate on many-to-one traffic pattern in WSNs. We first describe network-centric protocols, followed by application-specific protocols, and finally an approach that supports both network-centric as well as application-specific mechanisms.

### 8.5.1  Network-Centric Approaches

As described in the classification approach given in Sect. 8.4, network-centric congestion control approaches focus on traditional network centric performance metrics such as end-to-end throughput or reliability. In this section, we describe such approaches tailored toward WSNs.

#### 8.5.1.1 CODA: Energy Efficient Congestion Detection and Avoidance

Congestion detection and avoidance in sensor networks (CODA) [9] differs from classical approaches in attempting to conserve energy and in its focus on both transient and persistent hotspots and to regulate the individual sensor nodes generating higher data rates than others. As per the classification approach, CODA uses *local* congestion detection mechanism and *hop-by-hop backpressure* as well as *centralized, source control* as rate control mechanisms. More specifically, CODA consists of three mechanisms, receiver-based congestion detection, open-loop hop-by-hop backpressure, and closed-loop multi-source regulation. Receiver-based congestion detection mechanism can be used at sink as well as at intermediate nodes, and it is based on explicit congestion detection mechanism such as queue occupancy and channel state at the receiver. Once congestion is detected, if it is transient, the open-loop hop-by-hop backpressure technique is used to quickly mitigate it. However, if the congestion is persistent, rate control at the sources becomes necessary and the closed-loop multi-source regulation mechanism is applied. Thus, these two control mechanisms, when used in concert, complement each other. We now describe the three mechanisms in more detail.

1. *Receiver-based congestion detection*: Buffer occupancy has been extensively used in traditional congestion detection algorithms as a measure of congestion level. In this paper, the authors demonstrate that buffer occupancy alone is not a good measure of congestion in wireless networks because of the shared nature of the channel. In CODA, receivers not only monitor buffer occupancy, but also measure present and past channel utilization conditions to detect congestion. At the time of congestion, channel load generally increases at a much faster pace than buffer occupancy. In other words, the number of packets received successfully by a receiver is usually less than the number of packets transmitted by a sender, because of interference on the path. Thus, dropped buffer occupancy at the sender can give false information about the state of congestion in the network. However, continuous listening incurs high energy cost. Therefore, CODA uses a sampling scheme that activates local channel monitoring only under certain conditions, for example only when the send buffer is not empty, in order to save energy.
2. *Open-loop hop-by-hop backpressure*: When a receiver detects congestion, it sends backpressure signals toward the sources while the congestion state persists. The backpressure could propagate all the way to sources, or reach only intermediate nodes depending on their local congestion state. Routing protocols can also take advantage of the backpressure information to guide routing decisions and select better, non-congested paths for communication.
3. *Closed-loop multisource regulation*: CODA runs closed-loop congestion control mechanism on the sink to regulate multiple sources, in the case of persistent congestion. Essentially, when the transmission rate of a source exceeds maximum theoretical throughput ($S_{max}$), the source informs the sink by setting a bit in every packet that it transmits to the sink, as long as the transmission rate remains

higher than $S_{max}$. In response, sink starts sending ACKs to the source until the sink detects congestion. When the sink detects congestion, it stops sending ACKs until the congestion is mitigated, to implicitly notify the sender to drop its rate.

### 8.5.1.2   Fusion: Awareness of Packet Priority

Although CODA supports congestion mitigation, it does not provide fairness guarantees among sources. Fusion [3] addresses this problem and also supports prioritized MAC – a mechanism to drain queues of congested nodes quickly. Similar to CODA, Fusion also uses *hop-by-hop backpressure* mechanism for rate control. In addition, Fusion uses local congestion detection approach, however unlike CODA, the designers of the protocol show experimentally that congestion detection based on queue occupancy consistently outperforms that based on channel sampling. We now describe the approaches of Fusion in detail.

1. *Hop-by-hop flow control*: This mechanism is similar to that of CODA, with the exception that Fusion uses an implicit mechanism rather than the explicit messaging used by CODA. Specifically, nodes snoop on packets sent by their parent to check whether the congestion bit is set, in which case they throttle their transmissions, allowing the parent to come out of the congested state. This hop-by-hop backpressure can reach source nodes if the congestion is persistent. However, completely stopping children from transmitting packets, upon detection of congestion at parent node, can block further backpressure propagation toward source nodes. To avoid this problem, each child is allowed to transmit one packet having the congestion bit set, thereby allowing its children to overhear about the congestion.
2. *Limiting source rates*: This mechanism addresses an important problem, in which packets originated from distant sources get dropped near the sink due to congestion. To handle this problem, a passive snoop-based approach is used. Each sensor listens to the traffic its parent forwards to determine total number of nodes, $N$, transmitting packets through the parent. When parent transmits $N$ packets, each child takes one token. Each child is allowed to transmit as long as it has at least one token and each transmission costs one token. This simple token-based approach allows each sensor to match its transmission rate with the rate of its descendants. For this scheme to work, it is assumed that all sensors offer same traffic load, and routing tree is not significantly skewed.
3. *Prioritized MAC*: The CSMA MAC layer gives equal chances of transmissions to all sensor nodes. This is problematic especially in case of congestion scenarios, where a sensor node, serving as a parent to many source sensor nodes, tends to drop packets if its internal queues become full. Therefore, it becomes important to give preference to an already congested parent over its children in accessing the wireless medium. This problem is solved by using technique proposed by Aad and Castelluccia [14], in which the length of each sensor's randomized backoff becomes a function of its local congestion state. Therefore, backoff interval of a

congested sensor is set to one-fourth of the backoff interval of a non-congested sensor, thereby increasing the chances of congested sensor in gaining access to the wireless medium.

In general, the hop-by-hop flow control mechanism can throttle the transmissions at any link in the network, whereas the rate limiting technique can provide fairness to data generated by each source. In addition, the prioritized MAC also provides fairness by prioritizing the in-transit traffic. Together, these strategies complement each other, thereby achieving high level of efficiency even after network reaches saturation.

### 8.5.1.3 Distributed Rate-Control with Fairness

Ee et al. [15] present another approach that addresses the need to support fairness guarantees in addition to the congestion mitigation mechanism described in CODA. Similar to CODA, it uses *local* congestion detection mechanism, which is based on monitoring queue occupancy. In addition, the rate control algorithm is based on *hop-by-hop backpressure* mechanism. To employ *fairness*, each node assigns fair transmission rates to its upstream nodes. We now describe the basic rate-control and fairness mechanisms discussed in this work.

The basic congestion control mechanism is a closed-loop control algorithm, in which each node applies back pressure on its upstream nodes, when the node's queue is full or about to become full. Similarly, when the queue becomes empty, the node disseminates a higher reporting rate to its upstream nodes. To avoid interference incurred due to simultaneous transmissions of nodes at the same level, some jitter is introduced. The congestion control algorithm involves following steps:

1. *Measure average packet transmission rate r*: Assuming equal sized packets, the packet transmission rate can be estimated as the inverse of the time required to transmit one packet. The packet transmission time, $t$, is measured from the time when the transport layer first sends the packet to the network layer to the time when network layer notifies the transport layer that the packet has been transmitted. The estimated packet transmission time is tracked using an exponential moving average of $t$.
2. *Assign appropriate packet generation rate to upstream nodes*: The average packet transmission rate is divided among all upstream nodes, $n$, to assign data packet generation rate as

$$r_{\text{data}} = \frac{r}{n}. \tag{8.2}$$

To calculate $n$, a simple bottom-up propagation approach is used, in which each node embeds its subtree size in a packet and sends it to the parent. Parent retrieves the subtree counts of all children, adds one to it (if parent itself is generating data) and embeds the total count in the packet that it forwards further toward sink. When the queues are overflowing or about to overflow, the node assigns a lower packet generation rate to its upstream nodes.

3. *Compare the rate* $r_{data}$ *with the rate* $r_{data,parent}$ *obtained from parent, and propagate smaller rate to the upstream nodes:* $r_{data,parent}$ can either be piggy-backed by the parent in the packets it transmits or sent separately as a control message.

The fairness control mechanism uses either *probabilistic selection* or *epoch-based proportional selection*. In first, the child with a larger subtree size will get higher probability of its queue getting selected than the other children. While in second, within each epoch, the number of packets transmitted from each queue is a product of $n$ and the number of nodes serviced by that queue.

### 8.5.1.4 IFRC: Interference-Aware Fair Rate Control

IFRC [16] differs from the previous approaches in that it presents an interference aware approach to congestion control. IFRC uses *local* congestion detection algorithm, which is based on monitoring queue occupancy, and it also supports *fairness* in the network. Although, the work of Ee et al. [15] and IFRC focus on rate control with fairness guarantee, IFRC differs from the former in that in IFRC, a node controls the rates of all nodes, whose transmissions interfere with its transmissions, instead of controlling only its children. The main contributions of IFRC include identifying the set of all nodes that affect the rate at which the congested node is transmitting data (such flows may or may not be traversing through the congested node), and designing a low-overhead yet efficient mechanism to share congestion information to the set of such nodes. We now overview IFRC in more detail.

IFRC assumes that the CSMA-based MAC protocol is used and the MAC layer provides link-layer retransmissions to recover from packet losses. In addition, IFRC also assumes that a link-quality based routing protocol is used and it builds a tree structure having base station as a root. Finally, each node is assumed to be generating only one flow. IFRC considers that a node $n_j$ is a *potential interferer* of node $n_i$, if the flow originating at node $n_j$ uses a link that interferes with the traffic sent by node $n_i$. Thus, for many-to-one traffic, a set of potential interferers of node $n_i$ include neighbors of $n_i$, neighbors of parent of $n_i$, and subtrees of the parent as well as neighbors of the parent. We now present three main phases of the IFRC design:

1. *Congestion detection*: IFRC uses exponentially weighted moving average of queue occupancy as a measure of congestion.

$$q_{avg} = (1 - w_q) \, q_{avg} + w_q q_{inst} \tag{8.3}$$

IFRC detects oncoming congestion when the queue length crosses upper threshold, $U$. Upon detection of congestion, IFRC halves its data generation rate, $r_i$, then starts increasing it additively. Since, using a single upper threshold may still keep the node in congested state, even after halving its rate, the authors propose the use of multiple thresholds, $U(k)$, such that $U(k) - U(k-1)$ decreases as $k$ starts increasing. This allows the node to continue cutting its rate aggressively until its queue starts to drain.

2. *Congestion sharing*: In IFRC, a node transmitting data piggybacks the rate and congestion state of itself as well as its most congested child. This information is shared recursively in the network with the help of snooping. Thus, the IFRC's goal of assigning at least the most congested fair share rate to each flow can be achieved with the help of following rules:

   **Rule 1:** rate of a node should be less the rate of its parent.

   **Rule 2:** rate of a node should be less than the rate of its congested neighbor or the congested child of the neighbor.

3. *Rate adaptation*: Rate adaptation in IFRC is based on additive increase multiplicative decrease (AIMD) principle. At every inter-packet transmission time ($\frac{1}{rate}$) of node $n_i$, the rate is increased by $\frac{\delta}{rate}$, where $\delta$ represents intensity of additive increase. Similarly, the rate is halved (multiplicative decrease), when congestion is detected.

   To avoid *rate* jumping from $rate_{min}$ to $rate_{max}$ in one step, we require $\frac{\delta}{rate} \ll rate_{min}$ i.e.,

$$\delta = \varepsilon\, rate_{min}^2, \tag{8.4}$$

where $\varepsilon$ is a small positive number, whose value for small and sparse network is analytically determined to be:

$$\varepsilon < \frac{F_j}{8U_0} \tag{8.5}$$

and for large network as

$$\varepsilon < \frac{9U_1}{2s^{-2}F_j}, \tag{8.6}$$

where $U_0$ and $U_1$ are queue thresholds, $s$ is the average depth of the tree, and $F_j$ is a function of topology and the network size.

### 8.5.1.5  RCRT: Congestion Control with End-to-End Reliability

The congestion control approaches discussed so far do not support reliable data transport. Although 100% reliability is not a concern for some sensor network applications, especially in which colocated sensors collect redundant information [1], there is a class of sensor network applications that requires 100% reliability. For instance, in structural monitoring applications, structural mode shape estimation can be done by correlating readings taken from multiple sensors. The loss of samples can result in inaccurate estimations. To address this problem, RCRT [17] supports end-to-end reliability in addition to the rate control mechanism. RCRT focuses on applications involving *high-rate* data communications requiring 100% *reliability*. RCRT uses *global* congestion detection mechanism at sink, which is based on the variation in RTT between the sink and the source. In addition, RCRT's rate control mechanism is *centralized, source-control*-based approach. The rate control mechanism is implemented at the sink because the sink has a comprehensive view of the state of the network, which makes it possible to use more efficient source rate allocation. In comparison with IFRC, RCRT supports multiple concurrent streams from each sensor node as well as different rate allocation policies for different flows based on their demands. RCRT is shown to achieve more than twice the rate achieved by IFRC. We now explain different mechanisms supported by RCRT.

1. *End-to-end reliability*: RCRT supports 100% reliability using a standard end-to-end NACK-based feedback mechanism. Essentially, the sink maintains a list of missing packets per flow (here, flow represents the data transferred between a source and a sink), and sends a list of missing packets to each source for recovery. The sink also maintains out-of-order packets list per flow to support in-order delivery.

2. *Congestion detection*: RCRT uses an implicit congestion detection mechanism in which the sink assumes that the network is uncongested as long as the time taken for loss-repairs is less than a certain threshold. More specifically, number of RTTs required to recover a loss belonging to the flow from source $i$ is:

$$L_{norm,i} = \frac{L_i}{r_i RTT_i},  \tag{8.7}$$

where, $L_i$ is the length of source $i$'s out-of-order packet list, and $r_i$ is the rate assigned to source $i$. RCRT detects congestion if the exponentially weighted moving average of $L_{norm,i}$, denoted by $C_i$, is greater than an upper threshold $U$. Ideally, a loss should be repaired in approximately one RTT time ($Ci = 1$). Thus, when $C_i > 2$, the network is more likely congested. However, RCRT uses a more conservative value for the upper threshold ($U = 4$) because $C_i$ increases significantly when the network moves from uncongested to congested state. The lower threshold, $L$, is set to be 1.

3. *Rate adaptation*: RCRT uses an AIMD approach to control overall rate, $R(t)$ which is nothing but $\sum r_i(t)$. When network is not congested, it increases $R(t)$ additively:

$$R(t+1) = R(t) + A,  \tag{8.8}$$

where $A$ is a constant. Similarly, when network is congested, the rate is decreased multiplicatively:

$$R(t+1) = M(t)R(t),  \tag{8.9}$$

where $M(t)$ is a time-dependent multiplicative decrease factor. RCRT uses conservative approach for determining when to decrease the rate; after rate adjustment, sink waits for at least $2RTT_i$ time to get the feedback of the rate change. In addition, RCRT uses a better way to determine value of $M$ than just assuming it to be a constant as in $M = 0.5$ with TCP. Essentially, $M(t)$ is computed based on the loss rate experienced by $f_i$. If packet delivery ratio of $f_i$ is $p_i$ then the expected amount of traffic between source $i$ and sink is $\frac{r_i(1-p_i)}{p_i}$, including the traffic for losses. Thus, when a flow in the network is congested, $f_i$'s rate is adjusted such that the total amount of traffic from source $i$ is $r_i$. However, when a single flow is congested, RCRT conservatively adjusts the overall rate $R(t)$ by setting $M(t)$ to be:

$$M(t) = \frac{p_i(t)}{2 - p_i(t)}. \tag{8.10}$$

4. *Rate allocation*: RCRT allocates rates $r_i(t)$ to each flow based on the rate allocation policy $P$. RCRT currently supports three rate allocation policies: demand-proportional or weighted rate assignment, demand-limited in which overall rate is equally divided among all sources provided that no source gets more rate than it has demanded, and fair (equal) rate allocation policy.

## 8.5.2 Application-Specific Approaches

As described in Sect. 8.2.3, the reliability requirements in wireless sensor networks are application specific, and fundamentally different from network-centric reliability considerations present in typical transport protocols. WSNs involve application driven protocols where success is defined relative to the network's mission. Thus, in a typical data collection-based application, where events are detected by multiple colocated sensors and sent to the base station, the user is interested in reliably detecting all events, rather than reliably receiving every packet (including those with redundant information). The inherent redundancy typically present in sensor networks makes applications tolerant to some packet losses. Similarly, user is interested in knowing critical events as quickly as possible, when compared with the events with least importance. In this section, we describe WSN protocols that take into account these application-specific requirements.

### 8.5.2.1 ESRT: Application Specific, Centralized Rate Control

The event-to-sink reliable transport (ESRT) [18] protocol focuses on the problem of adjusting the reporting rates of sources to achieve required event reliability at the sink, with minimum resource utilization. According to the classification approach, ESRT uses *local* congestion detection mechanism. In addition, the rate control mechanism of ESRT is *centralized, source control* based approach. Essentially, ESRT uses a closed-loop congestion control mechanism, in which processing is done mainly at sink. ESRT assumes that sink nodes are powerful enough to reach all source nodes in the network by broadcast. The key idea in ESRT is that the sink instructs source nodes to adjust their reporting frequency based on the current reliability measure at the sink and the state of congestion in the network. On the one hand, when the reliability measure at the sink is lower than required, if there is no congestion in the network then the sink instructs sources to increase their reporting rate aggressively, whereas if there is congestion in the network, sink instructs sources to reduce their reporting rate conservatively. On the other hand, if the reliability measure at sink is higher than required then sink instructs sources to reduce their reporting rates to save energy.

ESRT tracks two parameters: (1) the reliability indicator, $\eta$, computed by sink; and (2) the current state of congestion. The sink computes $\eta_i$ for period $i$ as follows

$$\eta_i = \frac{r_i}{R_i}, \tag{8.11}$$

where $r_i$ is the observed event reliability and $R_i$ is the desired event reliability at the sink. To inform the sink about current state of congestion, each sensor node monitors its queue size and sets the congestion bit in the packet going to sink if the next chunk of source data is capable of causing buffer overflow at the node.

The actions in ESRT are based on the observation that increasing the data reporting rate under congestion may actually reduce the data quality. Specifically, Tilak et al. [1] make the observation that under congestion, packets are dropped without discrimination between which packets got dropped and which got forwarded. This may lead to lower data quality than a case with lower reporting rate where fewer packets are dropped and the sources receive better representation. Using these parameters, the sink node categorizes the current network state into five different states, as shown in Fig. 8.3, where the states have the following interpretations

1. *No Congestion, Low Reliability (NC, LR)*: Sources multiplicatively increase their reporting rates to raise reliability.
2. *No Congestion, High Reliability (NC, HR)*: Network is not congested, but observed reliability is higher than the desired reliability. Therefore, sink instructs source nodes to reduce the reporting rate cautiously, to always maintain the required reliability but lower overhead.

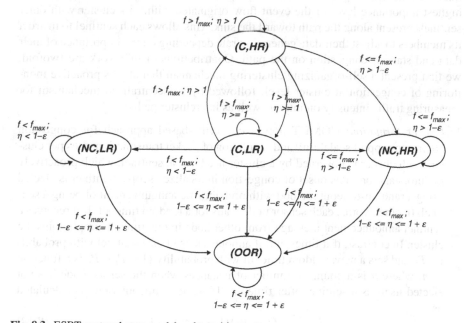

**Fig. 8.3** ESRT protocol state model and transitions

3. *Congestion, High Reliability (C, HR)*: Network is congested and reliability is higher than desired reliability. Therefore, reduce the reporting rates multiplicatively until congestion is resolved or reliability drops below the desired level.
4. *Congestion, Low Reliability (C, LR)*: This is the worst possible state, in which ESRT exponentially reduces the reporting frequency to relieve the congestion and potentially improve reliability.
5. *Optimal Operating Region (OOR)*: This is the optimal operating region where the reporting rate is just sufficient to meet the desired reliability. More precisely, $1 - \varepsilon \le \eta_i \le 1 + \varepsilon$, where $\varepsilon$ is a small error margin used to provide stability. The goal of ESRT is to always maintain the network state in OOR.

### 8.5.2.2   COMUT: Congestion Control Based on Importance of Data

In event-monitoring sensor networks, some events are of greater importance than others. Thus, event flows with different importance levels need different ways of handling to make sure that flows with greater importance will be delivered to the sink with higher fidelity and timeliness than the flows with lower importance. COMUT (congestion control for multi-class traffic) [19] takes into account this important observation and provides a cluster-based congestion control mechanism. The congestion detection mechanism of COMUT is *local*, which is based on the node's current queue occupancy. In addition, the rate control mechanism of COMUT is a *centralized, source-control* based approach, which is run at the cluster-head (sentinel) level. Essentially, each node notifies its local congestion level to the sentinel. Sentinel exchanges the collective congestion level of its cluster along with the highest importance level of the event-flow originated within its cluster with other sentinels present along the path toward the sink. This allows each sentinel to instruct its members to adjust their data generating rate depending on the importance of their data and state of congestion on the path. Contributions of this work are twofold, we first present a self-organizing clustering mechanism that allows proactive monitoring of congestion at cluster level, followed by a decentralized mechanism for measuring traffic intensity on intra as well as intercluster paths.

1. *Cluster formation*: COMUT uses a clustering-based approach for congestion control to support scalability and regulation of packet transmission rates per cluster. Each cluster is governed by a cluster head, a.k.a. sentinel, which proactively monitors and predicts onset of congestion in localized scope. Sentinel is elected using a random-timeout method with probabilistic announcement of being a sentinel. In other words, each sensor node waits for a random timeout. If it receives a sentinel announcement message from other node in that time, it simply joins the cluster. In contrast, if it times out, it announces itself as a sentinel with probability $P_n$ and sets a new, random timeout with probability $(1 - P_n)$. $P_n$ is a function of $n$, where $n$ is a count of number of instances when the sensor node has not elected itself as a sentinel after timeout. Thus, next timeout value is calculated as:

$$P_{n+1} = (1 - P_n)(1 - e^{-\alpha n}) + P_n, \tag{8.12}$$

where alpha represents effective degree of increase of $P_n$ with $n$. This is to make sure that with small number of iterative steps, each sensor node can either become a sentinel or a member of a cluster with neighboring node as a sentinel. To facilitate cluster formation, COMUT uses zone routing protocol (ZRP) [20].

2. *Traffic intensity estimation and rate regulation*: Each sensor monitors its local queue load and reports load to the sentinel in fixed time intervals. Sentinel calculates a collective estimate of load for the cluster by using the local readings from each cluster-member. These collective estimates are then communicated among sentinels over the entire path from the set of source nodes to sink to obtain aggregate estimate of the traffic intensity, a.k.a. congestion level, for that path. Sentinel periodically forwards the locally computed value of congestion level as well as the level of highest important flow observed in its cluster to other sentinels on the path toward source nodes. A threshold value can be calculated analytically, beyond which a cluster can be considered as congested. Thus, to save energy, congestion level can be forwarded only when it is above the threshold, instead of forwarding it periodically. This information is useful in adjusting packet generation rate at the source nodes. The proposed rate regulation policy is based on additive increase and multiplicative decrease (AIMD) policy. However, to support multiple classes of importance levels, the regulation policy is modified such that if the congestion level is above the threshold or if there exists a flow with higher importance along the path to the sink then the generation rate is dropped to some minimum rate for the less important flows, where the value of minimum rate can be measured experimentally or via simulations, and can be set before the deployment of sensors.

During the initialization phase (at network setup time), flooding event flows to sentinels can waste energy and create congestion, before giving a chance to sentinels to estimate the congestion level and allow sources to initiate rate regulation. To solve this problem, COMUT uses *slow start* mechanism, which assigns a number of Rate Regulation Epoch (RRE) intervals that a sensor should wait before it increases its rate. Thus, in addition to handling early congestion scenarios, COMUT favors flows with higher importance by assigning them a smaller waiting interval.

## 8.5.3  Hybrid Approach

In this section, we present Siphon [21], which provides a hybrid approach that supports traditional *network-centric* as well as *application-specific* congestion control mechanisms. Previously discussed congestion control schemes try to avoid or mitigate congestion either by limiting the data generation rates at sources, or by dropping packets, resulting in reduced overall application fidelity at sink. Thus, the main goal of Siphon is to maintain application fidelity measured at sink, even in case of congestion collapse, where application fidelity can be as simple as events/sec or complex (based on application specific metrics). Siphon uses *local* congestion detection mechanism similar to CODA to support network-centric congestion control initiated at node level, and *global* congestion detection mechanism to initiate application-specific congestion control at the base station. The main contribution of Siphon lies in its novel congestion control mechanism, which is based on the use of additional network capacity. Essentially, a small number of multi-radio virtual sinks (VS) are randomly distributed across the sensor field, which take the traffic load off the already loaded sensors, on the onset of congestion, and move it to the sink through their high capacity secondary radio network. Siphon uses stargate nodes as virtual sinks, which are equipped with a primary low-power mote radio, and a secondary long-range (e.g., IEEE 802.11) radio. Now we briefly describe the algorithms for design level policies of Siphon.

1. *VS discovery and visibility scope control*: Similar to the deployment of new services, Siphon may be deployed in an incremental fashion. As a result, the sink might not be always equipped with a secondary radio, consequently reducing the chances of having secondary network rooted at sink. In addition, there is no guarantee that VS will always be present near a congested node, making it imperative for a congested node to discover for a nearby VS. To facilitate VS discovery, sink broadcasts VS-discovery packet embedding signature byte, which contains VS-TTL (hop-count) set to either $l$ if sink is on the secondary network or NULL otherwise. Authors experimentally prove that an optimal value for $l$ is 2. When a VS receives VS-discovery message, it marks the forwarder of this packet as a next Siphon hop. If this packet is received from a secondary-radio, VS forwards the packet with signature byte embedded and VS-TTL set to $l$, over both the radio links. Thus, non-VS nodes receiving control packets with signature byte and VS-TTL greater than zero, add the VS in their neighbor list along with the path to the VS. Note that, if VS receives packet forwarded by a non-VS node then VS does not announce its presence to its neighbors, as there exist no path to the sink over secondary radio channel. In other words, such an VS ends up forwarding packets on the same path that would otherwise have been taken by original propagation funnels towards the sink, resulting in congestion near sink.

2. *Congestion detection*: Siphon uses mechanisms proposed in CODA [9] to detect local congestion level at a node. Thus, when congestion is detected, VS takes off the traffic from the overloaded nodes and redirects it over the secondary network to the sink. However, the important question is when to redirect traffic. If traffic is redirected as soon as there is a possibility of congestion, it diminishes the possibility of aggregation, which can be better performed later in the funnel, and vice versa. Thus, to strike a balance, it is most beneficial to redirect traffic just before the possibility of congestion to occur in the funnel. Another approach for detecting congestion in the network is a post-facto congestion detection, in which sink initiates VS-based redirection depending on measured application fidelity and event data quality. Although, this approach is not good for detecting transient congestion occurring deep in the network, it can detect congestion occurred close to the sink. In addition, it obviates the need for running congestion detection algorithms on low-power motes. Further, it has an advantage of avoiding early traffic redirection when network-wide aggregation is used.

3. *Traffic redirection*: A sensor forwards a packet to neighbor VS if the packet has redirection bit set. Siphon uses two approaches for setting the redirection bit: *on-demand redirection*, in which the redirection bit is set if congestion is detected, and *always-on redirection*, in which the redirection bit is always set. When VS receives the packet, it forwards it to the next hop toward the sink, where next hop could be either on the primary or the secondary-radio path. As a general guideline for traffic redirection, the next hop on the redirected path should have link estimation that is within 15% (lower-bound) of the link estimation of the next hop on regular path.

4. *Congestion in the secondary network*: Siphon uses congestion detection scheme for secondary network as proposed by Murty et al. [22]. A VS does not advertise its existence when it detects congestion on both primary and secondary networks. In such case, hop-by-hop backpressure mechanism proposed in CODA [9] is used. However, authors argue that there are less chances for VSs to become congested because they can communicate using two radios, on different channels, at the same time.

## 8.6 Reliability Requirements in WSNs

Typically, the transport protocol is also tasked with maintaining end-to-end reliability. In this section, we discuss the transport protocols that focus mainly on reliability ensuring issues in WSNs, with little or no attention given to the congestion control issues.

As discussed in Sect. 8.2, an important class of traffic in sensor networks is many-to-one flows; these flows typically arise when multiple sensor nodes monitoring for a phenomenon jointly, report their readings to a base station for further processing. However, other classes of traffic also arise in WSN, including one-to-one and one-to-many communication. For instance, in a data-centric storage system (for example GHT [23]), sensed data at one sensor node needs to be stored at some other sensor node, depending on the key associated with the data. In such a scenario, there is one-to-one communication between the sensing node and the node responsible for its storage. In such applications, if there is little redundancy in the collected samples, end-to-end reliability ensuring mechanisms [24, 25] are necessary. However, a standard protocol such as TCP, which supports end-to-end reliability, may not be a good choice for sensor networks because the large header of TCP can be an overhead for resource constrained sensor nodes. In addition, TCP assumes smart sender and simple receiver, which does not fit into the sensor network model, where the sender needs to be simple and the receiver (base station) can be complex. In addition, if data are redundant, then absolute reliability that is provided by TCP is wasteful, as long as sufficient information is received by the base station to achieve the desired monitoring quality.

Similarly, when reprogramming the sensor nodes from the sink, a one-to-many delay-tolerant communication pattern is observed. Reliability in such application is particularly important because even a loss of a single packet out of the whole program image can fail the purpose [26]. Queries in a sensor network often are generated from one node, but forwarded throughout the network (or some regions of it); again, this is an one-to-many pattern. Finally, to handle applications that may include all types of traffic flows, generic reliability guaranteeing mechanisms should be used.

### 8.6.1  RMST: A Transport Layer Over Directed Diffusion Routing Protocol [27]

RMST [24] is a transport layer protocol, which aims to provide guaranteed delivery and support for fragmentation/reassembly. To provide reliability, it uses a selective NACK-based protocol at either the receiving node or an intermediate node. Fast recovery in response to a NACK is achieved by using the local caches of the nodes along the path toward the source. If an intermediate node happens to have a copy of the packet required by the NACK, the retransmission request will be answered by the intermediate node itself, thereby reducing the overhead of end-to-end retransmissions.

### 8.6.2  RMBTS: Reliability for Block Data Transfers [25]

Reliability in WSNs is typically associated with the transfer of small amounts of data, such as low frequency sensor readings or event detections [28, 29]. However, there is a class of important applications in WSNs that requires bulk data transfer. For instance, in acoustic beamforming application, where raw data needs to be transferred to a centralized location, or applications involving a network of image sensors, where nonredundant image data are transferred to the base station, a service for reliable bulk data communication is required.

To provide reliability, the option of using redundant messages can help if the bandwidth is available. However, this approach degrades the throughput significantly in case of mass data transfer. In contrast, additional cost of using control messages to provide reliability is acceptable in case of longer packet bursts. For this reason, RMBTS uses a reliable MAC (using RTS/CTS and acknowledgments). Since a reliable MAC can greatly reduce packet loss due to transmission errors and collisions, a NACK-based end-to-end retransmission mechanism can be used to recover the remaining lost packets. In addition, a link monitor service is used on each node to collect statistics for the links connecting to the neighboring nodes by sending periodic ping messages and keeping track of the *reply counts* from each neighbor. This information can then be used by the node to select a better parent in the process of building a spanning tree having the base station as a root.

### 8.6.3   RBC: Reliability for Many-to-One, Bursty Traffic [30]

In this work, authors focus on bursty convergecast (many-to-one, bursty traffic) in wireless sensor networks, and show that the commonly used hop-by-hop packet recovery mechanisms, such as synchronous explicit ACK (SEA) and stop-and-wait implicit ACK (SWIA), do not result in performance improvement for this scenario. SEA, in which each packet's reception is informed by an immediate explicit ACK, suffers from the problems of channel contention due to unscheduled retransmission and ACK-losses, whereas SWIA, in which ACK is piggybacked on the packet that is transmitted by the next hop (snooping based approach), suffers from the problem of increased loss probability of piggybacked ACKs due to the larger sizes of the packets (resulting in unnecessary retransmissions), channel under-utilization due to in-order delivery constraint, and lack of retransmission scheduling.

To handle these problems, RBC uses window-less block acknowledgment scheme, which alleviates the need of in-order delivery with the assumption that the data packets are timestamped; at the same time, it reduces the ACK-loss probability by using block acknowledgments. To schedule retransmissions, RBC introduces differentiated contention control, which gives higher priority channel access to the packets that have been transmitted less number of times. When multiple nodes have packets of the same priority to transmit, preference is given to the node having more queued packets. Finally, RBC also uses fine-grained timer management to deal with varying ACK-delays resulting due to the changing network state and to expedite retransmissions of lost packets.

### 8.6.4   PSFQ: Reliability for One-to-Many Traffic Pattern [26]

PSFQ is a reliable transport protocol suitable for one-to-many routing. An important application in the context of sensor networks is reprogramming the sensor nodes based on application requirements. Reliability in such application is particularly important because even a loss of a single packet out of the whole program image can fail the purpose. The key idea behind the design of PSFQ is that the source node pumps data at slower pace, whereas receiver does the local recovery quickly by fetching the lost segments from neighbors aggressively.

The PSFQ algorithm consists of three steps: *(i) pump operation*: before transmitting data, the source waits for a period of time uniformly distributed within some bounds $t_{min}$ and $t_{max}$. Choosing an appropriate value of $t_{min}$ is important because it determines the local recovery time at the receiver and allows it to reduce redundant broadcasts. *(ii) fetch operation*: Upon detection of a sequence number gap, the receiver can initiate quick fetch operation by sending NACKs to the neighboring nodes. Intermediate nodes have local caching enabled to support quick recovery of the lost segments at receiver. *(iii) report operation*: it is necessary to inform the source about successful transmission in order to collect statistics of dissemination status as well as to allow sources to free up delivered segments. Reporting starts from the farthest target node, allowing the intermediate nodes to piggyback their reports, thereby achieving aggregation along the way.

### 8.6.5   STCP: Reliability for Hybrid Traffic Pattern [31]

STCP is a transport layer protocol, which supports heterogeneous applications, such as continuous flow or event-driven applications, and provides reliability and congestion control services for them. To provide reliability for continuous flow applications, it takes advantage of the base station's knowledge of the inter-arrival time between packets to implement NACK-based reliability. On the other hand, the base station is unaware of the expected arrival time of the next packet for event-based applications, for which STCP uses ACK-based reliability. In addition, STCP also supports handling different flows with different reliability requirements. For instance, the base station will not send NACKs for missing packets if the observed reliability is greater than the desired reliability value [1]. In case of data centric applications, since the number of sources detecting an event can be very high, STCP does not provide acknowledgment-based reliability support, since acknowledging each source can deplete network resources and energy. In contrast, they assume that the data from multiple sources are loss-tolerant due to the inherent redundancy or correlation in the data.

STCP also supports congestion control, in which intermediate nodes probabilistically set a congestion bit depending on their current buffer occupancy; this is a form of explicit congestion notification (ECN). Thus, upon receiving packets with congestion bit set, the sink informs the source(s) to take necessary measures, such as reducing the data reporting rate, or choosing another, noncongested path.

## 8.7   Other Related Works

In this section, we describe other related efforts that contribute to congestion control. Specifically, we describe protocols to avoid congestion in networks having one-to-one and one-to-many data delivery models, followed by a MAC protocol that supports rate control for many-to-one traffic in WSNs.

### 8.7.1   One-to-One Protocols

In this section, we describe Flush [32] – a pipelining-based approach to congestion control for multihop one-to-one communication. According to the classification scheme presented in Sect. 8.4, Flush is a *network-centric* approach to congestion avoidance and provides end-to-end *reliability*. In terms of target application model,

Flush is specially suited for *bulk data transfer* in WSNs. Similar to IFRC [16], Flush uses *local* congestion detection mechanism, which is based on the queue occupancy. However, unlike IFRC, which focuses on determining a set of potential interferers, Flush specifically focuses on intrapath interference (or self-interference). Essentially, *intrapath interference* occurs when transmission of the same packet by a successor node interferes with the reception of the next packet from the predecessor node, whereas *interpath interference* occurs when two or more flows interfere with each other. In addition, Flush proposes a novel rate control policy (based on stable rate estimates), which results in increased overall efficiency in comparison with IFRC, because the AIMD policy of IFRC can be less efficient due to the typical saw-tooth pattern of AIMD. We now describe the mechanisms presented in Flush in more detail.

Flush assumes that different flows do not interfere with each other, thereby ignoring the need to focus on interpath interference. In addition, Flush also assumes that: a node can snoop on the single-hop packets intended for other receivers, the link layer can provide efficient single-hop acknowledgments, and there exist underlying best-effort routing and delivery mechanisms to support packet-forwarding toward the data sink and from the data sink to the data sources, respectively. When the sink sends a query to a specific source, requesting for a data object, Flush moves through four phases: In *topology query phase*, sink computes timeout at the sink by measuring RTT with respect to the source. In *data transfer phase*, source sends data to the sink with maximum possible rate that will not cause congestion along the path. The *acknowledgment phase* begins after the source finishes its data transfer, in which the sink requests for retransmissions by supplying NACKs. Finally, in the *integrity-check phase*, i.e., upon receiving the whole data, the sink checks for the integrity of the data and sends a fresh request if the integrity-check fails. We now give an overview of contributions of Flush to achieve two important goals, reliable delivery and minimizing transfer time.

### 8.7.1.1  Reliability Protocol

During the data transfer from a source to a sink, some packets may be lost due to retransmission failures or queue overflows. When the sink believes that the source has finished sending data, the sink sends NACKs to the source, where each NACK packet contains at most three missing sequence numbers. Instead of sending a series of NACK packets containing all missing sequence number, Flush simplifies the algorithm by sending a single NACK packet every time, until all packets are successfully received by the sink.

### 8.7.1.2  Dynamic Rate Control Mechanism

Flush proposes a simple pipelining model for transferring data over a set of linearly connected nodes (from source to sink): when a node sends data to its successor, it should wait until (i) its successor forwards the packet and (ii) all nodes along the path, which can interfere with its successor, forward the packet. Thus, dynamic rate control algorithm is governed by two rules:

– *Rule1:* A node should transmit only when its successor is interference-free (to support pipelining model)

– *Rule2:* A node's sending rate should not be greater than the sending rate of its successor (to prevent rate-mismatching)

Therefore, after transmitting one packet, the delay for node $i$ to transmit next packet is

$$d_i = \delta_i + (\delta_{i-1} + f_{i-1}),$$ (8.13)

where $\delta_i$ is a time taken by node $i$ to send a packet, and $f_{i-1}$ is a time taken by other nodes to send packets that can interfere with the successor node, $(i - 1)$. To facilitate this calculation, each data packet transmitted from node $(i - 1)$ includes $\delta_{i-1}$ and $f_{i-1}$, which can be obtained by node $i$ by snooping. Node $i$ uses these values to determine the value of $f_i$, which is nothing but the sum of the $\delta$ s of all successors that the node $i$ can hear. As the values $\delta_{i-1}$ and $f_{i-1}$ can change over time, Flush continually estimates and updates $\delta_i$ and $f_i$. To prevent rate mismatch, each node calculates its sending interval as: $D_i = \max(d_i, D_{i-1})$. Again, each node simply includes $D_i$ in its data packet, so that the previous node can learn its value by snooping. To handle congestion at a node, it advertises sending interval by doubling the value of $\delta_i$, when its queue occupancy exceeds a specified threshold.

### 8.7.2 One-to-Many Protocols

One-to-many data delivery model can be typically observed in information/query dissemination applications. We briefly overview some of the protocols in this context.

Broadcast or flooding is a common approach for transferring data in one-to-many fashion. However, this simple approach involves high energy consumption of the network because of packet drops due to collisions at MAC layer and redundant transmissions to certain nodes through different paths (overlap problem). The former problem can be handled by using gossiping-based approaches [33], where the key idea is to forward the packet only to a randomly chosen neighbor, instead of all the neighbors. However, gossiping fails to handle the overlap problem. Thus, Heinzelman et al. [34] propose sensor protocols for information via negotiation (SPIN) protocols family, which solves the overlap problem in energy efficient manner by having nodes negotiate with each other in order to transfer only *useful* information.

Tilak et al. present a protocol for nonuniform information dissemination in sensor networks [12], where the underlying assumption is that information about an event is important to the nodes that are closer to the event. In other words, nodes that are farther away from the event source can tolerate some loss of information about that event, thus bringing nonuniformity in the information obtained by each node. To make sure that the network transfers only the enough amount of data that is required by the nodes, either deterministic or probabilistic approaches can be used. Essentially, the former approach applies filtering by forwarding only one packet out of $n$ packets, where $n$ is the protocol parameter such that $\frac{1}{n}$ represents the filtering frequency. The later approach probabilistically determines whether to transmit the data or not based on the value of a random number.

We also shed some light on another set of protocols, which have been proposed in the context of wireless ad hoc networks and mobile ad hoc networks (MANETs). In location based algorithm (LBA) [35], a node includes its location information in the packet, so that the receiving node can decide whether its broadcast provides sufficient coverage to be worth sending. In Ad Hoc Broadcast Protocol (ABHP) [36], nodes collect 2-hop neighbor information to explicitly select a set of 1-hop neighbors to rebroadcast the packet such that all 2-hop neighbors are covered. Similarly, in Scalable Broadcast Algorithm (SBA) [37], a node maintains 2-hop information to rebroadcast the packet only if this rebroadcast can cover additional nodes that were not covered by the sender of the packet. For the environments with high transmission error rate, scheme such as Double Covered Broadcast (DCB) [38] can be used. In DCB, forward nodes are selected such that sender's 2-hop neighbors and 1-hop forwarding neighbors are covered, whereas sender's 1-hop nonforwarding neighbors are covered by at least two forwarding neighbors. Hypergossiping [39] is a protocol specifically designed for sparse MANETs, which are more susceptible to network partitioning due to node movements. Hypergossiping tackles this problem by using gossiping algorithm inside partitions and performing broadcast repetition on partition joins.

### 8.7.3   ARC: A MAC Protocol with Support for Adaptive Rate Control

ARC [40] is an energy efficient MAC protocol with an adaptive rate control mechanism, aimed at providing channel access fairness for all the nodes in a WSN. ARC uses carrier sense multiple access (CSMA) but also attempts to reduce collisions by randomizing synchronized traffic (e.g., traffic due to event detection). Specifically, ARC introduces random transmission delays and uses phase shifting in backoff intervals. Here, backoff interval is a time period for which a node waits after contention with the hope of getting access to the channel after the end of the interval. Of more interest to our topic, ARC uses a simple adaptive rate control scheme, which is based on additive increase and multiplicative decrease (AIMD) policy. The rate adaptation decisions are taken independently by each node, depending on whether the parent node is able to successfully forward its packets. To achieve fairness with the self-generated as well as the route-through traffic, a node forwarding route-through traffic of $n$ children allocates the bandwidth for the packets generated by itself as $1/(n + 1)$. Thus, the adaptive rate control scheme coupled with the modified CSMA mechanism provides an effective and energy efficient solution that supports channel access fairness in WSNs.

## 8.8 Directions for Future Work

The survey presented in this chapter points to many interesting future research directions. For example, it would be interesting to implement the congestion management protocols discussed in this chapter on a real testbed. This would allow one to evaluate perfrormance of these protocols in a fair and consistent manner under a broad range of real-world scenarios. For example, one can compare the protocols using various metrics including robustness, energy-efficiency, etc. in real-world seetings where node failures, unfriendly nature of the physical environement are norms rather than exceptions. This will give researchers and end users insight into the behaviors of these protocols and would enable them to select the one that meets their requirements.

In addition, it will be useful to integrate congestion control mechanisms for different data delivery models, and approaches to support fairness and reliability, to evaluate the efficiency of the nearly complete transport protocol for sensor networks.

## 8.9 Summary and Concluding Remarks

Data centric nature and many-to-one data delivery model of WSNs have attracted many researchers to focus on congestion management related issues in WSNs. In this chapter, we have presented a survey of congestion and flow control approaches as well as mechanisms to ensure reliable data delivery in WSNs. A broad level comparison of the congestion control and reliability protocols discussed in this chapter is given in the Table 8.1.

We also take a note of a similar work on surveying transport layer protocols for WSNs [41], which compares existing protocols for congestion control and reliability based on different approaches used for congestion and loss detection, notification, and mitigation or recovery. In contrast, in this chapter, we present a survey of existing protocols for congestion control, flow control, and reliability and compare them based on various parameters such as mechanisms used for congestion detection and control, support for application specific design, target data delivery model, and support for fairness and reliability. In addition, we make this chapter more comprehensive by including some of the important recent contributions in this area [16, 17, 32].

**Table 8.1** Comparison of congestion control and reliability protocols

| | Congestion control support | Reliability guarantee | Congestion detection mechanism | Congestion control goal | Rate control mechanism | Fairness/QoS support | Target application model |
|---|---|---|---|---|---|---|---|
| CODA | Yes | No | Local | Network-centric | Hop-by-hop backpressure | No | Many-to-one |
| Fusion | Yes | No | Local | Network-centric | Hop-by-hop backpressure | No | Many-to-one |
| Ee et al. | Yes | No | Local | Network-centric | Hop-by-hop backpressure | Fairness | Many-to-one |
| IFRC | Yes | No | Local | Network-centric | AIMD | Fairness | Many-to-one |
| RCRT | Yes | Yes | Global | Network-centric | Centralized, source-control | Fairness | Many-to-one |
| ESRT | Yes | Yes | Local | Application-specific | Centralized, source-control | No | Many-to-one |
| COMUT | Yes | No | Local | Application-specific | Centralized, source-control | Event-importance based QoS | Many-to-one |
| Siphon | Yes | No | NA | Hybrid | Using additional network capacity | Constant reporting rate | Many-to-one |
| RMST | No | Yes | NA | NA | NA | No | One-to-one |
| RMBTS | No | Yes | NA | NA | NA | No | One-to-one |
| RBC | No | Yes | NA | NA | NA | No | Many-to-one |
| PSFQ | No | Yes | NA | NA | NA | No | One-to-Many |
| STCP | Yes | Yes | Local | Network-centric | Centralized, source-control | No | Hybrid |
| Flush | Yes | Yes | Local | Network-centric | Dynamic rate control | No | One-to-one |

# Questions

1. Why protocols for congestion control in WSNs need to be different than the protocols in wired or wireless ad hoc networks?
2. How can transient and persistent hot spots occur in a WSN? What mechanisms can be used to handle them?
3. What are the different types of unfairness that can happen in a sensor network? Discuss at least one mechanism that handles each type of unfairness.
4. Why TCP is not directly suited as a transport layer protocol for sensor networks?
5. What is a potential problem with ESRT, if different events should be given different transmission priorities? Why does ESRT reduce the overall throughput of the network?
6. Flush presents a pipelining based approach to congestion control, however it works only for a one-to-one connection. How can the scheme be extended to many-to-one communication with support for application level reliability?
7. Can we combine given approaches to create a protocol that supports reliable data delivery for one-to-many and many-to-one traffic, without loss of overall throughput?
8. Sink-oriented approaches are perfectly suitable for WSNs, since they have a complete view of the network as well as knowledge of application specific requirements. Is the argument valid? Discuss your reason(s). If the argument is invalid, can you suggest a possible approach to solve one of the problems?
9. In Siphon, the network is assumed to have a secondary network of a few resource rich nodes. One possible solution is to put minimal functionality on low-power sensor nodes: sensor nodes will sense the environment and report their readings to nearby VS. The VS then transfers this data to the base station over the secondary network. Discuss potential advantages and disadvantages of this approach.
10. COMUT supports rate allocation to sensor nodes based on the importance levels of the events they are reporting. However, COMUT can not be directly used to guarantee weighted fairness in the network. Discuss at least one additional factor that COMUT needs to consider to guarantee weighted fairness in the network. Can you suggest a possible direction for the solution?

# References

1. S. Tilak, N.B. Abu-Ghazaleh, and W. Heinzelman. Infrastructure tradeoffs for sensor networks. In WSNA '02: Proceedings of the 1st ACM international workshop on Wireless sensor networks and applications, ACM, New York, NY, USA, 2002. pages 49–58.
2. V. Jacobson. Congestion avoidance and control. In ACM SIGCOMM '88, Stanford, CA, August 1988, pages 314–329.
3. B. Hull, K. Jamieson, and H. Balakrishnan. Mitigating congestion in wireless sensor networks. In SenSys '04: Proceedings of the 2nd international conference on Embedded networked sensor systems, ACM, New York, NY, USA, 2004. pages 134–147.

4. K.K. Ramakrishnan and R. Jain. Congestion avoidance in computer networks with a connectionless network layer: Part iv: A selective binary feedback scheme for general topologies, August 1987.
5. S. Floyd. Random early detection gateways for congestion avoidance. IEEE/ACM Transactions on Networking (TON), 1(4):397–413, 1993.
6. M. Allman, V. Paxson, and W. Stevens. Tcp congestion control, 1999.
7. L.S. Brakmo, S.W. O'Malley, and L.L. Peterson. Tcp vegas: new techniques for congestion detection and avoidance. SIGCOMM Comput. Commun. Rev., 24(4):24–35, 1994.
8. J. Li, J. Jannotti, D.S.J. De Couto, D.R. Karger, and R. Morris. A scalable location service for geographic ad hoc routing. In MobiCom '00: Proceedings of the 6th annual international conference on Mobile computing and networking, New York, NY, USA, 2000. ACM, pages 120–130.
9. C.-Y. Wan, S.B. Eisenman, and A.T. Campbell. Coda: Congestion detection and avoidance in sensor networks. In SenSys '03: Proceedings of the 1st international conference on Embedded networked sensor systems, New York, NY, USA, 2003. ACM, pages 266–279.
10. G.-S. Ahn, S.G. Hong, E. Miluzzo, A.T. Campbell, and F. Cuomo. Funneling-mac: A localized, sink-oriented mac for boosting fidelity in sensor networks. In SenSys '06: Proceedings of the 4th international conference on Embedded networked sensor systems, ACM, New York, NY, USA, 2006. pages 293–306.
11. S. Tilak, N.B. Abu-Ghazaleh, and W. Heinzelman. A taxonomy of wireless micro-sensor network models. SIGMOBILE Mob. Comput. Commun. Rev., 6(2):28–36, 2002.
12. S. Tilak, A. Murphy, and W. Heinzelman. Non-uniform information dissemination for sensor networks. In ICNP '03: Proceedings of the 11th IEEE International Conference on Network Protocols, IEEE Computer Society, Washington, DC, USA, 2003. page 295.
13. O. Gnawali, K.Y. Jang, J. Paek, M. Vieira, R. Govindan, B. Greenstein, A. Joki, D. Estrin, and E. Kohler. The tenet architecture for tiered sensor networks. Proceedings of the 4th international conference on Embedded networked sensor systems, 2006. pages 153–166.
14. I. Aad, C. Castelluccia, and R.A. INRIA. Differentiation mechanisms for IEEE 802.11. INFOCOM 2001. Twentieth Annual Joint Conference of the IEEE Computer and Communications Societies. Proceedings. IEEE, 1, 2001.
15. C.T. Ee and R. Bajcsy. Congestion control and fairness for many-to-one routing in sensor networks. In SenSys '04: Proceedings of the 2nd international conference on Embedded networked sensor systems, ACM, New York, NY, USA, 2004. pages 148–161.
16. S. Rangwala, R. Gummadi, R. Govindan, and K. Psounis. Interference-aware fair rate control in wireless sensor networks. In SIGCOMM '06: Proceedings of the 2006 conference on Applications, technologies, architectures, and protocols for computer communications, ACM, New York, NY, USA, 2006. pages 63–74.
17. J. Paek and R. Govindan. Rcrt: Rate-controlled reliable transport for wireless sensor networks. In SenSys '07: Proceedings of the 5th international conference on Embedded networked sensor systems, ACM, New York, NY, USA, 2007. pages 305–319.
18. Y. Sankarasubramaniam, O. Akan, and I. Akyildiz. Esrt: Event-to-sink reliable transport in wireless sensor networks, 2003.
19. K. Karenos, V. Kalogeraki, and S.V. Krishnamurthy. Cluster-based congestion control for supporting multiple classes of traffic in sensor networks. In EmNets '05: Proceedings of the 2nd IEEE workshop on Embedded Networked Sensors, Washington, DC, USA, 2005. IEEE Computer Society, pages 107–114.
20. Z.J. Haas, M.R. Pearlman, et al. The Zone Routing Protocol (ZRP) for Ad Hoc Networks. TERNET DRAFT-Mobile Ad hoc Networking (MANET) Working Group of the bternet Engineering Task Force (ETF), November, 1997.
21. C.-Y. Wan, S.B. Eisenman, A.T. Campbell, and J. Crowcroft. Siphon: Overload traffic management using multiradio virtual sinks in sensor networks. In SenSys '05: Proceedings of the 3rd international conference on Embedded networked sensor systems, New York, NY, USA, 2005. ACM, pages 116–129.
22. D. Larson, T. Strategist, R. Murty, and E. Qi. An Adaptive Approach to Wireless Network Performance Optimization. Technology, page 1, 2004.

23. S. Ratnasamy, B. Karp, L. Yin, F. Yu, D. Estrin, R. Govindan, and S. Shenker. Ght: A geographic hash table for data-centric storage in sensornets, 2002.
24. F. Stann and J. Heidemann. Rmst: Reliable data transport in sensor networks. In Proceedings of the First International Workshop on Sensor Net Protocols and Applications, Anchorage, Alaska, USA, 2003. pages 102–112.
25. P. Volgyesi, A. Nadas, and A. Ledeczi. Reliable multihop bulk transfer service for wireless sensor networks. In ECBS '06: Proceedings of the 13th Annual IEEE International Symposium and Workshop on Engineering of Computer Based Systems (ECBS'06), IEEE Computer Society, Washington, DC, USA, 2006. pages 112–122.
26. C.-Y. Wan, A.T. Campbell, and L. Krishnamurthy. Psfq: A reliable transport protocol for wireless sensor networks. In WSNA '02: Proceedings of the 1st ACM international workshop on Wireless sensor networks and applications, New York, NY, USA, 2002. ACM, pages 1–11.
27. C. Intanagonwiwat, R. Govindan, and D. Estrin. Directed diffusion: A scalable and robust communication paradigm for sensor networks. In Mobile Computing and Networking, 2000. pages 56–67.
28. A. Mainwaring, D. Culler, J. Polastre, R. Szewczyk, and J. Anderson. Wireless sensor networks for habitat monitoring. Proceedings of the 1st ACM international workshop on Wireless sensor networks and applications, 2002, pages 88–97.
29. G. Simon, M. Maróti,' A. Lédeczi, G. Balogh, B. Kusy, A. Nádas, G. Pap, J. Sallai, and K. Frampton. Sensor network-based countersniper system. Proceedings of the 2nd international conference on Embedded networked sensor systems, 2004. pages 1–12.
30. H. Zhang, A. Arora, Y. Choi, and M.G. Gouda. Reliable bursty convergecast in wireless sensor networks. Proceedings of the 6th ACM international symposium on Mobile ad hoc networking and computing, 2005, pages 266–276.
31. Y.G. Iyer, S. Gandham, and S. Venkatesan. Stcp: A generic transport layer protocol for wireless sensor networks. In Proceedings. 14th International Conference on Computer Communications and Networks (ICCCN), 2005. pages 449–454.
32. S. Kim, R. Fonseca, P. Dutta, A. Tavakoli, D. Culler, P. Levis, S. Shenker, and I. Stoica. Flush: A reliable bulk transport protocol for multihop wireless networks. In To appear in Proceedings of the Fifth ACM Conference on Embedded Networked Sensor Systems (SenSys). ACM, 2007.
33. HEDETNIEMI-S. HEDETNIEMI, S. and A. LIESTMAN. A survey of gossiping and broadcasting in communication networks. In Networks 18, 1988.
34. W.R. Heinzelman, J. Kulik, and H. Balakrishnan. Adaptive protocols for information dissemination in wireless sensor networks. In MobiCom '99: Proceedings of the 5th annual ACM/IEEE international conference on Mobile computing and networking, ACM, New York, NY, USA, 1999. pages 174–185.
35. S.-Y. Ni, Y.-C. Tseng, Y.-S. Chen, and J.-P. Sheu. The broadcast storm problem in a mobile ad hoc network. In MobiCom '99: Proceedings of the 5th annual ACM/IEEE international conference on Mobile computing and networking, New York, NY, USA, 1999. ACM, pages 151–162.
36. W. Peng and X. Lu. Ahbp: An efficient broadcast protocol for mobile ad hoc networks. J. Comp. Sci. Tech., 16(2):114–125, 2001.
37. W. Peng and X.-C. Lu. On the reduction of broadcast redundancy in mobile ad hoc networks. In MobiHoc '00: Proceedings of the 1st ACM international symposium on Mobile ad hoc networking & computing, IEEE Press, Piscataway, NJ, USA, 2000. pages 129–130.
38. W. Lou and J. Wu. Double-covered broadcast (DCB): A simple reliable broadcast algorithm. In INFOCOM '04. Twenty-third Annual Joint Conference of the IEEE Computer and Communications Societies, 2004.
39. A. Khelil, P.J. Marrón, C. Becker, and K. Rothermel. Hypergossiping: A generalized broadcast strategy for mobile ad hoc networks. Ad Hoc Netw., 5(5):531–546, 2007.
40. A. Woo and D.E. Culler. A transmission control scheme for media access in sensor networks. In MobiCom '01: Proceedings of the 7th annual international conference on Mobile computing and networking, ACM, New York, NY, USA, 2001. pages 221–235.
41. S. Floyd. A survey of transport protocols for wireless sensor networks, 2006.

# Chapter 9
# Data Transport Control in Wireless Sensor Networks

**Hongwei Zhang and Vinayak Naik**

**Abstract** Dynamics of wireless communication, resource constraints, and application diversity pose significant challenges to data transport control in wireless sensor networks. In this chapter, we examine the issue of data transport control in the context of two typical communication patterns in wireless sensor networks: convergecast and broadcast. We study the similarity and differences of data transport control in convergecast and broadcast; we discuss existing convergecast and broadcast protocols, and we present open issues for data transport control in wireless sensor networks.

## 9.1 Introduction

Wireless sensor networks are increasingly innovating the way we interact with the physical world, and they tend to have a broad range of applications in science (e.g., ecology), engineering (e.g., industrial control), and our daily life (e.g., healthcare). Spatially distributed sensor nodes coordinate with one another through messaging passing. Two typical messaging passing tasks in wireless sensor networks are convergecast and broadcast. Convergecast enables a sink node to collect information (e.g., event detection) from multiple spatially distributed nodes, and broadcast enables a node to disseminate data (e.g., a new sensor node program) from itself to all the other nodes in the network.

Even though message passing has been studied extensively in traditional networks such as the Internet and wireless networks, wireless sensor networks bring unique challenges to the design of message passing services due to the

H. Zhang (✉)
Department of Computer Science, Wayne State University, Detroit, Michigan 48202, USA
e-mail: hzhang@cs.wayne.edu

S. Misra et al. (eds.), *Guide to Wireless Sensor Networks*, Computer Communications and Networks, DOI: 10.1007/978-1-84882-218-4_9,
© Springer-Verlag London Limited 2009

complex properties of wireless communication, resource constrains, and application diversity. Among other tasks, data transport control is an important, challenging task in sensor networks. Moreover, data transport control differs in different message passing tasks. For instance, fairness is an important issue in convergecast but not in broadcast; broadcast tends to require 100% reliability in most cases (e.g., in sensor network reprogramming), but reliability requirements may vary significantly in different convergecast scenarios; the source node of broadcast is a single node that may serve as a single-point-of-control in broadcast, yet the source nodes in convergecast are usually spatially distributed.

In this chapter, we examine in detail the data transport control issues in convergecast and broadcast in Sects. 9.2 and 9.3, respectively. The discussion of broadcast is from the perspective of sensor network reprogramming since it is one of the most commonly used broadcast services in sensor networks. We make concluding remarks in Sect. 9.4.

## 9.2 Data Transport Control in Convergecast

In this section, we first review the basic issues and approaches in data transport control for convergecast, and then give a detailed treatment of the protocol Reliable Bursty Convergecast (RBC) [1].

### 9.2.1 Introduction

In convergecast, multiple source nodes need to report data to a sink node, creating the funneling effect where the traffic load increases as the distance to the sink node decreases. One consequence of the funneling effect is network congestion where packet queues overflow because packets arrive at nodes faster than what the nodes can transmit. The funneling effect also increases channel contention and thus the probability of packet loss as a result of increased collision probability. To ensure reliable data transport in convergecast, therefore, two basic issues are congestion control and error control. Besides reliable data transport, another issue is to ensure fairness in delivering data from different source nodes. Fairness in data delivery is important because, otherwise, the sink node cannot detect or observe the phenomena happening in regions whose sensing packets experience significant loss. The research community has proposed different approaches to address the congestion control, error control, and fairness control issues in convergecast, and we discuss a few representative mechanisms in the next section.

### 9.2.2 Background

For congestion control, Wan et al. proposed the protocol CODA (for *Congestion Detection and Avoidance*) [2]. In CODA, a node monitors both its queue length and the channel load condition (e.g., the number of packets transmitted in a short interval) to detect any potential network congestion in its local neighborhood. A node declares the network as being congested when the queue length and/or channel load condition exceed certain threshold values. Once a node detects network congestion, it can use two complementary approaches to ameliorate congestion: open-loop, hop-by-hop congestion control, and closed-loop, end-to-end congestion control. In open-loop, hop-by-hop congestion control, a node having detected congestion will inform the corresponding transmitting nodes of the congestion; these transmitting nodes will reduce their transmitting rates accordingly and then propagate this "congestion" information backward along the direction toward the traffic sources, creating the diffusing "backpressure" so that the sources will eventually reduce their traffic generation rates too. In open-loop, end-to-end congestion control, the sink coordinates with the sources to regulate the traffic generation rates at different sources.

In CODA, Wan et al. did not differentiate between the congestion within a node and that in wireless transmission. To address this issue, Ee and Bajcsy [3] proposed a system where congestion within a node and congestion in wireless communication are treated separately. In [3], congestion within a node is detected by monitoring its queue length, and the congestion is dealt with by an open-loop, hop-by-hop control mechanism similar to that in CODA; congestion in wireless communication (also commonly referred to as *channel contention*) is addressed by letting nodes randomly backoff at the timescale of application transmission interval rather than at the timescale of radio transmission rate. Besides congestion control, Ee and Bajcsy also proposed a rate-based mechanism to ensure fairness in packet delivery. Corroborating several observations in [2] and [3], Hull et al. [4] studied the effectiveness of different congestion and fairness control mechanisms, and they found out that (1) hop-by-hop flow control is effective for all types of workloads and utilization levels, and (2) rate limiting is particularly effective in achieving fairness.

Focusing on the reliability of delivering information related to an event, ESRT [5] controls congestion based on the relationship between event reliability and source report frequency. More specifically, the sink node continuously measures the event reliability, and decides on the source report frequency accordingly; the sources will generate reports based on the frequency-feedback from the sink node to avoid congestion in the network.

For reliable packet delivery in sensor networks, Stann and Heidermann [6] studied the benefit of hop-by-hop error control and recovery compared with end-to-end error control. For instance, Figure 9.1 shows the number of transmissions required to send ten packets across ten hops in hop-by-hop and end-to-end error control, respectively. We see that hop-by-hop error control significantly reduces the number of transmissions required for reliable data delivery.

For reliable, real-time packet delivery in bursty convergecast where a huge burst of data need to be delivered from multiple source nodes to a sink node, Zhang et al. [1] proposed the protocol RBC. RBC addresses the challenge of reliable, real-time error control in the presence of high channel contention and collision. To improve channel utilization and to reduce ack-loss, RBC uses a windowless block acknowledgment scheme that guarantees continuous packet forwarding and

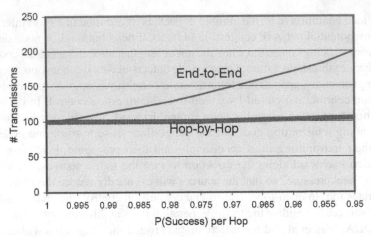

**Fig. 9.1** Number of transmissions required to send ten packets across ten hops

replicates the acknowledgment for a packet; to alleviate retransmission-incurred channel contention, RBC employs differentiated contention control. Moreover, RBC uses mechanisms to handle varying ack-delay and to reduce delay in timer-based retransmissions. We elaborate on the protocol RBC in the next section.

## 9.2.3 Reliable Bursty Convergecast

A typical application of wireless sensor networks is to monitor an environment (be it an agricultural field or a classified area) for events that are of interest to the users. Usually, the events are rare. Yet when an event occurs, a large burst of packets is often generated that needs to be transported reliably and in real time to a base station.

One exemplary event-driven application is demonstrated in the DARPA NEST field experiment "A Line in the Sand" (simply called *Lites* hereafter) [7]. In Lites, a typical event generates up to 100 packets within a few seconds and the packets need to be transported from different network locations to a base station, over multihop routes.

The high-volume bursty traffic in event-driven applications poses special challenges for reliable and real-time packet delivery. The large number of packets generated within a short period leads to high degree of channel contention and thus a high probability of packet collision. The situation is further exacerbated by the fact that packets travel over multihop routes: First, the total number of packets competing for channel access is increased by a factor of the average hop count of network routes; Second, the probability of packet collision increases in multihop networks due to problems such as hidden terminals. Consequently, packets are lost with high probability in bursty convergecast. For example, with the default radio stack of TinyOS [8], around 50% of packets are lost for most events in Lites.

For real-time packet delivery, hop-by-hop packet recovery is usually preferred over end-to-end recovery [Stann and Heidermann 2003], and this is especially the case when 100% packet delivery is not required (for instance, for bursty convergecast in sensor networks). Nevertheless, existing hop-by-hop control mechanisms do not work well in bursty convergecast. Via experiments with a testbed of 49 MICA2 motes and with traffic traces of Lites, Zhang et al. [1] observed that the commonly used link-layer error control mechanisms do not significantly improve and can even degenerate packet delivery reliability. For example, when packets are retransmitted up to twice at each hop, the overall packet delivery ratio increases by only 6.15%; when the number of retransmissions increases, the packet delivery ratio actually decreases, by 11.33%.

One issue with existing hop-by-hop control mechanisms is that they do not schedule packet retransmissions appropriately; as a result, retransmitted packets further increase the channel contention and cause more packet loss. Moreover, due to in-order packet delivery and conservative retransmission timers, packet delivery can be significantly delayed in existing hop-by-hop mechanisms, which leads to packet backlogging and reduction in network throughput.

On the other hand, the new network and application models of bursty convergecast in sensor networks offer unique opportunities for reliable and real-time transport control:

- First, the broadcast nature of wireless channels enables a node to determine, by snooping the channel, whether its packets are received and forwarded by its neighbors.
- Second, time synchronization and the fact that data packets are timestamped relieve transport layer from the constraint of in-order packet delivery, since applications can determine the order of packets by their timestamps.

Therefore, techniques that take advantage of these opportunities and meet the challenges of reliable and real-time bursty convergecast are desired.

Zhang et al. [1] studied the limitations of two commonly used hop-by-hop packet recovery schemes in bursty convergecast. They discovered that the lack of retransmission scheduling in both schemes makes retransmission-based packet recovery ineffective in the case of bursty convergecast. Moreover, in-order packet delivery makes the communication channel underutilized in the presence of packet- and ack-loss. To address the challenges, they designed protocol RBC (for *reliable bursty convergecast*). Taking advantage of the unique sensor network models, RBC features the following mechanisms:

- To improve channel utilization, RBC uses a windowless block acknowledgment scheme that enables continuous packet forwarding in the presence of packet- and ack-loss. The block acknowledgment also reduces the probability of ack-loss, by replicating the acknowledgment for a received packet.
- To ameliorate retransmission-incurred channel contention, RBC introduces differentiated contention control, which ranks nodes by their queuing conditions as well as the number of times that the enqueued packets have been transmitted. A node ranked the highest within its neighborhood accesses the channel first.

In the rest of this section, we examine in more detail the shortcomings of the existing error control mechanisms and how RBC addressed these shortcomings.

### 9.2.3.1   Performance with Existing Error Control Mechanisms

Two widely used hop-by-hop packet recovery mechanisms in sensor networks are synchronous explicit ack (SEA) and stop-and-wait implicit ack (SWIA). Zhang et al. [1] studied their performance in bursty convergecast, and we discuss their findings as follows.

Synchronous Explicit Ack

In SEA, a receiver switches to transmit mode and sends back the acknowledgment immediately after receiving a packet; the sender immediately retransmits a packet if the corresponding ack is not received after certain constant time. Zhang et al. studied the performance of SEA when used with B-MAC [9] and S-MAC [10]. B-MAC uses the mechanism of CSMA/CA (carrier sense multiple access with collision avoidance) to control channel access; S-MAC uses CSMA/CA too, but it also employs RTS-CTS handshake to reduce the impact of hidden terminals.

*SEA with B-MAC.* The event reliability, the average packet delivery delay, as well as the event goodput are shown in Table 9.1, where RT stands for the maximum number of retransmissions for each packet at each hop (e.g., RT = 0 means that packets are not retransmitted), ER stands for Event Reliability; PD stands for packet delivery latency; EG stands for event goodput [1].

Table 9.1 shows that when packets are retransmitted, the event reliability increases slightly (i.e., by up to 3.69%). Nevertheless, the maximum reliability is still only 54.74%, and, even worse, the event reliability as well as goodput decreases when the maximum number of retransmissions increases from 1 to 2.

*SEA with S-MAC.* Unlike B-MAC, S-MAC uses RTS-CTS handshake for unicast transmissions, which reduces packet collisions. The performance data for S-MAC is shown in Table 9.2.

**Table 9.1** SEA with B-MAC in Lites trace

| Metrics | RT = 0 | RT = 1 | RT = 2 |
|---|---|---|---|
| ER (%) | 51.05 | 54.74 | 54.63 |
| PD(s) | 0.21 | 0.25 | 0.26 |
| EG (packets/s) | 4.01 | 4.05 | 3.63 |

**Table 9.2** SEA with S-MAC in Lites trace

| Metrics | RT = 0 | RT = 1 | RT = 2 |
|---|---|---|---|
| ER (%) | 72.6 | 74.79 | 70.1 |
| PD (s) | 0.17 | 0.183 | 0.182 |
| EG (packets/s) | 5.01 | 4.68 | 4.37 |

Compared with B-MAC, RTS-CTS handshake improves the event reliability by about 20% in S-MAC. Yet packet retransmissions still do not significantly improve the event reliability and can even decrease the reliability.

*Analysis.* We can see that the reason why retransmission does not significantly improve – and can even degenerate – communication reliability is that, in SEA, lost packets are retransmitted while new packets are generated and forwarded, thus retransmissions, when not scheduled appropriately, only increase channel contention and cause more packet collision.[1] The situation is further exacerbated by ack-loss (with a probability as high as 10.29%), since ack-loss causes unnecessary retransmission of packets that have been received. To make retransmission effective in improving reliability, therefore, we need a retransmission scheduling mechanism that ameliorates retransmission-incurred channel contention.

## Stop-and-Wait Implicit Ack

Stop-and-wait implicit ack (SWIA) takes advantage of the fact that every node, except for the base station, forwards the packet it receives, and the forwarded packet can act as the acknowledgment to the sender at the previous hop [11]. In SWIA, the sender of a packet snoops the channel to check whether the packet is forwarded within certain constant threshold time; the sender regards the packet as received if it is forwarded within the threshold time, otherwise the packet is regarded as lost. The advantage of SWIA is that acknowledgment comes for free except for the limited control information piggybacked in data packets. The performance results for SWIA are shown in Table 9.3.

We can see that the maximum event reliability in SWIA is only 46.5%, and that the reliability decreases significantly when packets are retransmitted at most once at each hop. When packets are retransmitted up to twice at each hop, the packet delivery delay increases, and the event goodput decreases significantly despite the slightly increased reliability.

*Analysis.* The above phenomena are due to the following reasons. First, the length of data packets is increased by the piggybacked control information in SWIA, thus the ack-loss probability increases (as high as 18.39% in our experiments), which in turn increases unnecessary retransmissions. Second, most packets are queued upon reception and thus their forwarding is delayed. As a result, the piggybacked

**Table 9.3**  SEA with B-MAC in Lites trace

| Metrics | RT = 0 | RT = 1 | RT = 2 |
|---|---|---|---|
| ER (%) | 43.09 | 31.76 | 46.5 |
| PD (s) | 0.35 | 8.81 | 18.77 |
| EG (packets/s) | 3.48 | 2.58 | 1.41 |

---

[1] This is not the case in wireline networks and is due to the nature of wireless communications.

acknowledgments are delayed and the corresponding packets are retransmitted unnecessarily. Third, once a packet is waiting to be acknowledged, all the packets arriving later cannot be forwarded even if the communication channel is free. Therefore, channel utilization as well as system throughput decreases, and network queuing as well as packet delivery delay increases. Fourth, as in SEA, lack of retransmission scheduling allows retransmissions, be it necessary or unnecessary, to cause more channel contention and packet loss.

To address the limitations of SEA and SWIA in bursty convergecast, Zhang et al. [1] designed protocol RBC. In RBC, a windowless block acknowledgment scheme is designed to increase channel utilization and to reduce the probability of ack-loss; a distributed contention control scheme is also designed to schedule packet retransmissions and to reduce the contention between newly generated and retransmitted packets. Given that the number of packets competing for channel access is less in implicit-ack-based schemes than in explicit-ack-based schemes, Zhang et al. designed RBC based on the paradigm of implicit-ack (i.e., piggybacking control information in data packets).

### 9.2.3.2 Windowless Acknowledgment

In traditional block acknowledgment [12], a sliding window is used for both duplicate detection and in-order packet delivery.[2] The sliding window reduces network throughput once a packet is sent but remains unacknowledged (since the sender can only send up to its window size once a packet is unacknowledged), and in-order delivery increases packet delivery delay once a packet is lost (since the lost packet delays the delivery of every packet behind it). Therefore, the sliding-window-based block acknowledgment scheme does not apply to bursty convergecast, given the real-time requirement of the latter.

To address the constraints of traditional block acknowledgment in the presence of unreliable links, RBC takes advantage of the fact that in-order delivery is not required in bursty convergecast. Without considering the order of packet delivery, we only need to detect whether a sequence of packets is received without loss in the middle and whether a received packet is a duplicate of a previously received one. To this end, a *windowless block acknowledgment* scheme is designed to ensure continuous packet forwarding irrespective of the underlying link unreliability as well as the resulting packet- and ack-loss. For clarity of presentation, we consider an arbitrary pair of nodes S and R where S is the sender and R is the receiver.

*Windowless queue management.* The sender S organizes its packet queue as $(M + 2)$ linked lists, as shown in Fig. 9.2, where $M$ is the maximum number of retransmissions at each hop. For convenience, we call the linked lists *virtual queues*, denoted as $Q_0, \ldots, Q_M + 1$. The virtual queues are ranked such that a virtual queue $Q_k$ ranks higher than $Q_j$ if $k < j$.

---

[2] Note that SWIA is a special type of block acknowledgment where the window size is 1.

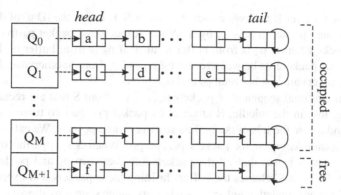

**Fig. 9.2** Virtual queues at a node

Virtual queues $Q_0$, $Q_1$, ..., and $Q_M$ buffer packets waiting to be sent or to be acknowledged, and $Q_{M+1}$ collects the list of free queue buffers. The virtual queues are maintained as follows:

- When a new packet arrives at S to be sent, S detaches the head buffer of $Q_{M+1}$, if any, stores the packet into the queue buffer, and attaches the queue buffer to the tail of $Q_0$.
- Packets stored in a virtual queue $Q_k$ ($k > 0$) will not be sent unless $Q_{k-1}$ is empty; packets in the same virtual queue are sent in FIFO order.
- After a packet in a virtual queue $Q_k$ ($k \geq 0$) is sent, the corresponding queue buffer is moved to the tail of $Q_{k+1}$, unless the packet has been retransmitted $M$ times in which case the queue buffer is moved to the tail of $Q_{M+1}$.
- When a packet is acknowledged to have been received, the buffer holding the packet is released and moved to the tail of $Q_{M+1}$.

The earlier rules help identify the relative freshness of packets at a node (which is used in the differentiated contention control in Sect. 9.2.3.3; they also help maintain without using sliding windows the order in which unacknowledged packets have been sent, providing the basis for windowless block acknowledgment. Moreover, newly arrived packets can be sent immediately without waiting for the previously sent packets to be acknowledged, which enables continuous packet forwarding in the presence of packet- and ack-loss.

*Block acknowledgment and reduced ack-loss.* Each queue buffer at S has an ID that is unique at S. When S sends a packet to the receiver R, S attaches the ID of the buffer holding the packet as well as the ID of the buffer holding the packet to be sent next. In Fig. 9.2, for example, when S sends the packet in buffer a, S attaches the values a and b. Given the queue maintenance procedure, if the buffer holding the packet being sent is the tail of $Q_0$ or the head of a virtual queue other than $Q_0$, S also attaches the ID of the head buffer of $Q_{M+1}$, if any, since one or more new packets may arrive before the next enqueued packet is sent in which case the newly arrived packet(s) will be sent first. For example, when the packet in buffer c of Fig. 9.2 is sent, S attaches the values $c$, $d$, and $f$.

When the receiver R receives a packet $p_0$ from S, R learns the ID $n'$ of the buffer holding the next packet to be sent by S. When R receives a packet $\mathbf{p}_n$ from S next time, R checks whether $\mathbf{p}_n$ is from buffer $n'$ at S: if $p_n$ is from buffer $n'$, R knows that there is no packet loss between receiving $p_0$ and $p_n$ from S; otherwise, R detects that some packets are lost between $p_0$ and $p_n$.

For each maximal sequence of packets $p_k, \ldots, p_{k'}$ from S that are received at R without any loss in the middle, R attaches to packet $p_{k'}$ the two tuple $<q_k, q_{k'}>$, where $q_k$ and $q'_k$ are the IDs of the buffers storing $p_k$ and $p_{k'}$ at S. We call $<q_k, q_{k'}>$ the *block acknowledgment* for packets $p_k, \ldots, p_{k'}$. When S snoops the forwarded packet $p_{k'}$ later, S learns that all the packets sent between $p_k$ and $p_{k'}$ have been received by R. Then S releases the buffers holding these packets. For example, if S snoops a block acknowledgment $<c, e>$ when its queue state is as shown in Fig. 9.2, S knows that all the packets in buffers between c and e in $Q_1$ have been received, and S releases buffers between c and e, including c and e.

One delicate detail in processing the block acknowledgment $<q_k, q_{k'}>$ is that after releasing buffer $q_k$, S will maintain a mapping $q_k \leftrightarrow q_{k''}$, where $q_{k''}$ is the buffer holding the packet sent (or to be sent next) after that in $q_{k'}$. When S snoops another block acknowledgment $<q_k, q_n>$ later, S knows, by $q_k \leftrightarrow q_{k''}$, that packets sent between those in buffers $q_{k''}$ and $q_n$ have been received by R; then S releases the buffers holding these packets, and S resets the mapping to $q_k \leftrightarrow q_{n''}$, where $q_{n''}$ is the buffer holding the packet sent (or to be sent next) after that in $q_n$. S maintains the mapping for $q_k$ until S receives a block-NACK $[n', n)$ or a block acknowledgment $<q, q'>$ where $q \neq q_k$, in which case S maintains the mapping for n or q, respectively. Via the buffer pointer mapped as above, the node S can process the incoming block acknowledgments and block-NACKs. For convenience, we call the buffer being mapped to the *anchor* of block acknowledgments. In the examples discussed earlier, buffers $q_{k''}$ and $q_{n''}$ have been anchors once. We also call the packet in an anchor buffer an *anchor packet*.

In the earlier block acknowledgment scheme, the acknowledgment for a received packet is piggybacked onto the packet itself as well as the packets that are received consecutively after the packet without any loss in the middle. Therefore, the acknowledgment is replicated and the probability for it to be lost decreases significantly.

*Duplicate detection and obsolete-ack filtering.* Since it is impossible to completely prevent ack-loss in lossy communication channels, packets whose acknowledgments are lost will be retransmitted unnecessarily. Therefore, it is necessary that duplicate packets be detected and dropped.

To enable duplicate detection, the sender S maintains a counter for each queue buffer, whose value is incremented by one each time a new packet is stored in the buffer. When S sends a packet, it attaches the current value of the corresponding buffer counter. For each buffer q at S, the receiver R maintains the counter value $c_q$ piggybacked in the last packet from the buffer. When R receives another packet from the buffer q later, R checks whether the counter value piggybacked in the packet equals to $c_q$: if they are equal, R knows that the packet is a duplicate and drops it; otherwise R regards the packet as a new one and accepts it. The duplicate detection is local in the sense that it only requires information local to each queue buffer instead of imposing any rule involving different buffers (such as in sliding window) that can degenerate system performance.

For the correctness of the earlier duplicate detection mechanism, we only need to choose the domain size $C$ for the counter value such that the probability of losing $C$ packets in succession is negligible. For example, for the high per-hop packet loss probability 22.7% in the case of Lites trace, $C$ could still be as small as 7, since the probability of losing seven packets in succession is only 0.003%. (Given the small domain size for the counter value as well as the usually small queue size at each node, the duplicate detection mechanism does not consume much memory. For example, it only takes 36 bytes in the case of Lites.)

In addition to duplicate detection, we also use buffer counter to filter out obsolete acknowledgment. Despite the low probability, packet forwarding at R may be severely delayed, such that the queue buffers signified in a block acknowledgment have been reused by S to hold packets arriving later. To deal with this, R attaches to each forwarded packet the ID as well as the counter value of the buffer holding the packet at S originally; when S snoops a packet forwarded by R, S checks whether the piggybacked counter value equals to the current value of the corresponding buffer: if they are equal, S regards as valid the piggybacked block acknowledgment; otherwise, S regards the block acknowledgment as obsolete and ignores it.

*Aggregated-ack at the base station.* In sensor networks, the base station usually forwards all the packets it receives to an external network. As a result, the children of the base station (i.e., the nodes that forward packets directly to the base station) are unable to snoop the packets the base station forwards, and the base station has to explicitly acknowledge the packets it receives. To reduce channel contention, the base station aggregates several acknowledgments, for packets received consecutively in a short period of time, into a single packet and broadcasts the packet to its children. Accordingly, the children of the base station adapt their control parameters to the way the base station handles acknowledgments.

### 9.2.3.3   Differentiated Contention Control

In wireless sensor networks where per-hop connectivity is reliable, most packet losses are due to collision in the presence of severe channel contention. To enable reliable packet delivery, lost packets need to be retransmitted. Nevertheless, packet retransmission may cause more channel contention and packet loss, thus degenerating communication reliability. Also, there exist unnecessary retransmissions due to ack-loss, which only increase channel contention and reduce communication reliability. Therefore, it is desirable to schedule packet retransmissions such that they do not interfere with transmissions of other packets.

The way the virtual queues are maintained in our windowless block acknowledgment scheme facilitates the retransmission scheduling, since packets are automatically grouped together by different virtual queues. Packets in higher ranked virtual queues have been transmitted less number of times, and the probability that the receiver has already received the packets in higher ranked virtual queues is lower (e.g., 0 for packets in $Q_0$). Therefore, we rank packets by the rank of the virtual queues holding the packets, and higher ranked packets have higher priority in accessing the communication channel. By this rule, packets that have been transmitted less number of times will be (re)transmitted earlier than those that have been transmitted more, and interference between packets of different ranks is reduced.

Windowless block acknowledgment already handles packet differentiation and scheduling within a node, thus we only need a mechanism that schedules packet transmission across different nodes. To reduce interference between packets of the same rank and to balance network queuing as well as channel contention across nodes, internode packet scheduling also takes into account the number of packets of a certain rank so that nodes having more such packets transmit earlier.

To implement the aforementioned concepts, we define the rank rank $(j)$ of a node $j$ as $<M-k, |Q_k|, \text{ID}(j)>$, where $Q_k$ is the highest-ranked nonempty virtual queue at $j$, $|Q_k|$ is the number of packets in $Q_k$, and $\text{ID}(j)$ is the ID of $j$. rank(j) is defined such that (1) the first field guarantees that packets having been transmitted fewer number of times will be (re)transmitted earlier, (2) the second field ensures that nodes having more packets enqueued get chances to transmit earlier, and (3) the third field is to break ties in the first two fields. A node with a larger rank value ranks higher. Then, the distributed transmission scheduling works as follows:

- Each node piggybacks its rank to the data packets it sends out.
- Upon snooping or receiving a packet, a node $j$ compares its rank with that of the packet sender $k$. $j$ will change its behavior only if $k$ ranks higher than $j$, in which case $j$ will not send any packet in the following $w(j, k) \times T_{\text{pkt}}$ time. $T_{\text{pkt}}$ is the time taken to transmit a packet at the MAC layer, and $w(j, k) = 4 - \text{i}$, when rank$(j)$ and rank$(k)$ differ at the i-th element of the three tuple ranks. $w(j, k)$ is defined such that the probability of all waiting nodes starting their transmissions simultaneously is reduced, and that higher ranked nodes tend to wait for shorter time. $T_{\text{pkt}}$ is estimated by the method of *exponentially weighted moving average* (EWMA).
- If a sending node $j$ detects that it will not send its next packet within $T_{\text{pkt}}$ time (i.e., when $j$ knows that, after the current packet transmission, it will rank lower than another node), $j$ signifies this by marking the packet being sent, so that the nodes overhearing the packet will skip $j$ in the contention control. (This mechanism reduces the probability of idle waiting, where the channel is free but no packet is sent.)

### 9.2.3.4 Experimental Results

Table 9.4 shows the performance results of RBC, and we can observe the following properties of RBC:

- The event reliability keeps increasing, in a significant manner, as the number of retransmissions increases. The increased reliability mainly attributes to reduced unnecessary retransmissions (by reduced ack loss and adaptive retransmission timer) and retransmission scheduling.

**Table 9.4** RBC in Lites trace

| Metrics | RT = 0 | RT = 1 | RT = 2 |
|---|---|---|---|
| ER (%) | 56.21 | 83.16 | 95.26 |
| PD (s) | 0.21 | 1.18 | 1.72 |
| EG (packets/s) | 4.28 | 5.72 | 6.37 |

- Compared with SWIA, which is also based on implicit-ack, RBC reduces packet delivery delay significantly. This mainly attributes to the ability of continuous packet forwarding in the presence of packet- and ack-loss and the reduction in timer-incurred delay.
- The rate of packet reception at the base station and the event goodput keep increasing as the number of retransmissions increases. When packets are retransmitted up to twice at each hop, the event goodput reaches 6.37 packets/s, quite close to the optimal goodput – 6.66 packets/second – for Lites trace.

Compared with SWIA, RBC improves reliability by a factor of 2.05 and reduces average packet delivery delay by a factor of 10.91. Compared with SEA with B-MAC (simply referred to as SEA hereafter), RBC improves reliability by a factor of 1.74, but the average packet delivery delay increases by a factor of 6.61 in RBC. Interestingly, however, RBC still improves the event goodput by a factor of 1.75 when compared with SEA. The reason is that, in RBC, lost packets are retransmitted and delivered after those packets that are generated later but transmitted less number of times. Therefore, the delivery delay for lost packets increases, which increases the average packet delivery delay, without degenerating the system goodput. The observation shows that, due to the unique application models in sensor networks, metrics evaluating aggregate system behaviors (such as the event goodput) tend to be of more relevance than metrics evaluating unit behaviors (such as the delay in delivering each individual packet).

## 9.3   Data Transport Control in Reprogramming

In this section, we discuss the basic issues and approaches in data transport control for the purpose of sensor network reprogramming.

### 9.3.1   Introduction

The large scale of deployments of the wireless networks of embedded devices demand an ability to reprogram the nodes in the field, possibly over multiple hops. Since a program has to reach in entirety, the reprogramming service has to deliver data with 100% reliability. Therefore, the need for a reprogramming service translates into a problem of reliable dissemination of bulk data in wireless networks of embedded devices. Designing such a service is a challenging problem due to the limited energy, memory, and lossy wireless links.

## 9.3.2 Background

Early work in wireless networks showed that simple retransmissions of broadcast messages lead to the broadcast storm problem, where redundancy, contention, and collision impair the ability to perform well. The naïve approach of simple retransmission is not reliable and fast. Hence, a more intricate handling of the transmissions in the space and time is needed.

The problem of sending a new program, which typically consists of many packets, is different from that of sending a command, which typically consists of a few packets. The probability of contention and collisions is more in the case of sending a new program. Further, optimizing latency of transmission is an important concern for the reprogramming case. Although broadcasting a few packets has its own research challenges, we will only focus on broadcasting a large number of packets due to the space constraints. Hence, the solutions for disseminating less data are not suitable for the reprogramming. In this section, we will look at the state-of-the-art reprogramming services for the wireless networks of embedded devices.

## 9.3.3 Challenges

The primary challenges in the problem of reprogramming are as follows:

- 100% reliability: The lossy links commonly found in the wireless networks of embedded devices make the problem of providing 100% reliability hard.
- Energy consumption: The battery-powered nature of the embedded devices necessitates that the energy consumption has to be minimized. The operations for Mica mote in the decreasing order of energy consumed are as follows: EEPROM write 16-bytes, transmit a packet, receive a packet, idle listen for 1 ms, and EEPROM read 16-bytes [13, 14].
- Time to reprogram the entire network: Since the primary objective of the networks is sensing, it is desirable to minimize the time required to reprogram the network.
- Memory consumption: Since the size of a program could be larger than that of the available RAM, the broadcast service must be scalable in terms of memory consumption.

All of the aforementioned challenges differentiate the problem of reliable dissemination in wireless embedded devices from that in wireless networks of PCs.

### 9.3.4  Techniques of Reprogramming

We enumerate the commonly used techniques to address the aforementioned challenges. These will help practitioners to understand the working of the state-of-the-art reprogramming services. The practitioners can use them to tune the performance of the existing reprogramming services or develop new services of their own.

- 100% reliability:

  - Hop-by-hop recovery: Given the lossy nature of the network, the recovery of the lost packets is done in a hop-by-hop manner. This reduces the number of transmissions as compared to that of end-to-end recovery.
  - Sender selection and suppression: The goal of the sender selection and suppression technique is to ensure that at most one node broadcasts the data in a radio range. An example of the selection criteria is to select a node that has larger number of potential receivers [15]. If a node sending data messages also overhears data messages from other nodes at the same time, it suppresses its transmissions based on any rule that uniquely determines an order among the nodes [14]. An example of such rule is the IDs of the nodes. The sender selection and suppression reduces the number of collisions.
  - Time division multiple access (TDMA): At the link layer, CSMA/CA protocol results in lower latency when the number of nodes, simultaneously transmitting in a neighborhood, is less. As the number of nodes increases, the number of backoffs increases. Also, the number of collisions due to the hidden terminal effect increases, thereby resulting in more retransmissions and hence more latency. One way to deal with the increased number of simultaneous transmissions is to use TDMA, where each node is allocated a time slot to transmit [16, 17]. The TDMA schedule is computed to guarantee that no two nodes within collision range from each other transmit at the same time. The collision range is approximately equal to twice the transmission range.
  - Use of implicit ACK and NACK-based explicit ACK: The use of TDMA creates a lower bound on the latency to hear from each of the transmitter in a node's range. In simpler words, a node knows when the other nodes in its neighborhood will transmit [16, 17]. This property enables a node to use implicit acknowledgment to detect message loss. The advantage of implicit acknowledgment over that of explicit is that the implicit acknowledgment reduces the number of message transmissions and hence is more energy efficient. However, implicit ACK requires the sender to maintain the state about the receivers. The amount of state grows linearly in terms of the number of receivers. Therefore, a more scalable approach is to use NACK-based recovery, where a receiver reports the sequence numbers of the lost packets to the sender and the sender rebroadcasts the requested packets [15].

- Energy consumption:
  - Sender selection and suppression: Use of sender selection and suppression reduces the number of concurrent transmissions in a neighborhood, thereby reducing the number of collisions.
  - Load balancing while selecting senders: An unfair sender selection process will tax a sender with transmissions and sapping its energy. A fair sender selection process takes into consideration the remaining energy at the node [15].
  - Duty cycling of radio: If a technique of sender selection and suppression is employed, a node that loses in the selection and suppression round can chose to switch off its radio [17]. The use of sender selection and suppression gives an opportunity to the unselected and suppressed nodes to switch off their radios while the selected sender is transmitting.
  - Minimum connected dominating set (MCDS) for selecting senders: The transmission of radio messages consumes significant amount of energy. Reducing the number of senders will save energy. A constraint to the problem of selecting senders is that all the nodes must receive the entire program. If we induce a graph over the wireless network, the problem of selecting a minimum set of senders is equivalent to that of finding a MCDS of the induced graph [16]. A formal definition of MCDS is given in the section titled "Terminologies"; here, we give an example as shown in Fig. 9.3.

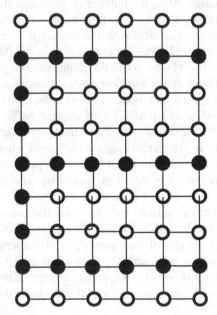

**Fig. 9.3** MCDS of a network. The circles represent nodes. A line between two nodes means that those two nodes can communicate with each other. The filled-in circles represent MCDS

- Time to reprogram the entire network:

  - Pipelining of messages over multiple hops: Although a node has not received an entire program, it can become a sender and start sending packets. This policy results in pipeline of transmissions, thereby reducing the latency [14, 16, 17]. However, the flipside of the policy can be hidden terminal effect if care is not taken to ensure that two nodes within colliding range are simultaneously transmitting.
  - Transmitting as fast as possible: The use of sender selection and suppression technique enables a sender to transmit data packets at the fastest rate [14].

- Memory consumption:

  - Forward/download phase in MNP: The size of a program could be far more than that of RAM. In fact, it could even be infeasible to save a bitmap, where a bit is allocated for each packet in the program, in the RAM. Two ways to deal with the small size of RAM are to either use a window-based recovery or save the information about the lost packets in the EEPROM instead of RAM. In window-based recovery, a sender does not transmit a new packet unless all the packets in the last window are successfully received. In the latter option of using EEPROM, although the size is not a problem, the latency becomes an issue since read and write access to the EEPROM is slower than that of RAM. One way to expedite access to the lost packets is to maintain a linked list connecting the slots of the lost packets [15]. This way, it is not necessary to traverse the entire list of packets to search for the lost packets.
  - Segmentation of the program: The entire program could be divided into packets and a fixed $N$ number of buffers could be allocated in the RAM, where the size of the buffer is same as that of a packet [14]. Since a node does have all of the packets in its RAM, the node can respond quickly only to those retransmit requests, for which the packets are in its RAM.

Although we have classified techniques depending upon which of the four distinct challenges they address, the classification is not disjoint. For example, sender selection technique not only reduces collisions and improves reliability, but it also saves energy by reducing the number of transmissions.

In Table 9.5, we summarize which of the aforementioned techniques the commonly used reprogramming services employ. The first row lists the four challenges and the first column lists the name of the services.

## 9.4   Thoughts for Practitioners

*Convergecast.* In RBC, the tolerance of out-of-order packet delivery enables the design of windowless block acknowledgment. In general, network protocol design tends to be application-specific in wireless sensor networks, and we should pay attention to the application properties in designing or choosing network protocols. For instance, open-loop, hop-by-hop control is more appropriate for transient

**Table 9.5** Techniques used by the well-known reprogramming services

|  | Reliability | Energy efficiency | Latency | Memory consumption |
|---|---|---|---|---|
| Deluge [14] | Sender selection and suppression, NACK-based recovery | Sender selection and suppression | Pipelining while forwarding packets | Dividing the program into pages and packets |
| Infuse [17] | TDMA, Implicit ACK | Duty cycling of radio | Pipelining while forwarding packets | Window-based recovery |
| MNP [15] | Sender selection and suppression, NACK-based recovery | Load balancing while selecting senders | – | Use of linked list in EEPROM to keep records of lost packets |
| Sprinkler [16] | TDMA, Implicit ACK | Use of MCDS while selecting senders | Pipelining while forwarding packets | – |

congestion, and closed-loop, end-to-end control is better for persistent congestion. Moreover, end-to-end error control may well be necessary to ensure 100% data delivery.

*Reprogramming.* The reprogramming services can be categorized into two broad classes, which are (a) ad-hoc and (b) structured, depending upon their approaches to selecting senders. The structured approach induces a graph over the network of nodes, where each node is a vertex and there is an edge between two vertices if the two corresponding nodes can communicate with each other. It then uses this graph to select senders. The ad-hoc services do not induce such a graph. While Deluge and MNP fall under the ad-hoc category, Infuse and Sprinkler fall under the structured category.

The benefit of using a structured approach is that computing a MCDS and a TDMA schedule requires fewer control messages than that of ad-hoc approach. An intuitive reason behind this is that position of a node in a graph can be used to decide whether the node becomes a sender or not. However, inducing a graph over nodes can be complex given the highly varying nature of the wireless links. For example, Infuse and Sprinkler assume that the distance between the nodes is an indication of the quality of the link between them and rely on location of the nodes to induce a graph. In practice, the assumption may not hold and could degrade the performance of the protocol.

## 9.5 Directions for Future Research

*Convergecast.* Despite the fact that many data transport control mechanisms have been proposed for convergecast in wireless sensor networks, how to effectively en-sure application-specific QoS remains an open issue. Much work is also needed to address the interaction between QoS provisioning and in-network processing in wireless sensor networks, since QoS provisioning affects the spatial and temporal data flow in the network, which in turns affects the effectiveness of in-network processing and thus messaging efficiency and reliability. Network coding tends to be an effective approach to improving the efficiency and QoS messaging in wireless networks, and it is worthwhile to explore how to apply network coding for the purpose of reliable, efficient data transport in sensor network convergecast.

*Reprogramming.* In the case of ad-hoc reprogramming services, the contention in the wireless medium may increase as the density of the network increases. For example, in Deluge the nodes advertise themselves as senders after they have received a page. In a dense network, these advertisements cause contention [14]. Although knowledge about density of the network will eliminate this problem, such knowledge needs partial information about the graphical topology of the network and hence violates the philosophy of the ad-hoc approach. Therefore, future research is needed to suppress the advertisements to avoid contention.

The structured approaches, such as Infuse and Sprinkler, assume location information at each node to induce a graph [16, 17]. Briefly, the idea is to use the position of nodes in the graph to decide whether two nodes are within contention range of each other and then compute MCDS and TDMA schedule. However, localization in wireless embedded networks is a hard problem in itself. Most of the state-of-the-art localization techniques demand special acoustic or ultrasonic hardware, which may not be available in all the embedded devices. Hence, computing MCDS and TDMA schedule without depending on a localization service demands further research.

## 9.6  Conclusions

We have reviewed the challenges and approaches for data transport control in sensor network convergecast and reprogramming-oriented broadcast. We have also seen the difference (in both challenges and solution methods) in data transport control for convergecast and broadcast. For guaranteed QoS and efficiency in convergecast and broadcast, we have also presented the important open problems in data transport control. In general, how to design application- and task-specific data transport control mechanisms remains an interesting, open problem.

## Terminologies

*Convergecast.* The transport of packets from multiple spatially distributed sensor nodes to a common sink node.

*Congestion control.* Control the packet generation rate at the sources and intermediate nodes to avoid overutilizing the network in terms of the node packet buffers and wireless channels.

*Error control.* Detect and recover transmission errors in the network, which are caused by packet transmission collision and other factors.

*Fairness.* Equality of different nodes in accessing network resources (e.g., wireless transmission bandwidth).

*Windowless block acknowledgment.* The block acknowledgment scheme that enables continuous data transmission irrespective of packet- and ack-loss without any constraint as imposed by the sliding window size in traditional window-based block acknowledgment mechanisms.

*MCDS.* A dominating set (DS) of a graph $G = (V, E)$ is a subset of $V'$ of $V$ such that every vertex $v \in V$ is either in $V'$ or adjacent to some member of $V'$. A minimum connected dominating set (MCDS) is a connected dominating set of minimum cardinality.

*TDMA-based transmission.* A time division multiple access (TDMA)-based transmission is a scheme where in each node is given a schedule, such that neither two adjacent nodes nor two nodes sharing a same adjacent node transmit at the same time.

*Pipelining of transmissions.* The process of pipelining of transmission is composed of simultaneous transmissions of packets by nodes in time.

*Implicit ACK.* Implicit ACK is an acknowledgment scheme, where a sender infers that (a) a packet has been successfully received if it overhears the forwarding of the packet by the sender's successor nodes and (b) otherwise if it does not overhear the forwarding.

*NACK-based recovery.* If a node recovers a lost packet by explicitly asking the sender to retransmit the lost packet then such a recovery is called as NACK-based recovery.

## Exercises

1. What are the basic issues in data transport control in sensor network converge-cast?
2. How is ESRT different from protocols such as CODA?
3. Study the paper on RBC [1], and discuss the respective roles of windowless block acknowledgment and distributed contention control in improving the reliability and goodput of convergecast?
4. Analyze the ack-loss probability in RBC.
5. RBC has focuses on windowless block acknowledgment and distributed contention control. But queue may still overflow without careful flow control. Please design a flow control mechanism to work with RBC.
6. What is the broadcast storm problem?
7. What are the challenges in the reliable broadcast in wireless networks of embedded devices?
8. When is TDMA faster than CSMA/CA in terms of latency?

9. What is MCDS and how is it useful for the reliable broadcast in wireless networks of embedded devices?
10. What are the two categories of the reliable reprogramming services and what are differences between them?

# References

1. Hongwei Zhang, Anish Arora, Young-Ri Choi, Mohamed Gouda (2007). Reliable Bursty Convergecast in Wireless Sensor Network. Computer Communications (Elsevier) 30(13):2560–2576
2. Chieh-Yih Wan, Shane B. Eisenman, Andrew Campbell (2003). CODA: Congestion Detection and Avoidance in Sensor Networks. ACM SenSys, Los Angeles, CA
3. Cheng Tien Ee and Ruzena Bajcsy (2004). Congestion Control and Fairness for Many-to-One Routing in Sensor Networks. ACM SenSys, Baltimore, MD
4. Bret Hull, Kyle Jamieson, Hari Balakrishnan (2004). Mitigating Congestion in Wireless Sensor Networks. ACM SenSys, Baltimore, MD
5. Yogesh Sankarasubramanjam, Ozgur B. Akan, Ian F. Akyildiz (2003). ESRT: Event-to-Sink Reliable Transport in Wireless Sensor Networks. ACM MobiHoc, Annapolis, MD
6. Fred Stann and John Heidemann (2003). RMST: Reliable Data Transport in Sensor Networks. International Workshop on Sensor Net Protocols and Applications
7. Anish Arora, Prabal Dutta et al. (2004). A Line in the Sand: A Wireless Sensor Network for Target Detection, Classification, and Tracking. Computer Networks (Elsevier) 46(5):605–634
8. TinyOS. http://www.tinyos.net/
9. Joseph Polastre, Jason Hill, David Culler (2004). Versatile Low Power Media Access for Wireless Sensor Networks. ACM SenSys, Baltimore, MD
10. Wei Ye, John Heidemann, Deborah. Estrin (2002). An Energy-Efficient MAC Protocol for Wireless Sensor Networks, IEEE INFOCOM, New York
11. Miklos Maroti (2004). The Directed Flood Routing Framework. Technical report, Vanderbilt University, Nashville, TN
12. Geoffrey Brown, Mohamed Gouda, Raymond Miller (1989). Block Acknowledgment: Redesigning the Window Protocol. ACM SIGCOMM, Austin, TX
13. Alan Mainwaring, David Culler, Joseph Polastre, Robert Szewczyk, John Anderson (2002). Wireless Sensor Networks for Habitat Monitoring. First ACM International Workshop on Wireless Sensor Networks and Applications, Atlanta, GA
14. Jonathan Hui, David Culler (2004). The Dynamic Behavior of a Data Dissemination Protocol for Network Programming at Scale. ACM SenSys, Baltimore, MD
15. Sandeep Kulkarni, Limin Wang (2005). MNP: Multihop Network Reprogramming Service for Sensor Networks, IEEE ICDCS, Columbus, OH
16. Vinayak Naik, Anish Arora, Prasun Sinha, Hongwei Zhang (2007). Sprinkler: A Reliable and Energy Efficient Data Dissemination Service for Extreme Scale Wireless Networks of Embedded Devices, IEEE Transactions on Mobile Computing, 6(7), 777–789
17. Sandeep Kulkarni, Mahesh Arumugam (2004). Infuse: A TDMA Based Data Dissemination Protocol for Sensor Networks. Technical Report MSU-CSE-04-46, Michigan State University, East Lanting, MI

# Chapter 10
# Fault-Tolerant Algorithms/Protocols in Wireless Sensor Networks

Hai Liu, Amiya Nayak, and Ivan Stojmenović

**Abstract** Wireless sensor networks (WSNs) have wide variety of applications and provide limitless future potentials. Nodes in WSNs are prone to be failure due to energy depletion, hardware failure, communication link errors, malicious attack, and so on. Therefore, fault tolerance is one of the critical issues in WSNs. The chapter investigates current research work on fault tolerance in WSNs. We study how fault tolerance is addressed in different applications of WSNs. Five categories of applications are discussed: node placement, topology control, target and event detection, data gathering and aggregation, and sensor surveillance. In each category, we focus on the representative research works that presented algorithms and approaches in application layer to achieve fault tolerance.

## 10.1 Introduction

### 10.1.1 Background

Wireless sensor networks (WSNs) have received significant attention in recent years due to their potential applications in military sensing, wildlife tracking, traffic surveillance, health care, environment monitoring, building structures monitoring, etc. WSNs can be treated as a special family of wireless ad hoc networks. A WSN is a self-organized network that consists of a large number of low-cost and low-powered sensor devices, called sensor nodes, which can be deployed on the ground, in the air, in vehicles, on bodies, under water, and inside buildings. Each sensor node is equipped with a sensing unit, which is used to capture events of interest, and a wireless transceiver, which is used to transform the captured events back to the base

H. Liu (✉)
Department of Computer Science, Hong Kong Baptist University, Hong Kong
e-mail: hliu@comp.hkbu.edu.hk

S. Misra et al. (eds.), *Guide to Wireless Sensor Networks*, Computer Communications and Networks, DOI: 10.1007/978-1-84882-218-4_10,
© Springer-Verlag London Limited 2009

station, called sink node. Sensor nodes collaborate with each other to perform tasks of data sensing, data communication, and data processing.

Nodes in WSNs are prone to failure due to energy depletion, hardware failure, communication link errors, malicious attack, and so on. Unlike the cellular networks and ad hoc networks where energy has no limits in base stations or batteries can be replaced as needed, nodes in sensor networks have very limited energy and their batteries cannot usually be recharged or replaced due to hostile or hazardous environments. So, one important characteristic of sensor networks is the stringent power budget of wireless sensor nodes. Two components of a sensor node, sensing unit and wireless transceiver, usually directly interact with the environment, which is subject to variety of physical, chemical, and biological factors. It results in low reliability of performance of sensor nodes. Even if condition of the hardware is good, the communication between sensor nodes is affected by many factors, such as signal strength, antenna angle, obstacles, weather conditions, and interference.

Fault tolerance is the ability of a system to deliver a desired level of functionality in the presence of faults [8]. Since the sensor nodes are prone to failure, fault tolerance should be seriously considered in many sensor network applications. Actually, extensive work has been done on fault tolerance and it has been one of the most important topics in WSNs. An early survey work can be found in [18]. However, its coverage is very limited and its references are outdated. The objective of the chapter is to investigate current research work on fault tolerance in WSNs. We study how fault tolerance is addressed in different applications of WSNs. More specifically, we address five categories of applications: node placement, topology control, target and event detection, data gathering and aggregation, and sensor surveillance. In each category, we focus on the representative research works that presented algorithms and approaches in application layer to achieve fault tolerance.

In the rest of this section, we discuss how faults happen in different levels of WSNs, and briefly introduce fault detection and recovery strategies. After that, organization of the chapter is as follows. Node placement in two-tired WSNs is discussed in Sect. 10.2. Fault tolerance in topology control is introduced in Sect. 10.3. Target detection and event detection are introduced are Sect. 10.4. Data gathering and data aggregation are discussed in Sect. 10.5. Sensor surveillance is studied in Sect. 10.6. We conclude the chapter in Sect. 10.7.

### 10.1.2 Fault Tolerance at Different Levels

Five levels of fault tolerance were discussed in [18]. They are physical layer, hardware layer, system software layer, middleware layer, and application layer. On the basis of study, we classify fault tolerance in WSNs into four levels from the system point of view. More specifically, fault tolerance in a WSN system may exist at hardware layer, software layer, network communication layer, and application layer.

#### 10.1.2.1   Hardware Layer

Faults at hardware layer can be caused by malfunction of any hardware component of a sensor node, such as memory, battery, microprocessor, sensing unit, and network interface (wireless radio). There are three main reasons that cause hardware failure of sensor nodes. The first is that sensor networks are usually for commercial use and sensor nodes are cost sensitive. Therefore, design of a sensor node will not always use the highest quality components. The second is that strict energy constraints restrict long and reliable performance of sensor nodes. For example, sensor readings may become incorrect when the battery of a sensor node reaches a certain level [26]. The third is that sensor networks are often deployed in harsh and hazardous environments, which affect normal operation of sensor nodes. The wireless radios of sensor nodes are severely affected by these environment factors.

#### 10.1.2.2   Software Layer

Software of a sensor node consists of two components: system software, such as operating system, and middleware, such as communication, routing, and aggregation. An important component of system software is to support distributed and simultaneous execution of localized algorithms. Software bugs are a common source of errors in WSNs. One promising method is through software diversity where each program is implemented in several different versions. Since it is difficult to provide fault tolerance in an economic way at hardware level of a sensor node, numerous fault-tolerant approaches are expected at the middleware level. The majority of current applications in WSNs are simple. To adapt the real-life applications, there is a need to develop much more complex middleware for WSNs.

#### 10.1.2.3   Network Communication Layer

Faults at network communication layer are the faults on wireless communication links. Assuming that there is no error on hardware, link faults in WSNs are usually related to surrounding environments. In addition, link faults can also be caused by radio interference of sensor nodes. For example, node $a$ can not successfully receive a message from node $b$ if node $a$ is within interference range of other nodes that are transmitting messages at the same time. The standard way to enhance the performance of wireless communication is to use aggressive error correction schemes and retransmission. These two methods may cause further delay of operation. It should be pointed out that there is always a trade-off between fault tolerance and efficiency.

#### 10.1.2.4   Application Layer

Fault tolerance can be addressed also at the application layer. For example, finding multiple node-disjoint paths provides fault tolerance in routing. The system can switch from an unavailable path with broken links to an available candidate path. However, an approach for fault tolerance in an application can not be directly applied to other applications. It requires proper addressing of fault tolerance in different applications, on a case by case basis. On the other side, fault tolerance in application level can be used to address faults in essentially any type of resource.

### 10.1.3  Fault Detection and Recovery

To tackle faults in a WSN, the system should follow two main steps. The first step is fault detection. It is to detect that a specific functionality is faulty, and to predict it will continue to function properly in the near future. After the system detects a fault, fault recovery is the second step to enable the system to recover from the faults.

Basically, there are two types of detection techniques: *self-diagnosis* and *cooperative diagnosis*. Some faults that can be determined by a sensor node itself can adopt self-diagnosis detection. For example, faults caused by depletion of battery can be detected by a sensor node itself. The remaining battery of the sensor node can be predicted by measuring current battery voltage. Another example is the detection of failure links. A sensor node may detect that some link to one of its neighbors is faulty if the node does not receive any message from the neighbor within a predetermined interval. However, there are some kinds of faults that require cooperative diagnosis among a set of sensor nodes. A large portion of faults in WSNs are in this category. For example, detection method proposed in [19] is to identify faulty sensor nodes in event detection application. The detection method is based on the assumption that sensor nodes in the same region should have similar sensed value unless a node is at the boundary of the event region. The method takes measurements of all neighbors of a node and uses the results to compute the probability of the node being faulty.

The most commonly used technique for fault recovery is replication or redundancy of components that are prone to be failure. For example, WSNs are usually used to periodically monitor a region and forward sensed data to a base station. When some nodes fail to provide data, the base station still gets sufficient data if redundant sensor nodes are deployed in the region. Multiple paths routing is another example. In the case of providing single route, a requested call can not be set up or be maintained if some nodes/links along the route fail. Keeping a set of candidate routes provides high reliability of the routes for routing. It requires $K$-connectivity of the network if it is able to tolerate failure of $K-1$ nodes.

Fault recovery mechanism in single-hop sensor networks was studied in [5]. The proposed fault recovery scheme is to deal with failure of sensor nodes, including the sink node. The basic idea is to partition the sensor memory into two parts, namely, data memory and redundant memory. The data memory is used to store sensed data and data recovered from failures of other sensor nodes. The redundant memory is used to store redundant data for future recovery. The recovered data is distributed in the memories of the nonfaulty sensors to be sent to the sink when it becomes available. It shows that the memory overhead is $(n+1)/n$, provided the total number of sensor nodes in the network is $n$.

## 10.2   Node Placement in Two-Tiered Wireless Sensor Networks

Since sensor nodes are prone to failure, one approach to improving reliability and prolonging lifetime of WSNs is the introduction of two-tiered network architecture. The architecture employs some powerful relay nodes whose main function is to gather information from sensor nodes and relay the information to the sink. That is, relay nodes serve as a backbone of the network. The relay nodes are more powerful than sensor nodes in terms of energy storage, computing, and communication capabilities. The network is partitioned into a set of clusters. The relay nodes act as cluster heads and they are connected with each other to perform the data forwarding task. Each cluster has only one cluster head and each sensor belongs to at least one cluster, such that sensor nodes can switch to backup cluster heads when current cluster head is not available. In each cluster, sensor nodes collect raw data and report to the cluster head. The cluster head analyzes the raw data, extracts useful information, and then generates outgoing packets with much smaller size to the sink via multihop paths.

A fault in transmitter can cause the relay nodes to stop transmitting tasks to the sensors as well as relaying the data to the sink. Data sent by the sensors will be lost if the receiver of a relay node fails. So, a communication link fault on a sensor requires the sensor to be reallocated to other cluster heads within communication range. If faults occur in intercluster heads, the two corresponding cluster heads should be reconnected by another multihop path. Therefore, in order to handle general communication faults, there should be at least two node-disjoint paths between each pair of relay nodes in the network.

An intuitive objective of relay node placement in two-tiered WSNs is to place the minimum number of relay nodes, such that some degree of fault tolerance can be achieved. A lot of work has been done on the minimum placement of relay nodes for fault tolerance in two-tiered WSNs [14, 15, 22, 28]. There are other works that study placement of sensor nodes to make a sensor network $k$-connected, such as [2]. It does not employ relay nodes and two-tiered architecture. However, it can be reduced to the same placement problem in two-tiered architecture by setting uniform communication ranges for both sensor nodes and relay nodes. So, we focus on relay node placement problem in two-tiered networks in this section.

There are variant definitions on the problem of minimum placement of relay nodes. Generally speaking, the problem can be described as follows. Given a set of sensor nodes that are randomly distributed in a region and their location, some relay nodes are needed to be placed on the region for forwarding data to the sink, such that each sensor node is covered by at least one relay node. The objective is to minimize the number of relay nodes that make the network $k$-connected (usually 2-connected is desirable).

Work in [15] assumes that the original sensor network is 2-connected and sensor nodes also participate in forwarding of the data. The objective is to guarantee that each sensor node is covered by at least two relay nodes and the network of relay nodes is 2-connected. The problem was shown to be an extension of Relay Node Double Cover problem, which has been proved to be NP-complete [11]. A polynomial time approximation algorithm was proposed. It was proved that performance of the proposed algorithm is bounded within $O(D \log n)$, where $n$ is the number of sensor nodes in the network and $D$ is the diameter of the network, which was defined in [15]. However, the assumption that the original sensor network is 2-connected is too strong to be applied in real applications. Moreover, it assumes that sensor nodes participate in forwarding task. Since sensor nodes usually have limited computing and communication capability, and especially very limited energy resource, it restricts application of the algorithm.

Work in [22] does not require earlier assumptions. Formal description of the problem is as follows: given a set of sensor nodes $S$ in a region and a uniform communication radius $d$, the problem is to place a set of relay nodes $R$, such that (1) the whole network $G$ is connected and (2) $G$ is 2-connected. The objective of the problem is to:

$$\text{Minimize } |R|,$$

where $|R|$ denotes the number of relay nodes in $R$.

The authors proposed a $(6 + \varepsilon)$-approximation solution for the case 1 of the minimum relay node placement problem (MRP-1 for short), and then proposed a $(24+\varepsilon)$-approximation solution for case 2 (MRP-2 for short), where $\varepsilon$ is an arbitrary positive number and running time is polynomial when $\varepsilon$ is fixed. The solutions were further extended to the scenario where communication radii of sensor nodes and relay nodes are different. The basic idea of the solutions is to partition the problem into two phases. The first phase is to place some relay nodes to cover all sensor nodes. The second phase is to add more relay nodes to make the whole network connected/2-connected.

The solution is based on two fundamental works. The first is the *covering with disks* problem. Given a set of points in the plane, the problem is to identify the minimum set of disks with prescribed radius to cover all the points. In [16], a polynomial time approximation scheme (PTAS) for this problem was proposed. That is, for any given error $\varepsilon \geq 0$, the ratio of the solution found by the scheme to the optimal solution is not larger than $(1 + \varepsilon)$. The running time is polynomial when $\varepsilon$ is fixed. The scheme was called *min-disk-cover scheme*.

The other fundamental work is the *Steiner tree problem with minimum number of Steiner points* (STP-MSP). Given a set of terminals in the Euclidean plane, the problem is to find a Steiner tree such that each edge in the tree has length at most $d$ and the number of Steiner points is minimized. Du et al. proposed a 2.5-approximation algorithm for the STP-MSP [10]. The algorithm was called *STP-MSP algorithm*. Note that sensor nodes do not participate in data forwarding. STP-MSP algorithm cannot be directly applied to the problem.

Based on earlier foundational works, the $(6 + \varepsilon)$-approximation algorithm for MRP-1 is as follows.

---

**Algorithm 1:**

---

Input: $S$, a set of sensor nodes with locations.

$\varepsilon$, any given error that is larger than 0.

$d$, the communication radius of sensor nodes and relay nodes.

Output: $G$, a connected network including sensor nodes and relay nodes.

1. Use the min-disk-cover scheme to place a set of relay nodes $R_1$, such that for$\forall s \in S$, $\exists r \in R_1$, and $r$ covers $s$.
2. Use $R_1$ as an input of the STP-MSP algorithm to place additional relay nodes $R_2$, such that $G$ is connected.
3. Output $G$ and the position of each relay node.

Since following theorem is the core technical part of the solution, introduction of the solution is incomplete without the theorem. We make a simple version on the original proof [22].

---

**Theorem 1.** *Let $R$ be the solution computed by algorithm 1 and $R^{opt}$ be the optimal solution to MRP-1. Then $\frac{|R|}{|R_{opt}|} \leq (6 + \varepsilon)$.*

*Proof.* Let $R_1^{opt}$ denote the minimum set of relay nodes that cover $S$. Since $R_1$ is the solution of PTAS, thus

$$|R_1| \leq (1 + \varepsilon)|R_1^{opt}|. \qquad (10.1)$$

Let $R_2^{opt}$ denote the minimum set of relay nodes that make $R_1$ connected. Since $R_2$ is the solution of the 2.5-approximation algorithm, thus

$$|R_2| \leq 2.5|R_2^{opt}|. \qquad (10.2)$$

For any $r \in R_1$, there must be at least one sensor node $s$, which is covered by $r$ (Otherwise $r$ can be removed from $R_1$). Consider the communication circle of $s$ (see Fig. 10.1), there exists $v \in R^{opt}$, such that both $r$ and $v$ can cover $s$. That is, relay nodes $r$ and $v$ are both in the communication circle of sensor $s$. Thus, $d(r, v) \leq 2d$. An additional relay node $u$ is placed in the middle point of line $v \to r$. Thus, $d(u, v) = d(r, u) \leq d$. It means node $v$ can communicate with node $r$ via node $u$. Therefore, for any relay node $r \in R_1$, an additional relay node is placed according to the earlier description, such that $R^{opt}$ can communicate with every relay

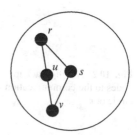

**Fig. 10.1** Communication circle of sensors

node in $R_1$. That is, $R^{\text{opt}}$ and these added relay nodes make $R_1$ connected. Note that $R_2^{\text{opt}}$ is the minimum set of relay nodes that make $R_1$ connected, and the number of added relay nodes is equal to $|R_1|$. Therefore,

$$|R_2^{\text{opt}}| \leq | \quad R^{\text{opt}}| + |R_1|. \tag{10.3}$$

According to (10.1)–(10.3), the total number of relay nodes placed by the algorithm to MRP-1 is as follows:

$$\begin{aligned} |R_1| + |R_2| &\leq (1 + \varepsilon)|R_1^{\text{opt}}| + 2.5(|R^{\text{opt}}| + |R_1|) \\ &\leq (1 + \varepsilon)|R^{\text{opt}}| + 2.5(|R^{\text{opt}}| + (1 + \varepsilon)|R^{\text{opt}}|) \\ &\leq 6|R^{\text{opt}}| + 3.5\varepsilon|R^{\text{opt}}| \end{aligned}$$

That is, $\frac{|R|}{|R^{\text{opt}}|} = \frac{|R_1| + |R_2|}{|R^{\text{opt}}|} \leq (6 + \varepsilon)$.

The $(24 + \varepsilon)$-approximation algorithm for MRP-2 is as follows. The basic idea is to add additional relay nodes to the connected network to make it 2-connected. □

The approximation ratio of algorithm 2 is $(24 + \varepsilon)$. Detailed proof can be found in [22].

---

**Algorithm 2**:

---

Input: $S$, a set of sensor nodes with locations.

   $\varepsilon$, any given error that is larger than 0.
   $d$, the communication radius of sensor nodes and relay nodes.

Output: $G$, a 2-connected network including sensor nodes and relay nodes.

1. Run algorithm 1 to get a set of relay nodes $R$, such that $S + R$ is connected.
2. Add three backup nodes in the communication circle of each $r \in R$ as in Fig. 10.2. The set of all backup nodes in this step is denoted by $R'$.
3. Output $G$ and positions of relay nodes in $R + R'$

---

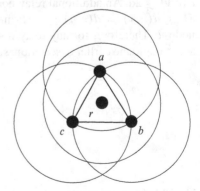

**Fig. 10.2** Adding backup nodes to the communication circle of $r$

The works in [15] and [22] were integrated and further extended in [28]. It studied four types of fault-tolerant relay node placement problems. They are single-tiered placement with/without base stations, and two-tiered node placement with/without base stations. In single-tiered model, an edge may exist between any two types of nodes. That is, both sensor nodes and relay nodes participate in packet forwarding in single-tiered model, while only relay nodes participate in packet forwarding in two-tiered model. For each problem, a polynomial constant approximation algorithm was proposed. The ratio of performance either is smaller than, or is the same as that of the previous-best. Four problems and their corresponding algorithms will be introduced one by one.

The basic technique is the use of *steinerization*. Suppose $x_i$ and $x_j$ are two sensor nodes. $R$ and $r$ are the communication ranges of relay nodes and sensors, respectively. If $d(x_i, x_j) \leq r$, $x_i$ and $x_j$ can directly communicate with each other. Otherwise, $x_i$ and $x_j$ can be connected by deploying a minimum number of relay nodes on the line segment $[x_i, x_j]$ in the following way (called steinerizing $[x_i, x_j]$).

- If $d(x_i, x_j) \in (r, 2r]$, place one relay node at the midpoint of line segment $[x_i, x_j]$.
- If $d(x_i, x_j) > 2r$, place $1 + \lceil (d(x_i, x_j) - 2r)/R \rceil$ relay nodes on the line segment $[x_i, x_j]$, such that two relay nodes, say $y_i$ and $y_j$, are at distance $r$ from $x_i$ and $x_j$, respectively, and the other $\lceil (d(x_i, x_j) - 2r)/R \rceil - 1$ relay nodes are evenly distributed on the line segment $[y_i, y_j]$.

Given $R \geq r > 0$ and a set of sensor nodes $X$, an edge weighted undirected complete graph $G^S(r, R, X)$, called the *steinerized graph* of $(r, R, X)$, consists of vertex set $V = X$ and edges with weight defined as follows:

$$c(x_i, x_j) \begin{cases} 0, & \text{if } d(x_i, x_j) \in [0, r]; \\ 1, & \text{if } d(x_i, x_j) \in (r, 2r]; \\ 1 + \lceil (d(x_i, x_j) - 2r)/R \rceil & \text{otherwise.} \end{cases} \quad (10.4)$$

Actually, the weight on each edge is the number of relay nodes needed to connected end nodes of the edge.

The approximation ratio of algorithm 3 is 14. Detailed proof can be found in [28]. The algorithm for single-tiered placement with base stations is similar to algorithm 3. The only difference is that it constructs the steinerized graph $G^S(r, R, B, X)$ in step 2, where $B$ is the set of base stations. The approximation ratio of the algorithm for single-tiered placement with base stations was proved to be 16.

The approximation ratio of algorithm 4 was proved to be $(10 + \varepsilon)$, which was claimed to improve $(24 + \varepsilon)$-approximation algorithm in [22]. However, it should be pointed out that both algorithm 4 and its integrated 5-approximation algorithm assume that nodes can be placed in the same position. It is not allowed in [22]. The basic idea of the 5-approximation algorithm in [23] is similar to algorithm 1. First, a minimum set of relay nodes, say set $A$, are placed to cover all sensor nodes by

---

**Algorithm 3**: (for single-tiered placement without base stations)

---

Input: $R \geq r > 0$ and set of sensor nodes $X = \{x_1, \ldots, x_n\}$.
Output: Set of relay nodes $Y = \{y_1, \ldots, y_l\}$.

1. Construct the steinerized graph $G^S(r, R, X)$.
2. Compute a 2-connected minimum weight spanning subgraph $G_A$ of $G^S(r, R, X)$ using the 2-approximation algorithm in [17]. ($G_A$ spans all sensor nodes in $X$.)
3. $l = 0$.
4. **for** each edge $(x_i, x_j) \in G_A$ s.t. $c(x_i, x_j) \geq 1$ **do**
5.     Steinerize edge $(x_i, x_j)$ with $c(x_i, x_j)$ relay nodes: $y_{l+1}, y_{l+2}, \ldots, y_{l+c(x_i,x_j)}$.
6.     $l = l + c(x_i, x_j)$
7. **endfor**

---

---

**Algorithm 4**: (for two-tiered placement without base stations)

---

Input: $R \geq r > 0$, $\varepsilon > 0$, and set of sensor nodes $X = \{x_1, \ldots, x_n\}$.
Output: Set of relay nodes $Y = \{y_1, \ldots, y_l\}$.

1. Apply 5-approximation algorithm in [23] to place set of relay nodes $Z = \{z_1, \ldots, z_k\}$, such that the resulting network is connected.
2. Duplicate each of the relay nodes in $Z$ to obtain $Y$.

---

using the min-disk-cover scheme. Second, it finds a subset of sensor nodes that are 1–1 mapped with nodes in $A$, and then places a set of relay nodes, say $B$, on the same locations of the subset. Finally, it calls STP-MSP algorithm to place a set of relay nodes, say $C$, to make the network connected. The set of required relay nodes is $A \cup B \cup C$.

The algorithm for two-tiered placement with base stations is similar to the 5-approximation algorithm in [23]. It was proved that the approximation ratio is $(20 + \varepsilon)$ [28].

Deploying relay nodes in heterogeneous WSNs was studied in [14]. It assumes that sensor nodes have different transmission ranges while relay nodes use the uniform transmission radius. The problem consists of two cases: (1) full fault-tolerance relay node placement, which aims to deploy a minimum number of relay nodes to establish $k$ vertex-disjoint paths between every pair of sensor and/or relay nodes; (2) partial fault-tolerance relay node placement, which aims to deploy a minimum number of relay nodes to establish $k$ vertex-disjoint paths only between every pair of sensor nodes. The basic idea of the proposed algorithm is similar to algorithm 3. It first constructs a steinerized weighted graph, and then applies existing algorithm for computing minimum $k$-vertex connected spanning graph on the weighted graph. The desired graph is achieved after steinerizing each edge in the spanning graph by placing relay nodes according to the weight of the edge.

All the aforementioned placement methods are deterministic. Stochastic node placement was discussed in [20]. It proved that even by random placement in a unit-area square region, the probability that the network $G(V, r_n)$ is $(k+1)$-connected is at least $e^{-e^{-\alpha}}$ when the transmission range $r_n$ satisfies $n\pi r_n^2 \geq \ln n + (2k-1)\ln\ln n - 2\ln k! + 2\alpha$ for $k > 0$ and $n$ sufficiently large, where $\alpha$ is any real number.

## 10.3   Topology Control

Although node placement provides a method to achieving fault tolerance in a WSN, the property of fault tolerance may be not valid due to movements and energy depletion of nodes. Therefore, topology control is required to construct and maintain the property of fault tolerance in WSNs.

A fault-tolerant topology control protocol was proposed in [4]. It first constructs a Connected Dominating Set (CDS) as a backbone of the network. For each node in the CDS, it adds necessary neighbors of the node to the backbone, such that it meets the required vertex connectivity degree. The power on/off model is adopted to turn on the nodes in the backbone to meet connectivity requirement, and other unnecessary nodes are off. Period rotation is used to keep the fairness among nodes. There are several selection metrics. One is power-based selection: the node in CDS selects nodes with more power one by one till the resulting graph is local $k$ vertex connected. Another metric is connection degree. The nodes with higher connection degree are first selected. It is because that the nodes with higher connection degree are supposed to have shorter delay. Some hybrid metrics are also discussed in [4]. Simulation results show the improvement of network lifetime with a desired vertex connectivity degree.

Most of existing works in fault-tolerant topology control aim to achieve $k$-vertex connectivity in the network. The objective is suitable for ad hoc networks where there is request to connect any two nodes in the network. However, data transmission in WSNs is usually in gathering and aggregation manner. It is rather important to have fault tolerance in the paths from sensors to sinks and gateway nodes, which are more powerful than sensor nodes.

The authors in [3] studied fault-tolerant topology control in heterogeneous WSN. The network consists of two types of wireless devices: a large number of sensor nodes and several resource-rich supernodes. The problem is to adjust each sensor's transmission range, such that there exist $k$-vertex disjoint communication paths from each sensor to the set of supernodes. The objective is to minimize the total power consumed by sensors. Three solutions were proposed. The first $k$-approximation algorithm consists of two steps. In the first step, a given graph is reduced to a direct graph where supernodes are merged as a root. In the second step, existing optimal solution for the Min-Weight $k$-OutConnectivity problem is adopted to compute the minimal transmission range of each sensor. The two steps are briefly introduced one by one.

The given graph is denoted by $G(V, E, c)$, where $V$ is the set of nodes, $E$ is the set of edges, and $c$ is the set of weight of the edge (indicating the power consumed in the edge). The reduced graph is constructed as follows. All supernodes in $V$ are merged into one node called the *root*. Edges between sensors remain the same, and an edge between a sensor and a supernode is replaced with an edge between the sensor and the root. The weight of the edge remains the same. It should be pointed out that if a sensor is connected to more than one supernode, only the edge to the closest supernode is kept. After that, every undirected edge between two sensors is replaced with two directed arcs that point to each of them. An undirected edge between a sensor and the root is replaced with one directed arc from the sensor to the root. The process of the step is illustrated in Fig. 10.3.

The algorithm in the second step is based on the reduced graph from the first step. It applies existing optimal solution for the Min-Weight $k$-OutConnectivity problem in the reduced graph. The final transmission range of each node is the transmission range used to meet the longest edge in final result. Detail of the algorithm is as follows.

---

**Algorithm 5**:

1. Construct the reduced graph of $G$.
2. Reverse the direction of each arc in the reduced graph and keep the weight of the arc the same.
3. Apply the optimal solution for the Min-Weight $k$-OutConnectivity problem.
4. Reverse back the direction of each arc.
5. **for** each sensor **do**
6.     Adjust transmission range to meet the longest arc in the graph.
7. **endfor**

---

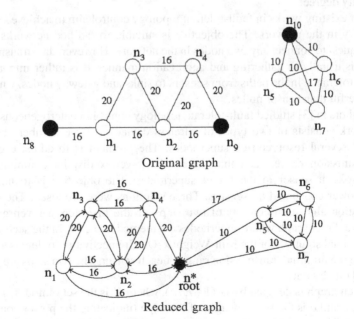

Original graph

Reduced graph

**Fig. 10.3** Original graph and reduced graph

The algorithm 5 was proved to achieve performance ratio $k$. A greedy algorithm that minimizes the maximum transmission power of sensors was further proposed. The basic idea is to sort all edges in the reduced graph in decreasing order. For each edge in the sorted order, the edge is discarded if the $k$-vertex connectivity to the root is still held without the edge. The process continues until all edges are computed. Similar to algorithm 5, the final transmission range of each node is the transmission range to meet the longest edge in final graph. Distributed version of the proposed algorithms can be found in [3].

Fault tolerance of wireless networks can also be achieved by movement control of nodes. Although work in [7] is for mobile robot networks, the proposed algorithm can be easily applied to WSNs without much modification. A localized movement control algorithm was proposed to form a fault tolerant biconnected network topology from a connected network. The objective is to minimize total distance of movement of nodes.

Each node is assumed to have information of its $p$-hop neighbors. The *p-hop subgraph* of a node is defined by the graph that contains all nodes that are within $p$-hops from the node and all corresponding links. A node is said to be a *p-hop critical node* if and only if its $p$-hop subgraph is disconnected without the node. The distributed algorithm is executed at each node and starts as follows. At initialization stage, each node checks whether it is a $p$-hop critical node. If a node finds itself a $p$-hop critical node, it broadcasts a critical announcement packet to all its direct neighbors.

To make the network biconnected, all critical nodes should become noncritical by movement of nodes. Note that the movement of a node may create new neighbors, but it may also break some existing links. Since the $p$-hop subgraph of a critical node is disconnected without the node, movement of a critical node may break some current links, which results in disconnection of the network. However, the network remains connected if all current links of a noncritical node are broken. The basic idea of movement control is to move noncritical nodes while keeping critical nodes static unless they become noncritical. According to the number of critical neighbors of a critical node, there are three cases: (1) critical node without critical neighbors, (2) critical node with one critical neighbor, and (3) critical node with several critical neighbors.

In case (1), a node finds itself a $p$-hop critical node and does not receive any critical announcement packet from its neighbors. The $p$-hop subgraph of any critical node can be divided into two disjointed components without the node. The basic idea of movement control is to select two close neighbors of the node from these two disjointed components, respectively, and then move them toward each other until they become connected. See the example shown in Fig. 10.4, where node 3 in black color is a critical node and other nodes in white color are noncritical nodes. Suppose $p = 2$ in this example. Since node 3 is critical, its 2-hop subgraph is divided into two disjointed sets $A = \{1, 2, 4, 5\}$ and $B = \{6, 7, 8\}$. Suppose distance of node 5 and 8 is the minimum among all possible pairs in these two sets (actually, other selection metrics can be applied). Node 3 computes new locations of node 5 and node 8 and asks them to move to become neighbors.

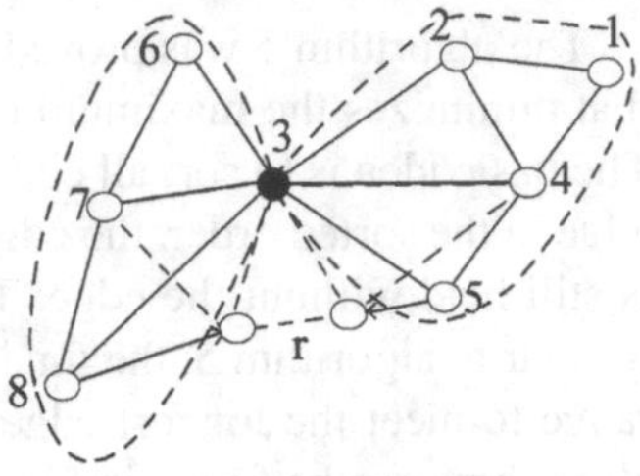

**Fig. 10.4** Critical nodes
without critical neighbors

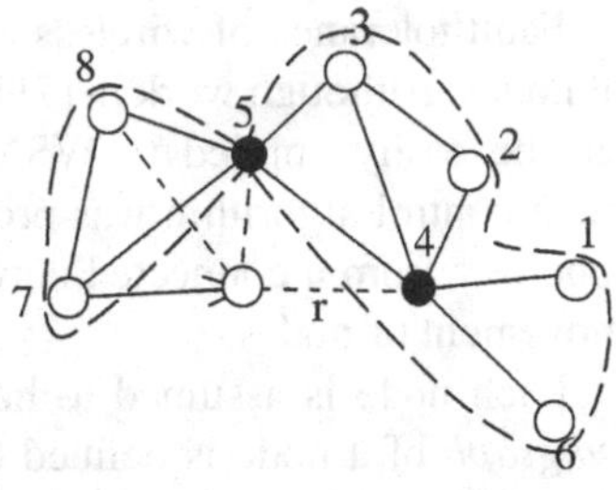

**Fig. 10.5** Critical node with
one critical neighbor

In case (2), the basic idea is to let the critical node with larger ID select one of
its noncritical neighbors to move toward the other critical node. See the example in
Fig. 10.5, where nodes 4 and 5 in black color are critical nodes and other nodes in
white color are noncritical nodes. Since ID of node 5 is larger than ID of node 4,
node 5 leads movement. Again $p$ is assumed to equal 2. Node 5 divides its 2-hop
subgraph into two disjoint sets $A = \{1, 2, 3, 4, 6\}$ and $B = \{7, 8\}$. Suppose distance
of nodes 4 and 7 is the minimum for all nodes in $B$. Node 5 computes new location
of node 7 and asks it to move toward node 4 until connected.

In case (3), some critical node has more than one critical neighbor. Note that each
node sends a critical announcement packet to all its direct neighbors if it finds itself
a $p$-hop critical node. After that, all nodes in the network know the status of their
neighbors. A critical node is said to be *available* if it has noncritical neighbors and
is *unavailable* otherwise. A critical node is available means that it has noncritical
neighbors that are able to move. An available/unavailable critical node broadcasts
an available/unavailable announcement packet to its neighbors. A critical node declares itself a *critical head* if and only if it is available and its ID is larger than
the ID of any available critical neighbor, or has no available critical neighbors. The
basic idea for this case is to use the pairwise merging strategy. Each critical head
dominates the pair merging and selects one of its critical neighbors to pair with. For
example, the available critical neighbor (if any) with largest ID is selected, or otherwise unavailable critical neighbor with the largest ID. Then the movement control
for case (2) is called for each pair to compute the new topology.

See the example in Fig. 10.6, where all black nodes are critical nodes (dashed
block with a node is subgraph of this node). Nodes 1, 5, and 6 are critical heads.
Node 1 becomes a critical head since node 3 is unavailable. Finally, it forms three
pairs: (1, 3), (5, 4), and (6, 4), dominated by nodes 1, 5, and 6, respectively. Each

**Fig. 10.6** Critical node with several critical neighbors

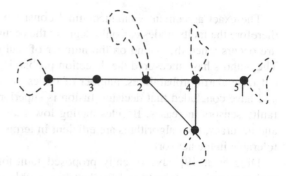

critical head in a pair calls the movement control algorithm for case (2) to merge the pair. Pairwise merging continues until all critical nodes become noncritical, i.e., the network is biconnected.

## 10.4 Target and Event Detection

One of the important sensor network applications is to detect, classify, and locate specific events, and track targets over a specific region. An example is to deploy a sensor network in battle field to detect tanks. Once a tank moves into a specific area, information of the tank, such as location and speed, will be gathered and reported by the sensors that are able to sense the tank. Another example is the use of WSNs as watchers to detect fire in forest. A lot of work has been done in fault tolerance on target and event detection. In this section, we introduce some representative works in target detection and event detection, respectively.

### 10.4.1 Target Detection

Clouqueur et al. [6] have proposed two fault-tolerant algorithms for collaborative target detection in sensor networks in which sensor nodes can either fail due to harsh environmental conditions or maliciously. Both algorithms are based on sensor nodes sharing information to reach consensus.

The first algorithm, called *value fusion*, works as follows. Each node obtains raw energy measurements from every node, computes an average by removing the largest $n$ and smallest $n$ values, and compares this average to a threshold for final decision for a given $n$. The second algorithm, called *decision fusion*, does not work on raw measurement but rather on local decision of each sensor node. It works in the same way as the value fusion algorithm. The authors mention that there is no need for dropping the data when all nodes are known to be fault-free.

The exact agreement guarantees final consistency among fault-free nodes and therefore the faulty nodes can only degrade the accuracy of the system. System failure occurs when the bound on the number of faulty nodes acceptable is violated. The authors have measured the detection probability of both algorithms for different false alarm probabilities, number of nodes, maximum power and decay factor and have concluded that decision fusion is superior to value fusion as the ratio of faulty sensors increases. Besides having low communication cost, high precision and accuracy, both algorithms are efficient in terms of the number of faulty sensors tolerable in the network.

Ding et al. [9] have recently proposed fault-tolerant algorithms to detect the region containing targets and to identify possible targets within the target region, while catering to a stingy sensor energy budget and faulty sensors. The basic idea behind their approach to target detection is that each sensor computes the median of signal measurements such that the disturbances of extreme measurements caused by faulty sensors are filtered out. If a median exceeds certain threshold then it implies that a possible target is present. It does not, however, tell how many targets exist and where they are. A target localization algorithm is used to compute the position of each target. The task of communicating with the base station and computing target positions is delegated to a particular sensor, called the root sensor, which is the one with the local maxima. The root sensor computes the geometric center of neighboring sensors with similar observations. Target position is further refined through the use of multiple epoch observations.

### Algorithm for target detection

1. Each sensor in a given neighborhood obtains its signal measurements.
2. Each sensor computes its median.
3. If the median exceeds a threshold the sensor becomes an event sensor.

### Algorithm for target localization

1. Obtain the estimated signal strength from all event sensors in a given neighborhood.
2. Compute the local event sensor that has the maximum signal strength in a given neighborhood and label them root sensors.
3. For each root sensor compute the location of a target based on the geometric center of a subset of event sensors.

### Algorithm for target identification

1. For each epoch, apply above Target Detection and Target Location algorithm.
2. After collecting raw data for $T$ epochs, the base station applies a clustering algorithm to group the estimates into a final target position computation. Each group is one target.

3. If the size of a group is less than half the number of epochs (i.e., $T/2$), then with high probability this group is a false alarm; otherwise, report a target and obtain the estimate of the position of the target using the geometric center of all raw data within the group.

The algorithms proposed by Ding et al. [9] work well in dense sensor networks as median is not robust in low density, and targets must be far apart in order to identify them as independent targets. The authors assume that each sensor can compute its physical location using either GPS or some GPS-less techniques, and there is no fault in processing and transmitting/receiving neighboring measurements as well as the proper execution of algorithms.

## 10.4.2  Event Detection

Krishnamachari and Iyengar [19] have proposed a distributed and localized fault-tolerant event detection method for WSNs. On the basis of the observation that the sensor faults are likely to be stochastically uncorrelated, while event measurements are likely to be spatially correlated, they propose an algorithm in which each sensor node communicates with its neighbors to collect their binary decisions to correct its own decision. A majority voting scheme is shown to be the optimal decision scheme for fault correction in their work. The algorithm can be described as follows:

Let $N_i$ be the neighbors of a sensor node $i$, each having a probability of failure $p$. Let the binary variables $T_i$ and $S_i$, represent the real situation and real output of $i$, respectively. That is, $T_i = 0$ if the node is a normal region, and $T_i = 1$ if the node is an event region. Similarly, $S_i = 0$ if the sensor measurement indicates a normal value, $S_i = 1$ otherwise. There are four possible scenarios: $(S_i = 0, T_i = 0), (S_i = 0, T_i = 1), (S_i = 1, T_i = 0)$, and $(S_i = 1, T_i = 1)$. It assumes that sensor fault probability is uncorrelated and symmetric, i.e.,

$$P(S_i = 0 \mid T_i = 1) = P(S_i = 1 \mid T_i = 0) = p.$$

The binary value is determined by introducing the threshold on the real-valued readings of sensors, $0.5(m_n + m_f)$, where $m_n$ is the mean value of normal reading and $m_f$ is the mean of event reading.

Let $E_i(a, k)$ be the evidence such that $k$ of its neighboring sensors report the same binary reading $a$ as node $i$ itself. The authors use a Bayesian fault recognition technique to determine an estimate $R_i$ of the true reading $T_i$. Since it assumes that the network is deployed with high density, nearby sensors are likely to have the similar event readings unless they are on the boundary of the event region. So, the following model is adopted:

$$P(R_i = a \mid E_i(a, k)) = k/N.$$

Therefore, the probability with which node $i$ can make a decision to accept its own reading $S_i$ in face of the evidence $E_i(a, k)$ is given by

$$P_{\text{aak}} = P(R_i = a | S_i = a, E_i(a, k)) = \frac{(1 - p)k}{(1 - p)k + p(N_i - k)}.$$

The probability of disregarding its own reading is simply $1 - P_{\text{aak}}$. Based on this Bayesian formulation, $P_{\text{aak}}$ and decision threshold $0 < \theta < 1$, the authors suggest three decision schemes:

**Algorithm using randomized decision scheme**

1. Obtain sensor readings $S_j$ of all $N_i$ neighbors of node $i$.
2. Determine $k_i$, the number of node $i$'s neighbors $j$ with $S_j = S_i$.
3. Calculate $P_{\text{aak}}$.
4. Generate a random number $u \in (0, 1)$.
5. If $u < P_{\text{aak}}$, set $R_i = S_i$, else set $R_i = \neg S_i$.
   ($S_i$ is a binary variable. $\neg S_i$ is opposite value of $S_i$.)

**Algorithm using threshold decision scheme**

1. Obtain sensor readings $S_j$ of all $N_i$ neighbors of node $i$.
2. Determine $k_i$, the number of node $i$'s neighbors $j$ with $S_j = S_i$.
3. Calculate $P_{\text{aak}}$.
4. If $P_{\text{aak}} > \theta$, set $R_i = S_i$, else set $R_i = \neg S_i$.

**Algorithm using optimal decision scheme**

1. Obtain sensor readings $S_j$ of all $N_i$ neighbors of node $i$.
2. Determine $k_i$, the number of node $i$'s neighbors $j$ with $S_j = S_i$.
3. If $k_i > 0.5 N_i$, set $R_i = S_i$, else set $R_i = \neg S_i$.

It proves that the best policy for each node is to accept its own reading if at least half of its neighbors have the same reading. Simulation results show that by using the optimal threshold decision scheme, faults can be reduced by as much as 85–95% with fault rates as high as 10%.

Krishnamachari and Iyengar's algorithm deals only with sensor faults and does not consider decision error caused by noisy measurements. Moreover, it is not known how large the neighborhood size should be.

In [24], Luo et al. have proposed a distributed approach that addresses both detection error and sensor fault simultaneously, while choosing a proper neighborhood size in order to be energy efficient and provide adequate error detection. Their approach is an improvement on the work of Krishnamachari and Iyengar [19], which provides a majority voting scheme to allow individual nodes to correct their binary decisions by communicating with their neighbors. The improvement is the additions of detection error in addition to sensor fault, as well as the determination of a proper neighborhood size to improve energy efficiency.

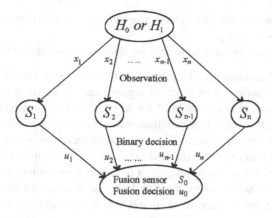

**Fig. 10.7** Decision Framework

In their model, each sensor node has $n$ neighbors and makes its binary decision independently based on its own measurement from the noisy environment. The faulty behavior of a node is considered as "event" by sensors at the location while sensor fault is treated as "no-event." The authors consider a two-layer detection system that consists of a fusion sensor and its $n$ neighbors. The fusion sensor makes a final decision whether an unknown hypothesis is $H_0$ or $H_1$ based on the decision from the $n$ sensors. Let $x_i$ denote the observation of the $i$th sensor, $i = 1, \ldots, n$. Let $u_i$ denote the binary decision (0 or 1) of the $i$th sensor, and let $\lambda$ be the decision threshold common to all nodes. Based on the received sensor decisions, the fusion sensor makes the final decision $u_0$ as shown in Fig. 10.7 with a $k$-out-of-$n$ rule for some majority $k$. Let $u_0 = 0$ if the fusion sensor decides $H_0$ and let $u_0 = 1$ if the fusion sensor decides $H_1$. That is, $u_0 = 1$ if $u_1 + \cdots + u_n \le k$, and $u_0 = 0$ if $u_1 + \cdots + u_n > k$.

The authors use a two-loop search algorithm to find the optimal solutions for $\tau g(= \ln \lambda)$, g, and $n$ for a given bound of detection error $P_e$ and a sensor fault probability $P_f$. Through optimization algorithm it is possible to find the optimal decision threshold $(\lambda)$ and decision majority $(k)$, such that the probability of a detection error is minimized. In the inner loop, the optimal $(\tau, k)$ pair is obtained through numerical optimization for a fixed neighborhood size $n$. In the outer loop, a binary search is employed to find the minimum $n$ that satisfies the given error bound. After optimizing the decision threshold, majority, and neighborhood size, each node then makes a decision based on the threshold and obtains the decisions of its neighbors and makes its final decision based on majority. The detection algorithm can be summarized as follows:

**Algorithm for distributed detection**

1. Set $\tau$, $k$, and $n$ in each sensor node. This can be done at the manufacturing time or after deployment.
2. Each sensor obtains its binary decision $u_i$ based on its measurement and $\tau(= \ln \lambda)$ with threshold test.

3. Each sensor obtains the binary decisions $u_1, u_2, \ldots, u_n$ of its $n$ neighbors and computes $u_1 + \cdots + u_n$.
4. Each sensor makes its final fault-tolerant decision based on the $k$-out-of-$n$ rule.

Luo et al. [24] assume that the ground truth is the same for all $n$ neighbors of nodes $i$; that is, if node $i$ is in an event region, then so are its neighbors and vice versa. This assumption does not hold for sensors at the event boundary. Experiments have shown that this causes confusion in sensors near the boundary and leads to a decrease in detection accuracy. In practical applications, some nodes may not have enough neighbors to meet the optimization criteria. Authors suggest keeping multiple alternative (less optimal) triples or to simply deploy more sensors. Communication errors from noisy communication links can affect event detection. Solution involves modeling these errors as additional sensor faults in the detection process.

## 10.5  Data Gathering and Aggregation

WSNs are usually used to monitor environments and collect information from sensor nodes to the sink or the querying node, which is responsible for further processing. Whatever be the application where a WSN is deployed, individual sensor readings of each sensor node needs to be collected and then reported to the sink. This is done by means of data gathering and data aggregation. Data gathering is to combine data coming from different sensor nodes, eliminate redundancy, and minimize the number of transmissions. In data aggregation, data sensed at neighboring nodes are either highly correlated or simply redundant and need to be aggregated before delivery to the upper layer. Since node failures and transmission failures are common in WSNs, fault tolerance should be considered in designing protocols of data gathering and data aggregation.

A general approach used for data gathering and data aggregation is to construct a spanning tree that is rooted at the sink and connects all nodes in the network. However, a tree topology is not robust against any node and transmission failure. An aggregation framework, called *synopsis diffusion*, that uses energy-efficient multi-path routing schemes was proposed in [25]. There are three basic functions on the synopsis.

- Synopsis generation function: $SG(.)$ takes a sensor reading (including its meta data) and generates a synopsis representing that data.
- Synopsis fusion function: $SF(.;.)$ takes two synopses and generates a new synopsis.
- Synopsis evaluation function: $SE(.)$ translates a synopsis into the final answer.

The synopsis diffusion algorithm consists of distribution phase and aggregation phase. In distribution phase, the aggregation query is flooded through the network and an aggregation topology is constructed. In aggregation phase, aggregation of individual reading of sensors is routed hop by hop toward the querying node. Details of the algorithm are as follows.

**Fig. 10.8** An example for synopsis diffusion

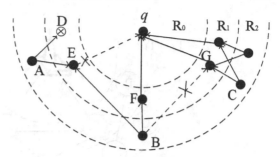

During query distribution phase, nodes in the network form a set of rings around the querying node, say $q$, according to their distance to $q$. $q$ is in ring $R_0$ and a node is in ring $R_i$ if $i$ hops away from $q$. The query aggregation period is divided into epochs and aggregation process is executed once at each epoch. It assumed that nodes in different rings are loosely time synchronized and are allocated specific time intervals when they should be awake to receive synopsis. For example, in Fig. 10.8, node $q$ is in $R_0$. There are five nodes in $R_1$ (including one fault node during the aggregation phase), and four nodes in $R_2$. At the beginning of each epoch, each node in the outermost ring ($R_2$ in the example) generates its local synopsis $s = SG(r)$, where $r$ is the sensor reading relevant to the query answer. The node broadcasts its synopsis to all neighbors. In general, a node in ring $R_i$ wakes up at its allocated time, generates its local synopsis $s = SG(.)$, and receives synopses from all nodes within transmission range in ring $R_{i+1}$. Upon receiving a synopsis $s'$, a node updates its local synopsis as $s = SF(s; s')$. The updated synopsis $s$ is broadcast at the end of allocated time of the node. Thus, the fused synopses propagate layer by layer toward node $q$, which returns $SE(s)$ as the answer to the query at the end of the epoch. We can see that synopsis of node $B$ can reach the querying node even if the transmissions $E \to q$ and $B \to G$ fail. However, since both nodes $D$ and transmission $E \to q$ are failure, synopsis of node $A$ can not be received by node $q$.

By dividing a network into a set of rings, data aggregation can be done over arbitrary topologies. Since multiple paths are used to route data to the querying node, another challenge of the aggregation algorithm is to support duplicate-sensitive aggregates. It exceeds the scope of the chapter. Interested readers can find more details in [25].

Two fault-tolerant schemes for duplicate-sensitive aggregation in WSNs were presented in [13]. The schemes use the available path redundancy to deliver a correct aggregate result to the sink. The basic idea is as follows. When a packet is lost between two sensors because of a link error, it is possible that one or more other sensors have correctly overheard the packet. If some of them have not yet transmitted their own values, they correct the error by aggregating the missing value into theirs. Since the lost packet is aggregated with another packet, error recovery does not cause extra overhead.

It assumes that a network is static in each query process. The track topology of the network is formed in layers, which is similar to Fig. 10.8. The only difference is that some edges may exist among nodes in the same track/layer (see Fig. 10.9).

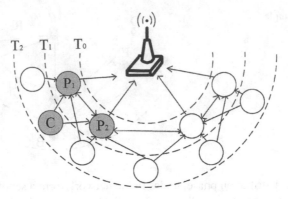

**Fig. 10.9**  Track topology

Edges are classified into three types: primary, backup, and side edges. Primary and backup edges are between adjacent layers (between a sensor node and its parents). Side edges are within the same layer (among parents). It assumes that errors in primary, backup, and/or side edges are independent. Each sensor selects one parent (and correspondingly one edge) as its primary parent and zero or more parents as backups. Primary edges form a spanning tree and are used as long as no communication error occurs. If an error occurs in a primary edge, data may be successfully delivered by some backup edges. The missing value is aggregated at most once by using side edges. Therefore, there is no duplicated value that is aggregated and reported to the sink. It should be noted that a sensor can be a primary parent for some children and at the same time a backup parent for some others.

Each parent attaches a bit vector to data messages it sends. The bit vector contains the IDs and bit positions of children whose values have been correctly received and aggregated. Each parent broadcast its bit vector. By overhearing the bit vectors over side edges, backup parents know link errors when one or more children are missing from the bit vector. Each parent determines the bit positions of its children inside the other parents' bit vectors during topology construction. The bit vector of a primary/backup parent contains two bits for each child. One is the *e-bit* that indicates error in the child's primary edge. The e-bit is set to 1 if the primary parent does not receive messages from a child. Overhearing the primary parent's signal, a backup parent sets its e-bit to 1 as well to propagate the error signal. The other is the *r-bit* that indicates that the sensor is correcting or helping correct the error.

See that in Fig. 10.9, there are three layers. Sensor $C$ in layer $T_2$ has two parents: primary parent $P_1$ and backup parent $P_2$. $C$ transmits a message to $P_1$ and $P_2$ overhears the message. If there is no error, only $P_1$ aggregates the message. Suppose there is a link error in primary edge $C \to P_1$, $P_2$ will receive the bit vector of $P_1$ over the side edge $P_1 \to P_2$. $P_2$ finds that $C$ is missing, and aggregates $C$'s value into its own to correct the error.

Data gathering is another important and fundamental operation in WSNs. A delay/fault-tolerant data gathering scheme for mobile sensor networks was studied in [27]. The scheme consists of two components: data transmission and queue management. Data transmission is to determine when and where to transmit data messages based on the *delivery probability*. Queue management is to determine which messages are to be transmitted or dropped based on the *fault tolerance*. These two important parameters are introduced in the following.

Delivery probability: It reflects the likelihood that a sensor can deliver data messages to the sink. The decision on data transmission is based on delivery probability. Let $\xi_i$ denote the delivery probability of a sensor $i$. $\xi_i$ is initialized with zero and updated upon an event of either message transmission or timer expiration. More specifically, if there is no message transmission within an interval of $\Delta$, the timer expires and it generates a timeout event. It means that the sensor could not transmit any data messages during $\Delta$. Thus, its delivery probability should be reduced. Whenever sensor $i$ successfully transmits a data message to another node $k$, $\xi_i$ should be updated to reflect its current ability in delivering data messages to the sinks. Since end-to-end acknowledgment is not employed in the scheme due to its low connectivity, sensor $i$ does not know whether the message transmitted to node $k$ will eventually reach the sink or not. Therefore, it estimates the probability of delivering the message to the sink by the delivery probability of node $k$, i.e., $\xi_k$. More specifically, $\xi_i$ is updated as follows,

$$\xi_i = \begin{cases} (1-\alpha)[\xi_i] + \alpha\xi_k, & \text{transmission;} \\ (1-\alpha)[\xi_i], & \text{timeout,} \end{cases} \qquad (10.5)$$

where $[\xi_i]$ is the delivery probability of sensor $i$ before it is updated, and $0 \leq \alpha \leq 1$ is a constant employed to keep partial memory of historic status. If $k$ is the sink, $\xi_k = 1$, because the message is already delivered to the sink. Otherwise, $\xi_k < 1$.

Fault tolerance: It indicates the importance of a message by duplicate copy of the message. Different from other gathering schemes where the packets are deleted after they are transmitted, sensors in the scheme may still keep a copy of the message after its transmission to other sensors. Therefore, multiple copies of the message may be created and maintained by different sensors in the network. Such redundancy for fault tolerance indicates the importance of a given message. Each message is assumed to carry a field that keeps its fault tolerance. Let $F_i^j$ denote the fault tolerance of message $j$ in the queue of sensor $i$. There are two approaches to define the fault tolerance of a message in [27].

The first is delivery probability-based approach. The fault tolerance of a message is defined to be the probability that at least one copy of the message is delivered to the sink by other sensors in the network. The fault tolerance of a message is initialized to be zero at the beginning. Suppose sensor $i$ multicasts a data message $j$ to $Z$ nearby sensors, denoted by $N_z, 1 \leq z \leq Z$. The multicast transmission essentially generates totally $Z + 1$ copies. Let $F_{N_z}^j$ denote the fault tolerance of message $j$ transmitted to sensor $N_z$. It can be computed as follows:

$$F_{N_z}^j = 1 - \left(1 - \left[F_i^j\right]\right)(1-\xi_i) \prod_{m=1,m\neq z}^{Z} (1 - \xi_{N_m}). \qquad (10.6)$$

The fault tolerance of the message at sensor $i$ is updated as

$$F_i^j = 1 - \left(1 - \left[F_i^j\right]\right) \prod_{m=1}^{Z} (1 - \xi_{N_m}), \qquad (10.7)$$

where $[F_i^j]$ is the fault tolerance of message $j$ at sensor $i$ before multicasting. The fault tolerance of each message is updated according to (10.2) and (10.3). In general, the more times a message has been forwarded, the more copies of the message are created. It will thus increase its delivery probability, which results in stronger fault tolerance.

The second approach is called message hop count-based approach, where fault tolerance is defined according to the hop count of the message. Let $h_j$ denote the number of times that the message $j$ has been forwarded. A message with larger $h_j$ usually has more copies in the network. Fault tolerance in this approach is defined with $F_i^j = h_j^2/H^2$, where $H$ is the maximum hop count. For a new message, $F_i^j = 0$. since $h_j = 0$. If a message has just been sent to the sink, $F_i^j = 1$. Simulation results in [27] show that the approach based on the message hop count is less accurate than the delivery probability-based approach.

Based on the delivery probability, process of data transmission is as follows. Suppose sensor $i$ has a message $j$ at the top of its data queue ready for transmission. When it moves into the communication ranges of a set of $Z$ sensors, sensor $i$ first learns their delivery probabilities and available buffer spaces via handshake. Message $j$ is transmitted to node $N_z, 1 \leq z \leq Z$, if $F_{N_z}^j > F_i^j$ and there are available buffers in $N_z$. Then, delivery probability of node $i$ is updated according to equation (10.3). The process continues until the updated $F_i^j$ is larger than a predetermined threshold.

Based on the fault tolerance, the process of queue management is as follows. Each sensor has a data queue that contains data messages ready for transmission. A message with smaller fault tolerance means that the message is forwarded less times and has less copy in other nodes. Therefore, the message is more important and should be transmitted with a higher priority. This is done by sorting the messages in the queue with an increasing order according to their fault tolerance. The message with the smallest fault tolerance is always at the top of the queue and is transmitted first. There are two cases that a message is dropped. First, if the queue is full and fault tolerance of a new message is larger than that of the message at the end of queue, the message is dropped. Otherwise, the message at the end of the queue is replaced with the new message Note that the new message should be inserted into the queue at appropriate position (not always at the end of the queue). Second, if the fault tolerance of a message is larger than a threshold, the message is dropped to reduce transmission overhead. It supposes that the message will be delivered to the sinks with a high probability by other sensors in the network. A message will be dropped immediately if it already reaches the sink.

## 10.6   Sensor Monitoring and Surveillance

Sensor monitoring/surveillance is different from target/event detection discussed in previous section. Target/event detection is usually to detect presence and status change of targets and events, while sensor monitoring/surveillance is usually to monitor static targets and interested area in applications. In sensor monitoring/surveillance application, a popular strategy for increasing the reliability is to deploy more nodes than strictly necessary. Since there are strict energy constraints on sensor nodes, usually only a portion of sensor nodes is active to maintain system operations while others go to sleep. Therefore, it requires a schedule that determines which ones remain active and which ones may go to sleep. To ensure proper operation of the network, sleeping nodes should frequently monitor active nodes. Those crashed nodes will be replaced once they are detected. On the other hand, nodes should remain sleeping as much as possible to save the energy. If the energy consumed in the monitoring process is too high, spare nodes may exhaust their batteries before they are needed.

The optimal monitoring period in fault-tolerant sensor networks was studied in [1]. A schedule algorithm called Sleep-Query-Active (SQA) was proposed. It is to ensure that the network remains connected and the lifetime of the network is maximized. It assumes that nodes are aware of their locations and uses the information to divide a two-dimensional space into grids. The distance of two farthest points in any two adjacent grids must be smaller than the communication range of sensor nodes $R$ (see Figure 10.10a). Thus, the cell side, $r$, should satisfy $r \leq R/\sqrt{5}$. It is to make sure of the connectivity of the network if more than one node exists in each grid.

It further assumes that each node in the network can only be in one of two states: either sleeping or active, as illustrated in Fig. 10.10b. The purpose of the wait state is to desynchronize nodes that start at the same time. Nodes in the network send *discovery messages* in following situations: (1) when they enter active state, (2) periodically when they are in the active state (to overcome the loss of messages), and (3) in active state when they receive a discovery message from a node with lower rank. Here, the ranks of nodes are determined by the estimated node active

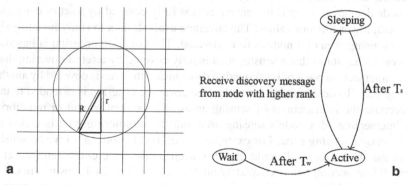

**a**    **b**

**Fig. 10.10**  Sleep-Query-Active algorithm

time, i.e., the remaining energy. That is, higher rank means longer expected lifetime. Whenever a node in active state receives a discovery message from a node with higher rank, it immediately sets up a timer to wake up and goes to sleeping state. The sleeping timeout, $T_s$, is treated as the monitoring period. Each time a node goes to sleep, it picks the value for $T_s$ from an interval with uniform probability. Selection of proper $T_s$ is done via extensive simulations since a theoretical approach to determine $T_s$ is a task of great difficulty [1].

Area coverage problem in sensor networks was addressed in [12]. The problem is to schedule sensor nodes to be active or sleeping, such that both connectivity and full area coverage are achieved. The objective is to achieve similar performance in terms of ratio of active sensors in a given round as the best existing localized solution, while significantly reducing the number of messages for making decision at each node. It assumes that sensing radio and communication radio are different, and sensor nodes are time synchronized. The authors proposed four variants of protocols, which relay on low communication overhead in order to be applied for highly dense networks. Overview of the proposed protocols is as follows.

Each node selects a random timeout and listens to messages sent by other nodes before the timeout expires. Once timeout ends, the neighbor table of a node contains every node that already made decision with shorter timeout. The node evaluates the status of coverage and connectivity and decides to be active if its sensing area is not fully covered or connectivity requirement is not satisfied. Otherwise, the node goes to sleep mode. The node announces its decision to its neighbors. Note that a node may hear from more active neighbors after it decides to be active. The node may change its mind by sending retreat message to its neighbors if its sensing area becomes fully covered or connectivity requirement becomes satisfied.

Framework of the proposed protocols consists of four components: timeout computation, coverage evaluation, connectivity conservation, and decision announcement. They are introduced one by one.

In timeout computation, it is assumed that any two neighboring nodes would select different random numbers, so that two nodes never attempt to send messages at the same time to avoid collisions. Actually, a node may make decision to go to sleep before the timeout expires once it receives enough information from neighbors and the sensing area is fully covered already. It saves computation on useless messages.

A node decides to sleep if its sensing area is fully covered by a set of connected nodes with lower timeout values. The coverage evaluation is to study how to judge that the sensing area of a node is fully covered. The covering criterion is based on the theorem that states that a sensing area is fully covered by a set of covering disks if every intersection point of two covering disks inside the area is covered by another covering disk. To deal with border nodes, the coverage criterion is extended to take into account the intersections of sensing areas and the deployment or monitoring area. Intersection of a node's sensing area and the monitoring area is called the node's *revised sensing area*. For example in Fig. 10.11, node $A'$s revised sensing area is the shaded area. $A$ could get into sleep mode since circle centered at $C$ covers all intersection points created by other circles and revised sensing area of $A$, while $C1$ and $C2$ are covered by circle centered at $D$.

**Fig. 10.11** Coverage
evaluation

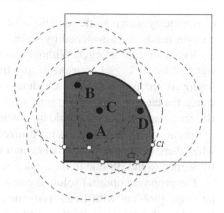

In connectivity conservation, a simple rule is that the connectivity is guaranteed
so long as the sensing area is fully covered and the communication range is at least
twice the sensing range. In the case that the communication range is less than double
of the sensing range, a node can decide to turn off if its sensing area is fully covered
by connected neighbors.

After verifying the coverage condition, each node decides whether or not to send
a message. Messages contain geographic position of nodes and activity status. Four
variants of the protocols are differentiated by the messages sent by nodes.

Positive-only: Each node that decides to be active sends exactly one message.
Nodes that decide to sleep do not send any message.

Positive and negative: Each node sends exactly one message. A positive or a
negative acknowledgment are for an active or sleep status, respectively.

Positive and retreat: Same as positive-only except that a node that has already
decided to be active can later switch to sleep status based on messages from newly
announced active nodes. Such nodes send one retreat message.

Positive, negative, and retreat: Any decision of a node will be transmitted. Each
node sends one message corresponding to the original decision on active or sleep
status. Nodes with originally positive decision may switch to sleep mode later and
send one retreat message.

Simulation results show that the last variant of the protocols has the best overall
performance among four variants, in terms of less percentage of active nodes and
longer lifetime of the network. However, it generates more messages than others.

The surveillance nature of sensor networks requires a long lifetime. A maximal
lifetime problem in sensor surveillance systems was studied in [21]. Given a set
of targets and sensors and a base station (BS) in an area, the sensors are used to
watch (or monitor) the targets and collect sensed data to the BS. Each sensor has an
initial energy reserve, and a fixed surveillance range and an adjustable transmission
range. A sensor can watch at most one target at a time. A target can be inside the
surveillance range of several sensors. A typical example is the use of camera to

continuously watch some targets, such as cargo containers. In some applications, a target needs to be watched by several sensors at any time for fault tolerance, or for localization of a target. Without loss of generality, each target is supposed to be watched by $k$, $k \geq 1$, sensors at any time. Since sensors are usually redundantly deployed, the problem is to schedule a subset of sensors to be active at a time to watch the targets and find the routes for the active sensors to send data back to the BS, such that each target should be watched by $k$ sensors at any time and the lifetime of the entire sensor network is maximized. The *lifetime* is the duration up to the time when there exists one target that cannot be watched by $k$ sensors or data can not be forwarded to the BS due to the depletion of energy of the sensor nodes.

The proposed optimal solution consists of three steps. In the first step, it formulated the problem with linear programming (LP) technique to compute the upper bound on the maximal lifetime and a workload matrix. The objective of the second step is to find the detailed schedule for sensors to watch targets based on the workload matrix. The basic idea is to represent the workload matrix as a bipartite graph and then apply the perfect matching technique to decompose the workload matrix into a sequence of schedule matrices. Each perfect matching corresponds to a schedule matrix. The process of finding perfect matching is continued until the workload matrix is completely decomposed. The last step to construct sensor surveillance trees for each session is based on the schedule matrices and data flow computed from the LP formulation.

## 10.7  Thoughts for Practitioners

Link fault is a common problem in real deployments of WSNs. To improve reliability of communications, cross-layer design could be adopted for specific applications of WSNs. For example, placement of redundant sensor nodes than necessary is a solution to achieve fault tolerance of the network. In MAC layer and communication layer, one standard solution is to increase reliability of wireless communication in adoption of error correction mechanisms and acknowledgement mechanism. In software layer, it requires to avoid software bugs in both OS and middleware. Based on difference in applications, techniques and algorithms that are introduced in the chapter could be employed.

However, there is a trade-off between fault tolerance and efficiency of the network. Many fault-tolerant techniques and algorithms may cause extra energy consumption, increase of overhead, transmission collision, and delay of operation. Therefore, balance point should be carefully analyzed and determined depending on the task of networks and objective of applications.

To apply WSNs in safety critical applications, security threats must be addressed during all operational phases of a fault-tolerant system. However, most current approaches do not include security measures.

## 10.8 Conclusions and Directions for Future Research

The goal of the chapter is to investigate current research work on fault tolerance in WSNs. We studied how fault-tolerant techniques were addressed in node placement, topology control, target and event detection, data gathering and aggregation, and sensor surveillance. We focused on the application layer and introduced representative works in each application. Actually, there are other applications where fault tolerance attracts attention, such as clustering, time synchronization, gateway assignment, etc.

Although extensive works have been done on fault tolerance in each layer of the WSN system, cross-layer solutions are expected in future. Use of the resource could be more efficient if resource can be properly integrated and scheduled in different layers. Therefore, cross-layer solutions are expected to have better performance than current solutions.

A new trend of WSNs is to cooperate or integrate with other wireless device/systems, such as actuator networks and RFID system. For example, there are an increasing number of applications that require the network system to interact with the physical system or environment via actuators. That is, it requires the use of sensor networks along with actuators to build wireless sensor and actuator networks (WSANs). Although fault tolerance techniques for WSNs could be reused in WSANs, there are new challenges that require new solutions. For example, when an actuator fails, the sensors that report their data to the actuator may either switch to another actuator or directly pass the data to the sink.

**Acknowledgments** This research is supported by NSERC Collaborative Research and Development Grant CRDPJ 319848-04 and the UK Royal Society Wolfson Research Merit Award.

## Terminologies

*Fault tolerance.* The ability of a system to deliver a desired level of functionality in the presence of faults.
*Self-diagnosis.* A sensor node determine faults by itself.
*Cooperative diagnosis.* Several sensor nodes determine faults by cooperation.
*Covering with disks.* Given a set of points in the plane, the problem is to identify the minimum set of disks with prescribed radius to cover all the points.
*Steiner tree problem with minimum number of Steiner points (STP-MSP).* Given a set of terminals in the Euclidean plane, the problem is to find a Steiner tree such that each edge in the tree has length at most $d$ and the number of Steiner points is minimized.
*Dominating Set (DS).* A subset of the vertices of a graph if every vertex in the graph is either in the subset or is adjacent to at least one vertex in the subset.
*Connected Dominating Set (CDS).* A connected DS.

*p-hop subgraph.* Of a node is defined by the graph that contains all nodes that are within $p$-hops from the node and all corresponding links.

*p-hop critical node.* The node is said to be $p$-hop critical node if and only if its $p$-hop subgraph is disconnected without the node.

*Available.* A critical node is said to be available if it has noncritical neighbors.

*Critical head.* A critical node is said to be a critical head if and only if it is available and its ID is larger than the ID of any available critical neighbor, or has no available critical neighbors.

## Questions

1. From system point of view, in which layers could faults happen in WSNs? And which layer does the chapter focus on?
2. What are the basic methods for fault detection and recovery in WSNs?
3. What is the basic idea of [22] to place the minimum number of relay nodes for WSNs?
4. What is the best known result for *covering with disks* problem?
5. What is the best known approximation ratio for STP-MSP?
6. Please find DS and CDS in Fig. 10.5.
7. Suppose $p = 1$, please find $p$-hop critical nodes in original graph of Fig. 10.3.
8. What are the basic methods in target/event detection?
9. What is the difference between data gathering and data aggregation?
10. What is the difference between sensor monitoring/surveillance and target/event detection?

## References

1. F. Araújo, and L. Rodrigues. "On the Monitoring Period for Fault-Tolerant Sensor Networks," *LADC 2005*, LNCS 3747, São Salvador da Bahia, Brazil, October 2005.
2. J.L. Brediny, E.D. Demainez, M.T. Hajiaghayiz, and D. Rus, "Deploying Sensor Networks with Guaranteed Capacity and Fault Tolerance," *MobiHoc* 2005, urbana-champaign, IL, 2005.
3. M. Cardei, S. Yang, and J. Wu, "Algorithms for Fault-Tolerant Topology in Heterogeneous Wireless Sensor Networks," *IEEE Transactions on Parallel and Distributed Systems*, vol. 19, no. 4, pp. 545–558, 2008.
4. Y. Chen, and S.H. Son, "A Fault Tolerant Topology Control in Wireless Sensor Networks," *Proceedings of the ACS/IEEE 2005 International Conference on Computer Systems and Applications*, 2005.
5. S. Chessa, and P. Maestrini, "Fault Recovery Mechanism in Single-Hop Sensor Networks," *Computer Communications*, vol. 28, issue 17, pp. 1877–1886, 2005.
6. T. Clouqueur, K.K. Saluja, and P. Ramanathan, "Fault Tolerance in Collaborative Sensor Networks for Target Detection," *IEEE Transactions on Computers*, vol. 53, no. 3, pp. 320–333, 2004.

7. S. Das, H. Liu, A. Kamath, A. Nayak, and I. Stojmenovic, "Localized Movement Control for Fault Tolerance of Mobile Robot Networks," *The First IFIP International Conference on Wireless Sensor and Actor Networks* (WSAN 2007), Albacete, Spain, 24–26 Sept. 2007.
8. M. Demirbas, "Scalable Design of Fault-Tolerance for Wireless Sensor Networks," PhD Dissertation, The Ohio State University, Columbus, OH, 2004.
9. M. Ding, F. Liu, A. Thaeler, D. Chen, and X. Cheng, "Fault-Tolerant Target Localization in Sensor Networks," *EURASIP Journal on Wireless Communications and Networking*, 2007.
10. D. Du, L. Wang, and B. Xu, "The Euclidean Bottleneck Steiner Tree and Steiner Tree with Minimum Number of Steiner Points," *Computing and Combinatorics, Seventh Annual International Conference COCOON 2001, Guilin, China, Aug. 2001 Proceedings.*
11. R.J. Fowler, M.S. Paterson, and S.L. Tanimoto, "Optimal Packing and Covering in the Plane are NP-complete," *Information Processing Letter*, vol. 12, pp. 133–137, 1981.
12. A. Gallais, J. Carle, D. Simplot-Ryl, and I. Stojmenovic, "Localized Sensor Area Coverage with Low Communication Overhead," *Proc. of IEEE Sensor'06*, Daegu, Korea, Oct. 2006.
13. S. Gobriel, S. Khattab, D. Mosse, J. Brustoloni, and R. Melhem, "Fault Tolerant Aggregation in Sensor Networks Using Corrective Actions," *Third Annual IEEE Communications Society on Sensor and Ad Hoc Communications and Networks*, vol. 2, pp. 595–604, 2006.
14. X. Han X. Cao E.L. Lloyd, and C.C. Shen, "Fault-Tolerant Relay Node Placement in Heterogeneous Wireless Sensor Networks," *INFOCOM* 2007.
15. B. Hao, H. Tang, and G.L. Xue, "Fault-Tolerant Relay Node Placement in Wireless Sensor Networks: Formulation and Approximation," *High Performance Switching and Routing* (HPSR 2004), pp. 246–250, 2004.
16. D.S. Hochbaum, and W. Maass, "Approximation Schemes for Covering and Packing in Image Processing and VLSI," *Journal of the ACM (JACM)*, vol. 32, issue 1, pp. 130–136, Jan. 1985.
17. S. Khuller, and B. Raghavachari, "Improved Approximation Algorithms for Uniform Connectivity Problems," *Journal of Algorithms*, vol. 21, pp. 214–235, 1996.
18. F. Koushanfar, M. Potkonjak, and A. Sangiovanni-Vincentelli, "Fault Tolerance in Wireless Sensor Networks," in *Handbook of Sensor Networks*, I. Mahgoub and M. Ilyas (eds.), CRC press, Section VIII, no. 36, 2004.
19. B. Krishnamachari, and S. Iyengar, "Distributed Bayesian Algorithms for Fault-Tolerant Event Region Detection in Wireless Sensor Networks," *IEEE Transactions on Computers*, vol. 53, no. 3, pp. 241–250, 2004.
20. X.Y. Li, P.J. Wan, Y. Wang, and C.W. Yi, "Fault Tolerant Deployment and Topology Control in Wireless Networks," *MobiHoc* 2003, Annapolis, MD, 2003.
21. H. Liu, P.J. Wan, and X. Jia, "Maximal Lifetime Scheduling for Sensor Surveillance Systems with *K* Sensors to 1 Target," *IEEE Transactions on Parallel and Distributed Systems*, vol. 17, no. 12, Dec. 2006.
22. H. Liu, P.J. Wan, and X. Jia, "On Optimal Placement of Relay Nodes for Reliable Connectivity in Wireless Sensor Networks," *Journal of Combinatorial Optimization*, vol. 11, pp. 249–260, 2006.
23. E. Lloyd, and G. Xue, "Relay Node Placement in Wireless Sensor Networks," *IEEE Transactions on Computer*, vol. 56, pp. 134–138, 2007.
24. X. Luo, M. Dong, and Y. Huang, "On Distributed Fault-Tolerant Detection in Wireless Sensor Networks," *IEEE Transactions on Computers*, vol. 55, no.1, pp. 58–70, 2006.
25. S. Nathy, P.B. Gibbons, S. Seshany, and Z.R. Anderson, "Synopsis Diffusion for Robust Aggregation in Sensor Networks," *SenSys'04*, November 3–5, 2004, Baltimore MD.
26. G. Tolle, J. Polastre, R. Szewczyk, D. Culler, N. Turner, K. Tu, S. Burgess, T. Dawson, P. Buonadonna, D. Gay, and W. Hong, "A macroscope in the redwoods," In *SenSys'05: Proceedings of the third International Conference on Embedded Networked Sensor Systems*, New York, 2005.
27. Y. Wang, and H. Wu, "DFT-MSN: The Delay/Fault-Tolerant Mobile Sensor Network for Pervasive Information Gathering," *INFOCOM* 2006, Barcelona, Spain, 2006.
28. W. Zhang, G.L. Xue, and S. Misra, "Fault-Tolerant Relay Node Placement in Wireless Sensor Networks: Problems and Algorithms," *INFOCOM* 2007, Anchorage, AL, 2007.

# Chapter 11
# Self-Organizing and Self-Healing Schemes in Wireless Sensor Networks

**Doina Bein**

**Abstract** The basis of sensor networks is the process of sensing, data processing, and information communication [1]. In this chapter we investigate methods for computing self-healing and self-configuring strategies that can be encoded into a sensor unit prior to deployment. A *strategy* is a set of rules that assigns actions to specific states; specifically, a sensor unit periodically checks whether its current state matches some rule, and in affirmative the corresponding action (or actions) is executed. Ideally, an optimization method evaluates the rules, refines them, and converges to a temporarily optimal strategy. We do not address reliability of the sensor node or the correctness of the data collected through the sensing unit. The design of self-organization is different among the wireless networks models (e.g., MANET, cellular networks, WSN) due to the performance objective of the network Clare et al. (Proc SPIE 3713: 229–237, 1999); (Sohrabi et al. IEEE Pers Comm Mag 7: 16–27, 2000); Sohrabi and Pottie (Proc IEEE Vehicular Tech Conf, 1222–1226, 1999).

## 11.1 Introduction

Started as a DARPA project, the Sensor Information Technology (SensIT) program focused on decentralized, highly redundant networks to be used in various aspects of daily life. The nodes in sensor networks are usually deployed for specific tasks: surveillance, reconnaissance, disaster relief operations, medical assistance, and the like. Increasing computing and wireless communication capabilities will expand the role of the sensors from mere information dissemination to more demanding tasks

D. Bein
Applied Research Laboratory, The Pennsylvania State University, University Park,
PA 16802-5018, USA
e-mail: siona@psu.edu

S. Misra et al. (eds.), *Guide to Wireless Sensor Networks*, Computer Communications and Networks, DOI: 10.1007/978-1-84882-218-4_11,
© Springer-Verlag London Limited 2009

as sensor fusion, classification, and collaborative target tracking. They may well be deployed in hostile environments and inaccessible terrains. Through a cooperative effort, proper information has to be passed to a higher level. The position of the sensor nodes is not predetermined, i.e., the network can start with an arbitrary topology.

The sensor nodes are generally densely scattered in a *sensor field*, and there are one or more nodes called *sinks* (or also *initiators*), capable of communicating with higher level networks (Internet, satellite, etc.) or applications. By a *dense* deployment of sensor nodes we mean that the nearest neighbor is at a distance much smaller than the transmission range of the node. In general, a spatial distribution of the sensors is a two-dimensional Poisson point process. The nodes must coordinate also as to exploit the redundancy provided by the high density of nodes to minimize the total energy consumption, thus extend the lifetime of the system overall, and to avoid collisions. Selecting fewer nodes saves energy, but the distance between neighboring active nodes could be too large, thus the packet loss rate could be so large that the energy required for transmission becomes prohibitive. Energy can be wasted by selecting more nodes, and shared channels will be congested with redundant messages, thus collision and subsequently loss of packets can occur.

Sensor nodes are equipped with a processor, but they have limited memory. They can carry out simple tasks and perform simple computations. The communication is wireless, e.g., radio, infrared, or optical media. Because of the large number of nodes and thus the amount of overhead, the sensor nodes may not have any global identification (ID). In some cases, they may carry a global positioning system (GPS). For most of the WSNs, nodes are not addressed by their IP addresses but by the data they generate. To be able to distinguish between neighbors, nodes may have local unique IDs. Examples of such identifiers are 802.11 MAC addresses [24] and Bluetooth cluster addresses [6].

Energy consumption occurs in three domains: sensing, data (signal) processing, and communication; communication is the major consumer of energy. This leads to the need for highly localized and distributed algorithms for networking. The required bandwidth for sensor data is relatively low, in the range of 1–100 kb/s [36]. Communication channels between sensor nodes must be established with few messages. This can be accomplished by allowing a node to decide when to invite another node to join a connection or when to drop a connection; thus, nodes must have full control of the connection process.

A self-configurable system must be able to extract the necessary information supporting its software intelligence from the data it collects. The need for automatic changes (self-configuring) should take the following design issues for a self-configuring network into account:

- *Ad hoc deployment.* The nodes might not be positioned in a regular pattern (grid, honeycomb, 3D grid, 3D honeycomb, etc.).
- *Limited resources and energy constraint.* A unit has limited resources (battery, memory, computational power). The number of actions a node executes and the time consumed by an action must be minimized, in order to prolong its battery lifetime.

- *No globally unique IDs*. The addressing scheme relies only on locally unique IDs.
- *Scalability*. The protocol for delivery of the data back to the initiator (or initiators) must scale not only according to the number of sensors but also according to the number of initiators and events.
- *Reliable delivery*. Reliable data delivery returned back to the initiator(s) must be ensured even if sensors are unreliable.
- *Error-prone wireless medium*. The wireless medium is more error-prone than the wired medium, and collisions can occur more frequently.

A good self-organizing protocol scales not only according to the number of nodes in WSN, but also according to the density of the nodes.

## 11.2  Background

A sensor communicates with its neighboring sensors over the shared wireless medium and has no global knowledge of the entire network. Since the number of sensors in a network is very large, any global knowledge is infeasible. Sensors are assumed to remain static throughout their lifetime. However, they may be highly unreliable, fail at any time without advance notice, and sometimes appear without any warning at all. For example, the latter situation can occur when, after the battery has failed, there is a rechargeable battery, and the battery is recharged.

We can summarize the characteristics of a WSN as follows:

- A sensor has limited power, computational capability, memory, and reduced transmission range (few meters).
- A large number of sensors are densely deployed.
- A sensor needs to work unattended, is prone to failure, can cooperate, and is application-specific.
- WSN is connected to the outside world through a *sink* node with high computational capability and unlimited power. The sink is also responsible for maintenance of the network.

A sensor node is made up of four basic components: a sensing unit (transducer), a power unit, a radio interface (transceiver), and a processing unit (see Fig. 11.1).

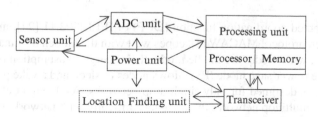

**Fig. 11.1**  Components of a Sensor Node

Smart sensors may have a location finding unit, responsible for locating the sensor position (using GPS or its European counterpart, Galileo). The transceiver unit connects the node to the rest of the network. A sensing unit is composed of a sensor and a signal converter, which converts from analog to digital (ADC). The observed phenomenon is captured by the sensor as an analog signal that is fed into the ADC unit that subsequently feeds it into the processing unit. The processing unit, equipped with a small memory, is responsible for signal processing, information communication, and managing other components. It also supervises a collaboration of the node with other nodes in accomplishing assigned sensing tasks.

There are two types of signal processing techniques, *coherent* and *noncoherent*. In *noncoherent processing*, the raw data collected from the sensor unit are preprocessed locally to extract a small set of values for given parameters to be forwarded to a *central node* (CN) for further processing. Ultimately it reaches the sink. In *coherent processing*, the raw sensor data collected from the sensor unit are tagged with a timestamp after some limited preprocessing and subsequently sent to a central node for more complicated computations. The advantage of noncoherent processing is that it has fairly small data size, while coherent processing may have large data size. The central node is elected from the neighboring sensors (near-field or far-field).

MANETs have been studied long before WSNs have. Naturally, the research for creating self-organizing WSN looked for inspiration from self-configuring and self-healing protocols in MANET, adapting them to the particularities of a WSN.

In Sect. 11.2.1 we discuss the drawbacks that MANET protocols have when they are used for a WSN and proposals that have solved some of the issues. Cluster-adapted protocols (e.g., sleep-and-awake, virtual clustering, and lightweight clustering) are presented in Sect. 11.2.2. Directed diffusion and other tree construction protocols specific to a WSN are discussed in Sect. 11.2.3. Recent approaches for designing self-organizing WSN imitate some aspects from natural life and we discuss research in that direction in Sect. 11.2.4.

## 11.2.1  Self-Organizing Protocols Inspired from MANETs

Previous work on MANET has focused on topology formation by discovering neighbors, on maintaining a neighborhood list, and on scheduling the communication among nodes. In many MAC protocols (IEEE 802.11, TDMA, or CDMA) nodes must monitor the channel all the time, thus these protocols are unsuitable for a WSN.

The distributed coordination function (DCF) of IEEE 802.11 [24], mainly built on the research protocol MACAW [5], copes well with the hidden terminal problem. Unfortunately, when nodes are in idle mode, their energy consumption is relatively high [41]. The power save mode (PS) allows nodes to sleep and awake periodically, but this mode is designed for only single-hop networks. As Tseng et al. [44] have observed, in a multihop network, PS mode has difficulty with network partitioning, neighborhood discovery, and clock synchronization. To overcome these drawbacks,

Tseng et al. [44] have proposed three schemes, but neither one synchronizes the sleep-and-awake periods of neighboring nodes.

A low duty cycle MAC scheme (IEEE 802.15.4) in connection with modified MAC protocols, such as clustering, spanning tree construction, can be used instead of IEEE 802.11, to reduce the energy consumed by a node. The protocol of Sohrabi et al. [40] (SMACS) combines neighborhood discovery and channel assignment phases, to let nodes form links on the fly. This means that the links last for the time the nodes are awake.

Piconet [4], an architecture designed for low-powered MANETs, sets the node to sleep, but there is no synchronization among the sleep-and-awake periods of neighboring nodes.

TDMA-based protocols require the nodes to form real clusters, and nodes communicate mainly within the cluster, thus interference can occur with a high probability. The TDMA schedule needs to be changed whenever the topology of a cluster changes. In the end, a relatively large number of control messages are generated for any change. Sohrabi and Pottie [39] propose a protocol based on a TDMA-like frame, called *superframe*. Here, a node talks with each neighbor, one at a time. The protocol does not prevent two nodes to access the medium at the same time, but the probability is lowered. One drawback is a poor utilization of the bandwidth, since the time slots for communicating with neighbors cannot be reutilized when there is no communication with certain neighbors.

A *pervasive sensor network* is a special type of sensor network that contains special type of sensor nodes called *parent nodes (PNs)*, with relatively high number of resources. The parent nodes are responsible for in-network data processing, routing the processed data to the initiator(s), and minimizing the communication delay. They can be dynamically added or removed from the network, and their functionality should be "plug-and-play." Periodically, the parent nodes exchange so-called *Network State Beacons (NSBs)* at a set Beacon Exchange Rate.

Iqbal et al. [26] propose a proactive self-configuration protocol by imposing bounds of the beacon exchange rate, and by dynamically updating the rate in response to changes of the network load.

Gupta et al. [19] propose a method of self-organizing a sensor network where nodes have unique IDs. Each time there is a request for data, in response to a query the network is reorganized such that only a small subset of sensors will process it: Only a small set of nodes process the query, thus a so-called logical partitioning of the network occurs. The number of messages used in this self-organization process is relatively small, thus the overhead introduced by this technique does not offset the expected benefit. The only drawback is that this method assumes unique IDs for the sensor nodes.

## 11.2.2   Cluster-Adapted Protocols

Early clustering algorithms proposed for MANET (e.g., highest or lowest ID, highest connectivity) cannot work for all WSNs, since the sensor nodes may not have unique IDs. Or these algorithms rely on current neighborhood information, which takes time and energy, while in a WSN the sensors are asleep most of the time. If the network density is high, the possibility of message collision is also high, which increases the energy cost and time requirements.

Still there is a conviction that better scalability can be achieved through structuring the network in a hierarchical manner [18]: Different nodes may have different roles. Hierarchical connected dominating set proposed by Wu [46, 47] is extended by Kochhal et al. [29] to a role-based hierarchical self-organization algorithm. Subramanian and Katz [42] propose a self-configuration architecture that is hierarchical. Their work is continued by Chevallay et al. [9]; the cluster head is elected based on energy and processing capabilities. Another approach for conserving energy is to construct multiple connected dominating sets for various transmission radii [13].

In *lightweight clustering* [2, 20] cluster heads are selected at random. Each node selects itself as a cluster head with a certain probability $p$. The value of $p$ is fixed, thus it is not adapted to the dynamics of the network. Unfortunately, the communication between the clusters is not fully addressed.

Ye et al. [49] propose a medium access control protocol (S-MAC), which enables sensors to form *virtual clusters* based on common sleep schedule.

Sleep-and-awake protocols schedule the nodes to sleep most of the time in order to save energy; they wake up either at random or at predefined moments. Using local neighbor information, nodes turn themselves on or off, depending on whether they are needed to cover a region that is either uncovered or is covered by other sensors. Slijepcevic and Potkonjak [38] propose a heuristic that selects a mutually exclusive set of sensor nodes in order to cover a monitored area. Tian and Georganas [43] use a node scheduling scheme to turn off redundant sensor nodes. Ye et al. propose the PEAS protocol [48] that divides the monitored area into disks of a given radius, and allows a single node inside of a disk to be active at any time, while all the other nodes are asleep. Olariu et al. [35] propose an energy-efficient self-organization protocol for a WSN that adapts to network or environment changes; to save energy, nodes are turned off as long as possible. Cerpa and Estrin [8] propose a self-configuring, adaptive algorithm (ASCENT) for a network of sensors and actuators: Adaptively, a set of nodes is selected to become "active" while the rest of the nodes remain "passive." Periodically the protocol checks each passive node whether it should become active. The active nodes are continuously on, performing packet routing and other operations. The protocol requires high computational power and memory on individual nodes.

## 11.2.3 Directed Diffusion and Other Spanning Tree Construction Protocols

Mirkovic et al. [34] propose a multiphase protocol, which has four phases for the tree construction and four phases for tree maintenance, for a sensor network with unique IDs. The network is organized into a structure that is self-optimizing, multicast tree-based. The self-optimizing concept is related to the minimization of the data forwarding paths between multiple sources and sinks. But the optimization heuristic is applied only when the tree is built for the first time. Subsequent actions, such as a node joining or leaving, have only a local effect. In case of joining, the closest neighbor becomes the "parent" of the new node in the tree; in case of a failure, attempts of reconnection are made to the nearer neighbors. From this point of view, the protocol proposed by Mirkovic et al. is more self-configuring than self-optimizing.

 *Directed diffusion* [25] is a data-centric protocol in which nodes are not identified by their ID, but by the data they generate as result of detecting (sensing). The data are organized as *attribute–value* pairs. An initiator node makes a request for a certain piece of data by broadcasting an *interest* for that piece of data throughout the sensor network. Different nodes may match the request on different degrees, and gradients are kept in order to point out the neighbor (or the neighbors) toward the initiator (or initiators) of that particular request. There is one gradient for each neighbor. The possible answers are forwarded back to the initiator, and intermediate nodes may perform a prefiltering of the answers. The initiator periodically broadcasts an interest to the rest of the network, to maintain the robustness and reliability of the network.

## 11.2.4  Bioinspired Self-Organizing Protocols

The connection between nature and self-organizing system was introduced by von Foerster [16] and later it was broadened by Eigen [15]. Bioinspired engineering currently relies on aspects of artificial immune systems (AISs), swarm intelligence, evolutionary (genetic) algorithms, molecular biology, and analysis of heartbeat rate.

 An AIS focuses on detecting changes in the environment or deviation form the normal behavior of a system and is used for virus and detection systems [21, 28]. It could also be used for self-protective systems. Swarm intelligence, observed at the level of large groups of interacting insects (as ants and bees), can be used for formation of clusters, or in search and exploration in self-organizing networks [7, 27]. Signaling pathways (intra- and intercellular) and diffuse communication from molecular biology could be effectively used in a WSN for efficient response from the WSN back to the sink, and shortening of routes [14, 30]. Analysis of heartbeat rate is used to achieve detection consensus in a fully connected sensor network [22] or globally connected sensor network [31].

 Recent research in self-organization in WSN and in sensor/actuators networks (SANET) using bioinspired mechanisms [12, 14, 17] has focused on bioinspired networking, or self-organization, and bioinspired data acquisition [11]. To reduce the number of messages sent to the sink, one solution is to identify areas where similar readings occur and allow only a single area to communicate data [37, 45]. Cuhna [11] proposes another solution that allows sensor nodes to identify patterns of the sensed phenomenon and report back to the sink node only abnormal variations of the pattern. This mechanism is similar to human and animal response to continually receiving stimuli. Barbarossa and Scutari [3] have proposed a self-synchronization mechanism to achieve global decision without a fusion center.

## 11.3  Thoughts for Practitioners

In accordance to the Moore's law, the cost of radio frequency integrated circuits (RFIC) and microcontrollers is falling rapidly. Currently, sensors have different design technologies, e.g., micro electrical-mechanical systems (MEMS), wireless integrated network Sensors (WINS), and Piconet.

If the sensor units are moved or changes in the environment occur, the network should be able to automatically reconfigure and readapt, i.e., self-adapt and self-configure. If the sensor units are malfunctioning, are underpowered, or external factors threaten their functionality, they should be able to reconfigure, i.e., self-healing and self-protection should occur. Given the hardware limitations, self-organizing algorithms must provide robust and energy-efficient communication.

A multihop sensor network must operate in both sensor-to-sink (convergecast) and sink-to-sensor modes (broadcast). When many nodes have failed, the medium access control (MAC) and the routing protocols must accommodate the formation of new links (by adjusting the transmission range) and the routes to the sink nodes, by rerouting packets through regions where nodes have more energy left.

## 11.4  Directions for Future Research

Lighting and HVAC (heating, ventilation, air conditioning) wireless systems [23], which cooperate with motion detectors and security sensors, are future contexts of WSNs. Such networks will learn to anticipate simple events and to flag abnormal events. Ideally, they should be "plug-and-play," i.e., upon deployment, units should automatically configure and optimize their performance based on local information (current environment, local density of sensor units, and network throughput), i.e., self-configure and self-optimize.

## 11.5  Conclusions

In this chapter we have investigated methods for computing self-healing and self-configuring strategies that can be encoded into a sensor unit prior to deployment. Because of the fact that sensors do not have global IDs, have nonrenewable battery, and are stationary, WSNs could not simply use existent self-organizing (self-configuring) and self-healing protocols developed for MANET. We have reasoned why, and we have presented a number of protocols that are inspired from MANET, and described protocols specific to WSNs. Recent research focuses on bio-inspiring communication protocols narrowing the gap between nature and an autonomous system.

# Terminologies

*Bioinspired protocol.* A network protocol that imitates a natural process.

*Coherent signal processing.* The raw sensor data collected from the sensor unit, after some limited preprocessing, are tagged with a timestamp and sent to a central node for deeper computation.

*Directed diffusion.* It is a data-centric protocol in which the nodes are not identified by their ID, but by the data they generate as result of detecting (sensing).

*Lightweight clustering.* A clustering protocol in which the cluster heads are selected at random (each node selects itself as a cluster head with a certain probability $p$). The value of $p$ is fixed.

*Noncoherent signal processing.* The raw data collected from the sensor unit are preprocessed locally to extract a small set of values for given parameters and are forwarded to a *central node* (CN) for further processing, ultimately reaching the sink.

*Pervasive sensor network.* A special type of sensor network that contains special type of sensor nodes with relatively high number of resources, so called *Parent Nodes (PNs)*.

*Sensor field.* The area where sensor nodes are generally (densely) scattered.

*Sensor node.* It is made of four basic components: a sensing unit (transducer), a power unit, a radio interface (transceiver), and a processing unit.

*Sink (initiator).* A node that is capable of communicating with higher level networks (Internet, satellite, etc.) or applications.

*Virtual clustering.* A protocol in which nodes are grouped into clusters based on their common sleep schedule.

# Questions

1. Exemplify two differences between MANET and WSN.
2. Exemplify two common characteristics of a WSN and MANET.
3. In a sensor node, what unit uses the most energy?
4. For which type of wireless ad hoc network, WSN or MANET, is the quality of service (QoS) the primary goal?
5. Discuss whether selecting fewer or larger number of nodes to service a query sent by a sink node is a good or a bad approach for designing a slef-organizing WSN.

6. Detail the types of signal processing and their advantages.
7. Present two drawbacks of the power save (PS) mode of IEEE 802.11. Which IEEE standard protocol has been designed later to overcome some of the drawbacks?
8. Present one cluster-adapted protocol for WSN.
9. What is directed diffusion and how has the data been organized?
10. Exemplify a case where biology can help designing a self-organizing WSN.

# References

1. Akyildiz IF, Su W, Sankarasubramanian Y, Cayirci E (2002) A survey on sensor networks. IEEE Commun Mag, 102–114
2. Bandyopadhyay S, Coyle E (2003) An efficient hierarchical clustering algorithm for wireless sensor networks. Proc INFOCOM, San Francisco
3. Barbarossa S, Scutari G (2007) Bio-inspired sensor network design. IEEE Signal Process Mag, 26–35
4. Bennett F, Clarke D, Evans JB, Hopper A, Jones A, Leask D (1997) Piconet: Embedded mobile networking. IEEE Pers Commun Mag, 4:8–15
5. Bharghavan V, Demers A, Shenker S, Zhang L (1994) MACAW: A media access protocol for wireless LANs. Proc ACM SIGCOMM, 212–225
6. Bluetooth Special Interest Group (1999) Bluetooth v1.0b specification. http://www.bluetooth. com. Accessed 20 February 2008
7. Bonabeau E, Dorigo M, Theraulaz G (1999) Swarm Intelligence: From Natural to Artificial Systems. Oxford University Press, New York
8. Cerpa A, Estrin D (2004) ASCENT: Adaptive self-configuring sensor networks topologies. IEEE Trans Mobile Computing, 3(3):272–285
9. Chevallay C, Van Dyck RE, Hall TA (2002) Self-organization protocols for wireless sensor networks. Thirty-sixth Conf Inf Sci Syst, Princeton, New Jersey
10. Clare LP, Pottie GJ, Agre JR (1999) Self-organizing distributed sensor networks. Proc SPIE, 3713:229–237
11. Cuhna DO (2005) Bio-Inspired Data Acquisition in Sensor Networks. Tech Report at Universidade Federal do Rio de Janeiro. http://www.gta.ufrj.br/ftp/gta/TechReports/CuDu05a.pdf. Accessed 20 February 2008
12. Das SK, Banerjee N, Roy A (2004) Solving optimization problems in wireless networks using genetic algorithms. Handbook of Bio-Inspired Algorithms, Chapman and Hall/CRC, London
13. Deb B, Bhatnagar S, Nath B (2003) Multi-resolution state retrieval in sensor networks. Proc IEEE Intl Workshop on Sensor Network Protocols and Applications, 19–29
14. Dressler F, Krueger B, Fuchs G, German R (2005) Self-organization in WSN using bio-inspired mechanisms. Proc 18th ACM/GI/ITG Intl Conf on Arch of Comput Syst (ARCS), Workshop on Self-Organization and Emergence, 139–144
15. Eigen M (1979) The Hypercycle: A Principle of Natural Self Organization. Springer, Berlin
16. Foester H (1960) On self-organizing systems and their environments. Proc Self-Organizing Syst, Yovitts MC, Cameron S (Editors), Pergamon Press, United Kingdam, 31–50
17. Gerhnson G, Heylighen F (2003) When can we call a system self-organizing? Proc 7th Euro Conf on Advances in Artificial Life (ECAL), 606–614
18. Giridhar A, Kumar PR (2006) Toward a Theory of in-network computation in wireless sensor networks. IEEE Commun Mag, 44:98–107
19. Gupta H, Zhou Z, Das SR, Gu Q (2006) Connected sensor cover: Self-organization of sensor networks for efficient query execution. IEEE/ACM Trans Networking, 14(1):55–67

20. Heinzelman W, Chandrakasan A, Balakrishnan H (2000) Energy-efficient communication protocols for wireless microsensor networks. Thirty-Third Hawaiian Intl Conf Syst Sci (HICSS), Hawaii

21. Hofmeyer S, Forrest S (2000) Architecture for an artificial immune system. Evol Comput, 8:443–473

22. Hong Y-W, Cheow LF, Scaglione A (2004) A simple method to reach detection consensus in massively distributed sensor networks. Proc Intl Symp Information Theory, 250

23. IEEE 802.15.4 Wireless Personal Area Network, http://www.ieee802.org/15/pub/TG4.html. Accessed February 2008

24. IEEE Computer Society LAN MAN Standards Committee (1997) Wireless LAN medium access control (MAC) and physical layer (PHY) specifications. Tech Report 802.11-1997, Inst of Electrical and Electronics Eng, New York

25. Intanagonwiwat C, Govindan R, Estrin D (2000) Directed diffusion: A scalable and robust communication paradigm for sensor networks. Proc 6th Annual Intl Conf on Mobile Computing and Networking, 56–67

26. Iqbal M, Gondal I, Dooley LS (2005) Distributed and load-adaptive self configuration in sensor networks. Proc Asia-Pacific Conf Commun, 554–558

27. Kennedy J, Eberhart RC, Shi Y (2001) Swarm Intelligence. Morgan Kaufmann, San Francisco

28. Kephart JO (1994) A biologically inspired immune system for computers. Proc Fourth Intl Workshop on Synthesis and Simulation of Living Systems, 130–139

29. Kochhal M, Schwiebert L, Gupta S (2003) Role-based hierarchical self organization for wireless ad hoc sensor networks. Proc ACM Workshop on Wireless Sensor Networks and Applications (WSNA), 98–107

30. Krueger B, Dressler F (2005) Molecular processes as a basis for autonomous networking. IPSI Trans Advances Research: Issues in Computer Sci and Eng 1:45–50

31. Lucarelli D, Wang I-J (2004) Decentralized synchronization protocols with nearest neighbor communication. Proc Second Intl Conf on Embedded Networked Sensor Syst (SenSys), 62–68

32. Meguerdichian S, Koushanfar F, Potkonjak M, Srivastava MB (2001) Coverage problem in wireless ad-hoc sensor networks. Proc INFOCOM 3:1380–1387

33. Meguerdichian S, Koushanfar F, Qu G, Potkonjak M (2001) Exposure in wireless ad hoc sensor networks. Proc Mobicom, 139–150

34. Mirkovic J, Venkataramani GP, Lu S, Zhang L (2001) A self-organizing approach to data forwarding in large-scale sensor networks. IEEE Intl Conf Commun (ICC), 1357–1361

35. Olariu S, Xu Q, Zomaya AY (2004) An energy efficient self-organization protocol for wireless sensor networks. Proc ISSNIP, 55–60

36. Pottie GJ, Kaiser WJ (2000) Wireless integrated network sensors. Commun ACM 43(5):51–58

37. Rahimi M, Pon R, Kaiser WJ, Sukhatme GS, Estrin D, Sirivastava M (2004) Adaptive sampling for environmental robotics. IEEE Intl Conf on Robotics and Automation, 3537–3544

38. Slijepcevic S, Potkonjak M (2001) Power efficient organization of wireless sensor networks. IEEE Intl Conf Commun (ICC), 472–476

39. Sohrabi K, Pottie G (1999) Performance of a novel self-organizing protocol for wireless ad hoc sensor networks. Proc IEEE Vehicular Tech Conf, 1222–1226

40. Sohrabi K, Gao J, Ailawadhi V, Pottie G (2000) Protocols for self organization of a wireless sensor network. IEEE Pers Commun Mag, 7:16–27

41. Stemm M, Katz RH (1997) Measuring and reducing energy consumption of network interfaces in hand-held devices. IEICE Trans Commun E80-B(8):1125–1131

42. Subramanian L, Katz RH (2000) An architecture for building self-configurable systems. IEEE/ACM Workshop on Mobile Ad Hoc Networking and Computing (MobiHoc), Boston, MA

43. Tian D, Georganas ND (2002) A coverage-preserving node scheduling scheme for large wireless sensor networks. Proc ACM Workshop on Wireless Sensor Networks and Applications (WSNA), Atlanta, GA

44. Tseng Y-C, Hsu C-S, Hsieh T-Y (2002) Power-saving protocols for IEEE 802.11-based multihop ad hoc networks. Proc INFOCOM, 200–209

45. Willet R, Martin A, Nowak R (2004) Backcasting: Adaptive sampling for sensor networks. Information Processing in Sensor Networks (ISPN), 124–133
46. Wu J (2002) Dominating set based routing in ad hoc wireless networks. In: Handbook of Wireless and Mobile Computing, Stojmenoic I (ed.) Wiley, New york, 425–450
47. Wu J, Li H (1999) On calculating connected dominating set for efficient routing in ad hoc wireless networks. Proc Third Intl Workshop on Discrete Algorithms and Methods for Mobile Computing and Commun, 7–14
48. Ye F, Zhong G, Cheng J, Lu S, Zhang L (2003) PEAS: A robust energy conserving protocol for long-lived sensor networks. Proc Intl Conf Distributed Comput Syst, 28–37
49. Ye W, Heidemann J, Estrin D (2004) Medium access control with coordinated adaptive sleeping for wireless sensor networks. IEEE/ACM Trans on Networking, 12(3):493–506

# Chapter 12
# Quality of Service in Wireless Sensor Networks

**Can Basaran and Kyoung-Don Kang**

**Abstract** Although well studied for traditional computer networks, quality of service (QoS) concepts have not been applied to wireless sensor networks (WSNs) until recently. QoS support is challenging due to severe energy and computational resource constrains of wireless sensors. Moreover, certain service properties such as the delay, reliability, network lifetime, and quality of data may conflict by nature. Multi-path routing, for example, can improve the reliability; however, it can increase the energy consumption and delay due to duplicate transmissions. Also, high resolution sensor readings incur more energy consumptions and delays. Modeling such relationships, measuring the provided quality, and providing means to control the balance is essential for QoS support. In this context, this chapter discusses existing approaches for QoS support in WSNs and suggests directions for further research.

## 12.1 Introduction

Quality of service management refers to systematic approaches to measuring and managing the quality of computational services. It has recently attracted a lot of interest, especially producing abundant research results in wired networks. A QoS study investigates the interplay between various service parameters such as bandwidth allocation and their impact on the provided service quality such as delay, jitter, and/or throughput. Reservation-based approaches such as IntServ [1] and reservation-less approaches such as DiffServ [2] are developed in order to provide bandwidth guarantees and service differentiation, respectively. IntServ architecture specifies a flow-based bandwidth reservation protocol to provide seamless data

C. Basaran (✉)
Department of Computer Science, Thomas J. Watson School of Engineering and Applied Science,
State University of New York at Binghamton, P.O. Box 6000, Binghamton, NY 13902-6000, USA
e-mail: cbasaran@cs.binghamton.edu

S. Misra et al. (eds.), *Guide to Wireless Sensor Networks*, Computer Communications
and Networks, DOI: 10.1007/978-1-84882-218-4_12,
© Springer-Verlag London Limited 2009

streams to users. On the other hand, DiffServ architecture does not maintain per-flow status. Instead, it supports differentiation between service classes to provide better QoS, e.g., shorter delay and smaller jitter, for a service class with a higher priority.

Mobile Ad hoc NETworks (MANETs) have different challenges for QoS support [3]. The differences stem from dynamic topologies, relatively low bandwidth, and shared wireless communication medium associated with MANETs. Despite all these differences, most QoS studies in MANETs focus on bandwidth allocation [4]. The fact that MANET topologies are more stable and nodes in such topologies are more capable than those of WSNs distinctly differentiates between the two network domains.

WSNs have an entirely different architecture. Individual nodes in a WSN have severe resource constraints in terms of energy, network bandwidth, memory, and CPU cycles. Also, they have unstable radio ranges, transient connectivity, and unidirectional links [5]. Despite these constraints, WSNs are often deployed for mission critical applications. These properties highlight the importance of QoS in WSNs [6]. Unfortunately, existing QoS approaches are not directly applicable to WSNs. For example, flow-based approaches such as IntServ need to establish end-to-end connections; however, an individual sensor node does not have sufficient resources to manage the state information per connection. Moreover, unstable connectivity between nodes makes it impossible to establish a persistent path between two distant ends.

A WSN acts as a collective unit to provide a sensor data service such as target tracking, fire detection, or habitat monitoring. Notably, QoS requirements, e.g., the required accuracy of sensor readings and the importance of a single reading, vary greatly from application to application. For instance, a routing protocol in a short-term target tracking application can be tuned to minimize the delay and maximize the reliability via real-time multipath routing for increased energy consumptions. For fire detection in a smart building, reliability is critical to ensure that important sensor readings are not lost. However, timeliness can be differentiated based on data values such that high temperature or pressure readings receive higher priority than normal readings. Also, redundant data indicating normal status can be aggressively aggregated to minimize energy consumptions. Further, a WSN deployed for long-term habitat monitoring may not need to support real-time data transmission. Data can be aggregated to minimize the energy consumption and stored at the base station to be sent to scientists every day, for example, via a satellite connection. Therefore, establishing a QoS model based on a specific application scenario allows us to identify key QoS requirements and metrics from which a feasible QoS management scheme potentially involving trade-offs can be derived. At the same time, it is important to identify key QoS requirements, if any, which apply to most WSN applications. Overall, QoS support in WSNs is a fairly new research problem with many remaining issues to investigate. In this section, we give a survey of well-known existing work, while discussing QoS issues in WSNs for future work.

## 12.2 Background

The quality of information provided by a WSN, e.g., accuracy, timeliness, or reliability, and the overall lifetime of the WSN are two major, conflicting properties. It has been reported that each link on a path increases the average packet loss ratio by approximately 5–10% depending on the node density and communication pattern [7], which results in the loss of approximately half of the packets along a path of 15 nodes. The same study [7] identifies that the shortest round-trip-time is approximately 600 ms, while the largest delay is approximately 5 s for a WSN covering an area of $1,200 \, m^2$. This is a clear indication of unstable behaviors of wireless communication media. Guaranteeing QoS in such a medium is certainly a challenging issue [8].

It is possible to craft tailored solutions for specific application needs and avoid generic approaches. But even so, an application often needs to execute in different states, while having flows with different priorities including control messages, periodic sensor readings, and alert messages. To provide network-wide QoS, each system component must comply with the desired QoS parameters. In this chapter, MAC (medium access control) layer, network layer, and in-network processing solutions for QoS support in WSNs are discussed. A number of these approaches are cross-layer solutions, since it is not always possible to divide system components into mutually exclusive modules. In fact, the reflections of the open systems interconnect (OSI) layers have a tendency to melt into each other in WSNs [9].

### 12.2.1 MAC Layer Solutions

The MAC layer provides channel access control services, which allow nodes to share the multiaccess wireless communication channel. Most network layer QoS solutions in the timeliness domain have MAC layer extensions. These extensions include, but are not limited to, modifications of the CSMA/CA protocol such that the back-off delay is inversely proportional to the priority of the packet being sent. Hence, upon a collision, a node with a high-priority packet to transmit waits for a shorter time interval before retrying to gain access to the wireless channel.

Since retransmissions are handled in the MAC layer in case of a transmission failure, upper layers may need to query the MAC layer in order to obtain information on the congestion state and the link quality. Most routing solutions examined in Sect. 12.2.2 utilize this information for delay estimation. Overall, MAC level QoS support is mostly limited to the policies implementing scheduling, channel allocation, buffer management, error control, and error recovery. The MAC layer QoS support in WSNs particularly focuses on scheduling and channel allocation to support upper layer services, such as routing and data aggregation, as discussed next.

QUIRE [10] is a cluster-based MAC protocol trying to form node clusters such that only one node in a cluster sends sensed data by communicating with a mobile agent flying over the deployment area. This approach focuses on WSNs with mobile agents that partition the region into hexagonal cells. Each agent broadcasts the cell radius, hovering over each cell. Each node in the cell receives this message and waits for a period inversely proportional to the quality of the received message. During the wait period, if a node hears a reply to the broadcast message from a node in the same cluster but with a better link quality, it cancels its own transmission. This approach aims at collecting sufficient data from the network such that the data distribution in the sensed field can be regenerated with a given probability $P_s$. At the same time, QUIRE ensures that the point of sensing can be estimated with a mean square error less than the *maximum distortion D*. Cell partitioning is accomplished taking into account the QoS metrics $P_s$ and $D$.

*Q-MAC* [11] is an energy-efficient, QoS-aware MAC protocol for WSNs designed to offer service differentiation between two classes via intra- and inter-node scheduling. Each node has an intra-node classifier which uses a separate FIFO queue for each priority level. Inter-node level classification gives the channel to the node with the highest urgency. The urgency of a node is evaluated by considering its packet priority, remaining number of hops to the destination, remaining energy of the node, and the queue lengths.

*CC-MAC* [12] exploits *spatial correlation* of sensor readings by pruning redundant data. Since sensor nodes within certain proximity generate similar data, based on the statistical information on node distribution, a *correlation radius* is calculated. This radius is used by CC-MAC to define *correlation regions* and perform filtering of messages belonging to the same correlation region. CC-MAC protocol is composed of two major components: Event MAC (E-MAC) and Network MAC (N-MAC). While E-MAC reduces the in-network traffic by dropping packets in the same proximity, N-MAC deals with forwarding filtered packets to the sink and prioritizes packets coming from foreign proximities. Although left as a future work, the QoS implications of proposed approach are evident, as the *correlation radius* may be tuned according to user-defined accuracy constraints.

An *Implicit Prioritized Access Protocol for WSNs* [13] defines a MAC layer protocol for cell-structured networks. The protocol provides delay guarantees for message delivery, while fully utilizing the available bandwidth via an earliest deadline first (EDF) scheduler, which exploits the periodic nature of WSN messages. In this approach, nodes are grouped into cells. All nodes within a cell are directly linked with each other. Intercell communication is handled by more capable *cluster heads* (CHs). A CH has two transceivers to transmit and receive packets at the same time. A total of seven radio channels are used in the whole network, which is modeled as a collection of hexagonal cells. A CH can communicate with the sensors in its cell, while communicating with the CHs of the (maximum six) neighboring cells using different channels. Intracell communication is based on a shared EDF schedule. The shared nature of this schedule allows all nodes to know precisely who should talk, and when. Moreover, some time slots are reserved for intercell communication. Using these time slots, CHs talk with each other based on another EDF schedule. If a node will not use the rest of the slots allocated to it, it broadcasts a yield message for the remaining slots. In this case, the next eligible node can take over the channel, increasing the bandwidth utilization.

## 12.2.2  Network Layer Solutions

MAC layer protocols can handle one-hop communication. For end-to-end QoS guarantees, network layer support is needed. The network layer QoS in WSNs encompasses end-to-end real-time service and reliability, which are fundamental requirements for mission-critical WSN applications. Because of energy demanding nature of radio transmissions [14, 15], QoS-aware routing protocols also have to use minimum number of control messages. It is challenging to support the desired QoS, while minimizing the number of control messages.

Given the scarce resources of sensor nodes [16], implementing an efficient routing protocol with no help from the underlying MAC layer is a daunting task. It follows that most of the approaches in this category are *cross-layer* solutions, which are built upon a QoS-aware MAC protocol to leverage lower level network information and service needed for QoS provision at the network layer.

In the reliability domain, using multipath routing [17–19] is a common approach. The idea behind this scheme is to exploit the high node density prevalent in WSNs. Because of the high density, there could be multiple paths between the source and the sink. If we assume that the packet delivery ratio for a link is 95%, then the delivery ratio for a 14-hop path is less than 50%. However, if there is a second, disjoint path with the same number of hops and link reliability, packets can be duplicated along the two paths to achieve a delivery ratio of 75%. The increase in reliability is achieved by sacrificing network lifetime, since energy consumption is roughly doubled if two paths rather than one are used. Alternatively, multipath routing can support load balancing to increase the network lifetime [20]. To transmit a packet, in this case, the routing algorithm only selects a single path among multiple paths for load balancing.

*RAP* [21] proposes a cross layer, i.e., network and MAC layers, architecture designed to support soft real-time requirements in WSNs. The architecture employs geographic forwarding (GF) [22, 23], in which a node forwards a packet to its one hop neighbor that is closer to the sink than the node is. When there are multiple one hop neighbors closer to the sink, it forwards the packet to the node that is closest to the sink. Thus, a node only has to keep the geographic locations of its one-hop neighbors.

In RAP, the sensing period and deadline are specified for each query. RAP applies Velocity Monotonic Scheduling, in which the priority of a data packet responding to a query is determined according to its requested *velocity* = *distance/deadline*. Specifically, in Static Velocity Monotonic (SVM), a permanent priority is assigned to a packet at the source according to the required velocity. On the other hand, Dynamic Velocity Monotonic (DVM) supports dynamic velocity adjustment at relay nodes. In DVM, when an intermediate node receives a packet, it assigns a new priority to the packet according to the remaining distance to the sink and time to the deadline. Therefore, when a packet suffers congestion, its priority, i.e., velocity, can be increased. On the contrary, if a packet moves faster than the requested velocity, its priority is decreased providing more bandwidth to others. This approach, however, requires either time synchronization [24] or MAC layer support for elapsed time calculation. In [21], DVM underperforms SVM possibly due to the lack of a reliable mechanism to measure the in-network delay and adjust the velocity accordingly. RAP requires prioritized MAC, which differentiates back-off delays according to the packet priority. Upon a collision, a node picks a random back-off delay in the interval [0,CW). The contention window $CW = CW_{prev} \times (2 + (PRIORITY - 1)/MAX\_PRIORITY)$ where $CW_{prev}$ is the previous contention window size and MAX_PRIORITY is the number of priority levels similar to [25]. As a result, a high-priority packet is likely to have a shorter back-off interval upon a collision.

Greedy GF is not always possible in the presence of a void, in which neither of the one hop neighbors is closer to the sink than the node that currently has the packet [22,23]. However, RAP simply assumes the constant availability of GF and it lacks void avoidance logic. It also has no congestion control mechanism. As a result, many packets with high-velocity requirements may miss their deadlines when the network is congested. Furthermore, the study does not consider the dynamic nature of WSN links and nodes in detail.

*SPEED* [26] is a routing protocol that aims to provide a uniform delivery speed across a WSN. Similar to RAP, SPEED relies on GF. Unlike RAP, it does not depend on MAC level real-time support. Each node keeps the location information of its one-hop neighbors and delay estimation for each link, which is computed using regular data messages and corresponding delays piggybacked on the acknowledgments (ACKs).

The QoS parameter *delivery speed* (*SetSpeed*) is supported in a best effort manner. Each node forwards a packet to a one-hop neighbor that is closer to the sink and to which it is connected via a wireless link that supports the *SetSpeed*. A neighbor supporting a higher speed is more likely to be selected. This forwarding approach is called the *stateless nondeterministic geographic forwarding*. If the required speed cannot be supported, packets are dropped with a given probability called *relay ratio*, which is calculated by the *neighborhood feedback loop* at the MAC layer based on the measured packet loss information indicating the severity of congestion or bad link quality.

When a node has no forwarding candidate, which is closer to the sink than itself or it cannot satisfy the desired speed to a specific destination, the node performs *backpressure* routing. The node issues a backpressure beacon to upstream nodes to notify them of the average delay suffered by the link on the path to the sink. Nodes receiving this information update their tables with the new information. If a node does not have the issuing node as a candidate to the destination it ignores this backpressure beacon. Voids are also avoided on the fly using backpressure beacons. A node identifying a void sets the average delay to infinity and informs upstream nodes.

*Multipath Multi-SPEED* (MMSPEED) [17] is a cross-layer protocol encompassing the network and MAC layers. It extends SPEED by providing *multiple network-wide speed levels for service differentiation*, while supporting QoS in the *reliability* domain at the same time. For scalability, MMSPEED relies on GF, similar to RAP and SPEED. The key idea is to have different speed layers over a single network. Hence, for $N$ speed layers, there are $N$ different *SetSpeeds*. Each virtual layer has its own FCFS queue. Different from SPEED, the MAC layer in MM-SPEED prioritizes packets belonging to higher speed layers over those of lower speed layers. Additionally, each node computes the remaining time to a packet's *deadline* and dynamically sets a new speed layer for the packet such that the new speed is the minimum speed able to satisfy the deadline requirement.

In the *reliability* domain, MMSPEED leverages the *loss rate* information provided by the MAC layer and *multipath routing*. Assuming a homogeneous loss rate across the network and a homogeneous hop distance, node $i$ locally estimates the end-to-end reachability (RP) of a packet to destination $d$ through one-hop neighbor node $j$ as follows:

$$RP^d_{i,j} = (1 - e_{i,j})(1 - e_{i,j})^{\text{[estimated number of hops]}}, \qquad (12.1)$$

ptwhere $e_{i,j}$ is the known one-hop loss rate for the link between nodes $i$ and $j$. The hop count in (12.1) is estimated by dividing the known distance to the final destination by the known one-hop distance to node $j$. Thus, the last part in (12.1), i.e., $(1 - e_{i,j})^{\text{[estimated number of hops]}}$, is a rough estimate for the rest of the network. Given a required reliability $P^{\text{req}}$, node $i$ can forward the packet to node $j$ if $RP^d_{i,j} \geq P^{\text{req}}$. Since decisions are based on estimates that can later prove to be incorrect, dynamic compensation logic is also implemented. When a node $s$ cannot find a single neighbor satisfying $P^{\text{req}}$, it can choose to forward the packet to two nodes ($j_1$ and $j_2$). In this example, if $P^{\text{req}} = 80\%$, $RP^d_{s,j_1} = 70\%$, and $RP^d_{s,j_2} = 60\%$, then node $s$ will calculate total reaching probability (TRP) as follows:

$$\begin{aligned} \text{TRP} &= 1 - (1 - RP^d_{s,j_1})(1 - RP^d_{s,j_2}), \\ &= 1 - (1 - 0.7)(1 - 0.6) = 0.88. \end{aligned}$$

where $(1 - RP^d_{s,j_1})$ is the probability that the path through node $j_1$ will fail and $(1 - RP^d_{s,j_2})$ is the probability that the path through $j_2$ will fail. Thus, TRP is the probability that at least one of them will deliver the packet toward the sink. Node $s$ can arbitrarily assign a new $P^{\text{req}}$ (e.g., 0.6 and 0.5) to each node because TRP $= 1-(1-0.6)(1-0.5) = 0.8$. Similar to the case of the *timeliness domain*, when a node suffers the existence of unreliable neighbors, it can use reliability *backpressure beacons* to decrease the expectations of upstream nodes. In this case, the node issuing a backpressure beacon will not be assigned a reliability level more than that specified in the beacon message. As the effect of a backpressure beacon only lasts for a limited time period, the impact of transient link problems on reliability assessment is limited.

By serving each level with just enough speed and reliability, MMSPPEED efficiently utilizes precious resources.

*JiTS* (Just-in-Time Scheduling) [27] is a network layer protocol for soft real-time packet delivery. JiTS only considers *timeliness* without considering reliability. It does not assume the underlying support of a QoS-aware MAC. Instead, it relies on the widely accepted nonprioritized IEEE 802.11 MAC. Unlike other routing protocols, JiTS aims at delaying a packet as much as its deadline allows. The idea of delaying packets is similar to the just-in-time delivery concept proposed in Mobicast [28] designed for mobile users of sensor data. Unlike Mobicast, JiTS does not assume user mobility. Also, it can be useful especially when in-network data aggregation is employed. Exploiting the slack-time increases the probability of similar data meeting at a relay node for aggregation.

JiTS forwarding logic uses a sorted queue where packets are inserted in nondecreasing order of target transmission time. The packet at the head of the queue is forwarded when the transmission time is reached. Target transmission time is calculated using the average one-hop delay estimated via ACK messages and estimated number of hops to the destination. The slack time determining the target transmission time (= current time + slack time) is estimated and uniformly distributed over all hops in the path:

$$\text{Slack time} = \frac{(\text{Deadline} - \text{EETD})}{\text{distance}(X, \text{sink})} \times a, \qquad (12.2)$$

where EETD is the estimated end-to-end transmission delay, which is equal to the product of the estimated average one-hop delay and the estimated number of hops to the destination. Also, variable $a$ in (12.2) is the *safety factor*. By setting it to a value less than 1 such as 0.7, JiTS can tolerate estimation errors.

JiTS has several variations. Especially, *nonlinear JiTS* shows the best performance among them. In a WSN, congestion may increase as packets approach the sink due to many-to-one communication patterns from sources to the sink. To reduce potential contention near the sink, nonlinear JiTS delays a packet more as the packet gets closer to the sink. Specifically, exponentially increasing portions of the estimated slack are allocated to the nodes closer to the sink:

$$\text{Slack time} = \frac{(\text{Deadline} - \text{EETD})}{2^{R/O}} \times a, \qquad (12.3)$$

where $R$ is the remaining distance to the sink and $O$ is the estimated one-hop distance.

Unlike RAP, SPEED, and MMSPEED, JiTS does not assume any specific routing protocol. In their simulation study, the shortest path routing protocol supported by many WSN systems such as TinyOS considerably outperforms GF in terms of deadline miss ratio and packet drop ratio.

*LESOP* (Low-Energy Self-Organizing Protocol) [9] is built based on a new two-layered network architecture called embedded wireless interconnect (EWI) replacing the OSI model. The design of EWI is justified by the fact that almost all solutions in WSNs require cross-layer implementations.

LESOP was specifically designed for target tracking applications, in which the first node detecting the target initiates the cooperation among nodes by broadcasting busy tones through a secondary wake-up radio channel. LESOP focuses on the accuracy of the target location by modeling trade-offs between QoS, i.e., the accuracy of the target location, and energy consumption. Increasing the idle time between sensing intervals decreases energy consumption, but it increases the delay for target detection. This is an evident trade-off between energy and QoS. Moreover, LESOP models the relationship between the target tracking error and the coverage to determine the minimum number of nodes that should be in sensing state. This minimum is calculated by using a QoS knob that is the minimum acceptable gain achieved by adding a new node to the sensing set. A drawback of LESOP is requiring a secondary radio, which can increase the cost.

## 12.2.3  In-Network Data Services

Since sensor readings are often redundant, sensor data can be aggregated in the network to reduce the number of packet transmissions and corresponding energy consumptions [29]. Such services can be implemented as a part of a user level application or a separate data service layer [30]. This domain deals with the quality and accuracy of the sensed data as well as minimization of in-network traffic, while conforming to a predetermined sensing accuracy. It is important to take into consideration that the process of aggregation may violate real-time constraints [48], because it requires relaying nodes to delay messages in order to aggregate them with data from different nodes. Thus, data aggregation should cooperate with QoS-aware routing and MAC to maximize the effect.

*Prediction-based monitoring in sensor networks* (PREMON) [31] applies the principles of MPEG [32] compression to the field of WSNs. In this approach, the sink node accumulates sufficient information to construct a *prediction model*. It then distributes this model to appropriate sensors along with the lifespan associated with this model. Sensor nodes receiving the prediction model change the mode of sensing to *update mode* and begin to send their sensor readings only when they differ from the predicted values by more than a predefined error margin. This is how a requested quality of monitoring (QoM) is provided.

PREMON is one of the first approaches to predicting sensor readings for the reduction of radio transmissions. The reduction depends on the *error tolerance*, i.e., the predefined QoM, and the correctness of the prediction model. Frequent distribution of a prediction model by a base station may consume significant amount of precious energy. Also, the scalability of PREMON is limited due to the centralized model construction.

*Temporal Coherency-Aware in-Network Aggregation* (TiNA) [33] proposes an approach to exploiting the *temporally coherent* nature of sensor readings. Each query has a *tct* (temporal coherency tolerance) value specified within the query itself. If the difference between the new reading and old one is smaller than the associated *tct* value, sensors do not report their readings. Parent nodes keep track of their child nodes while trying to aggregate their readings. If the parent does not receive an update from any of its children, the old reading from the child (or children) is used for aggregation. To distinguish a failed node from a node remaining silent because of the TiNA logic, a node sends periodic *heartbeat* messages to the parent. In this way, TiNAS's approach increases the quality of data (QoD) when network suffers severe congestion.

Romer et al. [34] reduce the amount of data to transmit based on the QoS requirements. A source and sink pair defines the tolerable error budget $e_{max}$. Rather than transmitting a complete stream of sensor data $\{x[k]\}$ from the source to sink, the source only sends a subset of data to the sink. More specifically, both the source and the sink run the same least-mean-square (LMS) predictor [35]. Using the LMS predictor, the source node computes the prediction error $e[k] = x'[k] - x[k]$, where $x'[k]$ is the sensor data predicted by the LMS method and $x[k]$ is the actual sensor reading. The source transmits $x[k]$ only if $e[k] > e_{max}$. Otherwise, $x[k]$ is simply dropped. Thus, a pair of the source and sink can support $e_{max}$.

## 12.3  Thoughts for Practitioners

WSNs with limited resources are often deployed for mission-critical applications. Thus, QoS-aware approaches are investigated to improve the cost–benefit ratio. Unfortunately, providing the desired QoS is not as straightforward as implementing a service. Services such as routing, MAC layer protocols, localization, time synchronization, and in-network data aggregation can be implemented in isolation, neglecting other system parameters. In the quality domain, all services share at least a common subset of interest such as network lifetime and delay. Essentially, the notion of *quality* denotes a black-box model, in which the end user expects a seamless integration of services that can be expressed in terms of inputs and outputs.

Is QoS a byproduct of a protocol, a result of a final skimming over a proposed solution, or the ultimate objective at the design phase? It is important to answer this question because, in the two former cases, QoS is only a minor issue in the application of interest. If QoS is the major concern, the integration and cooperation between the components and the resulting overall system performance matter. For example, a routing solution aiming to support soft real-time delivery guarantees may perform well in simulations; however, it might be useless in a real system, because it ignores global parameters of a real WSN system such as the MAC protocol and data aggregation, which can considerably affect the service delay.

Generally, QoS is a system-wide concept and it has to be handled as such. Because of severe resource constraints, most WSN protocols are optimized for specific applications of interest. Further, most solutions are cross-layered. As a result, QoS management becomes complex. Hence, a promising approach for QoS support is to analyze the application domain and design a QoS solution that conforms with the basic approaches employed by the intended set of target applications. A summary of key issues related to QoS-aware service design follows:

- QoS support must be an integrated part of the whole design/development process. Thus, a detailed analysis of the intended set of applications must be the starting point. The design should be compatible and cooperative with possible domain-specific system configurations unless it is generally applicable to different application domains.

- The QoS parameters to be supported must be decided considering their relevance to the target set of applications. Also the performance metrics and measurement methods, tools, and environments should be determined. If a cross-layer approach is taken, the required lower-layer services offered to the upper layers must be considered.
- Next, factors affecting the service performance for the selected QoS parameters must be investigated. For example, available wireless bandwidth and residual energy may affect data timeliness and reliability. In this way, the designers(s) of a QoS-aware WSN service can identify potential trade-offs between QoS parameters such as timeliness and reliability while considering a QoS model using the QoS parameters.
- Resource constraints such as memory and energy limitations should be considered. Because of severe resource constraints, a simple, lightweight approach is preferred. Also, one should take into account the nondeterministic, unreliable nature of wireless communication to let the QoS management scheme adapt to varying environments.
- Verifying models through simulations is important. If simulation results are convincing, before real-world deployment, the approach should be evaluated in a test bed composed of real battery-powered, wireless sensors such as MICA motes. Although there are many highly reliable simulation environments such as ns-2, virtually none of them can provide all the insights that can be gained from a testbed. On the other hand, a simulation study can cover a lager scope of experimental parameter settings and perform potentially intrusive or destructive experiments. Hence, simulation and testbed experiments are complementary to each other. From these experiments, issues that are important but were overlooked at design time can be newly identified to further improve the system design.
- If the previous steps are successful, the system can be deployed in a target environment starting from a relatively small-scale environment moving to a larger scale environment in a stepwise manner. The previous design steps may have to be revisited if new issues arise.

## 12.4 Directions for Future Research

Integration of various QoS functionalities within a system is an open research topic. We believe that future research efforts will follow this path and adopt a *holistic view* of QoS in WSNs. Definition of a holistic approach does not fall under the umbrella of a specific service category such as routing or channel allocation. Instead, it is a broader concept. A problem arises from the lack of established set of protocols even for specific application domains. Although there are groups trying to integrate existing research efforts [36] and proposals of complete systems [37], more work on service integration is required. In terms of QoS, neglecting *seemingly irrelevant* system functionalities is not a recommended approach, because they could affect each other. Thus, QoS-oriented approaches need to go beyond microscopic views and will embrace cross-cutting issues for seamless integration.

An appropriate place to start QoS integration is the operating system, which needs to provide necessary interfaces enabling access to information, such as link quality and delay, required by QoS management schemes at upper levels. If the QoS-centric operating system provides necessary information as well as low-level services for QoS management, QoS solutions in the upper layers can focus on certain aspects directly related to them. For these reasons, a QoS-centric operating system for WSNs needs to define rich and clean interfaces between system services. Such an operating system should also be modular so that only the necessary system components can be selectively integrated for a specific application. By being composable and providing basic low-level system information and services widely used for QoS management, QoS-centric operating system can provide a basis for a holistic QoS management in WSNs.

Furthermore, new languages or extensions over existing languages [38–42] are necessary to address WSN-specific challenges. For example, it is essential to compose event-driven services in WSNs. Compared with traditional programming languages, WSN programming languages, e.g., nesC, are relatively hard to understand and program. A new language is needed to directly support the event-driven nature of WSNs, while reducing the difficulty of programming.

Multimedia-based sensing in WSNs is also important [43, 44]. Multimedia data, in forms of snapshots, audio, and video require strict QoS support from the network. New QoS-compatible media formats [45], new collaborative distributed in-network data processing algorithms [46], and new real-time services [47] are required to support demanding multimedia services. In a near future, multimedia sensing may become one of the major research topics in WSNs. Currently the related work is scarce.

## 12.5 Conclusions

This chapter discusses state-of-the-art approaches for QoS support in WNSs. A number of existing approaches for MAC, routing, and data services are investigated. Most research effort in this specific field is devoted to routing services for timeliness. As WSN research is relatively new, QoS issues are not fully investigated in WSNs. Key services such as MAC, routing, and data aggregation can be further extended to support QoS. Moreover, seamless integration of available approaches for WSN QoS management at different layers is not studied in depth. A holistic view is required to thoroughly investigate QoS interactions between different layers. If these approaches are integrated without enough care, they can adversely affect each other causing undesirable results. For example, excessive data aggregation can significantly reduce the timeliness, while real-time multipath routing may consume too much energy when applied inappropriately. Thus, care should be taken to consider relevant QoS parameters in the context of an application of interest. At the same time, more research efforts are required to develop a general QoS model. QoS management mechanisms of the model could be composed to meet the needs of a specific application. In the future, bandwidth-demanding application scenarios such as multimedia sensing may further complicate QoS requirements in WSNs. This is another reason that a holistic approach is required for QoS management in WSNs.

**Acknowledgment**   This work was partly supported by a NSF grant CNS-0614771.

# Terminologies

*Quality of service.* QoS may employ many different meanings such as delay, jitter, throughput, reliability, sensor data accuracy, or network lifetime. A QoS study defines appropriate QoS metrics for an application of interest and develops approaches to supporting the desired QoS via trade-offs. It is also desired for QoS research to identify and develop common QoS metrics and QoS management schemes for a broad range of applications. In summary, QoS management is a systematic approach to measuring and managing the quality of computational services.

*Flow-based QoS.* Flow-based QoS supports the desired QoS such as the required bandwidth for an established end-to-end connection. It can support deterministic QoS guarantees by maintaining per-flow state information and reserving resources needed for QoS support. However, it has poor scalability due to large overheads for managing per-flow information. IntServ is a representative protocol providing flow-based QoS.

*Class-based QoS.* Class-based QoS is developed to address the scalability problem of flow-based QoS. Instead of supporting per-flow QoS, it provides QoS for aggregate traffic classes. DiffServ [2] is a representative example.

*Hard-QoS.* Service is subject to strict and deterministic quality guarantees. Resources are reserved for service guarantees. The system will reject requests that cannot be satisfied due to a resource shortage.

*Soft-QoS.* No hard guarantee but a probabilistic guarantee of QoS is provided. Because of unreliable wireless communication and severe resource constraints, it is infeasible to support hard QoS guarantees in WSNs. Rather, soft statistical QoS guarantees are needed in WSNs.

*Timeliness.* Timeliness measures the degree of timely delivery of data. This is a critical QoS metric in a number of WSN applications, requiring real-time sensing and control.

*Reliability.* Reliability measures the delivery ratio of requested data. This QoS metric is also very important to ensure the reliable delivery of sensor data to the sink at which more sophisticated data analysis can be performed.

*Multipath routing.* Multipath routing refers to a class of routing protocols utilizing more than one path for data communication. For reliability, a single data packet can be transmitted through multiple paths. Alternatively, for load balancing, one packet is forwarded through a single link at a time even if there are multiple links available.

*Quality of data (QoD).* QoD refers to the quality of information provided such as the data accuracy, resolution, and timeliness. Since WSNs are data centric, QoD is a broader concept than the traditional notion of QoS, which mainly focuses on the low level network performance such as the delay, jitter, or throughput.

*Temporal coherency.* Sensor data values do not largely oscillate within a given time interval. For example, temperature readings in a smart building may not change drastically from one sensing period to another. This property can be leveraged for data aggregation.

## Questions

1. Do you think reservation-based QoS provision is applicable to WSNs? Why or why not?
2. Give two specific application scenarios with different QoS expectations.
3. What are the main differences between wired networks, infrastructure-based wireless networks, mobile ad-hoc networks (MANets), and WSNs in terms of QoS?
4. What is the main focus of QoS efforts in the MAC layer?
5. What is the main focus of QoS efforts in the network layer?
6. What is data aggregation? How can it be utilized as a QoS tool?
7. How can temporal coherency of sensor readings be exploited to satisfy different QoD demands?
8. How can spatial coherency of sensor readings be exploited to satisfy different QoD demands?
9. What is the main difference between timeliness and reliability domains when served over multiple paths?
10. What is the most widely used approach to calculating path delay in routing? Why?

## References

1. R. Braden, D. Clark, and S. Shenker. Integrated Services in the Internet Architecture: An Overview. IETF RFC 1633, 1994
2. S. Blake, D. Black, M. Carlson, E. Davies, Z. Wang, and W. Weiss. An Architecture for Differentiated Services. IETF RFC 2475, 1998
3. P. Mahapatra, J. Li, and C. Gui. QoS in mobile ad hoc networks, IEEE Wireless Communication, vol. 10, no. 3, pp. 44–52, 2003

4. K. Wu and J. Harm, QoS support in mobile ad hoc networks, Crossing Boundaries, vol. 1, no. 1, Fall 2001

5. I.F. Akyildiz et al. Wireless sensor networks: A survey, Computer Networks, Elsevier Science, vol. 38, no. 4, pp. 393–422, 2002

6. Y. Wang, X. Liu, and J. Yin. Requirements of Quality of Service in Wireless Sensor Network. International Conference on Networking, International Conference on Systems and International Conference on Mobile Communications and Learning Technologies (ICNICON-SMCL'06), Mauritius, 2006

7. N. Ota, D. Hooks, P. Wright, D. Auslander, and T. Peffer. Poster Abstract: Wireless Sensor Network Characterization – Application to Demand Response Energy Pricing, In Proceedings of the First international conference on Embedded Networked Sensor Systems, November 05–07, 2003

8. M. Perillo and W. Heinzelman. Sensor Management, Wireless Sensor Networks, Kluwer Academic, 2004

9. L. Song and D. Hatzinakos. A cross-layer architecture of wireless sensor networks for target tracking, IEEE/ACM Transactions on Networking, vol. 15, no. 1, pp. 145–158, 2007

10. Q. Zao and L. Tong. QoS Specific Medium Access Control for Wireless Sensor Networks Fading, Eighth International Workshop on Signal Processing for Space Communications, Cataria, Italy, July 2003

11. Y. Liu, I. Elhanany, and H. Qi. An Energy-Efficient QoS-Aware Media Access Control Protocol for Wireless Sensor Networks. In Proceedings of the IEEE International Conference on Mobile Adhoc and Sensor Systems, November 2005

12. M.C. Vuran and I.F. Akyildiz. Spatial correlation-based collaborative medium access control in wireless sensor networks, IEEE/ACM Transactions on Networking, vol. 14, no. 2, pp. 316–329, 2006

13. M. Caccamo, L.Y. Zhang, L. Sha, and G. Buttazzo. An implicit prioritized access protocol for wireless sensor networks. In Proceedings of IEEE Real-Time Systems Symp., Dec. 2002, pp. 39–48

14. S. Madden, M.J. Franklin, J.M. Hellerstein, and W. Hong. "TinyDB: An Acquisitional Query Processing System for Sensor Networks," in ACM Transactions on Database Systems, 2005

15. W.S. Conner, J. Chhabra, M. Yarvis, and L. Krishnamurthy. Experimental Evaluation of Synchronization and Topology Control for In-Building Sensor Network Applications, in Proceedings of Wireless Sensor Networks and Applications, San Diego, CA, September 2003

16. J. Beutel. Metrics for sensor network platforms. In ACM RealWSN'6, Uppsala, Sweden, June 2006

17. E. Felemban, C. Lee, and E. Ekici. MMSPEED: Multipath multi-SPEED protocol for QoS guarantee of reliability and timeliness in wireless sensor networks, IEEE Transitions on Mobile Computing, vol. 5, no. 6, pp. 738–754, 2006

18. D. Ganesan et al. Highly resilient, energy efficient multipath routing in wireless sensor networks, Mobile Computing and Communications Review, vol. 5, no. 4, pp. 11–25, 2002

19. X. Huang and Y. Fang. Multiconstrained QoS multipath routing in wireless sensor networks, Wireless Networks Journal, vol. 14, no. 4, pp. 465–478, 2007

20. N. Jain, D. Madathil, and D. Agrawal. Energy aware multi-path routing for uniform resource utilization in sensor networks. In International Workshop on Information Processing in Sensor Networks (IPSN), April 2003

21. C. Lu et al., RAP: A Real-Time Communication Architecture for Large-Scale Wireless Sensor Networks, In Proceedings of the Eighth Real-Time and Embedded Technology and Applications Symposium, IEEE CS Press, Los Alamitos, CA, 2002

22. B. Karp and H.T. Kung. GPSR: Greedy Perimeter Stateless Routing for Wireless Networks, In Proceedings of the Sixth Annual International Conference on Mobile Computing and Networking, August 06–11, pp. 243–254, 2000

23. Y.-B. Ko and N.H. Vaidya. Location-aided routing (LAR) in mobile ad hoc networks. In Proceedings of the Fourth Annual ACM/IEEE International Conference on Mobile Computing and Networking, October 25–30, pp. 66–75, 1998

24. J. Elson. Deborah Estrin, Time Synchronization for Wireless Sensor Networks. In Proceedings of the 15th International Parallel & Distributed Processing Symposium, April 23–27, 2001, p. 186
25. I. Aad and C. Castelluccia. Differentiation Mechanisms for IEEE 802.11. IEEE INFOCOM 2001, Anchorage, Alaska, April 20
26. T. He, J.A. Stankovic, C. Lu, and T.F. Abdelzaher. SPEED: A Stateless Protocol for Real-Time Communication in Sensor Networks, In Proceedings of International Conference on. Distributed Computing Systems (ICDCS '03), Providence, RI, May 2003
27. K. Liu, N. Abu-Ghazaleh, and K.D. Kang. JiTS: Just-in-Time Scheduling for Real-Time Sensor Data Dissemination. In Proceedings of the Fourth Annual IEEE International Conference on Pervasive Computing and Communications (PERCOM'06), Washington, DC, IEEE Computer Society, Silver Spring, MD, 2006, pp. 42–46
28. Q. Huang, C. Lu, and G.-C. Roman. Mobicast: Just-in-time multicast for sensor networks under spatiotemporal constraints. International Workshop on Information Processing in Sensor Networks, Palo Alto, CA, April 2003
29. S. Madden, M.J. Franklin, J.M. Hellerstein, and W. Hong. TAG: A Tiny AGgregation service for Ad-Hoc sensor networks, In Proceedings of the Fifth symposium on Operating Systems Design and Implementation, December 09–11, 2002
30. R. Kumar, S. PalChaudhuri, D. Johnson, and U. Ramachandran. Network Stack Architecture for Future Sensors, Rice University, Computer Science, Technical Report, TR04-447, 2004
31. S. Goel and T. Imielinski. Prediction-based monitoring in sensor networks: Taking lessons from MPEG, ACM SIGCOMM Computer Communication Review, vol. 31, no. 5, October 2001
32. J. Watkinson, MPEG-2, Butterworth-Heinemann, Newton, MA, 1998
33. M.A. Sharaf, J. Beaver, A. Labrinidis, and P.K. Chrysanthis. TiNA: A scheme for temporal coherency-aware in-network aggregation. In Proceedings of the Third ACM International Workshop on Data Engineering for Wireless and Mobile Access, San Diego, CA, September 19–19, 2003
34. S. Santini and K. Romer. An Adaptive Strategy for Quality-Based Data Reduction in Wireless Sensor Networks. In Proceedings of the Third International Conference on Networked Sensing Systems (INSS'06), Chicago, IL, 2006
35. S. Haykin. Least-Mean-Square Adaptive Filters. Edited by S. Haykin, New York: Wiley-Interscience, 2003
36. Embedded WiSeNts. http://www.embedded-wisents.org/, December 2006
37. K. Sohrabi, J. Gao, V. Ailawadhi, and G. Pottie, Protocols for self-organization of a wireless sensor network, IEEE Personal Communications Magazine, vol. 7, no. 5, pp. 16–27, Oct. 2000
38. E. Cheong and J. Liu. galsC: A Language for Event-Driven Embedded Systems. In Proceedings of the Conference on Design, Automation and Test in Europe, March 07–11, 2005, pp. 1050–1055
39. D. Gay, P. Levis, R. von Behren, M. Welsh, E. Brewer, and D. Culler. The nesC Language: A Holistic Approach to Networked Embedded Systems. In Proceedings of Programming Language Design and Implementation (PLDI) 2003, San Diego, CA, June 2003
40. D. Janakiram and R. Venkateswarlu. A Distributed Compositional Language for Wireless Sensor Networks. In Proceedings of IEEE Conference on Enabling Technologies for Smart Appliances (ETSA), Hyderabad, India 2005
41. Srisathapornphat, C. Jaikaeo, and C. Chien-Chung Shen Sensor Information Networking Architecture. International Workshops on Parallel Processing, pp. 23–30, 2000
42. B. Greenstein, E. Kohler, and D. Estrin. A Sensor Network Application Construction Kit (SNACK), In Proceedings of the Second International Conference on Embedded Networked Sensor Systems, November 03–05, 2004
43. I.F. Akyildiz, T. Melodia, and K.R. Chowdhury. A survey on wireless multimedia sensor networks, Computer Networks, vol. 51, 921–960, 2007
44. Y. Gu, Y. Tian, and E. Ekinci. Real-time multimedia processing in video sensor networks, Image Communication, Elseiver Science, vol. 22, no. 3, 2007
45. Y. Wang, R.R. Reibman, and S. Lin. Multiple description coding for video delivery, In Proceedings of the IEEE, vol. 93, no. 1, pp 57–70, January 2005

46. M. Chu, J.E. Reich, and F. Zhao. Distributed attention for large video sensor networks. In Proceedings of the Institute of Defence and Strategic Studies (IDSS), London, UK, February 2004
47. S. Kompella, S. Mao, Y.T. Hou, and H.D. Sherali. Cross-layer optimized multipath routing for video communications in wireless networks, IEEE Journal on Selected Areas in Communications, vol. 25, no. 4, pp. 831–840, May 2007
48. B. Krishnamachari, D. Estrin, S.B. Wicker. The Impact of Data Aggregation in Wireless Sensor Networks. In Proceedings of the 22nd International Conference on Distributed Computing Systems, July 02–05, 2002, pp. 575–578

# Chapter 13
# Embedded Operating Systems in Wireless Sensor Networks

**Mohamed Moubarak and Mohamed K. Watfa**

**Abstract** Several operating systems (OS) for wireless sensor networks (WSNs) have been designed, implemented and are in the process of enhancement. However, early before implementation, designers face an important decision to make. The designer of an embedded operating system (EOS) has to conform to one of two completely different design philosophies and build his system according to that philosophy. This decision is crucial in the sense that the behavior and performance of each model differs, and those will be reflected on the WSN since the EOS is the core of the system, and any protocol built on top of it will drag with it the characteristics of the design model. Both models are investigated in this chapter by looking at the design and architectures of several EOSs built for WSNs.

## 13.1 Introduction

This chapter presents a survey and analysis of the current state of the art in the field of wireless sensor embedded systems, highlighting the open research challenges. It first introduces the importance of wireless sensor embedded systems and their crucial role in the performance of wireless sensor networks (WSNs). It then presents the essential characteristics an operating system should have to operate a WSN and shows why existing embedded operating systems (EOSs) are not a suitable choice. The design and architecture of state-of-the art EOSs are then discussed in a comparative manner. Important design issues are introduced and supported with examples. Finally, the latest research trends in the field are discussed. At the end of this chapter the reader should appreciate the effect of the underlying EOS on the performance of WSNs, the differences and trade-offs between different platforms and the design issues behind them, and develop a hands-on experience in developing applications under various platforms.

M. Moubarak (✉)
Computer Science Department, American University of Beirut, Beirut, Lebanon
e-mail: mam54, mw11@aub.edu.lb

S. Misra et al. (eds.), *Guide to Wireless Sensor Networks*, Computer Communications and Networks, DOI: 10.1007/978-1-84882-218-4_13,
© Springer-Verlag London Limited 2009

## 13.2 Background

An embedded system is one that completely engulfs the computer that controls it. This encapsulated computer in turn needs an operating system to manage its resources such as memory and energy. An operating system that manages an embedded system is called an EOS. EOSs are usually made for a particular function. The fields in which these EOSs are used range from industrial machines to consumer electronics. Some specific examples are industrial controllers, robots, networking, mobile phones and hand-held devices, calculators, military rocket launchers and even spacecrafts, all of which are designed to perform a unique set of tasks. This specialization gives the EOS its compact size in contrast to larger and more complex general purpose OSs. This compactness is also due to the small capabilities of microcontrollers that run EOS, in most cases at least. Since each field performs different tasks and provides different hardware capabilities, a single OS cannot manage all kinds of embedded systems. Hence, EOSs are specifically designed for the system in mind. While some systems require real-time operation and energy efficiency, other systems may require security or low production cost, all of which must be provided by the EOS.

Research in the field of EOSs is increasing due to the emergence of new embedded system domains, making them more ubiquitous and pervasive. One emerging domain that forms the main focus of this chapter is the domain of WSNs. The technology of WSNs made a great impact on the design and architecture of EOSs. Because of this technology, EOSs are gradually becoming fundamental elements of the infrastructure of our current world's ecosystem.

## 13.3 Wireless Sensor Operating Systems

A WSN consists of many sensor nodes with wireless communication modules. The characteristics of these sensor nodes make them a perfect example of an embedded system. For example, a wireless sensor node is designed to do a general task, which is to collect sensor readings and send them to a sink or base station to act upon them. This node has no user interface and is individually controlled by a set of buttons and uses LEDs as a form of display.

As a network, these sensors are expected to run a variety of sensing applications, reading in all types of data from acoustic to temperature values. To identify objects, a WSN will also need to do some pattern recognition, after which the sensors will diffuse the data on to an unmanageable network with low reliability. This requires the running of applications ranging from location-aware algorithms to energy-efficient routing [5]. At the same time, a node may act as a router, forwarding data toward their destinations, or even responding to queries issued from base stations far away. Hence, a new specially designed OS is needed to operate all these applications. The OS has to do so taking into consideration security, energy efficiency, and high amounts of concurrency. This sounds like a blend of three types of OSs that exist today: personal computers, distributed systems, and real-time systems. The required OS is also expected to run on a MMU-less hardware architecture, having a single 8-bit microcontroller running at 4 MHz with 8 KB of flash program memory and 512 bytes of system RAM [1]. Existing EOSs do not meet these requirements and hence the work on applicable OSs designed especially for WSNs has already begun.

Several EOSs for WSNs have been designed, implemented and are in the process of enhancement. However, early before implementation, designers face an important decision to make. The designer of an EOS has to conform to one of two completely different design philosophies and build his system according to that philosophy [2]. The two philosophies are called the event-driven and the thread-driven models. This decision is crucial in the sense that the behavior and performance of each model differs, and those will be reflected on the WSN since the EOS is the core of the system, and any protocol built on top of it will drag with it the characteristics of the design model. Even at the application level, each model has its unique programming structure that programmers have to follow. So, which model to choose and which programming structure is easier? To find out, both models are investigated in this chapter. The following sections will discuss the two models in more details highlighting the differences, advantages, and disadvantages of each to provide a better understanding of the underlying operating systems that are discussed later.

### 13.3.1    Event-Driven Model

Event-driven systems are based on a very simple mechanism and are more popular in the field of networking. That is because the model complements the way networking devices work. An event-driven system consists of one or more event handlers. Handlers basically wait for an event to occur and hence they are implemented as infinite loops. An event could be the availability of data from a sensor, the arrival of a packet, or the expiration of a timer. Each event could have a designated handler waiting for it to occur. When an event occurs, the associated event handler either starts processing the event accordingly or adds the event to a buffer for later execution. Events are removed from the buffer in a FIFO manner. Task preemption in this model may occur if an event that has a higher priority occurs. The execution model is therefore rather sequential. Figure 13.1 shows the execution model of event-driven systems.

### 13.3.2    Thread-Driven Model

The thread-driven model is process based. Processes run preemptively on the CPU in a seemingly parallel manner. That is each process is given a quantum, which is an amount of CPU time. When the quantum ends, the process must be preempted and another process is run. Preemption in thread-driven systems occurs more than

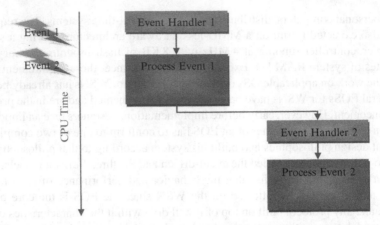

**Fig. 13.1** Event-driven execution model allows one process at a time. Event handler 1 polls for event 1 and processes it as it occurs. Even if event 2 occurred while processing event 1, it will only be processed when event handler 2 polls for it

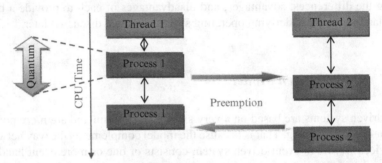

**Fig. 13.2** A thread-driven execution model allows simulates parallel execution on several CPUs. Each time process 1 spends its quantum on the CPU the scheduler preempts it and puts the next waiting process on the CPU for a quantum, in this case process 2

is strictly needed; however, this CPU sharing provides the virtualization of several CPUs existing instead of one real CPU. The main part of a thread-driven model, also called the heart of the system, is the kernel. The kernel provides all the system services such as resource allocation needed by the application level. The scheduler is the main controller of the system and it is built inside the kernel. It decides when to run a process and when to preempt it. Figure 13.2 later sketches the execution model of thread-driven systems.

**Table 13.1**  Advantages and disadvantages of the event-driven model

| Advantages | Disadvantages |
| --- | --- |
| Concurrency with low resources | Event-loop is in control |
| Complements the way networking protocols work | Program needs to be chopped to subprograms |
| Inexpensive scheduling technique | Bounded buffer producer–consumer problem |
| Highly portable | High learning curve |

**Table 13.2**  Advantages and disadvantages of the thread-driven model

| Advantages | Disadvantages |
| --- | --- |
| Eliminates bounded buffer problem | Complex shared memory |
| Programmer in control of program | Expensive context switches |
| Automatic scheduling | Complex stack analyses |
| Real-time performance | High memory footprint |
| Low learning curve | Not portable due to stack manipulation |
| Simulates parallel execution | Performs better on multiprocessors |

### 13.3.3  Event-Driven Versus Thread-Driven

Event-driven models are reputed by some researchers to provide more concurrency than thread-driven models do. However, other practitioners believe the opposite. To have a good idea about the trade-offs of each design, Tables 13.1 and 13.2 define the advantages and disadvantages of each model [3, 4].

The following section dwells into operating systems that are built using an event-driven model.

## 13.4  Event-Driven Embedded Operating Systems

Event-driven operating systems are built using the event-driven model discussed earlier. The OSs are discussed in detail in terms of their architecture and execution showing the impact of design on the real system.

### 13.4.1  TinyOS

To meet the tight constraints of WSNs, TinyOS adopted the event-driven approach as the concurrency model and is currently the standard OS for WSNs. TinyOS was designed to have a very small memory stamp, where the core OS could fit in less than 200 bytes of memory [1]. TinyOS' event-driven choice was based on the fact that it cuts down on stack sizes since one process could run at a time. Another fact it is that it eliminates unnecessary context switches, which are infamous for their energy inefficiency. TinyOS is entirely made of a set of reusable system components and an energy-efficient scheduler and hence has no kernel. Each component is made up of four parts: a set of commands, event handlers, a bundle of tasks, and a fixed size frame for storage. The commands and events a component supports must be predefined to enhance modularity.

### 13.4.1.1 TinyOS Components

Components in TinyOS are arranged hierarchically with low-level components closest to hardware and higher level components form the application layer. Components are of three types:

- Hardware abstraction components: These are the lowest level components that map the physical hardware to the TinyOS component model. One such component is the RFM radio component, which manipulates the pins connected to the RFM transceiver.
- Synthetic hardware components: These components simulate the behavior of hardware. For example, the Radio Byte component performs data encoding and decoding that can be performed by hardware. These components lie on top of the latter.
- High-level software components: These components form the application layer and are responsible for data management and routing. Data fusion applications fall into this category as well.

Since components are organized, some form of "wiring" or "binding" is required to make intercomponent protocols clear. This is provided by a component through its commands and events. As mentioned earlier, a TinyOS component is made up of commands, events, tasks, and a frame, as shown in Fig. 13.3. Commands are the set of function calls or services that a component will request from other components. Event handlers implement the handling of results returned from previous commands. Those results are triggered by the component that provided the service in a form of event to indicate completion of the service. Commands and events cannot block. Tasks on the other hand are a form of deferred computation. Most computational work is done through tasks. A component defines the tasks that it may post. When a task is posted, it is buffered until the scheduler runs it, which is a simple FIFO scheduler. When no tasks are pending, the scheduler puts the CPU in sleeping mode for energy efficiency. Only one task could run at a time and each runs to completion. Tasks may be preempted by commands or events. A task should not be long in order not to delay other tasks. Finally, the fixed size frame is used to depict the state of the component by storing parameters. The fixed size and static allocation of the frame allow for simpler memory management at compile time. Yet how do all these parts work together in a real life scenario? Example 1 provides a comprehensive answer to that question.

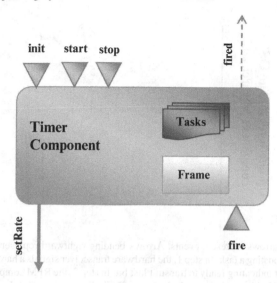

**Fig. 13.3** Visual representation of a TinyOS component. Upside-down triangles represent command handlers, triangles represent event handlers, upward dashed arcs represent signaled events, and downward solid arcs represent issued commands (adapted from [2])

**Table 13.3** The upward operation of events and the downward operation of commands suggest a fountain-like activity between components. To prevent a cycle in the flow, commands are not allowed to signal events

|  | Issue a command | Trigger an event | Post a task | Deposit to the frame |
|---|---|---|---|---|
| Commands | To lower components only | No | Yes | Yes |
| Events | To lower components only | To higher components only | Yes | Yes |
| Tasks | To lower components only | To higher components only | Yes | No |

*Example 1.* Component Interaction in TinyOS

To better understand the component structure of TinyOS and the way these components interact, this example provides a walk through inside various TinyOS components as it performs a simple send_message procedure. Specifically, this example illustrates how components interact during the transmission of the last bit of a packet. Before starting the walkthrough, Table 13.3 illustrates the operation of each of the four parts (commands, events, tasks, and frame) that make up a component.

Now that we have a complete idea on how components operate, we can visualize the fountain of interactions produced by TinyOS when trying to transmit the last bit of a message over the radio as part of the send_message command as shown in Fig. 13.4.

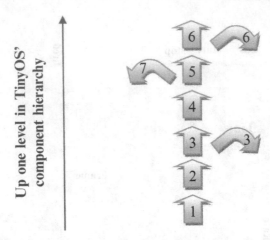

**Fig. 13.4** Upward arrows represent events. Arrows bending rightward represent commands and leftwards represent posting a task. In step 1, the hardware transceiver signals a hardware interrupt to the RFM component indicating ready to transmit last bit. In step 2, the RFM component transforms the interrupt into a TX_bit_evt that signals the upper Radio Byte component. In step 3, the Radio Byte component handles the TX_bit_evt by performing byte-level processing. It issues a command to the lower RFM component to transmit the final bit. It also triggers a TX_byte_ready event to the upper Radio Packet component indicating the end of the byte. In step 4, Radio Packet handles the TX_byte_ready event by issuing a TX_packet_done event to the upper AM component. In step 5, AM signals an event to the upper application layer to indicate that the send_message command has completed. In step 6, the application issues another send_message command down to AM. In step 7, AM posts a task to prepare the new packet

## 13.4.2  SOS

SOS is another event-driven OS targeting WSNs. Like TinyOS, SOS consists of components; however, these components or modules are dynamically reconfigurable [6]. To achieve such reconfiguration, SOS consists of a statically compiled kernel, and a set of dynamically loaded modules.

### 13.4.2.1  SOS Module

A module in SOS can be anything from a very low-level sensor driver to a high-level application. Each module is a position-independent binary that implements a specific task [6]. Modules communicate either by posting messages or by direct function calls. Each module is a message handler, which is similar in principle to event handlers in TinyOS.

### 13.4.2.2  SOS Kernel

The SOS kernel provides system services for the modules to use. This is done using the jump table mechanism. The jump table is stored in memory and acts as

Program Memory Layout

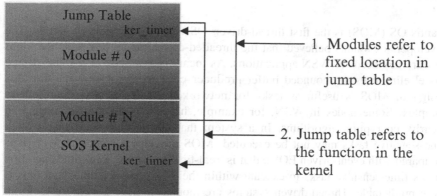

**Fig. 13.5** Jump table redirection for modules in SOS (adapted from [6])

an API for the dynamic modules. Figure 13.5 illustrates the layout of the jump table. Modules may also invoke functions on other modules. This functionality passes through the kernel. To support dynamic reconfiguration, the kernel provides a dynamic function registration and subscription service for modules. Using this service, any module can register its functions with the kernel making those functions available to other modules for subscription. A module registers its functions by providing the kernel with the relative address of the functions' implementations. The kernel stores these addresses and provides a handler for each function to the subscribing modules.

### 13.4.2.3 SOS Scheduler

The SOS scheduler is part of the kernel which operates in a FIFO manner with two priority queues. Memory allocation in SOS is also dynamic, unlike TinyOS' fixed size frames. In SOS, only modules can be dynamic, where the kernel cannot be upgraded the same way. Updating applications on other systems like TinyOS require a new system image; however SOS' approach is more energy efficient.

Opposing the event-driven model for designing OSs is the thread-driven model. This model's OSs are discussed in the following section.

## 13.5 Thread-Driven EOSs

The model discussed in Sect. 13.3.2 motivated WSN OS designers for its ease of programming and high concurrency model. OSs that followed that design are presented in the following sections.

## 13.5.1 Mantis OS

Mantis OS (MOS) is the first thread-driven OS targeting the field of WSNs. The developers of Mantis believed that the threaded-driven model best suits the high-concurrency needs of WSN applications. As mentioned in Sect. 13.3.3, this design model eliminates the bounded buffer producer–consumer problem. The threaded design of MOS is useful as tasks for networked sensors become increasingly complex. Some nodes in WSN, for example, have to perform time-consuming security encryption algorithms. In a system that allows only short tasks, other time-sensitive tasks may not be executed. MOS provides a unique characteristic compared with event-driven EOSs, that is, real-time operation. Real-time operation allows time-sensitive tasks to execute within their assigned deadlines and thus is more predictable. Thread-driven systems are thought to have a memory footprint that is large enough to render them useless in the field of WSNs; however, the developers of MOS were able to shrink a classic thread-driven OS into one that fits into 500 bytes of RAM [7]. So MOS' architecture is a traditional layered architecture. Figure 13.6 illustrates such architecture.

### 13.5.1.1 MOS API

MOS provides a portable and easy to use interface for building applications through the MOS API. Since MOS' programming language is C, porting existing protocols to MOS is easy. An application in MOS can read sensor data, toggle the LEDs, and transmit the data over the radio in as little as ten lines of code due to the rich API.

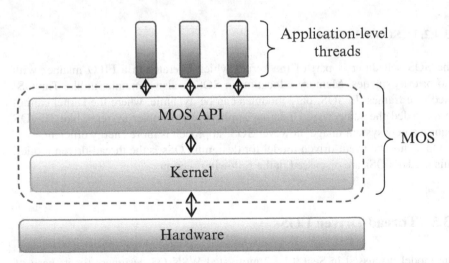

**Fig. 13.6** In a layered architecture, the kernel, scheduler, and device drivers lie just above the hardware. An application interface lies above the kernel and provides libraries and system calls for the application layer

To transmit a message, for example, MOS API provides the function com_send. For toggling the LEDs, the mos_led_toggle function is provided. APIs provided include functions to manipulate networking, onboard sensors, LEDs, and the scheduler. Such compact programs allow for a comparatively low learning curve. MOS API simplifies cross-platform support because it is preserved across both physical sensor nodes as well as virtual sensor nodes running on X86 platforms. That is, the same API is used for the real sensors and virtual sensors on X86 machines.

## 13.5.2 Kernel and Scheduler

The kernel in MOS is a classical kernel made up of a UNIX-like scheduler that is partially conforming to POSIX threads. The scheduler provides priority among threads using round-robin semantics within a priority level. For shared variables, binary and counting semaphores are supported. The goal behind the MOS kernel was to provide the conventional UNIX features for resource-constrained sensor nodes. Table 13.4 shows the compact sizes of structures in MOS.

RAM is managed as a heap, yet programmers cannot dynamically allocate heap in their applications. Each thread has space allocated for it in the heap. The space is recovered when the thread exits. The thread table forms the main data structure in the kernel. Each thread has a single entry in the table, just like a process table in a conventional OS. Since the table is allocated statically, there are a fixed number of allowable threads, which is 12. The context of a thread is not saved in the table entry; instead, it is saved on its stack. This significantly reduces the size of the thread table. Pointers are used to efficiently manipulate entries by the scheduler.

Context switches in MOS are triggered by a hardware interrupt, a system call, or a semaphore operation. The timer triggers an interrupt every 10 ms, which means that a quantum is 10-ms long. The timer interrupt is the only interrupt handled by the kernel. Other interrupts are directly handled by device drivers. Context switches are as expensive as 1,000 clock cycles [8]. At startup, the kernel creates an idle thread with the lowest priority, scheduled only when no other thread is running. The idle thread is used by the scheduler as a form of energy conservation. Whenever the idle thread is scheduled, MOS can enter a sleep mode to save on energy, since there are no tasks to perform.

**Table 13.4** MOS scheduler consists of a 120-byte thread table (12 threads, 10 bytes each) and a set of pointers, fields combined together to form 144 bytes

| Max. number of ready-list pointers | 4 |
| --- | --- |
| Size of each ready-list pointer | 5 bytes |
| Size of single current pointer | 2 bytes |
| Interrupt status field | 1 byte |
| Flags field | 1 byte |
| Max. number of threads | 12 |
| Size of each thread | 10 bytes |
| Scheduler overhead | 144 bytes |

## 13.5.3 RETOS

RETOS is another thread-driven OS specially designed for WSNs. It was designed with four objectives in mind: provide thread-driven interface, safe from erroneous applications, dynamic reconfiguration, and network abstraction. A powerful set of characteristics for constraint networked sensors. What is unique about RETOS is the optimization techniques used to cut down on energy consumption and space footprint [9]. RETOS developers intend to make the technology of WSNs more popular by providing an easy programming model, thus the thread-driven model was the choice. Yet they also believed that they have to optimize it to make it feasible.

### 13.5.3.1 System Resiliency

RETOS provides system resiliency by two techniques: dual mode operation and application code checking. Dual mode operation provides the logical separation of kernel and user execution [10]. Application code checking provides static analysis for compiled code and dynamic check for run-time behavior. Dual mode operation is implemented by switching from user stack to kernel stack and vice versa depending on the state of the thread. Application code checking prevents an application from accessing memory outside its boundary and from directly accessing hardware. To tackle the disadvantages of thread-driven systems, RETOS provides the following set of optimization techniques.

### 13.5.3.2 Single Kernel Stack

To reduce the memory stamp of RETOS, two techniques are used: single kernel stack and stack-size analysis. In a thread-driven system, each thread requires a user and kernel stack. In RETOS the kernel stacks are reduced to one shared kernel stack. This means that preemption at kernel level is not allowed. However, memory is significantly reduced. Stack-size analyses are used to determine the right amount of memory for each stack.

### 13.5.3.3 Variable Timer

Context switches are infamous for their high-energy consumption. A context switch occurs when the clock interrupt handler determines the end of a quantum. However, the clock will keep issuing interrupts at a certain rate. Since most of these interrupts are unhandled, a considerable amount of energy is wasted in triggering them. To overcome this problem, a variable timer can be implemented such that the rate at which interrupts occur depends on an upcoming time-out request (Fig 13.7). The variable timer in RETOS manages time-out requests from threads and sets the

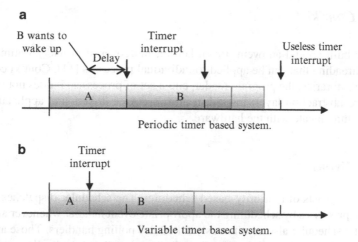

**Fig. 13.7** (adapted from [6]) in (**a**), thread B with higher priority arrives during the execution of A. But B has to wait until the timer issues an interrupt. In (**b**), thread B requests a timeout. A is preempted and B with higher priority is run. Along the execution there useless timer interrupts since the threads do not request any

clock-tick rate as such. Variable timers are not feasible in conventional OSs where the number of threads is very large. However, in networked nodes, the number of threads is small enough to allow for a variable timer.

#### 13.5.3.4 Event-Aware Thread Scheduling

The default scheduler in RETOS supports the POSIX real-time scheduling interface. However, RETOS also provides an optimized event-aware scheduler. This scheduler simply boosts the priority of the thread that is requesting to handle a certain event. This allows the event handler to preempt any running process. The boosted event loses its priority as it spends more time on the CPU.

We have seen so far various OSs designed using two different design models. Other OSs, however, try to provide both models at the same time. Such OS is presented in the following section.

## 13.6 Contiki Hybrid EOS

Operating systems supporting both the advantages of event-driven and thread-driven models of execution are highly desirable in WSNs. However, the most that has been achieved is merging both models in one OS, merging the disadvantages as well. An example of such OS is presented in this section.

## 13.6.1    Contiki

Contiki is built around an event-driven kernel; moreover, it provides optimal pre-emptive threading that can be applied to individual processes [11]. Contiki consists of a kernel, libraries, the program loader, and a set of processes. It does not provide a hardware abstraction layer; instead, it allows device drivers and applications to directly communicate with the hardware.

### 13.6.1.1    Kernel

The kernel consists of a priority-based scheduler. The scheduler dispatches events to running processes by scheduling the appropriate event handler whenever an event occurs. The scheduler also continuously schedules polling handlers. Those are handlers that continuously poll for an event. Polling handlers usually lie next to the hardware and poll for hardware updates. Process communication is done through posting events and always passes through the kernel. The kernel supports two kinds of events: synchronous and asynchronous. Synchronous events require the immediate running of the event handler and are required to run to completion. Asynchronous events on the other hand are a form of differed computation; hence, they could be dispatched at a later time. Unlike previous OSs, power management is not done by the scheduler. In Contiki, power management is left to the application programmer. Contiki provides the programmer with the size of the event queue. The programmer could use this to determine when the system could enter a sleeping mode.

### 13.6.1.2    Applications and Services

Contiki's programming language is C. Processes are either applications or services. Services are processes that implement functionality that can be used by other processes. Multiple applications may use the same service. Services and applications could be replaced at runtime using Contiki's program loader. An example of a service in Contiki is the communication protocol stack. Since services are replaceable, we could change the routing protocol during runtime. Services are managed by the service layer. The service layer monitors the running services and can be used to find the available services. Each service has an interface that indicates its id. Applications using the services do so by linking a stub library to the application. This library uses the service layer to find service processes. Once the service is located its id is cached, otherwise the request is aborted. Services may also call other services using the same mechanism. Figure 13.8 illustrates how the communication stack may split into two different services due to dynamic reprogramming. It also shows how service and event communication occurs.

What makes Contiki a hybrid then? Other than services, Contiki provides a set of libraries such as the stub library used in service communication. Another important library that Contiki provides is the threading library. This library provides the programmer with an API to implement preemptive threads. The library consists of two parts: a platform-independent part that interfaces with the kernel, and a platform-dependent part that deals with stack switching during preemption [11]. The library provides six functions, which are mt_yield(), mt_post(), mt_wait(), mt_exit, mt_start(), and mt_exec(). The first four are called by a running thread, while the last two are called to startup a thread.

*Example 2.* Dynamic programming, services and events in Contiki.

The presented OSs exploit the event-driven and thread-driven philosophies in constraint WSNs. A general comparison along with performance results of these OSs are presented in the following section.

## 13.7 Comparison and Analysis

We presented several OSs designed specifically for WSNs focusing on the execution model of each. Although some had similar models of execution yet they provided different services and functionalities required by WSNs. This section summarizes important features of WSN OSs in a comparative manner.

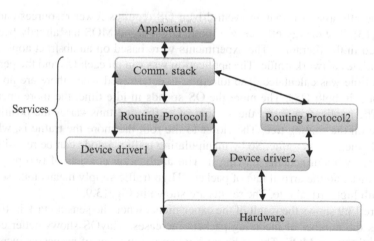

**Fig. 13.8** Because of dynamic programming at runtime, the communication stack is loosely coupled, with several routing protocols and device drivers. A device driver reads an incoming packet into a buffer and then calls the upper layer communication service using the service mechanism described earlier. The comm. stack processes the headers of the packet and posts a synchronous event to the application for which the packet is destined. The application processes the packet and buffers a reply before it returns control to the comm. stack (adapted from [11])

**Table 13.5** Comparison of WSN OSs [12]

| Features | TinyOS | MOS | SOS | RETOS | Contiki |
|---|---|---|---|---|---|
| Dynamic reprogramming | Yes | No | Yes | Yes | Yes |
| Priority scheduling | No | Yes | Yes | Yes | Yes |
| Real-time operation | No | Yes | No | Yes | Yes |
| Power-aware scheduler | Yes | Yes | No | Yes | Yes |
| Kernel based | No | Yes | Yes | Yes | Yes |
| Resiliency | Yes | No | No | Yes | No |
| Model | Event-driven components | Thread-driven | Event-driven modules | Thread-driven | Hybrid |

## 13.7.1 General Comparison

To preview the different characteristics of the aforementioned OSs, Table 13.5 presents a comparison of OSs for WSNs.

## 13.7.2 Thread-Driven Versus Event-Driven

It is generally assumed that an event-driven OS requires fewer resources and less energy [13]. The energy efficiency of both TinyOS and MOS has already been investigated in the literature. The experiments were based on an abstract application that simulates network traffic. The application was run on each OS and the percentage idle time was calculated. The idle time is determined when there are no tasks to perform in both OSs. The more the OS spends in idle time, the more energy it saves. The application varies the amount of traffic and thus varies the position of the node on the routing tree. The closer to the root the more the traffic is, while a leaf node means less traffic. So by manipulating traffic, a node can be repositioned on the tree without actually moving it. The application consists of two parts: the sensing task and the arrival rate of packets. High traffic simply means long sensing tasks with high arrival rate. The results are shown in Fig. 13.9.

Figure 13.9 shows the result of the experiments when the sensing task is 100-ms long. As the number of incoming packets increases, TinyOS shows better energy consumption than MOS. This is because when the number of processes increases, the scheduler will do a context switch more often. Context switches consume more CPU cycles than most operations. However, when traffic is low, both OSs have similar performance. More experiments have shown that although MOS is less energy efficient, it is much more predictable than TinyOS in real-time operation.

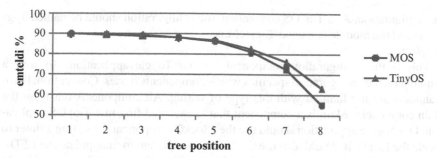

**Fig. 13.9**  As traffic increases, MOS tends to spend more energy than TinyOS due to the overhead of context switches (adapted from [13])

## 13.8  Thoughts for Practitioners

Practitioners or application developers may want to decide upon the easiest way to write their algorithms. However, other factors might be of concern such as energy efficiency. In this section we present practitioners with hands-on experience to learn how to start developing applications using the two design philosophies to guide them in their OS choice.

### 13.8.1  Implementing Your Own Program

After understanding the concepts behind different EOSs, let us get a hands-on experience with these concepts by developing our own WSN application. A practitioner who wants to write application programs for WSN will find this section useful not only as a starting point, but also as a reference for WSN programming concepts. We present two ways of writing the same application, one using the event-driven model on top of TinyOS and the other using the thread-driven model on top of MOS, in Examples 3 and 4, respectively. The application we intend to develop is a simple application that toggles one of the LEDs on a wireless sensor at a certain rate. Examples 3 and 4 are not intended for network programming; instead, they are an introduction to the programming principles of different OS models.

*Example 3.* Writing a Simple Event-Driven Program in TinyOS.

In this example we present component-based programming under TinyOS. The programming language used for TinyOS is called nesC, which is a C dialect. It has a powerful compiler that enforces the characteristics of TinyOS mentioned in Sect. 13.4.1. To learn more about nesC refer to [14, 15]. Let us start by naming our application. We shall call it "Toggle." An application in TinyOS consists of two components or files. One holds the implementation called the module and the other holds the "wiring," which tells TinyOS how to bind the components and is called the configuration. As a TinyOS convention, the configuration should be named Toggle.nc and the module is named ToggleM.nc.

*Toggle.nc*

This is the configuration component for our Toggle application. To wire a component, the configuration specifies which components it uses. Conventional programmers are not familiar with this type of wiring. All components must use the Main component. Main is a component that is executed first in a TinyOS application. Our Toggle application should use the ClockC component for setting a timer to toggle the LEDs. It should also use the LedsC component to manipulate the LEDs, and the ToggleM component that we will implement. So our configuration should look as follows:

```
1  configuration Toggle {
2  }
3  implementation {
4  components Main, ToggleM, ClockC, LedsC;
5  ... }
```

Our configuration is not complete yet. After defining the used components, we need to do the wiring. In nesC this is done using the ◊ operator. Before we describe the wiring, we need to understand a simple interface called "StdControl." StdControl is an interface that is used by all components to initialize and start.

That is, each component should implement the three functions of the StdControl interface, which are init(), start(), and stop(). The interface looks as follows:

```
1  interface StdControl {
2      command result_t init();
3      command result_t start();
4      command result_t stop();
5  }
```

The next part in our configuration wires the interfaces of other components used with our toggle application. The wiring is done inside the "implementation" clause as follows:

```
1  configuration Toggle {
2  }
3  implementation {
4      components Main, ToggleM, SingleTimer, LedsC;
5
6      Main.StdControl -> ToggleM.StdControl;
7      ToggleM.Clock -> ClockC;
8      ToggleM.Leds -> LedsC;
9  }
```

Line 6 wires Main.StdControl to ToggleM.StdControl, which means that ToggleM.StdControl.init() will be called by Main.StdControl.init(). The same applies for strat() and stop().

*ToggleM.nc*

```
1   module ToggleM {
2   provides {
3       interface StdControl;
4   }
5   uses {
6       interface Clock;
7       Interface Leds;
8   }
9   implementation {
10  bool state;/* state of the LED (on or off) */
11
12  /* implementation of StdControl interface */
13  command result_t StdControl.init() {
14      state = FALSE;
15      call Leds.init();
16      return SUCCESS;
17  }
18  command result_t StdControl.start() {
19          return call Clock.setRate(TOS_I1PS, TOS_S1PS);
20  }
21  command result_t StdControl.stop() {
22          return call Clock.setRate(TOS_I0PS, TOS_S0PS);
23  }
24  /* Implement the event handler for Clock.fire */
25  event result_t Clock.fire() {
26          state = !state;
27      if (state) {
28          call Leds.redOn();
29      } else {
30        Call Leds.redOff();
31      }
32      return SUCCESS;
33      }
34  }
```

In our module we have three clauses: the "provides" clause, which defines the set of provided interfaces, the "uses" clause, which defines the set of used interfaces and the "implementation" clause where we will find the implementation of both the commands of interfaces we provide and the event handlers of interfaces we use.

In lines 14 and 15 we are initializing the component. Line 19 creates a timer. As a result, when the timer stops the ClockC component will trigger an event upward to our application. This event is handled in lines 26–33. The event handler will toggle the red LED whenever the event occurs and thus at the rate we specified in line 19.

*Example 4.* Writing a simple thread-driven program in Mantis OS

In this example we implement the same program as before but under the thread-driven MOS. Programming with MOS is much simpler since it follows the conventional thread-driven approach and the highly portable C language. The entire aforementioned program could be written in three simple steps. More on MOS programming could be found in [16].

*Step 1*

First we include the headers needed to our Toggle.c file. The mos.h header should always be included. The msched.h header includes the Mantis scheduler. To manipulate the LEDs, we also include led.h.

```
1   #include ``mos.h''
2   #include ``mshced.h''
3   #include ``led.h''
```

*Step 2*

Now we implement the function that toggles one of the LEDs.

```
4   void toggle_thread(void){
5       while(1){
6       mos_led_toggle(0);
7       mos_thread_sleep(1000);
8       }
9   }...
```

The function called in line 6 takes a parameter, which indicates which LED to toggle. In our case it is the first LED with id 0. Line 7 calls a function that blocks the thread for 1s. This blocking simulates the timer in Example 3.

*Step 3*

Finally we implement the entry point of our application, the start() function. We convert our function into a thread; we give the thread 128 bytes of stack space and normal priority.

```
10   void start(void){
11   mos_thread_new(toggle_thread, 128,
     PRIORITY_NORMAL);
12   }
```

## 13.9   Directions for Future Research

Although several OSs exist for WSNs, yet most were developed as bases for future directions in WSN EOS design. Researchers are now interested in enhancing those EOSs for better energy consumption, space footprint, and real-time operation, focusing on the operation of single nodes. From the design point of view, as mentioned in Sect. 13.3, WSN OSs should have the characteristics of distributed systems, which are still not evident in present WSN OSs. More research needs to be conducted on the feasibility of sharing resources among WSN OSs. Moreover, WSNs in the future will consist of many thousands of nodes. Having so many different OS designs and execution models and different performances requires research on hybrid deployment to allow for the scalability of WSNs. Better yet, a global design for WSN OSs could be engineered. TinyOS, for example, is already noticed as the standard OS for WSNs and most research effort is done on top of it. To achieve a general design, more work should be done in eliminating the resource/accuracy trade-off between different OSs such as the one we saw in Sect. 13.7.2 where MOS provides more accuracy than TinyOS, which in turn provides better energy consumption than MOS. This could be done by optimizing preemption in thread-driven systems or by adding preemption to event-driven systems. Adding preemption to TinyOS is an ongoing research effort. Moreover, a more reliable comparison for WSN OSs is needed in order to pinpoint other tradeoffs and hence build a clearer picture of the intended general system. To do so, research on creating a benchmark for WSN OSs is indeed of great interest.

From the programming point of view, current programming models are too low-level. In Sect. 13.6.1, for example, we saw how Contiki allows the programmer to directly manipulate the hardware without a hardware abstraction layer. Although this may decrease the number of levels in the hierarchy, but it forces the developers to think about hardware details. This is reflected by the huge effort put in order to create demos [17]. We need a programming model that eliminates the irritable details of hardware. Moreover, current programming models are also node-centric such as nesC, for example. nesC focuses utterly on programming individual nodes. One area for research in this field is macroprogramming. The main purpose is to develop a high-level language to implement aggregate programs for a WSN. One such work called TinyDB has already taken this step [17]. More programming models that target an entire system are needed. Other programming tools for WSN programming are also at a stage that requires more work to be done such as tools for debugging and programming interfaces (IDEs).

## 13.10   Conclusions

In this chapter, we presented a survey of WSNs from an OS point of view. Different design models that form the core of WSN OSs were introduced exposing the trade-offs in performance and usability. We also showed how to optimize those design

models for WSNs by exploring different WSN OSs. For each OS, we explained
the design philosophy behind it and the motivation for choosing that philosophy.
We also presented the execution models of each OS with elaborative examples and
figures. Moreover, we presented a feature comparison between all the OSs and a
performance comparison between OSs of two different design models and showed
the trade-off in design. Later we provided a hands-on experience for practitioner in
the field on how programming for the two models differs allowing for a better understanding of all the previously mentioned concepts. Finally we presented directions
for future and open research areas in the field of WSN EOSs.

Although research in the field of WSN EOSs is actively growing, practitioners in
this field have very few choices in terms of OS and development tools. A practitioner
thus has to understand the principles of each OS and its programming model to make
a choice and present better results depending on his requirements. The main goal of
presenting different OSs was not to decide which is superior to the other, but to show
what conventional methods apply and what novel methods present as an alternative.
These were the same goals the developers of the discussed OSs had in mind. This is
because they were motivated by the resource challenges presented by WSNs. Now,
after all the efforts done to find out what is feasible and what is not, we may say that
the motivation for current research has partially shifted from challenging low-level
constraints of the nodes to high-level constraints of OSs.

## Terminologies

*Microcontroller.* A computer on a chip, which is more cost effective than a micro-
    processor.
*MMU.* The memory management unit is a hardware component that handles access
    to memory by translating virtual addresses to physical addresses.
*Bounded buffer producer–consumer problem.* Arises When items are removed
    from the buffer at a slower rate than being added. The buffer eventually be-
    comes full and does not allow more data.
*Flash memory.* Nonvolatile memory that can be electrically erased and repro-
    grammed.
*Hardware interrupt.* A signal from the hardware demonstrating the need for
    attention.
*RAM.* Random access memory is a type of computer storage.
*FIFO scheduler.* Processes are picked in a first in first out manner.
*Context switch.* When the scheduler replaces the running process on the CPU with
    another process.
*CPU cycles.* The unit of the CPU. The speed of the CPU is determined in terms of
    CPU cycles or clock ticks. A set of instructions usually require a fixed number
    of CPU cycles to execute. The faster the clock, the more instructions could be
    computed in time.

*Routing tree.* When networked sensors are connoted, they define a route that will be used to forward data. When nodes are placed hierarchically, the route will form a tree-like structure.

## Questions

1. Present two main differences between event-driven and thread-driven design models.
2. Is TinyOS an application-based OS? Why or why not?
3. Why is TinyOS more energy efficient than MOS?
4. How is the RETOS scheduler power aware?
5. How does the TinyOS scheduler determine that it should wake up from the sleep mode? What about the MOS scheduler? Is there a different way the MOS scheduler can wake up the system?
6. Assume a thread-driven model with variable sized tasks. Assume also that small tasks have higher priority and thus continuously preempt longer tasks causing extra context switches. Suggest a scheduling strategy to decrease the number of context switches. Hint: Lower context switches are more preferable than real-time operation.
7. Can an event-driven system provide preemption? Why or why not?
8. A context switch is as expensive as 1,000 clock cycles. Convert that into milliseconds on a 4-MHz processor.
9. Assume an event-driven system with a quantum of 5 ms with a 4-MHz processor. Also assume there are two running processes, each consuming 800,000 clock cycles and are started at the same time. Also assume that a context switch consumes 20,000 clock cycles. How many context switches will occur after

    (a) 100 ms
    (b) 200 ms

10. Give one advantage of providing several routing protocols as in Example 2 in Sect. 13.6.1.

## References

1. J. Hill, R. Szewczyk, A. Woo, S. Hollar, D. Culler, and K. Pister, "System architecture directions for networked sensors," *Proceedings of the Ninth International Conference on Architectural Support for Programming Languages and Operating Systems*, ACM Press, New York, USA, November 2000, pp. 93–104.
2. H. Lauer and R. Needham, "On the duality of operating system structures," *Proceedings of the Second International Symposium on Operating Systems*, IR1A, Rocquencourt, France, October 1978; reprinted in *Operating Systems Review*, April 1979, pp. 3–19.

3. R. Behren, J. Condit, and E. Brewer, "Why events are a bad idea (for high-concurrency servers)," *Proceedings of HotOS IX: The Ninth Workshop on Hot Topics in Operating Systems*, USENIX Association, Hawaii, USA, May 2003, pp. 19–24.
4. A. Gustafsson, "Threads without the pain," *Queue*, ACM Press, New York, USA, November 2005, pp. 34–41.
5. H. Karl and A. Willig, *Protocols and Architectures for Wireless Sensor Networks*, Wiley, April 2005, pp. 45–50.
6. C. Han, R. Kamur, R. Shea, E. Kohler, and M. Srivastava, "A dynamic operating system for sensor nodes," *Proceedings of the Third International Conference on Mobile Systems, Applications, and Services*, ACM Press, New York, USA, June 2005, pp. 163–176.
7. S. Bhatti, J. Carlson, H. Dai, J. Deng, J. Rose, A. Sheth, B. Shucker, C. Gruenwald, A. Torgerson, and R. Han, "MANTIS OS: An embedded multithreaded operating system for wireless micro sensor platforms," *ACMKluwer Mobile Networks and Applications Journal, Special Issue on Wireless Sensor Networks*, Kluer Academic, Hingham, USA, August 2005, pp. 563–579.
8. C. Duffy, U. Roedig, G. Herbert, and C. Sreenan, "An experimental comparison of event driven and multi-threaded sensor node operator systems," *Proceedings of the Fifth Annual IEEE International Conference on Pervasive Computing and Communications Workshops*, IEEE Computer Society, White Plains, New York, USA, March 2007, pp. 267–271.
9. H. Kim and H. Cha, "Multithreading optimization techniques for sensor network operating systems," *Wireless Sensor Networks*, Springer, Heidelberg, April 2007, pp. 293–308.
10. C. Hujumg, C. Sukwon, J. Inuk, K. Hyoseung, S. Hyojeong, Y. Jaehyun, and Y. Chanmin, "RETOS: Resilient, expandable, and threaded operating system for wireless sensor networks," *Proceedings of the Sixth International Conference on Information Processing in Sensor Networks*, Massachusetts, USA, April 2007, pp. 148–157.
11. A. Dunkels, B. Gronvall, and T. Voigt, "Contiki – A lightweight and flexible operating system for tiny networked sensors," *Proceedings of the Twenty Ninth Annual IEEE International Conference on Local Computer Networks*, November 2004, pp. 455–462.
12. S. Yi, H. Min, J. Heo, B. Boand, and E.F. Roberts, "Performance analysis of task schedulers in operating systems for wireless sensor networks," *Computational Science and Its Applications*, Springer, Heidelberg, May 2006, pp. 499–508.
13. C. Duffy, U. Roedig, G. Herbert, and C. Sreenan, "A performance analysis of mantis and tinyos," University College Cork, Ireland, Technical Report CS-2006-27-11, November 2006.
14. D. Gay, P. Levis, R. von Behren, M. Welsh, E. Brewer, and D. Culler, "The nesC language: A holistic approach to networked embedded systems," *Proceedings of Programming Language Design and Implementation*, California, USA, June 2003, pp. 1–11.
15. TinyOS. http://www.tinyos.net.
16. Multimodal Networks of In-situ Sensors. http://mantis.cs.colorado.edu.
17. M. Welsh and R. Newton, "Region streams: Functional macroprogramming for sensor networks," *Proceedings of the first workshop on data management for sensor networks*, Toronto, Canada, August 2004.

# Chapter 14
# Adaptive Distributed Resource Allocation for Sensor Networks

Hock Beng Lim, Di Ma, Cheng Fu, Bang Wang, and Meng Joo Er

**Abstract** A major research challenge in the field of sensor networks is the distributed resource allocation problem, which concerns how the limited resources in a sensor network should be allocated or scheduled to minimize costs and maximize the network capability. We survey the existing work on the distributed resource allocation problem. To address the drawbacks in the existing work, we propose the adaptive distributed resource allocation (ADRA) scheme, which specifies relatively simple local actions to be performed by individual sensor nodes in a sensor network for mode management. Each node adapts its operation over time in response to the status and feedback of its neighboring nodes. Desirable global behavior results from the local interactions between nodes.

We study the effectiveness of the general ADRA scheme for a realistic application scenario, namely, the sensor mode management for an acoustic wireless sensor network (WSN) to track vehicle movement. An enhanced version of ADRA, ADRA with node density compensator, is also proposed to improve the performance of the algorithm for randomly distributed sensor fields. We evaluated these algorithms via simulations and also prototyped the acoustic WSN scenario using the Crossbow MICA2 motes. Our simulation and hardware implementation results indicate that the ADRA scheme and its enhanced variant provide good trade-off between performance objectives such as coverage area, power consumption, and network lifetime.

## 14.1 Introduction

With the rapid advances in technologies such as MEMS sensor devices, low-power embedded processors, and wireless networking, it is now possible to deploy large-scale wireless sensor networks (WSNs) with hundreds or thousands of small, cheap,

H.B. Lim (✉)
Intelligent Systems Center, Nanyang Technological University, Research Techno Plaza,
BorderX Block, Level 7, 50 Nanyang Drive, Singapore 637553, Republic of Singapore
e-mail: limhb@ntu.edu.sg

S. Misra et al. (eds.), *Guide to Wireless Sensor Networks*, Computer Communications and Networks, DOI: 10.1007/978-1-84882-218-4_14,
© Springer-Verlag London Limited 2009

347

and smart sensor nodes. These sensor nodes can collaboratively sense the environment, collect and process environmental data, and guide intelligent decision making.

Sensor networks have generated much research interest in the academic and industry because they have a wide range of important potential applications, such as environmental and habitat monitoring, medical and healthcare monitoring, military and security surveillance, tracking of goods and manufacturing processes, smart homes and buildings, and many other applications we do not yet imagine [1, 9, 10, 15].

However, the design of large-scale sensor networks presents many challenging research issues. One of the most important open research issue is the distributed resource allocation problem – how to allocate, without a central coordinator, the limited sensing, processing, or communication resources in a sensor network to monitor the dynamically changing environment.

Distributed resource allocation in large-scale sensor networks is challenging for several reasons. First, there are a large number of decision makers. Second, there is limited communication among the decision makers. Thus, the information available for each decision maker is incomplete. Third, the environment is dynamically changing. Finally, the solution required is constrained by time.

Several approaches to tackle the distributed resource allocation problem have been proposed in the literature. In this chapter, we first survey the existing approaches for distributed resource allocation. We propose the adaptive distributed resource allocation (ADRA) scheme, which addresses the distributed resource allocation problem from a different angle compared to that of existing techniques. The ADRA scheme specifies relatively simple local actions to be performed by individual sensor nodes in a sensor network for mode management. Each node adapts its operation over time in response to the status and feedback of its neighboring nodes. Desirable global behavior results from the local interaction between nodes. The ADRA scheme is scalable since the coordination of the actions of neighboring nodes requires little communication. It is adaptive and robust with respect to the dynamic environment that the sensor network operates in.

Our scheme provides a general framework for efficient resource allocation in sensor networks. It can actually be applied to many sensor network applications and problems. In this chapter, to evaluate the effectiveness of the ADRA scheme, we apply it to a realistic application scenario. The scheme is used to perform sensor mode management in an acoustic WSN that tracks vehicle movement in an open terrain. A variant of the ADRA scheme, called the ADRA with density compensator (ADRA-dc) scheme, is also proposed to handle the scenario of random sensor node deployment more efficiently. We simulated both the schemes under the grid, random, and hotspot sensor node deployment scenarios, and also prototyped the ADRA scheme using the Crossbow MICA2 motes. Our simulation and hardware implementation results indicate that our proposed schemes provide a good trade-off between performance objectives such as coverage area, power consumption, and network lifetime.

The rest of this chapter is organized as follows. In Sect. 14.2, we review the related work and discuss our contributions. Section 14.3 presents the problem formulation, the general framework of ADRA scheme, and the enhanced ADRA-dc scheme. We describe the application scenario of the acoustic WSN for vehicle tracking in Sect. 14.4. Section 14.5 presents our algorithms to implement the ADRA scheme for three scenarios – grid, random, and hotspot deployment. The methodology and results of our simulation study are presented and discussed in Sect. 14.6. We discuss the hardware prototype implementation and the performance results in Sect. 14.7 for the benefit of practitioners. Finally, we conclude this chapter in Sect. 14.8 and outline the directions of future work.

## 14.2   Background

Different aspects of the distributed resource allocation problem have been investigated by researchers. In distributed real-time systems, which are widely utilized in critical control systems, high-speed communication systems, and various monitoring applications, an important research problem is how to allocate resources to applications and how to execute applications to maximize the performance according to some criteria with the real-time constraints. According to [2], there are two different approaches to solve this problem. In the static approach, the problem-solving process takes place before the operation of the system [5]. The dynamic approach is more flexible as it can handle unexpected situations that occur during the operation of the system, which may cause changes to the quality of service [16, 35]. However, previous research has shown that both approaches are in general NP-complete [11, 18]. Hence, heuristic techniques are usually developed to find the optimal solution of the resource allocation problem in real-time systems [2, 5, 22].

In [2], the system model is assumed to be a heterogeneous distributed real-time system continuously running applications. The resources are allocated for applications in a distributed real-time system by characterizing the specifications (e.g., hardware platform, quality of service constraint), and then developing appropriate heuristics to maximize the performance goal. An initial static allocation is determined to maximize the allowable workload increase, followed by the dynamic resource reallocation process to avoid QoS violation. Three greedy heuristics were developed for the problem-solving approach. The Most Critical Path First (MCPF) heuristic is suitable for latency-constrained real-time heterogeneous systems. The other two heuristics are the Most Critical Task First (MCTF) and the Tie-Breaking Two Phase Greedy (TB TPG) heuristics. They are designed to suit the requirements of throughput-constrained heterogeneous systems. Detailed discussions on greedy heuristics are provided in [3, 17].

A systematic formulation to map the distributed resource allocation problem into the distributed constraint satisfaction problem (DCSP) has been addressed by many researchers [19, 27, 28, 32, 38, 39]. Formally, the constraint satisfaction problem (CSP) consists of $n$ variables whose values are taken from finite and discrete domains, and a set of constraints on the values of the variables. To solve a CSP problem means to search for a consistent assignment of values to all the variables such that the constraints are satisfied. A good example of a CSP is the n-queens puzzle.

In the area of artificial intelligence, the research on CSP has a long and distinguished history [21]. When the variables and constraints become distributed among multi-agents, the CSP becomes a distributed CSP, i.e., the DCSP. The mapping of the distributed resource allocation problem into the DCSP is sufficiently generalized and reusable to tackle some specific difficulties such as ambiguity and dynamism. The problem is then solved by finding a solution to the DCSP, which is actually the assignment of values for distributed variables to satisfy all distributed constraints. A review of different algorithms for solving DCSP problems is given in [39] and a comparison of the performance of some algorithms is presented in [25]. The utilization of DCSP for resource allocation in wireless networks is discussed in [4, 20].

In market-based techniques [24, 36], the distributed system is modeled as the interaction between agents taking economic roles. Resources are allocated through buying and selling activities between agents. A seller seeks to maximize its earnings whereas a buyer seeks to minimize its spending. Resource requests and price notifications are communicated among the agents. Certain heuristics or strategies are used to control agent behaviors through the propagation of price information.

In [24], a market-based macroprogramming (MBM) paradigm was proposed to allocate resources in sensor networks. The sensor nodes receive profit for performing simple local actions in response to globally advertised price information, thus the sensor network forms a virtual market. A cost-evaluation function is implemented on each of the sensor nodes, and the global behavior is induced throughout the network by advertising price information that drives nodes to react. The system goals of lifetime, accuracy, or latency are met through the tuning of price information. The macroprogramming is therefore encoded in the process used to update price information in response to changing network conditions.

Another interesting approach to the multiagent resource allocation problem is the auction and bidding techniques for allocating resources to tasks, such as combinatorial auction [29, 30] and coalition formation [33].

In [30], a variant of the classic resource allocation problem, called the setting-based resource allocation problem was formulated in the context of WSN. This formulation reflects the challenges posed in domains in which sensor nodes have multiple settings, each of which could be useful to multiple tasks. Then, these tasks and resource allocations are translated into bids that can be solved by a modified combinatorial auction, where recent developments in the solution of such auctions can be utilized.

In [33], cooperative task execution in multiagent environments is discussed. Given a set of agents and a set of tasks that they have to satisfy, the situations where each task should be attached to a group of agents that will perform the task were considered. Task allocation to groups of agents is necessary when tasks cannot be performed by a single agent. However, it may also be beneficial when groups perform more efficiently than the single agent. Several solutions to the problem of task allocation among autonomous agents were proposed. It was shown that distributed agents perform tasks more efficiently in coalition. Though this scheme was proposed in the context of distributed agents, it could be applied to the context of WSNs without much alteration.

Much of the work in the multiagent systems community has focused on multiagent negotiation over the allocation of resources. In [6, 7], a "contract net" framework was proposed for communication and control in distributed problem solving. It functions as the common medium for contract negotiation, which is an essential form of task distribution. The protocol for the negotiation process should help determine the content of exchanged messages, and it is not just a means of physical communication.

In [23], the presented negotiation protocol is distributed and mediation-based. It takes advantage of the cooperative nature of multiple agents in the environment to maximize social utility. Each agent is able to act in a mediator capacity when resource conflicts are recognized. An agent gains local view of the global allocation problem and makes suggestions regarding the allocations. It employs a finite state machine as the heart of the negotiation protocol. The protocol also shares some common characteristics with distributed breakout algorithm [37]. In [12], a family of cooperative and adaptive algorithms for distributed resource allocation problems is presented in which the underlying strategy of combinatorial auction is utilized. The discussion in [8] takes a different angle of view for the negotiation process. It considers the problem that the agents are solving to construct a plan; hence, the agents need to solve planning problem as well as negotiate with each other. The agents work together by decomposing problems into subproblems, allocating and exchanging the subproblems as well as the solutions, and synthesizing the overall solution.

Our work differs from the aforementioned work and makes two important contributions. Firstly, our ADRA scheme is a scalable and adaptive distributed resource allocation scheme for sensor networks. It is scalable since it relies only on near-neighbor communications between nodes in a sensor network. It is adaptive because each node reacts to the environment (such as the presence of targets) as well as the status and feedback of its neighbors. These local node interactions produce desirable global system behavior.

Secondly, the ADRA scheme provides a general framework that is applicable to many WSN applications. For example, we have applied the ADRA scheme to a realistic application scenario of an acoustic WSN for target detection and tracking. Also, the performance evaluation in previous work is usually done via analytical modeling or simulation of a generic sensor network. In our work, apart from evaluating the ADRA scheme via simulation, we have implemented the scheme on a real sensor network platform and evaluated its effectiveness.

## 14.3  ADRA Scheme

### 14.3.1  Problem Formulation

The basic entity in a sensor network is a sensor node, which has sensing, processing, and communication capabilities. The sensor network has a set of stationary sensor nodes. Each node has a set of modes (or actions) that it is able to partake, and it can only choose one mode at any particular time instance.

During the operation of the sensor network, each sensor node's utility value is a function of several factors including the sensing coverage (or target detection), target localization, and target localization error minimization. For a sensor node, the rate of energy consumption affects its useful lifetime. When the sensor node is out of power, it is no longer able to sense, process, or communicate, and thus its utility value will be zero.

The sensor nodes have no prior information on the targets and their movement. They can sense and detect targets and are able to obtain directional information of the targets from sensor bearing measurements. It takes at least two sensor nodes to localize a detected target. Minimizing the error in localization requires more than two sensor nodes detecting the target simultaneously. The sensor nodes can communicate with their neighbors, which are within the communication range. However, from available communications and environment sensing, the nodes would not have full knowledge of the entire network.

The problem of resource allocation in a WSN can thus be defined as the maximization of some WSN objective functions subject to certain performance constraints. For example, in the mode management for the acoustic WSN to track vehicle movement, the objective function is to maximize the WSN lifetime by minimizing the resource usage, and the constraint is that the coverage ratio must be at least 50% of the absolute coverage achieved by turning on all the nodes.

### 14.3.2 The General ADRA Scheme

We propose the ADRA scheme as a framework or methodology to guide sensor nodes for efficient resource allocation. The ADRA scheme is shown in Algorithm 1.

Under the ADRA scheme, a sensor node goes through many operational cycles repetitively in its lifetime. An operational cycle represents a complete and self-contained activity period during which the node gathers sufficient information regarding the targets from the ambient environment and its neighbors for decision making. It determines the necessary actions to adapt itself to the environment while aiming for maximal performance of the whole network. Each cycle is split into three phases. Within each phase, the ADRA scheme specifies the necessary local actions to be performed to achieve efficient mode management and sensor resource allocation in the global context.

In Phase 1 (Initialization), each node initializes its internal states and prepares itself by querying its neighbors' mode status and the environment information such as the targets within range. At the end of Phase 1, each node shares the gathered preliminary information with its neighbors. During Phase 2 (Processing), each node collects all preliminary information from neighbors. The information would be analyzed and combined with its own information to yield its behavioral plan, i.e., the likely action to be executed. Again, the plan will be shared among neighbors.

Phase 3 (Decision) is the stage to make a final decision. With all necessary information and action plans from neighbors, a node is able to determine how it should act to maximize the overall performance of the network.

---

**Algorithm 1**: Adaptive Distributed Resource Allocation

**Phase 1: Initialization**
Query neighbors' mode status.
Get information about detected targets (if any).
Update local variables (e.g., utility, battery life).
Send information on detected targets to neighbors.

**Phase 2: Processing**
Receive information on targets from neighbors.
Fuse own detected target info with neighbors' detected target info.
Compute change in utility based on information from neighbors.
Compute own plan regarding sensor mode.
Optional: compute plan for neighbors.
Send information on the plan to neighbors.

**Phase 3: Decision**
Receive information on neighbors' plans.
Resolve own plan with neighbors' influence.
Execute the plan to change own sensor mode.

---

## 14.3.3   The Enhanced ADRA Scheme: ADRA with Density Compensator

The general ADRA scheme provides a framework for distributed resource allocation in WSNs. It works well for a WSN whose sensor nodes have a regular grid-like layout, but does not produce as good results for a WSN whose nodes are randomly distributed. This is because the general ADRA scheme does not consider the density, especially local density, of the distribution of sensor nodes in the entire WSN. It imposes the same computation rule for the sensor nodes' mode management regardless of sensor node distribution. This is fine for a WSN whose nodes are in a grid layout since the nodes' spacing is similar across the sensor field. However, in a random node distribution environment, it is possible that certain areas in the sensor field have more nodes than other areas. Thus, the nodes in these densely deployed areas should have a lower probability of being allocated a particular unit of resource compared with the nodes in those sparsely deployed areas. Although the concept is simple, it is difficult to partition a sensor field into subzones, which have different local densities. This is because without specifying the area for each subzone to be partitioned, there are infinite ways of partitioning the sensor field, which is an NP-hard problem.

We propose an enhancement of the general ADRA scheme, called the ADRA with density compensator (ADRA-dc) scheme, to handle the random node distribution case. ADRA-dc has a density compensator component compared with the general ADRA. Instead of partitioning the sensor field into different density zones, the density compensator calculates the relative density of a particular node with respect to the entire WSN. In other words, it quantifies the likelihood of a node being in a densely deployed zone. Thus, the mode management algorithm will take this density compensator into account to achieve a fairer resource allocation. The ADRA-dc scheme is shown in Algorithm 2.

---

**Algorithm 2**: Adaptive Distributed Resource Allocation – Density Compensator

---

**Phase 1: Initialization (same as in general ADRA)**
Query neighbors' mode status.
Get information about detected targets (if any).
Update local variables (e.g., utility, battery life).
Send information on detected targets to neighbors.

**Phase 2: Processing**
Receive information on targets from neighbors.
Fuse own detected target info with neighbors' detected target info.
Compute change in utility based on information from neighbors.
**Compute density compensator.**
**Compute own plan with density compensator regarding sensor mode.**
Optional: compute plan for neighbors.
Send information on the plan to neighbors.

**Phase 3: Decision (same as in general ADRA)**
Receive information on neighbors' plans.
Resolve own plan with neighbors' influence.
Execute the plan to change own sensor mode.

---

## 14.4   Mode Management in Acoustic Sensor Network

We consider an acoustic WSN deployed for the purpose of monitoring vehicle movement in an open terrain. In this network, the acoustic sensor nodes are powered by batteries. The scheme provides a controlled trade-off between the ability to provide coverage for the area of interest and to perform localization of the targets, and the battery power conservation to prolong the network lifetime.

Each acoustic sensor node has two modes: on and standby. When the acoustic sensor node is in the "on" mode, it has full sensing, processing, and communications functionalities. When the node is in the "standby" mode, it stops sensing the environment and has limited communications capabilities. The amount of battery life consumed by the node per time unit in this state is assumed to be ten times lesser than when the node is in the "on" mode. A node in "standby" mode can still communicate and exchange messages with its neighbors according to the ADRA scheme and switch to the "on" mode to sense a target when necessary.

The acoustic sensor's sensing capability is omnidirectional in nature, i.e., it can detect a target's acoustic signal from any direction, with an error variance of 1 rad. A target is considered to be detected when it is within the sensing range of the sensor. Sensor measurements or target detections are in the form of bearing (or angular) values of the target with respect to the sensors monitoring it, which are combined to form a positional fix of that target. The target bearing values and messages from the ADRA scheme will be transmitted among neighboring nodes.

## 14.5   Algorithm Description

### 14.5.1   Stansfield Algorithm

For the ADRA scheme, we adopt the Stansfield algorithm [34] to combine the bearing values of a target detected by multiple sensor nodes to localize the target, i.e., to obtain a positional fix of the target. The Stansfield algorithm computes the positional fix of the target in the form of the best point estimate of the target coordinates and an uncertainty ellipse that bounds the likely location of the target.

Figure 14.1 illustrates how the Stansfield algorithm works. The figure shows 4 sensor nodes and a target. The constraint is that each sensor node can only detect

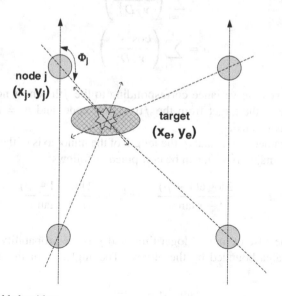

**Fig. 14.1** Stansfield algorithm

the bearing of the target, but not the exact position and the distance to the target. This constraint is applicable to a wide range of sensors. Each sensor node might contain an array of rudimentary sensors internally to provide direction finding capability. The Stansfield algorithm outputs two metrics for target localization: a best point estimate of the target coordinates, and an uncertainty ellipse that bounds the likely location of the target.

The coordinates of the best point estimate are computed using the following equation:

$$\begin{bmatrix} x_e \\ y_e \end{bmatrix} = \left( \sum_{j=1}^{n} \begin{bmatrix} \cos^2 \varPhi_j & -\sin \varPhi_j \cos \varPhi_j \\ -\sin \varPhi_j \cos \varPhi_j & \sin^2 \varPhi_j \end{bmatrix} \right)^{-1} \cdot$$
$$\sum_{j=1}^{n} \begin{bmatrix} x_j \cos^2 \varPhi_j & -y_j \sin \varPhi_j \cos \varPhi_j \\ -x_j \sin \varPhi_j \cos \varPhi_j & y_j \sin^2 \varPhi_j \end{bmatrix},$$

where $\varPhi_j$ = the bearing of the target from the $j$th sensor node with respect to true North, $x_j$ = the $x$-coordinate of the $j$th sensor node, and $y_j$ = the $y$-coordinate of the $j$th sensor node.

The uncertainty ellipse represents the accuracy tolerance of the algorithm. Its parameters are calculated from the following geometric equations:

$$s = \sum_{j=1}^{n} \left( \frac{\cos \varPhi_j \sin \varPhi_j}{v_j D_j^2} \right),$$

$$t = \sum_{j=1}^{n} \left( \frac{\sin^2 \varPhi_j}{v_j D_j^2} \right),$$

$$u = \sum_{j=1}^{n} \left( \frac{\cos^2 \varPhi_j}{v_j D_j^2} \right),$$

where $v_j$ = bearing variance corresponding to the $j$th sensor node, $D_j$ = estimated distance of the target from the $j$th sensor node, and $n$ = the number of bearing values associated.

Using the parameters $s$, $t$, and $u$, the length of the minor axis of the ellipse, $a$, and the length of the major axis, $b$, can be computed as follows:

$$a^2 = -\frac{2 \log e(1-p)}{t - s \tan \varphi}, \qquad b^2 = -\frac{2 \log e(1-p)}{u + s \tan \varphi},$$

where $e$ is the base of natural logarithm, and $p$ is the probability that the target lies within the area bounded by the ellipse. The angle, $\varphi$, of the ellipse is computed by:

$$\tan 2\varphi = -\frac{2s}{t - u}.$$

## 14.5.2   Mode Management in Acoustic Sensor Network Using General ADRA

Our algorithm to perform the ADRA scheme for the acoustic scenario is shown in Algorithm 3. In the first phase (initAndSend), each node obtains its own sensor measurements of the targets' bearing values, and computes the targets' positional fixes using the Stansfield Algorithm. It also updates its own potential, which is the utility value used for deciding the mode of the node (on or standby). Then, it sends the information on the detected targets and its own mode to the neighbors.

In the second phase (rcvProcessSend), each node receives the bearing values and positional fixes of target from its neighbors. Then, it fuses its own and the neighbors' bearing values to obtain the new positional fixes of the targets. The node updates its own potential and sends its potential and battery life to the neighbors.

In the third phase (rcvExe), each node receives the potential and battery life information from its neighbors. Based on the difference in battery life between itself and its neighbors, the node computes its new potential value. After computing its new potential, the node decides whether to be "on" or "standby" by comparing the potential with a threshold value.

---

**Algorithm 3**: Mode management in acoustic sensor network

---

1: **main**()
2: Constants : battPri, /* priority value for battery life conservation */
3:   covPri, /* priority value for coverage */
4:   locPri, /* priority value for localization */
5:   threshold /* threshold value */
6: Variables : potential, /* potential for on or standby mode */
7:   battLife, /* battery life of node */
8:   battLifeDiff /* battery life difference between self and neighbor */
9: **repeat**
10:   initAndSend();
11:   rcvProcessSend();
12:   rcvExe();
13: **until** termination of operation, or if node depletes its battery life.

14: **procedure initAndSend**()
15: Query neighbors' mode status.
16: Get own sensor measurement of target(s) bearing value(s).
17: Compute target(s) positional fix(es) using Stansfield Algorithm.
18: Update own potential.
19: Send to neighbors : target(s) bearing value(s) and existing positional fix(es), own mode (on or standby).

20: **procedure rcvProcessSend**()
21: Receive from neighbors : targets' bearing values and positional fixes, neighbors' modes.

---

(continued)

**Algorithm 3**: (continued)

22: Update potential.
23: Fuse and update current set of bearing value and positional fix with new values from self
      and neighbors.
24: **for** each bearing value from self and neighbors **do**
25:   Increase own potential (by covPri).
26: **end for**
27: **for** each positional fix from self and neighbors **do**
28:   Increase own potential (by locPri).
29: **end for**
30: Send to neighbors : own potential and battLife.

31: **procedure rcvExe()**
32: Receive from neighbors : potential values and battLife info.
33: **for** each neighbor **do**
34:   Compute battLifeDiff.
35:   **if** (neighbor_battLife > battLife) **then**
36:       Decrease potential by (battPri * battLifeDiff).
37:   **else**
38:       Increase potential by (battPri * battLifeDiff).
39:   **end if**
40: **end for**
41: **if** (potential < threshold) **then**
42:   Switch to "standby" mode.
43: **else**
44:   Switch to "on" mode.
45: **end if**

## 14.5.3   Mode Management in Acoustic Sensor Network Using ADRA-dc

As discussed earlier, the general ADRA scheme does not take the relative densities of the nodes in the WSN field into consideration. Thus, it imposes the same
computation rule for all the nodes in the field, even though nodes that are located
in a densely deployed subzone should have a lower probability of being allocated
a resource. ADRA-dc is designed to address this problem. Since it is difficult and
costly to partition the WSN field into absolute subzones with different densities, we
introduce a density compensator to compute the relative local density of a node with
respect to the global density of the WSN field. It is described in the following steps:

### 14.5.3.1   Step 1

After sensor nodes have been deployed in the field, we can compute the mean distance $m_i$ of each node to its neighbors based on localization information (we assume

known localization for each node), where $i$ is the node index and $i \in n$. The result of the computation will form a $1 \times n$ dimension matrix $[m_1, m_2, m_3, \ldots, m_n]$, while $n$ is the number of nodes in the field.

### 14.5.3.2  Step 2

We then search the matrix to find out its largest and the smallest item, denoted as $m_{\max}$ and $m_{\min}$. These two values are used for normalization of the matrix.

$$\left[ \frac{m_1 + a - m_{\min}}{m_{\max} - m_{\min}}, \frac{m_2 + a - m_{\min}}{m_{\max} - m_{\min}}, \ldots, \frac{m_n + a - m_{\min}}{m_{\max} - m_{\min}} \right],$$

where $a$ is a constant that prevents the normalized $m_i$ from becoming 0. The intuition here is that a node that is located in a densely deployed subzone will have a relatively smaller mean distance than nodes in a loosely deployed subzone. The normalized mean distance will give an indication of this tendency. The smaller the normalized mean, the more likely the nodes are located in a dense subzone. This normalized mean will be one of the basic components of the density compensator.

### 14.5.3.3  Step 3

However, it is not enough to consider just the normalized mean distance alone to determine the relative density of a node. Consider the case that two nodes are located near to each other but far away from the rest of the nodes, as depicted in Fig. 14.2.

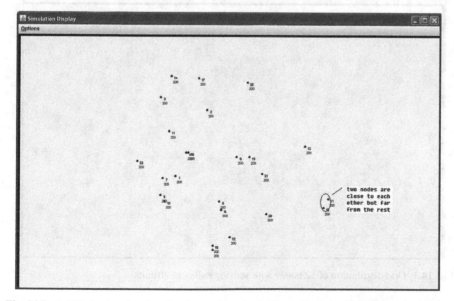

**Fig. 14.2** An illustration showing that the number of neighbors of a node also indicates its relative density

These two nodes will have a very small normalized mean distance because they do not have any other neighbors. However, it is obviously not true to say that these two nodes are located in a densely deployed zone.

Thus, the number of neighbors of a node also indicates how likely this node is located in a densely deployed subzone. The more neighbors a node has, the more likely this node is in a densely deployed zone. For example, consider the same case mentioned earlier; even though the normalized mean distances of the two nodes are small, it does not mean that they are in a densely deployed zone because they are just close to each other but far from the rest of the nodes. To take this factor into consideration, the number of neighbors of a node needs to be inserted into the density compensator.

### 14.5.3.4 Step 4

To further enhance the ADRA-dc algorithm, we also take the sensor sensing range (or radius) into consideration. Consider the two scenarios shown in Fig. 14.3 (case 1) and in Fig. 14.4 (case 2). The nodes in Fig. 14.3 have a sensing range of 20 units whereas the nodes in Fig. 14.4 have a sensing range of 10 units. Geographically, we can say that both case 1 and case 2 have the same node distribution pattern. However, the effective sensing coverage in case 2 is much smaller than in case 1. Thus, functionally, case 1 is denser than case 2. As a result, the sensing range of the node should be incorporated in the density compensator formula as well.

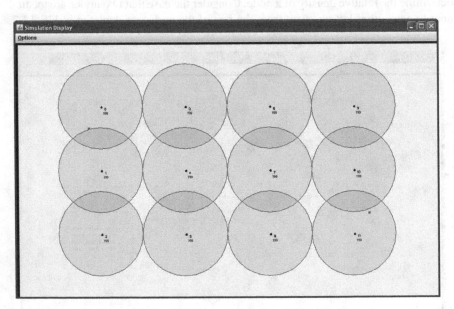

**Fig. 14.3** Grid distribution of 12 nodes with sensing radius of 20 units

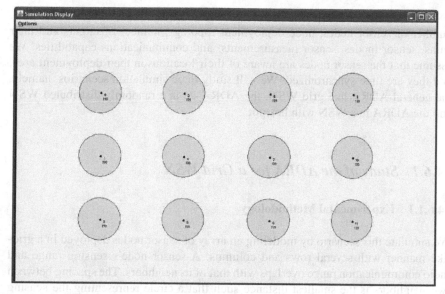

**Fig. 14.4** Grid distribution of 12 nodes with sensing radius of 10 units

### 14.5.3.5   Step 5

The final computation of the density compensator is shown here. This density compensator will then be used in the computation of node potential for node resource allocation.

$$\text{Density compensator} = \beta \frac{(\log sr)(H_i^r)}{m_i'},$$

where $\beta$ = scaling factor of the density compensator; it is set to 0.145 in the simulations done in this chapter, $sr$ = sensing radius; it has been set to 10, 20, and 30 in the simulations performed in this chapter, $H_i^\gamma$ = neighbor factor in the density compensator. $H_i$ is the number of neighbors for node $i$, $\gamma$ is a nonlinear scaling factor and is set to 1.65 in the simulations, and $m_i'$ = the normalized mean distance to its neighbors for node $i$.

## 14.6   Simulation Evaluation

We conducted simulation studies of the application scenario using the Recursive Porous Agent Simulation Toolkit (Repast 3.0) [31], an open source agent-based simulation and modeling toolkit. It is written in Java, and was originally developed at the University of Chicago.

In the simulation setup, we model attributes such as the simulation world size, number of sensor nodes, node deployment topology, number of targets and their paths, sensor modes, sensor measurements, and communications capabilities. We assume that the sensor nodes are aware of their locations in their deployment area, and they are time-synchronized. We will study three simulation scenarios, namely, the general ADRA in a grid WSN, the ADRA-dc in a randomly distributed WSN and the ADRA in a WSN with hotspot.

## 14.6.1 Study of the ADRA for a Grid WSN

### 14.6.1.1 Experimental Methodology

We simulate this scenario by modeling an array of sensor nodes deployed in a grid-like manner with several rows and columns. A sensor node's sensing range and radio communication range overlaps with that of its neighbors. The spacing between two neighbors is the smallest distance such that a circle representing the sensing coverage area of an internal node only intersects with those coverage circles of its four neighbors and no other nodes. By simple geometric rule, the node spacing $d$ is related to the sensing range sr by the equation: $d = sr\sqrt{2}$.

We investigate the general ADRA scheme with a fixed simulation setting: Net24 with 24 nodes ($6 \times 4$ grid). The sensing range is set to be 20 units, and so the spacing between nodes is $d = 28.3$ units. As each node needs to exchange messages with its neighbors, the radio communication range must be larger than the node spacing $d$. For this simulation study, the communication range is set to be 30 units. The corresponding dimensions of the grid areas for the simulation are 200 units $\times$ 120 units. We also model eight targets moving across the sensor field at a constant velocity of 1/3 unit per simulation tick.

We study three cases of the acoustic sensor network operation under such fixed simulation setting. In the baseline ("WithoutAlgo") case, the network does not use the ADRA scheme, i.e., all the nodes would be "on" until they exhaust their battery power. In the other two cases "WithAlgoWithoutTarget" and "WithAlgoWithTarget," the network uses the ADRA scheme to control its operation. There are no targets in the former case, while there are targets to be tracked in the latter case.

Figure 14.5 shows a screenshot of the Net24 simulation. The sensing coverage radius of an active node is delineated by a circle. The absence of such a circle indicates that a node is in standby mode.

We use the network coverage area and the sensor network lifetime as performance metrics. The coverage area is defined to be the largest area such that any inside point is covered by at least one circle, without double counting the regions where the circles overlap. Each sensor node starts off with a predefined battery life. As simulation time passes, each node consumes battery life at a varying rate according to the changes in its modes, until its battery life is depleted. We measure

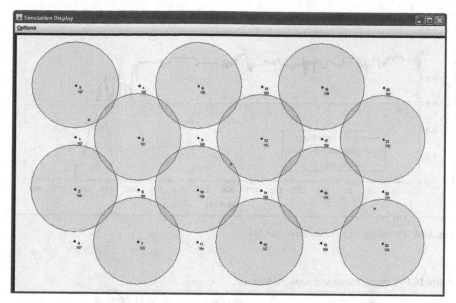

**Fig. 14.5** Screenshot of acoustic sensor network simulation (Net24)

the coverage area against time. As more and more nodes eventually use up their battery life, the trend is that the sensor network coverage area declines with time. We define the sensor network lifetime as the amount of time for the coverage area to drop to zero.

### 14.6.1.2   Simulation Result and Discussion

The coverage area against time for Net24 is shown in Fig. 14.6. Also, the results for the average coverage area and the network lifetime for this network under the three cases are shown in Table 14.1.

The baseline ("WithoutAlgo") case is the simplest to understand. As all the nodes are always "on," the maximum possible coverage area (100%) is provided until all nodes deplete their battery life. In our simulation, we set the nodes' initial battery life such that the network lifetime will be 200 s for the simulation. As a result, the WSN lifetime is 200 s in "WithoutAlgo" case.

In the "WithAlgoWithoutTarget" case, the network converges into two steady state configurations. In one configuration, the nodes at alternating diagonals are "on" and the rest are in "standby" mode. In the other configuration, the modes of the "on" and "standby" nodes are reversed. Triggered by the adaptive nature of the ADRA scheme, the network periodically switches back and forth between these two configurations by reversing the modes of the nodes. In this manner, the battery

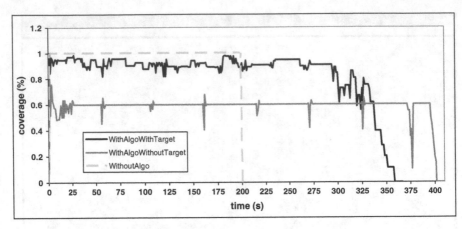

**Fig. 14.6** Coverage area versus time (Net24)

**Table 14.1** Coverage area and network lifetime

| Net24 cases | Avg coverage area (%) | Network lifetime (s) |
| --- | --- | --- |
| WithoutAlgo | 100 | 200 |
| WithAlgoWithoutTarget | 59.4 | 402 |
| WithAlgoWithTarget | 84.3 | 358 |

life consumption of the nodes is balanced as much as possible across the network as time progresses. Consequently, half of the nodes are "on" at steady state, and the network life time in this case is 402 s, double that of the "WithoutAlgo" case. However, as not all the nodes are "on" at all times, the tradeoff is that the coverage area for the Net24 has dropped to 59.4% of the maximal coverage in the baseline case, respectively.

The "WithAlgoWithTarget" case shows the effect of target tracking. The coverage area rises above the "WithAlgoWithoutTarget" case at the beginning since the ADRA scheme turns on more nodes to help track the targets. With more nodes turned on, the power consumption is higher too. Eventually, the coverage area starts to drop as more and more nodes deplete their battery life. Thus, the network lifetime in this case is shorter than that of the "WithAlgoWithoutTarget" case, but still longer than the "WithoutAlgo" case. The coverage area in this case is 84.3% of the maximal coverage provided by the baseline case; the WSN lifetime is 358 s.

In general, the ADRA scheme provides a significant improvement in network lifetime at the cost of a decrease in the coverage area in networks. Our results also show that the ADRA scheme is scalable and it can work well for larger networks too.

## 14.6.2    Study of the ADRA-dc for a Randomly Distributed WSN

### 14.6.2.1    Experimental Methodology

We simulate the randomly distributed WSN scenario by modeling a sensor field with sensor nodes deployed randomly. A sensor node's sensing range and radio communication range overlaps with that of its neighbors in random manner. First, we will investigate the effect of general ADRA scheme applied to this random WSN field. Second, we will apply ADRA-dc to the random WSN field. To demonstrate the effectiveness of the density compensator component, we will compare the results of both simulations.

The number of sensor nodes simulated is still 24 (Random – Net24). In order to achieve a repeatable randomness (i.e., to use the same random node distribution for both simulation runs), we fix the random seed (a seed for random number generator) to be 1,234 We also want the random coordinate generated to be within the simulation world (of size 200 × 120). Thus, we set the mean and standard deviation of the nodes' random $x$-coordinate to be 200/2 and $mean_x/1.5$, respectively, and the mean and standard deviation of the nodes' random $y$-coordinate to be 120/2 and $mean_y/1.5$, respectively. Effectively, we are generating 24 random Cartesian coordinates centered on the centre point (100, 60) of the simulation world, with a standard deviation of 66.7 for $x$-coordinate and 40 for $y$-coordinate. Fig. 14.7 shows the distribution of the nodes.

To demonstrate the effectiveness of the density compensator component, we apply the general ADRA and the ADRA-dc scheme to this randomly deployed WSN.

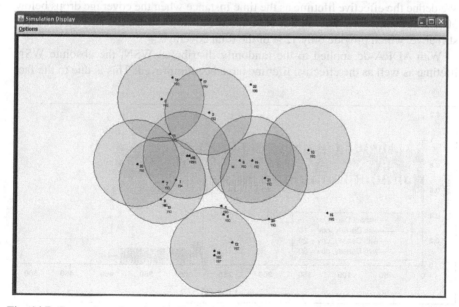

**Fig. 14.7** Screenshot of randomly distributed acoustic sensor network simulation (Random – Net24)

We do not model targets moving across the sensor field as the presence of targets only introduces a linear component in the sensor mode computation rule, and it does not affect the result of the density component.

We still use the network coverage area and the sensor network lifetime as performance metrics. We will compare the results generated by the general ADRA and the ADRA-dc, especially sensor network lifetime under different schemes. Since the sensor nodes' sensing radius should also be taken into consideration for the density compensator computation, we will investigate three sensing radius settings in the ADRA-dc simulation, i.e., the sensing radius is set to be 10, 20, and 30 units, respectively.

### 14.6.2.2   Simulation Results and Discussion

In Fig. 14.8, the "withoutDensity" case is the general ADRA applied to the randomly distributed WSN as shown in Fig. 14.7. The "withDensity" "cov = 10, 20, 30" are the three cases of ADRA-dc with the sensor nodes' sensing radius set to be 10, 20, and 30 units, respectively.

It can be seen from Fig. 14.8 that the absolute lifetime for the "withoutDensity" case is 401 s, which means that at that time, all the sensors deplete their battery power. However, we can also see from the plot that at time 233 s, the coverage ratio drops drastically. This clearly indicates that at time 233 s, most of the nodes have depleted their battery power and only a few nodes are still alive. Table 14.2 shows the coverage ratio and the number of sensor nodes alive before and after time 233 s. This characteristic is undesirable as the effective lifetime of the WSN is only 234 s. We define the effective lifetime as the time instance when the coverage drops below 50% of maximal coverage. In this case, after time 234 s, there are only two sensors still alive, which provide only 12% of the total coverage.

With ADRA-dc applied to the randomly distributed WSN, the absolute WSN lifetime as well as the effective lifetime have been improved. This is due to the fact

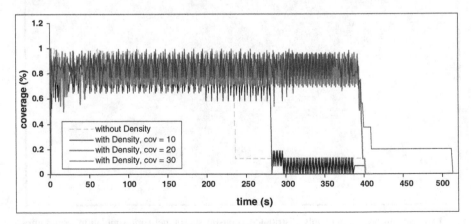

**Fig. 14.8** Coverage area versus time (Random – Net24)

**Table 14.2**  General ADRA, before and after effective lifetime

|                          | 0–233 s | 234 s | 235 s | 236 s |
|--------------------------|---------|-------|-------|-------|
| Avg coverage area (%)    | 90.4%   | 61.9% | 12.0% | 12.0% |
| Number of alive nodes    | 24      | 11    | 2     | 2     |

**Table 14.3**  ADRA-dc, before and after effective lifetime

| $Cov = 10$            | 0–280s  | 281 s | 282 s | 283 s |
|-----------------------|---------|-------|-------|-------|
| Avg coverage area (%) | 76.7%   | 31.3% | 18.8% | 18.8% |
| Number of alive nodes | 24      | 9     | 3     | 3     |
| $Cov = 20$            | 0–395 s | 396 s | 397 s | 398 s |
| Avg coverage area (%) | 81.9%   | 43.1% | 33.7% | 12.0% |
| Number of alive nodes | 17      | 16    | 10    | 4     |
| $Cov = 30$            | 0–398 s | 399 s | 400 s | 401 s |
| Avg coverage area (%) | 85.6%   | 37.2% | 37.2% | 37.2% |
| Number of alive nodes | 17      | 16    | 5     | 5     |

that density compensator component actually computes a node's relative density with respect to the rest, and subsequently updates a node's potential function using this component to achieve a fairer resource allocation.

The absolute lifetimes for the cases where the sensing radius is 10, 20, and 30 units are 401, 401, and 512 s, respectively. Table 14.3 shows the coverage ratio and effective lifetime for the three cases. As shown in Table 14.3, the effective lifetimes for the three cases are 280, 395, and 398 s, respectively. When the sensing radius is set to be 20 and 30 units, the effective lifetime under these two cases is almost double that of the general ADRA case. However, when the sensing radius is set to be 10 units, the effective lifetime only increases by 20%. This is because when the sensing radius is set to 10 units, the WSN is regarded as a sparsely distributed network regardless of the geographical density. The density compensator will have a smaller value to have more nodes to be turned on.

The average coverage ratios within the effective lifetime for the cases where the sensing radius is 10, 20, and 30 units are 76.7, 81.9, and 85.6%, respectively. Take note that the average coverage ratio increases almost linearly with the increase in sensing radius. Using standard least square regression method, we obtain a function that describes the linear relationship between average coverage ratio and the sensing radius for this random node distribution.

$$y = 72.5 + 0.455x,$$

where y is the average coverage ratio in percentage and x is the sensing radius. With this linear function, we can customize the network by balancing the trade-off between lifetimes, coverage, and sensing radius. We can also estimate the performance of a WSN before the nodes have been deployed.

### 14.6.3 Study of the ADRA and ADRA-dc for a WSN with Hotspot

#### 14.6.3.1 Experimental Methodology

In a typical WSN, it is likely that some areas in the sensor field may have more nodes than others. For example, in an acoustic sensor network for target tracking, we may deploy more nodes near a road junction. The area where more nodes have been deployed is called a hotspot in the WSN. By deploying more nodes near or within the hotspot, we actually allocate more physical resources for that area. We then study how the general ADRA and the ADRA-dc scheme can handle such cases. Note that it is possible that a hotspot may also exist in a randomly distributed WSN, but such a hotspot is randomly generated (as it is possible that some areas may have more nodes than others). Since the randomly generated hotspot serves no purpose functionally, it is not of interest to us. To study the hotspot scenario, we systematically generate a hotspot within the sensor field and investigate the effectiveness of the ADRA and the ADRA-dc scheme for such deployment.

The number of nodes simulated in this study is 24. As shown in Fig. 14.9, there are three hotspots in the WSN field. Each hotspot has eight nodes deployed nearby.

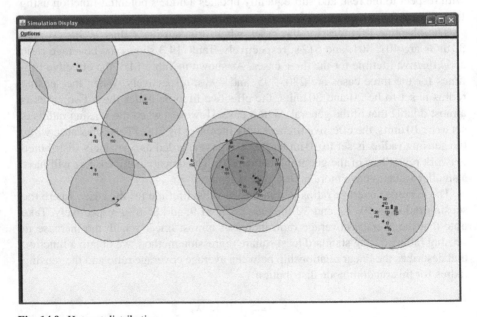

**Fig. 14.9** Hotspot distribution

**Table 14.4**   Configuration for three hotspots

| Hotspot | Center coordinate mean | Standard deviation $x$ | Standard deviation $y$ |
|---------|------------------------|------------------------|------------------------|
| 1 | (167,80) | 5 | 5 |
| 2 | (100,60) | 8 | 8 |
| 3 | (34,40) | 23 | 23 |

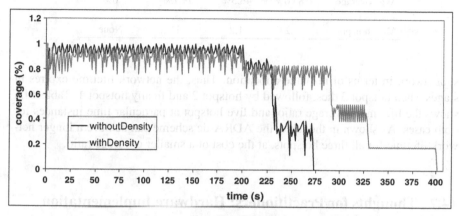

**Fig. 14.10**   Coverage area versus time for a WSN with hotspots

The first hotspot's coordinate is (167, 80); the eight randomly distributed nodes near the first hotspot have a mean coordinate of (167, 80) and standard deviation of 8 for both $x$ and $y$ coordinates. The second and third hotspots' nodes are deployed in a similar manner, except that the standard deviation is 13 and 18 for the second and third hotspots, respectively. The difference in standard deviation causes the hotspots generated to have different node density. As shown in Fig. 14.9, the smaller the standard deviation, the denser the node distribution around that hotspot. Table 14.4 shows the configuration for the three hotspots.

We investigate how the general ADRA and the ADRA-dc scheme perform in this scenario. The performance metrics are still coverage ratio and network lifetime. In this study, we do not investigate the effect of the sensor nodes' sensing radius as it has been studied in the previous simulation study.

### 14.6.3.2   Simulation Results and Discussion

In Fig. 14.10, there are two plots of coverage versus lifetime. The "withoutDensity" case is the general ADRA scheme applied to this scenario. The "withDensity" case is the ADRA-dc scheme with the sensing radius set to 20 applied to this scenario.

It can be seen from Fig. 14.10 that the absolute network lifetimes for the general ADRA and the ADRA-dc are 272 and 400 s, respectively. Also, the coverage ratio drops in a step manner for both of these cases. This is because the WSN field is unevenly distributed with three hotspots, each with eight nodes and function as a

**Table 14.5** General ADRA and ADRA-dc for a WSN with hotspots

| General ADRA | 0–200 s | 201–229 s | 230–272 s | 273 s onwards |
|---|---|---|---|---|
| Avg coverage area (%) | 94.4% | 82.8% | 33.3% | 0% |
| Alive hotspot | 1,23 | 1,2 | 1 | None |
| ADRA-dc | 0–290s | 291–325s | 325–400s | 400 s onwards |
| Avg coverage area (%) | 83.0% | 46.5% | 19.1% | 0% |
| Alive hotspot | 1,23 | 1,2 | 1 | None |

subnetwork in terms of resource allocation. Thus, the network lifetime expires in stages when hotspot 3 dies, followed by hotspot 2 and finally hotspot 1. Table 14.5 shows the lifetime, coverage ratio, and live hotspot at particular time instances for both cases. As shown in the table, the ADRA-dc scheme can achieve a longer network lifetime for all three hotspots, at the cost of a smaller coverage ratio.

## 14.7 Thoughts for Practitioners: Hardware Implementation

### 14.7.1 Hardware Prototype Implementation

To assess the actual performance of the ADRA scheme on a real sensor hardware platform, we prototyped the acoustic sensor network scenario using the Crossbow MICA2 motes [26]. The motes are programmed in nesC [13] under the TinyOS development environment [14]. nesC is an extension of the C programming language. TinyOS is an event-driven operating system designed for sensor nodes. Our hardware testbed deploys 16 MICA2 motes in a $4 \times 4$ grid. We also use the Crossbow MTS310CA sensor boards, which are plugged onto the MICA2 motes.

The total power consumption of a mote is an aggregation of the power consumption of its components, including the processor, radio, logger memory, and sensor board. Each component can operate in different functional modes. The power consumption of each component is different when operating in different modes. For example, the microcontroller draws around 8 mA during full operation but only 8 μA during sleep mode [26]. Therefore, the overall power consumption is the sum of all component-based consumptions, averaged by the duty cycles of operational modes for each component. In our testbed, we empirically measured the power consumption of a MICA2 mote as approximately 25 mA in active mode and 11 mA in standby mode.

Our testbed only aims to prototype the acoustic sensor network scenario to demonstrate a proof-of-concept hardware implementation of the ADRA scheme. We disable all sensors except the acoustic sensor for power saving. The acoustic sensor on the MTS310CA sensor board is a microphone capable only of providing the magnitude reading of an acoustic signal. It is unable to provide the direction of arrival of an acoustic signal. Thus, we simplify our implementation of Algorithm 3 so that it performs only target detection but not target localization. Fortunately, this simplification does not have any big impact on demonstrating the efficacy of the ADRA scheme because our key performance metrics of coverage area and network lifetime are still relevant.

We use the beeping sound of the MTS310CA sounder (at acoustic frequency of 4 KHz) to emulate the noise from a target. The spacing between two motes is related to the sensing range of the acoustic sensor, in a similar manner as in the simulation. From empirical measurements, we determine that a good spacing distance between the motes in our testbed is 50 cm, as it is a suitable distance for detecting the MTS310CA sounder signal with a reasonable internal threshold.

### 14.7.2   Results and Discussion

Figure 14.11 shows the coverage area against time for the three cases measured on our 16-node testbed. The unit of time in the $x$-axis is in terms of time cycles. A short time cycle duration makes packet collision reduction and power management difficult to control, whereas a long time cycle hampers the target detection. In our implementation, we empirically determined that a time cycle duration of 5 s gives acceptable performance. Each MICA2 mote is powered by a pair of AA batteries, which can last for days. To expedite the data collection and analysis process, we consider only the first 250 cycles as shown in Fig. 14.11

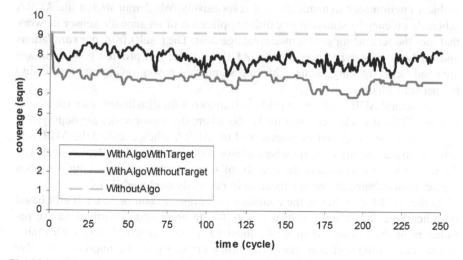

**Fig. 14.11** Coverage area versus time for 16-node MICA2 testbed

As expected, the baseline ("WithoutAlgo") case is very simple: all nodes are always "on" and hence the coverage area is constant at $9.1\,m^2$ over time. In the "WithAlgoWithoutTarget" case, the coverage area dropped to an average value of $6.7\,m^2$. When targets are introduced in the "WithAlgoWithTarget" case, more nodes are triggered to turn on and hence we get a higher coverage area than the case of "WithAlgoWithoutTarget." In Fig. 14.11, the graph representing "WithAlgo-WithTarget" is above that of the "WithAlgoWithoutTarget" case during the duration of 250 cycles. The average coverage area in the presence of targets is $7.9\,m^2$.

During the duration of 250 cycles, the coverage areas of the "WithAlgoWith-outTarget" case and the "WithAlgoWithTarget" case are 73.6% and 86.8% that of the"WithoutAlgo" case, respectively. However, if we were to run this experiment for a longer time, the coverage area graphs for both these cases should drop as more and more motes deplete their batteries, just like in the simulation.

## 14.8   Conclusions and Future Work

In WSNs, many sensor nodes are cooperating to perform various sensing, com-munication, and processing functionalities. For such large-scale distributed system, a key challenge is that how to efficiently allocate limited distributed resources in a dynamically changing environment. In this chapter, we have reviewed ex-isting literature addressing different aspects of the distributed resource allocation problem.

We have proposed the ADRA scheme, which specifies the coordination amongst neighboring nodes in a sensor network for action and decision making in mode management. The ADRA scheme helps sensor networks adapt to changes in the ambient environment dynamically and responsively. We demonstrated the ADRA scheme's efficacy by studying a realistic application of an acoustic sensor network that uses the scheme for sensor mode management. The results from our simulations and hardware prototype show that the ADRA scheme can provide good coverage area and target tracking, while achieving significant power saving and prolonging the network lifetime.

The general ADRA scheme provides a framework for distributed resource alloca-tion in WSNs. It works very well for WSNs where the sensor nodes are deployed in a grid layout. We proposed an extension of the ADRA scheme, called the ADRA-dc scheme, to address the scenario where sensor nodes are in random deployment. The ADRA-dc scheme considers the density of sensor nodes when making decisions regarding distributed resource allocation in randomly deployed WSNs.

In the ADRA-dc scheme, the computation of density compensator is still based on a heuristic. To achieve optimal results for different node distributions the pa-rameters of the density compensator must be tuned accordingly. As such tuning is time-consuming and may result in human error, a possible approach to solve this problem is to use genetic programming (GP). This issue can be addressed by focusing on how GP can be used to solve the parameter estimation problem. Finally, discovering further extensions of the ADRA scheme is also a good direction for future work.

# Terminologies

*ADRA.* ADRA stands for adaptive distributed resource allocation. It specifies relatively simple local actions to be performed by individual sensor nodes in a sensor network for mode management. The mode of each node is dynamically determined in a distributed manner based on local rules and near-neighbor interactions.

*ADRA-dc.* The ADRA with density compensator (ADRA-dc) is an extension of the general ADRA scheme. The ADRA-dc scheme takes into account the density of the nodes in a particular region of a sensor network. In a densely deployed area, the resource allocation strategy will be different from that in a sparsely deployed area.

*Agent-based simulation.* Agent-based simulation is an experimental framework that specifies a computational model for simulating the actions and interactions of autonomous agents in a network, with a view to assess the effects of the agents on the system as a whole.

*Multiagent system.* A multiagent system is a system composed of multiple software or hardware agents, collectively capable of achieving goals that are difficult to achieve by an individual agent.

*Stansfield algorithm.* Stansfield algorithm is a computational technique to fuse the bearing values of a target detected by multiple sensor nodes to localize the target. It computes the positional fix of the target in the form of a best point estimate of the target coordinates, and an uncertainty ellipse that bounds the likely location of the target.

*Sensor scheduling.* In WSNs, due to the limited resources such as battery power, processing power, bandwidth, etc., sensor nodes in the network must be scheduled properly to either save the energy consumption or to avoid conflicts in using a particular system resource. There are two general approaches to solve the sensor scheduling problem, namely, centralized and distributed. In centralized sensor scheduling, a central coordinator will compute the relevant cost function and determines the mod for each individual node. In distributed sensor scheduling, each node computes its own mode based on local rules and interactions with its neighbor nodes.

*QoS.* QoS stands for Quality of Service. In the field of communication engineering, Quality of Service is the ability to provide different priorities to different applications, users, or data flows to guarantee a certain level of performance. In the context of WSN, the QoS varies according to the specific application of the network deployed. For example, for a WSN tracking moving targets, the QoS can be metrics such as the tracking accuracy, confidence level, error covariance, etc.

*MCPF and MCTF.* MCPF stands for Most Critical Path First and MCTF stands for Most Critical Task First. These are greedy heuristic algorithms for solving the distributed resource allocation problem by determining the modes of sensor nodes based on predefined priority.

*CSP and DCSP.* The constraint satisfaction problem (CSP) consists of $n$ variables whose values are taken from finite and discrete domains, and a set of constraints on their values. Solving a CSP problem means to search for a consistent assignment of values to all variables such that the constraints are satisfied. The distributed constraint satisfaction problem (DCSP) is the extension of CSP into distributed system domain.

*MBM.* MBM stands for market-based macroprogramming, which is a paradigm for allocating resources in sensor networks. The individual sensor nodes receive profit for performing simple local actions in response to globally advertised price information, thus the sensor network forms a virtual market. A cost evaluation function is implemented on each of the sensor node, and the global behavior is induced throughout the network by advertising price information that drives nodes to react. The system goals of lifetime, accuracy, or latency are met through the tuning of price information. The macroprogramming is therefore encoded in the process used to update price information in response to changing network conditions.

## Questions

1. Explain the importance of distributed resource allocation in WSNs.
2. Briefly discuss two existing WSN resource allocation algorithms.
3. Explain what is market-based macroprogramming.
4. Explain what is the ADRA scheme.
5. Write down the logic flow for the general ADRA scheme.
6. Explain what is the ADRA-dc scheme.
7. Write down the logic flow for the ADRA scheme with density compensator.
8. Assuming there are four nodes deployed in a WSN, three of the nodes' coordinates are known, denoted as $(x_i, y_i)$. The node with unknown position can measure the distance (subject to noise) from itself to the three nodes with known positions. Let the unknown position be denoted as $(x_c, y_c)$ and the distance measured be denoted as $d_i$, where $i = 1, 2, 3$. Derive the least square sensor network localization formulation to determine $(x_c, y_c)$.
9. In a WSN target tracking application, given the noisy measurements as shown here, estimate the true target trajectory using least square method.

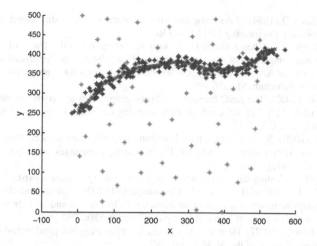

10. What is the disadvantage of fusing a target trajectory using the method in the previans question? What would be the fix?

# References

1. Akyildiz I, Su W, Sankarasubramaniam Y, Cayirci E (2002). Wireless sensor networks: a survey. Computer Networks 38(4):393–422.
2. Ali S, Kim J, Siegel H, Maciejewski A, Yu Y, Gundala S, Gertphol S, Prasanna V (2002) Greedy heuristics for resource allocation in dynamic distributed real-time heterogeneous computing systems. Proc. of the 2002 Intl. Conf. on Parallel and Distributed Processing Techniques and Applications (PDPTA 02), Las Vegas, NV, 519–530.
3. Armstrong R, Hensgen D, Kidd T (1998). The relative performance of various mapping algorithms is independent of sizable variances in run-time predictions. Proc. of the Seventh IEEE Heterogeneous Computing Workshop, 79–87.
4. Bejar R, Krishnamachari B, Gomes C, Selman B (2001). Distributed constraint satisfaction in a wireless sensor tracking system. Proc. of Workshop on Distributed Constraint Reasoning, Seattle, WA.
5. Braun TD, Siegel HJ, Beck N, Boloni LL, Maheswaran M, Reuther AI, Robertson JP, Theys MD, Yao B, Hensgen D, Freund RF (2001). A comparison of eleven static heuristics for mapping a class of independent tasks onto heterogeneous distributed computing systems. Journal of Parallel and Distributed Computing 6:810–837.
6. Davis R, Smith R (1983). Negotiation as a metaphor for distributed problem solving. Artificial Intelligence 20(1):63–109.
7. Davis R, Smith R (2003). Negotiation as a metaphor for distributed problem solving. Communication in Multiagent Systems: Agent Communication Languages and Conversation Policies 20:63–109.
8. Durfee E (2001). Distributed problem solving and planning. The Ninth ECCAI Advanced Course on Multi-Agent Systems and Applications, Prague, Czech Republic.
9. Estrin D, Govindan R, Heidemann J, Kumar S (1999). Next century challenges: scalable coordination in sensor networks. Proc of the Fifth ACM/IEEE Intl. Conf. on Mobile Computing and Networking 263–270.
10. Estrin D, Culler D, Pister K, Sukhatme G (2002). Connecting the physical world with pervasive networks. IEEE Pervasive Computing 1(1):59–69.

11. Fernandez-Baca D (1989). Allocating modules to processors in a distributed system. IEEE Trans. on Software Engineering 11:1427–1436.
12. Frank M, Bugacov A, Chen J, Dakin G, Szekely P, Neches B (2001). The marbles manifesto: a definition and comparison of cooperative negotiation schemes for distributed resource allocation. Proc. of the AAAI Fall Symp. on Negotation Methods for Autonomous Cooperative Systems, North Falmouth, MA, 36–45.
13. Gay D, et al. (2003). The nesC language: a holistic approach to networked embedded systems. Proc. of ACM SIGPLAN Conf. on Programming Language Design and Implementation (PLDI), San Diego, CA, 1–11.
14. Hill J, et al. (2000). System architecture directions for networked sensors. Proc. of the Ninth Intl. Conf. on Architectural Support for Programming Languages and Operating Systems (ASPLOS), Cambridge, MA, 93–104.
15. Huang G (2003). Casting the wireless sensor net. Technology Review 106(6):50–56.
16. Huh E, Welch LR, Shirazi BA, Tjaden B, Cavanaugh CD (2000). Accommodating QoS prediction in an adaptive resource management framework. In Parallel and Distributed Processing. Rolim J, et al. (eds.), Lecture Notes in Computer Science 1800:792–799.
17. Ibarra OH, Kim CE (1997). Heuristic algorithms for scheduling independent tasks on nonidentical processors. Journal of the ACM 2:280–289.
18. Islam KMJ, Shirazi BA, Welch LR, Tjaden BC, Cavanaugh C, Anwar S (2000). Network load monitoring in distributed systems. In Parallel and Distributed Processing. Rolim J, et al. (eds.), Lecture Notes in Computer Science 800–807.
19. Jung H, Tambe M, Kulkarni S (2001). Argumentation as distributed constraint satisfaction: applications and results. Proc. of the Fifth Intl. Conf. on Autonomous Agents 324–331.
20. Krishnamachari B, Bejar R, Wicker S (2002). Distributed problem solving and the boundaries of self-configuration in multi-hop wireless networks. Proc. of the 35th Intl. Conf. on System Sciences 3856–3865.
21. Mackworth A (1994). The logic of constraint satisfaction. Constraint-Based Reasoning 58: 3–20.
22. Maheswaran M, Ali S, Siegel HJ, Hensgen D, Freund RF (1999). Dynamic mapping of a class of independent tasks onto heterogeneous computing systems. Journal of Parallel and Distributed Computing 2:107–131.
23. Mailler R, Lesser V, Horling B (2003). Cooperative negotiation for soft real-time distributed resource allocation. Proc. of the Second Intl. Conf. on Autonomous Agents and Multiagent Systems, Melbourne, Australia 576–583.
24. Mainland G, Kang L, Lahaie S, Parkes D, Welsh M (2004). Using virtual markets to program global behavior in sensor networks. Proc. of the 11th ACM SIGOPS European Workshop, Leuven, Belgium.
25. Meisels A, Kaplansky E, Razgon I, Zivan R (2002). Comparing performance of distributed constraints processing algorithms. Proc. of AAMAS Workshop on Distributed Constraint Reasoning, Bolojna, Italy.
26. Mica2 user's manual. http://www.xbow.com/support/support_pdf_files/mts-mda_series_users_manual.pdf.
27. Modi P, Jung H, Shen WM, Tambe M, Kulkarni S (2001). A dynamic distributed constraint satisfaction approach to resource allocation. Proc. of the Seventh Intl. Conf. on Principles and Practice of Constraint Programming (CP 2001), Paphos, Cyprus, 685–700.
28. Modi P, Scerri P, Shen WM, Tambe M (2003). Distributed Resource Allocation: A Distributed Constraint Reasoning Approach. In Distributed Sensor Networks: A Multiagent Perspective. Kluwer Academic, New York.
29. Nisan N (2000). Bidding and allocation in combinatorial auctions. Proc. of the Second ACM Conf. on Electronic Commerce, Minneapolis, MN 1–12.
30. Ostwald J, Lesser V (2004). Combinatorial auctions for resource allocation in a distributed sensor network. Technical Report 04-72, University of Massachusetts at Amherst.
31. Repast 3.0 – Recursive Porous Agent Simulation Toolkit, http://repast.sourceforge.net.
32. Salido M, Barber F (2003). Distributed constraint satisfaction problems for resource allocation. Proc. of AAMAS Workshop on Decentralized Resource Allocation, Melbourne, Australia.

33. Shehory O, Kraus S (1998). Methods for task allocation via agent coalition formation. Artificial Intelligence 101(1–2):165–200.
34. Stansfield RG (1947). Statistical theory of DF fixing. Journal of the IEE, Part IIIA 94(15): 762–770.
35. Welch LR, Shirazi BA, Ravindran B, Bruggeman C (1999). DeSiDeRaTa: QoS management technology for dynamic, scalable, dependable, real-time systems. In Distributed Computer Control Systems, De Paoli F, MacLeod IM(eds). Elsevier Science, Kidlington, UK, 7–12.
36. Wellman M (1996). Market-oriented programming: some early lessons. In Market-Based Control: A Paradigm for Distributed Resource Allocation. World Scientific, River Edge, NJ.
37. Yokoo M, Hirayama K (1997). Distributed breakout algorithm for solving distributed constraint satisfaction problems. Report of Research Institute for Marine Cargo Transportation 8:43–50.
38. Yokoo M, Hirayama K (2000). Algorithms for distributed constraint satisfaction: a review. Autonomous Agents and Multi-Agent Systems 3(2):185–207.
39. Yokoo M, Durfee E, Ishida T, Kuwabara K (1998). The distributed constraint satisfaction problem: formalization and algorithms. IEEE Transactions on Knowledge and Data Engineering 10(5):673–685.

# Chapter 15
# Scheduling Activities in Wireless Sensor Networks

**Yu Chen and Eric Fleury**

**Abstract** We investigate scheduling activities in sensor networks; the materials covered are far beyond medium access control (MAC) protocols and the purpose is not to review specific or general purpose MAC approaches. Our purpose is more generic and we investigate scheduling strategies and techniques that could be applied to avoid interference, to prolong the network lifetime by reducing energy consumption, to optimize network performance by taking into account the underlying application communication patterns, to guarantee sensing coverage in monitoring tasks, and to achieve good levels of QoS. We examine scheduling under various interference models, including the traditional *channel separation constraints* model, the *protocol model*, and the *physical Signal-to-Interference-plus-Noise-Ratio* model. For each topic covered in this chapter, we survey the results and one or two representative works are examined in details as examples.

## 15.1 Introduction

In this chapter, we investigate the scheduling problem in sensor networks. This chapter concentrates on the scheduling strategies and techniques in various scenarios under various interference models; the materials covered are far beyond medium access control (MAC) protocols and the purpose is not to review specific or general purpose MAC approaches. Our purpose is more generic and we investigate scheduling techniques that could be applied to avoid interference, to prolong the network lifetime by reducing energy consumption, to optimize network performance by taking into account the underlying application communication patterns, to guarantee sensing coverage in monitoring tasks, and to achieve good levels of QoS. We

Y. Chen (✉)
ARES/INRIA, INSA Lyon Villeurbanne 69100, France
e-mail: chenyu@google.com

S. Misra et al. (eds.), *Guide to Wireless Sensor Networks*, Computer Communications and Networks, DOI: 10.1007/978-1-84882-218-4_15,
© Springer-Verlag London Limited 2009

examine scheduling techniques under various interference models, including the traditional *channel separation constraints* model, the *protocol model*, and the *physical Signal-to-Interference-plus-Noise-Ratio* (SINR) model [1].

Scheduling of sensor nodes' activities has been the topic of much interest over the past several decades. In wireless sensor networks, each sensor node is equipped with a wireless transceiver and they communicate via wireless radio over the shared communication medium or channel (we consider networks with a single shared channel if not explicitly stated). A typical sensor node can be in one of four types of modes: *transmit, receive, idle listen,* or *sleep.* A sensor is said to be idle listening if its radio is on and it is neither transmitting nor receiving; a idle listening node is able to switch into receive mode if it hears a transmission. When a sensor is transmitting, it transmits to the shared channel or, if the shared channel is divided into subchannels, one of the available subchannels. A schedule of sensor nodes' activities is to specify the state in which a sensor node may stay in each time slot. For example, a schedule might designate one of the three options for a given sensor node at a time (1) this sensor node's radio is turned off and it stays in sleep mode, (2) the radio is on and the sensor node is idle listening, thus it can switch to receive mode when it hears transmission, and (3) the radio is on and the sensor node is allowed to transmit if it has a packet to forward; if the shared communication channel is divided into subchannels, the subchannel to which it can transmit is also specified. Scheduling of sensor nodes' activities has been proved an important and effective mechanism in various aspects of wireless sensor networking, including interference avoidance, energy saving, and theoretical analyses on the best performance that can be achieved in a network.

One important application of scheduling is to avoid interference by scheduling sensor nodes' transmissions. Interference is caused by simultaneous transmissions in a sensor node's proximity, resulting in damaged useless received packets. A spectrum of MAC protocols [2–4] has been developed to handle interference. One class is *contention-based protocols*, which are prevalent in wireless networks due to their simplicity. However, as retransmissions are required to resolve contentions, these protocols are energy inefficient and they are not desirable for energy-constrained sensor networks. Another approach is called *allocation-based protocols*, or *scheduling protocols*, which guarantees interference-free receptions at the intended receivers by carefully scheduling sensor nodes' transmissions; since no energy is wasted due to channel contention, they suit well in sensor networks. The basic idea is first to divide the shared communication channel into subchannels and then allocate subchannels to sensor nodes. Various techniques have been developed for channel division. For example, the shared channel is divided into subchannels by *frequency bands* in frequency division multiple access (FDMA) schemes and by *orthogonal modulation codes* in code division multiple access (CDMA) schemes [3,5]. Given a set of subchannels, the amount of interference between two simultaneous transmissions depends on the spacing between the subchannels used by them and the distance between transmitters. The subchannel allocation should guarantee that the amounts of interference at the intended receivers are acceptable.

Scheduling can also be used to save energy in low traffic networks. Sensors are normally battery operated and energy efficiency is one of the most important constraints in sensor networking [6]. Studies have identified that idle listening is a significant consumer of power [7–10]. For example, WorldSens [11] sensor nodes are based on the Chipcon [12] CC1100 RF transceiver with current draw 16.2 mA for Rx mode and 15.1 mA for Tx mode (0 dbm output). It has been shown that periodic duty cycling of sensor nodes, that is, scheduling sensor nodes between active and sleep modes, can achieve better energy efficiency if the traffic load is light most of the time [8, 9, 13–16]; here by saying a sensor is active, we mean its radio is on, that is, it is either transmitting, receiving, or idle listening. One important issue in duty cycling is how to guarantee communication connectivity and small packet forwarding delay in the presence of sleeping nodes. Various scheduling schemes have been proposed for duty cycling [7–10, 17–26].

In addition to protocol design, scheduling is also an important tool in theoretical analyses on the best performance that can be achieved in a network. As a careful designed schedule represents an ideal situation of sensor nodes' activities, scheduling has been used in investigating the capacity of wireless networks [1, 27]. In [1], a constructive lower bound on the throughput capacity of a wireless network is obtained by spatially and temporally scheduling nodes' transmissions in arbitrary and random networks. Such a technique is also used in [27] to evaluate the capacity of networks with multiple transceivers and channels. Broadcast capacity and data aggregation capacity are examined by constructing appropriate schedules in [28] and [29], respectively. Scheduling complexity, that is, the time required to schedule all the requests in a given set of communication requests, is investigated in [30–32].

Scheduling strategies strongly rely on the interference model that describes the condition for a transmission to be received without interference. In this chapter, we investigate the scheduling problem under various interference models. One of the most common models is graph labelings with *channel separation constraints* [33–39], where constraints are defined on the minimum spacing between subchannels assigned to two nodes according to the distance between them. In [40], a geometrical distance is defined as *the interference range* for each sensor node and the transmissions from a node can interfere with all the nodes within this distance. In [1], the *protocol model* is proposed, which uses a guard zone parameter to ensure that other concurrently transmitting nodes are sufficiently away from the receiver. Interference between sensor nodes is described by a set of *interference links* in [41], where given two nodes $u$ and $v$, the link from $u$ to $v$ is an interference link if and only if the transmission from $u$ can interfere with node $v$. In [1], the *physical model* (the physical SINR model) is used to describe the accumulative effect of interference caused by different transmitters.

In this chapter, we investigate the scheduling strategies and techniques for various scenarios under different interference models. The following topics are covered; for each topic, we survey the results and one or two representative works are examined in detail as examples. We start in Sect. 15.3 with a scheduling problem that has been extensively studied – the channel allocation problem modeled by graph labelings with channel separation constraints. Schedules generated based on such labelings guarantee entirely interference-free communication. However, as indicated by the theoretical analyses, the span of subchannels required by such a schedule is large, especially in dense networks. In Sect. 15.4, we discuss a recently proposed scheduling strategy, called *light scheduling*, which reduces the span of required subchannels by only imposing channel separation constraints on communication links required by applications. Another form of scheduling, duty cycling, which is an important mechanism in energy saving, is introduced in Sect. 15.5. As sensor networks are usually deployed for specific applications, scheduling can be optimized by taking into account the underlying application communication pattern; such a strategy is examined in Sect. 15.6. In Sect. 15.7, our focus is on the protocol model and the SINR model; a representative scheduling is examined to illustrate how scheduling can be used in theoretical analyses.

## 15.2  Background

In this chapter, given a graph $G$, we denote the set of nodes (edges respectively) in graph $G$ by $V(G)$ ($E(G)$, respectively). As most works in sensor networking, we model a sensor network by a general graph $G$ if not explicitly stated. Each vertex in $V(G)$ corresponds to a sensor node, and for any two nodes $u, v \in V(G)$, there is an edge from node $u$ to node $v$ in $E(G)$ if and only if $v$ can receive the transmission from $u$ when $u$ is the only node that transmits in the network; we say $v$ is in the transmission range of $u$. Since some works consider sensor networks modeled by *unit disc graphs* (UDGs), we also present the definition here: UDGs refer to graphs in which nodes are associated with equal-sized discs and there is an edge between two nodes if and only if their discs intersect. The channel allocation problem is one of the most important problems on scheduling. It is usually modeled by graph labelings where distinct labels represent distinct subchannels. Given a graph $G$, a labeling is represented by a function $l() : V(G) \to \mathbf{N}$ that maps each node $u \in V(G)$ to a non-negative integer $l(u) \in \mathbf{N}$. Given a labeling $l()$, node $u$ is allocated the subchannel represented by the label $l(u)$.

Constraints defined in a graph labeling are subject to the purpose of scheduling the scheduling strategy and the interference model that describes the condition for interference-free communication. For each problem investigated in this chapter, we define the interference model and give the graph labeling definition. Since subchannels represent scarce system resources, e.g., frequencies in FDMA schemes and orthogonal codes in CDMA schemes, one critical performance metric of a graph labeling model or a labeling scheme is the number of labels that are required. We define the *span* of a labeling as the maximum label minus the minimum label; the span of a labeling corresponds to the bandwidth used by the corresponding channel allocation. Given a labeling definition, the minimum span of labels required by any labeling on a given graph that satisfies the defined constraints is called the *labeling number*. Note the labeling number is independent of the specific labeling schemes; it is decided only by the labeling definition and the network to be labeled. In this chapter, given $K \geq 1$, we use integers $\{0, \ldots, K\}$ to represent the set of labels with span $K$.

The following graph denotations are defined. Given a graph $G$ and two nodes, $x, y \in V(G)$, if $G$ is undirected, we denote the link between $x$ and $y$ by $x \leftrightarrow y$, the set of $x$'s neighbors by $N_G(x) \equiv \{y \in V(G) | x \leftrightarrow y \in E(G)\}$, $x$'s degree by $\delta_G(x) \equiv |N_G(x)|$, and the degree of $G$ by $\Delta_G \equiv \max\{\delta_G(x) | x \in V(G)\}$. Given $H \subseteq V(G)$, we define $N_G(H) \equiv \cup_{x \in H} N_G(x)$. If $G$ is directed, we denote the directed link from node $x$ to $y$ by $x \rightarrow y$, the sets of $x$'s outgoing and incoming neighbors by $N_G^+(x) \equiv \{y \in V(G) | \exists x \rightarrow y \in E(G)\}$ and $N_G^-(x) \equiv \{y \in V(G) | \exists y \rightarrow x \in E(G)\}$, respectively, and the outdegree and indegree of $x$ by $\delta_G^+(x) \equiv |N_G^+(x)|$ and $\delta_G^-(x) \equiv |N_G^-(x)|$, respectively. A path from node $x$ to node $y$ has the form of $z_1 \rightarrow z_2 \rightarrow \cdots \rightarrow z_k$, where $z_1 = x, z_k = y$, and $\forall i \in [1, k-1]$, $z_i \rightarrow z_{i+1} \in E(G)$. Given two nodes $x, y \in V(G)$, we define the distance $d_G(x, y)$ between $x$ and $y$ in graph $G$ as the number of edges in the shortest path starting at $x$ and ending at $y$. The diameter $D_G$ of graph $G$ is defined as the maximum distance between two nodes, $D_G \equiv \max\{d_G(x, y) | x, y \in V(G)\}$. The geometric distance between $x$ and $y$ is denoted by $||x, y||$.

## 15.3 Entirely Interference-Free Scheduling

A spectrum of works [33–39] formulates the channel allocation problem by graph labelings with channel separation constraints, where constraints are defined on the separation of subchannels used by two sensor nodes according to the distance between them. This is based on the observation that, given a set of subchannels, there are two major factors that affect the amount of interference between simultaneous transmissions. First is the *proximity* in the radio spectrum of the subchannels used by simultaneous transmissions; generally speaking, the amount of interference between subchannels close to each other is larger than that between subchannels which are far apart. Another factor is the *distance* between the intended receiver and other transmitters, as signal strength decays with distance. The definition of graph labeling with channel separation constraints is given in Definition 1, where a set of positive integers parameters $d_1, d_2, \ldots, d_k$ is used to describe the channel separation constraints; in particular, $d_i$ is the minimum spacing between the subchannels assigned to nodes that are distance $i$ from each other.

*Example 1.* IEEE 802.11a provisions for 12 channels [42]. Due to adjacent channel interference, guard bands are required in FDMA schemes (Fig. 15.1) and adjacent channels cannot be used simultaneously in a sensor node's proximity [43].

**Definition 1.** ($L(d_1, d_2, \ldots, d_k)$-**labeling**) Given a graph $G$ and a set of positive integers $d_1 \geq d_2 \geq \cdots \geq d_k > 0$, an $L(d_1, d_2, \ldots, d_k)$-labeling on $G$ is a function $l() : V(G) \rightarrow \mathbf{N}$ that satisfies: $\forall v, u \in V(G)$ such that $d_G(u, v) = i$, $|l(u) - l(v)| \geq d_i$. The labeling number of $L(d_1, d_2, \ldots, d_k)$-labelings on $G$ is denoted by $\lambda_{d_1, d_2, \ldots, d_k}(G)$.

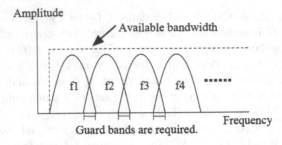

**Fig. 15.1** Adjacent channel interference

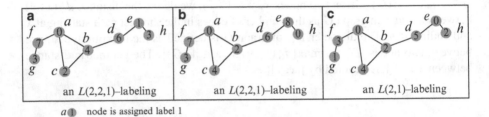

*a* ①   node is assigned label 1

**Fig. 15.2** Examples of $L(2,2,1)$-labeling and $L(2,1)$-labeling

*Example 2.* In Fig. 15.2a, b, $L(2, 2, 1)$-labelings are given, which guarantee the spacing of any two nodes with distance 1 (e.g., nodes $a$ and $c$) is at least 2, the spacing of any two nodes with distance 2 (e.g., nodes $b$ and $e$) is at least 2, and the spacing of any two nodes with distance 3 (e.g., nodes $a$ and $e$) is at least 1.

*Example 3.* In Fig. 15.2c, an $L(2, 1)$-labeling is given, which guarantees that the spacing of any two nodes with distance 1 (e.g., nodes $a$ and $c$) is at least 2, and the spacing of any two nodes with distance 2 (e.g., nodes $c$ and $d$) is at least 1.

The values of $d_i$, $i \in [1, k]$, should be defined appropriately to restrict the amount of interference between simultaneously transmissions from nodes with distance $i$. As signal strength decays with distance, if the distance between two nodes is sufficient large, the amount of interference caused by simultaneous transmissions from them is insignificant, even when they are transmitting to the same subchannel. We refer *reuse distance* to the minimum distance between two nodes that can use the same channel without interfering with each other, which is $k+1$ when the subchannel allocation is modeled by $L(d_1, d_2, \ldots, d_k)$-labelings.

It is worth pointing out that a subchannel allocation based on such a labeling is actually *entirely interference-free* in the sense that every transmission from each node is received by all the other nodes in its transmission range without interference. A fundamental case of $L(d_1, d_2, \ldots, d_k)$-labelings is the one where constraints are defined on distance two, that is, $L(d_1, d_2)$-labelings. The definition of $L(d_1, d_2)$-labelings was first proposed in [39] and since then it has attracted extensive research [33, 35, 37, 44]. In the sequel, we review the results in graph labelings for both general $L(d_1, d_2, \ldots, d_k)$-labelings and the special case $L(d_1, d_2)$-labelings.

*Example 4.* The reuse distance of an $L(2, 2, 1)$-labeling is 4. That is, any two nodes with distance at least 4 can be assigned the same label. In Fig. 15.2b, the distance between nodes $a$ and $h$ is 4 and they are assigned the same label 0.

### 15.3.1 Labeling on General Graphs

Graph labeling problem is well known to be NP-complete, even in drastically simplified cases. In [39], it is proved by a reduction to Hamiltonian paths that the $L(2, 1)$-labeling problem is NP-complete on graphs of diameter two. The special case $L(d_1)$-labeling is proved to be NP-complete on *finite induced subgraphs* of the *infinite triangular lattice* graph in [45] by a reduction to the coloring problem on planar graphs; the proof can be adapted for $L(d_1, d_2, \ldots, d_k)$-labelings for any given values of $d_1, d_2, \ldots, d_k$. Graph labeling on UDGs has also been investigated [46, 47] and it is shown that this problem remains NP-hard for such graphs. In [48], it is proved that the problem of approximating the minimum labeling span within any constant ratio is still NP-hard. As the problem of finding the optimal graph labeling is NP-complete, most research in this area focuses on either efficient heuristics which produce suboptimal but acceptable results, or on near optimal solutions on graphs with special properties. In the sequel, we will review the results in both aspects.

We present a very simple but commonly used labeling strategy – greedy labeling (Algorithm 1). In this strategy, nodes are labeled in a certain order. Each time an unlabeled node is examined, it is assigned the smallest label that does not invalidate the labeling constraints. Such an approach has been used by many works [33,49,50], in which different criterions are used in defining the label ordering, such as random ordering and increasing/decreasing number of neighbors. The resulting span of the used labels depends heavily on the order in which nodes are labeled, and it might not be optimal. However, such a heuristic finds a labeling in polynomial time and the extreme simplicity makes it attractive in practice.

---

**Algorithm 1**: Greedy graph labeling

---

**Input:** An ordering of sensor nodes: $v_1, v_2, \ldots, v_n$.
1: $l(v_1) = 0$;
2: **for** $i = 2$ to $n$ **do**
3:  $X = 0$;
4:  **for** $j = 1$ to $i - 1$ **do**
5:   **if** $d_G(i, j) <= k$ **then**
6:    $X = X \cup [\max\{0, l(j) - d_{dG(i,j)} + 1\}, l(j) + d_{dG(i,j)} - 1]$;
7:   **end if**
8:  **end for**
9:  $l(i) =$ the smallest label not in $X$;
10: **end for**

---

*Example 5.* We consider $L(2, 2, 1)$-labeling on the graph in Fig. 15.2. Let the labeling order be $a, b, c, d, e, f, g, h$. An $L(2, 2, 1)$-labeling generated by greedy labeling is given in Fig. 15.2b. For example, when $f$ is examined, nodes $a, b, c, d$, and $e$ have been labeled and we have $l(a) = 0, l(b) = 2, l(c) = 4, l(d) = 6, l(e) = 8$. Note $g$ is unlabeled at this point. We have (1) node $a$ is one hop from $f$, so labels 0, 1 cannot be picked for $f$ due to the $d_1 = 2$ constraint, (2) node $b$ is two hops from $f$, so labels 1, 2, 3 cannot be picked for $f$ due to the $d_2 = 2$ constraint, (3) node $c$ is two hops from $f$, so labels 3, 4, 5 cannot be picked for $f$ due to the $d_2 = 2$ constraint, and (4) node $d$ is three hops from $f$, so label 6 cannot be picked for $f$ due to the $d_3 = 1$ constraint. So the smallest label that can be assigned to $f$ is 7.

Now we consider bounds on the labeling number, that is, the minimum span of subchannels required by an $L(d_1, d_2, \ldots, d_k)$-labeling on a graph $G$. Most existing lower bound results are based on the size of cliques [51, 52]. Lower bounds on $\lambda_{d_1, d_2, \ldots, d_k}(G)$ can also be obtained by investigating the relation between graph labeling problem and the maximum independent set problem [53], traveling salesman problem [39, 53], and tile cover problem [54, 55]. Here we present the lower bound based on the size of cliques as an example. Given a graph $G$, an $i$-clique of $G$ is defined as a subset $\{u, v \in V(G) | d_G(u, v) \le i\}$ of $V(G)$. We have the following theorem, which is adapted from [51].

**Theorem 1.** [51] *For any graph $G$ and any set of integers $d_1 \ge \cdots \ge d_k > 0$, we have*

$$\lambda_{d_1, d_2, \ldots, d_k}(G) \ge \max_{i \in [1, k]} \{d_i(|C| - 1) | C \text{ is an } i\text{-clique of } G\}.$$

*Proof.* For any $i \in [1, k]$, we consider any $i$-clique $C$ of graph $G$. For any two nodes $u, v \in C$, by the definition of $i$-clique, we have $d_G(u, v) \le i$. By the definition of an $L(d_1, d_2, \ldots, d_k)$-labeling, we have $|l(u) - l(v)| \ge d_i$. Thus, the difference between the maximum label and the minimum label assigned to nodes in $C$ is $\max\{l(u) | u \in C\} - \min\{l(u) | u \in C\} \ge d_i(|C| - 1)$. $\qquad\square$

*Example 6.* In Fig. 15.2, $\{a, b, c, d\}$ is a 2-clique, $\{a, b, c, f, g\}$ and $\{f, a, b, c, d\}$ are 3-cliques, as indicated in Fig. 15.3.

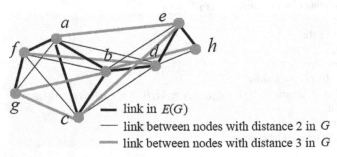

**Fig. 15.3** A lower bound on $\lambda_{2,2,1}$ for Fig 15.2

*Example 7.* In Fig. 15.2, the maximum size of 1-cliques is 3, the maximum size of 2-cliques is 4, and the maximum size of 3-cliques is 5. By Theorem 1, a lower bound is $\lambda_{2,2,1}(G) \geq \max\{2 \cdot (3-1), 2 \cdot (4-1), 1 \cdot (5-1)\} = 6$.

Now we discuss the special case $L(d_1, d_2)$-labeling. By applying Theorem 1 on $L(d_1, d_2)$-labelings, we have a lower bound $\lambda_{d_1,d_2}(G) \geq \max\{d_1(|C|-1), d_2 \Delta_G\}$, where $C$ is the maximum clique in $G$. Another lower bound can be obtained by examining nodes' one-hop neighborhood. We consider a node $u$ which has the maximum number $\Delta_G$ of neighbors. By the constraint on nodes with distance two, the difference in the labels assigned to nodes in $N_G(u)$ is at least $d_2(\Delta_G - 1)$; due to the constraint on nodes with distance one, a lower bound is $\lambda_{d_1,d_2}(G) \geq d_2(\Delta_G - 1) + d_1$. In particular, we have $\lambda_{1,1} \geq \Delta_G$ [56] and $\lambda_{2,1} \geq \Delta_G + 1$ [39]. As for the upper bounds for $L(d_1, d_2)$-labelings, by a greedy labeling we have $\lambda_{1,1} \leq \Delta_G^2$ for $L(1, 1)$-labelings. The special case $L(2, 1)$-labelings are investigated in [39] and it is showed that $\lambda_{2,1} \leq \Delta_G^2 + 2\Delta_G$. This upper bound is improved to $\lambda_{2,1} \leq \Delta_G^2 + \Delta_G$ in [44], to $\lambda_{2,1} \leq \Delta_G^2 + \Delta_G - 1$ in [57], and to the currently best bound $\lambda_{2,1} \leq \Delta_G^2 + \Delta_G - 2$ in [58]. More generally, an upper bound $\Delta_G^2 + (d-1)\Delta_G$ on $\lambda_{d,1}(G)$ is presented in [35]. We summarize the results below.

**Theorem 2.** [35, 39, 49, 56, 58] *Given any graph $G$ and any integers $d_1 \geq d_2 > 0$, we have (1) $\Delta_G \leq \lambda_{1,1}(G) \leq \Delta_G^2$, (2) $\Delta_G + 1 \leq \lambda_{2,1}(G) \leq \Delta_G^2 + \Delta_G - 2$, (3) $\Delta_G - 1 + d \leq \lambda_{d,1}(G) \leq \Delta_G^2 + (d-1)\Delta_G$, and (4) $d_2(\Delta_G - 1) + d_1 \leq \lambda_{d_1,d_2}(G)$.*

The lower bounds in Theorem 2 have been discussed; the proof for the upper bound $\lambda_{d,1}(G) \leq \Delta_G^2 + (d-1)\Delta_G$ is presented in the proof of Theorem 4 in Sect. 15.4, as a special case of $L_S(d_1, d_2)$-labelings which are defined in Sect. 15.4.

## 15.3.2  Labeling on Special Graphs

Graph labeling on special graphs has been extensively investigated. Since usually the positions of sensor nodes are not carefully designed and a sensor network does not necessarily have one of these specific topologies, here we present in Table 15.1 a summary of the results without further discussion on the proofs or algorithms; detailed surveys can be found in [62, 63]. For presentation simplicity, in Table 15.1 we omit the subscription "$G$" from the denotation $\Delta_G$.

## 15.4  Light Scheduling

Subchannel allocation based on the graph labelings defined in Sect. 15.3 guarantees entirely interference-free transmissions. However, as indicated by the analyses, such a schedule requires a large number of subchannels. For example, the upper bound given in Theorem 2 on the span of labels required by $L(d, 1)$-labelings has

**Table 15.1** Summary of results on $L(d_1, d_2, \ldots, d_k)$-labeling on special graphs

| Type of graph | Bounds on labeling number | Reference |
|---|---|---|
| Path | $\lambda_{1,1}(G) = 2$ | [33] |
| | $\lambda_{2,1}(G) = 2, 3, \text{or } 4$ | [39] |
| | $\lambda_{1,1,1}(G) = 3$ | [59] |
| | $\lambda_{2,1,1}(G) = 4$ | [34] |
| Hexagonal grids | $\lambda_{1,1}(G) = 3$ | [33] |
| | $\lambda_{2,1}(G) = 5$ | [34,39] |
| | $\lambda_{1,1,1}(G) = 5$ | [59] |
| | $\lambda_{2,1,1}(G) = 6$ | [34] |
| Tree $T$ | $\lambda_{2,1}(T) \in [\Delta + 1, \Delta + 2]$ | [39] |
| | $\lambda_{d,1}(T) \in [\Delta + d - 1, \min\{\Delta + 2d - 2, 2\Delta + d - 2\}]$ | [35] |
| | $\lambda_{d_1,d_2}(T) \in [d_1 + (\Delta - 1)d_2, d_1 + (2\Delta - 2)d_2]$, if $d_1/d_2 \geq \Delta$ | [60] |
| Complete binary tree $T$ | $\lambda_{1,1}(T) = 3$ | [33] |
| | $\lambda_{2,1}(T) = 4$ | [39] |
| with size $n \geq 31$ | $\lambda_{1,1,1}(T) = 5$ | [59] |
| | $\lambda_{2,1,1}(T) = 6$ | [34] |
| Cycle of order $n$, $Cn$ | $\lambda_{1,1}(Cn) = 2 \text{ or } 3$ | [33] |
| | $\lambda_{2,1}(Cn) = 4$ | [39] |
| | $\lambda_{d_1,d_2}(Cn) = \begin{cases} 2d_1 & \text{if } n \text{ is odd and } n \geq 3 \text{ and } d_1/d_2 > 2 \\ d_1 + 2d_2 & \text{if } n \equiv 0 \bmod 4 \text{ and } d_1/d_2 > 2 \\ 2d_1 & \text{if } n \equiv 2 \bmod 4 \text{ and } 3 \leq d_1/d_2 \leq 2 \\ d_1 + 3d_2 & \text{if } n \equiv 2 \bmod 4 \text{ and } d_1/d_2 > 3 \\ 2d_1 & \text{if } n \equiv 0 \bmod 3 \text{ and } d_1/d_2 \leq 2 \\ 4d_2 & \text{if } n \equiv 5 \text{ and } d_1/d_2 \leq 2 \\ d_1 + 2d_2 & \text{otherwise} \end{cases}$ | [60] |
| | $\lambda_{1,1,1}(Cn) = 3 \text{ or } 4$ | [59] |
| | $\lambda_{2,1,1}(Cn) = 4$ | [34] |
| Bidimensional grids | $\lambda_{1,1}(G) = 4$ | [33,59] |
| | $\lambda_{2,1}(G) = 6$ | [34] |
| | $\lambda_{1,1,1}(G) = 7$ | [59] |
| | $\lambda_{2,1,1}(G) = 8$ | [34] |
| Outer planar | $\lambda_{2,1}(G) \leq \begin{cases} \Delta + 2 & \text{if } \Delta \geq 8 \\ 10 & \text{otherwise} \end{cases}$ | [61] |
| Triangular outer planar | $\lambda_{2,1}(G) \leq \Delta + 6$ | [50] |
| Planar | $\lambda_{2,1}(G) \leq 3\Delta + 28$ | [50] |
| Triangular planar | $\lambda_{2,1}(G) \leq 3\Delta + 22$ | [50] |
| t-tree[a] | $\lambda_{d,1}(G) \leq (2d - 1 + \Delta - t)t$ | [35] |
| Chordal[b] | $\lambda_{d,1}(G) \leq (2d + \Delta - 1)^2/4$ | [35] |
| Graph of dimeter 2 | $\lambda_{2,1}(G) \leq \Delta^2$ | [39] |

[a] Given an integer $t > 0$, a *t-tree* is a graph of $n \geq t + 1$ vertices defined recursively as follows: (1) a clique of $(t + 1)$ vertices is a $t$-tree and (2) a $t$-tree with $(n + 1)$ vertices can be formed from a $t$-tree with $n$ vertices by making a new vertex adjacent to exactly all vertices of a $t$-clique in the $t$-tree with $n$ vertices

[b] A graph is *chordal* if and only if every cycle of length $\geq 4$ has a chord (i.e., there is no induced cycle of length $\geq 4$)

$O(\Delta^2_G)$ complexity and most labeling schemes use $O(\Delta^2_G)$ number of labels. As subchannels are scarce resources and sensor networks are usually densely deployed, an entirely interference-free schedule might not always be feasible or desirable. In this section, we present another scheduling strategy, called *light scheduling*, which aims to reduce the number of required subchannels while maintaining satisfactory communication connectivity. In particular, the span of subchannels can be reduced to $O(\Delta_G)$ in UDGs.

Light scheduling is based on the observation that an entirely interference-free schedule is not always necessary. Specific applications have their own communication patterns. For example, data gathering in monitoring tasks only requires that each node is connected to a sink by an interference-free path. Even when communication between any two nodes is required, it is sufficient to guarantee intererence-free communication along links that form a strongly connected component.

*Example 8.* In Fig. 15.4a, the gray lines form a directed tree rooted at the sink, with tree edges toward to the root; interference-free transmissions along these tree edges will be sufficient for data gathering. In (b), communication between any two nodes can be achieved by interference-free transmissions along the gray lines.

A definition that generalizes the traditional graph labelings, called $L_S(d_1, d_2)$-labelings [64], is given in Definition 2, where subgraph $S$ is a parameter that captures applications' communication pattern. The aim is to guarantee interference-free transmissions along each link in $S$. While networks under consideration are modeled by undirected graphs, the subgraph $S$ can be directed. Since the focus is on the transmissions along each of the links in $S$, the constraint is defined based on the *connectivity* in $S$. This is different from the traditional $L(d_1, d_2, \ldots, d_k)$-labelings, where the constraint is defined on the *distance* between two nodes.

**Definition 2. (A Light Labeling: $L_S(d_1, d_2)$-labeling [64])** Given a graph $G$, a subgraph $S$ such that $V(S) \subseteq V(G)$ and $E(S) \subseteq E(G)$, and integers $d_1 \geq d_2 > 0$, an $L_S(d_1, d_2)$-labeling on $G$ is a function $l():V(S) \rightarrow N$ that satisfies: $\forall x \rightarrow y \in E(S)$, we have $|l(x) - l(y)| \geq d_1$ and $|l(x) - l(z)| \geq d_2, \forall z \neq x, z \in N_G(y)$ (Fig. 15.5). The labeling number is denoted by $\lambda_{d_1, d_2}(G, S)$.

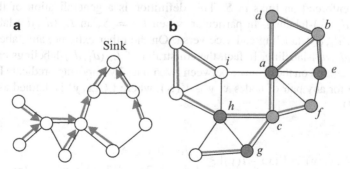

**Fig. 15.4** Entirely interference-free scheduling is not always necessary. (a) Communication to sink (b) Communication between any two nodes

**Fig. 15.5** $< x \to y > \in$
$E(S), z, z', z'' \in N_G(y)$

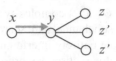

**Fig. 15.6** An example of
$L_S(1,1)$-labeling

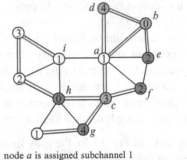

$a \, ①$     node $a$ is assigned subchannel 1

●—●    $a$ and $b$ are within the transmission of each other
$a$    $b$

*Example 9.* An example of $L_S(1, 1)$-labeling is given in Fig. 15.6, where links in
$S$ are marked by gray lines. When we consider link $\forall a \to b \in E(S)$, the spacing
between the label assigned to $a$ and that assigned to $b$ should be at least $d_1$, that is,
$|l(a) - l(b)| \geq d_1$, and the spacing between the label assigned to $a$ and that assigned
to $b$'s neighbors, that is, $d$ and $e$, should be at least $d_2$, that is, $|l(a) - l(d)| \geq d_2$
and $|l(a) - l(e)| \geq d_2$. Note the channel separation constraint is not enforced on
link $e \to a$, thus we can have $|l(e) - l(f)| = 0 \leq d_1$.

This definition implies that, in order to guarantee interference-free transmission from $x$ to $y$, the subchannel assigned to $x$ should be distinguished among
those assigned to $y$'s other neighboring nodes; the requirement on such a dis-
tinction is represented by channel separation constraints $d_1$ and $d_2$. Note different
from the traditional $L(d_1, d_2, \ldots, d_k)$-labelings, where channel separation constrains are enforced on all the links in $E$, in an $L_S(d_1, d_2)$-labeling these constraints
are only enforced on links in $S$. This definition is a generalization of the traditional $L(d_1, d_2)$-labeling. In particular, when $G = S$, an $L_G(d_1, d_2)$-labeling is
also an $L(d_1, d_2)$-labeling and vice versa. On the other extreme, any labeling is a
valid $L_\phi(d_1, d_2)$-labeling. In fact, the constraint of $L_S(d_1, d_2)$-labelings can be alternately defined on the distance between two nodes by imposing predicate $LC(x, y)$
to be true for any pair of nodes $x, y \in V(S)$, where $LC(x, y)$ is defined as follows
(Fig. 15.7):

$$LC(x, \, y) \equiv |l(x) - l(y)| \leq \begin{cases} d_1 & \text{if } y \in N_S^+(x) \\ d_2 & \text{if } \exists z \in N_S^+(x) \wedge y \in N_G(z). \end{cases}$$

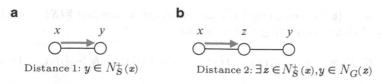

Fig. 15.7  Predicate $LC(x, y)$

It is easily to see that a labeling is an $L_S(d_1, d_2)$-labeling if and only if $LC(x, y)$ is true for all $x, y \in V(S)$, $x \neq y$. Different from the traditional $L(d_1, d_2, \ldots, d_k)$-labelings, in $LC(x, y)$ the first hop is defined by $S$ and the second hop is defined by $G$ – it is required at least $d_1$ spacing between two nodes with distance one (defined in $S$) and at least $d_2$ spacing between two nodes with distance two (the first hop is defined in $S$ and the second hop is defined in $G$).

*Example 10.* In the example in Fig. 15.6, the light gray nodes, (the dark gray nodes, respectively) are the first (second respectively) hop neighbors of node $a$.

The purpose to propose $L_S(d_1, d_2)$-labelings is to reduce the number of required subchannels. Below we give an example in which an $L_S(d_1, d_2)$-labeling requires less subchannels than an $L(d_1, d_2)$-labeling.

*Example 11.* A lower bound on the span of a traditional $L(1, 1)$-labeling on graph $G$ is $\Delta_G$. But the $L_S(1, 1)$-labeling in Fig. 15.6 indicates that the span of an $L_S(1, 1)$-labeling can be less than $\Delta_G$, which is $4 \leq \Delta_G = 6$ in this example.

## 15.4.1  Bounds on $\lambda_{d_1, d_2}(G, S)$

We have seen that the constraint of an $L_S(d_1, d_2)$-labeling defined on the connectivity in $S$ or the distance between nodes (with the first hop defined by $S$ and the second defined by $G$). Here we rephrase the definition by giving the constraint on each node's one-hop neighborhood; the equivalency can be easily obtained by comparing it to Definition 2.

$$\forall x \in V(G), \ \forall y \in N_S^-(x), \ |l(y) - l(x)| \geq d_1 \text{ and } |l(y) - l(z)| \geq d_2,$$
$$\forall z \neq y, z \in N_G(x).$$

This definition indicates a weaker requirement on the "uniqueness" of labels in a neighborhood than that of the traditional $L(d_1, d_2)$-labeling, and therefore a better lower bound. Let us take $L_S(1, 1)$-labelings as an example. While an $L(1, 1)$-labeling on $G$ requires unique labels in each node's neighborhood defined by $G$, an $L_S(1, 1)$-labeling only requires each node's incoming neighbors defined by $S$ should be assigned unique labels among neighbors defined by $G$. The following lower bound on $\lambda_{d_1, d_2}(G, S)$ can be derived from this definition.

**Theorem 3.** [64] *Given a graph $G$ and a subgraph $S$ such that $V(S) \subseteq V(G)$ and $E(S) \subseteq E(G)$, integers $d_1 \geq d_2 > 0$, we have*

$$\lambda_{d_1,d_2}(G, S) \geq d_2 \max\{\delta_S^-(x) + f(x) - 1 | x \in V(G)\} + d_1 \geq d_2(\Delta_s - 1) + d_1,$$

*where $f()$ is a binary function defined on $V(G)$ as follows: $\forall x \in V(G)$, $f(x) = 1$ if $(N_G(x) - N_S^-(x)) \neq \phi$, $f(x) = 0$ otherwise.*

*Note if $S = G$, we have $f(x) = 0$ for all $x$ and the above lower bound is $d_2(\Delta_G - 1) + d_1$. Thus we have $\lambda_{1,1}(G, G) \geq \Delta_G$ and $\lambda_{2,1}(G, G) \geq \Delta_G + 1$, which are consistent with the results stated in Theorem 2.*

*Example 12.* In an $L_S(1, 1)$-labeling on Fig. 15.8, nodes in $\{x, y_0, \ldots, y_4\}$ are required to have a unique label among those in $\{x, y_0, \ldots, y_4, z_0, \ldots, z_4\}$ and Seven labels are sufficient, while an $L(1, 1)$-labeling requires each node in $\{x, y_0, \ldots, y_4, z_0, \ldots, z_4\}$ to have a unique label and thus 11 labels are required.

Now we present an upper bound on $\lambda_{d,1}(G, S)$, where undirected $S$ is considered. This upper bound can be extended to directed $S$, since each valid $L_S(d, 1)$-labeling is also a valid $L_S(d, 1)$-labeling, where $S'$ is an undirected graph constructed from $S$ by ignoring the direction of each link. Predicate LC( ) indicates that, given any node $x$, $x$'s label is constrained by the label of nodes within distance two, that is, nodes in $N_S^+(x) \cup \{y | \exists z \in N_S^+(x), y \in N_{G(z)}\} - \{x\}$; we denote this set by $N_{S,G}(x)$. Since the labeling constraint is imposed on all the pairs of nodes, $x$'s label is not only constrained by nodes in $N_{S,G}(x)$, but also by every node $y$ such that $x \in N_{S,G}(y)$. We denote by $D_{S,G}(x)$ all the nodes that have impact on the label of node $x$: $D_{S,G}(x) \equiv N_{S,G}(x) \cup \{y | x \in N_{S,G}(y)\}$. Note if $x \in D_{S,G}(y)$ then $y \in D_{S,G}(x)$.

*Example 13.* In Fig. 15.9, we have $N_{S,G}(x) = \{z_2, y_2, y_3, y_4\}$ and $D_{S,G}(x)$ consists of all the gray nodes.

We consider an $L_S(d, 1)$-labeling scheme is Algorithm 2.

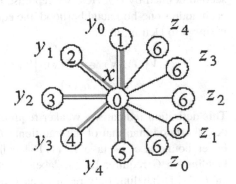

**Fig. 15.8** Example of
Theorem 3

**Fig. 15.9** Example of
$N_{S,G}(x)$ and $D_{S,G}(x)$

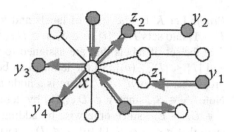

---

**Algorithm 2**: [64] An $L_S(d, 1)$-labeling scheme

---

1. Initially all the nodes are unlabeled.
2. In each step $k = 0, 1, \ldots,$
    2.1: The following denotations are defined:
       $\ldots U_k = V(G) - \bigcup_{i=0}^{k-1} S_i$, the set of unlabeled nodes,
       $\ldots P_k = \bigcup_{i=\max\{0, k-d+1\}}^{k-1} S_i$, the set of nodes that are assigned a lable $k'$ such that
           $k - k' < d$, and
       $\ldots E_k = N_S(P_k)$, the set of nodes adjacent in $S$ to a node with lable $k'$ such that
           $k - k' < d$.
    2.2: A set $S_k$ is computed as a *maximal* subset of $U_k - E_k$ such that $\forall x, y \in S_k$,
       $x \notin D_{S,G}(y)$.
    2.3: All the nodes in $S_k$ are assigned label $k$, that is, $\forall u \in S_k, l(u) = k$.
3. The computation of $S_k$ is repeated until all the nodes are labeled.

---

**Lemma 1.** **[64]** *Given an undirected graph $G$, a subgraph $S$ such that $V(S) \subseteq V(G)$ and $E(S) \subseteq E(G)$, and a positive integer $d$, Algorithm 2 generates an $L_S(d, 1)$-labeling on $G$.*

*Proof.* The lemma can be proved by showing that in each step $i$, LC$(x, y)$ is true for any nodes $x, y \in \bigcup_{j=0}^{i-1} S_j$. It is true for $i = 0$, as $\bigcup_{j=0}^{i-1} S_j = \phi$. Now we show that if it is true in step $k$, then it is true in step $(k + 1)$. Given any two nodes $x$, $y \in \bigcup_{j=0}^{i-1} S_j$, denoting the labels of $x$ and $y$ by $i_x$ and $i_y$, respectively, we have $x \in S_{i_x}, y \in S_{i_y}$, and $i_x, i_y \leq k$. If $i_x < k$ and $i_y < k$, LC$(x, y)$ is true by the induction assumption. If $i_x = k$ and $i_y = k$, by the selection of $S_k$ in line 2.2 we have $y \notin D_{S,G}(x)$, thus LC$(x, y)$ is true. Otherwise, without loss of generality, we assume $i_x = k$ and $i_y \neq k$. Since $i_x - i_y \geq 1$, we only need to consider the case $y \in N_S(x)$. In this case, we have $y \notin P_k$ since otherwise $x \in E_k$, and therefore $x \notin S_k$ by line 2.2, which contradicts to $x \in S_{i_x} = S_k$. By the definition of $P_k$, we have $k - i_y \geq d$, that is, $i_x - i_y \geq d$, so we prove LC$(x, y)$. $\square$

The span of labels used by this scheme is given below.

**Lemma 2.** **[64]** *The span of labels used by Algorithm 2 is no more than $\max_{u \in V(S)} |D_{S,G}(u)| + (d - 1)\Delta_S$.*

*Proof.* Let $K$ be the span of labels and $u$ be a node in $S_K$. We divide the set $[0, K-1]$ into sets $I = \{i | \exists v \in S_i, v \in D_{S,G}(u)\}$ and $I' = [0, K-1] - I$. Intuitively, $I$ is the set of labels that are assigned to nodes in $D_{S,G}(u)$. Note $K = |I| + |I'|$ and $|I| \leq |D_{S,G}(u)|$. So the lemma can be proved if $|I'| \leq (d-1)\delta_S(u)$. We first prove that for any $i' \in I'$, there is a node $u' \in N_S(u)$, such that $u' \in P_{i'}$ in step $i'$. Note $\forall v \in S_i$, since $v \notin D_{S,G}(u)$ by definition of $I'$, we have $u \notin D_{S,G}(v)$. Thus $u \notin U_{i'} - E_{i'}$, since otherwise by adding $u$ to $S_{i'}$, we have a larger set, $S_{i'} \cup \{u\}$ such that $\forall x, y \in S_i \cup \{u\}, x \notin D_{S,G}(y)$ and $y \notin D_{S,G}(x)$, which contradicts to the "maximal" property in line 2.2. Since $u \in S_K \subseteq U_{i'}$, we have $u \in E_{i'}$, that is, $\exists u' \in P_{i'}, u' \in N_S(u)$.

So we have $|I'| \leq |\cup_{u' \in N_S(u)} \{i | i' < K \wedge u' \in P_i\}|$. We have $|\{i | x \in P_i\}| \leq d-1$ since $\forall x, \{i | x \in P_i\} \subseteq [l(x) + 1, l(x) + d - 1]$. Thus we prove $|I'| \leq (d-1)\delta_S(u)$.                                                                                                                    □

Note $|D_{S,G}(x)| = |N_S(x) \cup \{y | \exists z \in N_S(x), y \in N_G(z)\} \cup \{y | \exists z \in N_G(x), y \in N_S(z)\} - \{x\}| \leq \Delta_S + \Delta_G \Delta_S + \Delta_G \Delta_S = \Delta_S + 2\Delta_G \Delta_S$. We also have $|D_{S,G}(x)| \leq \Delta_G^2$ since $D_{S,G}(x) \subseteq N_G(N_G(x))$. Thus $|D_{S,G(x)}| \leq \min\{\Delta_G^2, \Delta_S + 2\Delta_S\Delta_G\}$. So we have the following upper bound.

**Theorem 4.** [64] *Given an undirected graph $G$, a subgraph $S$ such that $V(S) \subseteq V(G)$ and $E(S) \subseteq E(G)$, and a positive integer $d$, we have $\lambda_{d,1}(G, S) \leq \min\{\Delta_G^2, \Delta_S + 2\Delta_S\Delta_G\} + (d-1)\Delta_S$.*

This upper bound is $\lambda_{d,1}(G, S) \leq \min\{\Delta_G^2, \Delta_S + 2\Delta_S\Delta_G\} + (d-1)\Delta_S = \Delta_G^2 + (d-1)\Delta_G$ in the special case $S = G$, thus the upper bound on $\lambda_{d,1}(G)$ in Theorem 2 in Sect. 15.3 is proved. For a general $G$, if $\Delta_S$ is bounded by a constant, then we have $\lambda_{d,1}(G, S) = O(\Delta_G + d)$. Given a connected graph, research has been done in generating spanning-connected subgraphs with constant bounded degrees. For example, given any connected UDG $G$, node degrees in a connected spanning subgraph called *local minimum spanning tree* (LMST) are bounded by six [65], which implies that $O(\Delta_G + d)$ labels are sufficient to guarantee that each pair of nodes are connected by an interference-free path.

**Theorem 5.** [64] *Given any connected UDG $G$, there is a labeling with the span of labels $O(\Delta_G + d)$ that guarantees that, given any two nodes $u, v \in V(G)$, there exists a path that connects node $u$ to node $v$, in which each link $\forall x \leftrightarrow y \in E(S)$, $|l(x) - l(y)| \geq d$, $|l(x) - l(z)| \geq 1$, $\forall z \neq x, z \in N_G(y)$.*

### 15.4.2 Heuristics of $L_S(d_1, d_2)$-Labeling

As an $L_G(1, 1)$-labeling is a special case of an $L_S(1, 1)$-labeling, the problem of minimizing the number of labels used by an $L_S(d_1, d_2)$-labeling is NP-complete. Heuristics of $L_S(d_1, d_2)$-labelings on general networks and subgraphs can be designed similarly to those of $L(d_1, d_2)$-labelings. Here we present a scheduling scheme (Algorithm 3) for data gathering applications, which require the existence of a (directed) interference-free path from each sensor node to the sink. In particular, interference-free communication is guaranteed along links in a directed breadth first search (BFS) tree rooted at the sink with all the edges toward the sink. In this scheme, the directed BFS tree is not explicitly constructed; it is implied in the token circulation by some algorithm that circulates a token in BFS order [66]. For presentation simplicity, here by saying a node labels a neighbor, we mean it sends a packet that contains the assigned label to that neighbor. We can prove the number of required labels is no more than $\Delta_G + 1$.

---

**Algorithm 3**: [64] An $L_S(1, 1)$-labeling for data gathering

---

Initially, all the nodes are unlabeled. Code on node $p$:

- **If $p$ is the sink**: $p$ labels itself by 0 and its children by $1, 2, ..., \delta_{G(p)}$ respectively, then $p$ initiates a token circulation in BFS order.
- **When $p$ receives the token for the first time**: We denote by $C$ the set of nodes in $N_G(p)$ that have been labeled. Before node $p$ forwards the token to the next node in the BFS token circulation, it labels nodes in $N_G(p) - C$ in such a way that all the nodes in $N_G(p) - C$ have unique labels among those in $N_G(p) \bigcup \{p\}$.

---

**Theorem 6.** [64] *Given a graph $G$ and a node $r$, the scheme in Algorithm 3 uses no more than $\Delta_G + 1$ labels to generate an $L_T(1, 1)$-labeling, where $T$ is a directed BFS tree rooted at $r$ with edges toward $r$.*

## 15.5 Duty Cycling

In sensor networks, sensor nodes are normally battery-operated and energy efficiency is one important constraint [6]. Based on the observation that many applications in wireless sensor networks are rather undemanding and can tolerate long end-to-end latency, a strategy to save energy is to trade off network performance, such as packet forwarding latency and throughput, for a reduction in energy consumption. In this section, we present an energy saving strategy, called *duty cycling*, which aims to reduce the amount of energy consumed by *idle listening*. In sensor networks, since a sensor node cannot tell when a message will be sent to it if no additional information is provided, in order not to miss packets, a sensor node must keep its radio active and listen to the shared communication channel; this is so-called *idle listening*. Idle listening has been identified a significant consumer of power in a sensor network, especially when the traffic is low; as indicated by many experimental results, the energy consumption rate of idle listening is 50–100% of that required by receiving [9, 67, 68]. Studies have shown that periodic duty cycling of sensor nodes can reduce energy consumption by idle listening and achieve better energy efficiency in networks where the traffic load is light most of the time [8, 9, 13–16].

*Example 14.* The Digitan 2 Mbps wireless LAN module (IEEE 802.11/2 Mbps) specification shows the ratios of energy consumption rates of idle listening, receiving, and transmitting are 1:2:2.5 [67]. If the data traffic is low and a node spends 80% of the time idle listening, 10% of the time receiving, and 10% of the time transmitting, then the ratios of the total amounts of energy consumed by idle listening, receiving, and transmitting are 8:2:2.5.

Although duty cycling can save energy, it also disrupts networks performance. Extra latency is introduced by duty cycling: the data sampled by a source node during its sleep period have to be queued until the active period, and an intermediate node has to wait until the next hop wakes up to forward the packet. Duty cycling might also introduce link disconnections, which can cause network partitions; for example, one end node of a link cannot communicate with the other end along this link if no slot is scheduled in which both end nodes are active. A well-designed duty cycling scheme should be able to cope with these problems.

Our focus in this section is on the algorithmic aspect of duty cycling. Many research works have been devoted to duty cycling and various techniques have been proposed [7–10, 17–26]. These works differ in their scheduling policy and their requirement on the collaboration among sensor nodes. Many works require an efficient synchronization mechanism [8,9], and on the other extreme, uncoordinated duty cycling schemes have also been investigated [7]. In the sequel, we first review existing duty cycling techniques. Then we concentrate on the algorithmic aspect of duty cycling. We present a graph–theoretical abstraction and theoretical analyses for duty cycling based on a synchronization mechanism [8].

### 15.5.1 An Overview on Duty Cycling

A spectrum of duty cycling schemes requires an efficient synchronization mechanism. Compared to TDMA, duty cycling requires a much looser synchronization (e.g., an active duration of 0.5 s is more than $10^5$ times longer than typical clock drift rates [9]), which enables energy saving by reducing message exchange for synchronization; furthermore, efficient synchronization protocols have been proposed particularly for duty cycling [69,70]. One example of such a duty cycling scheme is the classic S-MAC scheme [9]. In S-MAC scheme, sensor nodes broadcast the time stamp of their local clocks at the beginning of each active period, which enables them to adjust their local clocks, and therefore to follow a common sleep/active schedule. Since sensor nodes follow the same schedule, communication is always feasible during the active periods and link disconnection is avoided. Such a simple idea has been proved effective in reducing idle listening overhead. However, the latency will be as large as the number of hops times the duration of each period; furthermore, since the duty cycle is selected before network deployment, it does not deal well with traffic fluctuations. Timeout MAC (T-MAC) scheme [15] is proposed to adapt to the real traffic and to reduce latency. In T-MAC, an adaptive timeout phase is scheduled at the beginning of each period, in which each node listens to the shared channel – it goes into sleep mode if it hears no communication, otherwise it stays active until no communication has been observed. A similar idea, called *adaptive listening* is used in an extension of S-MAC [71]. Most of the existing works evaluate the performance of a duty cycled network through simulations. Theoretical aspects of duty cycling based on a synchronization mechanism has also been investigated; a graph–theoretical model is proposed in [8] and it is proved that minimizing the end-to-end communication latency is NP-hard. All these works either assume that interference is not a concern [8] or handle it by a contention-based scheme [9]. Based on the observation that a synchronization mechanism has been required by duty cycling, a scheduling scheme that integrates interference avoidance and duty cycling is proposed in [64].

Due to the complexity and the message overhead caused by synchronization, research is also done in asynchronous duty cycling schemes. Such a scheme eliminates the requirement of synchronization and it is more resilient to network dynamics, such as nodes joining and nodes failures. One technique in designing asynchronous duty cycling schemes is to employ a form of preamble sampling [72–74]. The basic idea is to prepend each packet with a long preamble. Sensor nodes sleep for most of the time and periodically wake up to sense the channel; if the channel is idle, it goes back to sleep, otherwise it stays active and continues to listen until the packet is received. If the time required to transmit the preamble is longer than the sleep interval, a sender is guaranteed to wake up the intended receivers. Another technique is to use randomization mechanism. In [7], a completely uncoordinated duty cycling scheme is examined, in which each node switches between active and sleep mode independently from each other and the duration of their active and sleep periods are two independent sequences of independently identical distributed random variables. Rigorous theoretical analyses are presented in [7] on the latency of such a scheme in random networks.

## 15.5.2 Duty Cycling Based on a Synchronization Mechanism

In this section, we concentrate on duty cycling based on a synchronization mechanism. We assume that time is organized into slots, and each slot is long enough that interference can be resolved in one slot by some random access scheme. The scenario under consideration is that every pair of sensor nodes is equally likely to communicate and the goal is to guarantee a small communication delay between any pair of nodes.

We present the graph–theoretical abstraction of the duty cycling problem given in [8]. In this model, a parameter $k$ is used to represent applications' requirement on energy efficiency – it is required that a sensor node is active in exactly one of the $k$ slots if it has no data to forward. Given a network modeled by a graph $G$, a duty cycling scheme is represented by a slot assignment function $f : V(G) \rightarrow [0, k-1]$, which assigns each node $u$ a slot, $f(u) \in [0, k-1]$. The schedule of each node $u$ is defined by the duty cycling schedule $f()$ as follows:

**Fig. 15.10** Examples of
duty cycled networks [8].
(a) Example 1 (b) Example 2

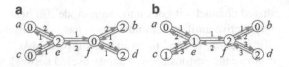

- Each sensor node $u$ is active only in slots $ik + f(u)$, $i = 0, 1, \ldots,$ if it has no data to transmit; the set of slots $\{ik + f(u), i = 0, 1, \ldots\}$ is called the active slots of $u$.
- If a sensor node has a packet to forward to a neighbor, it can wake up in the active slot of that neighbor and transmit the packet.

*Example 15.* Two duty cycling schedules [8] are given for the network in Fig. 15.10. Each node is assigned one of every $k = 3$ slots, indicated by the number in the circle; the gray arrows and their associated numbers will be explained later. In (a), node $a$ wakes up to listen to the channel in slots $\{3i \mid i = 0, 1, \ldots\}$. If it has a packet to forward to a neighbor, say, node $e$, it can wake up in one of the active slots of that neighbor to forward the packet, which are $\{3i + 2, k = 0, 1, \ldots\}$ for node $e$.

Given this model, for any two nodes $u, v \in V(G)$ such that $u \leftrightarrow v \in E(G)$, the delay of transmitting data from $u$ to $v$ is defined below; intuitively, node $u$ receives a data in one of its active slots, and it needs to wait until a later slot in which node $v$ is active to forward the packet to $v$.

$$d_G^f(u, v) = \begin{cases} k & \text{if } f(u) = f(v) \\ (f(v) - f(u)) \mod k & \text{otherwise} \end{cases}.$$

The delay of the communication along a path $P$ is defined as $d_G^f(P) = \sum_{u \to v \in P} d_G^f(u, v)$.

*Example 16.* In Fig. 15.10, the numbers with the gray arrows show the delay of a link in the corresponding direction. For example, the delay from node $c$ to $e$ in (a) is $f(e) - f(c) \mod 3 = 2$ and that in (b) is $k=3$ since $f(e) = f(c)$.

Similar to the definition of graph notations *distance* and *network diameter*, the *delay distance* from one node to another is defined as the minimum delay along all the paths that connect them and the *delay diameter* of a duty cycled network is defined as the largest delay distance between any two nodes.

**Definition 3. (Delay Distance $D_G^f(u, v)$ and Delay Diameter $D_G^f$)** Given a network $G$, a positive number $k$, and a slot assignment function $f : V(G) \to [0, k-1]$, for any two node $u, v \in V(G)$, the delay distance between $u$ and $v$ is defined as $D_G^f(u, v) \equiv \min\{d_G^f(P) \mid P \text{ is a path from } u \text{ to } v\}$ and the delay diameter is defined as $D_G^f \equiv \max_{u, v \in V(G)} D_G^f(u, v)$.

Thus the design goal is to give a duty cycling scheme $f()$ that minimizes the delay diameter. However, it is proved in [8] that minimizing the delay diameter is NP-Complete by a reduction from 3-Conjunctive Normal Form-Satisfiability (3-CNF-SAT). The corresponding decision problem is defined below.

**Theorem 7.** [8] *The decision problem of delay efficient sleep scheduling DESS* $(G, k, f, C)$ *is defined as "given a network $G$, a positive number $k$ and a slot assignment function $f : V(G) \rightarrow [0, k - 1]$, a positive integer $C$, is $D_G^f \leq C$?".* *The DESS($G, k, f, C$) problem is NP-Complete.*

*Proof.* Since there are a polynomial number pairs of nodes and the delay distance from one node to the other can be computed in polynomial time, the delay diameter can be computed and compared to $C$ in polynomial time. So DESS($G, k, f, C$) $\in$ NP. To prove that DESS($G, k, f, C$) is NP-hard, we present below a polynomial time reduction from 3-CNF-SAT to a special case, DESS($G, 2, f', 4$), where $G$ and $f'$ are defined as follows.

We consider a 3-CNF formula $F$ consisting of $n$ clauses and $m$ literals, that is, $F = c_1 \wedge c_2 \ldots \wedge c_n$, where $\forall i \in [1, n]$, $c_i = y_{i1} \vee y_{i2} \vee y_{i3}$, and $y_{ij} \in \{x_1, \bar{x}_1, \ldots, x_m, \bar{x}_m\}$, $j \in [1, 3]$; for nontriviality, it is assumed that $\forall i \in [1, n], \forall k \in [1, m]$, if $x_k \in \{y_{i1}, y_{i2}, y_{i3}\}$, then $\bar{x}_k \notin \{y_{i1}, y_{i2}, y_{i3}\}$. We give a reduction from $F$ to DESS($G, 2, f', 4$), where the graph $G$ is constructed as follows (Fig. 15.11); note the diameter of $G$ so constructed is 4. The set of vertices is $V(G) = \{S\} \cup \{X_i, X_{i1}, X_{i2}, i \in [1, m]\} \cup \{C_i, i \in [1, n]\}$, where (1) $S$ is a special node, (2) each literal $x_i$ has three corresponding nodes: $X_i$, $X_{i1}$ (representing $x_i$) and $X_{i2}$ (representing $\bar{x}_i$), and (3) each clause $c_i$ has one corresponding node $C_i$. The set of edges $E(G)$ is computed as follows. For each literal $x_i$, $i \in [1, m]$, edges $S \leftrightarrow X_{i1}$, $S \leftrightarrow X_{i2}$, $X_i \leftrightarrow X_{i1}$, and $X_i \leftrightarrow X_{i2}$ are added to $E(G)$ (black lines in Fig. 15.11). Then for each clause $c_j$, $j \in [1, n]$, we examine each literal $x_i$, $i \in [1, m]$: if $x_i$ appears in $c_j$, that is, $x_i \in \{y_{j1}, y_{j2}, y_{j3}\}$, $X_{i1} \leftrightarrow C_j$ is added to $E(G)$, and if $\bar{x}_i$ appears in $c_j$, that is, $\bar{x}_i \in \{y_{j1}, y_{j2}, y_{j3}\}$, $X_{i2} \leftrightarrow C_j$ is added to $E(G)$ (gray lines in Fig. 15.11).

A slot assignment function $f'$ is defined as follows: (1) $f'(v) = 1$ for $v \in \{S\} \cup \{X_i, i \in [1, m]\} \cup \{C_i, i \in [1, n]\}$, and (2) $f'(X_{i1}) = 0$ if $x_i$ is true, else $f'(X_{i1}) = 1$. Moreover, $f'(X_{i1}) + f'(X_{i2}) = 1$. Note we have $k = 2$ and given any two adjacent nodes $u$ and $v$, $d_G^{f'}(u, v) = d_G^{f'}(v, u) = 1$ if and only if $f'(u) \neq f'(v)$, otherwise $d_G^{f'}(u, v) = d_G^{f'}(v, u) = 2$. This reduction can be computed in polynomial time. Now we prove that a formula $F$ is satisfiable if and only if $D_G^{f'} \leq 4$.

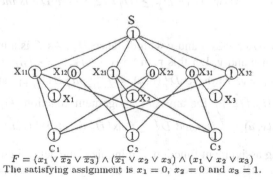

$$F = (x_1 \vee \overline{x_2} \vee \overline{x_3}) \wedge (\overline{x_1} \vee x_2 \vee x_3) \wedge (x_1 \vee x_2 \vee x_3)$$
The satisfying assignment is $x_1 = 0$, $x_2 = 0$ and $x_3 = 1$.

**Fig. 15.11** Reduction from 3CNFSAT to DESS [8]

We first prove $D_G^{f'} \leq 4$ if the formula $F$ is satisfiable by showing that for any node $u$, $D_G^{f'}(u, S) \leq 2$ and $D_G^{f'}(S, u) \leq 2$:

- $\forall i \in [1, m]$, we have (1) $D_G^{f'}(X_i, S) = D_G^{f'}(S, X_i) \leq 2$, since $\exists k \in [1, 2]$, such that $f'(X_{ik}) = 0$ due to $f'(X_{i1}) + f'(X_{i2}) = 1$, and thus $d_G^{f'}(X_i \to X_{ik} \to S) \leq 2$ and (2) $\max\{D_G^{f'}(X_{i1}, S), D_G^{f'}(X_{i2}, S)\} = 2$, since exactly one of $f'(X_{i1})$ or $f'(X_{i2})$ is 1 and $f'(S) = 1$.
- $\forall i \in [1, n]$, we consider the delay distance between $C_i$ and $S$. Since the clause $c_i = y_{i1} \vee y_{i2} \vee y_{i3}$ is true, there is $k \in [1, 3]$, such that $y_{ik} \in \{x_1, \bar{x}_1, \ldots, x_m, \bar{x}_m\}$ is true. Let $y_{ik}$ be $x_j$ or $\bar{x}_j, j \in [1, m]$. Then $\exists l \in [1, 2], f'(X_{jl}) = 0$ and $X_{jl} \leftrightarrow C_i \in E(G)$. Thus we have $D_G^{f'}(C_i, S) = D_G^{f'}(S, C_i) \leq 2$, since there is a path $C_i \to X_{jl} \to S$ (and vice versa) which has an alternating 1, 0 slots.

Now we show $D_G^{f'} > 4$ if $F$ is not satisfiable. Since $F$ is not satisfiable, there is a false clause $c_l$. We denote the neighbor of $c_l$ in $G$ by $N = \{X_{ik} | C_l \leftrightarrow X_{ik} \in E(G)\}$; note $C_l$ is only adjacent to nodes in $\{X_{ik}, i \in [1, m], k \in [1, 2]\}$. We first show $f'(X_{ik}) = 1$ for each $X_{ik} \in N$: if $k = 1$, we have $x_i \in \{y_{l1}, y_{l2}, y_{l3}\}$ by the construction of $G$, and $x_i$ is false since $c_l$ is false, so $f'(X_{i1}) = 1$ by the definition of $f'()$; if $k = 2$, we have $\bar{x}_i \in \{y_{l1}, y_{l2}, y_{l3}\}$ and $x_i$ is true since $c_l$ is false, so $f'(X_{i1}) = 0$ and $f'(X_{i2}) = 1$.

Given a node $X_{ip}$ such that $X_{ik} \in N$, where $k = 3 - p$, we can show that $D_G^{f'}(C_l, X_{ip}) > 4$. Every path from $C_l$ to $X_{ip}$ will reach a node in $N$, and then either go to $S$ or a node in $\{C_j, j \in [1, n]\}$ or a node in $\{X_j, j \in [1, m]\}$, then at least one hop is needed to reach $X_{ip}$. Note the delay of the first hop is 2 because, for all $X_{ik} \in N$, we have $d_G^{f'}(C_l, X_{jk}) = d_G^{f'}(X_{jk}, C_l) = 2$ since $f'(C_l) = f'(X_{jk}) = 1$. The second hop also has delay 2 because $f'(X_{ik}) = 1$ for all $X_{ik} \in N$ and $f'(u) = 1$ for all $u \in \{S\} \cup \{X_i, i \in [1, m]\} \cup \{C_i, i \in [1, n]\}$. Thus the proof is done.  □

Given the fact that minimizing the communication delay is NP-Complete, optimal algorithms have been designed on specific topologies. Here we give an optimal slot assignment function $f$ on tree [8]. First we present a lower bound.

**Theorem 8.** [8] *Given a tree $T$ and a positive number $k$, for every slot assignment $f : V(G) \to [0, k - 1]$, we have $D_T^f \geq D_T k/2$, where $D_T$ is the diameter of the tree $T$.*

*Proof.* Consider two nodes $u$ and $v$ with distance $D_T$. As $T$ is a tree, there is only one path between $u$ and $v$. Let this path be $i_1 \leftrightarrow i_2 \ldots \leftrightarrow i_{D_T} \leftrightarrow i_{D_T+1}$, where $i_1 = u$ and $i_{D_T+1} = v$. We have $D_T^f(u, v) = \sum_{j=1}^{D_T} d_T^f(i_j, i_{j+1})$ and $D_T^f(u, v) = \sum_{j=1}^{D_T} \left( k - d_T^f(i_j, i_{j+1}) \right)$ for any slot assignment function $f()$. So we show $D_T^f(u, v) + D_T^f(v, u) = kD_T$ and $D_T^f \geq \max\{D_T^f(u, v), D_T^f(v, u)\} \geq kD_T/2$.  □

The following assignment function $f$ minimizes the delay diameter of a tree $T$. Let $r$ be a node such that there is a node $u$ satisfying $d_T(r, u) = D_T$. For each node $u$, we denote by $p(u)$ the parent of $u$ in the tree rooted at $r$. The assignment function $f$ is defined as follow: (1) let $f(r) = 0$, (2) for each unassigned node $u$, if $f(p(u)) = 0$, then $f(u) = \lceil k/2 \rceil$, otherwise $f(u) = 0$.

*Example 17.* The duty cycling scheme in Fig. 15.10a is optimal with $k = 3$.

## 15.6  Application-Oriented Scheduling

Sensor networks are usually deployed for specific purposes, such as battlefield surveillance, monitoring of human physiological data, and vehicle tracking. Applications that support each of these specific purposes have their own communication patterns. Recently, the concept of application-oriented scheduling has been proposed, which takes advantage of this characteristics to reduce the communication latency, as well as the amount of resources required by scheduling. Efficient application-oriented scheduling protocols have been proposed for data gathering [75, 76] and backbone-based communication [41]. In these schemes, the schedules of sensor nodes' activities are optimized according to the specific communication pattern, and routing decisions are often jointly considered; since in a specific communication pattern, it is usually unnecessary to keep all the sensor nodes active, a certain level of duty cycling is often employed. In this section, we first review several representative works. Then we examine an asymptotically optimal scheduling scheme for data aggregation as an example.

### 15.6.1  An Overview

The Data gathering MAC (DMAC) in [73] addresses the latency for data gathering problem in duty cycled networks. It is designed for sensor networks where the communication is restricted to an established undirected data gathering tree rooted at the sink. The basic idea is to schedule the active periods of each sensor node according to its depth in the spanning tree. By scheduling sensor nodes going to sleep as soon as they finish forwarding the packets to the next level, and waking up just in time to receive the next round of packets, data packets can be forwarded level by level from the leaves of the tree toward the sink continuously.

The focus of [72] is on interference-free scheduling for data dissemination in sensor networks modeled by UDGs. In [72], a data aggregation tree is constructed from a BFS tree rooted at the sink; the technique of maximal independent sets is used in the tree construction. The transmissions of sensor nodes are scheduled based on this data aggregation tree and the designed schedule achieves an asymptotically optimal latency $O(\Delta + D)$ in UDGs, where $\Delta$ is the node degree and $D$ is the network diameter.

Backbone structures have been widely used in various aspects of wireless networking, but less work has been done on scheduling for communication based on such a structure. In [41] a scheduling scheme for backbone-based communication is proposed, which achieves asymptotically optimal latency in networks with certain properties. The basic idea is to divide the communication into three phases: transmissions from nonbackbone nodes to backbone nodes, transmissions from backbone nodes to nonbackbone nodes, and communication in the backbone structure. Schedules are designed for each phase to guarantee interference-free communication. Although the communication between nonbackbone nodes and backbone nodes might be expensive due to a large node degree, the number of such hops in the communication between any two nodes can be constant bounded. The schedule proposed in [41] guarantees an $O(\Delta^2 + D)$ latency for the communication between any pair of nodes if sensor nodes have bounded transmission range, where $\Delta$ is the node degree and $D$ is the network diameter.

## 15.6.2  Scheduling for Data Aggregation

We examine a scheduling scheme for data aggregation in sensor networks designed for monitoring tasks. In a monitoring task, each sensor node generates data packets and transmits them to a set of special stations, called *sinks*, which are responsible for collecting data. In particular, we consider *data aggregation*, in which when a sensor node receives a data packet from its neighbor, it can merge this packet with its own data packet and merged packet contains all the required information. Examples of such scenarios include environmental monitoring where the information of interest is the maximum temperature in the monitoring area; in this case, two (or more) pieces of data packets can be merged by taking their maximum.

Here we consider *time division scheduling* in sensor networks modeled by UDGs, where time is organized into *slots*. Given a network $G$, the scheduling scheme is represented by a labeling function $l : V(G) \to N$, which assigns each node $u$ a nonnegative integer $l(u)$. Each node $u$ is allowed to transmit only in slots $\{l(u) + i(K + 1)|  i = 0, 1, \ldots\}$, where $K$ is the span of labels used by labeling $l(\ )$. We assume that two nodes can interfere with each other if and only if they can communicate with each other. Thus a transmission can be received without interference by its intended receiver if and only if only one node transmits in the receiver's neighborhood. This can be formally described as follows: $\forall$ nodes $x, y \in V(G)$ such that $x \leftrightarrow y \in E(G)$, the transmission from node $x$ to $y$ is interference-free if and only if $l(x) \neq l(y)$ and $l(x) \neq l(z), \forall z \in N_G(y)$. Note this model is consistent with that in $L(1,1)$-labelings; in fact, by imposing the above condition on all the links in $E(G)$, we will have an $L(1, 1)$-labeling. If an $L(1, 1)$-labeling is applied for data aggregation, the latency will be $O(D\Delta^2)$, as most labeling algorithms use $O(\Delta^2)$ labels. In the sequel, we examine a scheduling [72] designed specifically for the communication in data aggregation and a latency of $O(\Delta + D)$ is achieved. In [72], a data aggregation tree rooted at the sink is first constructed and then transmissions of sensor nodes are scheduled based on this tree.

### 15.6.2.1 Data Aggregation Tree Construction

The data aggregation tree is constructed in three steps. In Step 1, a BFS tree rooted at the sink is constructed, which divides sensor nodes into layers – the sink is the only node in layer 0 and layer $i$ consists of the set of nodes, denoted by $Layer_i$, that are at distance $i$ to the sink. We number these layers by $0, 1, \ldots, L$, where $L$ is the maximum distance between a node and the sink.

*Example 18.* In Fig. 15.12a, a BFS tree is built and sensor nodes are divided into layers, as indicated by the numbers. For example, $Layer_0 = \{sink\}, Layer_1 = \{a, b, c\}$ and $Layer_2 = \{d, e, f, g, h\}$.

In Step 2, a maximal independent set, *Black*, is formed layer by layer as follows. Initially *Black* is an empty set. For each layer $i = 0, \ldots, L$, a maximal subset of nodes $Black_i \subseteq Layer_i$ that satisfied the following two conditions are added to *Black*: (1) $Black_i$ is an independent set and (2) nodes in $Black_i$ are independent of nodes that have been added to *Black*. We denote the parent of node $v$ in the BFS tree by $p_B(v)$. It is easy to see the following properties at the end of this step.

**Lemma 3.** *At the end of step 2, $\forall i \in [1, L]$, we have (1) $\forall v \in Black_i$, $p_B(v) \in Layer_{i-1}$ and $p_B(v) \notin Black_{i-1}$, and (2) $\forall w \in (Layer_i - Black_i)$, $\exists u \in Black_i \cup Black_{i-1}$ such that $w \leftrightarrow u \in E(G)$; we call $u$ a dominator of $w$.*

*Example 19.* In Fig. 15.12a, black nodes are marked. We have $Black_0 = \{sink\}$ and $Black_1 = \phi$. For layer 2, a maximal independent set $Black_2 = \{d, f, h\}$ of $Layer_2$ is selected, nodes in which are also independent of those in $Black_0 \cup Black_1 = \{sink\}$. Node $e \in Layer_2 - Black_2$ has dominators $d, f \subset Black_2$. Node $m \in Layer_3 - Black_3$ has dominator $d \in Black_2$.

In Step 3, a spanning data aggregation tree $T$ is constructed by letting $V(T) = V(G)$ and edges $E(T)$ computed as follows. Note that a node might have multiple dominators; here we pick one of the dominators for each node $u$ and denote the picked node by $d_u$.

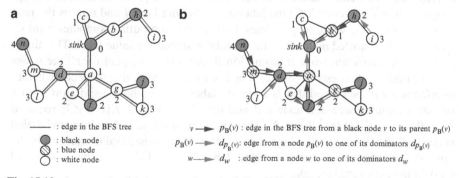

Fig. 15.12 An example of data aggregation scheduling [72]. (a) Step 2 (b) Step 3

- *Step 3.1.* $\forall i \in [1, L-1]$, $\forall v \in Black_{i+1}$, edge $v \leftrightarrow p_B(v)$ is added to $E(T)$ and edge $p_B(v) \leftrightarrow d_{p_B(v)}$ is added if it has not been added to $E(T)$. Note $d_{p_B(v)} \in Black_i \cup Black_{i-1}$ by Lemma 3. We denote by $Blue \equiv \{p_B(v)|v \in Black\}$ the set of parent nodes; in particular, the parent nodes in layer $i$ is denoted by $Blue_i \equiv Blue \cap Layer_i$. We define $White \equiv V(G) - Black - Blue$.
- *Step 3.2.* $\forall w \in White$, $w \leftrightarrow d_w$ is added to $E(T)$. Note $d_w \in Black$ by Lemma 3(2).

*Example 20.* We consider node $d \in Black_2$ in Fig. 15.12b as an example. Since $p_B(d) = a$ and $a$ has a dominator $d_a = sink$, edges $d \leftrightarrow a$ and $a \leftrightarrow sink$ are added to $E(T)$. As for node $e \in White$, edge $e \leftrightarrow f$ is added since $f$ is a dominator of $e$.

The tree topology of $T$ can be proved. Given a node $v \in V(G)$, we denote by $p_T(v)$ the parent of node $v$ defined in tree $T$; given a subset $S \subseteq V(G)$, we denote by $p_T(S) \equiv \cup_{u \in S} \{p_T(u)\}$. We can prove the following properties.

**Lemma 4.** *Given a network $G$, the subgraph $T$ so constructed is a tree with $p_T(White) \subseteq Black$, $p_T(Black_i) = Blue_{i-1}$ and $p_T(Blue_{i-1}) \subseteq Black_{i-1} \cup Black_{i-2}$.*

### 15.6.2.2  Data Aggregation Scheduling

Given the aggregation tree, data aggregation scheduling is quite straightforward. By the properties presented in Lemma 4, data aggregation can be achieved by aggregating data along tree edges bottom-up to the root as follows. First data are forwarded along tree edges from nodes in *White* to nodes in *Black*. For each layer $i$, if data have been collected by nodes in *Black* within distance $i$ to the root, then they can be aggregated as follows to nodes in *Black* within distance $i - 1$ to the root: data are forwarded from nodes in $Black_i$ to nodes in $p_T(Black_i)$, which is $Blue_{i-1}$, and then from nodes in $Blue_{i-1}$ to $p_T(Blue_{i-1}) \subseteq (Black_{i-1} \cup Black_{i-2})$. Thus by repeating this procedure for $i = L, \ldots, 1$, data packets can be aggregated layer by layer and going all the way to the root (the sink).

Based on this observation, an aggregation schedule is given in Algorithm 4, where $schedule(S, T, G, l)$ is a subroutine that labels nodes in $S$ to guarantee interference-free transmissions from nodes in $S$ to their parent defined by the aggregation tree $T$ in network $G$; the labeling starts with label $l$ and returns the next available label. The while-loop in lines 1–10 of this subroutine computes a maximal set $X$ of unlabeled nodes in $S$ that can be assigned the same label. The aim is to guarantee interference-free transmission from node $x$ to $p_T(x)$ for all the nodes $x \in X$. Initially $X$ is empty (line 2). For each node $s \in S$, line 4 checks whether interference will be caused if it is assigned label $l$: since the transmission from $s$ introduces interference at a node $u$ if and only if $s \leftrightarrow u \in E(G)$, it is required $s \leftrightarrow P_T(x) \notin E(G)$ for all $x \in X$. This checking is repeated for all the unlabeled nodes in $S$ and $X$ is a maximal set of nodes that can be scheduled to transmit in the same slot and they are assigned the same label (line 8). This procedure is repeated until all the nodes are labeled.

*Example 21.* We consider *schedule(White, T, G,* 0) for Fig. 15.12. We have *White* = {*l, e, k, c, i*}. In the first iteration of the while-loop, since nodes *l, k, c, i* can be assigned the same label, they are labeled 0. Node *e* is labeled 1 in the next iteration.

---

**Algorithm 4**: [72] Data aggregation scheduling

---

1: int $l$ = 0; // the next available lable ;
2: $l = schedule(White, T, G, l)$;
3: **for** $i = L$ to I **do**
4:    $l = schedule(Black, T, G, l)$;
5:    $l = schedule(Blue_{i-1}, T, G, l)$;
6: **end for**

**Subroutine** *schedule* $(S, T, G, l)$;
**Input:** a set of senders $S$, a data aggregation tree $T$. a network $G$ and the next available lablel $l$.
1: **while** $S \neq \emptyset$ **do**
2:    $X = \emptyset$
3:    **for all** $s \in S$ **do**
4:       **if** $(\forall x \in X, \langle s \rightarrow pr(x) \rangle \notin E(G))$ **then**
5:          Add $s$ to $X$ and remove $s$ from $S$
6:       **end if**
7:    **end for**
8:    $\forall x \in X, l(x) = l$;
9:    $l + +$;
10: **end while**
11: **return** $l$

---

   The correctness follows the explanations of the algorithm. We present the latency of data aggregation under this scheduling in the following theorem. Here we give the intuition of the proof; readers are referred to [72] for more information. Note the number of labels required by subroutine *schedule(S, T, G, l)* is polynomial in the degree (defined by the aggregation tree $T$) of nodes in $S$. The maximal independent set property in the tree construction and the geometric properties of UDGs guarantee that nodes in *Black* and *Blue* have constant-bounded degrees in $T$ and the diameter of $T$ is in the same order as that of $G$. Thus the latency of the first-hop forwarding, that is, from nodes in *White* to their parents in tree $T$, can be $O(\Delta)$ and the latency of each of the subsequence $O(D)$ hops is $O(1)$.

**Theorem 9.** [72] *Given a network modeled by a UDG $G$, data aggregation from sensor nodes to the sink can be completed in $O(D + \Delta)$ under the proposed schedul-ing, where $\Delta$ is the node degree and $D$ is the diameter of $G$.*

## 15.7  Scheduling Under the Protocol Model and SINR Model

Many works on the scheduling problem consider interference models based on graph–theoretic notions. In this section, we investigate the scheduling problem under two more models: the *protocol model* and the *physical model* [1]. In the protocol model, a parameter $\Gamma$ is used to model situations where a guard zone is specified by the protocol to prevent a neighboring node from transmitting to the same channel at the same time. In the physical model, also called the *physical* SINR model, the condition for interference-free communication is defined based on the accumulated effect of interference from different transmitters. These definitions are given below.

**Definition 4. (Protocol Model [1])** The transmission from node $x$ to node $y$ is interference-free if, for any other node $z$ simultaneously transmitting to the same channel, we have $||z, y|| \geq (1 + \Gamma)||x, y||$, where $||x, y||$ is the geometric distance between nodes $x$ and $y$.

**Definition 5. (Physical Model (SINR Model) [1])** Let $\{x_0, \ldots, x_K\}$ be the set of nodes simultaneously transmitting at some time instant over the same channel. Let $P_i$ be the power level chosen by node $x_i$, $i \in [0, K]$, to transmit. Then the transmission from a node $x_i$, $i \in [0, K]$, is interference-free received by a node $x$ if

$$\frac{\frac{P_i}{||x_i - x||^\alpha}}{N + \sum_{k \in [0, K], k \neq i} \frac{P_k}{||x_k - x||^\alpha}} \geq \beta.$$

Here $\beta$ is the minimum signal-to-interference ratio (SIR), $N$ is the ambient noise power level, and signal power decays with distance $r$ as $1/r^\alpha$, where we suppose $\alpha > 2$.

*Example 22.* In Fig. 15.13, let the distance between nodes $u$ and $v$ be $r = ||u, v||$. In order to guarantee an interference-free transmission from node $u$ to $v$, all the nodes

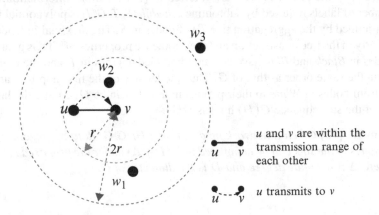

**Fig. 15.13**  Protocol model with $\Gamma = 1$

that are within $(1 + \Gamma)\|u, v\| = 2r$ distance from $v$, that is, $w_1$ and $w_2$, should not transmit at the same time, while the transmission from $w_3$ will not interfere with the transmission from $u$ to $v$.

These models have received extensive research interest since they were proposed. Research has been done on scheduling design and complexity analyses under these two models [1, 27–32]. In the sequel, we first give a review of these works and then examine an analysis, obtained by constructing schedules of nodes' transmissions, on the capacity of a multichannel wireless network under the protocol model [27].

### 15.7.1  An Overview of Scheduling Under the Protocol Model and the SINR Model

We present here the main results on the scheduling problem under the protocol model and the SINR model. We denote by $n$ the number of nodes and by $W$ the transmit rate of each node in bits/s. In these works, a schedule refers to an assignment of a transmission power to each node and, if multiple channels are available, the channel to which the node can transmit to, in each of a sequence of slots; if a node is not scheduled to transmit in a slot, the power assigned is 0.

In [1], the throughput capacity of a wireless network is investigated under the protocol model and the SINR model. Two types of networks are investigated: *arbitrary networks* and *random networks*. In the arbitrary network settings, network capacity is measured in terms of "bit-meter/second" – the network is said to transport one "bit-meter/second" when one bit has been transported across a distance of 1 m in 1 s. In random network setting, nodes are randomly located on the surface of a unit torus. Each node sets up a flow to a randomly selected destination. Network capacity is defined to be the sum of throughput over all the flows and measured in termed of "bits/s". In [1], it is shown that in arbitrary networks, if the nodes are optimal placed, the traffic patterns are optimally assigned, and each node's transmission range is optimally chosen, the transport capacity scales as $\Theta\left(W\sqrt{n}\right)$ bit-m/s under the protocol model; under the SINR model, $cW\sqrt{n}$ bit-m/s is feasible, while $c'Wn^{(\alpha-1)/\alpha}$ bit-m/s is not, for appropriate $c$ and $c'$. As for a random network, the throughput of each node scales as $\Theta\left(W/\sqrt{n\log n}\right)$ bits/s under the protocol model, and under the SINR model, a throughput for each node of $cW/\sqrt{n\log n}$ bit/s is feasible, while $c'W/\sqrt{n}$ bit/s is not, for appropriate $c$ and $c'$, both with probability approaching one as $n$ approaches $+\infty$.

In [27], the capacity of networks with multiple transceivers and channels is examined under the protocol model. Let $m$ be the number of transceivers and $c$ be the number of channels. Here the transmit rate of each node $W$ is the total data rate using all the channels, which is divided equally among the channels, and thus the data rate supported by each of the $c$ channels is $W/c$ bits/s. It is shown that in arbitrary networks, if $c/m = O(n)$, the network capacity is $\Theta\left(W\sqrt{mn/c}\right)$ bit-m/s, and if $c/m = \Omega(n)$, the network capacity is $\Theta(Wmn/c)$bit-m/s. In random networks, the network capacity is $\Theta\left(W\sqrt{n/\log n}\right)$ bits/s if $c/m = O(\log n)$, it is $\Theta\left(W\sqrt{mn/c}\right)$ bits/s if $c/m = \Omega(\log n)$ and also $c/m = O(n(\log\log n/\log n)^2)$, and it is $\Theta(Wnm\log\log n/(c\log n))$ bits/s if $c/m = \Omega(n(\log\log n/\log n)^2)$.

Scheduling complexity in wireless networks is discussed in [31] under the SINR model. Given a set of communication requests, the scheduling complexity is defined as the minimum amount of time required to successfully schedule all the requests. A scheduling algorithm that successfully schedules a strongly connected set of links in time $O(\log^4 n)$ is presented in [31], which indicates that the scheduling complexity of connectivity grows only polylogarithmically in the number of nodes in arbitrary networks.

In [29], the data aggregation capacity is investigated under both the protocol model and the SINR model, where the data aggregation capacity refers to the maximum sustainable rate (bits/s) at which each sensor node can continuously transmit data to the sink. It is shown that, under the SINR model, a rate of $\Omega(1/\log^2 n)$ can be achieved. In contrast, the best possible rate under the protocol model is shown to be $\Theta(1/n)$.

The protocol model is considered in [28] in the study of the broadcast capacity in multihop wireless networks, which is defined as the maximum rate at which broadcast packets can be generated in the network such that all the nodes receive the packets successfully in a finite time. It is proved that the broadcast capacity of a homogeneous dense network is $\Theta\left(W/\max\{1, \Gamma^d\}\right)$, where $\Gamma$ is the guard zone parameter used in the protocol model and $d$ is the number of dimensions of space in which the network lies.

## 15.7.2 A Scheduling Scheme Under the Protocol Model for Multichannel Networks

As a careful designed scheduling represents an ideal situation of sensor nodes' activities, scheduling has been used in theoretical analyses on the best performance that can be achieved in a network. In [1, 27], constructive lower bounds on the throughput capacity of wireless networks are obtained by spatially and temporally scheduling nodes' transmissions. Here we examine a scheduling scheme presented in [27] as an example. The schedule is designed for random networks with a set $V$ of $n$ nodes randomly located on the surface of a torus of unit area. The aim is to investigate the capacity of *single-transceiver multiple-channel* networks, that is, each node is equipped with a single transceiver and communication can happen via multiple channels. We denote the number of channels by $c$ and number the channels as $0, \ldots, c - 1$.

It is assumed that the transmission range of each node can be set by the scheduling scheme, and two nodes can communicate if and only if they are within the transmission range of each other. The protocol model is used in [27] to describe the condition for interference-free communication. Under this mode, given any two nodes $u$ and $v$, the transmission from $u$ to $v$ is successful if and only if (1) nodes $v$ is in the transmission range of $u$, (2) node $u$ is allowed to transmit to and node $v$ is listening to the same channel, and (3) the interference-free constraints defined by the protocol model are satisfied.

It is assumed the total data rate possible by using all the channels is $W$. The total data rate is divided equally among the channels, and the data rate supported by each of the $c$ channels is $W/c$ bits/s. The network capacity is defined under the traffic pattern such that each of the $n$ nodes sends $\gamma(n)$ bits/s to a randomly selected destination. We refer a "flow" to the traffic between a source and a destination pair. The per-node throughput is defined as the highest value of $\gamma(n)$ that can be supported with a high probability (w.h.p.) and the network capacity is defined to be $n\gamma(n)$, as there are a total of $n$ flows. The following lower bound on the network capacity is given in [27].

**Theorem 10.** [27] *The network capacity of a random single-transceiver multi-channel network is as follows, where $n$ is the number of nodes, $c$ is the number of available channels, and $W$ is the total data rate: (1) if $c = O(\log n)$, the network capacity is $\Theta(W\sqrt{n/\log n})$ bits/s; (2) if $c = \Omega(\log n)$ and also $O(n(\log\log n/\log n)^2)$, the network capacity is $\Theta(W\sqrt{n/c})$ bits/s; and (3) if $c = \Omega(n(\log\log n/\log n)^2)$, the network capacity is $\Theta(Wn\log\log n/(c\log n))$ bits/s.*

This lower bound is proved in [27] by constructing a routing scheme and a transmission schedule. Here a transmission schedule specifies for each node in each slot (1) the channel assigned to this node and (2) the action (transmitting or listening) it can take on the assigned channel. The proof consists of three phases (1) cell construction and transmission range assignment, (2) routing scheme, and (3) scheduling scheme. We present the basic idea of each phase below.

### 15.7.2.1 Cell Construction and Transmission Range Assignment

The surface of the unit torus is divided into square cells. The area of each cell is $a(n) \equiv \min\left\{\max\{(100\log n)/n, c/n\}, (1/D(n))^2\right\}$, where $D(n) \equiv \Theta$ $(\log n/\log\log n)$, and the transmission range of each node is set to be $r(n) \equiv \sqrt{8a(n)}$. The following properties can be proved for such a cell construction:

- The number of flows for which a node is a destination is no more than $D(n)$ w.h.p.; proof can be found in [27].
- The number of nodes in each cell is bounded; in particular, if $a(n) > (50\log n)/n$, each cell has $\Theta(na(n))$ nodes w.h.p.; readers are referred to [27] for the proof. Note $a(n) > (50\log n)/n$ when $n$ is suitably large.
- Each node in one cell can communicate with any node in its eight neighboring cells.

- The number of cells that can be interfered by the transmission from a node is $k_{inter} \leq 72(2 + \Gamma)^2$. This can be proved as follows. We consider the condition for a transmission from a node, say $w$, to interfere with another transmission, say, from node $u$ to $v$. Under the protocol model, $w$ is within distance $(1+\Gamma)r(n)$ to $v$. Since the distance between $u$ and $v$ is at most $r(n)$, the distance between the $w$ and $u$ is at most $(2+\Gamma)r(n)$. Therefore, nodes in a cell can be interfered with by only nodes in cells within a distance of $(2+\Gamma)r(n)$, which is completely enclosed in a larger square of side $3(2+\Gamma)\,r(n)$. So there are at most $(3(2 + \Gamma)r(n))^2/a(n) = 72(2 + \Gamma)^2$ cells that can be interfered by the transmission from $w$.

Given the assignment of nodes' transmission ranges, an interference graph $G_I$ can be built as follows: $V(G_I) = V$ and a link $u \leftrightarrow v \in E(G_I)$ if and only if nodes $u$ and $v$ can interfere with each other. Since each cell may interfere with no more than some constant $k_{inter}$ number of cells, and each cell has $\Theta(na(n))$ nodes, the degree is $\Delta_{G_I} = O(na(n))$.

### 15.7.2.2 Routing Scheme

We present the routing scheme for the $n$ flows, each of which corresponds to one of the $n$ source–destination pairs. For each source–destination pair, say node $S$ and node $D$, packets are routed through the cells that lie along the straight line joining $S$ and $D$, and for each of these cells, one node in the cell is "assigned" to the flow. The node in a cell that is assigned to a flow is the only node in that cell which forward data along that flow. The assignment is done by the *flow distribution procedure* [27] presented in Algorithm 5, in which Step 2 balances the assignment of flows to ensure that all nodes are assigned (nearly) the same number of flows. The node assigned to a flow will receive packets from a node in the previous cell and send the packet to a node in the next cell.

---

**Algorithm 5**: [27] Flow distribution procedure

- Step1: For any flow that originates in a cell, the source node $S$ is assigned to the flow. Similarly, for any flow that terminates in a cell, the destination node $D$ is assigned to the flow.
- Setp2: Each such remaining flow (passing through a cell)is assigned to the node in the cell that has the least mumber of flows assigned to it so far.

---

It can be proved as follows that the total number of flows assigned to each node is $O\left(1/\sqrt{a(n)}\right)$ w.h.p.

- Each node is the originator of one flow. Since each node is the destination of at most $D(n)$ flows w.h.p., in Step 1 in the flow distribution procedure, each node is assigned at most $1+D(n)$ flows w.h.p.

- Step 2 assigns to each node at most $O\left(1/\sqrt{a(n)}\right)$ flows, since (1) the assignment in Step 2 ensures that all nodes end up with nearly same number of flows, (2) each cell has $\Theta(na(n))$ nodes, and (3) it can be proved that the number of source–destination lines that intersect a cell, that is, the number of flows that pass through a cell, is $O\left(n\sqrt{a(n)}\right)$ (readers are referred to [77] for the proof).

- So the total flows assigned to each node is $O\left(D(n) + 1/\sqrt{a(n)}\right)$, which is $O\left(1/\sqrt{a(n)}\right)$ since the value chosen for $a(n)$ guarantees $a(n) \le (1/D(n))^2$.

Given such an assignment, a routing graph $G_R$ is built with $V(G_R) = V$ and $E(G_R)$ consists of links $x \leftrightarrow y$ for each pair of consecutive nodes in the route established for each flow. Since each flow through a node can result in at most two edges, one incoming and one outgoing, and each node is assigned to at most $O\left(1/\sqrt{a(n)}\right)$ flows, graph $G_R$ has degree $\Delta_{G_R} = O\left(1/\sqrt{a(n)}\right)$.

### 15.7.2.3 Scheduling Scheme

The transmissions of nodes is scheduled based on an edge-labeling on the routing graph $G_R$ and a vertex-labeling on the interference graph $G_I$.

- The edge labeling on $G_R$ guarantees that any two edges that share common end nodes are assigned different labels. It is represented by a function $l_e()$: $E(G_R) \rightarrow \mathbf{N}$, which satisfies $\forall x \leftrightarrow y, x' \leftrightarrow y' \in E(G_R)$, if $\{x, y\} \cup \{x', y'\} \ne \phi$, then $l_e(x \leftrightarrow y) \ne l_e(x' \leftrightarrow y')$. We denote the span of labels by $L_e$.
- The vertex labeling on $G_I$ guarantees that any two adjacent vertices are assigned different labels. It is represented by function $l_v() : V(G_I) \rightarrow \mathbf{N}$, such that $\forall x \leftrightarrow y \in E(G_I), l_v(x) \ne l_v(y)$. We denote the span of labels by $L_v$.

The scheduling scheme is given in Algorithm 6. The period of every second is divided into $(L_e + 1)M$ slots; each has length $1/((L_e + 1)M)$, where $M = \lceil (L_v + 1)/c \rceil$. Intuitively, the link labeling determines the set of links that can be scheduled in the same slot – links with the same label can be scheduled in the same slot, and the vertex labeling determines the channel assigned to a node. If the number of channels is $c \ge L_v + 1$, links with the same label can be scheduled in one slot – for each link $x \leftrightarrow y$, node $x$ and node $y$ are assigned channel $l_v(x)$, and $x$ is allowed to transmit to while node $y$ is listening to this channel. Since links with the same edge label do not share any common end node, the state of each node is specified unambiguously. The constraints on vertex labeling guarantees that interference will not occur. When the number of channels $c < L_v + 1$, links with the same label is divided into $M$ disjoint sets and each set is scheduled in one slot.

It is well known that given a graph $G$, there is an edge-labeling using at most $O(\Delta_G)$ labels and a vertex-labeling using at most $(\Delta_G + 1)$ labels [78]. Thus there is an edge labeling on $G_R$ with span of labels $L_e = O(\Delta_{G_R}) = O\left(1\Big/\sqrt{a(n)}\right)$ and a vertex-labeling on $G_I$ with span of labels $L_v = O(\Delta_{G_I}) = O(na(n))$ labels. So the length of each slot $1/((L_e + 1)M) = \Omega\left(\sqrt{a(n)}\big/\lceil na(n)/c\rceil\right)$. Each channel can transmit at the rate of $W/c$ bits/s. Thus in each slot,

$$\gamma(n) = \Omega\left(\left(W = \sqrt{a(n)}\right)\Big/\left(c\left\lceil\frac{na(n)}{c}\right\rceil\right)\right)$$

bits can be transported. So the lower bound on the total network capacity $n\gamma(n)$ can be obtained by substituting the value of $a(n)$.

---

**Algorithm 6**: [27] Scheduling of transmissions

---

1: for $k = 0, \ldots, L_e$ do
2:     $E_k$ = the set of links assigned edge-label $k$;
3:     **for all** links $\langle x \to y\rangle \in E_k$ **do**
4:         $t = l_y(x)\%c$; // note $t \in [0, c-1]$
5:         $s = \left[\frac{l_y(x)}{c}\right]$; // note $s \in [0, M-1]$
6:         node $x$ is allowed to transmit to channel $t$ in slot $kM + s$ in each second;
7:         node $y$ is listening to channel $t$ in slot $kM + s$ in each second;
8:     **end for**
9: **end for**

---

## 15.8 Thoughts for Practitioners

Scheduling has been proved an effective mechanism in wireless sensor networks. Since entirely interference-free scheduling requires a large number of subchannels, it would be interesting to jointly consider scheduling problem with routing scheme to optimize network performance based on the specific communication pattern required by applications.

## 15.9 Directions for Future Research

Energy efficiency is an important issue in sensor networking. Currently no many works have been done on integrating interference avoidance with duty cycling. Interference in wireless sensor networks is often described by graph–theoretic models, and therefore the scheduling problem usually boils down to variants of graph labeling problems, which have been extensively studied in the past several decades. However, these graph models does not capture the accumulated effect of interference from different transmitters. It will be promising to investigate the scheduling problem under more accurate models like the physical SINR model.

## 15.10  Conclusions

In this chapter, we examine the scheduling problem for various scenarios under different interference models. Topics investigated in this chapter include the channel allocation problem modeled by graph labeling with channel separation constraints, light scheduling that reduces the span of required subchannels by imposing constraints only on the communication links required by applications, duty cycling schemes which achieve energy efficiency by switching sensor nodes between active and sleep modes, and scheduling schemes optimized for specific applications. We also discuss scheduling under the protocol model and the SINR model. For each topic, we survey the results and one or two representative works are examined in details.

## Terminologies

*Interference.* Interference is caused by simultaneous transmissions to the same communication channel in a wireless receiver's proximity, resulting in damaged useless received packets.

*Scheduling.* A schedule of sensor nodes' activities is to specify the modes a sensor might stay at each of a sequence of time slots. Scheduling of sensor nodes' transmissions, that is, specifying whether a sensor node is allowed to transmit at a time slot, can be used to avoid interference. Another form of scheduling, called duty cycling, which switches sensor nodes between active and sleep modes, is an important mechanism to achieve energy efficiency in sensor networks.

*Allocation-based MAC protocols.* Allocation-based MAC protocols aim to avoid interference by scheduling nodes' transmissions. The basic idea is first to divide the shared communication channel into subchannels and then allocate the subchannels to sensor nodes.

*CDMA, FDMA.* CDMA and FDMA are allocation-based medium access technologies that enable multiple access to the shared communication channel by dividing the shared channel into subchannels using orthogonal modulation codes and carrier frequencies, respectively.

*Graph labeling.* A labeling on a graph is an assignment of integer values to the vertices so that certain constraints are satisfied. By letting distinct labels representing distinct subchannels, graph labeling can be used to model the channel allocation problem for an allocation-based MAC protocol. The constraints defined in a graph labeling are subject to the interference model that describes the condition for interference-free communication.

*Duty cycling.* Duty cycling is to switch sensor nodes between active and sleep modes. Such a mechanism can achieve energy efficiency in sensor networks where the traffic is low most of the time.

*Idle listening.* In sensor networks, a sensor node cannot tell when a message will be sent to it if no additional information is provided. In order not to miss packets, a sensor node must keep its radio active and listen to the shared communication channel; this is called as idle listening.

*Data aggregation.* In sensor networks designed for monitoring tasks, each sensor node generates data packets and transmits them to a set of special stations, called sinks, which are responsible for collecting data. In data aggregation, when a sensor node receives a data packet from its neighbor, it can merge this packet with its own data packet and the merged packet contains all the required information. Examples of such scenarios include environmental monitoring where the information of interest is the maximum temperature in the monitoring area; in this case, two (or more) pieces of data packets can be merged by taking their maximum.

*The protocol model.* In the protocol model, a parameter $\Gamma$ is used to model situations where a guard zone is required by the protocol to prevent a neighboring node from transmitting to the same channel at the same time. In this model, the transmission from a node, say $x$, to a node, say $y$, is interference free if, for any other node $z$ simultaneously transmitting to the same channel, we have $||z, y|| \geq (1 + \Gamma)||x, y||$, where $||x, y||$ is the geometric distance between nodes $x$ and $y$.

*The physical SINR model.* It is a model to describe the condition for interference-free communication in wireless networks. Let $\{x_0, \ldots, x_K\}$ be the set of nodes simultaneously transmitting at some time instant over the same channel. Let $P_i$ be the power level chosen by node $x_i$, $i \in [0, K]$, to transmit. Then the transmission from a node $x_i$, $i \in [0, K]$, is interference-free received by a node $x$ if

$$\frac{\frac{P_i}{||x_i - x||^\alpha}}{N + \sum_{k \in [0,k], k \neq i} \frac{P_k}{||x_k - x||^\alpha}} \geq \beta.$$

Here $\beta$ is the minimum SIR, $N$ is the ambient noise power level, and signal power decays with distance $r$ as $1/r^\alpha$.

## Questions

1. Design an $L(1,1)$-labeling on the graph in Fig. 15.2.
2. What is the lower bound on the labeling number of an $L(1,1)$-labeling on Fig. 15.2 according to Theorem 2?
3. Prove the labeling numbers $\lambda_{1,1}(G) = 2$ for a graph $G$ that has a path topology.
4. Give an $L_S(1,1)$-labelings on the graph in Fig. 15.4a, where $S$ is the directed tree routed at *sink* with tree edges marked by gray links.
5. Design a distributed greedy labeling scheme for $L_S(1,1)$-labelings on general graph $G$ for general subgraph $S$.
6. Prove that Algorithm 3 generates an $L_S(1, 1)$-labeling for data gathering, where $S$ is a directed breadth first search tree routed at the *sink* with edges toward the *sink*.

7. Prove the optimality of the slot assignment function for a tree presented in Sect. 15.5.2.
8. Prove Lemma 3 for the data aggregation scheduling presented in Sect. 15.6.2.
9. Prove Lemma 4 for the data aggregation scheduling presented in Sect. 15.6.2.
10. Design a scheduling for broadcasting problem, that is, propagating messages from a node, called *source*, to all the other nodes in a network. Use the interference model presented in Sect. 15.6.2.

# References

1. P. Gupta and P. Kumar. The capacity of wireless networks. IEEE Transactions on Information Theory, 46(2):388–404, 2000.
2. K. Kredo II and P. Mohapatra. Medium access control in wireless sensor networks. Computer Networks, 51(4):961–994, 2007.
3. T. Rappaport. Wirelss Communications, Principles and Practice. Prentice Hall, Upper Saddle River, NJ, 1996.
4. L. Wang and Y. Xiao. A survey of energy-efficient scheduling mechanisms in sensor networks. Mobile Networks and Applications, 11:723–740, 2006.
5. I. Katzela and M. Naghshineh. Channel assignment schemes for cellular mobile telecommunications: a comprehensive survey. IEEE Personal Communications, 3(3):10–31, 1996.
6. I. W. Akyildiz, Y. Su, Sankarasubramaniam, and F. Cayirci. Wireless sensor networks: a survey. Computer Networks, 38(4):393–422, 2002.
7. O. Dousse, P. Mannersalo, and P. Thiran. Latency of wireless sensor networks with uncoordinated power saving mechanisms. In Proceedings of Fifth ACM international symposium on mobile ad hoc networking & computing (MobiHoc), 2004.
8. G. Lu, N. Sadagopan, B. Krishnamachari, and A. Goel. Delay efficient sleep scheduling in wireless sensor networks. In Proceedings of IEEE INFOCOM 2005.
9. W. Ye, J. Heidemann, and D. Estrin. An energy-efficient MAC protocol for wireless sensor networks. In Proceedings of IEEE INFOCOM, 2002.
10. W. Ye, J. Heidemann, and D. Estrin. Medium access control with coordinated, adaptive sleeping for wireless sensor networks. Technical Report ISI-TR-567, USC, Jan. 2003.
11. Worldsens, http://worldsens.citi.insa-lyon.fr/.
12. Chipcon Inc. http://www.chipcon.com/.
13. E. Jung and N. Vaidya. An energy efficient MAC protocol for wireless LANs. In Proceedings IEEE INFOCOM, 2002.
14. S. Singh and C. Raghavendra. PAMAS: Power aware multi-access protocol with signalling for ad hoc networks. SIGCOMM Computer. Communication. Review., 28(3):5–26, 1998.
15. T. van Dam and K. Langendoen. An adaptive energy-efficient MAC protocol for wireless sensor networks. In ACM Sensys, 2003.
16. R. Zheng, J. Hou, and L. Sha. Asynchronous wakeup for ad hoc networks. In Proceedings of Fourth ACM International Symposium on Mobile Ad Hoc Networking and Computing (MobiHoc), 2003.
17. Q. Cao, T. Abdelzaher, T. He, and J. Stankovic. Towards optimal sleep scheduling in sensor networks for rare-event detection. In Proceedings of Fourth International Symposium on Information Processing in Sensor Networks (IPSN), 2005.
18. Y. Chen, E. Fleury, and V. Syrotiuk. Topology-transparent duty cycling for wireless sensor networks. In Proceedings of 21st IEEE International Parallel and Distributed Processing Symposium (IPDPS), 2007.
19. A. Keshavarzian, H. Lee, and L. Venkatraman. Wakeup scheduling in wireless sensor networks. In Proceedings of ACM International Symposium on Mobile Ad Hoc Networking and Computing (MobiHoc), 2006.

20. F. Koushanfar, N. Taft, and M. Potkonjak. Sleeping coordination for comprehensive sensing using isotonic regression and domatic partitions. In Proceedings of IEEE INFOCOM, 2006.
21. M. Miller and N. Vaidya. A MAC protocol to reduce sensor network energy consumption using a wakeup radio. IEEE Transactions on Mobile Computing, 4(3):228–242, 2005.
22. Ö. Akan and I. Akyildiz. Event-to-sink reliable transport in wireless sensor networks. IEEE/ACM Transactions on Networks., 13(5):1003–1016, 2005.
23. V. Rajendran, K. Obraczka, and J. Garcia-Luna-Aceves. Energy-efficient, collision-free medium access control for wireless sensor networks. Wireless Networks, 12(1):63–78, 2006.
24. M. Sichitiu. Corss-layer scheduling for power efficiency in wireless sensor networks. In Proceedings of IEEE INFOCOM, 2004.
25. R. Subramanian and F. Fekri. Sleep scheduling and lifetime maximization in sensor networks: fundamental limits and optimal solutions. In Proceedings of Fifth International Conference on Information Processing in Sensor Networks (IPSN), 2006.
26. N. Trigoni, Y. Yao, A. Demers, J. Gehrke, and R. Rajaraman. Wave scheduling and routing in sensor networks. ACM Transactions on Sensor Networks, 3(1):2, 2007.
27. P. Kyasanur and N. Vaidya. Capacity of multi-channel wireless networks: impact of number of channels and interfaces. In Proceedings of 11th Annual International Conference on Mobile Computing and Networking (MobiCom), 2005.
28. A. Keshavarz-Haddad, V. Ribeiro, and R. Riedi. Broadcast capacity in multihop wireless networks. In Proceedings of 12th Annual International Conference on Mobile Computing and Networking (MobiCom), 2006.
29. T. Moscibroda. The worst-case capacity of wireless sensor networks. In Proceedings of Sixth International Conference on Information Processing in Sensor Networks (IPSN), 2007.
30. T. Moscibroda, Y. Oswald, and R. Wattenhofer. How optimal are wireless scheduling protocols? In Proceedings of IEEE INFOCOM, 2007.
31. T. Moscibroda and R. Wattenhofer. The complexity of connectivity in wireless networks. In Proceedings of IEEE INFOCOM 2006.
32. T. Moscibroda, R. Wattenhofer, and A. Zollinger. Topology control meets SINR: the scheduling complexity of arbitrary topologies. In Proceedings of 12th Annual International Conference on Mobile Computing and Networking (MobiCom), 2006.
33. R. Battiti, A. Bertossi, and M. Bonuccelli. Assigning codes in wireless networks: bounds and scaling properties. Wirelen Networks, 5(3):195–209, 1999.
34. A. Bertossi, C. Pinotti, and R. Tan. Efficient use of radio spectrum in wireless networks with channel separation between close stations. In Proceedings of Fourth international Workshop on Discrete Algorithms and Methods for Mobile Computing and Communications (DIALM), 2000.
35. G. Chang, W. Ke, D. Kuo, D. Liu, and R. Yeh. On $L(d,1)$-labeling of graphs. Discrete Mathematics, 220:57–66, 2000.
36. I. Chlamtac and S. Pinter. Distributed nodes organization algorithm for channel access in a multihop dynamic radio network. IEEE Transactions on Computing, 36(6):728–737, 1987.
37. J. Georges and D. Mauro. Labeling trees with a condition at distance two. Discrete Mathematics, 269:127–148, 2003.
38. W. Hale. Frequency assignment: theory and application. In Proceedings of IEEE, volume 68, pp. 1497–1514, 1980.
39. J. Griggs and R.Yeh. Labeling graphs with a condition at distance 2. SIAM Journal on Discrete Mathematics, 5:586–595, 1992.
40. W. Wang, Y. Wang, X. Li, W. Song, and O. Frieder. Efficient interference-aware TDMA link scheduling for static wireless networks. In Proceedings of 12th Annual International Conference on Mobile Computing and Networking (MobiCom), 2006.
41. Y. Chen and E. Fleury. Backbone-based scheduling for data dissemination in wireless sensor networks with mobile sinks. In Proceedings of Fourth ACM SIGACT-SIGOPS International Workshop on Foundations of Mobile Computing (DIAL M-POMC), 2007.
42. IEEE Standard for Wireless LAN Medium Access Control and Physical Layer Specification, 802.11. 1999.

43. S. Kapp. 802.11a. more bandwidth without the wires. Internet Computing, IEEE, 6(4):75–79, 2002.
44. C. Chang and D. Kuo. The $L(2,1)$-labeling on graphs. SIAM Journal on Discrete Mathematics, 9:309–316, 1996.
45. C. McDiarmid and B. Reed. Channel assignment and weighted coloring. Networks, 36(2):114–117, 2000.
46. B. Clark, C. Colbourn, and D. Johnson. Unit disk graphs. Discrete Mathematics, 86(1-3):165–177, 1990.
47. A. Gräf, M. Stumpf, and G. Weißenfels. On coloring unit disk graphs. Algorithmica, 20(3):277–293, 1998.
48. C. Lund and M. Yannakakis. On the hardness of approximating minimization problems. Journal of the ACM, 41(5):960–981, 1994.
49. A. Bertossi and M. Bonuccelli. Code assignment for hidden terminal interference avoidance in multihop packet radio networks. IEEE/ACM Transactions on Networks, 3(4):441–449, 1995.
50. H. Bodlaender, T. Kloks, R. Tan, and J. Leeuwen. Approximations for λ-coloring of graphs. In Proceedings STACS, 2000.
51. A. Gamst. Some lower bounds for a class of frequency assignment problems. IEEE Transactions on Vehiculor Technology, 35(1):8–14, 1986.
52. C. Sung and W. Wong. A graph theoretic approach to the channel assignment problem in cellular systems. In Proceedings of IEEE 45th Vehicular Technology Conference, 1995.
53. D. Smith and S. Hurley. Bounds for the frequency assignment problem. Discrete Mathematics, 167–168:571–582, 1997.
54. J. Janssen and K. Kilakos. Tile covers, closed tours and the radio spectrum. Telecommunications Network Planning. Kluwer, Boston, MA, 1999.
55. J. Janssen, T. Wentzell, and S. Fitzpatrick. Lower bounds from tile covers for the channel assignment problem. SIAM Journal on Discrete Mathematics, 18(4):679–696, 2005.
56. R. Yeh. Labeling graphs with a condition at distance two. PhD Thesis, University of South Carolina, 1990.
57. D. Král and R. Skrekovski. A theorem about the channel assignment problem. SIAM Jorunal on Discrete Mathematics, 16(3):426–437, 2003.
58. D. Goncalves. On the l(p,1)-labeling of graphs. Discrete Mathematics and Theoretical Computer Science, AE:81–86, 2005.
59. A. Bertossi and C. Pinotti. Mappings for conflict-free access of paths in bidimensional arrays, circular lists, and complete trees. Journal of Paralled and Distibuted Computing, 62(8):1314–1333, 2002.
60. J. Georges and D. Mauro. Generalized vertex labelings with a condition at distance two. Congressus Numerantium, 109:141–159, 1995.
61. T. Calamoneri and R. Petreschi. $L(h, 1)$-labeling subclasses of planar graphs. Journal of Parallel and Distributed Computing, 64:414–426, 2004.
62. J. Janssen. Channel Assignment and Graph Labeling. Wiley., New York, NY, 2002.
63. R. Yeh. A survey on labeling graphs with a condition at distance two. Discrete Mathematics, 306:1217–1231, 2006.
64. Y. Chen and E. Fleury. A distributed policy scheduling for wireless sensor networks. In Proceedings of IEEE INFOCOM, 2007.
65. N. Li, J. Hou, and L. Sha. Design and analysis of an MST-based topology control algorithm. In Proceedings of IEEE INFOCOM, 2003.
66. B. Awerbuch and R. Gallager. A new distributed algorithm to find breadth first search trees. IEEE Transactions on Information Theory, 33(3):315–322, 1987.
67. O. Kasten. Energy Consumption, http://www.inf.ethz.ch/ kasten/research/bathtub/energy consumption.html.
68. M. Stemm and R. Katz. Measuring and reducing energy consumption of networks interfaces in hand-held devices. IEICE Transactions on Communications, E80-B(8):1125–1131, 1997.
69. S. Ganeriwal, D. Ganesan, H. Shim, V. Tsiatsis, and M. B. Srivastava. Estimating clock uncertainty for efficient duty-cycling in sensor networks. In Proceedings of Third International Conference on Embedded Networked Sensor Systems (SenSys), 2005.

70. G. Werner-Allen, G. Tewari, A. Patel, M. Welsh, and R. Nagpal. Firefly-inspired sensor network synchronicity with realistic radio effects. In Proceedings of Third International Conference on Embedded Networked Sensor Systems (SenSys), 2005.

71. W. Ye, J. Heidemann, and D. Estrin. Medium access control with coordinated adaptive sleeping for wireless sensor networks. IEEE/ACM Transactions on Networks, 12(3):493–506, 2004.

72. A. El-Hoiydi and J. Decotignie. WiseMAC: An ultra low power MAC protocol for multi-hop wireless sensor networks. In ALGOSENSORS, 2004.

73. J. Hill and D. Culler. MICA: A wireless platform for deeply embedded networks. IEEE Micro, 22(6):12–24, 2002.

74. J. Polastre, J. Hill, and D. Culler. Versatile low power media access for wireless sensor networks. In Proceedings of Second International Conference on Embedded Networked Sensor Systems (SenSys), 2004.

75. S. Huang, P. Wan, C. Vu, Y. Li, and F. Yao. Nearly constant approximation for data aggregation scheduling in wireless sensor networks. In Proceedings of IEEE INFOCOM, 2007.

76. G. Lu, B. Krishnamachari, and C. Raghavendra. An adaptive energy-efficient and low-latency MAC for data gathering in wireless sensor networks. In Proceedings of 18th International Parallel and Distributed Processing symposium (IPDPS), 2004.

77. A. Gamal, J. Mammen, B. Prabhakar, and D. Shah. Throughput-delay trade-off in wireless networks. In Proceedings of IEEE INFOCOM, 2004.

78. D. West. Introduction to Graph Theory. Prentice Hall, Upper Saddle River, NJ, 2001.

# Chapter 16
# Energy-Efficient Medium Access Control in Wireless Sensor Networks

**Gang Li and Robin Doss**

**Abstract** Medium access control for wireless sensor networks has been an active research area in the past decade. This chapter discusses a set of important medium access control (MAC) attributes and possible design trade-offs in protocol design, with an emphasis on energy efficiency. Then we categorize existing MAC protocols into five groups, outline the representative protocols, and compare their advantages and disadvantages in the context of wireless sensor network. Finally, thoughts for practitioners are presented and open research issues are also discussed.

## 16.1 Introduction

Advances in wireless communication, low-power electronics, and low-power radio frequency (RF) design have enabled the development of low-power sensor nodes with integrated sensing, processing, and wireless communication capabilities. These tiny sensor nodes self-organize into a network and collaborate to accomplish a common task such as industrial sensing, environment monitoring, or asset tracking. Typically, communication is achieved by multihop wireless communication between the sensor nodes and a central point commonly referred to as a *sink*. However, other communication patterns such as one-to-one communication between sensor nodes are also observed in many emerging applications [1]. One of the main challenges in sensor network communication relates to the energy-efficient sharing of the wireless channel. The broadcast nature of the wireless medium requires that singularity is established between a sender and receiver if successful communication is to be achieved. The medium access control (MAC) protocol has the responsibility of enforcing this singularity in a network where multiple nodes have data to transmit.

Gang Li (✉)
School of Information Technology, Deakin University, VIC 3125, Australia
e-mail: gang.li@deakin.edu.au

S. Misra et al. (eds.), *Guide to Wireless Sensor Networks*, Computer Communications and Networks, DOI: 10.1007/978-1-84882-218-4_16,

MAC protocols have been extensively investigated in wireless voice and data communications. In the 1970s, the ALOHA protocol [2] was developed for packet radio networks. The emergence of wireless local area networks (WLAN) in the 1990s saw renewed interest in the development of MAC protocols. A well-cited survey on MAC protocols for ad hoc networks can be found in [3]. Generally, we can observe two major strategies for MAC protocol design in wireless networks. They are schedule-based and contention-based protocols [4].

- *Schedule-based MACs.* This class of MAC protocols is based on reservation and scheduling. Typically a central authority (such as an access point) regulates the access to the medium by broadcasting a schedule that specifies when, and for how long, each node may transmit over the shared medium. An example of such a schedule-based MAC protocol is the time-division multiple access (TDMA) protocol. As in Fig. 16.1, TDMA divides the channel into individual time slots, which are then grouped into frames. In each time slot, only one node is allowed to transmit. The access point schedules which time slot is to be used by which node, and this decision can be made on a per frame or multiple frame basis. TDMA is frequently used in cellular wireless communication systems, and the mobile nodes communicate only with the base station; there is no peer-to-peer communications between nodes. One of the main advantages of TDMA is that TDMA is intrinsically energy-efficient. A sensor node can turn off its radio during all time slots in a frame during which it is not engaged in communication with the access point. However, TDMA requires the nodes to form clusters, analogous to the cells in the cellular communication systems. In addition, the accurate time synchronization between the sensor nodes and the access point requires that a node wakes up exactly at the start of "its" time slot. Moreover, when the number of nodes in a cluster changes, it is difficult for a TDMA protocol to dynamically change its frame length and time slot assignment.

- *Contention-based MACs.* This class of MAC protocols do not use channel division (in time or frequency), reservation, or scheduling. Instead, access to the

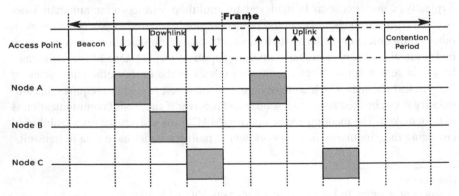

**Fig. 16.1** TDMA communication: Each node is allocated a fixed time slot within each frame. Communication is inherently collision free and energy-efficient

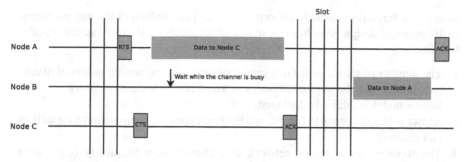

**Fig. 16.2** CSMA communication: Data transmission follows the RTS/CTS exchange with a high probability of the communication being collision-free

shared medium is contention-based and allocated to a node that has data to transmit in an on-demand fashion. An example of such a contention-based scheme is the carrier sense multiple access (CSMA) protocol. In CSMA-based protocols a transmitter senses the wireless channel before transmitting. If no on-going communication is detected, it assumes that the medium is clear and begins data transmission. There are several variants of CSMA, including nonpersistent, 1-persistent, and p-persistent CSMA [5]. When all nodes in the network can sense each other's transmission, CSMA performs efficiently. However, in a multihop wireless network, CSMA suffers from hidden terminal problems that lead to receiver collisions [5]. The CSMA/CA protocol, where CA stands for collision avoidance, was introduced to address the hidden terminal problem, and is adopted by the IEEE 802.11 wireless LAN standard [6]. CSMA/CA introduces a three-way handshake to make hidden nodes aware of the upcoming transmissions, thus preventing collisions. As in Fig. 16.2, the handshake starts with the sender initiating a short request-to-send (RTS) frame to the intended receiver. The receiver responds with a clear-to-send (CTS) frame, which informs its neighbors of the upcoming transfer. The reception of the CTS frame starts the data transmission at the sender. A successful RTS/CTS handshake thus ensures that there is a high probability that the communication is collision-free.

The unique characteristics and requirements of sensor networks require a rethink of current MAC techniques. First, traditional wireless MAC protocols were designed and optimized for single-hop wireless networks (e.g., WLANs), and their application to multihop communication is not straightforward. Second, and more important, the energy and complexity constraints of wireless sensor networks prevent the use of established wireless MAC protocols that do not aim to optimize energy and require the development of innovative energy-efficient solutions. Sensor nodes are battery powered, and in-network replenishment of battery is impossible. Third, a sensor network is composed of hundreds to several thousands of nodes spread throughout the sensor field and is often deployed in an ad hoc fashion rather than with careful preplanning. Consequently, the MAC protocol needs to lend itself effectively to network formation through self-organization. The challenges to energy-efficient MAC protocol design arise from these and other factors and can be summarized as follows:

1. The number of sensor nodes in a sensor network can be several orders of magnitude higher than the number of nodes in an ad hoc or wireless network.
2. Sensor nodes are densely deployed.
3. Sensor nodes are prone to failure, and limited in power, computational capacities, and memory.
4. The topology of a sensor network can change very frequently (e.g., when nodes die).
5. Sensor nodes may not have global identification because of the large amount of overhead and large number of sensors.
6. The sensing events can be sporadic leading to extremely bursty traffic in the sensor network.

In this chapter, we present a survey of MAC protocols for wireless sensor networks. Our aim is to provide an overall understanding of the present research issues in this field, and to summarize existing approaches to address these problems. We also provide some future directions on open research issues that have not been investigated adequately. For a more detailed technical specification of popular MAC protocols for sensor networks we refer the reader to [8] and [9].

The remainder of this chapter is organized as follows: In Sect. 16.2, we discuss the MAC layer-related sensor network properties including an analysis of major causes of energy waste in sensor networks. Section 16.3 reviews major MAC protocols for sensor networks, and finally in Sect. 16.4, we conclude the chapter with some insight into research issues for future MAC protocol design.

## 16.2  Background

The main function of the medium access control (MAC) layer is to ensure efficient usage of the physical medium by the nodes of a network, while providing error-free data transfer to the network layer above it. MAC protocols are influenced by a number of constraints. Consequently, in protocol design there is required a trade-off among several, often contradictory factors, such as quality of service, throughput, and energy efficiency. In wireless sensor networks, the MAC protocol must achieve the following two goals [7]:

1. Create the network infrastructure: Because thousands of sensor nodes are densely deployed, the MAC layer must establish the basic infrastructure needed for hop by hop communication and give the sensor network self-organizing ability.
2. Allow fair and efficient sharing of the wireless communication medium between sensor nodes.

The following sections discuss some important factors that need to be considered for MAC protocol design in order to meet the energy and complexity constraints of sensor networks and its applications.

## 16.2.1   Energy Efficiency

It is envisioned that sensor nodes will be cheap enough to be discarded rather than recharged and are efficient enough to operate with only ambient power sources (sunlight, vibrations, etc.). These energy constraints require energy efficiency to be a primary design goal for MAC protocols in sensor networks. The wireless sensor node, being a microelectronic device, can only be equipped with a limited power source, and run by battery. The operational lifetime of a sensor node is thus dependent on battery lifetime. The high level of redundancy observed in dense sensor network deployments ensures that the failure of one or even a few sensor nodes may not harm the overall functioning of a sensor network. However, a conservative estimate of network lifetime is defined as the time at which the first node in the network exhausts all its energy. In practice though, the lifetime of the network can be defined as the time at which the network experiences network partitioning. Irrespective of the definition of network lifetime the goal remains to optimize the lifetime of the network, and the role of the MAC protocol in achieving this energy efficiency is vital.

On many hardware platforms, the radio is a major energy consumer. The MAC layer directly controls radio activities, and its energy efficiency is thus a very important performance metric, directly influencing the network lifetime. Most sensor network applications involve data gathering. Consequently, the main task of a sensor node in the sensor network is to detect data/events, perform some preliminary data processing, and then transmit the data. The energy consumption can hence be attributed to data sensing, data processing, and data communication. Of the three factors a sensor node expends maximum energy in data communication. This involves both data transmission and reception with the ratio between energy consumption in data reception and transmission being 1:2.5 [10].

In [7] the following sources of energy waste have been identified for contention-based CSMA style MAC protocols.

- *Idle listening*. Since a node does not know when it will be the receiver of a message, it will need to keep its radio in receive mode. In sensor network applications where the traffic is light, this is a major source of energy waste, since typical radios consume two orders of magnitude more energy in receive mode than in standby mode, even though no receiving is happening.
- *Collisions*. When a sensor node receives more than one packet at the same time, these packets will be corrupted and must be discarded. Hence, the energy used during transmission and reception is wasted. If *Automatic Repeat reQuest* (ARQ) is employed, the follow-on retransmission of these packets will increase the energy consumption. Clearly, collision is a major problem in contention-based

MAC protocols The RTS-CTS handshake can resolve the collision problem for unicast messages, but this is at the expense of increased protocol overhead that will increase energy consumption.

- *Overhearing*. Since the radio channel is a shared medium, a sensor node may receive packets that are destined to other nodes. Overhearing can be a dominant factor of energy waste when traffic is heavy and sensor nodes are densely deployed. Adaptive power control that limits the interference range of sensor nodes needs to be considered.
- *Protocol overhead*. The MAC headers and control packets used for signaling do not directly convey data and are therefore considered as overhead; sending, receiving, and listening for these protocol overheads consume energy. In many applications where only a few byes of data are transmitted in each message, these overheads can be significant.
- *Traffic fluctuations*. A sensed event will lead to a sudden peak in the sensor network traffic and increase the probability of a collision. More importantly, the follow-on random back-off procedure will increase latency and consume energy. When the traffic load approaches the channel capacity, the performance can collapse with little data being delivered while the radio is consuming a lot of energy by repeatedly sensing to identify a clear channel.

It can be argued that since the radio consumes the most energy, an obvious means of energy conservation is to turn the radio off when it is not required. However this can prove to be suboptimal and increase rather than decrease energy consumption. Since sensor nodes communicate using short data packets the data communication energy is dominated by the radio startup energy in most deployments. Hence, turning the radio off during each idle period will result in negative energy gains. This requires that a well-designed MAC protocol should achieve energy efficiency by finding the right balance between smart radio control and efficient protocol design.

### 16.2.2   MAC Performance

The energy constraints of wireless sensor nodes require that energy efficiency is one of the primary design goals for MAC protocols in sensor networks, but it cannot be the only design goal. For efficient operation sensor networks will need to provide certain quality of service guarantees that satisfy traditional performance indicators. Some important factors that need to be considered and incorporated into the protocol design are as follows [11]:

- *Effective collision avoidance*. Collision avoidance is the core task for all MAC protocols. It determines when and how a sensor node can access the shared medium and send its data. Contention-based MAC protocols allow some level of collisions, but the avoidance of repeated collisions is an important factor that should be considered in protocol design.
- *Scalability and adaptivity*. In a sensor field, the number of sensor nodes may be of the order of hundreds or thousands. Since sensor networks consist of

low-power volatile nodes, it is likely that links will appear or disappear over time, and sensor nodes may join or leave the network, or the size of the neighborhood will change due to changes in the physical environment. A good MAC protocol should accommodate changes in network size, node density, and topology gracefully. Scalability and adaptivity are two important MAC attributes for sensor networks, which are deployed without preplanning, and often operate in uncertain environments.

- *Efficient channel utilization*. Channel utilization is a traditional MAC protocol metric that illustrates protocol efficiency, and it reflects how well the available bandwidth of the channel is utilized in communication. High channel utilization is critical for delivering a large number of packets with minimum latency.

- *Latency*. Latency is the delay from when a sensor node has a packet to send until the packet is successfully received by the receiving node. Different applications have varying emphasis on latency. Monitoring applications can usually tolerate some additional message latency, since the network speed is typically much faster than the object speed, and the object speed places a bound on how rapidly the network must react. Latency is usually considered as a less important attribute.

- *Throughput*. It refers to the amount of data successfully transferred from a sender to a receiver in a given time. Similar to latency, the importance of throughput depends on the application. Applications that demand long lifetime often accept longer latency and lower throughput values.

- *Fairness*. In traditional communication networks, fairness is an important attribute that reflects the ability of different nodes to share the medium equally. However, fairness is usually not a design goal, because all sensor nodes share a common task and cooperate toward it. Accordingly, in sensor networks, success is measured by the performance of the application as a whole rather than ensuring that each node receives a fair share of the medium.

The design of MAC protocols for wireless sensor networks is a nontrivial task as there are conflicting constraints that need to be satisfied. As noted earlier there are numerous contributing factors that need to be considered. However, for wireless sensor networks, the most important design goals can be identified as energy efficiency, effective collision avoidance, scalability, and adaptivity. Other attributes while important are not critical and hence can be considered as secondary goals.

## 16.3   MAC Protocols for Sensor Networks

We now proceed with describing briefly some of the MAC protocols developed for sensor networks. Similar to general MAC protocols, MAC protocols for sensor networks can also be broadly divided into two major groups: schedule-based and contention-based. However, most sensor network MAC protocols are hybrid protocols that seek to exploit the inherent advantages of both the schedule-based and the contention-based strategies. In this section, we review these protocols based on their intrinsic similarities and identify five classes of sensor network MAC protocols.

### 16.3.1 Contention-Based MAC Protocols

This group of MAC protocols is more close to the contention-based CSMA/CA protocol in which nodes can start a transmission at any random moment and must contend for the channel. Without having to maintain and share schedule, this kind of MAC protocol will consume less processing resources and be more flexible to network scale. The main challenge with contention-based protocols is to reduce the energy consumption caused by collisions, overhearing, and idle listening.

CSMA/CA features have been adopted in IEEE 802.11 standards for wireless ad hoc network. However, for wireless sensor networks, the main drawback is the energy wasted by idle listening. Preamble sampling [12] and low power listening [13] are two MAC protocols that aim to reduce this idle-listening energy waste. The common technique shared by these two protocols is a low-level carrier sense method, which effectively turns off the radio without losing any incoming data. This is shown in Fig. 16.3. It operates at the physical layer and starts off the PHY header with a preamble that is used to notify receivers of the transfer and allows receivers to adjust their circuitry to the current channel conditions. Following the preamble a startbyte is used to signal the beginning of the data transfer. This efficient carrier-sense technique allows the receiver nodes to periodically turn on the radio to detect if a preamble is present: if present, the node listens until the startbyte arrives and the message can be received; otherwise, the radio is turned off until the next sample. Using this technique, the receiver node can save a lot of energy since it is periodically in the sleep mode until the preamble is sensed, and the cost is mainly on the sending nodes, which have to increase the length of the preamble until data can be transmitted. It can also be applied to other contention-based MAC protocols. Preamble sampling [12] can be considered as a combination of this efficient carrier-sense technique and the ALOHA protocol, while low power listening [13] is a combination of this carrier-sense technique and the CSMA protocol.

WiseMAC extends the Preamble Sampling protocol by having sensor nodes that maintain the schedule offsets of their neighbors through piggybacked information on the ACK of the underlying CSMA protocol [14]. Based on the neighbor's sleep

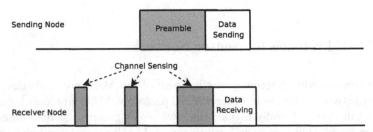

**Fig. 16.3** Low-level carrier sense technique: Preamble sampling and low power listening both use this common technique to effectively turn off the radio without losing any incoming data

schedule table, WiseMAC schedules transmissions so that the receiving node's sampling time corresponds to the middle of the sender's preamble. Another parameter affecting the choice of the preamble length is the potential clock drift between the sender node and the receiver node. When a node needs to send a message to a specific neighbor, it uses the neighbor's schedule offset table to decide when to start transmitting the preamble, and to account for any clock drift, the preamble is extended with a time proportional to the length of the interval since the last message exchange. The overall effect of these measures is that the WiseMAC protocol achieves better performance under variable traffic conditions: when traffic is light, WiseMAC uses long preambles and consumes low power since the cost of receiver nodes dominates the consumption; while when traffic is heavy, WiseMAC uses short preambles and operates energy efficiently since overheads are minimized. WiseMac operation is illustrated in Fig. 16.4.

Woo and Culler proposed a CSMA-based MAC protocol for convergecasting sensor network applications such as remote environmental monitoring [15]. It assumes that the base station tries to collect data equally from all sensor nodes in the field. Considering the fact that sensor nodes close to the base station carry more traffic because of forwarding more data from other nodes, the MAC protocol combines CSMA with an adaptive rate control mechanism: each sensor node dynamically adjusts its rate of injecting its original packets to the network, and linearly increases the rate if a packet is injected successfully, otherwise the rate is multiplicatively decreased. Using this measure, it achieves a fair bandwidth allocation to all nodes in the network.

A further improvement over the WiseMAC protocol is the CSMA with minimal preamble sampling (CSMA-MPS) that improves energy efficiency and reduces latency [16]. The sending nodes alternate between transmitting small control messages and listening for a response from the receiving nodes. Energy efficiency is achieved by using small control messages that allow the sending node to estimate the sampling offset of the receiving node. In doing so, a neighbor's sampling offset can be learnt without requiring extra fields in the ACK messages.

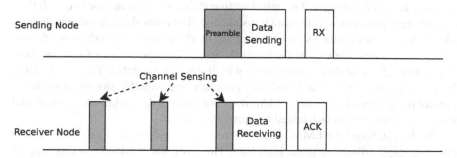

**Fig. 16.4** WiseMAC: Based on the neighbor's sleep schedule table, WiseMAC schedules transmissions so that the receiving node's sampling time corresponds to the middle of the sender's preamble

## 16.3.2 TDMA Variants

TDMA has attracted the interest of sensor network researchers as it is inherently collision free, and idle listening can be effectively eliminated since nodes are aware of when to expect incoming data. However, for sensor networks, the TDMA protocol has some limitations: one of the biggest challenges is to adapt TDMA to operate efficiently in sensor networks where there is no infrastructure. TDMA requires nodes to form clusters with one of the nodes in the cluster required to act as the base station (or cluster head). Further, TDMA usually assumes direct single-hop communication with the base station, rather than indirect multihop communication with peer nodes. Moreover, the scalability of TDMA is limited; when nodes leave the cluster or new nodes join, the frame length will have to be adjusted. More importantly, intercluster communications and intercluster interference are serious problems that need to be handled in sensor networks. Various solutions have been proposed to extend the basic TDMA protocol in different ways to accommodate the requirements of sensor networks.

van Hoesel et al. proposed a set of TDMA-based sensor network protocols including EMACS [17], LMAC [18], and AI-LMAC [19].

In the EMACS protocol, each time slot is divided into three sections: a contention phase, a traffic control section, and a data section [17]. Based on the assumption that only some nodes need to form a backbone network infrastructure, nodes are divided into active ones, passive ones, and dormant ones. An active node owns a slot and can transmit in its own slot, while a passive node does not own a slot and can only transmit after requesting a slot from an active node. Dormant nodes sleep until they wake up to participate in an active or passive role. Slot numbers are assigned uniquely within a two-hop neighborhood, which allows slot reuse at noninterfering distances.

LMAC improves EMACS by allowing all sensor nodes to own a slot, i.e., all nodes are active. Accordingly, the contention phase becomes unnecessary [18]. Energy efficiency is increased as nodes can turn off their radio during the data part if they are not the intended receiver. When a node wants to transmit, it broadcasts a message header detailing in the control section the destination and the length. It then immediately proceeds with the data transmission. The control section also includes a bit-set that specifies which slots are occupied by the sender's immediate neighbors. Thus, when a new node joins the network, it is sufficient to listen for a complete frame from all traffic flows to determine which slots are available. This is calculated by a simple OR operation and randomly claiming one available slot by transmitting control information in that slot. LMAC does not provide data acknowledgement and hence reliability has to be handled by upper layers.

The EMACS and the LMAC protocols provide simple time-slots allocation and can have high utilization under high load. However, the weakness of this method is that the network setup will take a lot of time, especially for large-scale applications where the setup starts from the access point and slot collisions may take several frames to resolve. Moreover, the sensor nodes can not adapt slot ownerships according to varying traffic conditions.

AI-LMAC improves upon LMAC by varying the number of slots a node owns based on traffic conditions [19]. Each node maintains a Data Distribution Table to record simple statistics on the data generated and forwarded by it. Based on this and other statistical information the number of slots assigned to a node can be increased or decreased. To further conserve energy, a node only transmits a control message in the first time slot it owns within a frame. The AI-LMAC control message also provides acknowledgements on the correct reception of the data. The drawbacks of AI-LMAC's adaptive mechanism lie in the maintenance of the Data Distribution Table as it consumes computational and energy resources. Further nodes must always listen to the control sections of all slots in a frame, even the unused ones, since new nodes may join the network.

Zebra-MAC (Z-MAC) is another TDMA variant, which assigns sensor nodes a time slot, but allows nodes to utilize slots they do not own through a CSMA mechanism with prioritized backoff times [20]. It can be considered as a hybrid of contention-based and schedule-based protocols. When the traffic in the network is light, the Z-MAC protocol performs similar to CSMA, and when the traffic is heavy, it approximates a strict TDMA scheme that can save a large amount of energy. The energy efficiency comes at the cost of protocol overhead, primarily caused by the TDMA structure. Similar to any TDMA protocol, synchronization has to be maintained, which introduces energy overheads.

The Traffic-Adaptive Medium Access (TRAMA) protocol is another TDMA-based protocol [21]. Similar to Z-MAC it also aims to increase the utilization of TDMA in an energy-efficient manner by combining the schedule-based approach and the contention-based approaches. Sensor nodes regularly broadcast the identities of their immediate neighbors and information about traffic flows routed through them. This ensures that each node can know its two-hop neighbors and the demands of its immediate neighbors. By means of a distributed hash function, it is possible to calculate the winning node for each slot based on the node identities and slot number. Nodes with little traffic may release their slots for the remainder of the frame for use by other nodes with heavy traffic. By trading-off latency and algorithm complexity, TRAMA achieves high channel utilization.

### 16.3.3  S-MAC and Its Variants

S-MAC is a MAC protocol specifically designed for wireless sensor networks [11]. It is perhaps the most studied scheduled MAC protocol for sensor networks and has been extended in different ways since it was proposed in 2004. S-MAC is based on the assumption that nodes are dedicated to a single application, which can tolerate some latency, and it has long idle periods. S-MAC introduces a technique called virtual clustering to achieve locally managed synchronizations and uses a coarse-grained sleep/wakeup cycle to allow nodes to spend most of their time in sleep mode. Nodes have the freedom to choose their own listen/sleep schedules, which is then shared with their neighbors so that the peer-to-peer communication will be possible. To prevent long-term clock drift, periodical SYNC packets are sent to immediate neighbors, so that nodes receiving these packets can adjust their clocks to compensate for any drift. The SYNC packets allow nodes to learn their neighbors' schedules so that they can wake up at the proper time to transmit a message. They also enable new nodes to join the network.

Data transmission occurs using the traditional RTS-CTS-DATA-ACK sequence to limit collisions and avoid the hidden terminal problem. Each packet also contains a duration field, which indicates the time needed for the current transmission. Consequently if a node overhears any packet, it knows how long it needs to keep silent. In this case, S-MAC puts the node into the sleep state for that amount of time so that energy waste due to overhearing can be avoided. Further, an important feature of S-MAC is the concept of message-passing, an optimization that allows multiple frames from a message to be sent in a burst. In message-passing, a single RTS and CTS exchange is used to reserve the medium for the entire time needed for transmitting all the fragments belonging to a message, thereby reducing the overhead. This reduction in the communication overhead results in energy savings. However, each fragment is separately acknowledged to enable selective retransmission if necessary.

To decrease the latency for delay-sensitive applications, the Dynamic Sensor-MAC (DSMAC) extends S-MAC by allowing sensor nodes to adopt dynamic duty cycles based on traffic and energy considerations [22]. Utilizing added fields in the SYNC period, all nodes share their one-hop latency values. All nodes start with the same duty cycle. However, when a receiving node notices that the average one-hop latency value is high, it decides to shorten its sleep time and announce it within the SYNC period. After the sender receives this decreased sleep period signal, it attempts to increase its duty cycle. To ensure that sensor nodes within the same virtual cluster remain synchronized, any increases to the duty cycle occur as multiplicative powers of 2.

Another extension to S-MAC is the Timeout-MAC (T-MAC) protocol that extends S-MAC to improve its performance with respect to latency and throughput under variable traffic load [23]. It uses a timer to indicate the end of the active period instead of relying on a fixed duty cycle schedule. This measure frees the application from the burden of selecting an appropriate duty cycle and also saves energy by lowering the amount of time spent in idle listening. It also adapts to changes in traffic conditions. The adaptive duty cycle allows T-MAC to automatically adjust to variations in network traffic. However, this has the drawback of desynchronization of the listen periods between nodes within the virtual clusters, which results in the early sleeping problem that limits the number of hops a message can travel in each frame time.

AC-MAC and MS-MAC are further variants of S-MAC. AC-MAC is an alternative approach to extend S-MAC [24]. Instead of using an adaptive duty cycle, AC-MAC allows sensor nodes with many buffered messages to introduce multiple data exchange periods per SYNC frame. The sender node includes a value proportional to its occupied buffer capacity in the first RTS message of a SYNC frame. Based on this value the receiving node calculates the duty cycle to use within the virtual cluster for the current SYNC period. MS-MAC extends S-MAC protocol into the mobile sensor networks [25]. Each sensor node records the received signal strength value for its neighbor and uses any changes as indications of movement. MS-MAC trades energy consumption for faster schedule synchronization: nodes with a high mobility and their neighbors look for additional schedules much more frequently and adopt schedules with a lower latency.

## 16.3.4   Self-Organizing MAC Protocols

Self-organizing features such as clustering or grouping allow protocols to scale more easily since the protocol can view the whole cluster as a single entity. In addition, intercluster and intracluster traffic can be differentiated contributing to further energy efficiency. S-MAC and its variants can also be considered as self-organizing protocols, as they self-organize into virtual clusters and synchronize the sleep schedules of neighboring sensor nodes.

   The Low-Energy Adaptive Clustering Hierarchy (LEACH) was designed for sensor network applications such as remote monitoring [26] where data from individual sensor nodes is sent to a central sink. To conserve energy, LEACH groups nodes into clusters where a special node called the cluster head coordinates the cluster and forwards data generated within the cluster. Once a cluster forms, the cluster head computes a schedule and distributes it to the nodes it controls. To increase network lifetime, the role of the cluster head is rotated among nodes within the cluster so that the energy of a single node is not exhausted. Nodes send data to the cluster head in their own time slot and rely on the cluster head to transmit the data to the base station. The cluster head can also perform some message aggregation (or remove any redundancy) so that each cluster produces traffic equivalent to a single node. The intracluster communication is done through direct sequence spread spectrum (DSSS) communication to alleviate interference with other clusters. Further, cluster heads use a reserved sequence for communication with the base station.

   LEACH requires that each cluster head can directly communicate with the base station. Further, since the role of the cluster head is rotated within the cluster, this in effect means that all sensor nodes should be able to communicate with the base station. The GANGS protocol avoids this problem by forcing the cluster heads into a routing backbone [27]. The clusters are formed in two steps: First, the initial cluster head is selected according to the available energy resources, and any node with remaining energy can declare itself a cluster head and transmit a message announcing it; the second step will connect those cluster heads together so that the whole network is connected. To assign slots, the cluster heads execute a distributed algorithm, which results in each cluster head having a slot to transmit in and learning the slots of each neighbor. The cluster heads communicate with each other based on a TDMA schedule. After the cluster heads determine the TDMA schedule, they disseminate the information within the cluster so that other nodes may use the available slots at the end of the frame to send their data. Similar to LEACH, the GANGS protocol needs extra time and energy for the cluster formation and cluster head rotation, but

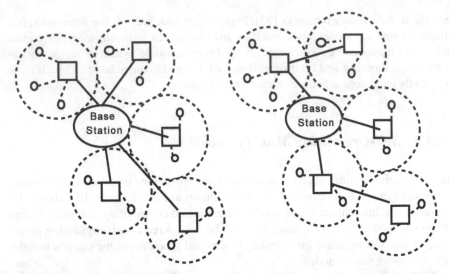

**Fig. 16.5** LEACH and GANGS mechanism: In LEACH all cluster heads are required to be able to communicate directly with the base station. GANGS allow cluster heads to be more than one hop away from the base station

it provides contention-free traffic flow for forwarded traffic while retaining the simplicity of a random access protocol within the clusters. The architecture adopted by LEACH and GANGS is shown in Fig. 16.5.

Group TDMA is an alternative protocol that aims to limit collisions and provide improved channel utilization by dividing nodes into groups [28]. Different from other self-organizing protocols, at each time, Group TDMA selects a subset of the nodes as receivers while the others can transmit data during their scheduled slot. Groups are built around the receivers. The receiver group formation occurs in a distributed manner based on random time-out values. The receiver selection and group formation is repeated until each sensor node is a receiver at least once. The protocol then assigns time slots to different groups so that collisions between groups can be avoided. Similar to other self-organizing protocols, the grouping phase in Group TDMA can consume significant time and energy. Hence, it may not be suitable for highly dynamic sensor network applications.

### 16.3.5 Mobile Sensor Network MAC Protocols

Most of the research into sensor network MAC protocols assumes that the sensor nodes are static, and hence fail to provide an acceptable performance when applied to networks with mobile sensor nodes. MS-MAC, as introduced in Sect. 16.3.3, considers the mobility at the MAC layer.

MMAC is one of the few MAC protocols explicitly designed for mobile sensor networks [29]. Following similar principles of TRAMA, MMAC introduces a mobility-adaptive frame time that enables the protocol to dynamically adapt to changes in mobility patterns, making it suitable for mobile sensor networks. The MMAC protocol assumes that all sensor nodes are aware of their current location, which can be used to predict the mobility pattern of the nodes according to the AR-1 model [30]. The traffic information and the mobility patterns are used to decide the particular time slot for transmission. MMAC is similar to a schedule-based protocol in this regard. Empirical results indicate that its performance is close to TRAMA for static sensor network, and for mobile scenarios, MMAC outperforms both S-MAC and TRAMA in terms of energy efficiency and latency. However, the mobility model used in MMAC is a very simple random model leaving much room for future research in this area.

## 16.4   Thoughts for Practitioners

As explained in Sect. 16.2, MAC protocols for wireless sensor networks are influenced by a number of constraints that force a protocol designer to make different choices regarding the trade-off between energy and performance. On the basis of a survey of 20 MAC protocols specifically designed for sensor networks Langendoen and Halkes specify three important design trade-off decisions [8]:

The number of the physical channels used: whether or not the radio should be capable of dividing the available bandwidth into multiple channels. Two common techniques are Frequency Division Multiple Access (FDMA) and Code-Division Multiple Access (CDMA). Multiple channels can allow the simultaneous and collision-free transmission of multiple messages. However, this requires sophisticated radio design and will consume considerably more energy.

The degree of organization between nodes: it refers if, and how much, the nodes in the sensor network should be organized to act together at the MAC layer. Is this organization completely random, or based on frames, or based on other structures like slots? The CSMA and the TDMA protocols can be considered as two extremes in terms of the degree of organization. In the CSMA protocol, the node organization is completely random, and it is easy to accommodate mobile nodes and traffic fluctuation, while the TDMA protocol is strictly organized based on frames providing inherent energy efficiency due to the lack of collisions, overhearing, and idle-listening overheads.

The way in which a node is notified of an incoming message: it relates to how the receiver is notified of an incoming message. In general, there are three typical methods: scheduled, wake-up, and listening. In schedule-based protocols such as TDMA, the receiving nodes know exactly when to turn on the radio because the data transfers are strictly scheduled, while in contention-based protocols, the nodes must listen continuously without any assistance from the sender and be prepared to handle an incoming transfer at any moment. To eliminate the resulting idle-listening overhead, senders may send a wake-up signal over a second radio to prepare an idle radio to receive an incoming message.

## 16.5   Directions for Future Research

In terms of future work, it is time to address other unsolved problems in the MAC layer, rather than solely focus on energy efficiency. Ali et al. present nine issues for future research [31], and here we note that at least the following directions should be addressed gracefully:

- Standardized Sensor Hardware: With a standardized radio, the sensor nodes can communicate with any number of devices while sharing the same physical layer. IEEE 802.15.4 is the emerging standard for lower layers, but its radio interfaces are packet-based, and its MAC specification is not suitable for all sensor network applications.
- Cross-Layer Protocol Design: Sharing information between protocol layers may allow protocols to cooperate and limit the resources needed for operation. A cooperative scheduled MAC and proactive routing layer could use a single message to share state information among sensor nodes and distribute the routing information resulting in reducing the amount of energy consumption. Developing such cross-layer protocol architectures would be a growing and organic process. Techniques such as diversity combining and cooperative relaying have been identified as having some potential in this area [32].
- Mobile Sensor Networks: As mentioned in Sect. 16.3.5, mobility in wireless sensor networks poses unique challenges to the MAC protocol design. As the interest in medical care and disaster response applications of sensor networks increases, it is becoming more and more important for MAC protocols to explicitly address the effects of mobility in the protocol design.
- New Optimization Criteria: Energy efficiency might be the most important design goal for MAC protocols, but it should not be the only one. Other performance attributes, such as latency, may gain importance in different applications. Achieving optimal trade-offs between multiple, conflicting criteria is a significant research challenge.
- Peaceful Coexistence: It is possible that a situation will emerge where sensor networks from different vendors operate at a common frequency band in the same physical environment. Integration of different MAC protocols and security issues needs to be resolved.

## 16.6   Conclusion

In the last few years, medium access control for wireless sensor networks has been a very active research area. Various MAC protocols have been proposed. Energy efficiency has been achieved in most protocols based on a trade-off between network lifetime and performance. Smart radio management and efficient protocol design are essential as they result in reducing the energy consumption due to idle listening, collisions, and overhearing.

In this chapter we have reviewed MAC protocols for wireless sensor networks and discussed important MAC attributes and possible trade-offs in protocol design. We categorized typical MAC protocols into five groups describing representative protocols in each group.

Although there are many proposed MAC protocols for sensor networks, none of them have been accepted as a standard. One of the reasons for this can be identified as the lack of standards at the physical layer and in sensor hardware. Moreover, MAC protocols are generally believed to be application dependent, and it remains an open question whether a flexible MAC protocols exists that can support different sensor network applications while providing acceptable transmission performance with minimal energy consumption.

# Terminologies

*Medium access control.* The broadcast nature of the wireless medium requires that singularity is established between a sender and receiver if successful communication is to be achieved. Medium access control (MAC) refers to the enforcing of this singularity in a network where multiple nodes have data to transmit.

*Idle listening.* Since a node does not know when it will be the receiver of a message, it will need to keep its radio in receive mode. In sensor network applications where the traffic is light, this is a major source of energy waste, since typical radios consume two orders of magnitude more energy in receive mode than in standby mode, even though no receiving is happening.

*Collisions.* When a wireless devise receives more than one packet at the same time, the packets "collide" to become corrupted and hence must be discarded. Collision is a major problem in pure contention-based MAC protocols.

*Efficient channel utilization.* Channel utilization is a traditional MAC protocol metric that illustrates protocol efficiency, and it reflects how well the available bandwidth of the channel is utilized in communication. High channel utilization is critical for delivering a large number of packets with minimum latency.

*Latency.* Latency is the delay from when a sensor node has a packet to send until the packet is successfully received by the receiving node. Different applications have varying emphasis on latency. Monitoring applications can usually tolerate some additional message latency, since the network speed is typically much faster than the object speed, and the object speed places a bound on how rapidly the network must react.

*Throughput.* It refers to the amount of data successfully transferred from a sender to a receiver in a given time. Similar to latency, the importance of throughput depends on the application. Applications that demand long lifetime often accept lower throughput values.

*Protocol overhead.* The MAC headers and control packets used for signaling do not directly convey data and are therefore considered as overhead; sending, receiving, and listening for these protocol overheads consume energy.

*Overhearing*. Since the radio channel is a shared medium, a sensor node may receive packets that are destined to other nodes. Overhearing can be a dominant factor of energy waste when traffic is heavy and sensor nodes are densely deployed.

*Schedule-based MACs*. This is a class of MAC protocols that is based on reservation and scheduling. An example is TDMA.

*Contention-based MACs*. This class of MAC protocols do not use channel division (in time or frequency), reservation, or scheduling. Instead, access to the shared medium is contention-based and allocated to a node that has data to transmit in an on-demand fashion. An example is CSMA.

# Questions

1. Describe the role of the MAC protocol.
2. What are the two main categories that MAC protocols can be categorized into?
3. Describe the challenges to energy-efficient MAC protocol design in wireless sensor networks.
4. List the factors that affect MAC protocol performance.
5. List some examples of contention-based MAC protocols for wireless sensor networks.
6. List some examples of schedule-based MAC protocols for wireless sensor networks.
7. Describe the operation of Z-MAC.
8. Give two examples of self-organizing MAC protocols for use in sensor networks.
9. Describe the operation of Group-TDMA.
10. Describe the message passing optimization used in S-MAC.

# References

1. P. Coronel, R. Doss, and W. Schott. Geographic routing with cooperative relaying and leapfrogging in WSNs. IEEE Global Telecommunications Conference (Globecom'07), USA, November 2007
2. N. Abramson. The ALOHA system – Another alternative for computer communications. In Fall Joint Computer Conference, Montvale, NJ, 37:281–285, 1970
3. R. Jurdak, C. Lopes, and P. Baldi. A Survey, classification and comparative analysis of medium access control protocols for ad hoc networks. IEEE Communications Surveys and Tutorials, 6(1):2–16, 2004
4. J.F. Kurose and K.W. Ross. Computer Networking: A Top–Down Approach Featuring the Internet. Reading, MA: Addison Wesley, Third edition, 2005
5. F. Tobagi and L. Kleinrock. Packet switching in radio channels: Part II – the hidden terminal problem in carrier sense multiple access and the busy-tone solution. IEEE Transactions on Communications, 23(12):1417–1433, 1975

6. Wireless LAN Medium Access Control (MAC) and Physical Layer (PHY) Specification, IEEE Std. 802.11

7. I.F. Akyildiz, W. Su, Y. Sankarasubramaniam, and E. Cayirci. Wireless sensor networks: A survey. Computer Networks, 38(4):393–422, March 2002

8. K. Langedoen and G. Halkes. Energy-efficient medium access control. In R. Zurawski (ed.), Embedded Systems Handbook, Boca Raton, FL: CRC Press, 2005

9. K. Kredo II and P. Mohapatra. Medium access control in wireless sensor networks. Computer Networks, 51(4):961–994, 2007

10. R. Doss, G. Li, V. Mak, S. Yu, and M. Chowdhury. The crossroads approach to information discovery in WSNs. Lecture Notes in Computer Science 4094, January 2008

11. W. Ye, J. Heidemann, and D. Estrin. Medium access control with coordinated adaptive sleeping for wireless sensor networks. IEEE/ACM Transactions on Networking, 12(3):493–506, 2004

12. A. El-Hoiydi. ALOHA with preamble sampling for sporadic traffic in ad hoc wireless sensor networks. In IEEE International Conference on Communications (ICC2002), New York, April 2002

13. J. Hill and D. Culler. MICA: A wireless platform for deeply embedded networks. IEEE Micro, 22(6):12–24, 2002

14. A. El-Hoiydi and J.-D. Decotignie. WiseMAC: An ultra low power MAC protocol for multi-hop wireless sensor networks. In Proceedings of the International Workshop on Algorithmic Aspects of Wireless Sensor Networks (Algosensors), pages 18–31, July 2004

15. A. Woo and D. Culler. A transmission control scheme for media access in sensor networks. In Proceedings of ACM/IEEE International Conference on Mobile Computing and Networking, Rome, Italy, pages 221–235, July 2001

16. S. Mahlknecht and M. Böck. CMSA-MPS: A minimum preamble sampling MAC protocol for low power wireless sensor networks. In Proceedings of the IEEE International Workshop on Factory Communication Systems, pages 73–80, September 2004

17. L.F.W. van Hoesel and P.J.M. Havinga. Poster abstract: A TDMA-based MAC protocol for WSNs. In Proceedings of the International Conference on Embedded Networked Sensor Systems (SenSys), pages 303–304, November 2004

18. L.F.W. van Hoesel and P.J.M. Havinga. A lightweight medium access protocol (LMAC) for wireless sensor networks: Reducing preamble transmissions and transceiver state switches. In Proceedings of the International Conference on Networked Sensing Systems (INSS), Tokyo, Japan, June 2004

19. S. Chatterjea, L.F.W. van Hoesel, and P.J.M. Havinga. AI-LMAC: An adaptive, information-centric and lightweight MAC protocol for wireless sensor networks. In Proceedings of the Intelligent Sensors, Sensor Networks, and Information Processing Conference, pages 381–388, December 2004

20. I. Rhee, A. Warrier, M. Aia, and J. Min. Z-MAC: A hybrid MAC for wireless sensor networks. In Proceedings of the International Conference on Embedded Networked Sensor Systems (SenSys), pages 90–101, November 2005

21. V. Rajendran, K. Obraczka, and J.J. Garcia-Luna-Aceves. Energy-efficient, collision-free medium access control for wireless sensor networks. In Proceedings of the International Conference on Embedded Networked Sensor Systems (SenSys), pages 181–192, November 2003

22. P. Lin, C. Qiao, and X. Wang. Medium access control with a dynamic duty cycle for sensor networks. In Proceedings of the IEEE Wireless Communications and Networking Conference (WCNC), volume 3, pages 1534–1539, March 2004

23. T. van Dam and K. Langendoen. An adaptive energy-efficient MAC protocol for wireless sensor networks. In Proceedings of the International Conference on Embedded Networked Sensor Systems (SenSys), pages 171–180, November 2003

24. J. Ai, J. Kong, and D. Turgut. An adaptive coordinated medium access control for wireless sensor networks. In Proceedings of the International Symposium on Computers and Communications, volume 1, pages 214–219, July 2004

25. H. Pham and S. Jha. An adaptive mobility-aware MAC protocol for sensor networks (MS-MAC). In Proceedings of the IEEE International Conference on Mobile Ad-hoc and Sensor Systems (MASS), pages 214–226, October 2004

26. W.B. Heinzelman, A.P. Chandrakasan, and H. Balakrishnan. An application-specific protocol architecture for wireless microsensor networks. IEEE Transactions on Wireless Communications, 1(4):660–670, October 2002

27. S. Biaz and Y.D. Barowski. GANGS: An energy efficient MAC protocol for sensor networks. In Proceedings of the Annual Southeast Regional Conference, pages 82–87, April 2004

28. Y.E. Sagduyu and A. Ephremides. The problem of medium access control in wireless sensor networks. IEEE Wireless Communications, 11(6):44–53, December 2004

29. M. Ali, T. Suleman, and Z.A. Uzmi. MMAC: A mobility-adaptive, collision-free MAC protocol for wireless sensor networks. In Proceedings of the 24th IEEE International Performance Computing and Communications Conference, Phoenix, Arizona, pages 401–407, USA, April 2005

30. Z.R. Zaidi and B.L. Mark, Mobility estimation for wireless networks based on an autoregressive model. In Proceedings of 2004 Global Telecommunications Conference, Dallas, TX, pages 3405–3409, December 2004

31. M. Ali, U. Saif, A. Dunkels, T. Voigt, K. Romer, and K. Langendoen, Medium access control issues in sensor networks. ACM SIGCOMM Computer Communication Review, 36(2): 33–36, April 2006

32. J. Polastre, J. Hui, P. Levis, J. Zhao, D. Culler, S. Shenker, and I. Stoica. A unifying link abstraction for wireless sensor networks. SenSys '05: Proceedings of the 3rd international conference on Embedded networked sensor systems, San Diego, California, USA, pages 76–89, 2005

# Chapter 17
# Energy-Efficient Resource Management Techniques in Wireless Sensor Networks

**Xiao-Hui Lin, Yu-Kwong Kwok, and Hui Wang**

**Abstract** Devices in a wireless sensor network are typically powered by limited and sometimes unchargeable batteries, which are supposed to sustain for months or even years. To enhance the lifetime of a sensor network, highly efficient energy management techniques are mandatory, in order to successfully achieve the missions of the network. These techniques, however, involve all levels of the sensor system hierarchy in data processing and transmitting. Thus, energy awareness should be incorporated into every level of the system design and operation to maximize the lifetime and connectivity as much as possible. In this chapter, state-of-the-art techniques at each layer for optimizing the energy usage proposed in literature are introduced. To illustrate the efficacies of the approaches, design examples in reducing energy expenditure are also given as well. In addition, new thoughts in energy conservation by exploiting the interactions between different layers are presented. These new ideas could be effective in reducing the energy consumption.

## 17.1 Introduction

Advances in MEMS (microelectronic-mechanical systems) based sensor technology, coupled with low-power, low-cost digital signal processors (DSPs) and radio frequency (RF) circuits, have spurred the proliferation of wireless sensor networks in a wide spectrum of civil and military applications, such as environment monitoring, battlefield surveillance, and home networking, for collecting, processing, and disseminating wide ranges of complex environmental data [1]. In most cases, such a sensor network is composed of hundreds or thousands of low-cost and battery-powered devices, for carrying out sensing, computing, and wireless communication

X.-H. Lin (✉)
Department of Communication Engineering, Shenzhen University, Guangdong, China
e-mail: xhlin@szu.edu.cn

S. Misra et al. (eds.), *Guide to Wireless Sensor Networks*, Computer Communications and Networks, DOI: 10.1007/978-1-84882-218-4_17,
© Springer-Verlag London Limited 2009

tasks in a hostile environment. In essence, a collection of tiny sensors are designed to form an autonomous and robust data computing and communication distributed system for automated information gathering and distributed sensing.

While the applications enabled by sensor networks are attractive, the real implementation of sensor networks is still restricted by many physical constraints such as limited battery power, memory, and computation capabilities. Among all these constraint factors, the most crucial one is energy consumption. Indeed, the replacement of the battery can be very difficult or impossible in many application scenarios, especially when the sensor network operates in hostile or remote environments. Consequently, the lifetime of the whole sensor network critically depends on individual node's battery lifetime. For a sensor node, energy consumed by the communication tasks using radio transmissions is the dominating component. Thus, to maximize the sensor node's lifetime, power awareness must be incorporated into the design of network protocol stacks used for data communications [2, 5].

In general, protocol layers that contribute to significant energy consumption in communications are physical layer, data link layer (DLL), and network layer. For physical layer, its major functions include modulation and coding so that data can be reliably transmitted in the presence of channel fading. At the same time, this layer also handles processor and radio parameter adjustments according to the working environment. DLL deals with packet fragmentation and error correction. In addition, medium access control is performed at this layer to harmonize channel access among multiple competing sensors. The network layer is responsible for establishing a path from the data collection point to the sink and forwarding data along this path.

In this chapter, we address the incorporation of energy saving into all these layers for sensor networks. Indeed, the energy efficiency of the network critically depends on all the system components. Thus, energy awareness must be incorporated into every level of the system design and operation, so as to maximize the battery usage and enhance network lifetime as much as possible [1, 9, 10]. We mainly discuss low-power strategies at each layer for reducing the energy expenditure.

The remainder of this chapter is organized as follows. In Sect. 17.2, we give some background on the sensor network, briefly demonstrating a trade-off between energy efficiency and performance. In Sect. 17.3, two representative energy conservation strategies – dynamic voltage scaling (DVS) and dynamic modulation scaling at physical layer – are introduced. In Sect. 17.4, we discuss low-power techniques for the DLL, and these techniques include adaptive packet fragmentation, forward error correction, and low-power MAC control. Their mechanisms are also presented in detail. Then in Sect. 17.5, from the perspective of energy efficiency, we describe several routing protocols for sensor networks, and their advantages and disadvantages are also discussed. Afterward, some cross-layer approaches to resource management are given in Sect. 17.6. Their mechanisms and gains achieved in energy efficiency are also presented. In Sect. 17.7, we discuss future directions for further research. Finally, we conclude this chapter by giving a summary in Sect. 17.6.

## 17.2   Background

With the advances in MEMS technologies, sensor nodes are getting smaller in size and more powerful in functions. However, despite all these successes achieved in communication and miniaturization technologies, the advance in developing long-life battery power is still far from satisfactory. This has imposed a constraint on the networking of tiny sensor nodes, as the energy available for processing and communication is still severely limited. Thus, to save precious battery resource, proper power management techniques are necessary.

As illustrated in Fig. 17.1, a sensor node is made up of four basic components: sensing unit, processing unit, transceiver unit, and power unit. When an incident happens and is detected by the sensing unit, the observed data are converted to digital signals by ADC (analog-to-digital convert) integrated in the sensing unit. Then the digital signals are fed to the processing unit, which performs data packing and encoding for robustness and security [4]. Afterward, the processed data are transmitted over the wireless channel to the end user by the transceiver unit. Therefore, with respect to system operation, the sources of power consumption in a sensor node at the physical layer can be divided into two types: computation related and communication related. Computation in a sensor mainly involves data processing, swapping, and CPU (central processing unit) usage, etc. Communication concerns packet transmission and reception by the radio. In the next section, by using DVS [12] and DMS (Dynamic Modulation Scaling) [10] as examples, we discuss how the output quality can be traded-off for computation and communication energy according to varying environmental conditions in the system operation.

Traditionally, the users of wireless networks might desire high performance (high data processing and transmission rate, low delay, and high throughput, etc.), and adopt a design philosophy that assumes that systems are working in the worst case operation state most of the time. For example, they might unnecessarily let the processor working at a full or fixed speed to achieve the lowest processing latency, at the expense of more energy consumption. Another example is that, to have low transmission delay, the highest modulation level is always used, and even much higher

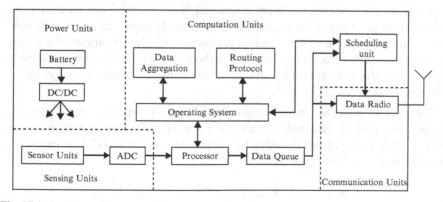

**Fig. 17.1** Structure of a sensor node

transmission power level is required. This is, obviously, in conflict with energy conservation criteria for wireless sensor network. In contrast to conventional cellular or ad hoc networks, in which, provision of high QoS (high network throughput, low packet delay) or terminal mobility management is of higher priorities, we are more concerned with the network lifetime in a sensor network. In fact, due to the high correlation of the sensing data, sensor networks are more fault-tolerant to the packet loss and delay. Therefore, network resiliency has provided us great flexibility to strike a balance between energy and system performance. In the physical level operation, the processor working state and the radio modulation level should be adaptively adjusted according to the time-varying computation and communication loads, to reduce the energy consumption and prolong the battery life.

## 17.3 Low-Power Techniques at the Physical Layer

### 17.3.1 Dynamic Voltage Scaling

Computation load varies with time, and peak system performance is not always required. By exploiting this fact, DVS is an effective technique for reducing CPU energy by dynamically adjusting the clock speed and supply voltage based on the instantaneous workload. Reducing the operation frequency during the period of reduced activity leads to linear decreases in energy consumption, and also means greater critical path delay and a compromise in the peak performance as well. However, some performance loss is endurable or negligible by the system. When the computation load of the embedded system is light, in view of energy efficiency, it is unnecessary to set the supply voltage (clock frequency) to the maximum value. The goal of DVS is to adapt the supply voltage and clock frequency to match the instantaneous workload.

*Example 1.* An SA-1100 DVS-enabled microprocessor can dynamically adjust the core supply power from 0.9 to 2.0 V, and clock speed from 59 to 206 MHz, in 30 discrete increments. It performs beamforming computation in an object tracking application. The computation latency required is 20 ms. We keep increasing the workload (more objects to be tracked) and dynamically adjust the core supply voltage to fulfill the computation. We adopt parameters in [16] and simulate the energy gain achieved with DVS. Simulation results show that, with DSV, energy consumption can be reduced by approximately up to 60% without performance degradation. The results are given in Fig. 17.2.

For a sensor node, the energy consumed by digital processing circuit goes into the processing of the gathered data and the protocol stack implementation. Let us see how DSV can yield energy saving for computation. Equation (17.1) is the energy consumption model for a processor.

$$E = E_{\text{switch}} + E_{\text{leakage}} = C V_{\text{DD}}^2 + (t V_{\text{DD}}) I_0 e^{\frac{V_{\text{DD}}}{n V_{\text{th}}}}, \tag{17.1}$$

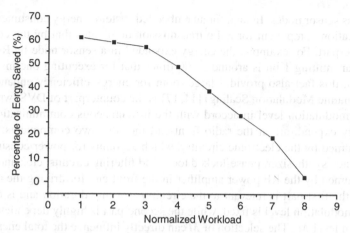

**Fig. 17.2**  Percentage of energy saved with workload

where $C$ is the switched capacitance, $V_{DD}$ is the supply voltage, $V_{th}$ is thermal voltage, and $I_0$ and $n$ are constants for the processor technology. In the equation, the total energy expenditure is composed of two parts: switching energy (the first term of the equation) that goes to energizing parasitic capacitors on the circuit from zero voltage (representing 0) to the supply voltage $V_{DD}$ (representing 1), and leakage energy (the second term of the equation) originating from the unavoidable leakage of the current from power to the ground when the processor is powered on. We can see that (1) for a given task, given a fixed supply voltage $V_{DD}$, switching energy is independent of time, while the leakage energy increases linearly with time $t$; and (2) switching and leakage energy have a quadratic and approximately exponential relationship with $V_{DD}$, respectively. Different computation applications have different performance requirements, and it is energy inefficient to set $V_{DD}$ to the maximum value (equal to the worst case workload) all the time. Thus, when the computation workload is high, the supply voltage should be raised in order to provide sufficient current drive and reduce the propagation delay in the integrated circuit to satisfy the demanding performance requirement. However, when the workload is low, the supply voltage should be scaled down to reduce the leakage energy, as the processor can be idle most of the time. In this manner, supply voltage serves as a controlling knob at the circuit level to realize an energy–performance trade off, and energy saving can be achieved by tuning supply voltage to deliver just the required performance. Thus, by properly adjusting supply voltage, no extra energy is wasted while the task can still be fulfilled as required.

## 17.3.2  Dynamic Modulation Scaling

Similarly, incorporating power awareness into the radio subsystem can also result in significant energy saving, since wireless communication is a major power consumer for wireless sensor nodes. In fact, for an embedded system, energy consumed by the communication component for radio transmission dominates that by the computation counterpart. For example, the energy expended in a sensor node by Rockwell Inc. for transmitting 1 bit is around 2,000 times that for executing one instruction [13]. Thus, this fact also provides large room for energy efficiency enhancement. DMS (Dynamic Modulation Scaling) [14, 17] is the counterpart of DVS, which can adapt the modulation level in accord with the instantaneous communication load. The energy expenditure at the radio front-end includes two components: (1) energy consumed by the electronic circuitry, which accounts for power consumed by the frequency synthesizer, phase-locked loop, and filtering circuits, etc.; and (2) energy consumed by the RF power amplifier in the front end for driving the antenna. Note that the first part is relevant to the circuitry working duration and is constant once the modulation level is fixed, while the second part is highly dependent on the modulation level $M$. The selection of $M$ can directly influence the total energy consumption and the communication latency, thus providing a control panel for striking a balance between energy and system performance. Mathematically, the energy cost for transmitting one information bit can be expressed as [17]:

$$E_{\text{bit}} = \frac{E_{\text{start}}}{L} + \frac{P_{\text{elec}}(M) + P_{\text{RF}}(M)}{R_s \times \log_2(M)} \times \left(1 + \frac{H}{L}\right), \qquad (17.2)$$

where $H$ and $L$ are packet payload size, and header size, respectively, $E_{\text{start}}$ is the energy consumed to start-up transmitter circuitry, $P_{\text{elec}}(M)$ is the power consumed by the electronic circuitry, $P_{\text{RF}}(M)$ is the power consumed by output radiation amplifier under the modulation level $M$, and $R_s$ is the symbol rate for the $M$-ary modulation scheme. In (17.2), variables $H, L, R_s$, and $E_{\text{start}}$ are fixed, thus the energy cost is the function of modulation level $M$. To reduce the energy consumption at electronic circuitry $P_{\text{elec}}(M)$, we should adopt higher modulation level $M$ to shorten the circuit working duration. However, once modulation level $M$ is upgraded, to keep the bit error rate (BER) within the acceptable level, we have to increase the transmission power $P_{\text{RF}}(M)$ to radiate more energy out. Therefore, there should be an optimal value for $M$, which can minimize the energy cost in (17.2).

*Example 2.* Table 17.1 shows the energy cost in transmission of one information bit versus the modulation level. The hardware and channel parameters are adopted from [17]. We can see that, when transmission distance is short, i.e., 10 m, the use of high modulation level can reduce energy consumption. However, when the transmission distance is increased to 100 m, M = 4 is the optimal level, i.e., constellation size of 2 bits/symbol is the most energy-efficient modulation mode. This is because

**Table 17.1** Energy consumed per useful bit (dB mJ)

| Modulation level | M = 2 | M = 4 | M = 8 | M = 16 |
|---|---|---|---|---|
| Distance 10 m | −20.2 | −22.4 | −25.3 | −27.6 |
| Distance 100 m | −15.1 | −17.2 | −12.0 | −10.5 |

Packet Size =1,000 bit, BER = $10^{-5}$, Path loss exponent = −3.5

now the energy consumed by the power amplifier dominates that by the circuitry. Although using high modulation level can shorten transmission time (and reduce circuitry energy consumption as well), to overcome the long distance attenuation, more energy must be radiated out, thus counteracting the gain achieved.

The earlier example reveals the dependence of optimal modulation level M on the working environment. With respect to energy efficiency, this provides us a practical guideline for adjusting the communication system: when the instantaneous communication load is low, to save energy, we should choose the optimal modulation level and transmit data at this "cruising" speed; however, when the communication load is high, to maintain an acceptable level of communication latency, we should increase the modulation level, at the expense of more energy consumption. This is the central idea of DMS: adapting the modulation level to match the instantaneous traffic load. Through modulation level dynamic adjustment, we can keep a balance between energy and performance.

## 17.4  Power-Aware Strategies at DLL

With increasing interest in battery-powered wireless sensor networks, energy efficiency has become one of the most important metrics of the system performance. Reducing the power consumption for sensor networks entails the design consideration of all levels of the network system. Similarly, DLL can also be tailored to optimize the energy utilization. For the DLL, the protocol stack components include (1) data fragmentation, (2) error control, and (3) medium access control. They together represent the main functions of the DLL and have direct influences on the energy expenditure for a sensor network.

### 17.4.1  Automatic Packet Fragmentation

Because of multipath fading and shadowing, wireless link is a highly unreliable channel characterized by high-error rate and bursty errors. Packet transmission over the air is inevitably prone to corruption and delivery failure, leading to retransmission of the entire packet, which is very energy consuming. Thus, how we should transmit packet under a hostile environment in an efficient manner is a challenging issue.

The fragmentation of data stream from the upper layer is performed at the DLL. The reliability of packet transmission over a lossy link is very sensitive to the fragmentation size. A Long packet is more prone to channel fading, and once one bit is corrupted, the whole packet has to be retransmitted. Thus, shortening the packet size can enhance the transmission reliability under the fading environment. However, too small a packet size is also energy inefficient because of the fixed overhead required per packet. Traditionally, designers of link layer protocol prefer a fixed packet size and choose a size for the worst working environment. Although this can leave enough redundancy for safe delivery, it is very inefficient in that a large portion of energy is spent on overhead transmission. Therefore, how to strike a balance between efficiency and reliability, thus maximizing the energy utilization is a complex problem. The difficulty is further aggravated by the fact that the wireless channel is a time-varying function. This requires a dynamic adjustment of packet length according to the channel variations. When no error protection is included, the optimal packet size adopted by the DLL can be written as [11]:

$$L_{\text{opt}} = \frac{-h \ln(1 - p) - \sqrt{-4h \ln(1 - p) + h^2 \ln(1 - p^2)}}{2 \ln(1 - p)}, \qquad (17.3)$$

where $h$ is the number of bits included in the packet head, and $p$ is the BER of the channel. With packet size chosen according to the equation, packet corruption probability can be minimized.

Figure 17.3 shows the optimal packet size versus channel BER. It can be seen that, when the channel quality deteriorates, we should decrease the payload size to lower the corruption probability of the transmitted packet. On the other hand, in case the received packet is in error, ARQ (automatic repeat request) is performed by retransmitting the corrupted packet. Figure 17.4 shows the percentage of energy saved after length optimization, assuming that ARQ is used. The reference packet size is fixed at 2,000 bits.

To implement the adaptive packet fragmentation in a resource-limited sensor network, the crux is to let the sending side know the instantaneous BER of the wireless link, and react to this fluctuation in a timely manner. In a wireless communication

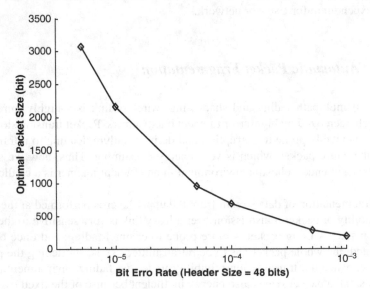

**Fig. 17.3** Optimal packet size with BER

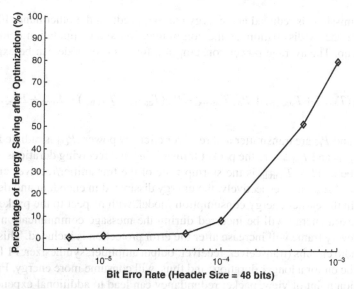

**Fig. 17.4**  Gain in energy saving with BER

system, when modulation and coding schemes are fixed, the BER is solely the function of the SNR (signal noise ratio) of the received signal. The SNR of the signal can be measured at the receiving side and mapped to the BER. Then, this calculated BER is sent back to the transmitting side through a feedback channel in a handshake manner like RTS-CTS in 802.11x. In this way, adaptive fragmentation can be implemented according to the equation given earlier. Another concern on this scheme is the channel variation speed. If channel quality changes too fast, the feedback channel state information (CSI) will have been outdated when it reaches the transmitting side, rendering an unsuitable packet size selected. Nevertheless, in general, the sensor node in the network is stationary or in low mobility, thus the channel varies sufficiently slowly. For example, at mobile speed of 1 m/s, the coherence time is approximately 122.88 ms for a center frequency of 2.4 GHz. Thus, since duration of one frame is around several milliseconds, it can be ensured that the channel remains approximately constant for the duration of at least several frames.

## 17.4.2   Forward Error Correction

Another important function of the DLL is error recovery of the received packet. Forward error correction (FEC) is such a recovering strategy. By incorporating error protection in the raw data, the transmitted packet is more robust to the channel fading, and delivery reliability is enhanced. This strategy has long been adopted in the cellular network system. However, for resource-constrained sensors, the effects of FEC are arguable: FEC can decrease the packet error probability, thus the number of retransmissions is reduced and energy can be saved; on the other hand, FEC can incur extra energy dissipation, as the communication and computation complexity also goes up. The average power consumption for a sensor node can be expressed as [16]:

$$E = P_{tx}(T_{on-tx} + T_{startup}) + P_{out}T_{on-tx} + P_{rx}(T_{on-rx} + T_{startup}) + E_{enc} + E_{dec}. \quad (17.4)$$

Here, $P_{tx}$ and $P_{rx}$ are transmitter and receiver circuitry power, $P_{out}$ is radiated output power, $T_{on-tx}$ and $T_{on-rx}$ are the packet transmitting and receiving durations, respectively, at both sides, $T_{startup}$ is the startup time of the transmitter/receiver circuitry, and $E_{enc}$ and $E_{dec}$ are, respectively, the energy dissipated in encoding and decoding the data. In the earlier energy consumption model, with respect to the packet transmission, extra energy will be incurred during the message communication, as the length of every frame will increase after the error protection is included. This means that the radio circuits (transceiver, receiver, output amplifier, synthesizer, PLL/VOC, etc.) will be on for a longer duration, and thus, will consume more energy. From the computation point of view, packet redundancy can lead to additional expended energy that goes into encoding and decoding data at two communication sides. The needed energy is drawn from the limited battery source, and thus, it should be taken into consideration in power management. Therefore, in view of energy efficiency, it is necessary to decide whether FEC scheme should be used.

In general, if convolutional code is adopted, the energy needed to encode data can be neglected. However, the energy required in Viterbi decoding at the receiving side is very energy consuming, and might exceed that expended in communication. Hence, if the associated energy expended in error protection is greater than the coding gain, then FEC strategy is energy inefficiency and the system is better off without coding. Let us look at an example [16].

*Example 3.* For a StrongARM SA-1100 microprocessor-driven wireless sensor, with path loss 70 dB, transmission rate 1 Mbps, and circuitry noise level 10 dB, we compare the energy efficiency of transmit schemes with and without forward error correction. Convolutional code is considered, with code rate 3/4 and constraint length 6. We vary the bit error probability requirement.

For desired error probability of $10^{-5}$ (voice data), the average required energy for the transmission of one useful bit is $1.2 \times 10^{-7}$ J if data is coded, while the corresponding energy required without coding is $1.35 \times 10^{-6}$ J. This illustrates that the decoding energy dominates the communication energy. In fact, in this case, to successfully deliver one useful bit, more than 90% of the energy is consumed by the encoding/decoding circuitry.

We further change the desired error probability to $10^{-8}$ (error-sensitive data), and compare the energy efficiency of both schemes again. For the coding scheme, the energy required is $1.4 \times 10^{-6}$ J, while the corresponding required energy of uncoding has a sharp increase to $1.3 \times 10^{-5}$ J. This is because to satisfy a higher communication quality, uncoding scheme has to increase radiated power level significantly, and the communication energy exceeds that consumed in encoding/decoding data. While the coding scheme is more robust against channel loss, it does not need to raise the transmission level greatly.

From the earlier example, it is shown that FEC is not always an energy-efficient strategy for data transmission. Adopting FEC or not depends on the communication quality requirement. If low error rate is needed, FEC is a suitable solution with respect to energy efficiency. However, if the error rate requirement is not so demanding, we can just transmit the packet without coding. As a matter of fact, in many application scenarios, the sensor network is very fault-tolerant due to large redundancy in collected data, thus the reliability requirement is not so rigid. Therefore, in these cases, to save energy, FEC is not recommended.

### 17.4.3 Energy-Efficient Medium Access Control

A typical sensor network is composed of hundreds or thousands of tiny sensor nodes connected by wireless links. In general, wireless channel is shared by multiple users, and packet transmission collision can happen due to the simultaneous transmission by two nearby terminals. This can incur retransmission and unnecessary energy and bandwidth waste. Therefore, harmonizing medium access among multiple competing users is the main task of medium access control (MAC) protocol. The functions of MAC protocol for wireless networks include (1) establishing communication links among users such that the sensor network can be set up in a self-organizing manner, and (2) efficiently resolving the channel access conflicts among senor nodes. However, as battery carried by sensor node is a precious resource, which is expected to sustain for years, the depletion of the battery energy means the failure of the node and partial partition of the network, resulting in the "blind area" of the corresponding location. Thus, different from MAC protocols for conventional wireless networks, in which high QoS (high throughput and low delay), high bandwidth efficiency, and fairness issues are of the first priorities, in contrast, for a sensor network, we are more interested in a power-aware MAC protocol that can prolong sensor lifetime.

#### 17.4.3.1 Traditional IEEE 802.11 MAC Protocol

For wireless LAN or ad hoc networks, IEEE 802.11 DCF (Distributed Coordination Function) is a standard MAC protocol. IEEE 802.11 is a fully distributed medium access control scheme based on CSMA/CA (carrier sense multiple access with collision avoidance). In the scheme, each mobile terminal accesses the medium on a contention basis. Before a data transmission begins, the sender and receiver must have a RTS-CTS signaling handshake to "reserve" the channel. The whole transmission sequence is a RTS-CTS-DATA-ACK four-way handshake as illustrated in Fig. 17.5. When a sender has a packet to transmit, it senses the channel

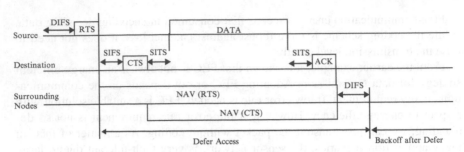

**Fig. 17.5** Handshake mechanism in IEEE 802.11

by detecting the air interface (in the physical layer) and looking up its NAV (net-work allocation vector). If the channel is busy, the terminal waits until the channel becomes free, in which case it sends a RTS to the destination terminal. On success-fully receiving the RTS, the destination replies the source with a CTS. The source can begin data transmission after the CTS is received. After the data are received at the destination, the destination sends an ACK to the source, confirming the success of a data reception. This is an ideal case of a four-way handshake. If the source fails to receive CTS or ACK (due to collision at source or destination), it backs off for a random period of time by doubly increasing its contention window (CW) size. Each packet, including RTS, CTS, DATA, and ACK, has a duration time in its header, which is used to specify the time that the wireless channel will still be occupied. The terminals in the neighborhood, on receiving these packets, adjust their NAVs as illustrated in Fig. 17.5. Thus, the wireless channel is deemed being occupied by a terminal if either its physical air interface or the NAV indicates so. The handshake and exponential back off mechanisms have made IEEE 802.11 a highly distributed, scalable, and robust medium access protocol that has long been commercialized and adopted in home networking.

Although IEEE 802.11 has been widely adopted as a standard MAC protocol for wireless LAN, with respect to energy efficiency, it is unsuitable for wireless sensor networks. The energy waste sources in IEEE 802.11 scheme include [20] the following: (1) Idle listening. In sensor networks, each terminal may possibly act as a router for any surrounding neighbor, and therefore, it has to keep sensing the channel as it does not know when the next transmission is going to take place. Even when there is no traffic on the air, the sensor has to keep the receive circuitry on most of the time. According to the measurement in [20], the energy consumed in idle state can be comparable to that of receiving sate, and thus is a main source of energy waste. (2) Unnecessary packet overhearing, i.e., receiving packets that in fact are not destined to or routed through it. (3) *Packet collision*. If the transmitted packet is corrupted, it has to be retransmitted, leading to extra energy expenditure. In fact, wireless sensor network is characterized by high node density, thus packet transmission on the air is more prone to collision than ad hoc network. (4) Control overhead. Before each packet transmission, both sides should finish a RTS-CTS handshake, even when there are large amount of packets to be sent. Before RTS-CTS handshake, there must be a channel sensing period to ensure that the channel is clear. This can incur latency and unnecessary energy waste.

### 17.4.3.2   S-MAC: A Sensor-MAC Protocol Design

To address the energy inefficiencies identified earlier, authors in [20] propose S-MAC to reduce the unnecessary waste from all sources. The power-aware strategies in S-MAC include periodic listen and sleep, collision and overhear avoidance, and message passing.

#### Periodic Listen and Sleep

During the period that no incident happens, there is little communication traffic among sensor nodes. Hence, most of the time, a sensor node is in idle state, which has been illustrated as a source of energy waste. Therefore, it is unnecessary to keep the sensor in listening mode all the time. To save battery energy, a feasible solution is to let the sensor enter sleep state and shut down the communication unit to reduce energy dissipation due to idle listening. In S-MAC, by periodically cycling the active and sleep state of sensor, the duty cycle and hereby the energy wasted in idle state are significantly reduced. This is a basic power-conserving strategy for S-MAC and is shown in Fig. 17.6. In the figure, the time domain is divided into continuous cycle periods, and each period consists of two states: listening and sleep states. In the listening state, the sensor is active and can communicate with the surrounding neighbors. In the sleep state, the sensor will power off its radio to conserve energy. The sensor alternates between these two states, and thus the active duty cycle is reduced, further decreasing energy wasted in idle listening.

Although S-MAC is effective in energy reduction, two challenging issues need to be properly handled. The first one is the communication latency, especially when the traffic must be routed through multiple sensors. Putting the sensor into a sleep state can incur delay, because its neighbors have to wait for the wake-up of the sensor to transmit the buffered packets. This is a trade-off between energy efficiency and performance. If higher communication performance is needed, the sleep duration in the period can be shortened at the sacrifice of more energy consumption. Therefore, the adjustment of the sleep duration depends on the applications running at the upper layer. The second issue is the synchronization between communication nodes. To enhance the communication and energy efficiency, it is ideal that the adjacent sensors can synchronize in listen and sleep period, i.e., they can sleep and wake up at exactly the same time to maximize the utilization of activity duration, thus reducing the communication latency. To synchronize with neighbors, each sensor node maintains a schedule table, recording the listening and sleeping time of its neighbors. The synchronization within the network is through the periodic exchange of SYNC packet among sensor nodes. The mechanism is detailed as follows.

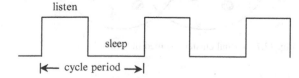

**Fig. 17.6**  Periodic listen and sleep in S-MAC

When a sensor node starts up, it must synchronize with a neighbor by following the neighbor's listening and sleeping time. Note that listening and sleep time indicates when this neighbor will enter listening and sleeping state. The new startup sensor firstly listens for a period of time, waiting for the broadcast of SYNC packet from one of its neighbors. If it receives a SYNC packet, the sensor synchronizes with this node, by choosing the same listening and sleeping time included in SYNC. Then after a random period, it generates a SYNC packet including its schedule, and broadcasts the packet out. In case the sensor cannot receive any SYNC, it schedules its own listening and sleeping time and then includes this information in a SYNC packet, which is also broadcast in the network. If a sensor receives a SYNC packet after it broadcast its own schedule, it adopts both schedules and wake-up accordingly.

As a SYNC packet is very short, and the broadcasting period is about 10 s [20], thus S-MAC is efficient in energy and bandwidth utilization. In this way, sensor nodes having the same schedule table will form a virtual cluster in that they can synchronize in communication activity. The whole network is divided into multiple virtual clusters, and a sensor node adopting more than two schedule tables will become a gateway sensor, as shown in Fig. 17.7. In sensor networks, nodes can die anytime due to the depletion of battery, and a new member can also join the group anytime. By periodically SYNC broadcasting, this synchronization scheme makes the network highly adaptive to the topology changes, and the cost is also very low.

### Collision and Overhear Avoidance

As in a virtual cluster, multiple sensors can contend the channel access in the listening period to avoid the packet collision. S-MAC also adopts the similar collision avoidance mechanism as in IEEE 802.11. Before sending the packets, the communication pair should have a RTS-CTS exchange. The RTS and CTS include duration the channel will be occupied. Thus, when a surrounding sensor receives RTS or CTS, it will set its NAV accordingly and keep silent during the specified time. In this way, the hidden collision problem is resolved. In addition, physical carrier sensing is also performed by the sending side to ensure that there is no transmission on

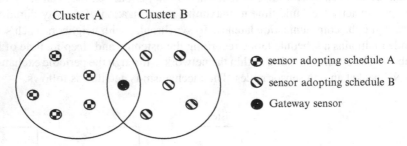

**Fig. 17.7** Virtual clusters formed in S-MAC

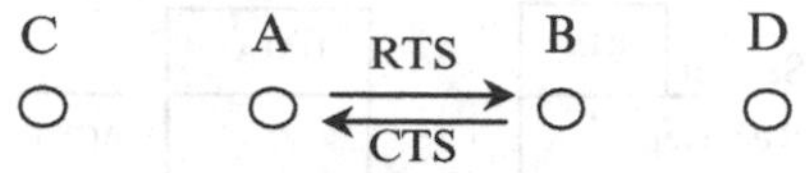

**Fig. 17.8** Collision and overhearing avoidance in S-MAC

C       A    RTS    B       D

O       O ⇄ O       O
              CTS

the air. Hence, the channel is deemed clear if both NAV and physical layer indicate so. As mentioned before, overhearing is another source of energy waste. To avoid receiving packets not destined itself, a sensor can enter sleep state on receiving RTS and CTS for the duration specified in RTS/CTS.

*Example 4.* Figure 17.8 is an example on how packet collision and overhearing are avoided. Sensor A wants to send a packet to B, and a RTS-CTS exchange is performed in between. The duration time the channel will be busy is also included in RTS-CTS signaling packets. For sensor C, on receiving RTS, it should enter sleep state to avoid receiving data packet destined to B. While for sensor D, on receiving CTS, it also enters sleep state and keeps silent to avoid packet collision at B. The sleeping durations for sensor C and D are specified in RTS and CTS, respectively.

### Message Passing

In a sensor network, sometimes the sensed data must be totally transmitted to a receiving node before data can be further processed or aggregated. The transmission of the raw data needs long packets. However, the long packet is more prone to corruption and once several bits are in error, the whole packet has to be retransmitted, which can incur latency and extra energy consumption. The solution to the problem is packet fragmentation, i.e., dividing a long message into multiple short packets and each time transmitting one packet. For IEEE 802.11, this method is inefficient in energy utilization and can result in carrier sensing delay due to the following reasons:

1. For each short packet transmission, there is a RTS-CTS-DATA-ACK exchange. No matter whether the transmission is successful or not, the sending side has to give up the channel and compete again for the next medium access to ensure the fairness among multiple sensors. This can incur contention latency.
2. In each contention, all the sensors must wake up, which means some energy is wasted in circuitry switching. Furthermore, in carrier sensing, all sensors simultaneously keep their radios *on* to probe the channel, and this is also a source of energy waste.

To solve these problems, S-MAC has made some revisions to the handshake mechanism of IEEE 802.11. Instead of performing one RTS-CTS handshake for each short packet, S-MAC has only one RTS-CTS exchange to reserve the channel for all the sequential short packet transmissions as shown in Fig. 17.9.

In the figure, there is only one RTS-CTS exchange, and the duration the channel will be busy is also included in RTS-CTS. On receiving this signaling exchange,

**Fig. 17.9** Message passing in S-MAC

the surrounding sensors will enter sleep state for the specified duration, thus avoiding frequent circuitry switching and unnecessary carrier sensing. For each arriving short packet, the receiving sensor sends an ACK back to the sender. If no ACK is received, the sender will retransmit the lost packet. ACK packet also bears the channel occupation time, thus the wake-up or newly startup sensor around the receiver can also enter sleep state. In this way, there will be no interference from competing sensors around sender and receiver, thus the communication latency is reduced. For these competing sensors, they are also exempted from frequent circuitry switching and carrier sensing, and thus reducing energy waste.

## 17.5 Energy-Efficient Packet Routing

A self-organizing wireless sensor network consists of hundreds or thousands of small, cheap, and battery-driven sensing devices densely scattering over the range of observation area. To collect information from zones of interest, sensor nodes must collaborate together in forwarding gathered data to the sink. This is the function of network layer. Wireless sensor network shares many similarities with ad hoc network, in that they are both battery powered, randomly deployed, and autonomous distributed systems. However, routing protocols designed for traditional ad hoc network do not necessarily fit the requirements of sensor network due to much harsher working constraints: (1) the senor device is smaller in size and is more prone to energy depletion, (2) The sensor network can be several orders of magnitude higher in number of nodes (3) Low data rate wireless radio, limited memory, and computation ability make network-wide cooperative computation difficult, thus many path setup mechanisms adopted in ad hoc network cannot be used. Moreover, as sensor network has attribute-based addressing and location awareness, the unique node address in ad hoc network is useless in sensor network. All these adverse factors, especially the physical constraints, have made the design of routing protocols for sensor networks very challenging.

In the design of routing protocols, these principles must be taken into consideration:

- Energy efficiency should be always put at the first priority.
- Sensor networks are data-centric and should have attribute-based addressing and location awareness.
- Internode data and filtering aggregation are useful in energy dissipation reduction.

## 17.5.1 Flooding

Flooding is the simplest method for message delivery from the observation site to the sink [1]. In this scheme, each sensor neither needs to maintain a routing table, nor needs to compute the next hop to send the message. On receiving the packet, a sensor just rebroadcasts it out if it is not the potential information consumer. Thus, flooding is robust to topology changes as each sensor does not need to keep a neighbor list. It has the advantages of low computing complexity and no memory needed in caching the paths.

Nevertheless, the scheme has some fatal deficiencies:

- *Implosion.* Implosion is the phenomenon that duplicated messages are received by the same node. For example, in Fig. 17.10, initially a message is broadcast from node A. Finally, sink node S receives three copies of the same message from its neighbors. Moreover, in rebroadcast, each node also receives several copies of the message from its own neighbors.
- *Overlay.* If two nodes are monitoring the same zone, and both detect the stimuli simultaneously, then overlapping pieces of sensing data are gathered. When overlapping data are flooded by these two nodes, network communication load will be doubled. Therefore, it is a waste of energy and bandwidth by sending two copies of data to the sink node. This phenomenon will get worse with the increase in network scale. An example is shown in Fig. 17.11; sensors A and B detect the event "E," thus they both report it to sink S, resulting in the overlay problem.
- *Resource blindness.* Each sensor blindly rebroadcasts messages without any consideration of the energy and bandwidth consumptions. Thus, flooding is

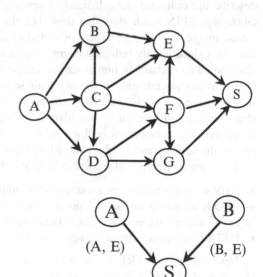

**Fig. 17.10** Implosion problem in flooding

**Fig. 17.11** Overlay problem in flooding

extremely inefficient in the resource utilization and can only be applied in small-scale networks in which complicated routing protocol is not a must and traffic load is not heavy either.

## 17.5.2 Gossiping

Gossiping is a slightly improved version of flooding [1]. Instead of broadcasting a received message to all its surrounding nodes, the sensor randomly selects one neighbor to send the message. This neighbor also randomly chooses a neighbor to pass the message. The process continues until the message finally reaches the destination. Therefore, the implosion problem in flooding is avoided. However, this scheme can cause long delay to propagate message to the destination, hence is not an efficient routing protocol.

## 17.5.3 SPIN: Sensor Protocols for Information via Negotiation

SPIN (sensor protocols for information via negotiation) is a negotiation-based information dissemination protocol suitable for wireless sensor networks [1]. Compared with Flooding and Gossiping, SPIN is efficient in that it uses metadata negotiations to eliminate redundant data transmission throughout the network. Moreover, SPIN enables the sensor node to base its communication decision on the available battery resource, thus consuming energy more efficiently and extending the sensor lifetime.

Specifically, in SPIN, the sensor uses metadata to concisely and completely describe the collected data. Instead of sending the whole data as in Flooding or Gossiping, SPIN sends data that describes the properties of collected data (e.g., sound, image, or video). The size of metadata is much smaller than that of the raw data, thus significantly reducing energy consumption. Before actual data transmission, metadata exchanges among sensors are performed via a data advertisement. On receiving this advertisement, the neighbor sensor checks whether it has this data. If it is interested in the data, it retrieves the data by sending a request message. Otherwise, it can just ignore the advertisement. In this way, metadata negotiation overcomes the deficiencies in flooding – redundant information passing, overlapping in the sensing area, and resource blindness.

There are three types of messages in SPIN: ADV, REQ, and DATA.

- ADV is an advertisement message containing meta-data of the actual data.
- REQ is a request message. Upon receiving ADV, if the senor node is interested in the actual data, it sends REQ to retrieve the data.
- DATA is the actual data message containing abundant information.

Note that ADV and REQ include only metadata, thus their sizes are much smaller than that of DATA. The precedent ADV-REQ exchange is cheaper in energy expenditure than the corresponding DATA. Therefore ADV-REQ exchange is a good

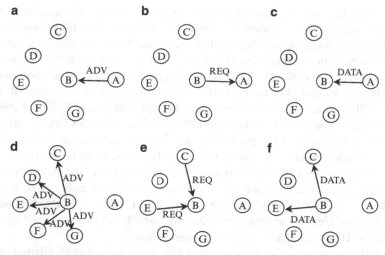

**Fig. 17.12**  Packet dissemination in SPIN

approach to avoiding energy wasted in redundant message passing. Let us use an example to illustrate the working mechanism of SPIN.

*Example 5.* In Figure 17.12, sensor A obtains new data and wants to disseminate it to the other sensors in the network. First, sensor A generates an ADV packet, including the metadata describing the actual data, and sends ADV to its neighbor sensor B (**a**). On receiving ADV, sensor B checks whether it really needs the data. If it is interested in these data, it sends a REQ back to A (**b**). When A receives REQ, it begins to transmit the whole actual data to B. Afterward, sensor B repeats the same process by performing the same ADV-REQ-DATA handshake with its surrounding sensor nodes (**d–f**).

Some simple energy conservation mechanism can also be integrated into SPIN. For example, if the energy level is high, a sensor node can participate in ADV-REQ-DATA as described earlier. When the sensor finds that its energy level is below a prescribed threshold, it can adaptively reduce its participation in message exchanges, thus to conserve its limited energy. Another advantage of SPIN is that it is robust to topology changes as a sensor only needs to know its one-hop neighbors. Thus, the computation complexity is low. However, there is no guarantee for successful data delivery. Assume the case that the interested sensors are multiple hops away from the originating sensor, while the sensors in between are not interested in the data, thus data cannot be relayed to the destinations.

### 17.5.4 *LEACH: Low-Energy Adaptive Clustering Hierarchy*

In wireless sensor networks, to facilitate remote monitoring and controlling, the collected raw data must be transmitted to a base station responsible for communications with the outside world. There are many routing protocols available, and they can be classified as either multihop routing or direct transmission. In multihop routing protocols (such as Gossiping and SPIN), data are forwarded in a hop-by-hop manner, while in direct transmission, data are directly sent to the base station. These protocols, however, are deficient in terms of energy efficiency. For multihop routing protocols, sensors closest to the base station must relay packets for sensors that are far away. Therefore, these sensors are unfairly exploited and will die out quickly due to energy depletion. For direct transmission, distant sensors have to raise the output power to ensure reliable packet delivery. Thus, on an average, in transmitting one packet, these sensors consume more energy than those close to the base station. The phenomenon described earlier is what we call energy load unbalance problem. In addition, as the data collected by sensors in vicinity are highly correlated, raw data transmission can be very resource consuming in terms of bandwidth and energy efficiency. Hence, the data should be processed locally to get rid of the data redundancy. This technique is called data aggregation.

LEACH (low-energy adaptive clustering hierarchy) is a clustering-based protocol that minimizes energy dissipation in sensor networks [3]. In LEACH, the whole network is divided into clusters, and in each cluster a sensor node is elected as the *cluster head*, which collects data from the other sensors within the cluster and performs local data aggregation and data forwarding to the base station. Therefore, compared with traditional routing protocols, short-range communication within the cluster and data aggregation at the cluster head significantly reduce the energy consumed. Moreover, to maintain fairness in energy utilization, all sensors rotate to take the responsibility of being a cluster head.

Specifically, key features of LEACH include the following:

- Network management is easy as no complicated routing protocol is needed.
- Inner data transmission between a sensor and the cluster head is much cheaper in energy consumption than a direct transmission.
- Communication burden is alleviated due to local data aggregation.
- Energy load is evenly distributed across the network, and no sensor is overexploited.

To achieve balance in energy consumption among sensors, the operation of LEACH is divided into "rounds," and in each round, sensors are regrouped to form new clusters. Thus, one round consists of two phases: a setup phase, during which new clusters are organized, and a steady phase, during which the sensors transmit data to the cluster head for data aggregation and forwarding.

### 17.5.4.1 Cluster Setup Phase

In the setup phase, each sensor decides whether it should take the duty of cluster head for the current round. The decision is based on the predetermined percentage of nodes that can become cluster head and the number of times this sensor has been cluster head thus far. Specifically, this sensor node $n$ generates a random number evenly distributed between [0, 1], and if the number is less than threshold, the node becomes the cluster head for the current round. The threshold is set as follows:

$$T(n) = \begin{cases} \frac{p}{1-p\times(r \bmod \frac{1}{p})} & \text{if } n \in G, \\ 0 & \text{otherwise.} \end{cases} \tag{17.5}$$

Here, $p(0 < p < 1)$ is the desired percentage of sensor node that can become cluster head, $r$ is the current round, and $G$ is the set of nodes that have not been cluster head in the past rounds.

If a sensor node has decided to act as cluster head for this round, it broadcasts a "cluster head advertisement" to surrounding nodes. Note that there can be multiple advertisements from multiple newly elected cluster heads. A non-cluster-head node thus must choose one to follow. To do this, it selects a cluster head from which it receives advertisement with the strongest signal strength, as this implies that the selected cluster head is the closest one in proximity.

### 17.5.4.2 Cluster Steady Phase

*Example 6.* Figure 17.13 shows how the whole network is divided into clusters. $P = 0.2$; 33 sensors are deployed in an area of 50 m × 50 m.

Once a sensor has selected its cluster, it must notify the corresponding cluster head. A cluster head should also keep a member list for sensors belonging to this cluster. As such, a new cluster is formed. To avoid communication inference among different clusters, each cluster is allocated a unique CDMA code. Likewise, to harmonize data transmission within cluster, TDMA MAC is used and each sensor is allocated a share of time slots. After a new cluster is formed, the TDMA schedule table is broadcast to all member nodes. Therefore, each sensor knows when it can

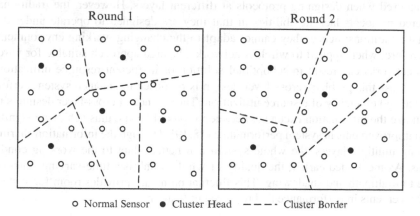

o Normal Sensor    • Cluster Head    −−−− Cluster Border

**Fig. 17.13** Cluster splitting in LEACH

send, and to save energy, it will also power off its radio during the slots allocated to other sensors. In each round, the cluster head must keep its radio on to receive data from its member sensors, thus consuming more energy than normal sensors. Nevertheless, the novel characteristic of LEACH is that each sensor takes turn to assume the responsibility of cluster head. Consequently, no sensor node is unfairly exploited. After network enters steady phase, cluster head begins to assume its duty: data gathering, aggregation, and forwarding to the base station.

## 17.6 Thoughts for Practitioners: Cross-Layer Design for Energy Efficiency

In traditional computer networks, the seven-layer open systems interconnect (OSI) layered hierarchy has been widely recognized and adopted by the industry as the design standard for communication systems. According to the OSI model, the overall networking tasks are divided into seven layers, and services provided by each layer are well defined. This layered hierarchy has been proved successful in that it provides modularity, transparency, and standardization in computer networks. Moreover, each layer must offer certain transparent services to its adjacent higher layer, and the higher layer is shielded from how these services are implemented [6, 15]. This approach has reduced the system design complexity, as designers can just focus their efforts on a particular subsystem without the concern that the overall system architecture will change. Such layered structure has also provided easy compatibility and interoperability among different networks and equipment from various manufacturers.

Different from wireline networks, in wireless networks, packets are transmitted through the wireless channel, which is highly unreliable and characterized by attenuation, shadowing, and multipath fading. In addition, the wireless channel fluctuates with time, and the throughput is a time-varying function. These factors cannot be ignored when designing protocols at different layers. However, the traditional layered protocols lack flexibilities in that they are designed to operate under the worst conditions; hence, they cannot adapt to the changing working environment. Therefore, when applied to wireless networks, layered approach suitable for wireline networks can result in suboptimal solution or inefficient resource utilization. To address this problem, cross-layer design is proposed to enhance system performance and efficiency of resource utilization. The essence of cross-layer design is to optimize the information exchange between protocol layers, thus achieving significant improvements in system performance [18,19]. Through the information sharing among multiple layers, the whole system can better adapt to the working conditions. As mentioned earlier, the quality of wireless link is a time-varying function due to multipath and shadowing. This fluctuation nature provides room for system improvements in performance and resource utilization.

*Example 7.* In wireless ad hoc networks, channel state information can be fed to the network layer for route computation. Thus, with this information from the physical layer, the network layer can discard paths in deep fading and select a route that can support higher data rate. This is the central idea of channel adaptive routing protocol of RICA (receiver initiated channel adaptive routing) [7]. Therefore, bandwidth and energy utilizations are optimized, and network performances are also enhanced [7]. The interaction of network layer and physical layer is illustrated in Fig. 17.14. The average node energy consumption speed and link throughput of the four protocols are given in Figs. 17.15 and 17.16, respectively. It is shown that the sharing of channel state information at the network layer can enable an intelligent routing adjustment according to the instantaneous channel variations.

Similarly, the time-varying property of the wireless channel can also be exploited at the MAC layer of sensor networks, and an interaction between physical layer and MAC layer is also feasible. One such example is adaptive packet fragmentation,

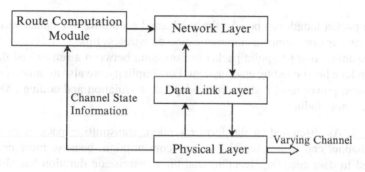

**Fig. 17.14**  Channel state information is fed to network layer

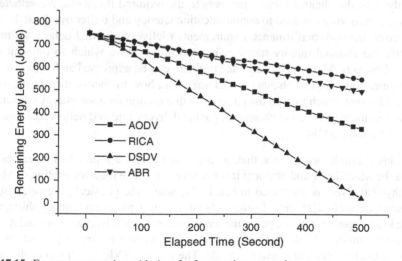

**Fig. 17.15**  Energy consumption with time for four routing protocols

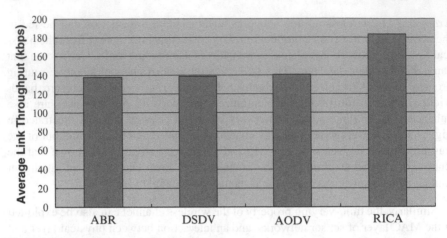

**Fig. 17.16** Link throughput for four routing protocols

in which packet length can be adaptively adjusted according to channel variations. Another case is opportunistic communication on wireless link.

We assume a point-to-point packet transmission between a sensor and the base station under a hostile fading environment. For simplicity, we also assume a constant transmission power level and resort to adaptive modulation and coding (AMC) to combat channel fading.

*Example 8.* As discussed in the former section, transmitting packets over deep fading channel can incur excessive energy consumption, because more energy is dissipated in data encoding/decoding and the transmission duration has also been prolonged. Thus, to save energy, an instinctive approach is to buffer the packet temporarily until the channel quality recovers to the required threshold. Nevertheless, the packet buffering can lead to communication latency and buffer overflow. Therefore, given network performance requirements (delivery rate and delay), we must quantify the channel quality transmission threshold above which to transmit the buffered packets. More specifically, the problem can be expressed as follows: when communication load and channel model are given, how to choose the transmission threshold above which to send the packets, so that communication energy consumption is minimized, while the prescribed packet delivery rate and delay requirements can still be fulfilled [8].

In this example, we can see that, to conserve energy, a joint cross-layer design across the MAC layer and physical layer is necessary. The joint integration of MAC and physical layers is illustrated in Fig. 17.17, where the physical parameters such as instantaneous CSI, channel fluctuations (signal strength and frequency shift), and traffic load, together with the performance requirements (delivery rate and delay) are fed into the mode selection unit, which performs queuing recursion [8] and selects the most energy-efficient operation mode. The output (SNR threshold) of this unit and the instantaneous CSI from the physical layer are provided to the MAC layer

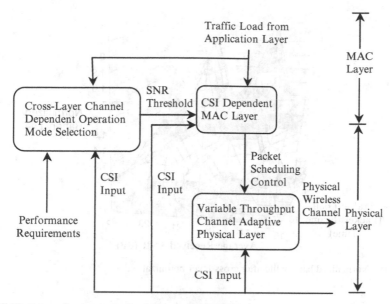

**Fig. 17.17**   Interaction between physical layer and MAC layer

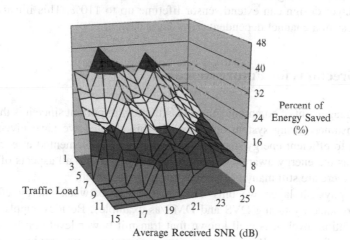

**Fig. 17.18**   Percentage of energy saved after cross-layer optimization

for packet scheduling: if the instantaneous channel quality is above the threshold, send the packet; otherwise, buffer them until the channel recovers. At the physical layer, the transmitter is under the control of MAC layer scheduler and performs a "burst-by-burst" throughput adaptation according to the current channel quality. The percentage of energy saved versus traffic load and average received SNR are presented in Fig. 17.18. It is observed that after optimization, we can save up to 40% of the energy consumption. Energy saving can also extend the sensor lifetime,

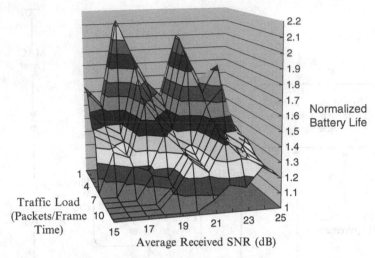

**Fig. 17.19** Normalized battery life after cross-layer optimization

and the corresponding gain in lifetime is shown in Fig. 17.19. We observe that using the cross-layer design can extend sensor lifetime up to 110%. This illustrates the effectiveness of a channel-dependent cross-layer design.

## 17.7 Directions for Future Research

Energy management of communications is extremely important since it is the major power consumer during system operation. In this chapter, we have given an introduction to efficient energy management techniques implemented at each layer. However, as the energy-awareness must be incorporated into all aspects of system hierarchy, there are still many open issues for future research.

At the physical layer, we only discuss trade-off between energy efficiency and performance by taking DVS and DMS as examples. Besides supply voltage and modulation level, it is well known that transmit power level can be adjusted according to distance, SNR, and performance requirement, thus providing us a control knob for energy efficiency. However, transmit power control in wireless sensor network is very challenging, as it requires complicated auxiliary circuits and fast-response adjusting algorithm, which might be difficult to implement in a tiny sensor network. In addition, harmonizing sensors using different power levels is also an interesting problem. On the other hand, to overcome packet corruption due to channel fading or congestion, dynamic and adaptive spectrum utilization is necessary, and it can significantly improve both performance and energy efficiency. Therefore, cognitive radio provides a solution to judicious spectrum management, and this topic is largely unexplored in wireless sensor networks. Some other research issues on energy efficiency at physical layer include hardware design, reducing switching power, and predicting workload in processor.

At DLL, besides energy-conserving schemes we discuss, there is large room for further exploration. Management of operation mode for power saving [1] is such an open issue. To reduce energy expenditure, a sensor node should power off its communication unit when it is not required, and turn it on when packets arrive. Though some work has been done on this topic, more research efforts are still needed. For example, switching off communication unit can incur packet latency and loss; how to strike a balance between energy efficiency and performance is a complicated problem, which entails the considerations of communication load, available battery, performance requirement, and startup overhead. In addition, adaptive packet fragmentation and FEC can be combined with ARQ at DLL to further enhance the energy efficiency, instead of using each alone. Finally, a dynamic power control and coding protocol for optimizing throughput and battery life is also possible, and it has not been addressed in the research literature of wireless sensor network.

Routing in wireless sensor networks is important and it has attracted a lot of attention. In this chapter, we have discussed several classical routing protocols designed for sensor networks. Different from traditional networks, sensor networks are data-centric, and the naming schemes might not be sufficient for complex queries and they usually depend on the specific applications. Therefore, efficient standard naming schemes are an interesting open issue. On the other hand, for cluster-based routing protocol such as LEACH, an unexplored topic is how to form the clusters so that the energy consumption and latency overhead are optimized. The communication among cluster heads (packet routing among cluster heads) is an open issue for future research. In addition, data aggregation and fusion among clusters is also an interesting problem to explore. Finally, location awareness can optimize energy consumption, thus another possible future research issue is how to intelligently utilize location information to provide energy-efficient routing.

## 17.8   Conclusions

The advances in wireless communications, MEMS, and computation technologies have spurred the proliferation of low-cost, low-power, and multifunctional wireless sensor networks. Driven by protocol stacks at each layer, sensor nodes can collaborate together, forming a fully distributed and automatic monitoring and controlling system. It has great application potentials in our daily life, including military, industry, and home networking. Thus, wireless sensor networks have been recognized as one of the most promising technologies in this century.

However, before wireless sensor networks can be widely implemented in practice, there are still a lot of physical problems to be tackled. One of such problems is low-energy design, which is also the focus of our discussion in this chapter. Specifically, due to the limited energy constraint, energy awareness must be integrated into protocol stacks at each layer. As communication consumes the most amount of energy, in this chapter, we have discussed all kinds of typical low-power techniques adopted at layers that are closely related to data communication. With these methods, we can significantly reduce the energy dissipation and extend the sensor lifetime to the best effort. In addition, we have also introduced some cross-layer approaches to energy saving for sensor network. Through interactions between layers, cross-layer design can promote adaptability at various layers based on information exchange. This joint design method has been shown to be effective in energy conservation. Techniques introduced in this chapter are mainly on the communication aspects. Nevertheless, efficient energy management involves all levels of the sensor system hierarchy, from hardware to software architecture, and from operating system to communication protocols. Indeed, all the system components can critically influence the energy dissipation, depending on the applications. Therefore, there is still large room for the future exploitation, and more research efforts need to be put into this topic.

**Acknowledgment** The work was jointly supported by research grant from Natural Science Foundation of China under project number 60602066 and 60773203, and grant from Guangdong Natural Science Foundation under project number 5010494. The work has also got support from Foundation of Shenzhen City under project number QK200601.

# Terminologies

*Energy management.* As battery carried by each sensor is limited, to prolong the battery life, energy resource must be properly utilized. Energy management entails all sorts of techniques at each layer to reduce the power consumption to the best effort.

*Power-aware design.* Limited battery power imposes constraints on the deployment and operation of sensor node. To maximize the usage of scarce energy and extend sensor lifetime as much as possible, power conservation must be incorporated into every level of the system design, from hardware to software architecture, and from operating system to communication protocol.

*DVS.* DVS is the abbreviation of "dynamic voltage scaling." It is an effective technique for reducing CPU energy by dynamically adjusting the clock speed and supply voltage based on the instantaneous workload.

*DMS.* DMS is the abbreviation of "dynamic modulation scaling." It can adapt the modulation level in accord with the instantaneous communication load, thus reducing the energy consumption.

*Packet fragmentation.* The data stream from the upper layer should be fragmented and encapsulated within a data link layer frame. The length of each frame at link layer can significantly influence the transmission reliability of packet on the air, and further the energy efficiency as well. Therefore, the frame length should be adaptively adjusted according to the instantaneous channel state.

*Forward error correction.* By adding redundant data into the message at the sender side, errors can be both detected and corrected at the receiver. This technique can enhance the transmission reliability and decrease the number of sender retransmission, and sometimes reduce the energy consumed in communications as well.

*Medium access control.* Medium access control is a part of the data link layer specified in the OSI model (layer 2). It provides channel access control mechanisms that make it possible for several network terminals to share the same channel. In wireless sensor network, a medium access control protocol is very important, in that it can harmonize the channel access among multiple competing sensor nodes, thus reducing energy wasted in packet collision.

*Data aggregation.* As the data collected by sensors in vicinity are highly correlated, raw data transmission can be very resource consuming in terms of bandwidth and energy efficiency. Hence, the data should be processed locally to get rid of the data redundancy. This technique is called data aggregation.

*Channel fading.* Channel fading refers to the distortion that a carrier-modulated signal experiences over certain propagation media. In wireless systems, fading is due to multipath propagation and is sometimes referred to as multipath-induced fading. Mathematically, fading is usually modeled as a time-varying random change in the amplitude and phase of the transmitted signal.

*Cross-layer optimization.* The central idea of cross-layer optimization is to optimize the control and exchange of information over two or more layers to achieve significant performance improvement by exploiting the interaction between various protocol layers.

## Questions

1. Compared with tr aditional wireless networks, what are the physical limitations restricting the deployment of sensor networks?
2. Briefly explain "energy-aware design."
3. How can system performance be traded-off for energy efficiency?
4. What are the functions of the data link layer in a wireless sensor network?
5. Given the bit error rate $p$ and packet header size $h$, derive the optimal payload size in (17.3).
6. How can S-MAC reduce the energy waste sources in IEEE 802.11?
7. Can routing protocols in ad hoc networks be applied in wireless sensor networks?
8. What are the advantages of LEACH in terms of energy utilization?
9. How can SPIN save energy?
10. Briefly explain "cross-layer design."

## References

1. Akyildiz IF, Su W, Sankarasubramaniam Y, Cayirci E (2002) A Survey on Sensor Networks. IEEE Communication Magazine 40:8, pp 102 114
2. Carle J, Ryl DS (2004) Energy-Efficient Area Monitoring for Sensor Networks. IEEE Transactions on Computer 37:2, pp 40–46

3. Heinzelman WR, Chandrakasan A, Balakrishnan H (2000) Energy Efficient Communication Protocol for Wireless Microsensor Networks. In: Proc. 33rd Annual Hawaii International Conference on System Sciences, pp 1–10
4. Kwok YK (2007) Key Management in Wireless Sensor Networks. In: Y Xiao and Y Pan (eds), Security in Distributed and Networking Systems. World Scientific, London
5. Kwok YK, Lau KN (2007) Wireless Internet and Mobile Computing: Interoperability and Performance. Wiley, New York
6. Lau KN, Kwok YK (2006) Channel Adaptive Technologies and Cross Layer Designs for Wireless Systems with Multiple Antennas: Theory and Application. Wiley, New York
7. Lin XH, Kwok YK, Lau KN (2005) A Quantitative Comparison of Ad Hoc Routing Protocols with and Without Channel Adaptation. IEEE Transactions on Mobile Computing 4:2, pp 111–128
8. Lin XH, Kwok YK, Wang H (2007) On Improving the Energy Efficiency of Wireless Sensor Networks Under Time-Varying Environment. In: Proc. of the 32nd IEEE Conference on Local Computer Networks (LCN), Dublin, Ireland
9. Lu G, Krishnamachari B, Raghavendra CS (2004) An Adaptive Energy-Efficient and Low-Latency MAC for Data Gathering in Wireless Sensor Networks. In: Proc. of IPDPS, pp 26–30
10. Min R, Bhardwaj M, Cho SH (2002) Energy-Centric Enabling Technologies for Wireless Sensor Networks. IEEE Transactions on Wireless Communications 9:4, pp 28–39
11. Modiana E (1999) An Adaptive Algorithm for Optimizing the Packet Size Used in Wireless ARQ Protocols. Wireless Networks 5:4, pp 279–286
12. Pering T, Burd T, Broderson R (1998) The Simulation and Evaluation of Dynamic Voltage Scaling Algorithm. In: Proc. of International Symposium on Low Power Electronic and Design, pp 76–81
13. Raghunathan V, Schurgers C, Parg S, Srivastava MB (2002) Energy-Aware Wireless Microsensor Networks. IEEE Transactions on Signal Processing 19:2, pp 40–50
14. Sinha A, Chandrakasan A (2001) Dynamic Power Management in Wireless Sensor Networks. IEEE Transactions on Design and Test of Computer 18:2, pp 62–74
15. Shakkottai S, Rappaprt TS (2003) Cross-Layer Design for Wireless Networks. IEEE Communication Magazine 41:10, pp 74–80
16. Shih E, Cho SH, Ickes N, Min R (2001) Physical Layer Driven Protocol and Algorithm design for Energy-Efficient Wireless Sensor Networks. In: Proc. of ACM. MOBICOM 2001, Rome, Italy
17. Shurgers C, Aberthorne O, Srivastava MB (2001) Modulation Scaling for Energy Aware Communication Systems. In: Proc. of ISLPED 2001, pp 96–99
18. Srivastava V, Motani M (2005) Cross-Layer Design: A Survey and the Road Ahead. IEEE Transactions on Wireless Communications 43:12, pp 112–119
19. Su W, Lim TL (2006) Cross-Layer Design and Optimization for Wireless Sensor Networks. In: Proc. of the Seventh ACIS International Conference on Software Engineering, Artificial Intelligence, Networking, and Parallel Computing, pp 23–29
20. Ye W, Heidemann J, Estrin D (2002) An Energy-Efficient MAC Protocol for Wireless Sensor Networks. In: Proc. of INFOCOM, pp 1567–1576

# Chapter 18
# Transmission Power Control Techniques in Ad Hoc Networks

Luiz Henrique Andrade Correia, Daniel Fernandes Macedo, Aldri Luiz dos Santos, and José Marcos Silva Nogueira

**Abstract** Communication is usually the most energy-consuming event on mobile ad hoc networks. Hence, medium access control (MAC) protocols use techniques to mitigate energy consumption on the transmission of data. This chapter presents one such technique, which consists of adjusting the transmission power of the packets. We discuss the fundamentals of transmission power control (TPC), showing their effects in wireless communication and the requirements of TPC solutions. Next, we examine the issues associated with their implementation and show the difficulties of implementing those techniques on real hardware, based on our experience with TPC-aware MAC protocols for wireless sensor networks. We close this chapter with a glimpse of future challenges of TPC-aware MAC protocols on MANETs and WSNs.

## 18.1 Introduction

Mobile ad hoc networks (MANETs) are used in situations where there is no preinstalled infrastructure, or the existing infrastructure cannot be used due to catastrophic events. Hence, the devices organize themselves to form a mobile wireless network, where nodes forward messages from one another, acting as data sources and forwarders. MANETs have more stringent requirements than wired networks, either due to the environment itself or due to hardware limitations. The most important restriction, from our point of view, is the limited amount of energy available to the nodes. This occurs because nodes should be small in order to be mobile, thus they must resort to batteries as their power source. The scarce energy availability,

L.H.A. Correia (✉)

Department of Computer Science, Campus Universitário – Caixa Postal 3037, 37200-000 Lavras –, Federal University of Lavras, Brazil
e-mail: lcorreia@ufla.br

S. Misra et al. (eds.), *Guide to Wireless Sensor Networks*, Computer Communications and Networks, DOI: 10.1007/978-1-84882-218-4_18,
© Springer-Verlag London Limited 2009

**Table 18.1** Energy consumption in Mica Motes2

| Device | Current |
|---|---|
| *Processor* | |
| Full operation | 8 mA |
| Sleep mode | 8 μA |
| Transceiver (0 dBm) | |
| Reception | 8 mA |
| Transmission | 12 mA |
| Sleep | 2 μA |
| *Flash memory* | |
| Write | 15 mA |
| Read | 4 mA |
| Sleep | 2 μA |
| *Sensor* | |
| Full operation | 5 mA |
| Sleep | 5 μA |

Source: [5]

in turn, impacts the design of hardware and software for MANETs, which should always be designed with energy consumption in mind.

Among the operations performed by MANETs, communication is usually the most energy consuming, as shown in Table 18.1[1] for a wireless sensor node. Thus, protocols developed for MANETs should be adapted to WSNs in order to reduce their resource usage and improve their scalability. Approaches to reduce the energy consumption of MANETs can be found on every communication layer [15, 18, 23]. However, the medium access control (MAC) layer is one of the most relevant on reducing the energy consumption involved in the communication, since it coordinates the transmission of every packet.

There are three sources of energy waste on the MAC layer, as we portray in Fig. 18.1. The first one occurs when radios listen to the medium but it is not necessary, instead of staying on an idle mode. To avoid idle listening, CSMA/CA protocols employ low duty cycles, where nodes sense the medium for a certain amount of time to check if there are ongoing transmissions. If a transmission is not found, the radios are disabled again until the next sensing period. One protocol that employs this strategy is B-MAC [18]. Another approach is to multiplex the access to the medium, allocating a certain amount of time, frequency, or codes for each node to transmit its data. One protocol that employs this technique is S-MAC [24].

Once rid of idle times, MAC protocols should optimize the energy consumption of the frames. One alternative is to reduce the amount of frames sent by turning off nodes that are not essential to the operation of the network. This task is commonly performed by topology control protocols, which identify the minimum set of nodes necessary to maintain an acceptable sensing coverage of the entire area and also

---

[1] Even though the flash consumes as much energy as the transceiver, these memories are scarcely accessed due to its high read and write time.

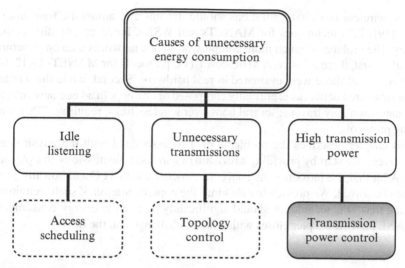

**Fig. 18.1** Reducing energy consumption on the MAC layer

build efficient routes [2]. Besides having fewer frames to transmit, the amount of contention on the medium will be reduced, as less nodes will compete to acquire the medium.

Once the problems above have been tackled, we look at the amount of energy required to send frames. To be correctly received, frames must arrive at the sender at a minimum power, so that they can be detected by the transceiver and at the same time be discernible from noise. Normally, MAC protocols employ a very high transmission power, thus wasting energy. Hence, we could reduce the transmission power so that the received signal is closer to the minimum power required for frames to be decoded. This technique, known as transmission power control (TPC), is the focus of this chapter.

There are several benefits on the use of TPC techniques on MANETs. Besides reducing energy consumption, TPC techniques reduce the number of collisions and the amount of packets lost due to interference. Further, it increases spatial reuse, allowing more nodes to transmit at the same time. As a consequence, the capacity and the throughput of the network can be increased [6].

In MANETs, TPC techniques have been implemented either in the MAC layer [4, 9, 14] or in the routing layer [11]. We argue that, since the information required to calculate the minimum transmission power is provided by the MAC and PHY layers, this operation should be done on the MAC layer. Routing protocols, on the other hand, would depend on the PHY and MAC layers to obtain the required data. Further, MAC protocols could adjust the transmission power of each frame, be it a RTS, CTS, data, or an ACK, while the routing layer would be limited to adjusting the transmission power on a coarser granularity.

TPC techniques have been extensively used on infrastructured networks [16, 20]. Recent standards, such as WiMax and IEEE 802.11h, mandate that equipment used on wireless broadband solutions should dynamically adjust the transmission power [19]. TPC techniques for MANETs and WSNs, however, are still on their infancy. The problem is much more complex on ad hoc networks than on structured networks. First, there are several MAC-level TPC protocols for MANETs [9,12,14]; however, few of those were evaluated in real hardware. Second, while the wireless part of structured networks is normally composed of one hop, in ad hoc networks the communication may traverse several hops over wireless links, requiring TPC-aware routing protocols.

This chapter describes the problems of TPC associated with the physical and MAC layers. We start by providing a revision of wireless transmissions on a physical level. After that, we introduce the concepts associated with TPC and how this problem can be solved. We provide insights into the implementation of such techniques, showing that TPC techniques should significantly reduce the energy consumption of MANETs. This chapter closes with a list of challenges on the field.

## 18.2 Background

To motivate the study of TPC techniques, this section describes the benefits of TPC on MANETs and the physical quantities that are employed to calculate the ideal transmission power. Next, the requirements for the implementation of TPC techniques are described; the causes of signal attenuation in the transmission medium and how to maintain a good link quality are also presented.

### 18.2.1 Benefits of TPC

TPC influences several characteristics of wireless communications, such as node connectivity, medium contention, delay, energy consumption, and data rate [11]. In the following we briefly discuss each of them.

Connectivity. Network connectivity is determined by the establishment of links among nodes, and depends on the correct reception and decoding of frames. TPC techniques influence this process, since the transmission power determines if the frame will overcome the interference, attenuation, and signal distortions imposed by the medium [7]. TPC techniques can be used to maintain a certain degree of connectivity, increasing the transmission power if link reliability falls below a certain threshold. Wireless links are usually asymmetric, that is, the characteristics of the transmission vary according to the sense of the communication. Link asymmetry may hinder the operation of several protocols that rely on messages coming on both ways. One example is frame acknowledgment on CSMA/CA MAC protocols. A node A might be able to send a frame to node B; however, A may not receive the acknowledgment frame, causing unnecessary retransmissions, collisions, and frame losses, hence leading to a reduced throughput [9]. TPC techniques can be used to mitigate asymmetric links by adjusting the transmission power in both senses of the communication.

Medium Contention. The transmission range is determined by signal strength. Thus, higher transmission powers will increase the number of nodes sharing the medium, and as a consequence the probability of collisions will be heightened. A more elevated contention

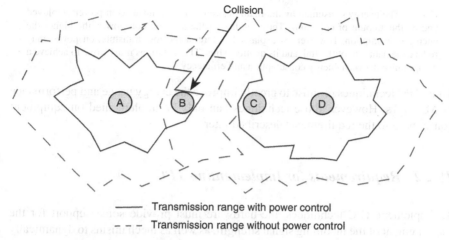

**Fig. 18.2**  Adjusting the transmission power to avoid collisions

on the medium also might increase latency and reduce the delivery rate. The amount of contention can be controlled using TPC techniques. By transmitting data at the exact power required to ensure a successful communication, only nodes that really must share the same "space" will contend to access the medium, as shown in Fig. 18.2. In this figure, the polygons represent the transmission range of nodes $A$ and $D$. The dashed lines represent the regular transmission range, while the solid lines represent the reduced transmission range. In this example, if nodes $A$ and $D$ concurrently transmit data at the regular transmission power to nodes $B$ and $C$, respectively, a collision will occur at node $B$. If the transmission power is reduced to the minimum necessary to reach the destination of the packet, no collisions occur, and $A$ can send data to $B$ without disrupting the transmission from $D$ to $C$. As a result, contention will be reduced, and the network will support more simultaneous transmissions, enhancing network utilization. Finally, a smaller transmission range will reduce the amount of hidden and exposed terminals [14], since fewer nodes overhear transmissions from others.

*Throughput.* The throughput of the links is directly influenced by the number of nodes within the transmission range, due to contention. The adjustment of the transmission power will reduce the amount of competing nodes, thus less retransmissions will be required to send data [8]. TPC may be used to increase medium reuse, since a reduced range allows more transmissions to be carried out simultaneously [6]. Further, TPC can be used to transmit frames using stronger coding and modulation, thus providing higher data rates. In several wireless standards, such as Wi-Fi and WiMax, the data rate of the link is dynamic, changing according to the conditions of the environment. For example, on an IEEE 802.11g network, you might get 54 Mbps on favorable conditions (i.e., a station near the AP, with no walls between them); however, the data rate may drop for distant stations or for stations suffering heavy interference.

This mechanism compensates the interference by using more resistant coding techniques; however, this comes with the cost of a reduced data rate. Using TPC, nodes could increase the transmission power on links with higher interference in order to maintain a high data rate.

*Latency.* Latency is a function of the number of hops traversed in the communication. TPC techniques may increase or decrease latency by changing the amount of hops on each route. Higher transmission powers allow shorter routes, which will have a smaller latency. Meanwhile, lower transmission powers will require longer routes.

*Energy.* The energy consumed by the radio depends on the transmission power employed and on the amount of collisions. Thus, the higher the transmission power, the higher the energy consumption. However, as explained before, a too low transmission power may reduce the data rate of the link and its quality. Thus, TPC solutions must strive to achieve a compromise between energy consumption and efficiency.

Thus, TPC techniques promise to greatly improve the energy usage and performance of MANETs. However, those techniques can only be implemented on equipment conforming to the requirements described later.

## 18.2.2  Requirements for Implementing TPC

To implement TPC techniques, the hardware must provide some support for the measurement of the incoming signal strength, as well as mechanisms to dynamically change the strength of outgoing transmissions. We describe those requirements here.

- *Fast transmission power switching.* The transceiver must be able to switch the transmission power on-line, without restarting the hardware. The modification of the transmission power should also be fast, allowing the use of different transmission powers from one packet to another. While in structured wireless networks packets are always forwarded to the same node (the access point), in MANETs flows may be forwarded to different nodes. Thus, the transmission power may change for every outgoing packet.
- *Incoming signal strength readings.* The radio must provide interfaces to measure the instantaneous strength of the signal picked up by the radio. Those measurements are used to calculate the average noise floor, as described later in this section, and also as input for the equations used to calculate the minimum transmission power.

Once the radio supports the measurement of link characteristics, we must quantify the condition of the link when idle, as well as the average signal loss incurred when the signal travels from source to destination, in order to implement efficient TPC algorithms. The following section gives an overview of those tasks.

## 18.2.3  Assessing Link Condition

To calculate the ideal transmission power, TPC algorithms require the knowledge of the signal strength when there are no transmissions, as well as the signal strength of each incoming frame. This is provided by readings coming from the *received signal strength indicator (RSSI)* port.

To distinguish data packets from garbage, wireless radios periodically sense the signal strength during the inactivity periods, which is called *noise floor (NF)*. Since the sources of electromagnetic radiation change constantly, the noise floor is an approximation that changes over time. The measurement of the noise floor is performed by *clear channel assessment* (CCA) algorithms and is implemented in hardware in some radios, while in others it is up to the host processor to run the CCA algorithm. Usually, the CCA calculation is used by MAC protocols to check if the medium is busy before transmitting frames. Thus, nodes only transmit data if the instantaneous strength of the signal is not higher than *NF*. If this is not the case, probably there is a transmission in place, and nodes wait a certain amount of time before retrying their transmission. The noise floor is also important to determine the minimum reception strength required to correctly decode a frame, as we will show in Sect. 18.3.2.

Another measurement necessary for the implementation of the TPC techniques is the reception strength, which is also provided by RSSI readings. The reception strength is compared against the transmission strength, in order to identify the loss of power suffered by the signal on its way from the sender to the receiver, which is called *signal attenuation*.

Signal attenuation determines the reach of the signal. Reijers et al. [21] and Lal et al. [13] demonstrated that signal range does not follow the concentric circle model. Also, the attenuation varies with humidity, temperature, node, and antenna positioning. Finally, the attenuation is influenced by the existence of obstacles and moving sources of interference (animals, devices transmitting at the same frequency, buildings, cars, etc.).

Consider the network topology in Fig. 18.3, where node *A* broadcasts a message to nodes *B* and *C*. Although node *B* is the closest to *A*, *C* will probably receive the signal from *A* with a higher signal strength than *B* (the strength is indicated by the thickness of the line), since there is an obstacle blocking the signal, increasing the attenuation of the link from *A* to *B*.

The first technique that comes to mind to determine the attenuation is the use of signal propagation models commonly used in simulations (e.g., Friis, Two ray

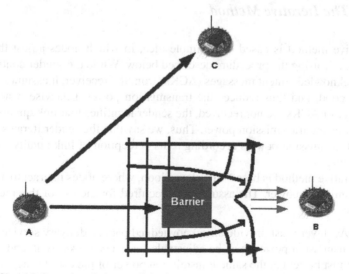

**Fig. 18.3** Signal fading when facing an obstacle

ground, Gilbert-Elliot, etc.). However, this approach does not overcome the existence of obstacles. Without providing nodes with a priori knowledge of existing obstacles and how the signal will degrade on each of them, which is impossible in most deployments, propagation models will yield unreliable results. Furthermore, more complex propagation methods that address the dynamics of the medium, such as Okumura, Hata, and logarithmic attenuation, are too computation intensive to be employed in the restricted hardware used in MANETs.

Because of those limitations, the estimation of the attenuation must be performed either by comparing the transmission and reception powers, or by using trial and error to identify the minimum transmission power that overcomes the attenuation imposed by the medium. Those two methods form the basis of existing TPC algorithms and will be described in the following section.

## 18.3 Thoughts for Practitioners

In this section we describe general approaches to adjust the transmission power on MANETs and WSNs. The first approach employs dynamic adjustments, where the transmission power is increased or decreased by small steps at a time, until the minimum transmission power is found. The second calculates the ideal transmission power according to signal attenuation in the link, thus finding the minimum transmission power using only one iteration. The third and fourth methods are improvements of the first two, based on issues brought up by the implementation of TPC-aware techniques on real hardware.

### 18.3.1 The Iterative Method

The iterative method is based on a simple idea, in which nodes adjust the transmission power using the procedure described below. While the sender continuously receives acknowledgement messages (ACKs) from the receiver, it assumes that link quality is good, and thus reduces the transmission power. Likewise, when a certain number of ACKs are not received, the sender identifies that link quality is poor and increases the transmission power. Thus, we say that the sender iterates over the available transmission powers according to its perception of link quality, hence the name *iterative*.

The iterative method is based on a closed loop, where nodes interact to assess the ideal transmission power. The assumptions required for the use of this method are as follows:

- The MAC layer must provide an acknowledged packed delivery service.
- The transmission power must be adjustable for every packet sent, and the ACK frame must be sent at the same transmission power of the data frame, in order to avoid asymmetric links.

- The transceiver must provide a finite number of discrete power values, in which the method will iterate over (in the Mica Motes2 platform for WSNs, for example, there are 22 power levels, separated by roughly 1 dBm [3]).
- The transmission of data occurs in a continuous flow, allowing the method enough time to converge. For rare transmissions, the characteristics of the medium change faster than the method can cope with, and in such situations the minimum transmission power may never converge.

The iterative method operates in two phases, as shown in Fig. 18.4. The first phase determines the ideal transmission power, while the second phase adapts the transmission power to changes in environmental conditions.

*First phase.* The first phase of the iterative method identifies the minimum transmission power required to correctly send packets to a given destination. The first packet is sent at the maximum power allowed by the radio ($P_{TXmax}$). If the reception is acknowledged, the sender will decrease the transmission power by one level ($P_{TX} = P_{TX} - 1$). This is repeated until a packet is not acknowledged, since the non-reception of a packet indicates that the transmission power employed was not sufficient to reach the destination. Thus, the ideal transmission power is the least value that allowed the correct reception of the packet. Next, the method enters its second phase.

*Second phase.* In this phase, the method adapts the transmission power to any variations that might occur in the environment, such as node mobility and interference. This reaction is determined by thresholds, which dictate when the transmission power must be decreased or increased. Those thresholds are adjusted according to

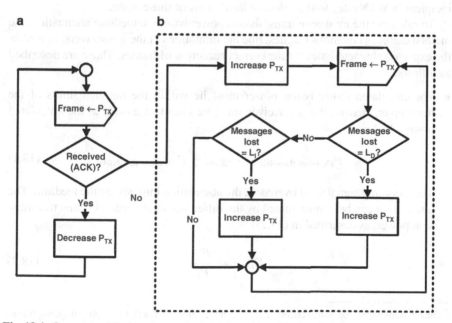

**Fig. 18.4** Operation of the iterative method: (**a**) first phase (**b**) second phase

the characteristics of each deployment. Thus, if a given number of successive transmissions are acknowledged, reaching a *decrease threshold* (*LD*), the transmission
power is decreased by one level. Similarly, if a predetermined amount of messages
are not acknowledged, reaching the *increase threshold* (*LI*), the transmission power
is increased by one level.

To avoid constant recalculation of the transmission power, this approach stores
the calculated value on a table. Thus, before sending a packet, nodes verify if an
entry referring to the destination exists on this table. In this case, nodes will use the
stored transmission power to send their packets. If no such entry exists, they will
use the maximum transmission power allowed by the radio ($P_{\text{TXmax}}$).

## 18.3.2 The Attenuation Method

Another approach for the calculation of the minimum transmission power is to quantify the attenuation suffered by the transmitted signal when traversing the medium.
Thus, whenever an incoming frame arrives at the transceiver, nodes compare the
value of the transmission power with the reception power, in order to determine the
amount of attenuation suffered by the signal. With this information, nodes adjust
the transmission power in a way that the reception power is the minimum required
to correctly decode frames, considering that the signal strength will be degraded by
the previously measured attenuation value.

This approach was initially proposed by Karn [10] in the context of WLANs and
has been extensively employed in MANETs [14, 17]. However, this technique is still
incipient in WSNs due to the resource limitations of those nodes.

To calculate the minimum transmission power based on medium attenuation, the
method employs equations that describe the limitations of the transceivers, as well as
the expected characteristics for the correct reception of frames. Those are described
as follows:

- The calculated transmission power must lie within the nominal limits of the
  transceiver, that is, the transceiver must be able to transmit at the calculated
  power (18.1)[2]

$$P_{\text{TX lower threshold}} \leq P_{\text{TXmin}} \leq P_{\text{TX upper threshold}}. \tag{18.1}$$

The incoming signal must overcome the attenuation imposed by the medium. The
attenuation can be approximated by the difference of the reception and transmission power, as described in (18.2)

$$G_{i \to j} = \frac{P_{\text{RX}j}}{P_{\text{TX}i}}. \tag{18.2}$$

---

[2] The relations expressed in this chapter are written for values in Watts, to simplify the comprehen-
sion of the method. A transformation is required to apply those equations to values in dBm.

- The strength of the signal arriving at the receiver must be such that the frame can be correctly decoded. To ensure that, the reception strength must be higher than an empirically defined threshold ($RX_{threshold}$), which guarantees a reception without errors. Thus, the signal must be transmitted in a power such that, after subtracting signal attenuation, its value will still be superior than $RX_{threshold}$ (18.3):

$$P_{TXmin(i \to j)} \geq \frac{RX_{threshold}}{G_{i \to j}}. \tag{18.3}$$

- The received signal must be discernible from ambient noise ($NF_j$). This can be accomplished by assuring that the received signal strength is greater than the noise by an amount called signal-to-noise ratio (SNR). Equation 18.4 describes this relation, rewriting the reception power so that it is a function of the transmission power.

$$P_{TX\,min(i \to j)} \geq \frac{SNR_{threshold} \times N_j}{G_{i \to j}}. \tag{18.4}$$

To implement this method, we use the signal strength values provided by RSSI readings and the CCA algorithm (described in Sect. 18.2.3) as input to the aforementioned equations. As the previous method, the attenuation method employs closed loop control. The operations used are presented in Fig. 18.5 and are described later.

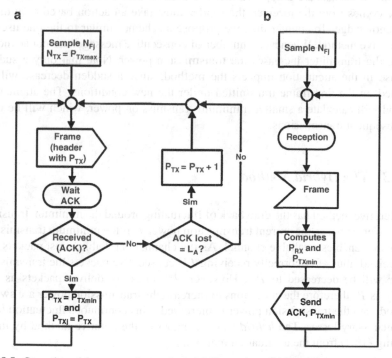

**Fig. 18.5** Operation of the attenuation method: (**a**) Transmitter, (**b**) Receiver

Nodes periodically sample the medium when there are no transmissions, in order to determine the noise floor. A node $i$, when transmitting a frame to a node $j$, informs the transmission power value in a header of the data frame. If this is the first transmission from node $i$ to $j$, the standard transmission power value is used.

Otherwise, the transmission power calculated on the last iteration of the attenuation algorithm is used. When node $j$ receives the frame coming from node $i$, it uses RSSI readings to assess the reception strength ($P_{RXj}$).

Using the average ambient noise and the reception and the transmission power values, node $j$ calculates the minimum transmission power. Equation 18.5, presented here, satisfies the restrictions of (18.1)–(18.4).

$$P_{TX\,min(i \to j)} = \max \left\{ \frac{RX_{threshold}}{G_{i \to j}}, \frac{SNR_{threshold} \times N_j}{G_{i \to j}} \right\}. \qquad (18.5)$$

After calculating the minimum transmission power, node $j$ confirms the reception of the message to node $i$ using an ACK frame, which is sent at the same transmission power employed by node $i$. The calculated transmission power ($P_{TXmin}(i \to j)$) is embedded in the ACK. Thus, at the end of only one transmission, node $i$ will know the minimum transmission power from itself to node $j$. Further, subsequent transmissions will employ this transmission power.

Since medium conditions change, the calculated transmission power should have a lifetime. Suppose, for example, that the condition of the medium gets worse very fast, and all packets are dropped. Since the method uses a control loop, and no response comes from the receiver, the sender must take an action based only on its local knowledge. To counter this, we propose a scheme similar to the one used in the iterative method. Whenever a number of consecutive messages are not acknowledged, the transmitter increases the transmission power. Note that only a sudden increase in the attenuation impacts the method, since a sudden decrease will be detected on the first frame transmitted under the new conditions. The attenuation method will calculate a smaller minimum transmission power, which will be used on subsequent transmissions.

### 18.3.3 The Hybrid Method

The Iterative method has the drawback of fluctuating around the minimum transmission power. Suppose the current transmission power $P$ is the minimum transmission power that enables a reliable channel. As the channel is reliable, most packets will be received, thus after correctly receiving $LD$ consecutive packets the transmission power will be decreased to $P^-$. However, $P^-$ does not deliver packets as efficiently as $P$, therefore the conditions to increase the transmission power are swiftly reached, and the transmission power is increased. This continuous fluctuation leads to unnecessary losses. The *hybrid method* improves the iterative method by incorporating (18.4) from the attenuation method.

To assure that the reception power does not decrease below a certain value, maintaining the quality of the link, the hybrid method controls the reception power of each frame. To do so, it employs (18.6), which is described below ensuring that the transmission power does not drop below a safety value that guarantees the correct reception of the packet.

$$P_{\text{RX}j} \geq \text{SNR}_{\text{threshold}} + \text{NF}_j. \tag{18.6}$$

Apart from the aforementioned modifications, the hybrid method works in the same way as the iterative method. It has two phases, where the first one determines the minimum transmission power. The second one, as in the iterative method, adjusts the ideal transmission power.

The hybrid method is presented in Fig. 18.6, and works as follows. In the first phase, the transmission power is decreased by one level whenever an acknowledgment is received. Upon the first frame loss, or if the receiver signals the sender that it must increase its transmission power, the hybrid method is switched to the second

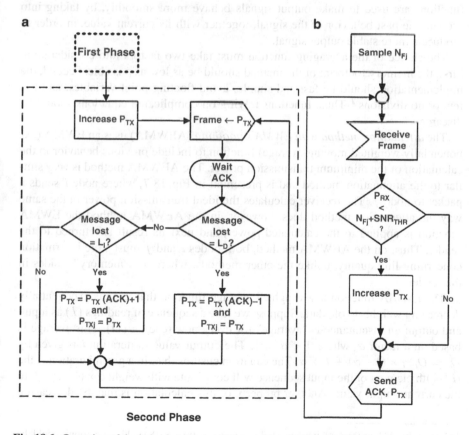

**Fig. 18.6** Operation of the hybrid method: (a) Transmitter, (b) Receiver

phase. In this phase, whenever a packet arrives at a reception power that is too close to the average noise, the receiver notifies the sender that the transmission power must be increased by one level. This notification, embedded in the ACK frame, makes the transmitter increase the transmission power by one level. As before, an increase in the transmission power can also be triggered by a predetermined amount of unacknowledged frames.

### 18.3.4  The AEWMA Method

The attenuation method suffers from fluctuations that lead to a considerable loss of packets, since the input values are always changing due to variations in the environment and the quality of the batteries. Because of those variations, the calculated transmission power varied significantly, thus prompting us to find ways to keep it more stable. To accomplish this, we decided to use signal filters, which are mathematical functions commonly used in the digital control of physical systems. Those functions are used to make output signals behave more smoothly, by taking into account the past behavior of the signal, together with its current value, in order to produce a more stable output signal.

The choice of the averaging function must take two factors into consideration. First, the memory footprint of the method should be as low as possible. Second, the implementation should be fast and simple, avoid floating point variables, and use few or no divisions.[3] Thus, functions relying on complicated equations should be discarded.

The *attenuation method with EWMA smoothing* (AEWMA) uses an EWMA (exponentially weighted moving-average) function to include previous behavior in the calculation of the minimum transmission power. The AEWMA method is very similar to the attenuation method and is presented in Fig. 18.7, where node $i$ sends a packet to node $j$. The receiver calculates the ideal transmission power in the same way the attenuation method does; however, in the AEWMA method the EWMA function is applied to the calculated power, and next its result is returned to the sender. Thus, on the AEWMA method, both nodes $i$ and $j$ must store information concerning link quality, unlike the other methods, where the "memory" resides in the sender.

The EWMA function is a weighted average function that assigns exponentially decreasing weights to old data. Suppose we take a sequence of readings ($I$) as input, and output an instantaneous "average" for every new reading ($O$). This average is based on a factor $\alpha$, where $0 < \alpha < 1$. The output value in iteration $i$ is given by $O_i = O_{i-1} \times (1 - \alpha) + I_i \times \alpha$. The equation ensures that, in a given iteration $i$, the $(i - k)$th element of the input sequence will contribute with weight $\alpha \times (1 - \alpha)^{i-k}$ to the current output value. Another property of this calculation is that, by decreasing

---

[3] Most embedded processors implement division on software, thus this operation should be avoided.

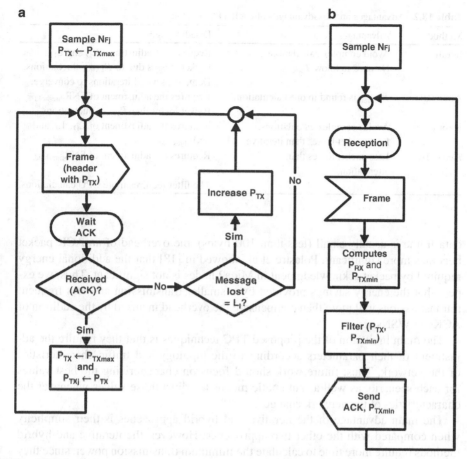

**Fig. 18.7** Operation of the AEWMA method: (**a**) Transmitter, (**b**) Receiver

the value of $\alpha$, more importance is given to past inputs over the recent ones in the final output. The value of $\alpha$ must be carefully chosen, since a high value leads to constant variations on the output, while a low value leads to very slow changes.

## 18.3.5 Analysis and Comparison of the Methods

The proposed TPC methods reduce energy consumption, as shown in [4]; however, they have some disadvantages, such as the necessity of packet acknowledgment and the adjustment of some parameters according to the characteristics of the network. Table 18.2 summarizes the advantages and disadvantages of each method.

Regarding 802.11 MANETs, the use of packet acknowledgement is a minor issue, since data packets are much larger than ACKs. However, in WSNs, where a

**Table 18.2** Advantages and disadvantages of each TPC method

| Method | Advantages | Disadvantages |
|--------|-----------|---------------|
| Iterative | Avoid complex calculations. No need to know $P_{RX}$. | Requires the adjustment $L_D$ and $L_I$. Packets losses due to $P_{TXmin}$ fluctuations. Demands several iterations to converge. |
| Attenuation | $P_{TXmin}$ is found in one calculation. | Requires the adjustment of $SNR_{threshold}$. Packet losses due $P_{TXmin}$ fluctuations. |
| Hybrid | Avoids complex calculations. Less packet losses than iterative. | Requires the adjustment of $L_D$, $L_I$, and $SNR_{threshold}$. |
| AEWMA | Less packet losses than attenuation. | Requires the adjustment of $SNR_{threshold}$ and $\alpha$. The filter requires more CPU calculations. |

data frame is usually small (less than 100 bytes), the overhead of an ACK packet becomes more significant. Polastre et al. showed in [18] that the additional energy required by packet acknowledgment in Mica2 nodes is not significant. Thus, we expect that the energy savings provided by a smaller transmission and less frequent retransmissions will more than compensate the overhead incurred by the addition of ACKs in WSNs.

The main limitation of the proposed TPC techniques is that they require the adjustment of their parameters according to the topology and traffic characteristics of the network. Thus, future work should focus on characterizing the best values for each scenario, as well as automatic means to adjust those values whenever the characteristics of the network change.

The main advantage of the iterative and hybrid approaches is their simplicity when compared with the other two approaches. However, the iterative and hybrid methods require more time to calculate the minimum transmission power, since they gradually reduce or increase the transmission power. The attenuation and AEWMA methods, on the other hand, are able to estimate the minimum transmission power with only one packet exchange.

## 18.4 Future Challenges in TPC

The introduction of TPC in wireless ad hoc and sensor networks provides new opportunities to increase the performance and lifetime of the network. However, the behavior of those protocols is still unclear in some situations. We identified several challenges in the adoption of TPC protocols and we discuss some of them in the following.

*Transmission power aware routing and topology control protocols.* With TPC, routing protocols are able to reach more nodes, as node neighborhood can be extended toward the maximum range of the radio. However, transmitting to a distant node over a multihop path might be more energy-efficient than using single-hop paths. Existing routing protocols traditionally are unaware of this fact, since the amount of energy required to transmit data to their neighbors is always the same. Hence, TPC requires routing decisions to consider the transmission power. Topology control protocols can also benefit from TPC. These protocols turn off redundant nodes while keeping sensing coverage of the area. To keep the network connected, topology control protocols sometimes have to let active some unproductive nodes just to maintain connectivity. Thus, by increasing the transmission power of productive nodes, the number of active but "redundant" nodes can be reduced.

*Node mobility.* Although WSNs are mostly static, some networks might employ mobile nodes. In this scenario, the network suffers partitions when nodes move out of the network, requiring adjustments in transmission power to reach the mobile nodes that leave the covered zone. On the other hand, when nodes move closer to the network, their transmission power can be lowered. The challenge in node mobility is how to cope with motion on different speeds, which dictate the frequency of transmission power updates.

*Multicast and broadcast messages.* Existing TPC techniques assume that the receiver acknowledges the reception of a packet. In multicast and broadcast transmissions, however, acknowledgments are not advisable because of the complexity that they bring to the protocols and of the potentially high number of messages sent by the receivers. There are a large number of publications focusing on calculating the ideal transmission power in order to provide a connected topology. Broadcast and multicast messages could be sent at this power, and multihop paths would be used to route the message to the remaining nodes not reachable within the transmission range. However, it is not proved yet that this is the most energy-efficient solution. That is, the transmission power could be higher, achieving a large percentage of the destination nodes with a single hop, and diminishing the number of message forwards needed to reach the more distant nodes.

*Reconfigurable radios.* Also known as software radios, this new kind of transceiver allows a fine adjustment of physical layer parameters such as operating frequency, bandwidth, number of channels, and modulation techniques [22]. By dynamically changing these parameters, WSNs will be more resilient to changes in environmental conditions. Such resilience can be provided by MAC protocols, if they are capable of negotiating radio parameters among nodes. Software radios will affect how TPC protocols operate, since nodes will be able to decrease the transmission power even further by employing modulation and encoding techniques more resilient to bit errors. The transmission power will encompass a much larger optimization space, since MAC protocols could trade-off energy consumption and bandwidth.

*Underwater networks.* Communication in underwater networks poses even more challenges than terrestrial wireless networks, since data are transmitted using acoustic waves [1]. In such conditions, signal propagation and attenuation differs according to the nature of the environment (shallow or deep ocean, salinity, temperature, depth, surface winds) and spatial orientation of the link (vertical or horizontal). As sound waves propagate five orders of magnitude slower than radio waves, protocols must also account for the delay of each data transmission. In such situations, TPC protocols must be completely redesigned to address those restrictions.

*Adaptive modulation and coding (AMC).* As described in Sect. 18.2.1, most wireless standards today use AMC. This technology influences the design of TPC techniques, since the minimum reception strength varies for each modulation and coding pair. Further, the data rate and the energy consumption of each link will influence the routing decisions, thus routing protocols should be TPC and AMC aware at the same time.

## 18.5 Conclusions

MANETs are composed of mobile nodes, which auto-organize themselves to build an infrastructure-less network. Being mobile, those nodes are usually powered by batteries, which have a reduced capacity due to the small sizes of the devices. Thus, in order to extend the lifetime of those networks, there have been several propositions that try to reduce the energy consumption of such networks. Since communication is usually the most energy-intensive operation in such networks, there is an extensive literature on energy-aware networking protocols for MANETs.

The adjustment of the transmission power, performed by TPC protocols, is a technique to diminish energy consumption on MANETs. TPC-aware protocols consume less energy per packet sent and increase the utilization of the medium, since concurrent transmissions interfere less with each other. This article examined some important topics related to the design and implementation of such protocols on the MAC level. The requirements to the establishment of reliable links have been described, as well as four approaches to adjust the transmission power. We showed that, due to limitations on the existing transceivers and mobile processors, TPC-aware techniques must be as simple as possible. Further, those techniques must cope with the imprecise nature of the data provided by the transceivers.

The use of TPC techniques brings new challenges to the design of WSNs, such as the awareness of the transmission power in routing and topology control protocols. Although TPC requires the coordination of both routing and MAC protocols, very little attention has been given to the routing aspects of TPC. Further, AMC, which is already implemented in most wireless standards, should be included in TPC-aware solutions.

## Terminologies

*Attenuation.* Is a quantity that represents the amount of strength lost when signals travel from their source to their destination. This is important in communications because transceivers can decode a signal only if it reaches the destination with a strength higher than a certain minimum value, called *sensitivity*. Transmissions arriving at lower values cannot be processed.

*Closed loop.* This expression, used in control theory, refers to controllers that use the current input value of the system as well as the last output to provide the next output. The use of the last output allows the controller to adapt to unexpected changes, which occur due to unknown conditions (e.g., an automatic speed controller in a car adapting to bumps in a road).

*Contention.* Indicates the amount of competing sources trying to access the medium. A higher contention can be caused by two things. First, if stations have a significant amount of data to transmit, they will frequently try to access the medium, increasing the contention. Second, when we add more stations, more packets will have to be transmitted, increasing the contention. In CSMA/CA MAC protocols, contention limits the amount of packets that can be sent, due to the increased probability of collisions.

*Duty cycle.* To avoid idle listening, some transceivers are turned off for a certain amount of time. A 60% duty cycle, for example, means that usually the radio will be turned off 60% of the time, while in the remaining amount of time it will be turned on, looking for data.

*Filter.* Are functions used in electronic circuits or computer programs to remove unwanted signal variations. Examples are the low pass, high pass, and PID filters, which remove values higher or lower than a certain value, or sudden variations, respectively.

*Idle listening.* Occurs when radios stay in the reception mode, even though there are no transmissions at the moment.

*Link asymmetry.* Occurs when, for a given pair of nodes, it is possible to send data in one direction, but not in the other. This situation is prejudicial in wireless transmissions, as several MAC protocols require symmetric links, required for a data message (data flows from sender to receiver) to be confirmed (data flows from receiver to sender).

*Modulation and coding.* Define the way data is conveyed on the physical me dium. Modulation defines how zeros and ones will be represented (e.g., turning the signal on and off, varying its strength, its amplitude, its frequency, among other possibilities). Once defined which modulation to use, we apply coding to define a sequence of events (e.g., transitions from zero to one and one to zero in NRZ and NRZI, pairs of values in Manchester) that will define symbols. The choice of the coding and modulation is defined by two factors: the resistance to noise and the achieved data rate. Usually, techniques more resistant to noise have a smaller data rate, because they require more resources to convey the same amount of information.

*Noise floor (NF).* Is an estimation of the average signal strength when there are no transmissions. It is used in MAC protocols to identify when a station can transmit without causing collisions.

*SNR (Signal to noise ratio).* An empirically determined amount, which defines the difference in power required to correctly decode data. If data arrives at a transmission power that is equal to or higher than $SNR + NF$, it has a very big probability of being received without errors.

## Questions

1. The MAC-level techniques employed to adjust the transmission power in MANETs differ from the techniques employed in structured networks in many aspects, due to the different configuration of each network. Which are the main differences among those networks that affect the implementation of TPC protocols?
2. What are the techniques that can be employed to reduce energy consumption in the MAC layer on MANETs? Describe each of them.
3. Describe how TPC techniques can enable the spatial reuse of the transmission medium.
4. Imagine how networks using TPC methods could reduce the latency in multihop paths by increasing energy consumption, or reduce the energy consumed by increasing latency. Suppose that a node $A$ transmits data to another node $B$. There are lots of nodes between them, and any of those could be used to build a route.
5. In wireless networks, the physical layer must know the difference among a data transmission and the noise level. Which are the methods implemented in hardware to distinguish signal from noise? Describe each of them.
6. The signal attenuation of the medium can be estimated using propagation models. However, these models cannot be employed in MANET or WSN due to hardware restrictions. Research in the Internet some propagation models (Okumura, Hata, logarithmic attenuation, and others) and verify the complexity of their equations. Could they be implemented without using floating point operations?
7. The attenuation method presented fluctuations in its calculated transmission power. Based on the description of this method, write a program that simulates the output of the equations. Test it in situations where the input values (the transmission power and the noise) vary frequently. Make a graphic to illustrate the output of your program.
8. Based on the definition of duty cycle, longer duty cycles would lead to more or less latency in packet delivery? Why?
9. Search on the Internet for closed loop signal filters, particularly the PID (proportional, integral, and differential) filter. Is it possible to implement such a filter without using floating point operations? Could we use a PID controller in restricted terminals, such as in nodes of a wireless sensor network?
10. Discuss the implications of using different transmission powers in the RTS/CTS dialogue.

## References

1. Akyildiz, I.F., Pompili, D., Melodia, T.: Underwater acoustic sensor networks: Research challenges. Ad Hoc Networks 3(3), 257–279 (2005)
2. Cardei, M., Wu, J.: Energy-efficient coverage problems in wireless ad hoc sensor networks. Elsevier Computer Communications, 29(4), 413–420 (2006)

3. CC1000: Chipcon products from texas instruments. CC1000 low power FSK transceiver. http://focus.ti.com/docs/prod/folders/print/cc1000.html (2007)
4. Correia, L.H.A., Macedo, D.F., dos Santos, A.L., Loureiro, A.A.F., Nogueira, J.M.S.: Transmission power control techniques for wireless sensor networks. Elsevier Computer Networks **51**(17), 4765–4779 (2007)
5. Crossbow: Mica2: Wireless Measurement System. http://www.xbow.com (2007)
6. Gomez, J., Campbell, A.T.: A case for variable-range transmission power control in wireless multihop networks. In: Proceedings of the IEEE Infocom, vol. 2, pp. 1425–1436 (2004)
7. Gupta, P., Kumar, P.: Critical power for asymptotic connectivity in wireless networks. In: Stochastic Analysis, Control, Optimization and Applications: A Volume in Honor of W.H. Fleming, W.M. McEneaney, G. Yin, and Q. Zhang (Eds.). Foundations and Applications Series, Boston, MA (1998)
8. Gupta, P., Kumar, P.: Capacity of wireless networks. IEEE Transactions of Information Theory **46**(2), 388–404 (2000)
9. Jung, E.S., Vaidya, N.H.: A power control MAC protocol for ad hoc networks. In: Mobi-Com '02: Proceedings of the 8th Annual International Conference on Mobile Computing and Networking, pp. 36–47. ACM Press, New York (2002)
10. Karn, P.: A new channel access protocol for packet radio. In: American Radio Relay League – Ninth Computer Networking Conference, London, ON (1990)
11. Kawadia, V., Kumar, P.R.: Principles and protocols for power control in wireless ad hoc networks. IEEE Journal on Selected Areas in Communications **23**(1), 76–88 (2005)
12. Kubisch, M., Karl, H., Wolisz, A., Zhong, L.C., Rabaey, J.: Distributed algorithms for transmission power control in wireless sensor networks. In: Proc. IEEE Wireless Communications and Networking Conference (WCNC'03), vol. 1, pp. 558–563 (2003)
13. Lal, D., Manjeshwar, A., Herrmann, F., Uysal-Biyikoglu, E., Keshavarzian, A.: Measurement and characterization of link quality metrics in energy constrained wireless sensor networks. In: IEEE GLOBECOM, pp. 172–187 (2003)
14. Monks, J.P.: Transmission Power Control for Enhancing the Performance of Wireless Packet Data Networks. Phd. Thesis, University of Illinois at Urbana-Champaign, Urbana, IL (2001)
15. Narayanaswamy, S., Kawadia, V., Sreenivas, R.S., Kumar, P.R.: The COMPOW protocol for power control in ad hoc networks: Theory, architecture, algorithm, implementation, and experimentation. In: European Wireless Conference, Florence, Italy (2002)
16. Oh, S.J., Wasserman, K.M.: Optimality of greedy power control and variable spreading gain in multi-class cdma mobile networks. In: MobiCom '99: Proceedings of the 5th Annual ACM/IEEE International Conference on Mobile computing and networking, pp. 102–112. ACM Press, New York (1999)
17. Pires, A.A., de Rezende, J.F., Cordeiro, C.: ALCA: A new scheme for power control on 802.11 ad hoc networks. In: IEEE International Symposium on a World of Wireless, Mobile and Multimedia Networks (WoWMoM), pp. 475–477 (2005)
18. Polastre, J., Hill, J., Culler, D.: Versatile low power media access for wireless sensor networks. In: Proceedings of the 2nd International Conference on Embedded Networked Sensor Systems, pp. 95–107. ACM Press, New York (2004)
19. Qiao, D., Choi, S., Jain, A., Shin, K.G.: Miser: An optimal low-energy transmission strategy for ieee 802.11a/h. In: MobiCom '03: Proceedings of the 9th Annual International Conference on Mobile Computing and Networking, pp. 161–175. ACM Press, New York (2003)
20. Rashid-Farrokhi, F., Liu, K., Tassiulas, L.: Downlink power control and base station assignment. IEEE Communications Letters **1**(4), 102–104 (1997)
21. Reijers, N., Halkes, G., Langendoen, K.: Link layer measurements in sensor networks. In: First IEEE Int. Conference on Mobile Ad Hoc and Sensor Systems (MASS '04), pp. 224–234 (2004)
22. Tuttlebee, W.H.W.: Software-defined radio: Facets of a developing technology. IEEE Personal Communications **6**(2), 38–44 (1999)
23. Wan, C.Y., Campbell, A.T., Krishnamurthy, L.: PSFQ: A reliable transport protocol for wireless sensor networks. In: Proceedings of the First ACM International Workshop on Wireless Sensor Networks and Applications, pp. 1–11. ACM Press, New York (2002)
24. Ye, W., Heidemann, J., Estrin, D.: An energy-efficient mac protocol for wireless sensor networks. In: Proceedings of the IEEE Infocom, pp. 1567–1576. New York (2002)

# Chapter 19
# Security in Wireless Sensor Networks

Eric Sabbah and Kyoung-Don Kang

**Abstract** Small, inexpensive, battery-powered wireless sensors can be easily deployed in places where human access could be difficult, dangerous, expensive, or even intrusive to the subject of sensing such as wild animals. Deployed wireless sensors can cooperate with each other to enhance the efficiency of, for example, scientific research, manufacturing, construction, transportation, or military operations. However, the open cooperative nature can expose wireless sensors to various attacks from a malicious adversary. No physical security is available to a wireless sensor network (WSN) deployed in an open environment. Most existing security solutions developed for wired networks are computationally too expensive for wireless sensors with limited energy, computational power, and communication bandwidth. Because of the cooperative nature, many sensor nodes can be affected even when a single node is compromised. This chapter discusses security challenges and vulnerabilities in WSNs. It gives a survey of representative security mechanisms designed to address known vulnerabilities. Finally, it highlights key research issues that remain to be tackled.

## 19.1 Introduction

A wireless sensor network (WSN) consists of small, inexpensive, battery-powered wireless sensors, which self-configure and collaborate with each other to support cost-effective sensing in situations where human observation or wired sytems deployment can be inefficient, expensive, dangerous, or otherwise untenable. Unfortunately, the desirable nature of a WSN introduces many security challenges.

E. Sabbah (✉)

Department of Computer Science, Thomas J. Watson School of Engineering and Applied Science, State University of New York at Binghamton, P.O. Box 6000, Binghamton, NY 13902-6000, USA
e-mail: esabbah@cs.binghamton.edu

S. Misra et al. (eds.), *Guide to Wireless Sensor Networks*, Computer Communications and Networks, DOI: 10.1007/978-1-84882-218-4_19,
© Springer-Verlag London Limited 2009

First, no physical security is available in WSNs. Wireless sensor nodes are often deployed in an open environment such as a battlefield. Thus, attackers can capture sensor nodes to steal secret data or reprogram them to execute malicious code. Because of the cooperative, self-configuring nature of wireless sensors, other uncompromised sensors can also be affected significantly. In the worst case, the entire WSN can fall in the attacker's hand. Wireless radio communication introduces another physical attack – radio jamming. An adversary may use higher energy signals to jam radio communications between sensors. To reduce the energy consumption and hardware cost, many existing wireless sensors, e.g., MICA motes, do not support frequency hopping or spread spectrum techniques to avoid jamming. As a result, parts of the network may become unreachable. Or, in severe cases, the entire system may become useless. Relatively little work has been done to address physical attacks in WSNs.

Because of the open, broadcast naure of wireless communication, attackers can easily eavesdrop on the communication channel. They can also inject false data or control packets into the system. Cryptography is the first line of defense for message confidentiality, integrity, and authentication. However, public key cryptography is computationally too expensive for small wireless sensors. It is known that public key encryption or message authentication takes several seconds in sensor nodes [1]. Public key systems consume more memory than symmetric key systems. Also, they considerably increase the message size. As a result, public key systems can significantly increase the consumption of precious energy and bandwidth. For these reasons, most existing cryptographic solutions in WSNs are based on symmetric key systems. In symmetric key systems, however, key distribution is a main challenge. A simplistic approach is to use a single, global shared key. Under such a scheme, however, the compromise of even a single node would render all cryptographic efforts void. A better solution may involve pairwise shared keys between neighbors. A more detailed discussion of key distribution is given later in this chapter.

WSN services heavily depend on self-configuring and highly distributed protocols for routing, localization, and time synchronization. Although a large number of routing protocols are developed, most of them are designed without considering security. Time and location information are critical, because they provide the spatial and temporal context information for sensor readings. Without the contextual information, the usefulness of sensor data can be greatly diminished. In-network data aggregation can reduce the energy and bandwidth consumption in WSNs by aggregating redundant sensor data. However, an adversary can intentionally inject fake data to disturb data aggregation. Further, end-to-end cryptography becomes implausible, since intermediate nodes need to be able to read the data being forwarded through them [2]. As a result, an adversary can see the sensor readings by extracting a cryptographic key of a single node on a communication path. Also, wrong decisions can be made based on false sensor data injected by an adversary. We will discuss security issues related to these underlying schemes and promising approaches for secure routing, localization, and data aggregation.

Most WSNs have unique traffic patterns, which are different from wired networks or other ad hoc networks [2]. Usually, queries are disseminated from the base station(s) to sensor nodes and sensor data are transmitted from nodes toward the base station(s). As a result, an adversary can observe heavier traffic near the base station. Also, the adversary can correlate message transmission intervals. In this way, he can figure out where the base station is located. Once that is determined, attacks can be concentrated on the base station or the nodes closest to it for maximal impact [3]. Corresponding results could be catastrophic because of the current WSN architecture, which mainly rely on the soundness and security of the base station and nodes near the base station.

Overall, in a typical WSN, sensor nodes with severe energy and other resource constraints are responsible for data collection, data processing, localization, time synchronization, and data forwarding upstream and downstream, while facing numerous security threats. Thus, security is critical to the success of WSNs. More research on WSN security backed by real-world deployment and experiments is required.

In the remainder of this chapter, security vulnerabilities of WSNs are described in detail. After that, a discussion of promising WSN security solutions is given, which is followed by a discussion of thoughts for practitioners, areas for future research, conclusions, definition of terms, and self-check questions.

## 19.2 Background

In this section, before describing security solutions, we clearly define known security vulnerabilities in WSNs. Notably, the following list is not exhaustive by its nature. New vulnerabilites can be found as WSN security research becomes more mature or WSNs are actually attacked in different ways.

*Node compromise* [4–6]. Compared to most computing systems, motes in a sensor network are highly susceptible to physical attack. They are generally placed in open areas, allowing attackers to capture them. This enables an adversary to steal cryptographic information, view and alter their programming, and damage or replace their hardware. Tamper-resistant packaging and camouflaging to prevent the attacker from easily locating motes are possibilities. However, these approaches would increase the financial cost of individual motes so as to make them impractical. Also, designers have found it challenging to accurately predict and thwart adversaries' potential for such attacks. Additionally, the open broadcast nature of radio communications makes it possible for an attacker to add malicious nodes. This attack is feasible, even if sensor nodes are tamper-proof. As a result, software solutions that can tolerate some compromised and/or malicious portion of the network are a must.

*Radio jamming* [5,6]. A denial of service (DoS) attack is broadly defined as any event that impairs or eradicates a network's aptitude for performing its expected function. Often the results of an attack can be similar in effect to service unavailability that results from other problems such as software bugs or power failures. However, for our discussion, we will focus on adversarial causes. The types of attack that constitute denial of service are somewhat broad, as there is much vulnerable functionality that could be subverted. The most basic type of DoS attack is radio jamming. In such a case, an attacker broadcasts a high-energy signal to prevent other nodes from communicating over the wireless channel. With a relatively small number of randomly distributed jamming nodes, the entire network can be rendered useless. It is possible for a node to determine that it is being jammed, since it can observe unusually high ambient energy levels. However, it is difficult for the node to do anything about it, since it cannot coordinate a response with other affected nodes or even report the attack to nodes outside the affected area, or to the base station.

*Compromising routing information* [5–8]. In the wired world, host machines send requests or responses to each other through intermediary routers. These routers are highly specialized machines with increased resources and protection. In contrast, sensor nodes in a typical WSN work as not only data sources but also routers. All these nodes with severe energy, communication, computation, and memory constraints are responsible for data collection, data processing, and data forwarding. Therefore, routing protocols need to be resilient enough to deal with failures that may occur anywhere in the network, while remaining simple enough to be scalable.

Unfortunately, many routing protocols suffer from security holes. An adversary could easily disrupt a network by communicating false routing updates. This could lead to routing loops, generation of false error conditions, an increase or decrease in path lengths, and many of the attacks that we will discuss shortly. As distinguishing a failed node from a malicious one is rather hard, some sort of prevention and/or tolerance is preferable. Basic message authentication protocols might protect against alteration or spoofing of routing packets, but there are other attacks that are hard to prevent by cryptographic approaches, such as authentication and encryption, as discussed next.

*Selective forwarding* [2, 6–8]. Selective forwarding attacks can take many forms. An adversary can drop packets arbitrarily, attempt to give unreasonable priority to its own messages, or misdirect traffic flows. Within the realm of misdirection, a malicious node can perpetrate an attack on a sender by diverting traffic away from the proper destination. Alternatively, messages from many nodes can be misdirected to a single receiver to overwhelm it with traffic.

*Sinkhole attacks* [2]. Sinkholes, also called black-holes, are created when an adversary advertises a very high-quality route to the base station. This may be a true advertisement or a faked one. The result of this announcement of a high-quality path is that neighboring nodes will choose to forward packets through the malicious or compromised node. The neighbors will also advertise to their neighbors that a good route has been found. Those neighbors will then advertise to theirs, and so on. The result is that a large amount of traffic can be diverted to the adversary giving it many opportunities to tamper with the data. In this way, sinkhole attacks can enable many other attacks such as selective forwarding.

*Sybil attacks* [7–9]. Another common attack is the Sybil attack in which an adversary pretends to be several different nodes. Sybil attack is relatively easy to launch in a WSN, since a node in a WSN does not usually have a unique, trusted identifier. Instead, for example, a sensor node is identified by its location augmented by its medium access control address. With no ID (identification) verification mechanism in WSNs, a malicious or compromised node can easily claim false identities. In this way, it can falsely convince many nodes that it is their neighbor. Hence, Sybil attack can make several forms of attacks such as sinkhole and selective forwarding attacks possible. It is also problematic for protocols that rely on voting/consensus schemes.

*Wormholes* [2]. Wormhole attacks commonly involve two distant malicious nodes colluding to understate their distance from each other by relaying packets along an out-of-band channel available only to the attackers. For example, a malicious node can be located near the base station and another distant malicious node can advertise a high-quality route to the base station. Also, a routing race condition typically arises when a node takes some action based on the first instance of a message it receives and subsequently ignores later instances of that message. Wormholes can cause nodes to receive desired routing information before it would otherwise reach them through multiple hops. Thus, an adversary can ensure that it always "wins" the race and can manipulate the topology that is derived by such routing protocols.

*HELLO flood attacks* [2]. This is a vulnerability that is particularly notable in TinyOS, since it uses beaconing for route discovery. The base station sends out periodic hello messages to its one-hop neighbors. Hearing these messages, the receiving motes assume that they are in normal radio range of the sink, and therefore, they mark it as their parent in a tree-like topology. They then rebroadcast the beacon to their neighbors. Downstream nodes accept the first beacon they receive, mark its sender as their parent, and rebroadcast the beacon with their own IDs. This continues recursively until every mote has marked a parent. This is a simple algorithm with little overhead, but unfortunately it opens the system up to much vulnerability. An adversary could replay or spoof one of the hello beacons, thus making itself the root of the tree, and potentially excluding the base station from data flow altogether.

Also, an attacker with greater transmission power such as a laptop-class attacker can potentially convince every node in the network that the adversary is within its normal radio range. When the nodes later attempt to send messages to this false neighbor, many will be sufficiently far away from the adversary as to be sending packets into voids.

*Acknowledgment spoofing* [2]. Acknowledgment spoofing can be quite detrimental and it can involve little more than simply retransmitting an ACK that had previously been sent legitimately. This type of attack can convince the sender that a weak link is strong or that a dead or disabled node is alive. This can lead to a type of DOS or energy exhaustion attack, since packets sent along weak or dead links will get lost. At the same time, precious energy is wasted.

*False data injection to disrupt data aggregation* [7, 8, 10]. The most limited resource in WSNs is energy. Especially, the most energy-demanding task is radio transmission. On the other hand, much of the data collected by sensors is redundant. Thus, reducing the number of transmitted packets via clever data aggregation schemes is a must. However, data aggregation makes end-to-end cryptography infeasible as discussed before. Also, an attacker can have an intense impact on an application by manipulating data without having to attack other aspects of the system such as routing or localization. For example, an adversary can inject a very high temperature reading to create a false fire alarm.

Attack on an aggregation point may allow an adversary to steal or corrupt not only the data that the node collects, but the data from all the downstream nodes. It also allows an attacker to potentially alter the overall aggregated result, which is reported to the base station. In the worst case, by attacking a single node, the adversary could cause equivalent damage to the WSN as if he had captured a large portion of the network. In the previous fire detection example, an adversary may compromise the whole WSN by injecting a single false temperature reading to make the base station misbelieve that there is a fire when the reality is the opposite.

A good solution to this difficult problem must provide error correction at the base station such that a certain level of sensing accuracy is supported even in the presence of false data injections. Furthermore, intermediate nodes doing in-network processing should quickly eliminate any injected false data to avoid spreading false data and consuming energy due to unnecessary forwarding.

*Traffic analysis* [3, 10]. Traffic analysis, sometimes called homing, is caused by the many-to-one traffic pattern prevalent in WSNs. Sensor nodes sense the environment and forward data toward the base station. In this current WSN architecture, compromising certain nodes would be more detrimental than compromising others. One representative example is the base station. If the base station is compromised, the whole WSN falls in the adversary's hands. Other important nodes include data aggregation points and cluster heads elected for special coordination activities. The simplest step that must be taken is to encrypt all information within the packets that could expose routing information. However, this is insufficient. Because of many-to-one traffic patterns, traffic intensity increases near the base station. An adversary can track the packet sending rate of nearby nodes and move toward those with higher rates, until reaching the base station. An attacker can also observe a node sending a message and then the subsequent forwarding of the message by the receiving node. In this way, a time correlation study can allow the adversary to figure out the path to the base station by following the propagation of a packet. To reduce the risk of traffic analysis, one can randomize the traffic pattern or apply data aggregation to avoid rate monitoring and time correlation in WSNs.

*Spoofing Location* [2, 7, 8]. Location information is critical to many sensor network applications. Most sensor readings can lose their meaning without appropriate location information. For example, sensor readings for fire detection or tracking of enemy tanks can be meaningless with missing location information or wrong/compromised location information. Under a collaborative localization scheme, a large number of sensor nodes may get wrong location estimates due to a few malicious or compromised nodes providing false location information. Although each sensor node can be equipped with a GPS (global positioning system), a GPS is too expensive and energy-consuming. To address this problem, much research has been done to discover low-cost distributed localization schemes requiring no expensive hardware such as a GPS [11–18]. Unfortunately, these algorithms are not designed with security in mind. Therefore, an adversary can report false location information to its neighbors without being detected. In the worst case, all sensor readings could become useless due to compromised location information.

If geographic routing is used, an adversary can orchestrate a selective forwarding attack by falsely claiming that it is closest to the base station. It can claim several different locations, if necessary, to attract as many packets as possible. The effect of this Sybil attack combined with selective forwarding is maximized when nodes have no location verification mechanism, but simply believe the location claims of other nodes. In geographic routing, it is also possible to forge location advertisements to create routing loops in data flows without having to actively participate in packet forwarding. This could cause a situation where $node_1$ sends to $node_2$ and then the attacker advertises $node_1$ as being closer to the final destination. Hearing this, $node_2$ forwards the packet to $node_1$, which then forwards to $node_2$, and so on. Such an attack would drain the energy of the attacked nodes, and prevent the packet from being sent where it needs to go, while costing the adversary little resources.

## 19.3  Existing Security Solutions

In this section, we turn to a discussion of well-known existing security solutions developed to address the WSN security challenges described in the previous section. It would be impossible for us to address every algorithm that has been proposed. Also, many of the ones described in this section still have vulnerabilities. However, this section is written to give readers general understandings of WSN security solutions and where to begin further reading.

*Jamming.* A simple response to a radio jamming attack is suggested by Akyildiz et al. [19]. When nodes sense continuous, ambient high-energy readings, they suspect jamming and switch to infrared, optical, or any other available form of communication rather than radio. This approach assumes that the adversary has not jammed these alternative communication media and that the devices being used have such alternatives. However, adding them to a sensor node will increase the cost and energy consumption. Also, infrared or optical communication suffers from the limitation of line-of-sight transmission. Thus, the applicability of [19] to WSNs is limited.

Another effective approach is to use spread spectrum radios [20] where the originally narrow band data are spread to the broadband. A popular approach is frequency hopping in which a sender periodically changes the frequency of the carrier signal in a pseudo random sequence. A receiver knowing the pseudo random sequence can decode the received signal. As a result, an adversary has to either adjust its frequency to follow suit or it has to continuously jam the whole band. However, spread spectrum requires broadband radio hardware and necessitates more complicated transmission algorithms, which cost more energy, memory, and computation.

In large networks, it is possible that jamming can only affect a portion of the network. On the basis of this observation, Wood and Stankovic [21] propose a novel approach to mapping a jammed area. If the jammed area is mapped, unaffected nodes can route around the area. A jammed node blindly sends a jam message with higher energy. Nodes on the border of the jammed region may receive the jam message and collaboratively derive the boundary of the region. Once the jammed area is successfully mapped, nodes outside the region can route around the region. Wood et al. [22] propose several approaches to fighting jamming attack by mote class attackers, which have the same power as WSN nodes. They implement and evaluate their approaches using MicaZ motes, which support frequency hopping. Jamming attack by more powerful attackers is studied in depth in [23].

*Cryptography* [7,8]. Eavesdropping has the goal not of slowing down or stopping the network from doing its business, but of gaining access to the information that is being collected. This may not be a problem in many systems, for example, if sensors are collecting climatic data. However, it can be a huge problem if sensors need to collect medical information or track military activities such as the movements of friendly tanks. The basic solution to this is encryption. In addition to protecting data confidentiality, encryption has additional benefits in protecting routing and other control information, which can help prevent many types of attack. As discussed before, public key cryptography is the mainstay of the wired network security world, but it is simply too expensive to be feasible in WSNs. Also, end-to-end encryption is ineffective due to a strong need for in-network processing of data, such as aggregation. Thus, link-layer symmetric key solutions are often utilized in WSNs. However, if a single global shared key is used, the compromise of even one node would nullify all encryption efforts as discussed before. Thus, key distribution is critical, for example, to support the use of cluster-based shared keys and/or pairwise shared keys between neighbors.

*Key distribution.* Eschenauer and Gligor [24] were one of the first to address the key distribution problem in WSNs. It is difficult to know which nodes will be one-hop neighbors in advance, since sensors are generally deployed randomly. Since it is not known which nodes need to be able to directly communicate with one another, it is impossible to completely set up pairwise key sharing in the predeployment phase. Thus, they developed a scheme where predeployment consists of each node randomly choosing $m$ keys from a predefined key pool that has a total of $n(n \gg m)$ keys. Immediately after deployment, a node communicates with its neighbors to discover which, if any, keys it shares with them by exchanging key IDs without actually giving up cryptographic secrets. The variable, $m$, is a tunable parameter, which can be set so as to support a desired probability of two neighboring nodes sharing at least one key. Nodes that discover that they share a key can verify whether or not their neighbor actually holds the key through challenge/response. The shared key then becomes the key for that link. If two nodes do not have a shared key a *path key* can be established through the neighbors they share keys with. Also, neighbors without shared keys can generate one and pass them to each other via an already secured path. If the network does not have complete secure connectivity after completing this process, *range extension* must be performed, in which nodes temporarily increase transmission power, if hardware permits. Otherwise, they could request that neighbors broadcast the key IDs a few hops until a node with a shared key can be found. Note that this approach is fully distributed and does not require the base station to set up secure paths between two sensor nodes unlike SNEP [25] or TinySec [26].

Chen et al. [27] describe three potential improvements to [24]. The first is the q-composite random key predistribution scheme. In this approach, q common keys are hashed together to compute a shared key between two sensor devices. The benefit of this approach is that, since there are more possible variations of shared keys, an adversary needs to capture and compromise more nodes to compromise the same fraction of communications as [24]. However, to maintain the same probability of connectivity requires a larger $m$. As a result, a compromised single node exposes more keys belonging to the entire set of the $n$ keys. If an adversary succeeds in compromising multiple nodes, the resilience of the algorithm becomes significantly worse than that of [24]. They propose the multipath key reinforcement scheme where a message is partitioned into several fragments and each fragment is routed through a separate secure path. Thus, an adversary needs to compromise at least one node in each path to retrieve the original data. This is more secure, but its overhead is prohibitively high when compared with [24]. Finally, they propose a random pairwise key distribution scheme that provides node-to-node authentication and is resilient to node capture. In the predeployment phase, the key distribution scheme generates $n$ unique node IDs. This $n$ may be larger than the number of nodes in the network, and therefore, more nodes can be added later. Each node ID is matched up with $m$ other randomly selected distinct node IDs and a unique pairwise key is generated for each pair offline. The key and the paired IDs are stored in both key rings. After the deployment, each node broadcasts its identity (node ID, not a key ID) to its neighbors and searches for received IDs in its key ring. It can then initiate challenge/response to verify keys as in the basic scheme.

*Authentication and secure broadcast.* A great number of authentication related-protocols are based on the μTesla algorithm for authenticated broadcast [25]. In the general computing world, authentication is done using asymmetric digital signatures, but this is too expensive for sensor nodes. μTesla introduces the needed asymmetry through a delayed disclosure of symmetric keys. To send an authenticated packet, the base station computes a MAC (message authentication code) on the packet using a secret key and sends it to all nodes. A MAC can be thought of as a secure checksum. No adversary could have altered the packet in transit, because only the base station knows the key at this point in time. Receiving nodes store the packet in a buffer. After a certain length of time that is determined based on the estimated time for communication to all nodes, the base station broadcasts the key to all receivers. When a node receives the disclosed key, it can easily verify it using the previous key, because each key is part of a key chain that was generated by a public one-way function. Before deployment, the base station had chosen the last key of the chain randomly. It then derived the other keys in reverse order through repeated invocation of the hash function. Each key in the chain is the result of hashing its successor key. The last key that is derived is given to all the nodes as a commitment, along with the hash function, before any authenticated broadcast. When a receiver wishes to authenticate a disclosed key, it applies the one-way hash function to it. If the hashing result is equal to the previous key, the disclosed key is valid. If the disclosed key is valid, the node can verify the MAC using the key. If the MAC is correctly verified, it can conclude that the stored packet actually came from the base station and it is unaltered.

*Secure, trust-based routing.* Ideally, a network would be able to detect errors and attacks, distinguish between them, and initiate appropriate responses to recover fully. However, such intrusion detection is difficult in WSNs due to the noisy, dynamic environment involved [28]. Also, the WSN should not be shut down while the source of the problem is eliminated. Therefore, routing protocols must be resilient to such adversarial actions [29–31].

INSENS (Intrusion-tolerant routing protocol for wireless Sensor Networks) [29] can limit the region of effectiveness of an attack. It works in three phases: predeployment, route discovery, and data forwarding. The predeployment phase is similar to that of μTesla. The route discovery phase takes place immediately following deployment and reruns periodically. Route discovery itself is made up of three rounds. In the first round, the base station uses authenticated broadcast to flood a route discovery request to sensor nodes. When a node receives a particular request for the first time, it adds the sender's ID to its neighbor list. It also appends its own ID to the path in the request message as well as a MAC of the entire path so far. After that, it broadcasts the extended request message to its neighbors. Upon receiving a request with a previously received sequence number, it will update its neighbor list but will not forward the request. In the second round of route discovery, a feedback message containing the list of neighbors' IDs and a path from itself to the base station is sent back from each node. This message includes all the MACs generated in round 1 and it is authenticated by a MAC generated using the unique key shared between the node and the base station. In the third round, the base station verifies the received neighborhood information to use it to determine the topology of the network. Based on this topology, it can figure out each node's routing table, which it securely sends to the node using the symmetric key that it shares with the receiving node. Even if captured, a node only reveals its individual key; therefore, the adversary cannot spoof enough MACs to provide a fake path in a feedback message. An attacker can replay discovery requests during route discovery. This could fool downstream nodes into believing an inaccurate topology, but would have no effect on upstream nodes. Unfortunately, this approach has two serious drawbacks: (1) Secure route discovery has a high overhead and (2) the algorithm relies heavily on the base station, which decreases scalability and aggravates the problem of a single point of failure/attack.

ARRIVE [30] is a routing protocol, which is resilient when utilized for sensor networks with a tree topology. In this algorithm, each node listens to and records the historic behavior of neighboring nodes to form a reputation metric. It then forwards packets not only to its parent, but along redundant paths to compensate for possible unreliability of a single forwarding path. For redundant forwarding, each node probabilistically forwards packets to its parent's neighbors that have exceeded a threshold value of the reputation metric.

Abu-Ghazaleh et al. [31] present a trust-based geographic routing protocol that is robust in the presence of packet dropping attacks. This algorithm also keeps track of observed behavior to formulate reputation results of one-hop neighbors and forwards along multiple paths. Once a neighboring node is determined untrustworthy, new paths to the base station are deduced from reputation statistics in conjunction with verified location information. Geographic routing is highly scalable since a node only has to keep the geographic locations of its one-hop neighbors. Reputation information can also be stored without creating large overhead. Additionally, mutually trusting nodes can swap this reputation information to gain knowledge beyond their immediate neighborhood.

A trust management system is a comprehensive scheme comprising of security policies, accreditations, and trust relations designed to tell which nodes are trustworthy and which are not. Given trust values, for example, a node can forward packets to the most trusted node among its neighbors. Also, nodes can elect the most trusted node as the cluster head, while excluding a node misbehaving due to either malice or failure. As WSNs operate in noisy open environments, it could be hard to distinguish a fault from malice. However, a compromised or malfunctioning node may not be cooperative, acquiring a low trust level. Given severe energy and resource constraints, it is challenging to develop an efficient, highly accurate distributed solution for trust management in WSNs. A number of trust management schemes, including [32–41], exist for ad hoc and P2P (peer to peer) networks. However, they do not directly consider WSN constraints. (Fernández-Gago et al. [42] discuss the applicability of the trust management approaches developed for ad-hoc and P2P networks to WSNs.)

A WSN often consists of hundreds or thousands of sensors. Hence, a centralized approach may not scale. Neither is it desirable from the security perspective. An adversary can add colluding malicious nodes to a WSN, which can defeat trust-based routing protocols based on snooping [30, 43]. Further, a malicious node may falsely accuse innocent nodes. A few protocols are developed for trust management in WSNs [30, 31, 44–46]; however, these approaches are still preliminary. They can only partly address the trust management issues in WSNs discussed before. Overall, relatively little work has been done for trust management in WSNs despite the importance. Significant research efforts are needed to clearly identify requirements and constraints for trust management in WSNs and develop solutions to meet the requirements.

*Secure localization and location verification.* Much research has been done regarding localization [11–18]. However, none of these address security issues. Lazos et al. [46] put forth an interesting scheme for secure range-independent localization. Their protocol considers attacks on the localization mechanism to ensure that nodes can determine their own location in a secure manner. However, it does nothing to stop a malicious node from claiming whatever location it wishes.

Sastry et al. [48] propose a scheme for secure verification of location claims. Suppose a node, called a prover, wishes to claim its location. The prover has to send the location claim to the other node, called a verifier, via a RF signal. The verifier sends back a nonce, i.e., a random bit string, to the prover, also via RF. The prover then has to send back the nonce via an ultrasonic channel. The verifier now knows the time it took from sending the nonce to receiving its own location. This location claim is used to derive the claimed distance to the prover. Also, both the speed of light and sound are known in advance. Thus, it can perform coarse grained location verification by calculating the claimed distance divided by the speed of sound to verify whether or not this result matches with the time taken from nonce transmittal to receipt. Unfortunately, this approach requires an additional ultrasonic channel, which can increase the cost of a sensor node. In fact, an adversary can change its location without changing its distance to the verifier; therefore, this approach cannot support location verification but distance verification. Additionally, in the presence of overloads, packet losses, or environmental interference, the prover may respond late. In this way, an innocent node can be misbelieved to transmit a false distance claim.

Abu-Ghazaleh et al. [31] present an approach for location verification that does not require additional hardware. It also provides enhanced location verification, since it does not solely depend on the distance to a single verifier unlike [48]. In this approach, the triangulation process, which is at the heart of localization, is reversed. Rather than listening for localization beacons being continually sent from anchor nodes, a mote sends out requests for localization to anchors. Trusted proximal anchors localize the requester and pass it location information cryptographically certified via the MAC.

A method for combating malicious anchors is discussed in [49]. Nonmalicious beacon nodes test for compromised ones by sending them a location request and estimating their distance from the amount of time it takes to receive a response. This estimated distance is compared with the distance implied by the received location claim. If the distances differ by more than a given threshold parameter, the testing beacon reports a suspected adversary to the base station. The base station decides which anchors it wants to issue revocation alerts based on the number of beacons that have recommended revocation. How many revocation recommendations each anchor node makes must also be considered to prevent an adversary from causing legitimate nodes to be revoked due to a DOS attack.

Another option for localization in the presence of malicious beacon nodes is to attempt to tolerate their presence rather than weed them out [50]. This can save overhead by eliminating extra messages that were used for both testing and revocation in [49]. Within this general methodology there are two options: (1) "Attack-Resistant Minimum Mean Square Estimation," which weeds out outlying location estimates, assuming that they were determined by malicious beacon signals, and (2) "Voting-Based Location Estimation," which divides the total area of the network into cells and treats various location estimates from various beacons as votes for a cell. Finally, the center of the cell with the most votes is the location estimate. Both approaches involve iterative refinement and tolerate malicious beacon signals well as long as the benign beacon signals make up the majority of the "consistent" beacon signals.

*Secure aggregation.* When using aggregation, an attack on an aggregation point allows an adversary to compromise the effect of a large portion of the network on the final result at the base station. By attacking a single node, the adversary can do as much damage to data collection as if he had subverted many individual data nodes. Wagner [51] shows that the minimum, maximum, sum, and mean are all insecure aggregation functions, because they can be affected to any desired degree with a single malicious value. For example, if an adversary wants the mean to increase by 100, it simply needs to add 100 times the number of values being averaged to its true result. The average can be made more resilient through truncation that places bounds on sensor reading's range or trimming, which ignores a certain number of the highest and lowest values. Also, the count is an acceptable aggregation function, as a compromised node can only change 1 to 0 or 0 to 1; that is, its impact is at most 1. The most resilient aggregation function discussed is the median, which is basically an extreme form of trimming (with all but the middle value being ignored). Moreover, Wagner proposes the notion of "approximate integrity" in which an adversary, which has compromised some intermediate node (not the base station or those immediately near it), has a limited effect on the aggregated result. An aggregation system has approximate integrity, if an upper bound can be determined for the effect on the result and the effect can be limited to the same order of magnitude as inherent errors due to noise in the physical world.

Pryzatek et al. [52] propose more specific secure data aggregation protocols, which involve repeatedly going through a three-stage process of aggregate, commit, and prove. In the first stage, the aggregator collects data from sensing nodes and calculates an aggregated result. Next, the aggregator commits to the collected data using a Merkle hash-tree [53, 54]. The leaves of the tree are plaintext data and each internal node in the Merkle tree is the hash value of the concatenation of its two children. The hash function in use is resilient to collisions. Thus, having used the root of the tree as a commitment, a malicious aggregator would be easily caught if he tried to alter any of the values of the leaves. In the final phase, the aggregator sends the aggregated result and the commitment to a verifier. This begins a process of interactive proofs to confirm its correctness. First, the verifying node checks whether the committed data are a good representation of that being reported by the WSN through random sampling. Then, it ensures that the aggregation result is close to the proper result, which should be obtained from the committed values.

*Thwarting data injection attacks.* Data in some WSN applications such as temperature readings are not secret, and thus, eavesdropping may not be an issue of concern. In any sensor network, however, it is important to ensure that the information the base station receives is accurate. This is challenging because compromised nodes can inject false information as discussed before. Two main approaches have been developed to deal with this type of attack: (1) statistical analysis can be done at the base station or aggregation points in order to come up with an approximation of the correct answer [52], or (2) false data can be eliminated through some consensus process. A good solution should not only provide accuracy at the base station, but also should eliminate the injected data as soon as possible to avoid unnecessary forwarding, thus reducing the network lifetime.

One solution of the detection/elimination variety is presented in [10]. This scheme presumes the existence of key distribution and broadcast authentication techniques, such as those discussed in the trust-based routing section. A summary of the five-phase solution follows:

1. Node initialization and deployment:

   (a) Initialization: Every node is loaded with unique ID and information needed to establish pairwise keys.
   (b) Deployment: Nodes establish pair-wise keys with each of their one-hop neighbors.

2. Association discovery:

   (a) Each node discovers the IDs of its associated nodes, i.e., the nodes that are $t+1$ hops away, where $t$ is a threshold parameter equal to the minimum number of cluster nodes that must agree on data along with the cluster head.
   (b) This is done periodically, or upon failure of a neighbor node.

3. Report endorsement:

   (a) Each endorsing node computes two MACs (message authenticating codes) for the data that is being endorsed.
   • One using its key shared with the base station
   • One second with its pairwise key shared with its upper associated node (the one that is upstream of it)
   (b) The nodes send the MACs to their cluster head, which wraps the MACs into a report, and forward report toward the base station.

4. En-route filtering:

   (a) Before forwarding the report:
   • Each node first verifies the authenticity of the one-hop neighbor from which the report was received, using the pairwise key shared with it.
   • Then, they confirm that there are $t + 1$ MACs present (or the number of hops to the base station if it is less than $t + 1$) in the report.
   • Finally, they authenticate the MAC from their lower association node.
   (b) If any of these three tests fail, they drop the report.
   (c) Otherwise, the forwarding node replaces its lower associated node's MAC with a new MAC based on a pairwise key that it shares with its upper associated node, and forwards the new report to the next node toward the base station.

5. Base station verification

   (a) If the base station detects $t + 1$ correct endorsements, it accepts the report; otherwise, it is discarded.

Using this heuristic, if at most $t$ nodes are compromised, then the system can filter out an injected false data after forwarding it to at most $t + 1$ uncompromised nodes. Notably, the more secure the system, the less flexible the deployment structure, since each cluster must have $t + 1$ nodes including the CH, and thus, network density is constrained. Obviously, a small $t$ is more secure, but it requires to have a small cluster size and more clusters in the network, which may limit the flexibility of cluster management. It is required that $t + 1$ nodes simultaneously detect an event of interest and collaboratively generate a report, which may not always be possible when events are transient and sensor nodes are subject to failures and susceptible to environmental noises. Further, a smaller $t$ requires a larger number of upper and lower associated node pairs, which increases the computational cost. Thus, $t$ is a tunable parameter considering the criticality of the appilication of interest and the cost to support a certain $t$. A drawback of this approach is that the bidirectionality of wireless links is required for sensor nodes to find upper associated nodes and report data toward the base station. Unfortunately, wireless links are often unreliable and only unidirectional. Thus, this approach may not work in noisy wireless environments common in the real world.

*Antitraffic analysis.* As disucssed before, an adversary can apply rate monitoring and time correlation for traffic analysis to identify the location of the base station or nodes near the base station. Deng et al. [3] propose four techniques to reduce these vulnerabilities. The first technique is to allow nodes to forward packets to one of a set of parent nodes to make routing patterns less obvious. Secondly, a random walk can be injected into the packet's path. This would serve to distribute data flow, which would reduce the possibility of successful rate monitoring. Another technique is for a random set of forwarding nodes to send bogus packets along fake paths at arbitrary time intervals. This would diminish what could be accomplished by time correlation observations. In the final technique, random areas of artificially high traffic can be induced. This would trick traffic analyzers into thinking that the base station is in the bogus location.

Conner et al. [55] propose a novel approach to antitraffic analysis in WSNs. In their model, sensors' data are first forwarded to a fake sink node, called decoy sink, at which data are aggregated and forwarded toward the real sink. In this way, the network traffic to the real sink is reduced, while the traffic to the decoy sinks increases. As a result, sinks can be hidden from an adversary. However, this approach is vulnerable to a relatively long-term traffic analysis, in which an adversary may find that information flows from decoy sinks to a real sink. Overall, very little prior work has considered antitraffic analysis and additional work is necessary.

## 19.4  Thoughts for Practitioners [7, 8]

Now that we have a firm idea as to the building blocks that are utilized in designing a secure WSN, the time has come to consider which of the tools to use and under which circumstances. One approach would be to attempt to secure a system from any and all possible attacks, and thus to use as many of the proceeding algorithms as possible. However, this is often impractical. Some algorithms may be inherently incompatible, but beyond that there is always a balance that must be struck between the perceived level of threat and the cost system designers are willing to incur to attempt to thwart the threat. This level of threat is a function of both the opportunity afforded to the adversary as well as the motivation level of potential attackers. Motivation level is in turn a function of the potential benefit that can be derived from either stealing the data or interfering in the mission of the network. The application centered security context formed from the combination of attacker motivation and vulnerability factors should be strongly considered when deciding on the comprehensive security policy to deploy.

To illustrate these points, we will outline their applicability to two types of sensor network applications. The first such application is that of habitat monitoring. WSNs are ideal for this work due to their ability to obtain high-resolution information about their surroundings. Additionally, they are unobtrusive compared to direct human observation, and thus will not suffer from the "observer effect" [56].

With regard to such an application, it is hard to see any potential benefit to be gained from attacking such a network other than random mischief. The remote area of many such networks would also increase the cost of attacks. This limited motivation comparative to cost indicates that only moderate security mechanisms are needed. Specifically, there is little reason to implement algorithms to prevent eavesdropping, physical compromise, or traffic analysis and subsequent attack. However, since the main purpose of such an application is to collect accurate data, the main concern is to prevent corruption of this data. Hence, there is a need to provide authentication mechanisms and to combat false data-injection attacks. In the case of data injection, application-specific knowledge in regard to expected ranges of values can aid in statistical methods used to weed out false packets. Also, as discussed, data are often useless without proper context information, so secure localization should also be implemented.

On the other hand, let us consider the case of a battlefield monitoring application. The potential gains from attacking such a system are very high from the adversary's perspective. Successful attacks can turn the tide of war and result in large casualties. Thus, there is high motivation for the adversary to commit resources to detecting the network, accessing its data, and/or disrupting its proper functionality. Also, they are by their very nature deployed in an area where the adversary may already have resources. As a result, it is imperative for a WSN with this type of application to support confidentiality, integrity, authenticity, correct routing, correct location information, and time stamps. Also, it must adhere to real-time constraints and avoid a traffic analysis. Therefore, designers of such systems have quite a challenge and have to be willing to compensate with higher energy and financial costs.

Several optimizations can be utilized if the application's context information is taken into account. At times of high enemy activity or increased proximity, the urgency of both speed and security increases. It may be necessary, for example, to temporarily forgo the benefits of aggregation and instead send high-priority messages using end-to-end encryption. During periods of little activity, these requirements wane, and energy saving approaches can be utilized.

The two cases illustrated are, of course, two extremes but are meant as general guideposts for practitioners to use when inferring the exact requirements and constraints of their own systems.

## 19.5  Directions for Future Research

Security in WSNs is an emerging area with many remaining issues. In this chapter, we have discussed various issues including (but not limited to) node compromise, jamming, key distribution, authentication, secure/trust-based routing, secure localization, and location verification, secure data aggregation, and antitraffic analysis. Each of these issues (and others) deserves further research. For example, relatively little work has been done to tackle jamming problems in WSNs.

Cryptographic approaches in WSNs mainly rely on shared key systems, resulting in challenges for key distribution. Although key distribution is well studied, most existing approaches have certain security drawbacks compared to a public key system. Thus, efficient public key systems such as the elliptic curve algorithm deserve further research. If an efficient public key system can be supported in resource-constrained wireless sensors, key distribution will become much easier. As a result, designing key security mechanisms such as pairwise shared key establishment, secure clustering, and authentication may become much easier, reducing potential mistakes, errors, or vulnerabilities in secure protocol design.

As there are numerous security threats and system failures in WSNs, secure trust-based routing is a must. However, the related work is scarce. Research efforts for defining trust requirements and building trust management systems are needed. A good starting point could be adapting trust management schemes developed for P2P and ad hoc networks to consider resource constraints and security issues unique in WSNs.

Location and time play an important role in WSNs. Without this information, sensor readings could be meaningless. Thus, secure localization in the presence of malicious beacon nodes and verification of location claims by nodes are required, deserving further research.

Secure aggregation and antitraffic analysis is important to securely support in-network processing and protect the base station, respectively. Relatively little prior work has been done for secure data aggregation and antitraffic analysis.

Another important issue that is often ignored is redesigning the sensor network architecture to enhance security. The current architecture requires individual low-end sensor nodes to work as both data sources and routers. By compromising a few sensor nodes, an adversary could control the whole network. Wireless communication in sensor networks is subject to jamming. Many-to-one traffic pattern, which is subject to many DOS attacks and traffic analysis, is rooted in the current architecture where sensors report to a single (or a few) base station(s). If the network architecture is less vulnerable, more secure sensing and control can be achieved. Thus, developing an alternative, more secure different from the current WSN architecture is called for.

As illustrated in Sect. 19.4, since there are many different types of WSN applications with different security requirements, it is reasonable to consider application-specific security context and optimize security solutions based on the attacker motivation and opportunities for attacks [7, 8]. As a result, an application designer could optimize the cost–benefit ratio between security and its impacts on the consumption of energy and other resources as well as performance implications. However, very little prior work has been done in this respect.

Moreover, most existing WSN security protocols are evaluated via simulation or controlled laboratory experiments. Thus, it is necessary to evaluate them in a real-world setting to evaluate the applicability of them to large-scale secure sensor networks deployed in noisy, unpredictable real-world environments. In general, security in WSNs is a fairly new research area with a lot of remaining work to do.

## 19.6 Conclusions

WSNs are a field that presents much promise due to its inexpensive, easily deployable, and self-confirguring nature. As we have seen this very nature, however, creates many security vulnerabilities. Securing a WSN is a challenging problem because of the limited energy, bandwidth, and computational complexity of motes as well as the typically open environment they are deployed in, which negates any notion of physical security.

While facing numerous threats, these severely resource-constrained nodes are responsible for data collection, data processing, localization, time synchronization, aggregation, and data forwarding. Most of this functionality is either not required or handled by specialized, relatively well protected, high-performance machines in traditional systems. Thus, addressing security issues is tremendously important to the success of WSNs. In this chapter, we have discussed security problems in WSNs and outlined well-known existing security solutions in WSNs. Generally, a great deal of further research is required, since many WSN security challenges remain unsolved.

**Acknowledgment** This work was supported, in part, by a NSF grant CNS-0614771.

## Terminologies

*DoS attack.* A denial of service attack is any attack that lessens or eliminates a network's ability to perform its proper duties. This can include low-level attacks such as radio jamming or energy exhaustion attacks. Also, they involve more sophisticated attacks on the routing or higher level protocols. Most attacks discussed in this chapter are DoS attacks.

*HELLO flood attacks.* Attacks where the adversary sends high-power HELLO-messages to convince every node in the network that it is their neighbor.

*Wormhole attacks.* Attacks where two adversaries communicate with each other using a powerful transmitter to pretend to be closer than they are in actuality.

*Sinkhole attacks.* Attacks aiming to draw in as much of the traffic as possible through manipulation of the routing processes. They can be a setup for any selective forwarding attacks.

*Sybil attacks.* Attacks where the adversary pretends to be multiple nodes. This is a vulnerability due to the problem of having no unique identifier like IP address.

*Confidentiality.* Supporting secrecy against eavesdropping attacks. Because of open broadcast nature of wireless communication, confidentiality requires encryption.

*Message Authentication.* Cryptographic process of verifying the identity of a sender, which is accomplished using a MAC (message authentication code). Via a MAC, a receiver can verify whether or not a received message has been altered or spoofed in transit.

*Aggregation.* The process of combining redundant data to reduce the number of packets that need to be transmitted to save precious energy and wireless network bandwidth. Secure aggregation needs to ensure that aggregation results are not skewed due to false data injections.

*Localization.* The process by which a node determines its own location often aided by anchors that already know their locations and broadcast them. Secure localization enables sensor nodes to localize despite compromised anchors potentially broadcasting false beacon messages. On the other hand, verifiers in localization verification check whether or not location claims by sensor nodes are valid. Secure localization and location verification are important because sensor readings in many WSN applications such as target tracking are meaningless without location information. Also, some routing protocols rely on highly accurate location estimates.

*Traffic analysis.* Traffic analysis is a process of examining packet transmission patterns. In WSNs, an adversary may analyze the traffic pattern to find the location of the base station or the nodes near the base station. It can be combined with other attacks to cripple the network by launching, for example, a focused DoS attack on the base station.

## Questions

1. Why must node compromise be tolerated in WSNs?
2. Why are we concerned with the adversary's ability to perform traffic analysis?
3. Why do we use link-layer symmetric key solutions to prevent eavesdropping?
4. What is the main challenge for applying symmetric key cryptosystems to WSNs?
5. What are the main vulnerabilities that should be addressed in a secure localization scheme?
6. Why is median a much more secure aggregation function than mean?
7. How are traffic analysis attacks usually accomplished?
8. What is a trust management system and why would it be beneficial?

9. Why is it undesirable to utilize security algorithms that heavily rely on the base station?
10. Describe the trade-off involved in tuning the five-phase solution [10] for thwarting data injection attacks.

# References

1. D. J. Malan, M. Welsh, and M. Smith, A public-key infrastructure for key distribution in TinyOS based on elliptic curve cryptography, In IEEE SECON, Santa Clard, CA, 2004.
2. C. Karlof and D. Wagner, Secure routing in wireless sensor networks: Attacks and countermeasures, Sensor Network Protocols and Applications, 2003.
3. J. Deng, R. Han, and S. Mishra, Countermeasures against traffic analysis attacks in wireless sensor networks. Technical report, CU-CS-987-04, 2004.
4. R. Anderson and M. Kuhn, Tamper resistance—A cautionary note, Proc. Second Usenix Workshop Electronic Commerce, Usenix, Berkeley, CA, 1996, pp. 1–11.
5. A. Perrig, J. Stankovic, and D. Wagner, Security in wireless sensor networks, Communications of the ACM, vol. 47, no. 6, June 2004.
6. A. Wood and J. Stankovic, Denial of service in sensor networks, IEEE Computer, pp. 54–62, September 2002.
7. E. Sabbah, K. D. Kang, A. Majeed, K. Liu, and N. AbuGhazaleh, An application driven perspective on wireless sensor network security. In Q2SWinet'06, October 2, 2006.
8. E. Sabbah, K. D. Kang, N. AbuGhazaleh, A. Majeed, and K. Liu, An application-driven approach to designing secure wireless sensor networks. Wireless Communications and Mobile Computing, Special Issue on Resources and Mobility Management in Wireless Networks, vol. 8, no. 3, pp. 369–384, March 2008.
9. J. Newsome, E. Shi, D. Song, and A. Perrig, The Sybil attack in sensor networks: Analysis and defenses, In IPSN '04, Berkeley, CA, 2004.
10. S. Zhu, S. Setia, S. Jajodia, and P. Ning, An interleaved hop-by-hop authentication scheme for filtering of injected false data in sensor networks, In IEEE Symposium on Security and Privacy, Berkeley, CA, 2004.
11. P. Bahl and V. N. Padmanabhan. RADAR: An in-building RF-based user location and tracking system, In Proceedings of the IEEE INFOCOM '00, March 2000, Tel Avice, Israel.
12. N. Bulusu, J. Heidemann, and D. Estrin. GPS-less low-cost outdoor localization for very small devices, IEEE Personal Communication, 2000.
13. T. He, C. Huang, B. Blum, J. Stankovic, and T. Abdelzaher. Range-free localization schemes for large scale sensor networks, In MobiCom'03, San Diego, CA 2003.
14. R. Nagpal. Organizing a global coordinate system from local information on an amorphous computer. Technical Report A.I. Memo 1666, MIT A.I. Laboratory, August 1999.
15. D. Niculescu and B. Nath. Ad hoc positioning system (APS) using AoA, In INFOCOM '03, San Francisco, 2003.
16. D. Niculescu and B. Nath. DV based positioning in ad hoc networks, Journal of Telecommunication Systems, vol. 22, nos. 1–4, pp. 267–280, January 2003.
17. A. Savvides, C. C. Han, and M. B. Srivastava, Dynamic fine-grained localization in ad-hoc networks of sensors, In MOBICOM '01, Rome, Italy, July 2001.
18. B. H. Wellenhoff, H. Lichtenegger, and J. Collins, Global Positions System: Theory and Practice, Fourth Edition, Springer Verlag, 1997.
19. I.F. Akyildiz et al., Wireless Sensor Networks: A Survey, Computer Networks, Elsevier Science, vol. 38, no. 4, 2002, pp. 393–422.
20. R. Anderson, Security Engineering: A Guide to Building Dependable Distributed Systems, Wiley Computer Publishing, New York, 2001, pp. 326–331.

21. A. D. Wood, J. A. Stankovic, and S. H. Son, JAM: A jammed-area mapping service for sensor networks, In Real-Time Systems Symposium (RTSS), Cancun, Mexico, 2003.
22. A. D. Wood, J. A. Stankovic, and G. Zhou, DEEJAM: Defeating energy-efficient jamming in IEEE 802.15.4-based wireless networks, in the Fourth Annual IEEE Communications Society Conference on Sensor, Mesh and Ad Hoc Communications and Networks (SECON), San Diego, CA, June 2007.
23. W. Xu, W. Trappe, Y. Zhang, and T. Wood, The feasibility of launching and detecting jamming attacks in wireless networks, in Proc. of MobiHoc. ACM Press, 2005, pp. 46–57.
24. L. Eschenauer and V. D. Gligor, A key-management scheme for distributed sensor networks, In the Ninth ACM conference on Computer and Communications Security, Washington, DC, 2002.
25. A. Perrig, R. Szewczyk, V. Wen, D. Culler, and J.D. Tygar, SPINS: Security protocols for sensor networks, In MobiCom, Rome, Italy, 2001.
26. C. Karlof, N. Sastry, and D. Wagner, TinySec: A link layer security architecture for wireless sensor networks, In ACM SenSys, Baltimore, MD 2004.
27. H. Chan, A. Perrig, and D. Song, Random key predistribution schemes for sensor networks, In IEEE Symposium on Security and Privacy, May 2003.
28. C. Baslie, M. Gupta, Z. Kalbarczyk, and R. K. Iyer, An approach for detecting and distinguishing errors versus attacks in sensor networks. In Performance and Dependability Symposium, International Conference on Dependable Systems and Networks, 2006, Philadelphia, PA.
29. J. Deng, R. Han, and S. Mishra, A performance evaluation of intrusion-tolerant routing in wireless sensor networks, In Second International Workshop on Information Processing in Sensor Networks (IPSN 03), Palo Alto, CA, April 2003.
30. C. Karlof, Y. Li, and J. Polastre, ARRIVE: Algorithm for robust routing in volatile environments. Technical Report UCB//CSD-03-1233, University of California at Berkeley, Berkeley, CA, 2003.
31. K. Liu, N. Abu-Ghazaleh, and K. D. Kang, Location verification and trust management for resilient geographic routing, Journal of Parallel and Distributed Computing, Vol. 67, pp. 215 228, 2007.
32. K. Aberer and Z. Despotovic, Managing trust in a peer-2-peer information system, Proceedings of the Tenth International Conference on Information and Knowledge Management (CIKM01), 2001, pp. 310–317.
33. E. Aivaloglou, S. Gritzalis, and C. Skianis. Trust establishment in ad hoc and sensor networks. In the First International Workshop on Critical Information Infrastructure Security, 2006 (CRITIS'06), Samos Island, Greece.
34. T. Bearly and V. Kumar, Expanding trust beyond reputation in peer to peer systems, In 15th International workshop on Database and Expert Systems Applications (DEXA'04), IEEE Computer Society, 2004.
35. F. Cornelli, E. Damiani, S. Paraboschi, and P. Samarati, Choosing reputable servents in a P2P Network, In 11th International World Wide Web Conference, Honolulu, HI, May 2002.
36. Z. Liang and W. Shi, PET: A personalized trust model with reputation and risk evaluation for P2P resource sharing, In 38th Hawaii International Conference on System Sciences, Hilton Waikoloa Village, Island of Hawaii, 2005.
37. Z. Liu, A. W. Joy, and R. A. Thompson. A dynamic trust model for mobile ad-hoc networks. In Tenth IEEE International Workshop on Future Trends of Distributed Computing Systems, pp. 80–85, Suzhou, China, May 2004.
38. A. Singh and L. Liu. TrustMe: Anonymous management of trust relationships in decentralized P2P systems, In Third International Conference on Peer-to-Peer Computing (P2P'03), IEEE, 2003.
39. N. Stakhanove, S. Basu, J. Wong, and O. Stakhanov. Trust framework for P2P networks using peer-profile based anomaly technique, In 25th IEEE International Conference on Distributed Computing Systems Workshops (ICDCSW'05), IEEE, 2005.
40. Y. Wang and J. Vassileva, Trust and reputation model in peer-to-peer networks, In Third International Conference on Peer-to-Peer Computing (P2P'03), 2003.

41. Z. Yan, P. Zhang, and T. Virtanen, Trust evaluation based security solutions in ad-hoc networks, In NordSec 2003, Proceedings of the Seventh Nordic Workshop on Security IT Systems, Norway, 2003.
42. M. Fernandez-Gago, R. Roman, and J. Lopez, A survey on the applicability of trust management systems for wireless sensor networks, In Proceedings of Third International Workshop on Security, Privacy and Trust in Pervasive and Ubiquitous Computing (SecPerU 2007), pp. 25–30, 2007.
43. S. Marti, T. J. Giuli, K. Lai, and M. Baker, Mitigating routing misbehavior in mobile ad hoc networks, In Mobile Computing and Networking, Atlanta, GA, 2000.
44. S. Ganeriwal and M. B. Srivastava, Reputation-based framework for high integrity sensor networks. In Proceedings of SASN '04: Proceedings of the Second ACM Workshop on Security of Ad Hoc and Sensor Networks, pp. 66–77, New York, 2004.
45. S. Tanachaiwiwat, P. Dave, R. Bhindwale, and A. Helmy, Location-centric isolation of misbehavior and trust routing in energy-constrained sensor networks, In IEEE Conference on Performance, Computing and Communications, pp. 463–469, 2003.
46. Z. Yao, D. Kim, I. Lee, K. Kim, and J. Jang, A Security Framework with Trust Management for Sensor Networks. In Workshop of the First International Conference on Security and Privacy for Emerging Areas in Communication Networks, pp. 190–198, 2005.
47. L. Lazos and R. Poovendran. SeRLoc: Secure range-independent localization for wireless sensor networks, In the ACM Workshop on Wireless Security, San Diego, CA, 2003.
48. N. Sastry, U. Shankar, and D. Wagner, Secure verification of location claims, In the ACM Workshop on Wireless Security, San Diego, CA, 2003.
49. D. Liu, P. Ning, and W. Du. Detecting malicious beacon nodes for secure location discovery in wireless sensor networks, In The 25th International Conference on Distributed Computing Systems, Columbus, OH, June 2005.
50. D. Liu, P. Ning, and W. Du, Attack-resistant location estimation in sensor networks, In IPSN '05, Los Angeles, CA, April 2005.
51. D. Wagner. Resilient aggregation in sensor networks, In SASN'04, Washington, DC, October 2004.
52. B. Przydatek, D. Song, and A. Perrig. SIA: Secure information aggregation in sensor networks, In Proceedings of ACM Sen-Sys, Los Angeles, CA, 2003.
53. R. C. Merkle, Protocols for public key cryptosystems. In Proceedings of the IEEE Symposium on Research in Security and Privacy, pp. 122–134, April 1980.
54. R. C. Merkle, A certified digital signature, In Proc. Crypto'89, pp. 218–238, 1989.
55. W. Conner, T. Abdelzaher, and K. Nahrstedt, Using data aggregation to prevent traffic analysis in wireless sensor networks, *DCoSS*, San Francisco, CA, June 2006.
56. A. Mainwaring, J. Polastre, R. Szewczyk, D. Culler, and J. Anderson, Wireless sensor networks for habitat monitoring, In WSNA, Atlanta, GA, 2002.

# Chapter 20
# Key Management in Wireless Sensor Networks

**Yee Wei Law and Marimuthu Palaniswami**

**Abstract** In wireless sensor networks, cryptography is the means to achieve data confidentiality, integrity, and authentication. To use cryptography effectively, however, the cryptographic keys need to be managed properly. First of all, the necessary keys need to be distributed to the sensor nodes before the nodes are deployed in the field, in such a way that any two or more nodes that need to communicate securely can establish a session key. Then, the session keys need to be refreshed from time to time to prevent birthday attacks. Finally, in case any of the nodes is found to be compromised, the key ring of the compromised node needs to be revoked and some or all of the compromised keys might need to be replaced. These processes, together with the policies and techniques needed to support them, are called key management. In this chapter, we explore different key management schemes with their respective advantages and disadvantages.

## 20.1 Introduction

Information assurance (IA) is a set of measures that protect and defend information and information systems by ensuring their availability, integrity, authentication, confidentiality, and nonrepudiation. These measures include providing for restoration of information systems by incorporating *protection*, *detection*, and *reaction* capabilities [4]. Cryptography lies at the heart of protection capabilities, but to use cryptography effectively, the cryptographic keys need to be managed properly. We can break down a Wireless sensor network (WSN) key management architecture into three main components (1) key establishment; (2) key refreshment; and (3) key revocation. Key establishment is about creating a *session key* between the parties

Y.W. Law (✉)
Department of Electrical and Electronic Engineering, The University of Melbourne,
Parkville, Victoria 3052, Australia
e-mail: y.law@ee.unimelb.edu.au

S. Misra et al. (eds.), *Guide to Wireless Sensor Networks*, Computer Communications
and Networks, DOI: 10.1007/978-1-84882-218-4_20,
© Springer-Verlag London Limited 2009

that need to communicate securely with each other. Key refreshment prolongs the effective lifetime of a cryptographic key, whereas key revocation ensures that an evicted node is no longer able to decipher the sensitive messages that are transmitted in the network. Each of these components must exist in a complete key management framework. A thorough understanding of what role these components play and how they integrate with each other is crucial to the design of a key management framework.

## 20.2  Background

The challenges to designing a key management framework for WSNs lie in the constraints that are unique to WSNs (Table 20.1).

When designing a key management framework, we are essentially meeting both the security objectives and the challenges in Table 20.1 at the same time. Table 20.1 allows us to distill a few design principles:

---

*Design Principle 1.* Favor computation over communication:
In general, we do not mind doing a little bit more computation just to save a few transmissions, as communication is three orders of magnitude more expensive than computation.

*Design Principle 2.* Minimal public-key cryptography:
Public-key algorithms remain prohibitively expensive on sensor nodes in terms of both storage and energy. The use of public-key cryptography should be kept to a minimum, if necessary at all.

*Design Principle 3.* Resilience:
Severe hardware and energy constraints suggest that security should never be overdone – on the contrary, tolerance is generally preferred to overaggressive prevention. This reasoning leads us to design key management schemes that, instead of trying to be perfectly secure, aim to be resilient.

---

Our goal for this chapter is to identify and introduce, based on these guidelines and the state of the art in the literature, key management building blocks for WSNs.

An aspect of key management that is often overlooked in the WSN literature is the *formal verification* of cryptographic protocols, that is, the use of formal methods in mathematics to prove or disprove the correctness of these protocols. In protocol verification, the two most important properties to verify are *secrecy* and *authentication.* However, these problems are well known to be *undecidable* (there is no way to tell whether the property is valid), if we assume the intruder can construct an infinite number of messages, or if there can be an unbounded number of parallel sessions (i.e., parallel executions of the same protocol). One approach to make the problem decidable is to limit the number of parallel sessions. Much of the

**Table 20.1** Constraints of WSNs and their security implications

| No. | Constraint | Implication |
| --- | --- | --- |
| 1 | A node has severe hardware and resource constraints | A node cannot execute cryptographic algorithms that either consume too much energy or occupy too much storage |
| 2 | Sensor nodes operate unattended and hence are susceptible to *node capture* attacks | An adversary can compromise any node |
| 3 | Sensor nodes are generally not tamper resistant | An adversary can compromise all of a node's keys once the adversary captures the node itself |
| 4 | There is no fixed infrastructure | A node cannot generally assume a special-purpose node exists in its vicinity |
| 5 | There is no preconfigured topology | A node does not know in advance who its neighbors are |
| 6 | Sensor nodes communicate in an open medium | All communications are world-readable and world-writable by default |

work that uses this strategy is based on constraint solving. Our secondary goal for this chapter is to give a primer on protocol verification via constraint solving, in the hope that protocol verification will become an integrated step in the design of key management schemes for WSNs in the future.

The rest of this chapter is organized as follows. As preliminaries, we will first introduce the notation for specifying cryptographic protocols. We will then discuss protocol verification by using constraint solving. We will then introduce the building blocks in the three areas: key establishment, key refreshment, and key revocation. All protocols mentioned in the course of discussion will be verified using constraint solving. Finally, we will give a brief conclusion.

## 20.3 Notation for Protocol Specification

The notation in Table 20.2 is used to specify cryptographic protocols for the rest of this chapter.

It is important to note that when both $E(K, M)$ and $\text{MAC}(K, M)$ appear in the same message, the encryption actually uses a subkey generated from $K$, and the MAC uses another subkey, also generated from $K$. For example, given a pseudorandom function $\text{PRF}(\cdot, \cdot)$ and a master key $K$, the encryption subkey can be derived as $\text{PRF}(K, 1)$, whereas the MAC subkey can be derived as $\text{PRF}(K, 2)$. The reason for not using $K$ directly is that some cipher operation modes like the popular CBC are susceptible to *birthday attacks*: if we use the same key to transform more than $O(2^{m/2})$ plaintexts (Exercises 1 and 2), it becomes likely that two or more of these

**Table 20.2** Notation for specifying cryptographic protocols

| Symbols | Meaning |
|---------|---------|
| $A, B, \ldots$ | Sensor nodes $A, B, \ldots$ |
| $N_A, N_B, \ldots$ | Nonces (random numbers) generated by $A, B, \ldots$ |
| $K_{AB}$ | A key shared between $A$ and $B$ |
| $E(K, M)$ | Encryption of message $M$ using key $K$ |
| $MAC(K, M)$ | Message authentication code (MAC) of message $M$ using key $K$ |
| $PRF(K, M)$ | Pseudorandom function with key $K$ applied to plaintext $M$ |
| $\|$ | Concatenation operator |
| $K'$ | New key for replacing $K$ during re-keying |

plaintexts might map to the same ciphertext, allowing data forgery to occur. We say $O(2^{m/2})$ is the *birthday threshold*. Using different subkeys for encryption and for authentication allows us to process more plaintexts before reaching the birthday threshold. Also, unforeseen problems may arise if the same key is used for both encryption and authentication.

## 20.4  Protocol Verification

A number of formal methods can be used for protocol verification, depending on the restriction we impose on the attacker model. If we limit the number of parallel sessions, we can model a protocol using the *strand space model* [8], and use constraint solving to verify its security properties efficiently. The strand space model can be understood informally as a mapping of the notions in the first column of Table 20.3, to the notions in the second column of Table 20.3.

Basically, a protocol consists of *roles* that exchange messages with each other, and the messages that "fly" back and forth between the roles can be visualized as *strands*. A *bundle* is basically a bunch of interleaving strands. A *system scenario* is a hypothetical instantiation of the protocol between some specified principals with a specified outcome. For example, we can specify a system scenario where the principals include one initiator, one responder, one server; we can then define their roles, and specify the outcome as the attacker getting the session key – all of these are our constraints. If we can find a bundle that satisfies these constraints, then we can say the protocol does *not* satisfy the secrecy requirement. Note the fact that a bundle cannot be infinite means we cannot model infinite number of parallel sessions. In WSNs, we are mainly after these three security requirements:

- *Secrecy*. A session key must only be known to the communicating nodes.
- *Authentication* (implies integrity). A key establishment protocol must end with every party properly authenticating the other parties it is communicating with. In other words, it must be impossible for any intruder $M$ to impersonate another node $A$ whose keys (used in the key establishment protocol) $M$ does not have.

**Table 20.3** The strand space model.

| Protocol | Strand space model | Example |
|---|---|---|
| Role: What a principal does in the protocol | Strand: A sequence of events | Initiator, responder, server |
| Complete run: A complete iteration of the protocol | Bundle: A set of strands – legitimate or otherwise – hooked together where one strand sends a message and another receives that same message, that represents a full protocol exchange | 1. Initiator → Attacker: ... <br> 2. Attacker → Responder: ... <br> 3. Responder → Attacker: ... <br> 4. Attacker → Initiator: ... |

- *Replay resistance.* The meaning of replay attack on a role $R$ is the possibility of unauthenticated parties to cause $R$ to run, i.e., for $R$ to process replayed messages. If $R$ happens to maintain the states of every run, then it would be maintaining the incorrect states.

The beauty of this approach is that it can easily be implemented using Prolog. One example is CoProVe.[1] All the protocols that are given in this chapter in standard notation have been verified using CoProVe.

## 20.5  Key Establishment

We start with the first component of key management: key establishment. In precise terms, key establishment is a process or protocol whereby a shared secret key becomes available to two or more parties, for subsequent cryptographic use. There are two types of key establishment protocols:

1. *Key transport*, where one party creates or otherwise obtains a secret value, and securely transfers it to the other(s)
2. *Key agreement*, where two or more parties derive a shared secret as a function of information contributed by, or associated with, each of the parties (ideally), such that no party can predetermine the resulting value.

A *key predistribution* protocol is a key agreement protocol whereby the resulting established keys are completely determined a priori by initial keying material. Key predistribution is essential to WSNs because (1) it minimizes the exchange of information, i.e., communication and (2) it does not require any key distribution

---

[1] http://wwwhome.cs.utwente.nl/~etalle/protocol_verification_TN/coprove/

center (KDC). However, as we shall see, key predistribution is not the only key establishment technique used in WSNs, because due to the resource constraints of sensor nodes, we can rarely predistribute enough keying material such that any pair of nodes would be able to establish a session key. We will look at some key predistribution schemes later.

In WSNs, key establishment is required to support these basic communication modes (1) global broadcast, or flooding; (2) local broadcast; and (3) unicast. Hence, we will discuss the key establishment protocols in the context of supporting these communication modes. Note that for each mode, we can in theory either use symmetric-key cryptography or public-key cryptography, but we are honoring Design Principle 2 by restricting ourselves to using symmetric-key cryptography. The following discusses how key establishment can be done to support the three basic communication modes.

### 20.5.1  Global Broadcast

In doing a global broadcast, a node (sender) intends to broadcast a message to all other nodes (receivers) in the network. The security objective is to ensure the integrity, authenticity, and optionally the confidentiality of the messages from the sender to the receivers. The sender cannot share a key with all the receivers, because any of the receivers can forge messages. The sender also cannot share a different key with each of the receivers, because this solution is not scalable – the sender would have to broadcast a large message consisting of multiple encryptions of the same information but with different keys. Instead, the standard solution for integrity and authentication is µTESLA (the "micro" version of the Timed, Efficient, Streaming, Loss-tolerant Authentication Protocol) [14].

To bootstrap the protocol, the sender first generates a *one-way key chain* $(K_1, K_2, \ldots, K_n)$, where $K_i = H(K_{i+1})$, $i = 1, \ldots, n-1$ and $H()$ is a *collision-resistant hash function*; and distributes the root of the key chain $K_1$ to the receivers securely. $K_1$ is called the *commitment* of the key chain. For this protocol to work, the sender and the receivers must synchronize their clocks. The sender and the receivers divide time into intervals. Whenever the sender broadcasts a message $M_i$ during time interval $i$, the sender always appends $MAC(K_i, M_i)$ to $M_i$. The receivers cannot authenticate $M_i$ until $\delta$ intervals later, when the sender would broadcast $K_i$ (Fig. 20.1). The receivers successfully authenticate the sender if (1) there exists a

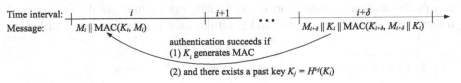

**Fig. 20.1** Keys are released according to a schedule in SPINS

past key $K_j = H^{-1}(K_i)$, $1 \leq j < i$ and (2) $K_i$ generates $MAC(K_i, M_i)$. Note that the security of µTESLA lies in the fact that an attacker cannot regenerate the key chain in ascending order of the index $(i, i+1, i+2, \ldots)$ due to the "one-wayness" of collision-resistant hash functions. The following are the finer points regarding µTESLA:

- How well should the sender's clock and the receiver's clock be synchronized?
  If the sender and the receivers know the maximum time synchronization error, they only need to be loosely synchronized.
- How long should a time interval be?
  A time interval should be long enough for a message to reach all receivers in the network.
- How much should δ be?
  The recommended minimum is 2, because it takes one interval for a message to be fully propagated to all receivers, an interval to serve as a time buffer (this time buffer has to be larger than two times the maximum time synchronization error [7]), and another interval for the corresponding key to be transmitted. In practice, δ is usually small anyway because sensor nodes have little storage to buffer messages while they wait for the authenticating key.
- How long should the key chain be?
  While the key chain needs to be as long as possible, the sensor nodes have limited storage. The obvious strategy is to store some intermediate keys of the key chain, and generate the remaining keys on demand by applying the hash function successively on the stored keys. The best algorithm so far requires $(m+1)(n^{\left(\frac{1}{m+1}\right)} - 1)$ memory units to store the intermediate keys, where $n$ is the length of the key chain and $m$ is number of hash function evaluations to generate one key [9]. However, note the assumption behind this algorithm is $n \leq m^{m+1}$, so $m$ cannot be too small lest $n$ becomes too small, but $m$ cannot be too large either since that would increase energy consumption and latency.

Since keys are distributed along with messages, µTESLA by itself cannot provide confidentiality. In this respect, a global key is usually used alongside µTESLA to provide data confidentiality.

## 20.5.2   Local Broadcast

In doing a local broadcast, a node (sender) intends to broadcast a message to all its neighbors (receivers). The security objective is to ensure the integrity, authenticity, and optionally the confidentiality of the messages from the sender to the receivers. As before, we may use µTESLA to provide integrity and authentication, and a *cluster key* (a key shared between a node and its neighbors) to provide confidentiality. Alternatively, we may relax the time synchronization requirement, because the receivers are just one hop away from the sender. The following protocol to be described is originally designed for *passive participation* – a data communication paradigm in which a node would suppress its own transmission if it overhears its neighbor(s) transmitting similar data. This alternative protocol is essentially μTESLA, used with a cluster key, but without a key disclosure schedule. In this protocol, the sender distributes, as in μTESLA, a commitment of its key chain, and additionally a cluster key to the receivers (which are also the sender's neighbors) [15]. The rationale behind using this key combination is as follows:

- If only the key chain is used, the keys in the key chain would have to be broadcast in the clear, and in the absence of time interval differentiation, a cluster-outsider would be able to forge messages using these keys.
- If only the cluster key is used, authentication of the sender cannot be achieved.
- But if used together, the cluster key can be used to encrypt messages as well as to hide the key chain keys from cluster-outsiders; and at the same time, the key chain keys can be used for authentication.

The disadvantage of this protocol is that an insider attacker would still be able to forge messages to other receivers. Note that this protocol is not suitable for global broadcasts because a global broadcast travels more than one hop, and the lack of time intervals allows a malicious upstream receiver to forge messages to downstream receivers.

### 20.5.3  Unicast

Unicast is one-to-one communication. The security objective is to ensure the integrity, authenticity, and optionally confidentiality of the messages exchanged between two communicating nodes. Denote the two nodes by $A$ and $B$. We only deal with the case where $A$ and $B$ are neighbors, because when $A$ and $B$ are multiple hops away, we can usually secure one hop at a time. Our goal is to establish a session key between $A$ and $B$, which in the WSN literature is called a *pairwise key*.

#### 20.5.3.1  Random Key Predistribution

The prevalent strategy for establishing pairwise keys is *random key pre-distribution* (RKP) (aka *probabilistic key sharing*). The general idea is to prepare a pool of keying material, called the *key pool*; and to each sensor node, distribute a fixed-size subset of keying material randomly chosen from the key pool. The keying material belonging to a node is called the node's *key ring*. Denote the key pool size by $P$ and key ring size by $K$. Having potentially different subsets of the key pool, two neighboring nodes can only establish a pairwise key at a certain probability that is related to $P$ and $K$; that is, a node may not be *securely* connected to all its neighbors. However by adjusting $P$ and $K$, it is possible to make a network securely connected with high probability.

In RKP, this is how two nodes establish a session key: When a node is added to the network, the node initiates *shared-key discovery*, by broadcasting a list of identifiers that identify the keys it has. The neighbors reply with their lists of key identifiers. By comparing the lists, the new node and its neighbors discover what key(s) they share. Session keys are then derived from the shared key(s), for example, by applying a PRF on the XOR of the shared key(s). The disadvantage of this approach is that it allows an attacker to find out which keys a node is holding, giving room to the attacker to attack strategically. An alternative approach is, instead of picking keying material randomly for a node, to pick the keying material according to the result of a PRF. For example, the index of node $A$'s $i$th key is given by $PRF(K_s, A||i) \mod P$, where $K_s$ is a secret key shared by all nodes. Using this approach, a node can, by just knowing the ID of its neighbor, determine the indexes of its neighbor's keys.

Different variants of RKP can be instantiated depending on what we use as "keying material":

- Symmetric keys [6]: The simplest case is to use symmetric keys as keying material. In this case, every node is imprinted with $K$ keys chosen at random from a key pool of size $P$. When a node $A$ is compromised, $A$'s keys may be used to compromise other secure channels that do not involve $A$, since the keys might be stored in some other nodes outside $A$'s communication range as well.

- Polynomials [11]: In this case, the key pool consists of $P$ symmetric $t$-degree bivariate polynomials over a Galois field $GF(q)$, i.e., a pool of polynomials of the form $f(x, y) = \sum_{i,j=0}^{t} a_{ij} x^i y^j$ with $a_{ij} = a_{ji}$; $a_{ij}, x, y \in GF(q)$; and $q$ is a prime chosen to be much larger than the number of nodes (Exercise 9) and at least as long as the desired key length. Denote this set of polynomials by $\{f_1(x, y), \ldots, f_P(x, y)\}$. Every node $A$ is then imprinted with $K$ polynomial shares $f_{i_1}(A, y), \ldots, f_{i_K}(A, y)$, by choosing $i_1, i_2, \ldots, i_K$ randomly from $\{1, \ldots, P\}$. By shared-key discovery, $A$ and $B$ can find out which polynomials they have in common. If that polynomial is $f_1(x, y)$, then without loss of generality, $A$ and $B$ can establish a session key with each other by calculating the key as $f_1(A, B)$, and as $f_1(B, A)$, respectively. When a node $A$ is compromised, $A$'s polynomial shares $f_{i_1}(A, y), \ldots, f_{i_K}(A, y)$ can only be used to compromise secure channels that involve $A$, unless the attacker manages to compromise $(t + 1)$ shares of one of the shared polynomials.

- Matrices [5]: In this case, the key pool consists of $P$ matrices $M_1, M_2, \ldots, M_P$ of size $N \times (t + 1)$ over Galois field $GF(q)$, where $N$ is the expected total number of nodes in the network; $t$ is a security parameter; and $q$ is a prime chosen to be much larger than $N$ and at least as long as the desired key length. The matrices are generated in three steps: First, a Vandermonde-like matrix $G$ of size $(t + 1) \times N$ over Galois field $GF(q)$ is generated using a primitive element $s$ of $GF(q)$:

$$G = \begin{bmatrix} 1 & 1 & 1 & \cdots & 1 \\ s & s^2 & s^3 & \cdots & s^N \\ s^2 & (s^2)^2 & (s^3)^2 & \cdots & (s^N)^2 \\ \vdots & \vdots & \vdots & \ddots & \vdots \\ s^t & (s^2)^t & (s^3)^t & \cdots & (s^N)^t \end{bmatrix}$$

At the second step, $P$ random symmetric matrices $D_1, D_2, \ldots, D_P$ of size $(t+1) \times (t+1)$ are generated. Thirdly and finally, the final matrices are calculated as $M_1 = (D_1 \cdot G)^T$, $M_2 = (D_2 \cdot G)^T, \ldots, M_P = (D_P \cdot G)^T$. $G$ has the following useful properties: (1) since $s$ is a primitive element and $N < q$, $s$, $s^2$, $\ldots$, $s^N$ are all unique and can be used as sensor IDs; (2) any $(t + 1)$ columns of $G$ are linearly independent. The following are what get distributed to the $j$th node: (1) the $j$th column of $G$, denoted $G(j)$; (2) the $j$th row from each of $M_{i_1}, M_{i_2}, \ldots, M_{i_K}$, denoted $M_{i_1}(j), M_{i_2}(j), \ldots, M_{i_K}(j)$, where $i_1, i_2, \ldots, i_K$ are randomly chosen from $\{1, \ldots, P\}$. Therefore, in theory, each node has to store 1 matrix column and $K$ matrix rows; but in practice, each node only has to store the 2nd element of its assigned column and $K$ matrix rows, because all elements of the same column are just different powers of the 2nd element of the column. For example, the 1st node stores $s$, the 2nd node stores $s^2$, and so on. By shared-key discovery, nodes $i$ and $j$ can find out which matrices they have in common. If that matrix is $M_1$, then without loss of generality, $i$ and $j$ can establish a session key with each other by first disclosing the column $G(i)$ and $G(j)$, respectively, to each other, and then calculating the key as $M_1(i)G(j)$ and as $M_1(j)G(i)$, respectively. Note that $M_1 G = G^T D_1^T G$ is symmetric, hence $M_1(i)G(j) = (M_1 G)_{ij} = (M_1 G)_{ji} = M_1(j)G(i)$, i.e., node $i$ and node $j$ are able to derive the same session key. When node $j$ is compromised, node $j$'s matrix rows $M_{i_1}(j), M_{i_2}(j), \ldots, M_{i_K}(j)$ can only be used to compromise secure channels that involve node $j$, unless the attacker manages to compromise $(t + 1)$ rows of $M_{i_1}$ or $M_{i_2}$ or $\ldots$ or $M_{i_K}$, because any $(t + 1)$ rows of $M_{i_1}$ or $M_{i_2}$ or $\ldots$ or $M_{i_K}$ are linearly independent.

We now consider the case where $A$ and $B$ do not share any key, but each has a secure link to a common neighbor $S$. In this case, $A$ and $B$ can still establish a session key through $S$ acting as a trusted third party. The following key transport protocol can be used to establish a session key $K_{AB}$ between $A$ and $B$ via $S$ [14]:

---

**Protocol 1**

$A \to B : N_A || A$

$B \to S : N_A || N_B || A || B || \mathrm{MAC}(K_{BS}, N_A || N_B || A || B)$

$S \to A : E(K_{AS}, K_{AB}) || \mathrm{MAC}(K_{AS}, N_A || B || E(K_{AS}, K_{AB}))$

$S \to B : E(K_{BS}, K_{AB}) || \mathrm{MAC}(K_{BS}, N_B || A || E(K_{BS}, K_{AB}))$

$A \to B : \mathrm{Ack} || \mathrm{MAC}(K_{AB}, \mathrm{Ack})$

---

Protocol 1 has been verified with CoProVe to be (1) secure with respect to the secrecy of $K_{AB}$, (2) secure in the mutual authentication between $A$ and $B$, and (3) secure against replay attacks on $S$ [10].

### 20.5.3.2   LEAP+

An alternative scheme to RKP, as part of LEAP+ [15], is as follows:

1. First, embed an initial key $K_{IN}$ in every node.
2. Upon bootstrapping, every node $A$ derives its own master key as $K_A = \text{PRF}$ $(K_{IN}, A)$, and set its timer to fire at time $T_{min}$ later. $T_{min}$ is the estimated minimum amount of time for an attacker to compromise a node. $A$ sends out a HELLO message containing its ID.
3. As long as the timer has not fired, if $A$ hears a HELLO message from a neighbor $B$, it will derive the pairwise key as $K_{BA} = \text{PRF}(\text{PRF}(K_{IN}, A), B)$ and acknowledge $B$. If on the contrary, $B$ receives and replies to $A$'s HELLO message first, then the pairwise key would be $K_{AB} = \text{PRF}(\text{PRF}(K_{IN}, B), A)$ instead.
4. When the timer fires, $K_{IN}$ is erased from memory.

This scheme is, however, only useful for static networks, since after $K_{IN}$ is erased, a node can no longer derive pairwise keys.

### 20.5.3.3   EBS

Exclusion Basis Systems (EBS) [12] is a variation of the symmetric-key-based RKP. Basically, instead of choosing $K$ out of $P$ keys at random, EBS chooses $K$ out of $P$ keys *uniquely* for each node, so there are only $P!/[K!(P-K)!]$ ways of choosing, and there can only be a maximum of $P!/[K!(P-K)!]$ nodes. By picking $K > P/2$, EBS makes sure every pair of nodes share at least one key, hence guaranteeing the network is connected. The drawback of this scheme is that, when a node is compromised, only $P - K$ keys, or less than half of the keys in the key pool remain intact. Because of this, a WSN that uses EBS is most often compartmentalized into clusters, with a different key pool assigned to each cluster, to make the whole system more resilient to node capture.

## 20.6   Key Refreshment

As mentioned, different subkeys are used for encryption and for authentication because that would allow the birthday threshold to be reached more slowly, but the birthday threshold will eventually be reached. The standard solution to further delaying the birthday threshold from being reached is key refreshment, i.e., the process of refreshing shared secrets periodically as a means to increase the birthday threshold (the cryptography literature generally uses "key refreshment" and "rekeying" interchangeably, but we reserve "rekeying" for the process that follows key revocation). There are two mainstream approaches [1]:

1. *Parallel rekeying.* We start with keys $K_{enc,0}$ and $K_{mac,0}$. The $i$th ($i = 1, 2, \ldots$) refreshed keys are $\mathrm{PRF}(K_{enc,0}, i)$ and $\mathrm{PRF}(K_{mac,0}, i)$. Note: $K_{enc,0}$ and $K_{mac,0}$ can be generated from the same master key $K_0$ via $\mathrm{PRF}(K_0, 1)$ and $\mathrm{PRF}(K_0, 2)$.
2. *Serial rekeying.* We start with key $K_0$. The 1st refreshed keys are $\mathrm{PRF}(K_0, 1)$ and $\mathrm{PRF}(K_0, 2)$, respectively, for encryption and MAC. The $i$th ($i = 2, 3, \ldots$) refreshed keys are $PRF(\underbrace{PRF(\ldots PRF}_{i-1 \text{ times}}(K_0, \underbrace{0)\ldots, 0}_{i-1 \text{ times}}), 1)$ and $\mathrm{PRF}(\underbrace{\mathrm{PRF}(\ldots \mathrm{PRF}}_{i-1 \text{ times}}(K_0, \underbrace{0)\ldots, 0}_{i-1 \text{ times}}), 2)$, again respectively for encryption and MAC.

The advantage of using these approaches is as follows. Suppose the key length is $k$. If the session key is not refreshed, the birthday threshold is $2^{k/2}$. If the session key is refreshed every $2^{k/3}$ function invocations (where "function" is either encryption or MAC), the session key can be refreshed $2^{k/3}$ times before birthday attack is likely to succeed. In other words, the birthday threshold is increased from $2^{k/2}$ to $2^{2k/3}$.

For WSNs, serial rekeying is preferred, because in parallel rekeying, the counter $i$ and the key $K_0$ have to be stored, and if a node is compromised, these information would allow an attacker to generate all past keys *in addition to* future keys. In other words, parallel rekeying does not provide *forward security*. On the other hand, in serial rekeying, only the term $\mathrm{PRF}(\underbrace{\ldots \mathrm{PRF}(K_0, }_{i-1 \text{ times}}\underbrace{0)\ldots, 0}_{i-1 \text{ times}})$ needs to be stored, and this does not allow any past key to be generated due to the noninvertibility of PRFs.

## 20.7 Key Revocation and Rekeying

Key revocation is the process of removing keys from operational use prior to their originally scheduled expiry, for reasons such as node capture. When a node is found to be compromised, a key revocation list is constructed and broadcast using µTESLA to the whole network. The list contains the ID of the compromised node, and optionally the indexes of the nodes' keys – these keys are keys from the key pool, and there is a mechanism for calculating these indexes based on the node IDs as described previously, so making the key indexes known is optional.

The process of removing keys is usually accompanied by rekeying. Because of this, the main challenge for doing key revocation efficiently is to do rekeying efficiently. Let us consider the types of keys that need to be replaced in case of a node capture (see Fig. 20.2 for example):

- *Global broadcast keys.* If the system uses µTESLA, then the compromised node $A$ must have stored the key chain commitment $K_S{}^{\text{chain}}$ distributed by the base station, but an attacker cannot recover the key chain corresponding to $K_S{}^{\text{chain}}$, so

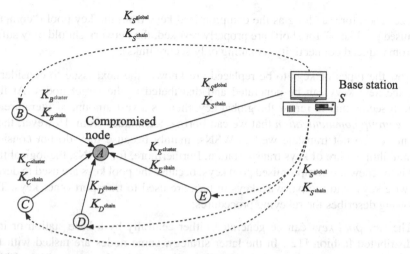

**Fig. 20.2** The keys that have been distributed before node $A$ is compromised. $K_v^{global}$, $K_v^{chain}$, $K_v^{cluster}$ represent the global key, the key chain commitment, and the cluster key distributed by node $v$, respectively

the uncompromised nodes do not need to replace their copies of $K_S^{chain}$. Recall that a µTESLA key chain only supports the authentication of the base station's messages. The global key $K_S^{global}$ is needed to provide confidentiality to $S$'s broadcast messages. If the compromised node has a copy of $K_S^{global}$, then the uncompromised nodes need to have their copies of $K_S^{global}$ replaced.

- *Local broadcast keys.* If the system uses passive participation, then the compromised node $A$ must have distributed its cluster key $K_A^{cluster}$ and key chain commitment $K_A^{chain}$ to its neighbors, as well as collected the cluster keys and key chain commitments from its neighbors, denoted by $K_v^{cluster}$ and $K_v^{chain}$, respectively ($v \in A$'s neighbors). Each of $A$'s neighbors, $v$, should purge $K_A^{cluster}$ and $K_A^{chain}$, as well as regenerate and redistribute $K_v^{cluster}$ and $K_v^{chain}$ to its neighbors. Note that unlike $K_S^{chain}$, $K_v^{chain}$ needs to be refreshed because without a key release schedule, an attacker with a compromised $K_v^{cluster}$ can forge messages using $K_v^{cluster}$ (this weakness has already been mentioned).
- *Unicast keys.* The pairwise keys are the unicast keys. There are two scenarios: either the compromised pairwise keys are only used for the secure channels that involve the evicted node, or the keys might actually be used elsewhere in the network for the secure channels that do not involve the evicted node at all. The first scenario applies to polynomial-based RKP, matrix-based RKP, and LEAP+; whereas the second scenario applies to symmetric-key-based RKP and EBS, because a key in these latter schemes is potentially shared by nodes distributed all over the network (all these schemes are mentioned in the last section). For EBS, rekeying is essential, because every node contains more than half of the keys from the key pool. For symmetric-key-based RKP, rekeying is less urgent, because in this case, a node's key ring is typically much smaller than the key pool

size. Therefore, as long as the compromised keys from the key pool ("compromised pool keys" for short) are properly revoked, the network should only suffer from reduced connectivity (counting only secure links).

Now the types of keys to be replaced are known, the next issue to consider is how the new keys can be generated and distributed to the target nodes. At first sight, it seems that to renew the global key, there is a vast amount of literature on *secure group communication* that we can borrow techniques from. However, these techniques do not translate well to WSNs, mainly because they do not consider the multihop nature of keys transportation. Furthermore, for WSNs, the logical first step is to renew the compromised pool keys, because the pool keys are used to derive pairwise keys, and the pairwise keys in turn are used to transport other keys. The following describes the rekeying procedures:

- The *new pool keys* can be generated either centrally by a base station or in a distributed fashion [12]. In the latter strategy, some nodes are tasked with the generation of certain keys, e.g., the $i$th node generates the $i$th pool key. Either way, the problem is getting the new keys to the right nodes. As mentioned, rekeying is essential for EBS. When a node is compromised, $P-K$ out of $P$ keys in the key pool remain secure, and all uncompromised nodes must have at least one of these $P-K$ keys (this is not the case for RKP schemes!). Suppose without loss of generality the compromised keys are $K_1, \ldots, K_m$. For each intact key $K_i (i = m + 1, \ldots, P)$, the message $E(K_i, E(K_1, K'_1)|| \ldots || E(K_m, K'_m))$ is generated and µTESLA-broadcast to the network. Every node will then be able to replace its compromised keys and derive new pairwise keys with its neighbors. On the other hand, for symmetric-key-based RKP, replacing the compromised pool keys is not crucial, and is actually not efficient to execute. In this scheme, the compromised node's neighbors purge the compromised keys from their system and re-establish pairwise keys with their respective neighbors.
- The *new cluster keys* are generated by the compromised node's neighbors. For example in Fig. 20.2, node $B$ would generate a new $K_B{}^{cluster}$ and a new $K_B{}^{chain}$ and send them to its neighbors, encrypted with new pairwise keys.
- The *new global key* is generated centrally by a base station and subsequently broadcast to the network. The base station $S$ does the following [15]:

1. $S$ generates the new global key as $K_g'$ (shorthand for a new $K_S{}^{global}$ in Fig. 20.2).
2. $S$ broadcasts the hash of $K_g'$, $h(K_g')$, to the network using µTESLA. Every node in the network caches $h(K_g')$. This hash will be used later to verify $K_g'$.
3. $S$ broadcasts $K_g'$ to its neighbors encrypted with its cluster key. $S$'s neighbors individually verify $K_g'$ with the hash $h(K_g')$ they have received earlier. $S$'s neighbors then re-encrypt $K_g'$ with their own cluster keys and broadcast the re-encrypted $K_g'$ to their respective neighbors. The process continues until $K_g'$ reaches every node in the network. This flooding process can be made more efficient, by optimizing the underlying routing protocol, but the principle remains the same.

## 20.8 Thoughts for Practitioners

Securing local broadcasts is generally too expensive for current generation of nodes. The priority is to secure query broadcasts, data convergecasts, and neighbor-to-neighbor unicasts. This means a node should minimally store a unique key shared with the base station, a µTESLA commitment distributed by the base station, a global key, and a set of pairwise keys, each of which is shared with a different neighbor.

## 20.9 Directions for Future Research

The most challenging task is to integrate key management with other components of a WSN. For example, an energy-efficient key management architecture should be optimized for the underlying routing protocol and vice versa. Secure data aggregation also needs to be taken into account. Meanwhile, existing building blocks can be further improved. In fact, polynomial-based and matrix-based RKP can be further generalized [13]; and re-keying for symmetric-key-based RKP is actually a difficult problem. Most importantly, the perpetual quest is to lower the resource requirements of key management.

## 20.10 Conclusions

Key management is one of the core areas in WSN security. Our approach is to break down a key management architecture into three components – key establishment, key refreshment, and key revocation – and introduce the building blocks for each of these components. Additionally, we also introduce constraint solving as a tool to verify the security of key management protocols. Future key management architectures can be designed based on these building blocks, and using constraint solving as a verification tool.

## Terminologies

Birthday attack:
    A cryptographic attack based on the observation that after $O(n^{1/2})$ evaluations of a function $H(\ )$ with uniformly distributed outputs, there is more than 50% chance of producing a collision, i.e., finding two arguments $x_1$ and $x_2$ with $H(x_1) = H(x_2)$.
Cipher-block chaining (CBC):
    A cipher operation mode defined by the encryption algorithm:

1. Divide message $M$ into $n$ $l$-sized blocks $M_1, M_2, \ldots, M_n$
2. Select $C_0 \in \{0, 1\}^l$ at random ($C_0$ is the *initialization vector*)
3. For $i \leftarrow 1$ to $n$ do
   $C_i = E(K, M_i \oplus C_{i-1})$
4. Return $C_0 C_1 \ldots C_n$

Cipher operation mode:
   A paradigm for encrypting multiple blocks of messages. The paradigm of encrypting blocks $M_1, M_2, \ldots, M_n$ to $E(K, M_1), E(K, M_2), \ldots, E(K, M_n)$ independently is called the electronic codebook mode (ECB) mode. ECB is insecure since an adversary can construct valid ciphertexts from the original ciphertext by arbitrarily rearranging, repeating, and/or omitting blocks from the original ciphertext. More secure operation modes add randomness into each block depending on previous blocks so that two identical blocks in the stream are encrypted differently.

Cluster key:
   A key shared between a node and its neighbors.

Collision-resistant hash function:
   A one-way hash function $H(\ )$ that also satisfies the property: it is computationally infeasible to find any $x$, $x'$ such that $H(x) = H(x')$ (the freedom of choice makes finding collision easier than finding preimages, so a function that satisfies this property is more secure than a function that does not).

Convergecast:
   A traffic pattern where, as opposed to multicast, message flows converge at a central point.

Finite field:
   See Galois field.

Galois field:
   A field of finite order. A field is a set with two binary operations $+$ and $\cdot$ that satisfies the field axioms:

| Property | $+$ | $\cdot$ |
|---|---|---|
| Identity | Exists, denoted $0$ | Exists, denoted $1$ |
| Inverse | Exists, denoted $-a$ | Exists, denoted $a^{-1}$ |
| Associativity | $(a + b) + c = a + (b + c)$ | $(a \cdot b) \cdot c = a \cdot (b \cdot c)$ |
| Commutivity | $a + b = b + a$ | $a \cdot b = b \cdot a$ |
| Distributivity | $a \cdot (b + c) = (a \cdot b) + (a \cdot c)$ | |

Examples are the integers modulo a prime, the real numbers, the rational numbers, etc.

Global key:
   A key shared by all the nodes in the network.

Message authentication code (MAC):
   A code generated from a key and a message, that lets the receiver of the message to authenticate the message as tamper-free and as coming from the sender, if the receiver shares the key with the sender.

One-way hash function:
   A hash function $H(\ )$ that satisfies the properties:

1. Preimage resistance: given $y$, it is computationally infeasible to compute $x$ such that $H(x) = y$.
2. Second preimage resistance: given $x$, it is computationally infeasible to compute $x'$ such that $H(x) = H(x')$.

Pairwise key:
   A key shared between a pair of in-range nodes.
Primitive element:
   A number $g$ is a primitive root of $m$ if the smallest $i$ such that $g^i \equiv 1 \bmod m$ is $m$.
Primitive root:
   See primitive element.
Pseudorandom function (PRF):
   A mapping $\{0, 1\}^k \times \{0, 1\}^l \to \{0, 1\}^m$ where $k$ is the length of the key, $l$ is the length of the input and $m$ is the length of the output, that satisfies the requirement that no probabilistic polynomial-time algorithm can distinguish with significant advantage between a function chosen from the pseudorandom function family and a random oracle (i.e., a truly random function).

## Questions

1. Birthday attacks result from the *birthday paradox*, which says that in a group of at least 23 people, there is a more than 0.5 chance that some pair of them will have the same birthday. Show that in a mapping from $n$ preimages to $r$ images, the probability that two or more preimages map to the same image is larger than $1 - e^{-(n^2-n)/(2r)}$.
2. Denote the block size of a PRF by $m$, so that the total number of possible output blocks of the PRF is $2^m$. Substitute $2^m$ for $r$ in Exercise 1, then show that when $n > 2^{m/2}\sqrt{2\ln 2}$, the probability that two or more plaintexts map to the same ciphertext $> 0.5$. Thus, the birthday threshold is $O(2^{m/2})$.
3. Suppose a base station does an authenticated broadcast every minute, and the network is to last for 3 years. If we use Kim's algorithm to implement μTESLA, what is the minimum amount of memory needed?
4. In symmetric-key-based RKP, two nodes can only establish a session key if they share at least a key. Given key pool size $P$ and key ring size $K$, calculate the probability $p_s$ of two nodes being able to establish a session key.
5. Continuing from the previous exercise, given the network size is $n$ and every node has $d$ neighbors, what is the approximate probability that the network is securely connected? Hint: When a network is connected, a node is securely connected to at least one neighbor.
6. Extending symmetric-key-based RKP, we can enforce that two nodes are only allowed to establish a session key if they share at least $q$ keys. This scheme

is called $q$-composite RKP [3]. Given key pool size $P$ and key ring size $K$, calculate the probability $p_s$ of two nodes being able to establish a session key.

7. Continuing from the previous exercise, if $x$ nodes are compromised, what is the probability a secure link between two uncompromised nodes is compromised?

8. In many proposals of RKP schemes, WSNs are modeled as random graphs. A random graph $G(n, p_s)$ is a graph of $n$ vertices (sensor nodes) for which the probability that an edge (a secure link) exists between any two vertices is independently determined by a coin flip of probability $p_s$ [2]. If $p_s$ is zero, then the graph is disconnected, and if $p_s$ is one, the graph is fully connected, so there must exist a certain value of $p_s$ such that the graph is almost surely connected. While the random graph-inspired intuition about connectivity is correct, the random graph model $G(n, p_s)$ does not capture the real nature of a WSN, why?

9. In polynomial-based RKP, $q$ must be much larger than the number of sensors, why?

10. In the original proposal [14], Protocol 1 does not include the last transmission, but why is the last transmission necessary? Protocol 1 is originally proposed as a general key establishment protocol between $A$ and $B$ with a trusted third party $S$, in the sense that $A$ and $B$ do not need to be neighbors. What potential problem might there be in this general setting?

# References

1. Abdalla M, Bellare M (2000) Increasing the lifetime of a key: A comparative analysis of the security of rekeying techniques. In: Advances in Cryptology – ASIACRYPT 2000, volume 1976 of LNCS, pp. 546–565. Springer, New York, NY
2. Bollobás B (1985) Random Graphs. Academic, New York, NY
3. Chan H, Perrig A, Song D (2003). Random key predistribution schemes for sensor networks. In: Proceedings of the 2003 IEEE Symposium on Security and Privacy. IEEE Computer Society, Washington, DC
4. CNSS (2006) National Information Assurance (IA) Glossary, CNSS Instruction No. 4009, revised June 2006. http://www.cnss.gov/Assets/pdf/cnssi_4009.pdf
5. Du W, Deng J, Han YS, Varshney PK, Katz J, Khalili A (2005) A pairwise key predistribution scheme for wireless sensor networks. ACM Transactions on Information and System Security, 8(2):228–258
6. Eschenauer L, Gligor VD (2002) A key-management scheme for distributed sensor networks. In: Proceedings of the Ninth ACM conference on Computer and communications Security, pp. 41–47. ACM, New York, NY
7. Hu Y-C, Perrig A, Johnson D (2002) Ariadne: a secure on-demand routing protocol for ad hoc networks. In: Proceedings of the Annual International Conference on Mobile Computing and Networking, pp. 12–23. ACM, New York, NY
8. Javier Thayer Fabrega F, Herzog JC, Guttman JD (1998) Strand spaces: Why is a security protocol correct? In: Proceedings of The 1998 IEEE Symposium on Security and Privacy, pp. 160–171. IEEE Computer Society, Washington, DC
9. Kim S-R (2005) Scalable hash chain traversal for mobile devices. In: Computational Science and Its Applications – ICCSA 2005, volume 3480 of LNCS, pp. 359–367. Springer, New York, NY
10. Law YW, Corin R, Etalle S, Hartel PH (2003) A formally verified decentralized key management architecture for wireless sensor networks. In: Proceedings of the Fourth IFIP TC6/WG6.8

International Conference on Personal Wireless Communications (PWC 2003), volume 2775 of LNCS, pp. 27–39. Springer, New York, NY

11. Liu D, Ning P, Li R (2005) Establishing pairwise keys in distributed sensor networks. ACM Transactions on Information and System Security, 8(1):41–77

12. Moharrum M, Eltoweissy M, Mukkamala R (2006) Dynamic combinatorial key management scheme for sensor networks. Wireless Communications and Mobile Computing, 6(7):1017–1035. Wiley, New York, NY

13. Padró C, Gracia I, Molleví SM, Morillo P (2002) Linear key predistribution schemes. Design, Codes and Cryptography, 25:281–298

14. Perrig A, Szewczyk R, Wen V, Culler D, Tygar JD (2001) SPINS: Security protocols for sensor networks. In: Proceedings of the Seventh Annual International Conference on Mobile Computing and Networking, pp. 189–199. ACM, New York, NY

15. Zhu S, Setia S, Jajodia S (2006) LEAP+: Efficient security mechanisms for large-scale distributed sensor networks. ACM Transactions on Sensor Networks, 2(4):500–528

# Chapter 21
# Secure Data Aggregation in Wireless Sensor Networks

**Yee Wei Law, Marimuthu Palaniswami, and Raphael Chung-Wei Phan**

**Abstract** The biggest advantage of building "intelligence" into a sensor is that the sensor can process data before sending them to a data consumer. The kind of processing that is often needed is to aggregate the data into a more compact representation called an aggregate, and send the aggregate to the data consumer instead. The main security challenges to such a process are (1) to prevent Byzantine-corrupted data from rendering the final aggregate totally meaningless and (2) to provide end-to-end confidentiality between the data providers and the data consumer. This chapter surveys the state of the art in techniques for addressing these challenges.

## 21.1 Introduction

The primary use of wireless sensor networks (WSNs) is to collect and process data. There are two reasons why we do not want to send all raw data samples (henceforth *samples* for short) directly to the sink node (the word "node" is sometimes omitted). The first reason is that the data rate of sensors is limited. The second reason is that communication is three orders of magnitude more energy consuming than computation. Instead, what is usually done is that the samples are aggregated along the way from the sources to the sink. The simplest example is when two samples from two neighboring sensors are nearly identical, we only need to send one of the samples to the sink. In general, spatial and/or temporal correlation may exist in the data, i.e., data collected by neighboring sensors and/or data collected by a sensor at different time instants maybe correlated. Therefore, instead of sending highly correlated data to the sink all at once, it maybe more energy-efficient to use some

Y.W. Law (✉)
Department of Electrical and Electronic Engineering, The University of Melbourne,
Parkville, VIC 3052, Australia
e-mail: y.law@ee.unimelb.edu.au

S. Misra et al. (eds.), *Guide to Wireless Sensor Networks*, Computer Communications and Networks, DOI: 10.1007/978-1-84882-218-4_21,

533

intermediate sensor node(s) to aggregate the data into a single comprehensive digest, and then send the digest toward the sink. This strategy is called *data aggregation*, or *data fusion*, or *in-network processing*.

The essence of data aggregation lies in how the data digest is generated. A function that takes raw data as input and produces a digest as output is called an *aggregation function*. Clearly, the kind of aggregation function used depends on the type of statistics we want to derive from the raw data. For example, we might be interested in the minimum, or the maximum, or the sum, or the average of the raw data. However, in the presence of malicious nodes, the aggregation functions mentioned just now are insecure, in the sense that a malicious sensor can arbitrarily bias the aggregation result by submitting just one false data. The search for secure aggregation functions is the primary objective of *secure data aggregation*. For convenience, we call this primary objective the *robustness objective*.

The secondary objective of secure data aggregation is to ensure that other than the sink and the sources, no node should have knowledge of the raw data or the aggregation result. For convenience, we call this secondary objective the *confidentiality objective*.

Our goal in this chapter is to explore the techniques to achieve these two objectives of secure data aggregation. A common theme in security is to set up multiple layers of defense. To achieve the robustness objective, our first line of defense is to prevent outsider attackers (i.e., attackers that do not possess our system's cryptographic keys) from sending wrong data to our network. Our second line of defense is to minimize the effect of insider attackers (i.e., compromised devices in our network) that try to invalidate our aggregation result by submitting wrong data. For our first line of defense, we can rely on cryptography, in particular neighbor-to-neighbor authentication; whereas for our second line of defense, we have to rely on an assortment of techniques ranging from *resilient aggregation* to *voting*. Our final line of defense is *result verification*, that is, to double check the aggregation result by verifying the result with some or all of the sources. To achieve the confidentiality objective, we need a special cryptographic construct called *privacy homomorphism* (which can be understood as "privacy-preserving homomorphism").

Thus, this chapter is organized as follows. First, we will give some background information on the data aggregation model and mechanisms. Then, the techniques of resilient aggregation, voting, result verification, and privacy homomorphism will be introduced.

## 21.2 Background

A secure data aggregation process can be divided into three phases:

1. *Query dissemination phase* (Fig. 21.1a): A sink (usually a base station) floods the network with an SQL-style query such as

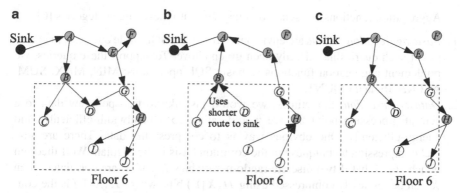

**Fig. 21.1** Phases of secure data aggregation: (**a**) query dissemination, (**b**) data aggregation, and (**c**) result verification

> SELECT AVERAGE (temperature) FROM sensors
> WHERE floor = 6
> EPOCH DURATION 30

In this sample query, AVERAGE is specified as the aggregation function.

2. *Data aggregation phase* (Fig. 21.1b): The nodes that satisfy the query self-elect as sources. For example, in response to the sample query above, only the nodes on the 6th floor, as indicated by the **WHERE** clause, would become sources; and as indicated by the **EPOCH DURATION** clause, the sources would send their samples toward the sink every 30 s. All nonleaf nodes aggregate the samples from their child nodes using an aggregation function that depends on the query. For example, for the sample query above, the aggregation function is AVERAGE. In Fig. 21.1b, nodes $A, B, G$, and $H$ aggregate data from their child nodes. Some aggregators only aggregate data (e.g., node $A, B, H$). Some aggregators double as sources, in that they also produce data (e.g., node $G$). Some nodes only forward data toward the sink and this type of nodes are called forwarders (e.g., node $E$). The biggest difference between an aggregator and a forwarder is that a forwarder forwards data straightaway without waiting for more data to aggregate. However, a node does not know, upon receiving the query, whether or not it should ideally become an aggregator or a forwarder. For example, upon receiving the query, node $E$ would not know whether it would receive any data from $F$ alone, or from $G$ alone, or from both. Later, we will discuss a scheme called SDAP that helps us decide whether a node should become a forwarder or an aggregator. What arrives at the sink is the final aggregation result.

3. *Result verification phase* (Fig. 21.1c): To get maximum assurance of the dependability of the aggregation result, the sink verifies the aggregation result with all the source nodes. Another term for this process is *attestation* because in effect the source nodes are requested to attest to the validity of the aggregation result. In view of the communication complexity, this step is reserved for aggregation results that are mission-critical, hence this step might not always be performed.

Aggregation functions are generally classified into these three categories [6]:

1. *Basic aggregation.* WSNs are envisioned to support SQL-style queries. An example of such queries has already been given above. To support these queries, we implement aggregation functions as basic SQL operations: MIN, MAX, SUM, AVERAGE, and COUNT.
2. *Data compression.* Sometimes, we would just like to transport raw data to a central processing point but we would rather do it bandwidth-efficiently and energy-efficiently. One obvious way is to compress the data. There are several compression techniques but the common basis is the Slepian–Wolf theorem which states that if two discrete random variables $X$ and $Y$ are correlated, then $X$ can be losslessly compressed using $H(X|Y)$ bits, where $H(X|Y)$ is the conditional entropy of $X$ conditioned on $Y$.
3. *Parameter estimation.* Often, the parameters of the sample data's underlying probability distribution function are of interest, and we can estimate these parameters in a distributed fashion. The problem of parameter estimation can be formulated as an optimization problem [19]. Traditionally, the cost function (i.e., the function to be minimized) is the mean square error. For example, if the parameter to be estimated is the mean and the samples are $x_1, \ldots, x_n$, then the cost function is

$$\sum_{i=1}^{n} (x_i - \mu)^2$$

There are several challenges to the above aggregation mechanisms, however. First, some sensors may be wrongly calibrated, or they may malfunction for example in a harsh environment. Additionally, some sensors may be captured, compromised, and reprogrammed to inject false data into the network. The result of these problems is that the aggregation result becomes invalid. The approaches we take toward solving this type of problems include resilient aggregation and voting. Using resilient aggregation, we can construct aggregation functions that are resilient against data contamination. Using voting, we can minimize the risk of accepting false data. Resilient aggregation and voting are complementary approaches toward thwarting malicious source nodes. Result verification is a countermeasure against malicious aggregators. In this step, the sources are allowed the final chance to verify if their contribution to the final aggregation result has really been accounted for.

Some applications might require the data to be encrypted during their passage from the sources to the sink. The approach we take toward solving this type of problem is privacy homomorphism. Private homomorphism, however, only works when there is no malicious source node in the network. To thwart attacks by malicious source nodes, techniques like resilient aggregation and voting are again needed. Figure 21.2 summarizes the techniques for securing data aggregation.

In a realistic threat model, any of the sources, aggregators, and sink might be compromised. We do not care, however, if the sink is compromised because the sink is the ultimate data consumer; and if it is compromised, there is no reason to consider other cases anyway.

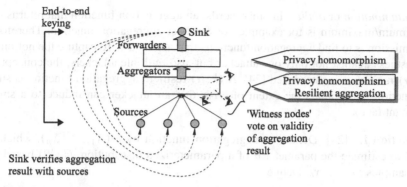

**Fig. 21.2**  Securing data aggregation

**Table 21.1**  Notation used in this chapter

| Symbols | Meaning |
| --- | --- |
| $f()$ | A generic multiparameter aggregation function |
| $rms()$ | The root mean square function |
| $E[]$ | The expectation function |
| $X_{(i)}$ | The $i$th order statistics of random variable $X$ |
| $\text{Med}(x_1, \ldots, x_n)$ | The sample median of $x_1, \ldots, x_n$ |
| $E(K, M)$ | An encryption of message $M$ using key $K$ |
| $h(M)$ | A collision-resistant hash of message $M$ |
| $\text{MAC}(K, M)$ | A message authentication code (MAC) of message $M$ using key $K$ |
| $\|$ | Concatenation operator |
| $q$ | Query ID |
| $x \in_R X$ | Choose $x$ from set $X$ in a uniform and random manner |
| $\mathbb{Z}_p$ or $(\mathbb{Z}/p\mathbb{Z})$ | The group of integers modulo $p$ |
| $\mathbb{Z}_p^*$ or $(\mathbb{Z}/p\mathbb{Z})^*$ | The multiplicative group of units (i.e., invertible integers) modulo $p$ |
| $GF(p)$ | The Galois field (finite field) of order $p$ |
| $GF(p)^*$ | The multiplicative group of nonzero elements of the Galois field of order $p$ |

For the rest of this chapter, the notation in Table 21.1 will be used. Protocol listings always abstract away the data fields specific to the underlying communication protocols.

## 21.3  Resilient Aggregation

Either because a sensor might malfunction, or because it might be compromised to inject false data, we must be careful when aggregating raw data. For example, if our aggregation function is to calculate the minimum/maximum value of a data set, a malicious sensor can bias the aggregate by injecting a false value that is many times lower/higher than the actual minimum/maximum – we call this value a *contamination* or *outlier*. In other words, an aggregation function that returns the minimum/maximum is for example not a robust aggregation function. Therefore, our mission is to find aggregation functions that are robust. To capture the notion of robustness against not only one attacker but also multiple attackers, the concept of $(k, \alpha)$-*resilience* is introduced [24]. A $(k, \alpha)$-resilient aggregation function is such a function that bounds the extent of error that an attacker can induce to a small constant factor.

**Definition 1.** [24]: Denote an aggregation function by $\hat{\Theta}(X_1, \ldots, X_n)$, which is used to estimate the parameter $\theta$ of a parameterized distribution $f(x|\theta)$ based on the samples $x_1, \ldots, x_n$. Define

$$\text{rms}(\hat{\Theta}) \triangleq \sqrt{\text{E}[(\hat{\Theta} - \theta)^2|\theta]}. \tag{21.1}$$

Denote $\hat{\Theta}^*$ as the outcome of $\hat{\Theta}$ when $k$ out of $n$ of the samples $x_1, \ldots, x_n$ are contaminated. Define the function $\text{rms}^*()$ as

$$\text{rms}^*(\hat{\Theta}, k) \triangleq \sqrt{E[(\hat{\Theta}^* - \theta)^2|\theta]}. \tag{21.2}$$

Then, the aggregation function $\hat{\Theta}(X_1, \ldots, X_n)$ is $(k, \alpha)$-resilient with respect to $f(x|\theta)$ if

$$\text{rms}^*(\hat{\Theta}, k) \leq \alpha \cdot \text{rms}(\hat{\Theta}) \tag{21.3}$$

Since $\text{rms}(\hat{\Theta})$ is dependent on the distribution parameterized by $\theta$, the notion of resilience is with respect to that distribution. Using this notion, it is easy to see why sum and sample median are not resilient: for any $\alpha$, if the attacker contaminates a value from $x$ to $x + \alpha \, \text{rms}(f)$, $\text{rms}^*(f, 1)/\text{rms}(f) > \alpha$ (Exercise 1).

We can calculate the $(k, \alpha)$-resilience of the sample median as follows. If the population distribution is a normal distribution $N(\mu, \sigma^2)$, then the asymptotic mean and asymptotic variance of the $p$th sample quantile (or equivalently the $pn$th order statistics) are $\mu + (z_p\sigma/\sqrt{n})$ and $(p(1 - p)\sigma^2)/(n\phi(z_p)^2)$, respectively, where $z_p = \Phi^{-1}(p)$, $\Phi$ and $\phi$ are the cdf and pdf of $N(0, 1)$, respectively ([17], p. 262). The sample median is the 0.5th sample quantile (or the $0.5n$th order statistics) with a standard deviation of

$$\text{rms}(f) = \sqrt{\text{Var}(X_{(0.5n)})} = \frac{\sigma}{\sqrt{n}} \sqrt{\frac{1}{4\phi(\Phi^{-1}(0.5))^2}} \approx \frac{\sigma}{\sqrt{n}} \sqrt{\frac{\pi}{2}}. \tag{21.4}$$

A contamination of $k$ values can increase or decrease the median to at most the $(0.5 + k/n)$th or the $(0.5 - k/n)$th sample quantile. Setting $p = 0.5 + k/n$ (the case with $p = 0.5 - k/n$ is symmetrical), we have

$$E\left[\left(X_{(pn)} - \left(\mu + \tfrac{z_p\sigma}{\sqrt{n}}\right)\right)^2\right] = \mathrm{Var}\left(X_{(pn)}\right) = \tfrac{p(1-p)\sigma^2}{n\phi(z_p)^2}$$

$$\Rightarrow E\left[\left(\left(X_{(pn)} - \mu\right) - \tfrac{z_p\sigma}{\sqrt{n}}\right)^2\right] = \tfrac{p(1-p)\sigma^2}{n\phi(z_p)^2}$$

$$\Rightarrow E\left[\left(X_{(pn)} - \mu\right)^2\right] - 2\tfrac{z_p\sigma}{\sqrt{n}}E\left[X_{(pn)} - \mu\right] = \tfrac{p(1-p)\sigma^2}{n\phi(z_p)^2} - \tfrac{z_p^2\sigma^2}{n} \cdot$$

$$\Rightarrow \mathrm{rms}^*(f,k) = \sqrt{E\left[\left(X_{(pn)} - \mu\right)^2\right]} = \tfrac{\sigma}{\sqrt{n}}\sqrt{\tfrac{p(1-p)}{\phi(z_p)^2} + z_p^2}.$$

(21.5)

What remain to be solved in (21.5) are $z_p$ and $\phi(z_p)$:

$$z_p = \Phi^{-1}\left(0.5 + k/n\right) = \sqrt{2}\mathrm{erf}^{-1}\left(2(0.5 + k/n) - 1\right) \approx \sqrt{2\pi}(k/n)$$
$$\Rightarrow \phi(z_p) \approx \tfrac{1}{\sqrt{2\pi}}\exp\left(-\pi(k/n)^2\right).$$

(21.6)

Combining (21.4), (21.5), and (21.6), and setting $a = 2k/n$, we have

$$\alpha = \sqrt{\frac{2}{\pi}\left(\frac{p(1-p)}{\phi(z_p)^2} + z_p^2\right)} \approx \sqrt{\frac{2}{\pi}\left[\frac{\frac{1}{4}(1-a^2)}{\frac{1}{2\pi}\exp\left(-\frac{\pi}{2}a^2\right)} + \frac{\pi}{2}a^2\right]}$$

$$= \sqrt{(1-a^2)\exp\left(\frac{\pi}{2}a^2\right) + a^2} = \sqrt{(1-a^2)\left(1 + \frac{\pi}{2}a^2 + \frac{\pi^2}{4\cdot2!}a^4 + \cdots\right) + a^2}$$

$$\approx \sqrt{1 + 2\pi(k/n)^2}.$$

(21.7)

Hence, when $k \ll n$, the sample median is $(k, \sqrt{1 + 2\pi(k/n)^2})$-resilient.

Another measure of robustness is the *breakdown point* $\varepsilon^*$, defined as the largest fraction of contamination that a data set may contain before the estimate of a parameter deviates arbitrarily from the true estimate. Mathematically,

$$\varepsilon^* = \sup\{k/n : \mathrm{rms}^*(\widehat{\Theta}, k) < \infty\}.$$

(21.8)

The largest possible breakdown point is 0.5, since obviously if more than half of the data are contaminated, there is no way at all to estimate a parameter correctly. Research in robust statistics tells us that the sample median gives us a breakdown point of 0.5. Table 21.2 lists the $(k, \alpha)$-resilience and breakdown point of other aggregation functions.

### 21.3.1 Quantiles Aggregation

The previous calculation of the resilience of the sample median is based on the model of a single aggregator exactly one hop away from the source nodes.

**Table 21.2** $(k, \alpha)$-Resilience and breakdown points of some aggregation functions.

| Aggregation function | Resilience $\alpha$ | Breakdown point $\varepsilon^*$ |
|---|---|---|
| Sample median wrt Gaussian distribution | $\approx \sqrt{1 + 2\pi(k/n)^2}$, if $k \ll n$ | 0.5 |
| 5%-trimmed average wrt Gaussian distribution | $\approx 1 + 6.278k/n$, if $k < 0.05n$ | 0.05 |
| $[l, u]$-truncated average wrt Gaussian distribution | $1 + (u - l)/\sigma \cdot k/\sqrt{n}$ | Not applicable |
| Count wrt Bernoulli distribution with parameter $p$ | $\sqrt{1 + k^2/[np(1 - p)]}$ | Not applicable |

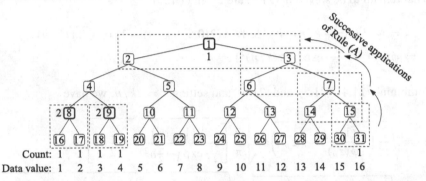

**Fig. 21.3** Producing a q-digest of the data set $\{1, 2, 3, 4, 16\}$ with compression parameter $k = 2$. A number beside or under a node is a count. The final q-digest comprises of the nodes with thickened edges, and is represented by $\{< 8, 2 >, < 9, 2 >, < 1, 1 >\}$

If there are multiple aggregators in the network, the above result is not applicable. Note that the median of medians is *not* resilient. For example, the data set $\{1, 2, 3, 4, 16\}$ has a median of 3, while the median of $\{\text{Med}(1, 2, 3), \text{Med}(4,16)\}$ is 6. Had we taken the median of the set $\{\text{Med}(1, 2, 3), 4, 16\}$, we would have gotten 4, which is closer to the actual median 3. The idea we can learn from this later approach, is that before the final median is calculated, some intermediate subset can be compressed (e.g., the median 2 is a reasonable compression of the set $\{1, 2, 3\}$), but some should not be compressed (e.g., the elements 4 and 16 are far enough apart from each other that their median is not representative).

An idea similar to the above can be found in Shrivastava et al.'s quantiles aggregation technique [22]. In this technique, raw data are aggregated into something called a q-digest, and multiple q-digests can be further aggregated into a new q-digest. The technique is best illustrated using an example (Fig. 21.3). Suppose a data value ranges from 1 to 16. We build a binary tree with 16 leaves (the leaf nodes are also called *buckets* in this context), and label the nodes starting from 1 to $2^{\log 16+1} - 1$ (note that "nodes" in this subsection mean the logical nodes in Fig. 21.3 and *not* sensor nodes). For the data set $\{1, 2, 3, 4, 16\}$, we assign a count of 1 to each of node 16, 17, 18, 19, and 31. All other nodes have a count of 0. Two rules, based on the compression parameter $k$, are used for compressing this tree:

> Rule (A): count(node) + count(parent) + count(siblings) $\geq \lfloor n/k \rfloor + 1$
> Rule (B): count(node) $\leq \lfloor n/k \rfloor$

These rules are applied to all nodes except the root and leaf nodes. Rule (A) stipulates that a node together with its parent and siblings should have a combined count of at least $\lfloor n/k \rfloor + 1$. This rule encourages us to combine the counts of a node and its siblings, if the combined count is low. For example,

- The cluster 8–16–17 has an initial combined count of $2 < \lfloor 5/2 \rfloor + 1$, so we combine the counts of nodes 16 and 17 into the count of node 8, that is, we assign a count of $1 + 1 = 2$ to the parent node 8. The cluster 8–16–17 has a combined count of $2 + 1 + 1 = 4 > \lfloor 5/2 \rfloor + 1$, so we do not elevate to the cluster 4–8–9 for further compression.
- The cluster 9–18–19 is processed similarly as above.
- The cluster 15–30–31 has an initial combined count of $1 < \lfloor 5/2 \rfloor + 1$, so we combine the counts of nodes 30 and 31 into the count of node 15, resulting in a new combined count of $1 + 1 = 2$. However, the combined count of the cluster still does not satisfy Rule (A), so we elevate to the cluster 7–14–15, and then the cluster 3–6–7, and finally to the cluster 1–2–3. In the end, node 1 is assigned a count of 1.

Rule (B) stipulates that a node should have a count of at most $\lfloor n/k \rfloor$. This rule prevents us from compressing the tree too much. For example,

- Nodes 8 and 9 have a count of $2 = \lfloor 5/2 \rfloor$, so they satisfy Rule (B).
- Node 1 has a count of $1 < \lfloor 5/2 \rfloor$, so it satisfies Rule (B) too.

To summarize this example, we only need node 8, 9, and 1 to represent the q-digest. A q-digest is expressed as a set of tuples of the form <node representation, count>. So for our example, the q-digest is $\{< 8, 2 >, < 9, 2 >, < 1, 1 >\}$. The median corresponding to this q-digest resides in node 9, and node 9 corresponds to data value 3 and 4, so we can calculate the median as 3.5.

The aggregation of two q-digests requires just an extra step. Given two q-digests $Q_1$ and $Q_2$,

1. The count of the $i$th node in the new q-digest ← the count of the $i$th node in $Q_1$ + the count of the $i$th node in $Q_2$.
2. Apply Rule (A) and Rule (B) to the new q-digest.

What arrives at the sink is a q-digest that represents a lossy histogram. From this histogram, statistics like minimum, maximum, and average can be estimated.

## 21.3.2 RANBAR

Another approach to estimating model parameters is to use, at the start, as few samples as possible to determine a preliminary model, and then use the preliminary

model to identify samples that are consistent with the model, and subsequently refine the model with all the samples that are found to be consistent. The model is
refined iteratively until the fraction of consistent data is above a certain threshold.
This approach is called RANSAC (RANdom SAmple Consensus), and the set of
data that are found to be consistent with the inferred model is called the *consensus
set*. RANSAC originates in the computer vision literature. Buttyán et al. [2] adapt
the RANSAC paradigm to secure data aggregation and call it RANBAR. If the target
distribution is Gaussian, the corresponding RANBAR algorithm is as follows:

1. Let the initial set size be $s$. $s \leftarrow 2$, because at least two samples are needed to
   estimate the variance, and the variance is a parameter for a Gaussian distribution.
2. Let the consensus set size be $t$. $t \leftarrow n/2$, because we assume in the worst case,
   half of the source nodes might be compromised.
3. Let the maximum number of iterations be $i$. $i \leftarrow 15$, based on Buttyán et al.'s
   empirical analysis [2].
4. Let the error tolerance be $\delta$. $\delta \leftarrow 0.3$, based on Buttyán et al.'s experimental
   result [2].
5. Build a histogram hist() of the samples after discarding the upper 0.5% and lower
   0.5% of the samples.
6. While number of trials $\leq i$, do
   Randomly select $s$ samples
   $\mu \leftarrow$ sample mean of the $s$ samples
   $\sigma^2 \leftarrow$ sample variance of the $s$ samples
   Instantiate the Gaussian model $N(\mu, \sigma^2)$
   For every sample $x$, $d \leftarrow |\text{pdf}_{\mu,\sigma}(x)-\text{hist}(x)|$
   Discard samples with $d > \delta$
   Consensus set $\leftarrow$ all remaining samples
   If the size of the consensus set $> t$,
      $\mu \leftarrow$ sample mean of the consensus set
      $\sigma^2 \leftarrow$ sample variance of the consensus set
      Return the Gaussian model $N(\mu, \sigma^2)$
      Break loop
   End if
   End while
7. Return failure.

To summarize this section, we have introduced the concept of resilient aggregation, and described two concrete approaches in *multihop* resilient aggregation.
The first approach is by aggregating quantiles and then calculating the sample median from the quantiles. The second approach is by the RANSAC paradigm, which
builds a preliminary model first based on as few data as possible, and then iteratively refines the model based on the samples that are found to be consistent with
the preliminary model. This approach only works when the system knows a priori
how the data are distributed (i.e., whether they are Gaussian, Poisson, etc.).

## 21.4   Voting

Some assumptions are in order before we describe how voting works.

---
**Assumption 1** Two nodes that are physically close are likely to sense the same phenomenon and hence return readings that are close to each other.

**Assumption 2** Among the neighbors of any single node, malicious nodes are the minority.

---

The general idea of voting is as follows. First, we consider the case of malicious sources. When a node $A$ sends its data to an aggregator, this data might be false. To get some assurance that this data is valid, by Assumption 1, we can ask the nodes that are close to $A$ to examine the data. Let us call these nearby nodes the "witness nodes" [9]. For example, a witness node for $A$ can be one of $A$'s neighbors. Since some of the witness nodes themselves might be malicious, the intuition is to ask the witness nodes to vote on the validity of the data. By Assumption 2, majority positive votes should convince us that the data is valid. However, it is very inefficient to ask the witness nodes to vote on every data that $A$ possibly wants to send. Instead, we can ask the witness nodes to vote on the aggregation result. The new scheme is as such: we ask the witness nodes of an aggregator to vote on the aggregation result, before the aggregator sends out the aggregation result toward the sink. An aggregator cannot forward the aggregation result without having majority positive votes from the witness nodes. The problem is how exactly a witness node should vote on a data. The standard approach in cryptography is to sign the data, that is, by either calculating a message authentication code (MAC) or a digital signature of the data.

A scheme that uses MAC is as follows [9]. Assume every node shares a unique key with the sink (not the aggregator). Before sending its aggregation result toward the sink, an aggregator consults its witness nodes about the validity of its aggregation result. Each of the witness nodes that approve the aggregation result generates a MAC based on the aggregation result, and sends its MAC to the aggregator. After collecting enough MAC's (the definition of "enough" is a system parameter), the aggregator sends the aggregation result and the MACs generated by the witness nodes ("witness MACs" for short) toward the sink. At the sink, if $k$ of the witness MACs are found to be consistent with the aggregation result, and the fraction $k/$(total number of witness nodes) is above a system-defined threshold, then the aggregation result is considered valid. The disadvantage of this scheme is that, as is, this scheme is only suitable for single-level aggregation trees.

A scheme that uses digital signature is as follows [13]. Assume (1) the network is divided into clusters, (2) each cluster has an associated public–private key pair, and (3) cluster heads work as aggregators. Every cluster member has a unique share of the cluster private key, generated using verifiable $(t, n)$-threshold secret sharing. It takes at least $t/n$ of the cluster members to generate a valid signature. Before sending its aggregation result toward the sink, an aggregator consults its witness nodes about the validity of its aggregation result. Each of the witness nodes that approve the aggregation result generates a partial signature based on the aggregation result, and sends the partial signature to the aggregator. After collecting enough partial signatures to assemble a valid full signature, the aggregator sends the aggregation result together with the signature toward the sink. The disadvantages of this scheme are that (1) digital signatures are more costly to generate than MACs and (2) the cluster structure of the network has to remain fixed.

Voting is generally costly to implement due to the communication complexity and the key management requirement, and hence is best reserved for small-scale networks.

## 21.5 Result Verification

Result verification completes the last line of defense against data tampering attackers. The purpose of the sink's performing result verification is to double check with the source nodes that the intermediate aggregators have not tampered with the intermediate aggregation results. For simplicity, we will begin our discussion with single-aggregator WSNs. Then, we will extend the ideas we learn from this simple-aggregator case to the more general and common multiaggregator case. For the schemes we are going to describe, there are two common key pre-distribution requirements (1) the sink must share a unique key with every other node and (2) every node can establish a pairwise key with each of its neighbors. These requirements are satisfiable in practice. As before, examples are used heavily in the following discussion.

### 21.5.1 The Single-Aggregator Case

In Fig. 21.4a, aggregator $A$ collects samples $x_1$, $x_2, \ldots, x_n$ from node $1, 2, \ldots, n$, respectively; and aggregates the samples into one result $y = f(x_1, x_2, \ldots, x_n)$. Suppose the aggregation function calculates the median of the set $X = \{x_1, x_2, \ldots, x_n\}$, the sink $S$ can verify $y$ by sampling just a subset of $X$, using an interactive proof algorithm [18]. For this algorithm to work, $A$ must have $X$ sorted

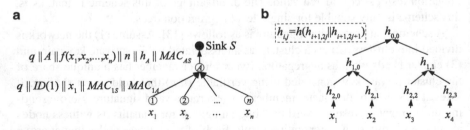

**Fig. 21.4** **a** A single-aggregator WSN. **b** A commitment tree for the case $n = 4$

in ascending order in the first place, so without loss of generality, we assume $X$ is sorted. The algorithm, which checks whether $y$ approximates well the real median of $X$, is as such:

1. $n \leftarrow$ number of data samples
2. if $n$ is odd
   request $x_{(n+1)/2}$ from $A$
   $z \leftarrow x_{(n+1)/2}$
   else if $n$ is even,
   request $x_{n/2}$ and $x_{(n+2)/2}$ from $A$
   $z \leftarrow (x_{n/2} + x_{(n+2)/2})/2$
3. if $y \neq z$ then reject $y$
4. for $i = 1$ to $(1/\varepsilon)$ do
   if $n$ is odd, pick $j \in_R \{1 \ldots n\} \backslash \{(n+1)/2\}$ without replacement
   else if $n$ is even, pick $j \in_R \{1 \ldots n\}$ without replacement
   request $x_j$ from $A$
   $t \leftarrow x_j$
   if $j \leq n/2$ and $t > y$ then reject $y$
   if $j > n/2$ and $t < y$ then reject $y$
5. accept $y$

Following this algorithm, $S$ makes $1/\varepsilon$ requests in the for-loop (which can be done in one go) to $A$, and with this many requests, $S$ can be sure that if $y$ is not present in $X$, or if the index of $y$ in $X$ is at least $\varepsilon n$ ($\varepsilon > 0$) away from the true index of the sample median, then with probability $1 - (1 - \varepsilon)^{1/\varepsilon}$, $S$ will correctly reject $y$ (Exercise 8). To use this algorithm, the following is the information $S$ requires from $A$ in the data aggregation phase:

- Aggregation result $y$
- The number of data samples $n$
- A commitment of the data samples $h_A$

The commitment is necessary because when $S$ requests a value from $A$ in the result verification phase, we must make sure $S$ gets a value that has previously been committed to by some source node in the data aggregation phase; or in other words, a value that the aggregator $A$ cannot simply make up. One way to construct $h_A$ is to calculate $h_A$ as the root of a binary *commitment tree*, i.e., when $n = 4$, the commitment $h_A$ in Fig. 21.4a is calculated as the root $h_{0,0}$ in Fig. 21.4b and $h_{0,0} = h(h(h(x_1)||h(x_2))||h(h(x_3)||h(x_4)))$. Suppose $S$ requests only the first and second element, i.e., $x_1$ and $x_2$, from $A$, $S$ can already reconstruct the commitment if $A$ also sends $h_{1,1}$ to $S$. If the length of a hash is the same as the length of a sample, sending $h_{1,1}$ instead $x_3$ and $x_4$ cuts the size of transmission in half. If the length of a hash is shorter than the length of a sample, the saving in bandwidth is even more.

At this point, we know how $S$ can check if the aggregation result more or less represents the sample median; and we also know what information $S$ needs to collect from $A$. We now show in detail how the protocol works. In the data aggregation phase, the source node 1 sends the following to the aggregator $A$:

$q||ID(1)||x_1||\text{MAC}_{1S}||\text{MAC}_{1A}$,
where $\text{MAC}_{1S} = \text{MAC}(K_{1S}, q||\text{ID}(1)||x_1)$
and $\text{MAC}_{1A} = \text{MAC}(K_{1A}, q||\text{ID}(1)||x_1||\text{MAC}_{1S})$

The aggregator $A$ in turn sends the following to the sink:

$q||A||f(x_1, x_2, \ldots, x_n)||n||h_A||\text{MAC}_{AS}$,
where $\text{MAC}_{AS} = \text{MAC}(K_{AS}, q||A||f(x_1, x_2, \ldots, x_n)||n||h_A)$

In the result verification phase, when $S$ requests for $x_1$ and $x_2$, for example, $A$ replies with

$M||\text{MAC}(K_{AS}, M)$,
where $M = q||\text{ID}(1)||x_1||\text{MAC}_{1S}||\text{ID}(2)||x_2||\text{MAC}_{2S}||h_{1,1}$ .

$S$ then (1) authenticates $x_1$ and $x_2$ using $\text{MAC}_{1S}$ and $\text{MAC}_{2S}$, respectively, (2) checks if $x_1$ and $x_2$ are acceptable using Przydatek et al.'s algorithm, and (3) checks if $x_1$, $x_2$ and $h_{1,1}$ allow the commitment to be reconstructed.

### 21.5.2 The Multiaggregator Case

Let us now consider the case of multiple, nested aggregators. First, we will discuss Chan et al.'s hierarchical in-network aggregation [5], and then we will look at Yang et al.'s secure hop-by-hop data aggregation protocol (SDAP) [26].

#### 21.5.2.1 Chan et al.'s [5] Hierarchical In-Network Aggregation

To describe this scheme, we need two binary operators:

- AGG(msg1, msg2): Let $\text{msg1} = q||v_1||c_1$ and $\text{msg2} = q||v_2||c_2$, then AGG(msg1, msg2) $= q||f(v_1, v_2)||c_1 + c_2$.
- COMB(msg1, msg2): Let $\text{msg1} = q||v_1||c_1$ and $\text{msg2} = q||v_2||c_2$, then COMB(msg1, msg2) $= q||v_1||c_1||v_2||c_2$.

Figure 21.5 illustrates the example we use for our ensuing discussion, which is derived from Fig. 21.2. In this example, the sensors in white – $C, D, G, I, J$ – are source nodes. We begin our discussion with the data aggregation phase. In this phase, every sensor sends a message of the following format to its parent:

query ID || value || complement || count || commitment || MAC,

where "query ID" is a number that identifies the query; "value" is the raw data or aggregation result computed by the sender; "complement" is the complement of value; "count" is the total number of samples from which the aggregate is derived;

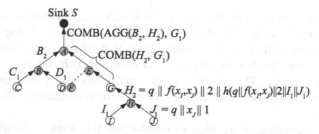

**Fig. 21.5** Chan et al.'s secure hierarchical in-network aggregation [5]. In this example, only sensors in white source data. Pairwise MACs are omitted for clarity. Messages are labeled following the pattern "nodeID$_{count}$"

"commitment" is defined as in the single-aggregator case; and MAC is a message authentication code signed with the pairwise key shared between the sender and the receiver. The "complement" field is for validating the "value" field, and for clarity, it will be omitted in the following discussion. Every message is assumed to be protected by a pairwise MAC which is not made explicit in our protocol description.

In the data aggregation phase, $I$ and $J$ each sends a data message to $H$, which aggregates the data into message $H_2$ (our naming scheme of the messages follows the pattern "nodeID$_{count}$"). Note the commitment in $H_2$ is dependent on $I_1$ and $J_1$. When $H_2$ arrives at $G$, $G$ does *not* aggregate $f(x_I, x_J)$ and $x_G$. Instead, $G$ sees $H_2$ as a three-node tree (strictly speaking, the root of a three-node tree) but $G_1$ as a single-node tree, and decides to combine $H_2$ with $G_1$ using the COMB operator. On receiving both $B_2$ and COMB$(H_2, G_1)$, $A$ sees $B_2$ and $H_2$ as both three-node trees but $G_1$ as a single-node tree, and decides to aggregate $B_2$ and $H_2$ into AGG$(B_2, H_2)$, and send COMB$(\text{AGG}(B_2, H_2), G_1)$ to $S$. The reason for aggregating only trees of the same size is to create balanced binary trees. The advantage of creating only balanced binary trees is that edge congestion (congestion on a link) is only $O(\log^2 n)$, where $n$ is the number of samples. In the end, $S$ only receives a forest of balanced binary trees – this is called the *commitment forest* – consisting of the seven-node tree AGG$(B_2, H_2)$ and the single-node tree $G_1$, from which $S$ calculates the aggregation result as AGG$(\text{AGG}(B_2, H_2), G_1)$.

In the result verification phase, $S$ broadcasts COMB$(\text{AGG}(B_2, H_2), G_1)$ to the network, for example, using μTESLA. Next, the following transmissions take place:

| | |
|---|---|
| $A \to B : H_2$ | $A \to E : \text{COMB}(B_2, G_1)$ |
| $B \to C : \text{COMB}(H_2, D_1)$ | $B \to D : \text{COMB}(H_2, C_1)$ |
| $E \to G : \text{COMB}(B_2, G_1)$ | $G \to H : B_2$ |
| $H \to I : \text{COMB}(B_2, J_1)$ | $H \to J : \text{COMB}(B_2, I_1).$ |

After these transmissions, both $C$ and $D$ can then independently reconstruct AGG$(B_2, H_2)$; $G$ can trivially reconstruct $G_1$; and both $I$ and $J$ can reconstruct AGG$(B_2, H_2)$. A difference between this and the single-aggregator case is that instead of reconstructing the commitment at the sink, we are now reconstructing the commitment at each individual source node. A source node that successfully reconstructs the commitment will send a confirmation message $q||\text{nodeID}||\text{OK}||\text{MAC}(K, q||\text{nodeID}||\text{OK})$ to the sink, where $K$ is the unique key the sensor shares with the sink. A source node that fails to reconstruct the commitment will immediately raise an alarm, for example, by broadcasting a negative confirmation message. The protocol ends successfully when enough confirmation messages are received and there is no negative confirmation; otherwise, the protocol fails.

Negative confirmations are used to combat the following scenario: since the sink only knows how many source nodes there are, but does not know which the source nodes are, any insider attacker can forge a confirmation message, but as long as one source node fails to reconstruct the commitment, it will be able to alert the entire network by sending a negative confirmation. However, using negative confirmations presents another risk, in that an insider attacker can forge a negative confirmation. The source of these problems lies in the fact that instead of at the sink, the commitment is reconstructed at the source nodes themselves.

### 21.5.2.2 SDAP

From Chan et al.'s proposal [5], we learn that reconstruction of the commitment is better done at the sink than at the source nodes themselves. To do this, we do not need to involve all source nodes to attest to their submitted data; on the contrary, we only need to watch out for suspicious sensors. This is the motivation behind SDAP. In the previous scheme, whenever $S$ needs to verify an aggregation result, it has to engage the whole subnetwork rooted at $A$ (Fig. 21.5), since it does not know who the source sensors are. However, if we divide the subnetwork into groups, we only need to check the groups which look suspicious (Fig. 21.6). This is more efficient and is the main innovation behind SDAP.

The essence of SDAP is the grouping algorithm. In the data aggregation phase, a sensor decides whether it would become a group leader by checking whether $h(q||\text{nodeID}) < F_g(c)$, where $F_g(c)$ is a function that increases with the data count $c$, i.e., the higher the data count, the more likely the inequality is satisfied. Assuming

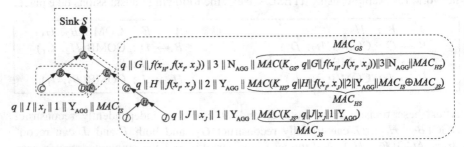

**Fig. 21.6** SDAP divides a network into groups (three groups in this case). In this example, only sensors in white source data. Pairwise MACs are omitted for clarity

$h(\ )$ and $F_g(\ )$ has a the same range $[0,\ M]$, one candidate for $F_g(\ )$ is $M(1-e^{-\beta c})^\gamma$, where $\beta$ is used to control the gradient of the curve and $\gamma$ is to control the shape of the curve of $F_g(\ )$. The role of a group leader is to set a Boolean flag in a message to $N_{AGG}$ to indicate the message needs only be forwarded. Nongroup leaders set the flag to $Y_{AGG}$ to indicate that the message should be aggregated as normal. The example in Fig. 21.6 shows that there are three group leaders $B$, $G$, and $S$. All the sensors that do not have a group leader among their ancestors fall under the group led by the sink.

Figure 21.6 illustrates the example we use for our ensuing discussion, which is derived from Fig. 21.2. In this example, the sensors in white — $C$, $D$, $G$, $I$, $J$ — are source nodes. We begin our discussion with the data aggregation phase. In this phase, every sensor sends a message of the following format to its parent:

---

query ID || node ID || value || count || aggregate-flag || commitment

---

The format is similar to Chan et al.'s scheme [5], except for the "aggregate-flag" which is discussed in the previous paragraph. Every message is assumed to be protected by a pairwise MAC which is not made explicit in our protocol description. In the data aggregation phase, $J$ sends the following message to its parent:

---

$J \rightarrow H$ : $q||J||x_J||1||Y_{AGG}||MAC_{JS}$,
where $MAC_{JS} = MAC(K_{JS},\ q||J||x_J||1||Y_{AGG})$.

---

$H$ aggregates $I$ and $J$'s data in the following message:

---

$H \rightarrow G$ : $q||H||x_H||2||Y_{AGG}||MAC_{HS}$,
where $x_H = f(x_I,\ x_J)$
$MAC_{HS} = MAC(K_{HS},\ q||H||x_H||2||Y_{AGG}||MAC_{IS} \oplus MAC_{JS})$.

---

Note both $MAC_{IS}$ and $MAC_{JS}$ contribute to $MAC_{HS}$. The XOR operator instead of the concatenation operator is used to combine $MAC_{IS}$ and $MAC_{JS}$ because the XOR operator is commutative. Since $G$ is a source node as well as a group leader, it not only increases the count by 1, but also sets the flag to $N_{AGG}$:

---

$G \rightarrow E$ : $q||G||x_G||3||N_{AGG}||MAC_{GS}$,
where $x_G = f(x_G,\ (x_I,\ x_J))$
$MAC_{GS} = MAC(K_{GS},\ q||G||x_G||3||N_{AGG}||MAC_{HS})$.

---

The above message is never aggregated with any other message, on its way from $G$ to $S$, because of the $N_{AGG}$ flag. At $S$, every group's aggregate is saved in the format:

(leader's node ID, value, count, commitment). After receiving the aggregates from all groups, for each group aggregate:

- $S$ tests if $h(q||\text{leader's nodeID}) < F_g(\text{count})$. If false, $S$ discards the group aggregate. Otherwise, $S$ proceeds with the next test.
- $S$ tests if the group aggregate represents an outlier, for example, by using Grubbs' test (and of course $S$ can also test for more than one outlier). Once an outlier is found, result verification begins. Note that an outlier is not necessarily invalid data. In fact, if result verification succeeds, in an anomaly-seeking application, this outlier will be regarded as interesting data. In other words, in such an application, outlier detection is only to confirm abnormal values are genuine.

To kick-start the result verification phase, suppose the suspicious group is found to be the group led by $G$, the following transmission would take place:

$$S \rightarrow A: \ G||q||q_a,$$

where $q_a$ is the ID that identifies this attestation round and is used to choose an *attestation path*. When the request reaches $G$, $G$ chooses an attestation path as follows. Suppose $G$ has $d$ children with counts $c_1, c_2, \ldots, c_d$. For each child with ID $ID$, G calculates $h(q_a||ID) \sum_{i=1}^{d} c_i$. $G$ picks the $j$th child if the result lies in the interval $[M \sum_{i=1}^{j-1} c_i, M \sum_{i=1}^{j} c_i)$, where $M$ is the maximum value of $h( )$. The result of this policy is that the bigger the contribution of a child's count is to $\sum_{i=1}^{d} c_i$, the more likely the child would be selected as the next node in the attestation path (Exercise 9). In this example, $G$ has only one child $H$, so $G$ has to choose $H$ anyhow. Suppose the final attestation path turns out to be $G\,H\,J$, the following messages are sent back to $S$:

$$G \rightarrow\rightarrow S: \ q_a||G||x_G||3||\text{MAC}_{GS},$$
$$H \rightarrow\rightarrow S: \ q_a||H||x_H||2||\text{MAC}_{HS},$$
$$J \rightarrow\rightarrow S: \ q_a||J||x_J||1||\text{MAC}_{JS},$$
$$I \rightarrow\rightarrow S: \ q_a||I||x_I||1||\text{MAC}_{IS}.$$

These messages must be sent in the correct order (Exercise 10). Upon receiving these messages, $S$ performs the following checks:

- $x_G$ is correctly derived from $f(x_G, f(x_J, x_I))$.
- $\text{MAC}_{GS}$ is correctly reconstructed in the following steps:

$$\text{MAC}_{IS} = \text{MAC}(K_{IS}, q||I||x_I||1),$$
$$\text{MAC}_{JS} = \text{MAC}(K_{JS}, q||J||x_J||1),$$
$$\text{MAC}_{HS} = \text{MAC}(K_{HS}, q||H||f(x_J, x_I)||2||\text{MAC}_{IS} \oplus \text{MAC}_{JS}),$$
$$\text{MAC}_{GS} = \text{MAC}(K_{GS}, q||G||f(x_G, f(x_J, x_I))||3||\text{MAC}_{HS}).$$

If all checks succeed, $S$ accepts $G$'s group aggregate and computes the final aggregation result. Otherwise, $S$ computes the final aggregation result regardless of $G$'s group aggregate.

## 21.6  Privacy Homomorphism

So far we have talked about the techniques to achieve the *robustness* objectives: resilient aggregation, voting, and result verification. Now we will talk about the technique for achieving the *confidentiality* objective: privacy homomorphism (PH). The purpose of using PH is to allow aggregators to aggregate encrypted data directly. PH is defined as follows. A function is $(\otimes, \oplus)$-homomorphic if $f(x) \otimes f(y) = f(x \oplus y)$, where "$\otimes$" is an operator in the range and "$\oplus$" is an operator in the domain. If $f$ is an encryption function and the inverse function $f^{-1}$ is the corresponding decryption function, then we have a PH.

PHs were first introduced [21] to allow processing on encrypted data. To the best of our knowledge, PHs were first applied to WSNs only recently [4, 10].

PHs are different from conventional ciphers in the sense that the highest attainable security [27] for PHs is *semantic security under nonadaptive chosen-ciphertext attacks* (IND-CCA1) [15]. Very loosely speaking, this means the best security that a PH can achieve is against an attacker that has limited access to an encryption–decryption black box. This is also to say that *no* PH is semantically secure under *adaptive chosen-ciphertext attacks* (IND-CCA2) [20]; or in other words when an attacker has unlimited access to an encryption–decryption black box, any PH can be defeated.

In practice, we only look for PHs that are semantically secure against *chosen-plaintext attacks* (IND-CPA) [11] (which is weaker than IND-CCA1). This is acceptable for data aggregation since we consider an attacker that can manipulate the inputs to a source node (has access to an encryption black box) but cannot see the output of a sink node (has no access to a decryption black box).

PHs are also by definition *malleable*, in the sense they cannot achieve the *nonmalleability* (NM-CCA) notion [7], because an attacker can derive a new ciphertext simply by applying the operator $\otimes$ to two known ciphertexts. In WSNs, we go around this problem by requiring the samples to be propagated to the sink hop by hop, using a secure channel for each hop. As such, a sample is first encrypted by a PH, and then signed with the session key of a secure channel, thereby preventing an outsider attacker from modifying the message.

There are three main approaches to PHs in WSNs so far (1) PHs that are based on *polynomial rings* [8, 25], (2) PHs that are based on *one-time pads* [4, 16], and (3) homomorphic *public-key* cryptosystems [14].

The first approach is unfortunately insecure under only *known-plaintext attacks* [23] which are much weaker than chosen-plaintext attacks discussed in preceding paragraphs. Thus, we are left with the second and third approach.

The second approach of using one-time pads has actually been long known [1]. In this approach, the encryption function simply adds a message $m$ with a key stream $k$ that simulates a one-time pad, i.e., $E(k, m) = m + k \pmod{p}$. When $n$ ciphertexts are added together, the resulting ciphertext $C$ becomes

$$C = \sum_{i=1}^{n} E(k_i, m_i) = \left( \sum_{i=1}^{n} m_i + \sum_{i=1}^{n} k_i \right) \bmod p.$$

To decrypt the resulting ciphertext $C$, the decryptor has to keep track of the keys used, and retrieve the plaintext sum as

$$\sum_{i=1}^{n} m_i \bmod p = C - \sum_{i=1}^{n} k_i \bmod p.$$

The security of this scheme lies in the randomness of the key stream. The key stream can be generated using a standard block cipher in a stream cipher mode, for example, using AES in counter mode; or if a lower security level is acceptable, a dedicated stream cipher. The drawbacks of this approach are:

1. The *security* of the scheme has not been rigorously proven, especially with respect to the use of the addition operator in place of the XOR operator in the plaintext space.
2. The sink has to *synchronize* its key streams with those of the sources. For example, if counter mode is used, then the sink will have to synchronize as many counters as there are sources, with the sources. This will become a scalability issue when there are a lot of sources.
3. A more significant *scalability* problem arises in schemes where a sensor has to share keys and synchronize counters with sensors from one to multiple hops away [16].
4. This approach, being symmetric in the sense that secret keys are shared by the sink with the sources, is not *intrusion-resilient*. Some schemes try to increase the resilience by requiring a sensor to share keys with sensors from one to multiple hops away [16], but scalability becomes an issue, as mentioned earlier.

The third approach is based on homomorphic public-key cryptosystems. The advantage of this approach over using key streams is that key management becomes easier.

Based on energy- and bandwidth-efficiency, two candidate cryptosystems have been identified [14]: (1) ElGamal on elliptic curves (EC-EG) and (2) Okamoto–Uchiyama (OU). These two cryptosystems are described in detail in Sects. 21.6.1.1 and 21.6.1.2.

### 21.6.1.1  ElGamal on Elliptic Curves

Key setup  Choose a large prime $p$ (163-bit is the standard minimum length).
Choose an elliptic curve $E$ defined over $GF(p)$.
Choose a point $G$ on $E$ with prime order $n$.
$x \in_R [1, n-1]$.
Calculate public key $Y$ as $Y = xG$.
Find a homomorphic encoding function $\varphi\,()$ that maps a message $m$ to a point $M$ on $E$.

Encryption  $M = \varphi(m)$.
$k \in_R [1, n-1]$.

The encryption of $M$ is the pair $(C_1, C_2) = (kG, kY + M)$.

Decryption Decrypt the pair $(C_1, C_2)$ as $M = -xC_1 + C_2$.
$$m = \varphi^{-1}(M).$$

ElGamal on Elliptic Curves EC-EG is $(+, +)$-homomorphic because if $M_1$ is encrypted to $(C_{11}, C_{12}) = (k_1G, k_1Y + M_1)$ and $M_2$ is encrypted to $(C_{21}, C_{22}) = (k_2G, k_2Y + M_2)$, then the component-wise sum of $(C_{11}, C_{12})$ and $(C_{21}, C_{22}) = ((k_1 + k_2)G, (k_1 + k_2)Y + (M_1 + M_2))$ decrypts to $M_1 + M_2 = \varphi(m_1 + m_2)$. The maximum number of aggregations is determined to be less than (order of elliptic curve $E$)÷(maximum value of $m$). In practice, care must be taken to ensure that $\varphi$ does map to a valid point on $E$ and likewise $\varphi^{-1}$ does map to a valid message.

The semantic security of EC-EG depends on the assumption that, given the public key $Y = xG$ and $kG$, it is impossible to differentiate between $xkG$ and $rkG$ where $r$ is a random number. This is the so-called Decisional Diffie-Hellman (DDH) assumption, and it is related to the discrete logarithm problem on elliptic curves. Currently, this is considered a computationally intractable problem and is the security basis of many cryptographic schemes.

#### 21.6.1.2 Okamoto–Uchiyama

Key setup  Choose two large $k$-bit primes $p$, $q$ and set $n = p^2q$.
Define subgroup $\Gamma = \{x | x \in (\mathbb{Z}/p^2\mathbb{Z})^*, x \equiv 1 \bmod p\}$.
Define function $L : \Gamma \rightarrow GF(p)$ as $L(x) = (x - 1)/p$. $L$ is $(\times, +)$-homomorphic, i.e.,

$$L(ab) = L(a) + L(b) \bmod p$$
$$L(a^b) = bL(a) \bmod p$$

Choose $g$ randomly from $\mathbb{Z}_p^*$ such that the order of $g^{p-1} \equiv 1 \bmod n$ is $p$.
Set $h = g^n \bmod n$.
Set private key $= (p, q)$.
Set public key $= (n, g, h, k)$.

Encryption  $r \in_R \mathbb{Z}_n$.
Ciphertext corresponding to plaintext $m$ $(0 < m < 2^{k-1})$, $C = g^m h^r \bmod n$

Decryption  Plaintext corresponding to ciphertext $C$
$$= L(C^{p-1} \bmod p^2)/L(g^{p-1} \bmod p^2) \bmod p$$
$$= L(g^{(m+nr)(p-1)} \bmod p^2)/L(g^{p-1} \bmod p^2) \bmod p$$
$$= L(g^{m(p-1)} \bmod p^2)/L(g^{p-1} \bmod p^2) \bmod p$$
$$= m$$

Okamoto–Uchiyama (OU) is $(+, \times)$-homomorphic because if $m_1$ is encrypted to $C_1$ and $m_2$ is encrypted to $C_2$, then the decryption of the ciphertext product $C_1C_2 = L((C_1C_2)^{p-1})/L(g^{p-1}) \bmod p = (m_1 + m_2)L(g^{p-1})/L(g^{p-1}) \bmod p = m_1 + m_2$, is the sum of the original messages. Note that since the maximum message value is $2^{k-1} - 1$, we have to make sure the maximum number of aggregation operations is less than $p/(2^{k-1} - 1)$.

The semantic security of OU depends on the intractability of factoring $n = p^2 q$. Factoring $n$ allows the attacker to know $p$ and $q$, the private key. Like ElGamal, OU can be constructed on elliptic curves, however, doing so turns out to be less energy efficient [14].

By using elliptic curves, EC-EG does not require as many bits to represent a ciphertext as OU does. However, the $\varphi^{-1}$ function of EC-EG requires an amount of computation that increases with the value of the aggregation result. Therefore, the general recommendation is to use EC-EG when the aggregation result is small, but OU if otherwise [14].

## 21.7 Thoughts for Practitioners

Among the techniques introduced so far, voting, result verification, and PH all require a lot of resources. Only resilient aggregation is the most practically implementable. If all data are only aggregated once, then a simple resilient aggregation function such as one of those in Table 21.2, or RANBAR, can be used directly. Otherwise, quantiles aggregation can be used to compress the data at each aggregation point, so that at the final sink node, the compressed data are still good to derive various statistics from.

## 21.8 Directions for Future Research

The first layer of defense against malicious source nodes is resilient aggregation. While sample median is a robust measure of location (the location is the center of the data set in this case), we sometimes need to collect statistics like the minimum or the maximum of the samples. When the data distribution is not known a priori, there is currently no $(k, \alpha)$-resilient way to aggregate these statistics. Further research is needed in this area. Result verification is a communication-intensive process. Some work is needed in quantifying and optimizing the overhead involved. Apart from that, existing schemes are designed with the assumption that the whole network is involved in data aggregation, which is not necessarily true. This fact can be used to improve existing schemes. Privacy homomorphism is still costly to implement, but substantial optimization work can be done in this area. It is also worthwhile here to note a recent work [12] that for the first time in the context of data aggregation in WSNs, considers how to achieve the additional objective of *privacy*. This is an interesting line of research.

## 21.9 Conclusions

There are two objectives in secure data aggregation: *robustness* and *confidentiality*. The robustness objective is to ensure that malicious source nodes or aggregators cannot arbitrarily bias aggregated data statistics from its true value. The techniques include resilient aggregation, voting, and result verification. Resilient aggregation is used to bind the potential distortion of the data caused by malicious source nodes. When there is a high degree of spatial correlation in the data, before an aggregator accepts a sensor's sample, the aggregator first consults the sensor's neighbors, which then vote on the data – this mechanism is called voting. Due to the resource requirement though, voting is reserved for small-scale networks. We use result verification to ensure intermediate aggregators and forwarders have not tampered with the aggregation results. As for the confidentiality objective, the requirement is that the intermediate sensors between the sink and the sources can aggregate encrypted samples but cannot know the actual aggregated values. To achieve this, the standard technique is privacy homomorphism.

## Terminologies

*Adaptive chosen-ciphertext attack.* This is the same as a chosen-ciphertext attack except that the decryption oracle is permanently available to the attacker.

*Chosen-ciphertext attack.* An attack against a cryptosystem that is divided into two phases: in the first phase, the attacker feeds a decryption oracle with carefully chosen ciphertexts and analyzes the resultant plaintexts in relation to the ciphertexts; in the second phase, the attacker loses the decryption oracle and tries to infer information about the plaintext corresponding to a target ciphertext.

*Chosen-plaintext attack.* An attack against a cryptosystem where the attacker feeds an encryption oracle with carefully chosen plaintexts, and analyzes the resultant ciphertexts in relation to the plaintexts. For public-key cryptosystems, this kind of attack is trivially doable by the attacker since the encryption key is public.

*Decryption oracle.* An abstraction that is independent of the adversary but decrypts ciphertexts for the adversary on request.

*Elliptic curve.* An elliptic curve over finite field $GF(p)$ is defined by an equation of the form $y^2 = x^3 + ax + b$, where $a, b \in GF(p)$ satisfy $4a^3 + 27b^2 \neq 0$ mod $p$.

*Encryption oracle.* An abstraction that is independent of the adversary but encrypts plaintexts for the adversary on request.

*Entropy.* The entropy of a random variable $X$ is a measure of the "uncertainty" of $X$. If $X$ takes on $n$ values with probabilities $p_1, p_2, \ldots, p_n$, then the entropy of $X$ is $-\sum_{i=1}^{n} p_i \log p_i$.

*Finite field.* See Galois field.

*Galois field.* A field of finite order. A field is a set with two binary operations $+$ and $\cdot$ that satisfies the field axioms:

Examples are the integers modulo a prime, the real numbers, the rational numbers, etc.

| Property | $+$ | $\cdot$ |
|---|---|---|
| Identity | Exists, denoted 0 | Exists, denoted 1 |
| Inverse | Exists, denoted $-a$ | Exists, denoted $a^{-1}$ |
| Associativity | $(a + b) + c = a + (b + c)$ | $(a \cdot b) \cdot c = a \cdot (b \cdot c)$ |
| Commutivity | $a + b = b + a$ | $a \cdot b = b \cdot a$ |
| Distributivity | $a \cdot (b + c) = (a \cdot b) + (a \cdot c)$ | |

*Group.* In abstract algebra, group is a (finite or infinite) set with a binary operation $+$ that satisfies the group axioms:

1. *Closure.* $a + b \in G, \forall a, b \in G$
2. *Identity.* There exists an identity element, denoted 0, s.t. $a + 0 = a, \forall a \in G$.
3. *Inverse.* There exists an inverse, denoted $-a$, s.t. $a + (-a) = 0, \forall a \in G$.
4. *Associativity.* $(a + b) + c = a + (b + c), \forall a, b, c \in G$.

When the operation is written in multiplicative notation, it is called a multiplicative group.

*Malleable.* If a cryptosystem message is malleable, an attacker can modify a plaintext message in a meaningfully controllable manner by modifying the corresponding ciphertext. Nonmalleability is now understood to be a desirable property for cryptosystems.

*Message authentication code (MAC).* A code generated from a secret key and a message, that lets the receiver of the message to authenticate the message as tamper-free and as coming from the sender, if the receiver shares the key with the sender.

*Order (of an element).* The order of an element $x \in G$ is the smallest positive integer $n$ such that $x^n = 1$, where 1 is the identity element of $G$.

*Pairwise key.* A key shared between a pair of in-range sensors. The MACs created using this key are informally called pairwise MACs in this chapter.

*Privacy homomorphism.* A function $f$ is $(\otimes, \oplus)$-homomorphic if $f(x) \otimes f(y) = f(x \oplus y)$, where "$\otimes$" is an operator in its range and "$\oplus$" is an operator in its domain. If $f$ is an encryption function and the inverse function $f^{-1}$ is the corresponding decryption function, then we have a privacy homomorphism.

*Semantic security.* A requirement of a cryptosystem that whatever is efficiently computable about the plaintext given the ciphertext is also efficiently computable without the ciphertext. In other words, an attacker should not learn anything about the plaintext from the ciphertext except the trivial information about its length.

## Questions

1. Show that "sum" is not a resilient aggregation function.
2. Show that "count" is a resilient aggregation function, given $\Pr\{$a sensor reports $1\} = p$, and $\Pr\{$a sensor reports $0\} = 1 - p$.

3. If we are collecting samples that are normally distributed with a variance of 1, we can estimate the average in the following manner [3]. Let the samples be $x_1, \ldots, x_n$ where $n$ is even. We divide the samples into two groups $Z_1 = X_1 + \cdots + X_{n/2}$ and $Z_2 = X_{n/2+1} + \cdots + X_n$, and take their difference $W = Z_1 - Z_2$. Intuitively, $W$ will have a sample mean of around 0 unless the samples are contaminated. In other words, if $|W| > h_\alpha$, for some threshold $h_\alpha$, then we can say the samples have been tampered with. What should $h_\alpha$ be to limit the probability of false detection to 0.05 for 50 samples?

4. Show that a q-digest $Q$ constructed with compression parameter $k$ has a size of at most $3k$ (q-digest nodes).

5. Show that in a q-digest created using the compression parameter $k$, the maximum error in count of any node is $\log \sigma \lfloor n/k \rfloor$, where $\sigma$ is the number of possible values a data sample can take.

6. In a quantile query, the aim is to find the value $x$ whose rank (i.e., order in a sorted sequence of the samples) is $pn$, e.g., to find the sample median, $p = 0.5$. Define the error $\varepsilon$ as

$$\varepsilon \triangleq \frac{|\text{true rank of } x - pn|}{n}.$$

Using the result from Exercises 4 and 5, show that given $m$ memory units to build a q-digest, it is possible to answer any quantile query with $\varepsilon < (3 \log \sigma)/m$.

7. In RANBAR, the maximum number of iterations is set at $i = 15$ by empirical analysis. What happens if we set $i$ too low or too high?

8. Prove that in Przydatek et al.'s algorithm, the probability of rejecting a false median, i.e., a sample that is $\varepsilon n (\varepsilon > 0)$ or more away from the true median, is at least $1 - e^{-1}$.

9. In the result verification phase of SDAP, a parent chooses the next node in the attestation path by seeing where $h(q_a || ID) \sum_{i=1}^{d} c_i$ lies. If the result lies in the interval

$$[M \sum_{i=1}^{j-1} c_i, M \sum_{i=1}^{j} c_i),$$

then the parent pick its $j$th child. Assuming the hash function produces uniformly distributed output, prove that the probability that this parent selects the $j$th child is $c_j / \sum_{i=1}^{d} c_i$.

10. In the result verification phase of SDAP, why is it important for the sensors to send their messages back to $S$ in order?

# References

1. Ahituv N, Lapid Y, Neumann S (1987) Processing encrypted data. Communications of the ACM, 30(9):777–780
2. Buttyán L, Schaffer P, Vajda I (2006a) RANBAR: RANSAC-based resilient aggregation in sensor networks. In: Proceedings of the Fourth ACM Workshop on Security of Ad Hoc and Sensor Networks (SASN '06), pp. 83–90. ACM, New York, NY

3. Buttyán L, Schaffer P, Vajda I (2006b) Resilient aggregation with attack detection in sensor networks. In: Proceedings of the Fourth Annual IEEE International Conference on Pervasive Computing and Communications Workshops (PERCOMW'06). IEEE, New York, NY
4. Castelluccia C, Mykletun E, Tsudik G (2005) Efficient aggregation of encrypted data in wireless sensor networks. In: Mobile and Ubiquitous Systems: Networking and Services (MobiQuitous '05). IEEE, New York, NY
5. Chan H, Perrig A, Song D (2006) Secure hierarchical in-network aggregation in sensor networks. In: Proceedings of the 13th ACM Conference on Computer and Communications Security (CCS '06), pp. 278–287. ACM, New York, NY
6. Chen W-P, Hou JC (2005) Chapter 15: Data gathering and fusion in sensor networks. In: Handbook of Sensor Networks: Algorithms and Architectures. Wiley, New York, NY
7. Dolev D, Dwork C, Naor M (1991) Non-malleable cryptography. In: Proceedings of the 23rd Annual ACM Symposium on Theory of Computing (STOC '91), pp. 542–552. ACM, New York, NY
8. Domingo-Ferrer J (2002) A provably secure additive and multiplicative privacy homomorphism. In: Information Security: Proceedings of the Fifth International Conference (ISC '02), Sao Paulo, Brazil, September 30–October 2, volume 2433 of LNCS, pp. 471–483. Springer, New York, NY
9. Du W, Deng J, Han YS, Varshney PK (2003) A witness-based approach for data fusion assurance in wireless sensor networks. In: IEEE Global Telecommunications Conference (GLOBECOM '03), volume 3, pp. 1435–1439. IEEE, New York, NY
10. Girao J, Westhoff D, Schneider M (2005) CDA: Concealed data aggregation for reverse multicast traffic in wireless sensor networks. In: IEEE International Conference on Communications (ICC '05), pp. 3044–3049, Seoul, Korea, May 2005. IEEE, New York, NY
11. Goldwasser S, Micali S (1984) Probabilistic encryption. Journal of Computer and System Sciences, 28:270–299
12. He W, Liu X, Nguyen H, Nahrstedt K, Abdelzaher T (2007) PDA: privacy-preserving data aggregation in wireless sensor networks. In: Proceedings of the IEEE Conference on Computer Communications (INFOCOM '07), pp. 2045–2053, Anchorage, Alaska, USA, 6–12 May 2007. IEEE, New York, NY
13. Mahimkar A, Rappaport TS (2004) SecureDAV: a secure data aggregation and verification protocol for sensor networks. In: IEEE Global Tele-communications Conference (GLOBECOM '04), volume 4, pp. 2175–2179. IEEE, New York, NY
14. Mykletun E, Girao J, Westhoff D (2006) Public key based crypto-schemes for data concealment in wireless sensor networks. In: IEEE International Conference on Communications (ICC '06), volume 5, pp. 2288–2295. IEEE, New York, NY
15. Naor M, Yung M (1990) Public-key cryptosystems provably secure against chosen ciphertext attacks. In: Proceedings of the Twenty-Second Annual ACM Symposium on Theory of Computing (STOC '90). ACM, New York, NY
16. Önen M, Molva R (2007) Secure data aggregation with multiple encryption. In: Wireless Sensor Networks, volume 4373 of LNCS, pp. 117–132. Springer, New York, NY
17. Patel JK, Read CB (1982) Handbook of the Normal Distribution, 1st edn. Marcel Dekker, New York, NY
18. Przydatek B, Song D, Perrig A (2003) SIA: secure information aggregation in sensor networks. In: Proceedings of the First International Conference on Embedded Networked Sensor Systems, pp. 255–265. ACM, New York, NY
19. Rabbat M, Nowak R (2004) Distributed optimization in sensor networks. In: IPSN '04: Proceedings of the Third International Symposium on Information Processing in Sensor Networks, pp. 20–27. ACM, New York, NY
20. Rackoff C, Simon DR (1991) Non-interactive zero-knowledge proof of knowledge and the chosen ciphertext attack. In: Advances in Cryptology: Proceedings of the 11th Annual International Cryptology Conference (CRYPTO '91), Santa Barbara, CA, USA, August 11–15, 1991, volume 576 of LNCS, pp. 433–444. Springer, New York, NY
21. Rivest RL, Adleman L, Dertouzos ML (1978) On data banks and privacy homomorphisms. In: Proceedings of Foundations of Secure Computation, pp. 169–179. Academic, New York, NY

22. Shrivastava N, Buragohain C, Agrawal D, Suri S (2004) Medians and beyond: new aggregation techniques for sensor networks. In: Proceedings of the Second International Conference on Embedded Networked Sensor Systems (SenSys '04), pp. 239–249. ACM, New York, NY
23. Wagner D (2003) Cryptanalysis of an algebraic privacy homomorphism. In: Information Security: Proceedings of the Sixth International Conference (ISC '03), Bristol, UK, October 1–3, 2003, volume 2851 of LNCS, pp. 234–239. Springer, New York, NY
24. Wagner D (2004) Resilient aggregation in sensor networks. In: Proceedings of the Second ACM Workshop on Security of Ad Hoc and Sensor Networks (SASN '04), pp. 78–87. ACM, New York, NY
25. Westhoff D, Girao J, Acharya M (2006) Concealed data aggregation for reverse multicast traffic in sensor networks: Encryption, key distribution, and routing adaptation. IEEE Transactions on Mobile Computing, 5(10):1417–1431
26. Yang Y, Wang X, Zhu S, Cao G (2006) SDAP: a secure hop-by-Hop data aggregation protocol for sensor networks. In: Proceedings of the Seventh ACM International Symposium on Mobile Ad Hoc Networking and Computing (MobiHoc '06), pp. 356–367. ACM, New York, NY
27. Yu Y, Leiwo J, Premkumar B (2008) A study on the security of privacy homomorphism. International Journal of Network Security, 6(1):33–39. Preliminary version appeared in Proceedings of the Third International Conference on Information Technology – New Generations (ITNG '06), pp. 470–475, Las Vegas, Nevada, USA, 2006. IEEE, New York, NY

# Chapter 22
# Wireless Multimedia Sensor Networks

Ivan Lee, William Shaw, and Xiaoming Fan

**Abstract** The emergence of low-cost and mature technologies in wireless communication, visual sensor devices, and digital signal processing facilitate of wireless multimedia sensor networks (WMSN). Like sensor networks which respond to sensory information such as temperature and humidity, WMSN interconnects autonomous devices for capturing and processing video and audio sensory information. This survey highlights the following topics (1) a summary of applications and challenges of WMSN; (2) an overview of advanced coding techniques for WMSN, including video and audio source coding, and distributed coding techniques; (3) a survey of WMSN communication protocols, including routing techniques and physical layer standards; and (4) a summary of Quality-of-Service (QoS) and security aspects of WMSN.

## 22.1 Introduction

With recent advances in microelectronics, development of tiny wireless sensors has drawn a lot of attentions both in the academia and in the industry. Given their low cost, low power, and small footprint characteristics, wireless sensors can be used in various applications, such as battlefield surveillance, distance monitoring, product inspection, inventory management, virtual keyboard, and smart office. Most of these applications are spanned over an extended physical area, and a large order of wireless sensors may be required for an adequate coverage. Wireless sensor network (WSN) is a system that coordinates sensors autonomously to facilitate the above-mentioned applications.

I. Lee (✉)
School of computer and Information Science, University of South Australia,
Mawson Lakes, SA 5095, Australia
e-mail: Ivan.Lee@unisa.edu

S. Misra et al. (eds.), *Guide to Wireless Sensor Networks*, Computer Communications
and Networks, DOI: 10.1007/978-1-84882-218-4_22,
© Springer-Verlag London Limited 2009

In general, WSN can be classified as either a homogeneous sensor network or a heterogeneous sensor network [1]. In a homogeneous sensor network, sensors are identical in terms of their processing capabilities, battery energy, and hardware complexity. A heterogeneous sensor network consists of wireless sensor nodes equipped with different battery capacities, which may serve for different applications. Heterogeneous WSNs are usually operated in an open space such as battlefield and wild environment, without human supervision. In addition, these sensor nodes are capable of self-organization [2], event-detection, and event-activation.

Sensors may be used for various applications: some sensors record the temperature and humidity; others may detect the noise level. For the ease of installation, these sensors may integrate a wireless network module for control signaling or data exchange. Major components of a typical wireless sensor node are illustrated in Fig. 22.1 [3], where the shaded components are further discussed throughout the rest of this chapter. The sensing unit usually consists of sensory devices and analog-to-digital converters (ADC). It is responsible for sensory data capturing, which is fed into the processing unit. The role of the processing unit is to process the data captured by the sensing unit, encapsulates, and forwards to other sensors or base station. Multimedia sensors extend sensors' audio and video sensing capabilities, thus able to perform media-rich applications such as video surveillance and traffic monitoring. The multimedia enhancement comes with costs in shaded units in Fig. 22.1: (1) an audio and/or a video sensor unit are required, which is typically costly; (2) a higher power consumption for processing and compressing multimedia sensory data. Typically, digital signal processors (DSPs) are used instead of microcontrollers; (3) a larger space for random access memory or fixed storage device is required; (4) a larger communication bandwidth may be required due to the excessive multimedia

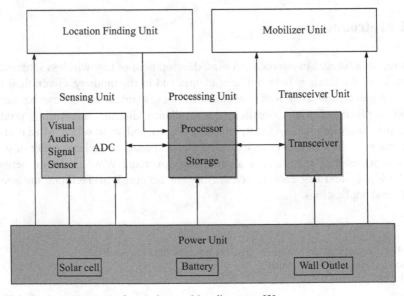

**Fig. 22.1** Basic components of a wireless multimedia sensor [3]

content; and (5) a large power unit is required to support the excessive computational complexity for processing multimedia data. There are two optional units in sensors: mobilizer unit and location finding unit. Although most multimedia sensors are placed in fixed locations, it is possible to embed a mobilizer unit for tracking moving objects, for example, location finding unit such as Global Positioning System (GPS) can provide neighborhood locations which can be used to compute routing metric (e.g., distance vector routing), thus improving the performance of the WMSN.

## 22.2  Background

WSN is sometimes considered as a subset of wireless ad hoc networks [4]. Although WSN carries many wireless ad hoc network properties, there are several subtle differences [3] (1) the number of sensor nodes in a sensor network could be several orders of magnitude higher than that of an ad hoc network; (2) the topology of a sensor network changes much more frequently due to the low power supply; (3) sensor nodes mainly use a broadcast communication paradigm, whereas, most ad hoc networks are based on point-to-point communications; (4) sensor nodes are constrained by energy supply, computational capabilities, and memory size; and (5) sensor nodes may not have global ID because of the large amount of overhead and large number of sensors. WSN serves many applications, which are divided into five categories: military applications, environmental monitoring, logistics support, human-centric applications, and robotics applications [5]. Sensory data can be relayed to base station directly or through multiple hops, as illustrated in Fig. 22.2.

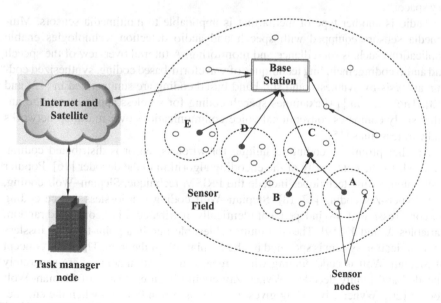

**Fig. 22.2**  WSN architecture

## 22.2.1 Source Coding

Because of limited power and processing capability, there are three main challenges for video and audio transmitting in WMSN: First of all, the radio spectrum is a limited resource in wireless communication [6]. A solution for this challenge is to use an advanced source coding technique to drop the payload size, hence reduce the bandwidth requirement. Second, the processing unit and the power unit are scarce resources for multimedia sensors. Therefore, using a complex source coding technique can significantly increase the power consumption and delay. Last, unlike wired networks, wireless communication is subject to a high error rate due to channel interference, multipath fading, and other loss mechanisms. Therefore, efficient and effective source and channel coding techniques are crucial to resolve the issues raised by the above-mentioned challenges.

Image and video are popular types of media used on multimedia sensors. Several image/video coding technologies are studied in video sensors or surveillance systems such as JPEG, differential JPEG [7, 8], and H.264 [9]. Comparisons of JPEG and JPEG2000 over ZigBee networks [7], and JPEG2000 have shown a superior error-resilience in terms of PSNR because of its improved multiple layer coding. JPEG and differential JPEG are applied to construct a video sensor platform which delivers a high quality video over 802.11 networks with a power consumption of less than 5 W [8]. For reducing the traffic load, H.264 [9] may be applied to yield a high compression ratio. Additional feature such as object tracking may apply a median filter algorithm to improve the computational speed [10]. However, its encoder complexity raises the power consumption issue [11], which may be inappropriate for WMSN applications since the sensors have limited processing power and memory space.

Audio is another type of media that is applicable to multimedia sensors. Multimedia sensors equipped with speech and audio detection technologies enable applications such as surveillance and monitoring. A tutorial overview of the speech and audio coding, including techniques of waveform-based coding, synthesized coding, analysis-by-synthesis, and sub-band based coding are summarized in [12] and [13]. The study in [14] examines speech coding for wireless applications, and another study compares conventional voice communications over multihop networks and wireless links [15].

Another promising coding technique for wireless sensor is distributed coding, which shifts the complexity of the encoding algorithm to the decoder [16]. Popular distributed coding techniques include the PRISM technique, Slepian–Wolf coding, and Wyner–Ziv coding [17, 18]. Slepian–Wolf coding is a lossless source coding for compression of two independent identically distributed (i.i.d) correlated random variables $X$ and $Y$ [19]. The minimum achievable rate in a point-to-point lossless communication system is specified by the Slepian–Wolf theorem. The basic concept of Slepian–Wolf is code binning which means information sources are separately encoded and jointly decoded. Wyner–Ziv coding is an extension of Slepian–Wolf coding [20]. Wyner–Ziv coding gives the rate distortion function when the encoder and the decoder both access to the side information. In addition, Wyner–Ziv coding can provide an error-resilient transmission [21], which is suitable for many-to-one uplink video communication systems such as wireless video and distributed sensor networks [22]. Paper [23] presents a WMSN system which maximizes the audio compression ratio while minimizing the energy consumption using a wavelet-based distributed audio coding (WDAC) technique.

## 22.2.2  Routing Protocols

Because WSN shares some similarities with ad hoc wireless networks, some researchers attempt to modify the existing routing protocols in ad hoc wireless networks for WSN. Aditya Mohan et al. test the routing protocols which include low power, reliable time sync, and mint standard [24]. By evaluating of routing protocols in a test bed where the sensors are densely spaced, HSN DSDV and Reliable Time Sync can achieve a higher performance. In addition, Nam N. Pham et al. draw comparisons between four WSN protocols: Multi Hop Router, Ad-hoc On-demand Distance-Vector (TinyAODV), Greedy Forwarding (GF), and Greedy Forward with Received Signal Strength Indication (GF-RSSI) [25]. Performance measurement was conducted on a WSN test bed for medical applications research. Based on their results, GF-RSSI has better performance in various operating conditions than others. It is also shown a higher success rate of packet delivery with moderate energy consumption in GF-RSSI. Furthermore, Jamal N. Al-Karaki et al. review routing techniques in WSN and summarize the routing protocols into three different network structures: flat (SPIN, rumor routing, etc.), hierarchical (LEACH, sensor aggregate, etc.), and location-based (SPAN,GAF, etc.) [26].

Unlike ad hoc wireless networks, typically, WMSN consists of hundred or thousand camera nodes. These nodes are capable of sensing the environment and transmitting data to the destination. Thus, the traditional routing protocols used in wireless local area networks (WLAN) and local area networks (LAN) cannot be directly employed into WMSN due to the following reasons. First, because of the large number of sensors, the traditional IP-based protocols such as IPv4 and IPv6 may create the overhead of ID for sensors which need to be monitored via Internet. Second, in contrast to the traditional wireless communication networks, WMSN is constrained by the energy supply, processing capabilities, and storage capacities. Complex routing algorithms may not be suitable because the massive number of the sensors can create huge routing table. Furthermore, the shortest path routing in WMSN may not always be the best solution. The balance of data traffic can also be an open research issue in WMSN. Third, the location information is also a key characteristic of WMSNs, since the data collection is normally based on the location. GPS can be a solution for localization technique in WMSN; however, in some environment, GPS is not accessible. In [27], a positioning algorithm based on WLAN is presented. A system estimated the user's position based on the received signal strength from three transmitters, and a pre–post cursor multipath mitigator is used for multipath interference cancellation. In [28], analysis of signal property can improve the design of better algorithms. It is observed that the signal strength cannot be specified by a single distribution (i.e., log-normal distribution). Statistical property is shown that signal strength is stationary under certain circumstances. Finally, sensor networks may be designed for application-specific such as battlefield monitoring and robot design. The quality of service routing may be considered in WMSN. With such design aspect, new routing protocols have been proposed for WSN. In the following paragraph, several famous routing techniques such as Low Energy Adaptive Clustering Hierarchy (LEACH), Sensor Protocol for Information via Negotiation (SPIN), and SPEED will be discussed.

LEACH is a cluster-based protocol proposed by Heinzelman et al. [29]. It is a self-organizing, adaptive clustering protocol that minimizes energy dissipation in sensor nodes. In LEACH, sensor nodes will organize themselves into local cluster based on the minimum communication energy. Once all the nodes were grouped into the corresponding cluster, sensor nodes started collecting the data and then transmitted to their cluster head. Since the cluster head receives all the data from its cluster, it aggregates the data and then transmits them to the base station. The key features of LEACH are localized coordination and control for cluster setup, rotation of the cluster head to enhance the system lifetime, and local aggregation and compression to reduce the overhead of communication. Cluster-based protocols, prolong the lifetime of farthest node to base station and improve sense manageability. In addition, in order to spread this energy usage over multiple nodes, LEACH also provided dynamic routing mechanism. The cluster head in LEACH will not be fixed and the position will rotate at different interval based on the cost function. In [29], authors discover based on their simulation model, only 5% of the nodes need to act as cluster heads. However, there are still some drawbacks of LEACH such as the extra overhead for cluster head changing [26] and the delay due to the in-networking process such as aggregation and encryption.

SPIN is an adaptive protocol proposed by Heinzelman et al. [30, 31]. Unlike conventional protocols such as flooding and gossiping which transmit the duplicate information and resource blindness, SPIN integrated negotiation and resource-adaptation mechanisms to overcome duplicate information. To efficiently disseminate information at each node in WSN, sensors use meta-data to describe the data they collect. In SPIN, meta-data is used to negotiate between nodes and thus the redundant data transmission can be avoided. This mechanism saves the energy and the bandwidth with duplicate packet transmission. In addition, nodes running SPIN can poll their resources before data transmission. Another advantage of SPIN is that each node only needs to know its neighbor. Thus, the logical change can be localized and routing table can be diminished. According to the study published in [32], the main disadvantage in SPIN is that nodes around a sink could deplete their battery quickly if the sink is interested in too many events.

In [33, 34], a spatiotemporal communication protocol for sensor network called SPEED was introduced. SPEED provides three types of real-time communication services: real-time unicast, real-time area multicast and real-time area any-cast. SPEED can maintain a desired delivery speed and reroute the traffic during the

**Table 22.1** Routing protocol comparisons

|  | LEACH [29] | SPIN [30, 31] | SPEED [33, 34] | Application-aware protocol [36] |
|---|---|---|---|---|
| Protocol base | Hierarchical network | Flat network | QoS-based | Application-based |
| Network lifetime | Very good | Good | N/A | N/A |
| Multimedia support | N/A | N/A | QoS support | Yes |
| Feature | Dynamic clustering, data aggregation, meta-data | Data aggregation, meta-data, maintain global topology | Real-time service, congestion avoid | New cost metric |
| Drawbacks | Traffic delay, extra overhead of dynamic clustering | Not scalable, sink could easily drain out the power | Not support differentiating various traffic | Only suitable for video sensor networks |

congestion by employing neighborhood feedback loop (NFL) and backpressure routing. In addition, the delay estimation mechanism can determine whether the congestion has occurred. In SPEED, each node keeps a neighbor table to maintain the routing information. Compared with dynamic source routing (DSR) and ad hoc on-demand vector routing (AODV), SPEED shows an improved performance in terms of a lower end-to-end delay and a lower miss ratio. On the other hand, the study in [35] shows that SPEED protocol provides only one network-wide speed, which is not suitable for differentiating traffic with different deadlines. SPEED is also limited to provide any guarantee in terms of the reliability domain. In addition, comparing with other energy-aware routing protocols, SPEED does not support further energy metric for routing; hence its energy consumption could be a potential concern.

In [36], the coverage and routing cost are integrated into an application-aware routing protocol. The study shows that the lifetime of video sensor networks can be prolonged because of the unique way that cameras capture data. Unlike the traditional sensor networks, the position of the cameras' filed of views (FoVs) is unpredictable. Thus, the application-aware protocol behaves different in video sensor. Furthermore, based on their result, the new cost function achieves a marginal improvement over energy-aware routing in wireless video-based sensors. Table 22.1 summarizes the protocols we discuss above.

## 22.2.3  Physical Layer

Unlike wired networks, packets in wireless networks are easily interfered by noise and lost during the transmission. In WMSN, since the nodes may run out of energy or might be damaged, it should provide certain robustness mechanisms to guarantee that end users can still receive information. For instance, redundant nodes in the networks can solve the case of damaged nodes or node run out of energy. Moreover, the packet loss and collision should also be considered in WM-SNs. The packet losses are inevitable due to the wireless characteristics and limited bandwidth; interference and collision error are easily occurred during transmission which result in an increase of damaged data. In [37], the authors classified existing MAC protocols of WSN into four categories: scheduling based, collision free, contention based, and hybrid schemes. They also summarize the challenges for designing MAC in WSN. By eliminating collisions during transmission, the power for retransmission and end-to-end delay will be reduced. In addition, current wireless MAC protocols are not feasible for WMSN because they only focus on throughputs instead of power consumptions. In [38], the authors studied the drawbacks of collision-free medium access in WSN such as Time Division Multiple Access (TDMA), Carrier Sense Multiple Access (CSMA), and Frequency Division Multiple Access (FDMA). The difficulties of TDMA systems are node synchronization, topology change adaptation, and throughput maximum issue. On the other hand, CSMA needs to employ additional collision detection mechanisms to achieve collision free and FDMA requires the additional circuitry to dynamically communicate with different channels. The study in [39] uses seven different channels to avoid cochannel interferences. However, this method needs extra hardware to support the collision free feature and will increase the cost of production.

WMSN targets to handle rich media data such as video and audio using sensors. In this section, a number of popular wireless communication standards will be discussed; these standards include IEEE 802.11, Bluetooth, ultra wide band (UWB), and Zigbee.

Bluetooth is a standard for Wireless Personal Area Network (WPAN), and it provides a universal short range wireless capability by using the 2.4-GHz spectrum. Bluetooth can be implemented between devices such as headphone sets, video game controllers, and printers where data can be exchange over a secure radiofrequency. In Bluetooth, nodes are organized into a piconet, consisting of a master and up to seven active slaves. The master can make the determination of the hopping sequence. However, two drawbacks of Bluetooth are provided [40]. First, Bluetooth needs to constantly have a master node, spending much energy on polling his slaves. Second, Bluetooth is limited by the number of active slave per piconet and some important data will be dropped during inactive time. Because of the two factors above, Bluetooth is less favorable for WMSN applications.

The IEEE 802.11 is another well-known standard using in WLAN. There are three physical media defined in the original 802.11 standard: direct sequence spread spectrum, frequency-hopping spread spectrum, and infrared. In [4], 802.11b has the maximum data rate at 11 Mbit/s. 802.11b is normally used in a point-to-multipoint configuration and an access point communicates via an omnidirectional antenna with one or more clients that are located in a coverage area around the access point. 802.11-based WMSN benefits from simple hardware requirement, high date rate, and the use of direct sequence means to avoid the problems of frequency hopping systems [41]. However, the cost and power consumption of 802.11 systems is far beyond the feasibility of wireless multimedia sensor networks [42]. IEEE also defined 802.11e to support the LAN applications with quality of services, including voice and video over WLAN. By deploying enhanced distributed channel access (EDCA) and hybrid coordination function (HCF) into 802.11e, traffic can be delivered based on predefined priorities.

Zigbee is a standard suite of a high level communication protocols that aims at small, low-power digital radio based on IEEE 802.15.4. ZigBee fulfills most of WSN application requirements, such as a low data rate, a long battery life, and a secure networking. Zigbee operates in the ISM radio bands: 868 MHz in Europe, 915 MHz in the USA, and 2.4 GHz in worldwide. This technology is intended to be simpler and cheaper than the competing WPAN standards such as Bluetooth. Although Zigbee is intended to be deployed in embedded applications that demand low data rate, low cost, and low power consumption, it is not feasible for WMSN: the best effort multihop transmission of JPEG and JPEG2000 images over Zigbee networks is examined in [43, 44]. The result shows that multihop transmissions of JPEG2000 images are unfortunately not completed due to adverse environment with interference from uncontrolled IEEE 802.15.4 and IEEE 802.11 wireless devices [44]. In addition, Kim et al. [45] implemented the face recognition applications in wireless image sensor networks. The image transmission speed and power consumption is considered in this paper. The result shows that most power is consumed on the radio transceiver, and transmission speed is reasonably low for systems that are not demanding a high frame rate. In [46], the wireless cameras networks over Zigbee are developed. The interaction between different subsystems of wireless cameras is proposed to reduce the overhead for interlayer communication and thus increases the performance.

UWB [47] aims at facilitating high-speed, low-power, low-cost multimedia applications. Two differences between UWB and other traditional narrow band are specified in [47]. First, the bandwidth of UWB system is more than 25% of an arithmetic centric frequency. UWB is intended to provide an efficient bandwidth for multimedia transmission. Second, UWB is typically implemented in a carrier-less fashion. UWB is different from the conventional "narrowband" system which uses the radio frequency (RF) carriers to move the signal [48].

In [49], a practical example of UWB WSNs is investigated. By applying UWB for WSNs, it ensures the low-power, low-cost, and wide-deployment sensor networks. Although the UWB transmission has been discussed for several years, the IEEE 802.15.3a task group still cannot reach the consensus. In [50], the open research issue was presented. The cross-layer communication based on UWB with the objective of delivering the QoS to WMSNs should be designed. A comparison between the above-mentioned standards is summarized in Table 22.2.

## 22.2.4  Security

Wireless multimedia sensors target to be one solution for the next generation mobile applications with their advantages of low cost, small size, and ease of deployment. However, WMSN in general is more vulnerable to malicious attacks in comparison

**Table 22.2** Physical layer protocol comparisons

| Protocol | Frequency range | Coverage | Data rate | QoS support |
|---|---|---|---|---|
| 802.11b | 2.4-GHz DSSS | Up to 110 m | Up to 11 Mbit/s | Yes with 802.11e |
| 802.15.1(Bluetooth) | 2.4-GHz FHSS | Up to 10 m | 1 Mbit/s | Yes |
| 802.15.3 | 2.4 GHz | 30–50 m | 10–55 Mbit/s | Yes |
| 802.15.3a (UWB) | 3.1–10.6 GHz | Up to 10 m | 100–500 Mbit/s | Yes |
| 802.15.4 (Zigbee) | 868–868.6 MHz, 902–928 MHz, 2,400–2,483.5 MHz | 10–75 m | 20 kbps, 40 kbps, 120 kbps | No |

with wired solutions due to the lack of physical medium protections [51,52]. For applications such as battlefield monitoring and surveillance, insecure communications may potentially be a catastrophic issue. Thus, delivery of sensitive multimedia data using an efficient and effective encryption technique while sustaining low-power consumption should be treated as a major design aspect for developing WMSNs. With such design aspect, applying the security mechanisms in popular WLAN standard such as IEEE 802.11e in WMSN may not be feasible.

Apart from confidentiality, another security issue of WMSN is related to the integrity and legitimacy of the multimedia content. Techniques such as digital watermarking and multimedia fingerprint may be applied to resolve these issues. In [53], a security mechanism using digital right management (DRM) is presented for video sensor networks. If the video content is transmitted over the Internet instead of private leased lines, the DRM technique serves two major benefits (1) DRM can effectively shield its protected data from unauthorized access. For example, for patient monitoring applications, the access to patient's private profile such as patient's appearance could be limited to certain individuals. (2) Some sensor contents may hold significant commercial values such as highway traffic monitoring, airport surveillance, and industry control monitoring. DRM provides a set of solutions for abuse of access account for trading, accounting, and transaction processing of digital contents as commodities [53].

In WMSN, sensors are limited by power, computing resource, storage, and transmission range; attackers can use a powerful laptop computer with high energy and long distance communication to perform attacks by these design aspects. Furthermore, several design challenges of secure WSN are raised [54]:

- In order to make sensor networks economically viable, sensor nodes are constrained by their energy computation and low processing capability. With this aspect, the complicated encryption method cannot be implemented in WMSN due to the computational overhead [54]. Moreover, for sensors equipped with batteries, energy management could be a main concern for a highly reliable solution to multimedia applications.
- Physical attacks will be a risk for WMSN due to the fact that sensor nodes are accessible in open areas. Thus, adversaries can easily locate and destroy the sensor nodes. Since nodes could be physically captured, attackers may compromise the

cryptographic keys from the sensor nodes and then install malicious node into the network to perform attacks such as sinkhole attack [55]. In addition, some sensors equipped with solar cell may rejoin the network later. Thus, a mechanism to guarantee that new joining sensors are not malicious nodes should be provided in WMSN

- The key establishment will be more difficult due to the large scale of sensor nodes. For example, it is undesirable to deploy the public key algorithms such as Diffie-Hellman key agreement [56] or RSA [57] because of their computational complexities. In addition, using different keys for each individual sensor will enlarge the memory size and increase the cost of production. Shared keys, on the other hand, has less overhead than public keys. The drawback of shared key is once attackers compromise a single node in a network would reveal the secret key and then the network traffic can be easily decrypted [54].

- Unlike traditional wireless applications, WMSN facilitates a distributed processing mechanism in which the data can be aggregated locally to reduce both the communication bandwidth and the energy consumption [58]. However, for scenarios such as ad hoc WSNs [59] and multiple layer architecture WMSN [36,58], sensors transmit the encrypted data to cluster heads; cluster heads must decrypt the payload before performing the aggregation task. The extra delay introduced by this aggregation task is not feasible for real-time WMSN applications such as surveillance and battlefield monitoring. In addition, some security protocols require key exchanges [54], and the overhand of the protocol handshakes should also be considered. It is also important to consider the denial of service (DoS) attacks which impact the communication between end devices and even the routing table updates. For example, routing protocols that use HELLO messages can be vulnerable to the HELLO flood attacks. With insecure routing protocols, the route of data transmission can have routing loop problem [60,61].

- WMSN can be considered as a specialized WSN for multimedia applications. Because of carrying rich multimedia content, the energy consumption and end-to-end transmission delay should be minimized in WMSN. In [62], link layer security architecture is studied. By detecting unauthorized packets when they first inject, the authors presented authenticity, integrity, and confidentiality of message exchanges between neighboring nodes. In [63], network layer attacks such as sinkhole attacks in sensor networks are discussed. In addition, by discussing these attacks, the possible solutions are also provided.

As indicated previously, the key establishment and trusted connection setup can deeply affect the WMSN security. Public key and share key approaches are not suitable for large scale WMSN due to resource constraints [51]. Therefore, key distributions have been an active research topic in WSN. Cheng et al. [64] classifies enhanced key predistribution mechanisms into three categories (1) random key predistribution schemes, (2) polynomial-key predistribution schemes, and (3) location based key predistribution schemes. Random key predistribution [65] allows nodes deploying at later time joining the network securely. By picking a random pool of keys from the total possible key space, nodes perform key discovery to find out a common key within their respective subsets, and it is used as their shared secret

key to initiate a secure link. Random key predistribution solve the computational
overhead in public key and provide a secure link between two nodes; however, it
also increases the overhead of communication during key discovery. Polynomial-
key predistribution, on the other hand, not only requires a lower communication
overhead, but also have less sufficient security than random key predistribution.
Location-based key predistribution [61] uses the location deployment of sensor node
to improve the networks' performance. As indicated earlier, WMSN is vulnerable to
various types of attacks because of hardware constraints of sensors. Attacks in the
network layer and possible solutions for these attacks are summarized below:

- *Replay attack.* In [66,67], replay attacks in WMSN is studied. An adversary that
  eavesdrops on a legitimate message sent between two authorized nodes and replays it at same later time. For surveillance applications, adversary can spoof
  the system by simply placing pictures in front of the camera and play recording.
  In [62], a common defense includes a monotonically increasing counter with
  message and reject message with old counter value. However, the main drawback is the cost of maintaining the neighbor table and extra device memories are
  required.
- *Cut and paste attack.* In [68], the cut and paste attack is investigated. By breaking
  apart an unauthenticated encrypted message and constructing another message
  which decrypts to something meaningful, cut and paste attack is a type of message modification attack that attackers removes a message from network traffic,
  alters message, and reinserts message into the network. The possible solution
  of cut and paste attack is integrated watermarking scheme into image [69].
  Main drawbacks of this solution include the increase in distortion and energy
  consumption.
- *Selective forwarding.* In multihop WMSN, sensors are based on the assumption
  that participating nodes will faithfully forward the received messages. However, an attacker may create malicious nodes that receive the data from the
  neighbor and refuse to forward any further. In WMSN applications such as battlefield surveillance, sensed data can be easily corrupted and caused a catastrophic
  problem. In [70], the detection method of selective forwarding is provided. By
  deploying a multihop acknowledgement technique to launch alarms, the responses from intermediate nodes can be obtained. An intermediate node can
  report abnormal packet loss and suspect nodes to both the base station and the
  source node. The main drawback of this approach is that extra processing demand
  will consume more energy.
- *Sinkhole attack.* An attacker can send unfaithful routing information to the neighbors, and then perform selective forwarding or alter the data passing through. To
  resolve the issue of this problem, a two-step algorithm for detecting sinkhole attack is presented in [71]. First, it locates a list of suspected nodes by checking the
  data consistency, and then identifies the intruder in the list through analyzing the
  network traffic flow. The drawback of this approach is its extra processing demand which also causes additional delay.
- *Sybil attack.* Analysis of Sybil attacks and its defense strategies are studied
  in [72]. An attacker can employ a physical device with multiple identities to

generate Sybil attack during the data aggregation, voting, and resource allocation. Furthermore, Sybil attack can reduce the effectiveness of fault-tolerant schemes significantly. The possible defense is to validate an identity to the corresponding physical devices. Newsome et al. [72] provided methods such as radio resource testing, random key predistribution, registration, position verification, and code attestation. The random key predistribution is the most promising technique to prevent Sybil attack without additional overhead [72].

- *Wormholes.* An attacker records a packet at one location in the network, tunnels the data to another location, and replays the packet there. The attacker can perform the attack even if the attacker does not have any cryptographic keys [73]. A malicious node could announce a shortest path through this node and create a black hole in this region. Temporal leashes are possible solutions for the wormhole attack, and the TIK provides an instant authentication of the received packets. A MAC using TIK can efficiently protect against reply, spoofing, and wormhole attacks without additional processing demand at the MAC layer [73].

- *HELLO flood attack.* For protocols that require HELLO messages to build the association and to announce their presence to their neighbors, an adversary may perform the HELLO flood attack [60]. In HELLO flood attack, a malicious node can transmit a message with an abnormal high power so as to make all nodes to believe that it is their neighbor. When normal sensors hear the HELLO message from this malicious node, they will treat the malicious node as the next hop and then a routing loop may be created. In this paper, the authors also presented the suspicious node information dissemination protocol (SNIDP) to deal with this problem. The concept of SNIDP is that node A detects a suspicious through the signal strength. Once suspicious node S is detected, node A will be the identity of node S to its neighbors and then perform a suspicious vote. With this process, malicious node can be detected. However, the drawback of this approach is the energy consumption associated with additional message checking, message transmissions and receptions incurred by the execution of SNIDP.

- *Node capture attack.* The node capture attacks are studied in [55]. Because of the physical constraint of sensor nodes, an attacker may physically capture some sensor nodes and compromise their data and communication keys. In [65, 74], instead of requiring sensors to store all assigned keys, the authors deploy the random key predistribution into WMSNs. Although attacker can compromise the cryptographic, only partial information will be decrypted.

## 22.2.5   Quality of Service (QoS)

QoS is a performance measure of the network to ensure efficient and reliable data transmissions. Delay, jitter, bandwidth, and packet loss are typical QoS for WMSN. However, some WMSN application may rely on other QoS indicators. Video surveillance applications, for example, require good visual coverage. Existence of unclear picture qualities or blind spots should be considered low QoS or even treated as the security breach. Another important QoS measure for WMSN is the network lifetime and its coverage area.

For real-time video applications such as battlefields monitoring and delays in video and audio transmission may cause severe or even catastrophic impacts on critical decisions. Factors causing latencies in WMSN include in-network processing, queuing delay, and transmission delay [75]. In general, the transmission delay is relatively small in comparison with the delay for data processing; therefore, the latency of processing time should be minimized. In addition, multihop transmission helps reducing the power consumption with reduced node-to-node distances, and hence improving the lifetime for WMSN [3]. However, the drawbacks of this approach include the extended end-to-end delay, security vulnerability, and difficulties in queue scheduling [41]. Furthermore, most delay is due to the waiting period for collecting the image data in the similar area and processing time of decrypting, uncompressing, and aggregating image data. Hence, reducing the delay is a crucial task for time-critical WMSN applications.

Reducing power consumption is an important topic for ensuring an improved network lifetime. Aggregating the redundant video data captured by sensory devices helps eliminating excessive bandwidth usage and consequently reducing the transmission power. However, the complex aggregation algorithms processed by cluster head sensors take additional processing power and memory. Another strategy for reducing energy consumption is by putting sensors into sleep mode, which might, however, introduce additional latencies such as sleep delay [76]. Exploring the optimal energy-latency trade-offs is therefore an important issue in QoS.

QoS-based routing is another important topic in WMSN. Conventional QoS routing protocols used in wireless ad hoc networks [77–79] may be inadequate for WMSN due to its severe resource constraints. Sequential assignment routing (SAR) [34] is one of the first QoS-based routing protocols in WSN. To balance between the energy consumption and the image quality, SAR uses three factors for making routing decisions: energy resource, QoS on the each route, and the priority level of packets. Furthermore, SAR also employs the multipath transmission. SPEED is another well-known routing protocol used in WSNs which guarantees a soft real-time requirement [34]. In addition, to solve the scarce resource issues such as limited bandwidth and energy in WMSNs, SPEED uses a nondeterministic forwarding to balance the flow among multiple routes. Although SPEED uses a novel back-pressure rerouting to overcome packet congestion, it does not have a packet prioritization scheme. MMSPEED [35] is a protocol based on SPEED, and it is designed for handling multimedia traffic with embedded scalability and adaptability. The energy-aware QoS routing mechanism proposed in [80] deals with real-time traffic in WMSN. By finding a least cost, delay-constrained path in terms of link cost, it captures nodes' energy reserve, transmission energy, and other parameters as routing metric. With this method, traffics will be divided into two classes: nonreal-time and real-time. The drawback for this method is that it does not support multiple priorities for real-time traffics [37].

While WMSN carries multimedia sensory data, it is important to examine the perceived quality of audio and/or video as part of the quality measurement, and such measurement could be referred to as the perceived quality of service or PQoS. In [81], the quality measurement considers the relationship between image quality and energy consumption. Others suggest that the trade-offs between power, rate, and distortion [82] play an important role for designing WSNs. Another study also suggests that the balance between complexity, rate, and distortion should be examined [8]. To study the impact of system resources to the overall system performance, resource-distortion analysis was proposed as an alternative measure of the conventional rate-distortion analysis [83]. In addition, the concept "accumulative visual information" (AVI) was introduced to measure the amount of visual information collected in wireless video sensors [8], and it jointly evaluates entropy, image distortion, encoding efficiency, and energy consumption as a measure of the system's quality measure [84].

While WMSN may be used to facilitate surveillance or monitoring applications, visual coverage is another important QoS parameter. Since sensors may be randomly placed in an open space, there may be different overlapping density covered by different sensors. This overprovision of sensors will make the energy consumption inefficient [36]. By coordinating sensors' activation status using a cost function and alternatively switching sensors into sleep mode, the overall network lifetime can be prolonged and the bandwidth demand is reduced [36]. Similar results were found for scenarios extended to three dimensions [85]. In [86], the trade-off between the network lifetime of WMSN and image distortion is discussed. By deploying the hybrid or adaptive camera selection into WMSN, the optimal lifetime-distortion trade-off will be provided. Although the overlapping coverage increases energy consumption, it provides more information to the end users. For example, with more sensor nodes in action, the tracked object can be viewed from different angles. Malicious attacks such as sending unrealistic images through the wireless transmission can be prevented.

## 22.3 Thoughts for Practitioners

WMSN not only maintains WSN properties such as a low power supply and a short transmission range, but also captures high bit rate sensory data such as video and audio. Thus, these networks can support a wide range of applications such as military and environmental monitoring. Below is a list of WMSN applications:

- *Security.* For scenarios such as temporary indoor exhibitions, installing traditional surveillance cameras is expensive and difficult to remove. WMSN can act as temporary surveillance systems or online tour guide. In addition, multimedia sensor nodes can be placed close to the entrance, and they can monitor and record customers, and then transmit the video and audio data to the base station.
- *Wild animal tracking.* In a national park, it is difficult to track the habitual behaviors of wild animals by using fixed surveillance system due to the high cost

of installation. Wireless sensors can provide mobility for moving object. Furthermore, in the case of direct transmission from sensors to base station is blocked; a multiple hop scheme embedded in WMSN can be utilized to carry over the transmission. Therefore, an information loss or delay can be avoided.

- *Traffic monitoring and environmental measurement.* Multimedia sensors can be applied into downtown area to monitor rush hour traffic and help drivers avoiding the congested roads. In addition, sensors can measure the noise level and air quality for research purposes.
- *Remote medicine.* In a desolate area, sensors can transmit video and audio of the patient to the doctors who resides in metropolitan areas. With these data, doctors can provide the first aid to someone is injured, for applications such as telemedicine, prescription, and monitoring the patient continually on heart beat, pulse, body temperature, and blood pressure [87].
- *Climate and sea shore monitoring.* Sensors can continually record the video data on the climate, such as the cloud, sunrise, sunset, moon, and temperature. In addition, for sea shore monitoring [88], WMSN can provide more flexibility and low cost devices.
- *Battlefield surveillance.* For battlefield surveillance, visual sensors can be used to monitor the enemy remotely. In addition, WMSN can be coupled with actor nodes to launch reactive missile attacks once certain sensory events are triggered [89].
- *Airport surveillance.* After the 9/11 terrorist attacks, attentions to airport security have been raised significantly. Low cost sensors can be used for monitoring the registration desk, luggage delivery, plane arrival or departure, and customer density.
- *Fire alarm and control.* Sensor and actor networks can be integrated into fire alarm systems. Water sprinkler actors can be activated before a fire becomes uncontrollable based on sensory information [90].

## 22.4 Directions for Future Research

Although WMSN can retrieve multimedia content such as video and audio, some challenges in WMSN still need to be studied. (1) Because WMSN provides more sensory information, high data rate and high bandwidth support are required in WMSN [50, 91]. Thus, different coding methods such as source coding and distributed coding can prolong the system lifetime and reduce the bandwidth usage [17]. (2) Multimedia sensors capture video and audio instead of sensing simple text data and the energy consumption for in-network processing such as image fusion can be huge. Although power scavenging technology can provide a temporary solution, the specialized processing technology should be developed. (3) Security in WMSN becomes a problem because of a lack of physical protection in wireless channels. Conventional security mechanisms in 802.11 wireless LANs are not suitable for WMSN. Therefore, a lightweight security mechanism should be designed for WMSN. (4) Most routing protocols for WMSN focus on energy saving rather than QoS. For applications such as remote medicine, high packet loss, and delay can cause a catastrophic issue. Thus, specific routing techniques and physical layer protocol should be designed for WMSN.

## 22.5   Conclusions

With a promising outlook of its potential applications, WMSN has been experiencing a rapid development in recent years. WMSN extends the multimedia capabilities from WSN, to facilitate media-rich applications such as environment monitoring, battlefield surveillance, structural health monitoring, robotics, video surveillance, and human motion capture. However, constrained by limited energy, some design aspects such as communication protocols, security mechanisms, QoS, and advanced source and channel coding techniques should be considered.

Sensors are in general powered by limited energy resources such as batteries, and tradeoffs between energy reduction and quality represent a major design challenge. WMSN captures, processes, and transmits video and audio sensory information, and conventional WSN standard may be inappropriate for WMSN. In particular, topics such as efficient and effective source coding technique, distributed coding technique, security, cross-layer design for routing and switching, and different aspects of quality of service present the main challenges of the future of WMSN.

## Terminologies

*Wireless sensor network (WSN)*. A wireless network consists of low cost, low power consumption, autonomous sensors spatially distributed for monitoring sensory data.

*Wireless multimedia sensors*. Small and cheap devices, which capture and process multimedia sensory signals such as video and audio. Wireless multimedia sensors consists of sensing units for video and audio capture and analog-to-digital conversion, processing unit such as DSP, volatile or nonvolatile memory, power unit, and wireless transceiver unit.

*Wireless multimedia sensor network (WMSN)*. A special type of WSNs with the support of rich multimedia sensory data such as video and audio.

*Heterogeneous sensor network*. A heterogeneous sensor network consists of wireless sensor nodes equipped with different battery capacities, which may serve for different applications.

*Homogeneous sensor network*. In a homogeneous sensor network, sensors are identical in terms of their processing capabilities, battery energy, and hardware complexity.

*WMSN source coding.* Compression techniques for encoding multimedia sensory data in WMSN.

*Distributed coding.* A new coding technique which aims to shifts the complexity of the encoding algorithm to the decoder, which is useful for WMSN.

*Cluster head.* It is the node which is responsible for collecting and aggregating sensory data from sensors within the cluster, and the aggregated data to the base station or the sink. It is used in Low Energy Adaptive Clustering Hierarchy (LEACH) routing protocol which aims to lower the communication energy in WMSN.

*Digital right management (DRM).* To ensure the integrity and the legitimacy of the multimedia sensory data, techniques such as digital watermarking and multimedia fingerprint may be applied. This is known as digital right management for WMSN.

*PQoS.* Perceived Quality of Service examines the perceptual quality of the received data at the destination. For example, the perceptual quality of video and audio transmitted over WMSN.

## Questions

1. What is the difference between sensor networks and multimedia sensor networks?
2. What are the major issues in multimedia sensor networks?
3. Describe the differences between heterogeneous sensor networks and homogeneous sensor networks.
4. What is the layered clustering network and what is advantage to use the clustering architecture?
5. Describe the difference between traditional source coding and distributing coding for multimedia signals.
6. Energy management is a critical issue in WMSNs. Please provide three current proposed methods to prolong the network lifetime.
7. Describe the security design challenges in WMSNs.
8. Point out the different between the view of traditional QoS and PQoS in WMSNs.
9. Describe the difference between WSN and ad hoc network.
10. Describe applications of WMSN.

## References

1. V. Mhatre, C. Rosenberg, Homogeneous vs heterogeneous clustered networks: A comparative study, Proceedings of IEEE International Conference on Communications, June 2004.
2. K. Sohrabi, J. Gao, V. Ailawadhi, and G. J. Pottie, Protocols for self-organization of a wireless sensor network, IEEE Wireless Communications, 7(5), 16–27, 2000.

3. I. F. Akyildiz, W. Su, Y. Sankarasubramaniam, and E. Cayirci, A survey on sensor networks, IEEE Communications Magazine, 40(8), 102–114, 2002.
4. H. Karl and A. Willig, Protocols and Architectures for Wireless Sensor Networks, Chichester: Wiley, 2005.
5. T. Arampatzis, J. Lygeros, and S. Manesis, A survey of applications of wireless sensors and wireless sensor networks, Proceedings of the IEEE International Symposium on Intelligent Control, Mediterrean Conference on Control and Automation, pp. 719–724, 2005.
6. J. Zander, Radio resource management – an overview, IEEE Vehicular Technology Conference, vol. 1, pp. 16–20, May 1996.
7. G. Pekhteryev, Z. Sahinoglu, P. Orlik, and G. Bhatti, Image transmission over IEEE 802.15.4 and ZigBee networks, IEEE International Symposium on Circuits and Systems, vol. 4, pp. 3539–3542, May 2005.
8. W. C. Feng, E. Kaiser, W. C. Feng, and M. Le Baillif, Panoptes: scalable low-power video sensor networking technologies, ACM Transactions on Multimedia Computing, Communications, and Applications, 1, 151–167, 2005.
9. Advanced video coding for generic audiovisual services, ITU-T Recommendation H.264.
10. Y. Zhao, and G. Taubin, Real-time median filtering for embedded smart cameras, IEEE International Conference on Computer Vision Systems, 2006.
11. T. Wiegand, G. J. Sullivan, G. Bjntegaard, and A. Luthra, Overview of the H.264/AVC video coding standard, IEEE Transactions on Circuits and Systems for Video Technology, 13(7), 560–576, 2003.
12. A. S. Spanias, Speech coding: a tutorial review, Proceedings of the IEEE, 82(10), 1541–1582, 1994.
13. A. Gersho, Advances in speech and audio compression, Proceedings of the IEEE, 82(6), 900–918, 1994.
14. M. Budagavi and J. D. Gibson, Speech coding in mobile radio communications, Proceedings of the IEEE, 86(7), pp. 1402–1412, 1998.
15. J. D. Gibson, Speech coding methods, standards, and applications, IEEE Circuits and Systems Magazine, 5(4), 30–49, 2005.
16. B. Girod, A. M. Aaron, S. Rane, and D. Rebollo-Monedero, Distributed video coding, Proceedings of the IEEE, 93(1), 71–83, 2005.
17. R. Puri, A. Majumbar, P. Ishwar, and K. Ramchandran, Distributed source coding for sensor networks, IEEE Signal Processing Magazine, 21(5), 80–94, 2004.
18. Z. Xiong, A. D. Liveris, and S. Cheng, Distributed source coding for sensor networks, IEEE Signal Processing Magazine, 21(5), 80–94, 2004.
19. D. Slepian and J. Wolf, Noiseless coding of correlated information sources, IEEE Transactions on Information Theory, 19(4), 471–480, 1973.
20. A. Wyner and J. Ziv, The rate-distortion function for source coding with side information at the decoder, IEEE Transactions on Information Theory, 22(1), 1–10, 1976.
21. A. Aaron, S. Rane, R. Zhang, and B. Girod, Wyner-Ziv coding for video: Applications to compression and error resilience, Proceedings of the Conference on Data Compression, 2003.
22. J. Garcia-Frias and Z. Xiong, Distributed source and joint source-channel coding: from theory to practice, Proceedings of IEEE International Conference on Acoustics, Speech, and Signal Processing, vol. 5, pp. 1093–1096, March 2005.
23. H. Dong, J. Lu, and Y. Sun, Distributed audio coding in wireless sensor networks, International Conference on Computational Intelligence and Security, vol. 2, pp. 1695–1699, Nov 2006.
24. A. Mohan and V. Kalogeraki, Speculative routing and update propagation: a kundali centric approach, IEEE International Conference on Communication, vol. 1, pp. 343–347, May 2003.
25. N. Pham, J. Youn, and W. Chulho, A comparison of wireless sensor network routing protocols on an experimental testbed, IEEE International Conference on Sensor Networks, Ubiquitous, and Trustworthy Computing, vol. 2, pp. 276–281, 2006.
26. J. N. Al-Karaki and A. E. Kamal, Routing techniques in wireless sensor networks: a survey, IEEE Wireless Communications, 11, 6–28, 2004.

27. R. Singh, M. Gandetto, M. Guainazzo, D. Angiati, and C. S. Ragazzoni, A novel positioning system for static location estimation employing WLAN in indoor environment, IEEE International Symposium on Personal, Indoor and Mobile Radio Communications, vol. 3, pp. 1762–1766, Sept 2004.
28. K. Kaemarungsi and P. Krishnamurthy, Properties of indoor received signal strength for WLAN location fingerprinting, International Conference on Mobile and Ubiquitous Systems: Networking and Services, pp. 14–23, Aug 2004.
29. W. R. Heinzelman, A. Chandrakasan, and H. Balakrishnan, Energy-efficient communication protocol for wireless microsensor networks, Proceedings of the Hawaii International Conference on System Sciences, vol. 2, p. 10, Jan 2000.
30. J. Kulik, W. Heinzelman, and H. Balakrishnan, Negotiation-based protocols for disseminating information in wireless sensor networks, Wireless Networks, 8, 169–185, 2002.
31. W. R. Heinzelman, J. Kulik, and H. Balakrishnan, Adaptive protocols for information dissemination in wireless sensor networks, Proceedings of the ACM/IEEE International Conference on Mobile Computing and Networking, pp. 174–185, 1999.
32. Q. Jiang and D. Manivannan, Routing protocols for sensor networks, IEEE Consumer Communications and Networking Conference, pp. 93–98, Jan 2004.
33. T. He, J. A. Stankovic, T. F. Abdelzaher, and C. Lu, A spatiotemporal communication protocol for wireless sensor networks, IEEE Transactions on Parallel and Distributed Systems, 16(10), 995–1006, 2005.
34. T. He, J. A. Stankovic, C. Lu, and T. Abdelzaher, SPEED: a stateless protocol for real-time communication in sensor networks, Proceedings of International Conference on Distributed Computing Systems, pp. 46–55, May 2003.
35. E. Felemban, C. G. Lee, and E. Ekici, MMSPEED: multipath multi-SPEED protocol for QoS guarantee of reliability and timeliness in wireless sensor networks, IEEE Transactions on Mobile Computing, 5(6), 738–754, 2006.
36. S. Soro and W. B. Heinzelman, On the coverage problem in video-based wireless sensor networks, Second International Conference on Broadband Networks, vol. 2, pp. 932–939, Oct 2005.
37. J. A. Stankovic, T. F. Abdelzaher, C. Lu, L. Sha, and J. C. Hou, Real-time communication and coordination in embedded sensor networks, Proceedings of the IEEE, vol. 91, pp. 1002–1022, 2003.
38. I. Demirkol, C. Ersoy, and F. Alagoz, MAC protocols for wireless sensor networks: a survey, IEEE Communications Magazine, 44(4), 115–121, 2006.
39. M. Caccamo, L. Y. Zhang, L. Sha, and G. Buttazzo, An implicit prioritized access protocol for wireless sensor networks, IEEE Real-Time Systems Symposium, pp. 39–48, 2002.
40. R. Nusser and R. M. Pelz, Bluetooth-based wireless connectivity in an automotive environment, IEEE Vehicular Technology Conference, vol. 4, pp.1935–1942, 2000.
41. R. Benkoczi, H. Hassanein, S. Akl, and S. Tai, QoS for data relaying in hierarchical wireless sensor networks, Proceedings of the First ACM International Workshop on Quality of Service and Security in Wireless and Mobile Networks, pp. 47–54, 2005.
42. E. H. Callaway, Jr., Wireless Sensor Networks, Architecture and Protocols, 2004. Boca Raton, FL: Auerbach.
43. L. Zheng, ZigBee Wireless Sensor Network in Industrial Applications, SICE-ICASE, pp. 1067–1070, Oct 2006.
44. G. Pekhteryev, Z. Sahinoglu, P. Orlik, and G. Bhatti, Image transmission over IEEE 802.15.4 and ZigBee networks, IEEE International Symposium on Circuits and Systems, vol. 4, pp. 3539–3542, May 2005.
45. I. Kim, J. Shim, J. Schlessman, and W. Wolf, Remote wireless face recognition employing zigbee, Workshop on Distributed Smart Cameras (DSC), Oct 2006.
46. E. Ljung, E. Simmons, A. Danilin, R. Kleihorst, and B. Schueler, 802.15.4 Powered distributed wireless smart camera network, Workshop on Distributed Smart Cameras, Boulder, CO, Oct 2006.

47. K. Mandke, H. Nam, L. Yerramneni, C. Zuniga, and T. Rappaport, The evolution of ultra wide band radio for wireless personal area networks, High Frequency Electron, pp. 22–32, Sept 2003.
48. J. Foerster, E. Green, S. Somayazulu, and D. Leeper, Ultra-wideband technology for short- or medium-range wireless communications, Intel Technology Journal, 2, 1–11, 2001.
49. I. Oppermann, L. Stoica, A. Rabbachin, Z. Shelby, and J. Haapola, UWB wireless sensor networks: UWEN – a practical example, IEEE Communications Magazine, 42(12), 27–32, 2004.
50. I. F. Akyildiz, T. Melodia, and K. R. Chowdhury, A survey on wireless multimedia sensor networks, Computer Networks, 51, 921–960, 2007.
51. D. Djenouri, L. Khelladi, and AN Badache, A survey of security issues in mobile ad hoc and sensor networks, Communications Surveys and Tutorials, 7, 2–28, 2005.
52. Y. Wang, G. Attebury, and B. Ramamurthy, A survey of security issues in mobile ad hoc and sensor networks, IEEE Communications Surveys and Tutorials, 7(4), 2–28, 2005.
53. T. Wu, L. Dai, Y. Xue, and Y. Cui, Digital rights management for video sensor network, Proceedings of the IEEE International Symposium on Multimedia, pp. 131–138, 2006.
54. A. Perrig, J. Stankovic, and D. Wagner, Security in wireless sensor networks, Communications of the ACM, 47, 53–57, 2004.
55. P. Tague and R. Poovendran, Modeling adaptive node capture attacks in multi-hop wireless networks, Ad Hoc Networks, 5, 801–814, 2007.
56. W. Diffie and M. Hellman, New directions in cryptography, IEEE Transactions on Information Theory, 22(6), 644–654, 1976.
57. R. L. Rivest, A. Shamir, and L. Adleman, A method for obtaining digital signatures and public-key cryptosystems, Communications of the ACM, 21(2), 120–126, 1978.
58. X. Fan, W. Shaw, and I. Lee, Layered clustering for solar powered wireless visual sensor networks, Proceedings of IEEE International Symposium on Multimedia, Dec 2007.
59. P. Biswas and Y. Ye, Semidefinite programming for ad hoc wireless sensor network localization, Proceedings of International Symposium on Information Processing in Sensor Networks, pp. 46–54, 2004.
60. W. R. Pires, T. H. P. Figueiredo, H. C. Wong, and A. A. F. Loureiro, Malicious node detection in wireless sensor networks, International Parallel and Distributed Processing Symposiums, 2004.
61. Y. Zhang, W. Liu, W. Lou, and Y. Fang, Securing sensor networks with location-based keys, IEEE Wireless Communications and Networking Conference, vol. 4, pp. 1909–1914, March 2005.
62. C. Karlof, N. Sastry, and D. Wagner, TinySec: a link layer security architecture for wireless sensor networks, Proceedings of the Second International Conference on Embedded Networked Sensor Systems, pp. 162–175, 2004.
63. C. Karlof and D. Wagner, Secure routing in wireless sensor networks: attacks and countermeasures, Proceedings of the First IEEE International Workshop on Sensor Network Protocols and Applications, pp. 113–127, May 2003.
64. Y. Cheng and D. P. Agrawal, An improved key distribution mechanism for large-scale hierarchical wireless sensor networks, Ad Hoc Networks, 5, 35–48, 2007.
65. H. Chan, A. Perrig, and D. Song, Random key predistribution schemes for sensor networks, Proceedings of Symposium on Security and Privacy, pp. 197–213, May 2003.
66. H. Bredin, A. Miguel, I. H. Witten, and G. Chollet, Detecting replay attacks in audiovisual identity verification, Proceedings of IEEE International Conference on Acoustics, Speech, and Signal Processing, 2006.
67. Y. C. Hu, D. B. Johnson, and A. Perrig, SEAD: secure efficient distance vector routing for mobile wireless ad hoc networks, Proceedings of IEEE Workshop on Mobile Computing Systems and Applications, pp. 3–13, 2002.
68. P. Barreto, H. Y. Kim, and V. Rijmen, Toward secure public-key blockwise fragile authentication watermarking, Proceedings of IEE Vision, Image and Signal Processing, 149(2), 57–62, 2002.
69. C. T. Li and H. Si, Wavelet-based fragile watermarking scheme for image authentication, Journal of Electronic Imaging, 16(1), 2007.

70. B. Yu and B. Xiao, Detecting selective forwarding attacks in wireless sensor networks, Proceedings of the International Workshop on Security in Systems and Networks, 2006.
71. E. C. H. Ngai, J. Liu, and M. R. Lyu, On the intruder detection for sinkhole attack in wireless sensor networks, Proceedings of the IEEE International Conference on Communications, 2006.
72. J. Newsome, E. Shi, D. Song, and A. Perrig, The Sybil attack in sensor networks: analysis and defences, Third International Symposium on Information Processing in Sensor Networks, pp. 259–268, 2004.
73. Y. C. Hu, A. Perrig, and D. B. Johnson, Packet leashes: A defense against wormhole attacks in wireless ad hoc networks, Proceedings of Annual Joint Conference of the IEEE Computer and Communications Societies, 2003.
74. K. Ren, K. Zeng, and W. Lou, On efficient key pre-distribution in large scale wireless sensor networks, IEEE Military Communications Conference, vol. 1, pp. 20–26, Oct 2005.
75. D. Chen and P. K. Varshney, QoS support in wireless sensor networks: A survey, Proceedings of International Conference on Wireless Networks, 2004.
76. W. Ye, J. Heidemann, and D. Estrin, Medium access control with coordinated adaptive sleeping for wireless sensor networks, IEEE/ACM Transactions on Networking, 12(3), 493–506, 2004.
77. R. Sivakumar, P. Sinha, and V. Bharghavan, CEDAR: a core-extraction distributed ad hoc routing algorithm, IEEE Journal on Selected Areas in Communications, 17(8), 1454–1465, 1999.
78. C. R. Lin, On-demand QoS routing in multihop mobile networks, Proceedings of Twentieth Annual Joint Conference of the IEEE Computer and Communications Societies, vol. 3, pp. 1735–1744, 2001.
79. C. Zhu, M. S. Corson, F. Technol, and N. J. Bedminster, QoS routing for mobile ad hoc networks, Proceedings of Joint Conference of the IEEE Computer and Communications Societies, vol. 2, pp. 958–967, 2002.
80. K. Akkaya and M. Younis, An energy-aware QoS routing protocol for wireless sensor networks, Proceedings of International Conference on Distributed Computing Systems Workshops, pp. 710–715, May 2003.
81. K. Chow, K. Lui, and E.Y. Lam, Balancing image quality and energy consumption in visual sensor networks, International Symposium on Wireless Pervasive Computing, p. 5, Jan 2006.
82. Z. He and D. Wu, Resource allocation and performance analysis of wireless video sensors, IEEE Transactions on Circuits and Systems for Video Technology, 16, 590–599, 2006.
83. Z. He, and D. Wu, Accumulative visual information in wireless video sensor network: Definition and analysis, IEEE International Conference on Communications, vol. 2, pp. 1205–1208, May 2005.
84. Z. He and C. W. Chen, From rate-distortion analysis to resource-distortion analysis, IEEE Circuits and Systems Magazine, 5(3), 6–18, 2005.
85. S. Soro and W. Heinzelman, Camera selection in visual sensor networks, Proceedings of IEEE International Conference on Advanced Video and Signal based Surveillance, Sep 2007.
86. C. Yu, S. Soro, G. Sharma, and W. Heinzelman, Coverage-distortion in image sensor networks, Proceedings of IEEE International Conference on Image Processing, San Antonio, CA, Sep 2007.
87. D. Malan, T. Fulford-Jones, M. Welsh, and S. Moulton, CodeBlue: An ad hoc sensor network infrastructure for emergency medical care, Proceedings of the Workshop on Applications of Mobile Embedded Systems, 2004.
88. R. Holman, J. Stanley, and T. Ozkan-Haller, Applying video sensor networks to nearshore environment monitoring, IEEE Pervasive Computing, 2(4), 14–21, 2003.
89. R. Vedantham, Z. Zhuang, and R. Sivakumar, Addressing hazards in wireless sensor and actor networks, Proceedings of the International Conference on Mobile Computing and Communications, 10, 20–21, 2006.
90. I. F. Akyildiz and I. H. Kasimoglu, Wireless sensor and actor networks: research challenges, Article Ad Hoc Networks, 2(4), 351–367, 2004.
91. E. Gurses and O. B. Akan, Multimedia communication in wireless sensor networks, Annals of Telecommunication, 60, 799–827, 2005.

# Chapter 23
# Middleware for Wireless Sensor Networks: The Comfortable Way of Application Development

Kirsten Terfloth, Mesut Güneş, and Jochen H. Schiller

**Abstract** Application development for wireless sensor networks (WSNs) demands for expertise in distributed as well as embedded programming. To ease the task of application development and make this area more accessible to nonexperts, middleware abstractions are commonly employed. Middleware is defined as software which is located in between software applications. Similar to operating systems, middleware systems provide applications with additional services to implement their functionality in a more abstract manner. Since devices forming a WSN have only little capabilities in terms of processing power and memory, their corresponding operating systems only provide very basic support for application development. At the same time various kinds of applications do have additional requirements to simplify their implementation. A multitude of middleware approaches are available to fill in this gap, thus provide support for comfortable application development. We will discuss common application building blocks in this domain, discuss a selection of middleware approaches available, and provide an evaluation of their applicability by mapping application needs to middleware services.

## 23.1 Introduction

In the past years, progress in miniaturization of electronic devices has led to the development of a multitude of embedded devices that are able to enhance our day-to-day life. Exemplary areas include automation and control of buildings, fine-tuning of vehicular performance, monitoring of personal health, or supervision of large areas, all of which depend on the measurement of some physical phenomena with the help of appropriate sensors. Often, more than one device is involved in the data

K. Terfloth (✉)
Institute of Computer Science, Computer Systems and Telematics (CST), Freie Universität Berlin,
Takustr. 9, 14195 Berlin, Germany
e-mail: terfloth@inf.fu-berlin.de

S. Misra et al. (eds.), *Guide to Wireless Sensor Networks*, Computer Communications and Networks, DOI: 10.1007/978-1-84882-218-4_23,
© Springer-Verlag London Limited 2009

acquisition process, which is, for example, the case when phenomena are spread over a certain geographical area or rely on multiple or different types of measurements at dedicated spots. Applications running on top of such distributed devices may observe the progression of sampled data in time and stream it to a central entity for further analysis, and/or may react to detected events in the physical world. In case a wireless medium is used for communication among participating devices, commonly entitled *sensor nodes*, the collection of such nodes is called a *wireless sensor network* (WSN).

Like all embedded devices, sensor nodes suffer from restrictions in terms of available resources. Bounds on energy consumption, memory constraints and limited processing capabilities have a great impact on the software stack that can be installed on a sensor node. This imposes a hardware-oriented point of view on the programmer when developing applications that are executed with a WSN. While this calls for detailed knowledge of a domain-expert, it is contradictory to the problem-oriented point of view an application programmer is used to, making software development a tedious task.

An abstraction layer that is situated in-between the bare system software and the application can help developers to focus on application needs. Up to now, many different solutions to provide a suitable abstraction and support applications in the domain of sensor networks have been proposed. Their characteristics can be evaluated from two orthogonal angles: On the one hand it is interesting to look at how good an approach serves at abstracting from system-related challenges. System-imposed criteria that hinder application development will therefore be briefly reviewed in Sect. 23.1.1. On the other hand, approaches target to facilitate application development by providing support on common application building blocks. To point out such application commonalities, we rely on a use case that envisions possible deployments of sensor networks in the near future and extract widely used operations and tasks. With these at hand, a catalog of parameters will be compiled to enable a qualitative evaluation of current middleware systems. For the sake of simplicity, we will refer to all of the approaches presented in this chapter as middleware approaches, although we are aware that the term middleware as used in the context of distributed systems may not exactly apply to all of them in the same manner. To alleviate this problem of using a well-established term in a wider connotation, we will clarify criteria for approaches to be denoted, before surveying representative middleware approaches for WSNs.

### 23.1.1 WSN-Specific Challenges

Application development for WSNs is a time-consuming task due to several reasons that are directly related to the general setup of these networks. These include:

- Constrained resources (energy, memory, processing capabilities)
- Distributed applications

- Unreliable communication
- Real-time requirements

Foremost, to enable their operation in application scenarios that were not easily accessible to technology beforehand, their physical size is envisioned to be arbitrarily small. The downside of this vision of ubiquitous sensing and reporting is that software will have to be programmed on *embedded devices*, calling for expertise knowledge in this domain of the programmer. Memory and energy scarcity, e.g., imply sensitive usage patterns on the software side and thus require a carefully crafted software stack.

A second challenge that naturally arises when the completion of a task involves more than a single device is the management of its *distribution*. This is, for example, the case when multiple nodes are questioned to take a data sample, calculate the median of their respective values and only report the result if a significant change has been observed among a majority of participating nodes. Here, shared data has to be prevented from being subject to race conditions, thus global state has to be synchronized among sensor nodes.

Depending on the type of application, unreliable communication may not be tolerable. In a wireless context, reflection, refraction, fading, and multipath propagation can lead to erroneous transmission of signals and asymmetric links between communication partners. To achieve the required robustness, corresponding protocols have to be developed to ensure communication quality as demanded by the application.

WSNs that are utilized to immediately report an event, such as an alerting system in a health monitoring application, may be subject to real-time requirements. In this case, both, the time needed for event processing as well as the time necessary for data transmission are critical parameters that decide upon application success. Hence, all software components that are involved in these time-critical operations have to be carefully designed to meet the required thresholds.

### 23.1.2  Use Case

The networking research community makes huge efforts to realize the vision of the *Internet-of-Things*, which will allow the communication of any daily *thing* with any others. This may simplify and provide additional comforts in daily living of humans. WSN applications can be set up to implement a multitude of different applications.

In the following, we will sketch an example day in the future with respect to sensor network applications and refer to Fig. 23.1.

People will live in *intelligent* or *smart* homes which will provide many value-added services to the inhabitants. There will be different computer and sensor systems installed and integrated into the home, e.g., air conditioner, heating systems, lighting system, and alarm systems.

When Alice wakes up in the morning depending on the sunshine the lighting system will adapt the luminosity in the rooms to support her to being in good mood and

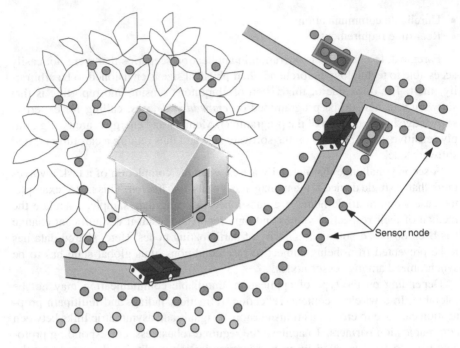

**Fig. 23.1** Scenario with sensor network applications

the heating system will adapt the temperature in the rooms to support her immune system. During the breakfast she may get a report of the status of the home, the garden, and the car. Upon the report she may trigger important tasks in the house, in the garden, and the car, e.g., the heating system of the car. When she leaves to work, she is reminded about important tasks during the day, the energy control system will shut down all devices which will not be used until she returns to save energy, and the alarm system will switch on.

Alice has various plants in her garden, which have different requirements on water, light, and nitrogen. To observe and take care of her plants she has installed a sensor network. Additionally, Alice is able to track her cat Bob over the sensor network in the house and in the garden.

On the way, Alice may communicate over the telecommunication means in the car without using her smartphone and checks the status of the car, e.g., the status of fuel, brakes, and the next service time. Additionally, information is gathered from the road infrastructure which may have various sensor data. For example, information about the air pollution is delivered to the air conditioning system in the car which switches to a suitable filter. The brake system of the car is informed about the wetness of the street and the car remembers the locations of potholes and adapts the suspensions of the car. Furthermore, the road infrastructure collects information about the cars to provide them with additional traffic information, e.g., to smooth the traffic flow by proposing speed intervals to the cars to avoid traffic jams.

### 23.1.3 Common Application Building Blocks

In this section, we discuss the common building blocks required to build applications on top of WSNs. For this we refer to the use case as well as other realized prototypical sensor network applications.

#### 23.1.3.1 Data Streaming Pattern (DSP)

Streaming of data from multiple nodes to a sink for the sake of data collection is a very common task performed in sensor networks. In the use case sketched above, Alices request for a status report on her belongings is a data streaming operation. This *read-only* request may either be applied to all of her devices or *filtered* in case she is only interested in a certain subset or *group* of nodes, such as the current water level of the plants in her garden. Periodic measurements for continuous supervision as necessary to guarantee correct automatic water supply for the plants over the course of a day or year may be required as well as one-time snapshots.

Real-world examples following a data streaming paradigm that have been build with WSNs include predominantly environmental deployments such as [1–4]. Here, the network has been used to learn more about the climate, eruptions of volcanoes, or the behavior of birds. Especially in this domain, infrastructure is often not directly accessible and phenomena are usually spread over a certain region, making WSNs a valuable tool for data acquisition. Other examples that rely on data streaming include personal health monitoring that allows for supervision of key health parameters with a Body Area Network or tracking applications that strive for providing fresh information on the geographic position on certain persons, animals such as Alice's cat, or vehicle.

#### 23.1.3.2 Sense-And-React Pattern

Another usage pattern for wireless sensor networks can be summed up under the keyword *Sense-And-React Pattern* (SARP). Alice's automatic light control is a nice example illustrating the altered scope of data in sense-and-react applications compared to data streaming: Data samples from luminosity sensors directly effects the control of both the lights in her bedroom if it is too dark in the morning and the shades in the summertime to provide optimal lighting. On the other side of the house though, the acquired sensor data is meaningless since the local lighting state is different. Therefore, data in SARP usually resembles a limited, local scope of validity and triggers local actions denoted by predefined control laws. These control laws can manifest themselves in simple events such as a value passing a certain threshold but can also be complex conjunction of multiple spatiotemporal conditions. To assure robustness, thus avoid triggering a wrong action based on a faulty sensor reading or due to the spatial distribution of an event, values are often collected from multiple nodes physically close to one another, referred to in as a *neighborhood*. Coordination and control of localized interaction and a meaningful and rich language to express events are major challenges in SARP.

Heating, ventilation, and air-conditioning (HVAC) applications [5] or event detection and classification [6] are real-world examples that follow a SARP paradigm.

### 23.1.3.3 Read–Write Pattern

While in a data streaming context, status information is predominantly transmitted from a data source to a sink, applications such as Alices garden network or the vehicle-to-street communication require an uplink to the sensor network. This uplink can either be used to tweak parameters such as the sampling frequency, as may be necessary when extreme climatic states are reached to prevent Alices plants from drying or freezing, or to provide software updates on deployed sensor nodes. If Alice puts new flowers into her garden, they certainly have a different nurturing scheme that needs to be supplied dynamically via software updates.

Deployments such as the monitoring of a fence [6] or the volcano monitoring [2] make use of a read–write pattern by adapting the sampling frequency on event recognition to assure a good trade-off between data granularity on the one side and energy exposure on the other. An emergency deployment for road tunnel supervision has been proposed in [7] that explicitly enables dynamic loading of new software components onto deployed sensor nodes to reconfigure usage scenes.

### 23.1.3.4 Group-Processing Pattern

Unlike Internet-scale networks, the goal of a single node in a WSN is often of minor importance. Instead, a network is set up to serve as an entity rather than individual nodes pursuing their own determination, an understanding that calls for new addressing and data manipulation schemes.

Looking at the use case above, one can observe the need for a group-processing pattern to enable tracking Alice's cat. One option to implement this is to ask each individual node in the network whether the cat is currently in its vicinity, thus to stream the complete status information of each node to a sink. Obviously, with increasing size of the network, the communication overhead is immense. A better approach is to push the application logic into the network and specify node behavior dependent on its membership to a certain group of nodes. Criteria for group formation can be functional, such as the availability of a specific sensor on a node, application-dependent, such as the detection of Alice's cat, or based on network connectivity thus contain all nodes in an $n$-hop neighborhood of an event or node. This way, coordination among nodes can be limited to group members, thus network load decreased.

Group addressing schemes will also be of great value when it comes to retasking a selection of nodes that share common functional or application-dependent attribute. This may be, for example, the case when Alice wants to change the sampling frequency of all sensor nodes that are currently able to spot the cat.

Group processing has for example been used in a redwood tree monitoring deployment [8].

### 23.1.3.5 Heterogeneity and Internetworking

While the patterns presented so far purely address network internal activity, Internetworking and heterogeneity become important issues to realize ubiquitous computing schemes.

Heterogeneity is a problem that is commonly addressed with the help of middleware and can be found throughout the depicted use case. The communication of a roads sensor layer with a vehicles control to signal slippery road parts or warn about upcoming road construction or congestion requires the ability for Internetworking. Participants in a network may be of different manufacturers and capabilities, as well as stemming from different networks.

Experiments that have been carried out with heterogeneous hardware include a small setup simulating a tunnel monitoring application [7].

## 23.2 Background: Middleware – Definition and Classification

Traditionally, middleware is defined to be situated in between operating system and applications in a distributed and networking context, see Fig. 23.2, a rather imprecise definition that may be attributed to libraries, software components, or frameworks as well. Bernstein [9] specifies a set of criteria commonly addressed by middleware services. These include independence of the chosen platform; hence middleware services have to be portable to a variety of system architectures with modest and predictable effort, supply of a functionality that meets the need of a wide variety

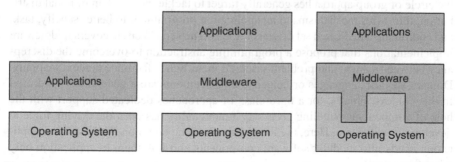

**Fig. 23.2** Relationship of operating system, middleware, and applications

of applications and distribution of the service itself. From a developer's point of view, a middleware implementation offers a platform-independent API to mask the complexity of distributed processing and underlying networks.

Well-known, classical middleware systems are the CORBA specifications [10], which enable multiple software components, potentially written in different languages and running on various computers to work together, or Message-Oriented-Middleware such as MQSeries that depends on a message queuing or message passing concept to interconnect applications in heterogeneous environments.

Looking at middleware utilized in WSNs, the boundary between operating system functionalities and middleware starts to blur and functions mingle as depicted in Fig. 23.2. A range of functions that are classically implemented as part of the operating system are now shifted to a middleware and may therefore not necessarily be available to an application programmer throughout a variety of middleware implementations. This observation can be explained by the origins of middleware, whose development has mainly been motivated to support a certain class of dedicated applications. Limitations in terms of available memory led to implementations that export a tailored API rather than establishing a common ground of middleware functionalities to serve a general-purpose setup. To characterize and eventually grasp the term middleware in a WSN context, two different options are available. Either, a minimal set of core functions that software has to provide to call it a middleware can be defined or the definition above can be widened to point out the diversity of approaches and provide insight on differences and commonalities.

Due to our understanding that there will be no such thing as a general-purpose sensor network, hence usage patterns will determine the need, nature, and value of proposed abstractions, we choose the latter. It is important to point out that all applications share system-related challenges, but their impact and therefore the need for addressing the arising problems in a middleware layer will differ among deployments.

In the following, we will therefore rely on a simple, but meaningful model to classify middleware approaches currently available as depicted in Fig. 23.3. The idea is be able to quickly comprehend at what end a solution operates.

Each of the three circles symbolizes a basic abstraction mechanism to address a core problem that application development for WSNs faces. Approaches unified in the circle of group approaches generally target to tackle the problem of nodal distribution, thus offer mechanisms to an application programmer to better specify, task, and coordinate a dedicated set of sensor nodes. The second circle covers middleware implementations that propose a programming abstraction to overcome the discrepancy between system and problem-oriented view, hence facilitate context mapping. Domain-specific languages or language enhancements are a general keyword used in this context. Finally, for a third class of approaches dedicated support with the help of components bundling necessary functionality has been the driving force to develop a middleware. Here, the idea is to outsource common WSN-specific application needs into modular software components and link them to application only when necessary.

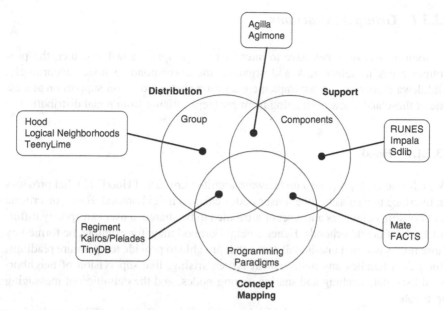

**Fig. 23.3** Classification of middleware systems

Clearly, these circles overlap. *Macroprogramming*, e.g., refers to abstractions that allow for programming the network as a whole, thus to specify a global program which is compiled down to node-level code. These approaches can be situated in the intersection area of Group Abstractions and Programming Paradigms. Approaches that feature both a programming abstraction and allow to encapsulate sets of functions in components can be summarized under the keyword *Composite Programming*. Finally, *Dispersed Structuring* refers to those approaches that lay in the intersection area of Group Abstraction and Component-based middleware systems.

The assignment of middleware approaches to abstraction categories has been accomplished based on their primary design goals. A more fine-grained evaluation of the approaches discussed in the following that will incorporate and discuss common application building blocks will be supplied and discussed at the end of this chapter.

## 23.3   Middleware Approaches for Wireless Sensor Networks

Magnitudes of abstractions that serve as a middleware have been proposed for WSNs over the past years. Due to space limitations, only a small set of approaches will be presented in the following. The selection has been motivated by the urge to illustrate the wide range of different abstraction mechanisms.

## 23.3.1  Group Abstractions

As soon as sets of nodes have to interact to accomplish a task together, the programmer has to select, task and organize the corresponding nodes accordingly. Middleware approaches that implement a group abstraction offer support on at least one of these and allow abstracting from problems arising from nodal distribution.

### 23.3.1.1  Hood

Whitehouse et al. propose a middleware architecture called Hood [11] that provides an interface to a subset of the sensor nodes called a neighborhood. Based on criteria for choosing neighbors and a set of attributes to be shared, a user can specify different kinds of neighborhoods. Hence a neighborhood may, for example, be formed by those nodes within a one-hop distance that are able to provide temperature readings. Hood then handles any management issues arising, like supervision of neighborhood lists, data caching and sharing among nodes, and the definition of messaging protocols.

Communication within a neighborhood is based on a broadcast/filter mechanism. If a node wants to share one of its attributes, it simply broadcasts the value. Incoming packets are filtered, and nodes can determine whether or not the received attribute is of interest and cache it. There is no feedback to the node that sent the value, so in contrast to concepts building upon reliable networking, Hood only needs asymmetric links between nodes.

Looking at the programming part of the approach, a neighborhood becomes a programming primitive. To create a new neighborhood and allow its individual parameterization, a code generation tool has to be invoked by the developer. The system itself builds upon TinyOS [12], an operating system widely deployed on sensor nodes. Several interfaces offered by Hood provide handles to access neighborhood attributes and define sharing and updating strategies. Furthermore, values of neighbors stored locally on a node can be annotated with so-called scribbles. These simply note extra information, for example the quality of the link to the mirrored node.

Overall, a programmer who builds applications upon Hood is given the possibility to address and control functional parts of the network together, instead of issuing single nodes. Thus, this project alleviates effort for maintenance and low-level concerns, and takes the burden of dealing with distribution in a one-hop neighborhood from the programmer.

### 23.3.1.2  Logical Neighborhoods

Logical Neighborhoods, proposed by Mottola et al. in [13] expand the idea of forming neighborhoods from a set of nodes in one-hop vicinity as proposed in Hood to a logical partition of a sensor network that operates in a multihop manner. To accomplish this task, the Logical Neighborhoods framework provides two basic components.

First of all, a declarative language is utilized to specify a node's exported attributes in a templates, and to define neighborhoods with the help of predicates over such node templates. The actual membership of nodes to a specific neighborhood is determined at instantiation time, and requires an application programmer to declare the starting point of neighborhood construction. A nice feature to bound energy consumption on sensor nodes in terms of messages sent is that the framework implements a credit-based cost function, which may be used to restrict the scope of neighborhoods and thus enable application-level control of resource consumption.

The second part of the Logical Neighborhood framework is a neighborhood routing primitive that enables nodes to logically broadcast values to other members of the neighborhood. To this end, a structureless routing mechanism has been implemented that enables message delivery even in dynamic environments.

From an application programmers' point of view, distributed applications may be build in a cost-sensitive manner without having to explicitly develop underlying network protocols when relying on Logical Neighborhoods.

### 23.3.1.3   TeenyLime

TeenyLime, proposed by Costa et al. [14], is the latest addition to the Lime family of middleware platforms. Specifically designed to enable sophisticated sense-and-react applications in WSNs, it allows for data sharing among neighboring nodes with its tuple space implementation. Stateful coordination with the possibility to reliably share data, the ability to specify multiple tasks, and support for reactive interactions are the main benefits that TeenyLime offers to an application programmer.

Central to the TeenyLime implementation is the tuple space, a shared data repository which can be accessed with read and write commands to insert and retract data tuples via pattern matching. The local tuple space of a node is automatically shared with its one-hop neighbors and therefore serves as a communication primitive. For instance, nodes can publish their ability to provide sensor data by putting a special tuple that indicates this ability into the tuple space. When another node wants to invoke a reading, it matches the pattern of this *capability tuple* and will be automatically provided with the data sample requested. The necessary communication between nodes is shielded from the developer and handled by the TeenyLime middleware Furthermore, the TeenyLime API specifies commands to add and remove reactions whose action parts will be triggered upon the emergence of a tuple, whereas stateful coordination is transparently supported by introducing reliable operations. The TeenyLime middleware is implemented in nesC on top of TinyOS.

Since data sharing in a local, one-hop context of a node is facilitated by the tuple space interface and a simple API that TeenyLime offers, a developer can benefit from relying on this middleware implementation when localized interactions have to be coordinated.

## 23.3.2 Macroprogramming

In case supportive functions to overcome distribution issues are directly incorporated in a programming language so that it is possible for an application programmer to write a network program, approaches are commonly summarized under the keyword of macroprogramming. Node-level code generation and coordination of nodes are subject to the middleware framework.

### 23.3.2.1 TinyDB

TinyDB [15] is a suggestion to alleviate network programming from application programmers. The idea of a middleware using a database abstraction is to enable the utilization of a well-known, declarative programming language upon the distributed nodes without having to deal with network issues. Therefore, the network established by the sensor nodes is understood as a distributed database which can be queried using a subset of SQL. The authors added some essential key-words specific to the sensor network domain to enhance the language, respectively.

In this concept, each node contributes one row to a single, virtual table, and each column represents one of the attributes that can be queried. A query processor is run on every node to handle and possibly aggregate the sensor data questioned by the query specification. Thus to obtain values from the network, a user issues a query, which is then automatically routed to all nodes. TinyDB maintains a spanning tree from the node where the request has been initialized, so that resulting data can be sent back in reverse direction. Queries may be marked to be evaluated periodically, or values to be aggregated, summed up, or grouped on their way back through the network. Any maintenance concerning bootstrapping or failure of nodes or routing issues will be handled by TinyDB without any interaction with the programmer.

TinyDB contributes with its SQL-style programming manner an interface that is already widely accepted. Since distribution issues are transparent to the user, even unexperienced users are able to task the network appropriately.

### 23.3.2.2 Regiment

The Regiment System as proposed in [16] and its actual implementation evaluated in [17] supports application programmers by providing a functional domain-specific language. Instead of specifying node-level behavior, Regiment offers macroprogramming primitives and thus enables the specification of global programs that are compiled into an intermediate language to be interpreted on individual nodes.

From the author's point of view, the core elements that WSN application scenarios feature are data streams. Originating from a set of spatially distributed sensor nodes, Regiment, therefore, provides language basics to process these streams, transparently aggregates data from multiple nodes forming neighborhoods, and abstracts from data acquisition details and storage. The result of processed data streams, e.g., detected events such as a sensor value exceeding a threshold within a certain region of nodes, is automatically forwarded to a predefined base station.

Regiment programs are composed of functions that manipulate so-called *signals*, which represent streams of data samples at a given time, and of functions that can act upon *regions*, which represent collections of signals. To construct a region, a programmer may either form it starting at a certain node in via a spanning tree algorithm, or he can rely on two different gossip-based primitives. With the help of these constructs, programs that include filters on data streams of nodes, their organization into neighborhoods, and their automatic delivery at a central entity can be easily defined. Note that it is not possible to alter node-local state since the Regiment language is side-effect free.

Source code is compiled to an intermediate language called Token Machine Language (TML), which is interpreted on the nodes accordingly. A token is akin to an active message, encapsulating a payload of private data and triggering the execution of an associated token-handler upon reception.

The main benefit of utilizing Regiment is that it offers a very high level of abstraction to organize data streams, which makes it especially appealing for rapid prototyping.

### 23.3.2.3 Kairos and Pleiades

Kairos [18] and its successor Pleiades [19] share the same idea of annotating sequential code with macroprogramming statements. They both provide language primitives to access node-local state and iterate over a set of nodes, but differ in the way they support serializability, concurrency, and code migration.

Three simple extensions to C source code have been introduced by Kairos: The *node* data type allows for logical naming of sensor nodes and exports a set of common operations on nodes, a call of *get_neighbors()* returns a list of one-hop neighbors that may be manipulated iteratively and the ability to access data remotely is supplied for named nodes. To put into effect the annotations mentioned above, a preprocessor filters the program for these enhancements. A compiler then generates node-level code, with Kairos commands being translated into calls to the Kairos runtime that has to be installed on every node. Any variable that is subject to remote access, so-called *managed objects*, or referenced by a node but resides at a remote node, so-called *cached objects*, are managed by the runtime environment. Program execution that involves remote access follows a synchronous execution model, and is internally dispatched to asynchronous message passing between participating sensor nodes by the Kairos runtime.

Pleiades basically explores a similar approach, thus augments C with language primitives, but furthermore offers support for reliable concurrent execution of code on multiple sensor nodes. The Pleiades runtime takes care of synchronized access to shared variables, guarantees serializability, and locks resources accordingly. Both, the compiler and the runtime system support a distributed deadlock detection and recovery algorithm to avoid potential deadlocks when invoking a concurrent iteration over a set of nodes, a feature made available by introducing a so-called *cfor* loop. Furthermore, the program will automatically be partitioned by the compiler into *nodecuts*, node-level programs, whose control flow is managed and migrated between nodes by the runtime environment in order to minimize communication costs. Pleiades programs are translated into nesC code, with the Pleiades runtime being a collection of TinyOS modules.

By injecting new statements that guard shared variables and allow for simple manipulation of a set of neighbors, a programmer relying on these macroprogramming approaches is relieved from having to explicitly address shared state.

### 23.3.3  Composite Programming

Difficulties in application development in an embedded context are often a result of a low level of abstraction concerning the underlying hardware. Domain-specific languages that are furthermore designed to enable modular composition of middleware functions can be classified under the term of Composite Programming approaches.

#### 23.3.3.1  Maté

To tackle the problem of retasking a network at runtime, Lewis et al. [20] have developed a virtual machine, Maté, and a specific byte-code it can interpret. This way, software updates may solely implicate transmission of the new application and not a complete binary image to flash.

Using Maté on sensor nodes presumes that applications have to be expressed in special Maté instructions. The design of a suitable language is therefore crucial for the ability to express applications the network is supposed to accomplish, and thus for the success of the virtual machine. Mandatory goals include the need for concise instructions and dense byte-code to save energy on transmission, and an expressive but simple language to enable the envisioned wide spectrum of possible applications. The authors decided on a stack-based architecture and an instruction set programming style. Since Maté is highly dependent on TinyOS and makes use of its messaging infrastructure, the complete system architecture is optimized for a symbiotic relationship. The sizes of Maté instructions are for example customized to perfectly fit into TinyOS packages. Programs can thus be segmented into equal-sized chunks, so-called capsules, which is advantageous for on-the-fly software installations.

The instruction set of Maté combines low- and high-level instructions, and allows three possible operand types to be used: values, sensor readings, and messages. Besides basic instructions for arithmetic computations, halting, and branches, sensor network specific commands are available and offer a convenient abstraction for an application programmer. A build-in routing algorithm can, for example, be called by issuing a single instruction, which is in charge of sending the specified package to its destination. Also, packets can forward themselves to install new applications in the network within a single instruction call. Furthermore, the instruction set includes eight instructions that may be implemented by the application programmer, and is thus tailored to the needs of a special application scenario.

A safe execution environment as provided by a virtual machine hides the complexity of the hardware or, in this case, TinyOS's complex, asynchronous execution model, and prevents system crashes. The instruction set design is especially targeted to the sensor network domain, enabling a programmer to express his application needs easily.

### 23.3.3.2   FACTS

FACTS [21] has been inspired by the event-centric nature of many sensor network applications and provides a solution to elegantly express asynchronous behavior. Modular pieces of processing instructions and rules encapsulate knowledge about when and how to handle data occurring in the system. This allows a programmer to specify tasks such as filtering, data aggregation, or more complex data processing schemes with the help of a couple of interacting rules. Since FACTS is designed to only react to changes relevant to an application, the node may go to low-power mode in case no event is triggering rule execution.

FACTS therefore provides a new language for rule specification, called the ruleset definition language (RDL). This enables grasping necessary commands for working with WSNs directly at the language level. Not only is it convenient to have a small and precise set of programming primitives at hand, but also it allows for optimizing the language toward the target domain. Sets of rules are compiled to a dense bytecode that is interpreted on the node at runtime, shielding a programmer from underlying hardware and networking issues. Data for instance is specified in a special format called a fact, a programming primitive that is transparently exchanged over the network. Local interaction between nodes becomes easy to express; global network behavior may therefore be the key focus of application developers.

Application programmers that utilize FACTS can benefit from its high-level data abstraction facility on the one hand, but also from a concise set of operations needed to express event-centricity and manipulate sensor data on the other. Sets of interacting rules may implement applications as well as middleware functionalities, hence FACTS provides a modular framework to task sensor networks according to application needs.

## 23.3.4   Dispersed Structuring

Approaches that can be subsumed under the label of Dispersed Structuring combine a component-oriented approach with the ability to transparently program groups of devices.

#### 23.3.4.1   Agilla

Agilla [22] provides an abstraction for WSNs that relies on mobile agents. A special runtime environment featuring a tuple space for asynchronous communication between multiple agents residing on a host and an execution platform for agents is the core features of Agilla. The general idea is to be able to deploy a vanilla sensor network that only features the Agilla runtime environment. Later on, different applications may be inserted into the network with the help of agents encapsulating application logic. These agents autonomously move around the network to gather data or coordinate local tasks.

Agent specification relies on the Mate instruction set, enhanced with specialized instructions to support agent migration, agent cloning, and tuple space modifications. The Agilla runtime environment is implemented in TinyOS.

Using Agilla to implement sensor network applications can be useful when multiple applications have to utilize the same network that is not known at deployment time.

#### 23.3.4.2   Agimone

Agimone [23] has been designed to allow for coupling several sensor networks using an IP-network in between. Therefore, the authors utilize two mobile agent platforms, Agilla, an implementation for WSNs and Limone, which has been proposed for more elaborate devices, and integrate them to achieve cross-network interaction. Each WSN is associated with one dedicated gateway to enable its advertisement to other participating networks. For an Agilla agent to migrate to distant WSNs, it has to be wrapped into a Limone agent at the gateway, transferred across the IP-network relying on Limone migration and unwrapped and reinjected into the target network.

When already relying on a mobile agent middleware, Agimone offers a straightforward approach to transfer information across network borders.

### 23.3.5   Component-Based Abstraction

In contrast to group and language abstractions, component-based approaches do not explicitly address a single WSN-specific problem, but rather provide an infrastructure to encapsulate support for applications. The main idea is to elegantly provide software as needed and enable its reuse by different applications.

#### 23.3.5.1   Runes Middleware

RUNES [7] is an approach that fulfills the classical requirements of a middleware. Designed to alleviate problems arising from heterogeneity of interacting devices, both in terms of manufacturer, operating system and system capabilities, and dynamic network settings, the authors propose a supporting two-level middleware. Self-contained components feature necessary middleware functionality and can be individually deployed at runtime according to application needs. To enable this, a component model serves as a basis to specify basic runtime units and their corresponding interfaces in a language-independent manner.

A variety of platform-dependent implementation of the component model have been developed, including a Java-virtual-machine-based implementation, a C/Unix-based implementation as well as an implementation running on the Contiki operating system to ensure the applicability of the model on heterogeneous systems.

The clear separation of platform-dependent and middleware concerns provides an application programmer with a nice tool to develop dedicated components.

### 23.3.5.2 Impala

Impala [24] is an architecture implemented within the ZebraNet project [25]. The primary design goal has been to build a modular, lightweight runtime environment for applications that manages both devices and events. Hence, Impala splits the field of duty into two layers, one to encapsulate the application protocols and programs for ZebraNet, and an underlying layer that contains functions for application updates, adaptation, and event filtering.

Application programming follows an event-based programming paradigm, thus any application deployed upon the nodes has to implement a set of event and data handlers to respond to different types of events, including timer, packet, data, and device events. Besides supplying event filter mechanisms, Impala emphasizes the need for integrating adaptation and updates of applications at runtime within the system architecture.

Adaptation of an application or an application-level protocol can become necessary due to changes of the system, e.g., failure of certain sensors or low battery level, as well as application specific modifications, e.g., a sudden drop of successfully delivered packages. A middleware agent, the Application Adapter, checks the overall state of the system on a regular basis and selects the most suitable configuration according to the present circumstances. Dynamic software updates may be mandatory during execution, but since the devices are not reprogrammable at the same time (ZebraNet equips wildlife animals with sensor nodes, which results in a highly dynamic topology, so software can be received in incomplete bundles of packets) the Application Updater serves as a management component for available versions and code bundles.

Overall, it is interesting to point out that Impala is a middleware which has been especially crafted for the ZebraNet project, a fact that reflects the application-oriented development process of middleware abstractions in the sensor network context.

### 23.3.6 Sdlib

The idea of sdlib [26] is to provide a standard library for operations commonly found in WSNs, encapsulate basic patterns, and, since then functionalities may be shared by multiple applications, reduce the overhead implied. As an example, implementations of basic services such as data collection or data dissemination are proposed and discussed in the paper.

Sdlib has been designed for TinyOS and therefore uses the nesC programming language and its wiring concept. The sdlib runtime engine serves as a core management entity with auxiliary components such as a dataflow component, a simple memory management component, or a component to guarantee reliable transmission of messages. These services may then be invoked by an application programmer, with the promise to alleviate problems arising from manual flow or data management.

## 23.4 Thoughts for Practitioners

The goal of middleware is to provide the application developer with comfortable means to implement applications. We discuss in this section the presented middleware approaches in respect to their appropriateness for the development of heterogeneous, distributed sensor network applications. This qualitative evaluation will cover application- as well as system-dependent criteria.

### 23.4.1 Application-Oriented Selection of Middleware Approaches

An overview of all presented middleware approaches is depicted in Table 23.1. Here, we analyze the middleware approaches based on the patterns compiled in the first section. The focus is on typical requirements for WSN applications and their support by the middleware approaches. We distinguish the provision of patterns in two classes. patterns explicitly supported by a middleware approach, whereas o patterns which can be possibly implemented with a middleware approach, but are not in the focus of the approach. This twofold evaluation of the middleware approaches is somehow arbitrary, but represents the experience and the understanding of the authors.

Most approaches explicitly support the DSP which is not astonishing, since the first bigger deployments of WSNs featured this pattern. However, the SARP is usually not implemented at the same time, although this becomes especially interesting when advancing from pure monitoring to sensor-actor networks. Only two discussed middleware approaches support both patterns, namely Runes and Facts.

**Table 23.1** Overview of middleware approaches

| Approach | DSP | SARP | RO | RW | Inet | EPP | GPP |
|---|---|---|---|---|---|---|---|
| Hood | O | | | O | | O | ● |
| Logical neighborhoods | ● | | | O | | O | ● |
| Regiment | ● | | ● | | | | ● |
| Kairos/Pleiades | O | | | ● | | O | ● |
| Maté | O | | | ● | | ● | ● |
| TinyDB | ● | | ● | O | | O | ● |
| TeenyLime | | ● | | ● | | O | ● |
| Agilla | | O | | O | | ● | |
| Aginome | | O | | O | ● | ● | |
| Runes | O | O | | O | ● | O | O |
| Impala | O | | | | O | ● | O |
| Facts | O | ● | | ● | | ● | |
| Sdlib | ● | | O | | | ● | |

● Explicitly supported by the middleware approach, O Middleware approach may be used, but is not in the focus, *DSP* Data streaming pattern, *SARP* Sense-and-react pattern, *RO* Read only, *RW* Read–Write, *Inet* Internetworking, *EPP* Entity processing pattern, *GPP* Group processing pattern

From the overview we can also derive that nearly all approaches support the *read–write* pattern (RW). To implement a read–write pattern *downlink* and *uplink* communication has to be provided on the lower layers.

Addressing schemes, entity processing pattern (EPP) and group processing pattern (GPP), differ among approaches. We distinguish approaches with an explicit group-addressing scheme from approaches that allow for entity-addressing only. Although a group may consist of only one sensor node, a sensor node is not individually addressed in a group processing scheme, but instead filtering is used to select that sensor node.

Among all middleware approaches only three support the Internetworking pattern of different sensor networks, thus restricting the set from which can be selected in case it is needed.

Summing up all characteristics, only one middleware, namely Runes, supports nearly all patterns. However, some attention has to be paid: Runes provides a framework architecture with service components to be implemented by the application developer.

## 23.4.2 System-Oriented Selection of Middleware Approaches

Middleware approaches provide powerful tools to accelerate application development for WSNs. Since each approach contributes to achieve a higher level of system abstraction by means of hiding domain-specific complexity, programming on top of a middleware will most likely be less erroneous, faster, and more focused on the actual application logic.

Nevertheless, the choice of which middleware to utilize should not only depend on the functional aspects as discussed above, but also include aspects relevant to the life-cycle of application development: While in a first state of prototyping the demands for low-level tuning of system parameters may not be relevant, it can be crucial for deployments later on. Energy efficiency and low duty cycling of nodes can become mandatory parameters that will decide about application success. Therefore, attention has to be paid whether an approach supplies access to these parameters or already includes corresponding support or whether these options are completely shielded from the developer.

## 23.5  Directions for Future Research

Unlike the usual progress we observe in system development that shows a trend to integrate a multitude of features into a single approach, middleware development for sensor networks will rather target a diversification. New deployments and applications will reveal new usage patterns and thus demand for new middleware solutions in the long run. In this context, especially the shift from pure sensing to more sophisticated sensor-actor networks will probably reveal new challenges to overcome.

A problem that will definitely be discussed in more detail in the near future will be bridging WSNs with other more mature networks to achieve a better interaction between different classes of devices. Efficient and robust mesh networking is the key challenge that has to be addressed to enable broader usage scenarios as those discussed in the introductory use case.

## 23.6  Conclusions

The development of application software for distributed sensor networks which consist of tiny computers with restricted resources is a complex task. Additional challenge into the software development process is brought by the fact that these systems cannot provide a highly sophisticated operating system, thus system-oriented issues have to be addressed on the application layer. An approach to alleviate this situation is by using middleware solutions appropriate for the sensor network domain.

In this chapter, we introduced a variety of middleware approaches for WSNs, motivated their deployment, derived requirements for general purpose middleware approaches, and classified them in respect to applications. We also discussed the middleware approaches from the point of view of an application developer who seeks for comfortable support. The discussion shows that middleware approaches for WSNs are biased according to particular applications.

# Terminologies

*Wireless sensor network.* A WSN consists of spatially distributed devices that communicate over a wireless medium and cooperatively monitor physical phenomena with the help of sensors.

*Middleware.* Middleware is software situated in between operating system and applications in a distributed system and networking context.

*Neighborhood.* In the WSN domain, a neighborhood describes a set of nodes sharing at least one common attribute, usually but not necessarily network proximity.

*Data streaming.* Data streaming refers to (periodically) sending acquired data samples to one or more dedicated data sinks, possibly relying on multihop routing.

*Sense-and-react.* In sense-and-react applications, data samples are processed to events that eventually trigger reactions. The scope of reactions range from node-local actions to global effects on the network.

*Macroprogramming.* Macroprogramming is defined as programming a sensor network in the large where programming systems do not contain explicit abstractions for nodes, but rather express a distributed computation in a network independent way.

*Domain-specific language.* In contrast to general-purpose languages, a domain-specific programming language is tailored to one specific kind of task or operational domain.

*Composite programming.* Composite programming incorporates a component-based software development approach into a domain-specific programming language.

*Dispersed structuring.* Dispersed structuring refers to approaches that integrate component-based software development with group-level tasking of sensor nodes.

*Routing.* Routing is the process of selecting paths along which data is send in a network of nodes.

# Questions

1. What are the main benefits of utilizing a middleware to write sensor network applications?
2. Describe three exemplary services a middleware implementation for WSNs offers to an application programmer, and point out the corresponding approaches.
3. What is the difference between approaches that offer a pure group abstraction approach to those that supply a macroprogramming abstraction?
4. Give an example for an application that is not suitable for macroprogramming.
5. Given a network of 50 sensor nodes, set up to monitor the consumption of coffee in a university building. Therefore, each node records the amount of water that is used to brew the coffee by means of measuring the water flowing through

the coffee machines pipes. Every hour, the attached node sends a report toward a data sink that calculates the average coffee consumption of the university per day.

To build this application, which middleware would you preferably use and why?

6. In case the network is supposed to only report high peaks of coffee consumption, instead of periodical status information, which middleware is able to map such application behavior better?

7. Consider you have an application that has to deal with group management of nodes in an event-centric manner and you want to realize this in a component-based implementation. Which middleware do you think is best suited to be extendible to meet all three abstraction concerns?

8. A TinyDB network can be programmed by specifying SQL queries that are evaluated on the nodes with results being streamed back to a data sink. Aggregation of values and grouping of results is possible during the routing process. Consider a TinyDB network executing a query as depicted below. *Sensors* refers to the table of all sensor nodes, thus to the complete network. Assume the GROUP BY statement to rely on the following average light groups.

Group 1: $0 < \text{temp} \leq 10$
Group 2: $10 < \text{temp} \leq 20$
Group 3: $20 < \text{temp} \leq 30$
Fill in the values of nodes 1, 3, 4.

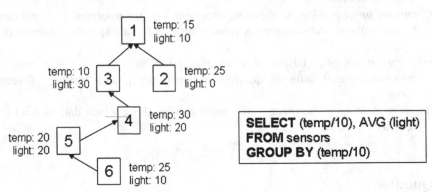

9. Three approaches, namely TeenyLime, Agilla, and FACTS share with the usage of "tuples" (called facts in FACTS) a common data abstraction model. Give a definition of the term tuple and name reasons why the authors may have chosen to use tuples.

10. Discuss whether you think current middleware approaches provide sufficient services to application developers.

# References

1. K. Martinez, P. Padhy, A. Riddoch, R. Ong, and J. Hart. Glacial environment monitoring using sensor networks. In: Real-World Wireless Sensor Networks (REALWSN 2005), June 20–21 2005, Stockholm, Sweden.
2. G. Werner-Allen, K. Lorincz, J. Johnson, J. Lees, and M. Welsh. Fidelity and yield in a volcano monitoring sensor network. In USENIX'06: Proceedings of the Seventh Conference on USENIX Symposium on Operating Systems Design and Implementation, pp. 381–396, Berkeley, CA, USA, 2006. Berkeley, CA: USENIX Association.
3. R. Freeman. Autonomous monitoring of vulnerable habitats. Available at: http://research. microsoft.com/ero/. Last access: 06.03.2008.
4. A. Mainwaring, J. Polastre, R. Szewczyk, D. Culler, and J. Anderson. Wireless sensor networks for habitat monitoring. In ACM International Workshop on Wireless Sensor Networks and Applications (WSNA'02), Atlanta, GA, September 2002.
5. A. Deshpande, C. Guestrin, and S. Madden. Resource-aware wireless sensor actuator networks. IEEE Data Engineering Bulletin, 28(1):40–47, 2005.
6. G. Wittenburg, K. Terfloth, F. L. Villafuerte, T. Naumowicz, H. Ritter, and J. Schiller. Fence monitoring – Experimental evaluation of a use case for wireless sensor networks. In Proceedings of the Fourth European Conference on Wireless Sensor Networks (EWSN '07), pp. 163–178, Delft, The Netherlands, January 2007.
7. P. Costa, G. Coulson, C. Mascolo, G. P. Picco, and S. Zachariadis. The RUNES middleware: A reconfigurable component-based approach to networked embedded systems. In Proceedings of the 16th Annual IEEE International Symposium on Personal Indoor and Mobile Radio Communications (PIMRC'05), Berlin (Germany), September 2005.
8. G. Tolle, J. Polastre, R. Szewczyk, D. Culler, N. Turner, K. Tu, S. Burgess, T. Dawson, P. Buonadonna, D. Gay, and W. Hong. A macroscope in the redwoods. In SenSys '05: Proceedings of the Third International Conference on Embedded Networked Sensor Systems, pp. 51–63, New York, NY: ACM, 2005.
9. P. A. Bernstein. Middleware: A model for distributed system services. Communications of the ACM, 39(2):86–98, 1996.
10. OMG Specification: The Common Object Request Broker: Architecture and Specification, Revision 2.0, OMG Document.
11. K. Whitehouse, C. Sharp, E. Brewer, and D. Culler. Hood: a neighborhood abstraction for sensor networks. In MobiSys '04: Proceedings of the Second International Conference on Mobile Systems, Applications, and Services, pp. 99–110, New York, NY: ACM Press, 2004.
12. D. Gay, P. Levis, R. von Behren, M. Welsh, E. Brewer, and D. Culler. The nesc language: A holistic approach to networked embedded systems, In Proceedings of the ACM SIGPLAN Conference on Programming Language Design and Implementation, 2003.
13. L. Mottola and G. P. Picco. Logical neighborhoods: A programming abstraction for wireless sensor networks. In Proceedings of the Second International Conference on Distributed Computing in Sensor Systems (DCOSS), Number 4026 in Lecture Notes on Computer Science, pp. 150–167, San Francisco, CA, June 2006.
14. P. Costa, L. Mottola, A. L. Murphy, and G. P. Picco. Programming wireless sensor networks with the teenylime middleware. In Proceedings of the Eighth ACM/IFIP/USENIX International Middleware Conference (Middleware 2007), Newport Beach, CA, November 2007.
15. S. R. Madden, M. J. Franklin, J. M. Hellerstein, and W. Hong, 2005. TinyDB: an acquisitional query processing system for sensor networks. ACM Transactions on Database System 30:1, 2005.
16. R. Newton and M. Welsh. Region streams: functional macroprogramming for sensor networks. In DMSN '04: Proceedings of the First International Workshop on Data Management for Sensor Networks, pp. 78–87, New York, NY: ACM, 2004.
17. R. Newton, G. Morrisett, and M. Welsh. The regiment macroprogramming system. In Proceedings of IPSN, pp. 489–498, New York, NY: ACM, 2007.

18. R. Gummadi, O. Gnawali, and R. Govindan, Macro-programming wireless sensor networks using kairos. In Proceedings of the First International Conference on Distributed Computing in Sensor Systems (DCOSS), 2005.
19. N. Kothari, R. Gummadi, T. Millstein, and R. Govindan. Reliable and efficient programming abstractions for wireless sensor networks. In PLDI '07: Proceedings of the 2007 ACM SIG-PLAN Conference on Programming Language Design and Implementation, pp. 200–210, New York, NY: ACM, 2007.
20. P. Levis and D. Culler. Mate: A tiny virtual machine for sensor networks. In International Conference on Architectural Support for Programming Languages and Operating Systems, San Jose, CA, October 2002.
21. K. Terfloth, G. Wittenburg, and J. Schiller. FACTS – A rule-based middleware architecture for wireless sensor networks. In Proceedings of the First IEEE/ACM International Conference on COMmunication System softWAre and MiddlewaRE (COMSWARE'06), New Delhi, India, January 2006.
22. C.-L. Fok, G.-C. Roman, and C. Lu. Mobile agent middleware for sensor networks: An application case study. In Proceedings of the Fourth International Conference on Information Processing in Sensor Networks (IPSN'05), pp. 382–387, IEEE, 2005.
23. G. Hackmann, C.-L. Fok, G.-C. Roman, and C. Lu. Agimone: Middleware support for seamless integration of sensor and IP networks. In Lecture Notes in Computer Science, vol. 4026, pp. 101–118, 2006.
24. T. Liu and M. Martonosi. Impala: A middleware system for managing autonomic, parallel sensor systems, In ACM SIGPLAN Symposium on Principles and Practice of Parallel Programming, June 2003.
25. P. Juang, H. Oki, Y. Wang, M. Martonosi, L. S. Peh, and D. Rubenstein. Energy-efficient computing for wildlife tracking: design tradeoffs and early experiences with zebranet. In Proceedings of the 10th International Conference on Architectural Support for Programming Languages and Operating Systems, vol. 37, pp. 96–107, New York, NY: ACM, 2002.
26. D. Chu, K. Lin, A. Linares, G. Nguyen, and J. M. Hellerstein. Sdlib: a sensor network data and communications library for rapid and robust application development. In IPSN '06: Proceedings of the Fifth International Conference on Information Processing in Sensor Networks, pp. 432–440, New York, NY: ACM, 2006.

# Chapter 24
# Wireless Mobile Sensor Networks: Protocols and Mobility Strategies

**Jung Hyun Jun, Bin Xie, and Dharma P. Agrawal**

**Abstract** In the last few years, tremendous efforts have been made to enhance the performance of stationary wireless sensor networks (WSNs). However, such improvements are constrained by the limitations of being a stationary network. Recent advances in robotic and the potential usage of naturally moving objects such as vehicle, animal, and even human, enable some of the sensors in the network to be mobile, and such a network is so called a Mobile WSN (MWSN). In this chapter, we study how mobility can improve the network performance such as the network lifetime, coverage, and connectivity. For example, the lifetime of a WSN can be improved by additionally deploying some mobile sensors in the hot spot around the Base Stations (BSs). The coverage is further enhanced by allowing some or all sensors to reposition themselves or move continuously. Furthermore, high connectivity along with coverage is maintained by replacing the broken links or adding extra sensors to reconnect the partitioned networks through the use of mobile relay units. To provide a complete understanding of these aspects, we perform a comprehensive examination of existing approaches in designing a MWSN.

## 24.1 Introduction

A Wireless Sensor Network (WSN) is a collection of spatially distributed sensors with which to collaboratively monitor various environmental changes (e.g., detecting forest fires, sensing the variation of soil temperature, and tracking the movement of intruders). Over the past few years, a number of research efforts have been made to develop sensor hardware, which is small, light weighted, and using ultra low power, to effectively deploy WSNs for a variety of applications. Many of these

J.-H. Jun (✉)
Department of Computer Science, OBR Center of Distributed and Mobile Computing,
University of Cincinnati, Cincinnati, OH 45221 0030, USA
e-mail: junjn@ececs.uc.edu

S. Misra et al. (eds.), *Guide to Wireless Sensor Networks*, Computer Communications
and Networks, DOI: 10.1007/978-1-84882-218-4_24,
© Springer-Verlag London Limited 2009

works focus on how to efficiently improve the network performance from the aspects of network lifetime, connectivity, coverage, detection time, and others. For example, CrossBow, a famous producer of sensor network devices and operating systems like tinyOS, recently developed the hardware called Telos motes, which allow a selectable power state to support sleep scheduling algorithms like SMAC (sensor-MAC (Medium Access Control)) [1] and DMAC (MAC for Data gathering) [2]. Therefore, the network lifetime can be extended by rendering the sensor to sleep or wakeup appropriately. Typically, Telos motes only use $2\,\mu W$ in the sleep mode, which is much lower than in the wakeup state. On the other hand, algorithms like ASCENT [3] and SPAN [4] have been developed to effectively maintain the network connectivity by adaptively reconfiguring the network topology. Dhillon [5] develops an optimal sensor placement algorithm to maximize the coverage using a grid topology. On the other hand, Coverage Configuration Protocol (CCP) [6] aims to provide different degrees of coverage desired by applications. To support these protocols, node localization and time synchronization [7, 8] have to be carefully designed.

Most of these works consider a WSN to be static, and we call this as a Stationary Wireless Sensor Network (SWSN). However, many researchers have realized that the network performance of a WSN is restricted by the nature of stationary and are envisioning the design of Mobile WSN (MWSN) to improve the WSN performance by using mobile entities. The MWSN technology can be practical and efficient due to recent advance in robotic technology [9, 10] or realizing the potential usage of naturally moving objects such as vehicles, animals, and even humans. Intel Corp. is also involved in developing mobile entities for sensors and advocates [11] the sensor network that could employ the mobile robot as the base station (BS), relay nodes, or sensors. For example, static sensors could be used to monitor the microclimate while the mobile robot could act as a mobile BS to collect the information from these static sensors. For this purpose, the BS has the ability of moving around the vineyard in a way to directly collect the sensed information from the static sensors in an order, which significantly reduces the number of multihop transmission and enhances the sensors' lifetime. CarTel project [12], developed by MIT (Massachusetts Institute of Technology) under the supervision of Professors Balakrishnon and Madden, is investigating a mobile distributed sensor system where sensors are located on the automobiles. In this system, a mobile sensor is designed to cover a larger area than a static sensor. The mobile sensor can also act as a dataMule, which is capable of collecting a volume of data from static sensors in a sequence, and later delivers them to the administrative center once it is available. The large-scale WSN like Cane-toad monitoring at the Kakadu national park in Australia [13] is vulnerable to network partitioning due to natural disasters like forest fires. To address this problem, the WSN is placed with a small set of mobile sensors which can used to reconnect the partitioned network.

In this chapter, we focus on the design of a MWSN and show how mobility can improve the network performance. In a MWSN, the mobile entity could be acted as a mobile BS, mobile sensor, mobile relay, or mobile clusterhead, according to the desirable role. We further categorize these mobile units into three types, depending on whether their trajectories are deterministic, predictable, or unpredictable. To provide a complete understanding of MWSN, we explore MWSN algorithms that have been developed to improve WSN lifetime, coverage, and connectivity. For completeness of the text, we also study the existing protocols for sensor mobility and data collection in a WSN. For example, the Joint Mobility and Routing Strategy is an algorithm based on the use of a controllable mobile BS. On the other hand, the Mobile Relay Strategy uses controllable mobile relay units to prolong the lifetime of a WSN. By reducing their workload of the bottlenecked sensors, in contrast, the Predictable Observer Strategy improves the network lifetime by using predictable but uncontrollable mobile units. We further look at the MWSN design for achieving enhanced coverage and connectivity.

The remainder of this chapter is structured as follows. Section 24.2 describes the background of MWSN. In Sect. 24.3, we identify mobile entities in MWSN and categorize them into three groups according to their mobility patterns. In Sect. 24.4, we present the most well-known strategies that improve the network lifetime in MWSN. In Sect. 24.5, we elaborate the deployment technique of mobile sensors that are designed to improve the network coverage. Section 24.6 further studies the technique to improve the network connectivity by using small set of relay nodes. Finally, the future direction and concluding remarks are included in Sects. 24.7–24.9.

## 24.2  Backgrounds

We first give the background of the WSN, its system architecture, the roles of each component in the system, and some key performance issues.

A WSN system [14] typically consists of spatially distributed sensors and the BS as shown in Fig. 24.1. Sensors are denoted by a set $N = \{0, 1, \ldots, n\}$ and $\lambda$ represents the node density. As shown in Fig. 24.1, nodes $X$ and $Y$ are two sensors deployed in the WSN while B is the BS. The network domain is covered by a large number of sensors, equipped with various components that perform the functionalities such as data sensing, computation, and networking, etc. Therefore, the sensor has the autonomous capabilities of sensing the environment, processing the sensed data, and wirelessly transmitting the data to its intended destination on the network. A WSN has one or a small set of BSs which act as the gateway to the administrative center. The BS is usually equipped with a high-gain antenna, a large data storage space, and a rich energy supply, and thus the BS can perform some sophisticated functionalities such as data computing, result analysis, and storing the data.

The sensors are generally equipped with limited power and deployed randomly in a given area, e.g., by dropping from airplanes, helicopters, or unmanned aerial or ground vehicle. A WSN may be expected to be deployed in a harsh or hostile environment which renders the manual deployment to be impractical. To monitor the environmental changes, the sensor needs to sense the environment and collaboratively relay the data to the BS through single- or multihop transmissions. Due to

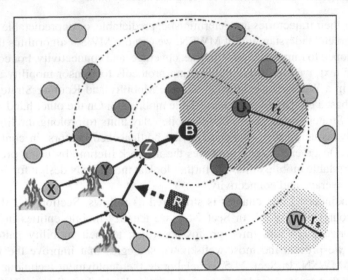

**Fig. 24.1** A WSN for forest fire monitoring application. BS B is located at the center of the network and each annulus represents the maximum transmission range for the sensors belonging to that annulus

limited power of a sensor, networking protocols have to be designed in such a way to consume less power used for sensing and transmission. For the same purpose, the BS should be located at the optimal location that conserves the sensor energy. It is because of the sending data of sensors is typically forwarded to the BS and the BS may disseminate control messages or a query to the network. Figure 24.1 shows a WSN that sensors are uniformly distributed and the BS is situated at the center of the network domain. In order to facilitate the analysis, the network can be divided into several annuli based on the maximum sensor transmission range, e.g., $r_t$ as shown in Fig. 24.1. Suppose the annulus is indexed in an increasing order (1 to $k$) from the center of network, the sensor in the $i$th annulus at least requires $i$ hop(s) to reach the BS. For example, when an environmental phenomenon like forest fire is sensed by node $X$ and $Y$, the alert messages of them have to be forwarded to the BS in a multihop fashion through node $Z$. Due to these randomly deployed sensors with scarce resources and multihop wireless communication, the designing of WSN protocols is very complicated and some of the critical design aspects include network lifetime, coverage, and connectivity as follows:

- *Network lifetime.* Network must remain operational for a long period of time as possible. The design of an environmental monitoring WSN should target to maintain a satisfactory sensing coverage and collect the required sensing data (e.g., the forest fire) to the BS, ranging from months to years.
- *Network coverage.* The network domain has to be well covered by sensors since it is unpredictable where an event occurs in the concerning area.
- *Network connectivity.* All sensors in the WSN must be connected so that the sensing data (e.g., an alert message) at any sensor can be wirelessly transmitted

to the BS once an event is sensed. Therefore, the network connectivity is needed to ensure that sensors can communicate with each other either for collaboration or for reporting the sensed data to the BS. For example, a process of data aggregation may be locally performed among sensors, and then the compressed data is reported to the administrative center (i.e., BS). If the network connectivity is not assured, it is meaningless, even a sensor detects the presence of an event.

In the following, we further illustrate the necessary background with respect to the network lifetime, coverage, and connectivity.

### 24.2.1   Network Lifetime

There are few variations in the meaning of the network lifetime of a WSN and is primarily due to application specific nature and dynamically changing of network topology. For example, the lifetime of a WSN can be defined as the time for the first sensor to die in data mining application where sensors are sparsely deployed [15]. In such a sparse network, the loss of one sensor could disconnect a large portion of the network. Conversely, the lifetime of a WSN can be defined as the time for a certain percentage of sensors to die in a densely deployed WSN. This is because the surviving sensors may still remain connected and the network functionality can be still performed well by them. Alternatively, the network lifetime can also be defined in terms of the packet delivery rate or the number of flows alive so that the quality of communication in the WSN could achieve. However, above definitions do not impair the network functionality which is important from the application perspective. Bhardwaj et al. [16] defines the network lifetime as the time for the first loss of coverage. The limitation of this definition is that it does not take the network connectivity into account. The most general definition of lifetime is defined by Blough and Santi [17] as follows. Given $G(t) = (V(t), E(t))$ is the communication graph of the sensor network at time $t$, and $V(t)$ is the set of living nodes at time $t$. Assume that $G(0)$ is connected and $V(0)$ covers the $d$-dimensional region $R = [0, l]^d$ at $t = 0$, where $l$ is the side length of the network domain, it denotes $n(t)$ as the cardinality of $V(t)$ with $n = n(0)$. The network lifetime therefore is defined as the minimum time among $t_1$, $t_2$, and $t_3$, where $t_1$ is the time it takes for the cardinality of the largest connected components of $G(t)$ to drop below $c_1 \cdot n(t)$, $t_2$ is the time it takes for $n(t)$ to drop below $c_2 \cdot n$, $t_3$ is the time it takes for the volume covered to drop below $c_3 \cdot l^d$, and $0 < c_1, c_2, c_3 \leq 1$. This definition can be reduced to most of the existing definitions by appropriately choosing the values for $c_1$, $c_2$, and $c_3$. For example, setting $c_1 = 0$ and $c_2 = 1$, it corresponds to the lifetime as the time of first sensor to die.

In this chapter, we separately deal with the coverage issue from the lifetime. Thus, the lifetime is defined as the time for the first sensor to die. This is a weak definition for densely deployed network. This is because the loss of a sensor quickly adds extra loads to its neighboring sensors. Therefore, they are also likely to die out very soon and this effect spreads like a chain reaction. Consequently, this definition could serve as an indication of the near end of the sensor lifetime in the dense network.

Under this definition, the lifetime of a SWSN is bounded by sensors closer to the BS as they have to forward more data packets from or to the BS. For example in Fig. 24.1, sensor node $Z$ is receiving at least three times of packets from the other sensors and forward them to the BS than the node in the outmost annulus. Therefore, no matter how optimal routing protocol is, such a limitation will always be present in a SWSN [18]. We often identify such a set of nodes as the hotspot.

### 24.2.2 Network Coverage

The network domain can be partitioned into two regions in terms of coverage: covered and uncovered. The covered region means that any point in the region is within the coverage of at least one sensor. The uncovered region is the complement of the covered region. Every sensor has a sensing range and the union of the sensing ranges of all sensors is called the network coverage. Let us consider an ideal sensor model of a sensor that is a circular sensing range represented by the sensing radius $r_s$. For example, the sensing range of node $W$ in Fig. 24.1 is shown as the shaded region centered at the node $W$. If an event occurs within the sensing range, it can be detected by the sensor. Equivalently, an event can be sensed if it has a sensor within the distance of $r_s$ away from it. For example, the network shown in Fig. 24.1 detects the fire since it is within the distance $r_s$ from node $X$ such that the heat and light radiating from the forest fire is strong enough to be sensed by sensor $X$. Let us consider a randomly deployed SWSN that the locations of sensors are uniformly and independently distributed in the region $R$ (i.e., network domain). The distribution of sensors can be modeled by a stationary two-dimensional Poisson point process. Denote the density of the underlying Poisson point process as $\lambda$. The number of sensors located in a region $R$, $N(R)$, follows a Poisson distribution of parameter $\lambda \| R \|$, where $\| R \|$ represent area of the region. The probability that $k$ number of sensors is located in $R$ is given by (24.1):

$$P(N(R) = k) = \frac{e^{-\lambda \| R \|} (\lambda \| R \|)^k}{k!}. \tag{24.1}$$

Suppose each sensor covers a disk of radius $r_s$, then the initial configuration of the sensor network can be described by a Poisson Boolean model $B(\lambda, r_s)$. The fraction of the geographical area covered by at least one sensor can be given by (24.2):

$$C(\lambda, r_s) = 1 - e^{-\lambda \pi r_s^2}. \tag{24.2}$$

It shows that the network coverage is determined by the initial network configuration: the sensing range $r_s$ and the sensor density $\lambda$. If we deploy more sensors or if we use the sensor which can sense a larger area, this will results in a better coverage.

However, the problem is that a dense deployment is not always feasible due to high cost (even though each sensor could be cheap, deploying a large number of them is still expensive) so that the random deployment may not guarantee a full coverage. It is also noted that the coverage holes may be created by the failure of sensors.

### 24.2.3 Network Connectivity

Every deployed sensor ideally has a wireless transmission range $r_t$ by operating on the shared wireless channel. The network topology can be modeled as a graph $G = G(V, E)$, where $V$ is set of vertices representing sensors and the BS and $E$ is a set of edges representing the communication links. For example in Fig. 24.1, node $U$ has the edge to the BS if the Euclidean distance between $U$ and the BS is less than the maximum transmission radius $r_t$. We ignore asynchronous communication between any pair of nodes and represent link to be bidirectional. It means that if node $U$ has a wireless link to the BS, this link can be used for communication from the BS to $U$. A network graph $G$ is connected if and only if there is at least one path between every pair of vertices in $V$, and otherwise it is disconnected. In other words, the network is disconnected if it contains one or more isolated nodes so that the network graph $G$ is divided into two or more disjoined graphs. The nonexistence of isolated nodes is consequently a necessary but not the sufficient condition for network connectivity. In a randomly deployed WSN, this existence of isolated nodes could occur frequently. This is based on the phenomenon in the random graph process, in which links between nodes are added uniformly at random [19], the network gets connected at the moment when we add the link connecting the last isolated node with high probability. The probability that a WSN with $n \gg 1$ sensors and node density $\lambda$ is connected, is

$$p(G \text{ is connected}) = \left(1 - \sum_{N=0}^{n_0-1} \frac{(\lambda \pi r_t^2)^N}{N!} e^{-\lambda \pi r_t^2}\right)^n \qquad (24.3)$$

with high probability, where $n_0$ is the minimum node degree in network graph $G$ [20]. For example, if $n_0$ is 2, then every node $v \in V$ has at least two neighbors. We refer to it as a network with at least 2-connected with high probability. Clearly, the network has high resistance to the link failure when $n_0$ is high. Also (24.3) shows that network is likely be connected if more sensors are deployed and transmission range $r_t$ is high.

## 24.3 Functionalities and Mobility in the MWSN

In a MWSN, there is at least one mobile entity and the remaining sensors are static. According to the design objective, the mobile entities are able to communicate with its neighboring sensors if required. In addition, if there are multiple mobile entities, they are capable of forming a local network like a MANET, which is self-configurable, adaptive to the changing environment, robust, and scalable. In accordance with the role performed, the mobile entities can either be mobile BS, which acts as the data sink of network, or mobile sensors that sense the environmental changes or serves as data relaying nodes. The mobility patterns of the mobile BS or mobile sensors could be used to boost the network performance such as network lifetime. In addition, a mobile entity could be a relaying node or a clusterhead in the network, depending on the deployment strategy, network architecture, and application. In Fig. 24.1, it is important for fire fighters to know the information about the fire such as the spreading direction of the fire, the area of the fire, current oxygen level, and the temperature. Therefore, the maintenance of continuous data flow from sensors to the BS is a key requirement in successfully monitoring the fire. However, the WSN may be partitioned due to hardware failure of sensors, resulting in disconnection of the flow of some sensors (e.g., nodes $Y$ and $Z$ in Fig. 24.1) to the BS. On the contrary, if a mobile relay node $R$ can be deployed to inherit the role of sensor nodes such as $Y$ and $Z$ in Fig. 24.1, the network connectivity can be continuously maintained. Before illustrating the detail of mobility and deploying strategies, we describe the underlying functionalities of these mobile entities followed by their mobility trajectories.

- *Mobile BSs* [18, 21, 22]. The basic role of the BS is to collect the data generated from various sensors. Additionally the mobile BS mounted on the mobile unit can effectively enhance the lifetime by periodically or continuously changing its locations according to a predefined strategy. The increase of lifetime is due to two reasons. At first, there are no fixed set of sensors close to the BS when the BS is moving. This helps to disperse the bottleneck sensors around the network which again evenly dissipates energy. Second, the number of transmission hop from the sensor to the BS could be reduced with an efficient data transmission scheduling. For example, the sensor in a delay-tolerant application can transmit its collected data to the mobile BS until the mobile BS has moved to its direct transmission range. Let us suppose node $R$ in Fig. 24.1 is a mobile BS and it is periodically circulating around the second annulus where node $Y$ is located. Therefore, it is possible to design a schedule that allows node $Y$ to send its packets to node $R$ when it is close by, and thus the long transmission path can be avoided.
- *Mobile sensors*. The basic functionality of a sensor is to sense the environment and transmit its data to the BS periodically. On the other hand, the mobile sensor can be used for increasing the efficiency of data relaying and controlling the network topology. Mobile sensors can be distributed in the network domain with minimum human intervention like other stationary sensors. In a SWSN, the coverage and connectivity are fixed once the deployment stage is performed. On the other hand, the mobile sensor could be used to form an ideal topology which could improve the coverage and connectivity, or release the relaying load for some bottleneck nodes.
- *Mobile relaying nodes* [23–25]. Mobile relaying nodes are special mobile sensors designed for releasing the relaying load of some sensors in the network. The mobile relaying node is able to move in a way to serve as a substitute relaying

node for the sensors. For example, they could be Mobile Ubiquitous LAN Extensions (MULEs) [24] or Message Ferry [23] which are mobile devices specifically designed to roam around to collect data from nearby sensors and deliver them to the BS.

- *Mobile cluster heads* [26]. A mobile clusterhead can be one of the mobile sensors through an election process or a special node placed manually. They form clusters in the network and forward the information collected within their own cluster to the BS. Unlike stationary clusterheads, mobile clusterheads can increase the energy efficiency and intelligently form the cluster topology adaptively according to the environment or changes in the network mission.

These mobile units can be introduced naturally or placed artificially. The mobility pattern of each mobile entity is typically determined based on specific application and the WSN size. Based on mobility patterns, mobile units can be categorized into three types:

- *Controllable mobile units.* They follow some predefined trajectories such as mobile robots that the network planner can program them to meet the requirements. An example of this is a TagBot of Fig. 24.2a [9], which is designed by the Carnegie Mellon University. TagBot is an advanced robot that can communicate with sensor motes like MicaZ or Telos. It can move both forward and backward, and turn in any direction by a controlling program, which is resided in an Intel's board. Three Robomotes [27] of Fig. 24.2b are small in size. Robomote in Fig. 24.2c is equipped with a solar panel for recharging the battery and move controllably as designed by a programmer.
- *Uncontrollable and unpredictable mobile units.* They move in a random fashion such that the next movement cannot be predicted. For example, the movement of an animal or human, which carries a sensor, is generally considered in this category. For example, if the sensor is mounted on an elephant in Africa in finding its group behavior, the mobility of sensor is random as the elephant moves.
- *Uncontrollable but predictable mobile units.* They are like bus or train that move according to a predefined schedule. Therefore, the movement of the sensor carried in the bus or train is usually not random and follows a predetermined path.

**Fig. 24.2** (a) TagBot, (b) Robomotes, (c) Robomote with solar panel for recharging the battery

However, they cannot be controlled by the sensor itself. For example, the movement path of a bus to collect sensor data may not be the best routine for WSN performance.

As a matter of fact, the mobility and the deployment design of a MWSN is a complex problem that involves design requirements, mobility capacity of mobile sensors, network environment, and application constraints such as delay requirements. For example, real-time requirement renders full connectivity from sensors to the mobile BS. According to these design constraints, mobility strategy, collaborative scheme, data packet, and routing schedule should be carefully addressed in terms of network performance. In the following sections, we investigate some approaches that improve the network lifetime, coverage, and connectivity by using mobile entities.

## 24.4  Network Lifetime Enhancement in the MWSN

There are three most recognized algorithms proposed to enhance the lifetime of WSN using mobile units: Joint Mobility and Routing Strategy [18], Predictable Observer Strategy [23], and Mobile Relay [25]. The first two algorithms use a mobile BS to improve the network lifetime. The former uses a controllable mobile unit while the later uses an uncontrollable but predictable mobile unit. The third algorithm also uses controllable mobile units but it differs from the first two since the mobile units are relay entities instead of a BS.

### 24.4.1  Joint Mobility and Routing Strategy

Since the number of BSs is generally smaller as compared to the number of sensors in the network, using a few mobile BSs instead of many mobile nodes is one of the most efficient ways to improve the lifetime. Joint Mobility and Routing Strategy is a mobility strategy for the BS to maximize the network lifetime while the data traffic is continuously flowing from every part of the network to the mobile BS. Due to mobility of the BS, it has to design a multihop routing protocol that enables the static sensor in the network to forward its data to the BS wherever it moves. As shown in Fig. 24.3, the multihop path is determined by the relative location of the BS and the sensor. Mobile BS B moves in a circle while the data traffic is delivered to the BS from different parts of the network. Node $C$ in the inner circle of radius $R_m$ transmits the data using the shortest path. On the other hand, node $A$ uses a round route toward BS. The latency is not significantly affected compared to stationary BS since the connection to the BS can be immediately established if they have packets to send.

The basic idea of using Joint Mobility and Routing Strategy is that the mobile BS periodically circulates around the network and broadcasts its mobility pattern.

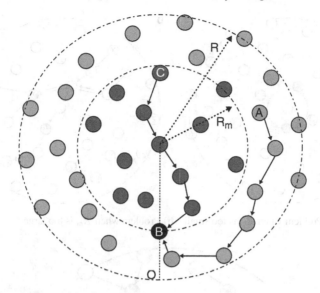

**Fig. 24.3**  Joint mobility and routing strategy

In the beginning, sensors are aware of their own locations by a localization scheme. They again learn the mobility pattern of the BS so that they know where the BS is at a given time. Therefore, the data packet is forwarded without occurring heavy overhead such as computing for the location of the mobile BS and the route toward the BS.

For example, the BS in Fig. 24.3 is moving on the circle of radius $R_m$. Sensor $C$, which is situated inside of the inner circle of radius $R_m$, transmits its data through the shortest path to BS as shown in Fig. 24.3. Whereas, sensor $A$ in the same annulus as the BS performs two-step routing: the path first circles around the center $O$ until it reaches line $OB$, and then it follows a short path to the BS. The direction of the round routing is decided by the location of a sensor: clockwise on one side of the diameter $OB$ and counterclockwise on the other side. For example, sensor $A$ has a clockwise round route because it is the direction where the round route is closer to BS. The heuristic behind this routing strategy is to reduce the size of the sub-network within the inner circle which uses the shortest path. However, the BS being too close to the center-point of the network is still inefficient since the hot spot problem is still there as shown in Fig. 24.4a. Also being too far from the center-point is also not desirable in terms of the network performance. This is because a large fraction of the data traffic is going through the center of network which creates a hot spot near the center-point. As shown in Fig. 24.4b, node $D$ is utilized three times as the relaying node when data traffic is flowing from $X$ to $B$ at time $t_0$, $Y$ to $B$ at $t_1$, and $Z$ to $B$ at $t_2$. Therefore, it is critical to choose right $R_m$ which provides the best energy efficiency: an efficient rule of thumb for choosing $R_m = 9R/10$ as studied by Luo and Hubaux [18], where $R$ is the radius of the network domain.

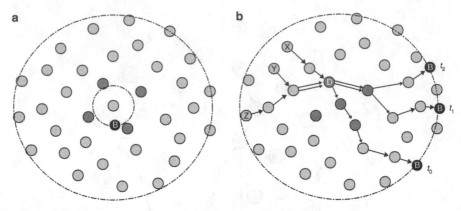

**Fig. 24.4** (**a**) Problem when $R_m$ is too small; (**b**) Problem when $R_m$ is too large

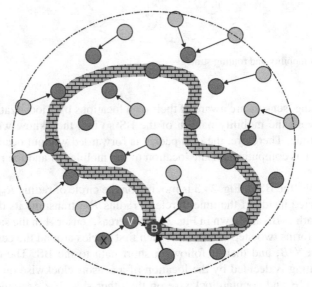

**Fig. 24.5** Sensor network with predictable but uncontrollable BS B

## 24.4.2 Predictable Observer Strategy

The Predictable Observer Strategy involves a data forward scheduling for improving the network lifetime. Rather than using a controllable MS, this approach takes advantage of a mobile BS whose movement is predictable but uncontrollable by the sensor network. For example, the mobile BS, denoted by B as shown in Fig. 24.5, is mounted on the bus traveling on the path with brick pattern. The dark gray circles represent the nodes within the vicinity of the BS. In Fig. 24.5, sensor node $X$ which is not in the vicinity of the BS first forward its data to the closest node $V$ which may be in the direction transmission range of the mobile BS. Node $V$ will forward the received data including its own data to the mobile BS when the mobile BS has moved close to it. Therefore, the long multihop route is no longer necessary in this approach and the energy used for packet transmission is accordingly conserved. In this approach, the observer is referred to the BS since it roams around the network, looking for some events like data transmission.

In this approach, due to uncontrollable mobility of the BS, it is critical to schedule the transmission from each sensor such that it can forward the packet to the observer in a cost-efficient manner. For example, the sensor out of the direct transmission range of the observer can complete its data transmission to a close by sensor that is directly connected to observer before observer moves away. For this purpose, Predictable Observer Strategy should have a correct setting of transmission power to minimize the outage and a transmission schedule of each sensor to minimize the collision from the concurrent transmissions. The outage is defined as a scenario that the observer moves out of the transmission range of the sensor before time $T$ which is the time required to complete sensor's data transmission to the observer.

There are certain conditions in selecting vicinity nodes that must be satisfied in order to reduce the outage. Let us consider a perpendicular distance $D$ from the closest tangent point on the path of the observer. $R$ is the maximum transmission radius of each sensor and the observer. For successful communication between the sensor and the observer, it is not enough for sensors to be within distance $R$ from the observer. As shown in Fig. 24.6a, for example, node $S$ only communicates with observer $B$ for a short period of time which is not sufficient to complete its data transmission. If they require time $T$ or more for completing their transmission, the relationship between $R$ and $D$ must satisfy (24.4) that assumes a straight-line path:

$$R \geq \sqrt{D^2 + (vT/2)^2}, \tag{24.4}$$

where $v$ is the velocity of the observer that is assumed to be a constant. Otherwise, it should be the maximum velocity of the observer for guaranteeing the needed transmission time.

In a randomly deployed WSN, there is a possibility that several nodes being located closely. If these sensors send their data to the observer at same time it may result in packets collision at the BS. One possible solution used in SWSN is to allow each sensor to listen to the channel and transmit when the channel is idle or backup otherwise. However, this can reduce network lifetime since listening to channel also consumes energy. Thus, an alternative protocol is designed in such

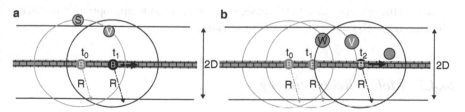

**Fig. 24.6** (a) An example of an outage occurrences at node $S$. (b) An example of the case where node $V$ needs to wait until $W$ finishes its transmission which require time $T = t_1 - t_0$

a way that individual sensor could maximally save its energy. Alternatively, the medium access control, resolving contention, dealing with collisions, and various kinds of failures will be handled by the observer rather than the sensor. For instance, the observer listens to one of the sensors at a given time and the rest of them must wait until the current transmission of a sensor completes. For example, as shown as Fig. 24.6b, the BS is in vicinity of sensor $V$ at time $t_1$ so if $V$ is unaware of transmission between sensors $W$ to the BS at $t_1$ it start its transmission to BS and it will result in packet collision. In order to avoid this problem, it is important for the BS to schedule $V$ to start its transmission after sensor $W$ has finished its transmission at time $t_2$. On the other hand, any node cannot just wait for ever, and thus it defines a maximum waiting time $t$ for each sensor. The maximum waiting time is a function of distance $D$ from the path of observer which is given by (24.5):

$$t = \frac{2\left(\sqrt{R^2 - D^2}\right) - vT}{v}. \tag{24.5}$$

Equation (24.5) indicates the waiting time required for the sensor to begin the communication with the observer. If the communication starts after $t$, it will be impossible for the sensor to send all its packets to the observer and it will result in an outage.

In the Predictable Observer Strategy, a communication protocol is designed to ascertain this waiting time and it has three phases – *startup, steady, and failure.*

1. *Startup phase.* The observer broadcast its beacon messages while it is moving on the predefined path. The beacon messages are collected by the sensor to estimate the observer cycle and the duration of the observer staying in its range. When the estimation is finished, the sensor reports the result back to the observer which decides the priority among multiple sensors waiting to transfer data during the next steady phase.
2. *Steady phase.* The observer initiates the communication by sending a wakeup signal to the sensor that it knows to be within the transmission range. When there are several sensors within this range, it assigns a higher priority to the sensor that has a smaller maximum waiting time. Each sensor can predict when the observer is likely to be in its transmission range by using the information collected from the startup phase. To do this efficiently, they monitor the channel only when the observer is expected to be nearby.
3. *Failure detection.* The observer can detect node failures if a sensor is unable to respond the wakeup calls. In this case, the observer has to appropriately reschedule the transmission of the remaining sensors.

### 24.4.3  Mobile Relay Strategy

Mobile Relay is another technique used in a MWSN to prolong the lifetime of WSN that uses the controllable mobile relaying nodes. This approach is developed for the

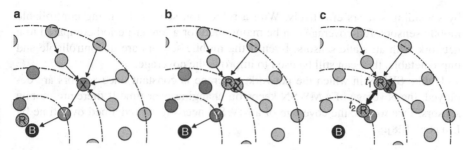

**Fig. 24.7** A mobile relay node (i.e., $R$). (**a**) The initial state of the network. (**b**) A relay node inherits a bottleneck node $X$. (**c**) Both node $X$ and $Y$ become bottleneck nodes

scenario where the BS and sensors are stationary. The capabilities of mobile relay nodes are identical to usual mobile sensors, but they have enough power to move. The mobile relay node moves and inherits the responsibilities of any bottleneck sensors for a while or permanently. The responsibilities include sensing environment, processing data, and forwarding the sensed or received data to the BS.

Figure 24.7 shows a strategy how a mobile relay node replaces the bottleneck sensor. The mobile relay node (i.e., $R$ in Fig. 24.7) initially stays around the BS as shown in Fig. 24.7a. When the BS receives the notification from a sensor (i.e., $X$ in Fig. 24.7) that it is about to die, the BS requests the relay node to inherit the roles of $X$ as shown in Fig. 24.7b. It can be seen that the use of the mobile relay node is similar to the scenario that a static redundant sensor has awakened by the request of node $X$. The redundant sensors are deployed around the network and sleep until it is requested by active sensors. In addition, the use of relay node can be more beneficial if a relay node can take the roles of two or more bottleneck nodes. Considering the scenario in Fig. 24.7c, nodes $X$ and $Y$ are both bottleneck nodes. In this case, a single redundant sensor would not be enough to handle two bottleneck nodes. On the contrary, the mobile relay node can perform the role of these two sensors (e.g., $X$ and $Y$ in Fig. 24.7) as follows. At first, the relay node performs the functionalities of node $X$ at time $t_1$, and then moves to the location where node $Y$ is. Therefore, the relay node take over the role of node $Y$ at time $t_2$ and hereafter returns back to node $X$ again.

With this simple example, we can see that one mobile relay node could double the network lifetime with a proper scheduling for the mobile relay node. In fact, the lifetime of WSN could be improved up to four times with a single mobile relay node as studied by Wang et al. [25].

## 24.5  Network Coverage Improvement

In MWSN, the coverage is not only determined by the initial network configuration but also depends on the mobility behavior of sensors. With some mobile sensors that have controllable mobile units, the network coverage can be significantly improved by spreading sensors effectively. With a relocation scheme by using controllable mobile sensors, full coverage can be maintained for a longer period compared to a network with all static sensors. Even if the mobile sensors are uncontrollable and unpredictable, they can still be used to improve the coverage.

Unlike SWSN in which the network coverage is constant once sensors are deployed, the coverage of a MWSN keeps on changing over time if there are mobile sensors. Due to this, the coverage of a MWSN needs to be redefined over time by Liu et al. [28].

**Definition 1.** *Area coverage*: The area coverage of a sensor network at time $t$, $f_a(t)$, is the fraction of the geographical area covered by one or more sensors at time $t$.

**Definition 2.** *Area coverage over a time interval*: The area coverage of a sensor network during time interval $[0, t]$, $f_i(t)$, is the fraction of the geographical area covered by at least one sensor within the time interval.

In the following part, we first look at two deployment strategies that improve the area coverage by using the controllable sensor units and later study the strategy that improve the area coverage over a time interval by using uncontrollable and unpredictable mobile sensor units:

- *Deployment using potential fields* [29]. In this approach, mobile sensors deploy themselves autonomously to enhance the area coverage. To achieve this purpose, the movement of the sensor in the robotic is controlled by potential fields to evenly distribute the senor in a domain while avoiding the obstacles.
- *Movement assisted deployment* [30]. The coverage hole is the place without the cover of any sensor. In this strategy, Voronoi polygon is used to detect the coverage hole locally and remove the coverage hole by moving a sensor this uncovered area.
- *Random mobility strategy*. This strategy takes advantage of random mobility that every sensor continuously moves in a random fashion so that the area coverage is improved over a time interval.

### 24.5.1 Deployment Using Potential Fields

The purpose of the sensor deployment is to place the sensors in the network such that the area "covered" by sensors is maximized and all sensors are connected to the network. To achieve this, a potential-field-based approach is proposed by Howard et al. [29] that considers every sensor has the locomotion capability. The basic idea of this approach is that the fields are constructed such that each node is repelled by both obstacles and other sensors. Therefore, it automatically forces the sensor to spread itself throughout the network domain. In other words, sensors and obstacles (e.g., walls or tables in office environment) have the potential field which repels each other, until it comes to the state of static equilibrium where all the sensors are stationary. The equilibrium state has the characteristics of spreading the sensors evenly around the region and achieving the maximum area coverage with a connected WSN. Assuming each sensor node has holonomic drive mechanisms (i.e., it can move equally well in any direction), the deployment steps using the potential field include:

1. *Determine the force*. The force is defined as the gradient of a scalar potential field by (24.6) and (24.7).
2. *Estimate the trajectory*. The node trajectory of movement is estimated by (24.8) using the force, and gets updated by (24.9).
3. *Reaching static equilibrium*. Repeat steps (1) and (2) until the network reaches a state of equilibrium.

### 24.5.1.1  Determine the Force

Each sensor is subject to a force $F$. The force $F$ is divided into two parts $F_o$ and $F_n$, where $F = F_o + F_n$, $F_o$ represents the exerted force due to obstacles, and $F_n$ is the exerted force due to sensors. Let $x$ denotes the position of the sensor and $x_i$ denotes the position of other objects (an object can be either a sensor or an obstacle). These forces could be modeled by using their relative distance $\widehat{r}_i = x_i - x$ and Euclidean distance $r_i = |x_i - x|$ from the sensor to object $i = \{0, 1, \ldots, u\}$ as described in (24.6) and (24.7):

$$F_o = -k_o \sum_{\forall i} \frac{1}{r_i^2} \frac{\widehat{r}_i}{r_i},  \tag{24.6}$$

$$F_n = -k_n \sum_{\forall i} \frac{1}{r_i^2} \frac{\widehat{r}_i}{r_i},  \tag{24.7}$$

where $k_o$ and $k_n$ are constants describing the strength of each type of fields and there are $(u + 1)$ objects.

### 24.5.1.2  Estimate Trajectory

Based on the force generated from the potential, each node determines its trajectory at every iteration until the network reaches the state of equilibrium. The trajectory could be determined by this algorithmic expression of (24.8) and (24.9).

$$\Delta v \longleftarrow \frac{F - \upsilon v}{m} \Delta t.  \tag{24.8}$$

The right side of (24.8) is an approximation of equation of motion, which will be assigned to variable $\Delta v$ in (24.8). In (24.8), $v$ is the velocity of a node, $\Delta v$ is the change in velocity between time $t$ to $t + \Delta t$, $\upsilon$ is a viscous coefficient, and $m$ is the mass of a sensor node. Then, the velocity $v$ gets updated by using (24.9).

$$v \longleftarrow v + \Delta v.  \tag{24.9}$$

On the other hand, $\Delta v$ and $v$ is constrained by the maximum acceleration and maximum velocity of the mobile sensor node. If they exceed the constraint, they will be clipped to their maximum values accordingly. Also when $v$ is small (close to zero velocity), viscous friction tend to produce oscillation rather than asymptotic convergence to zero velocity. This kind of behavior is typical in the discrete control system and can be eliminated by introducing a velocity "dead-band." Dead-band means a band where no action occurs (i.e., velocity is zero).

### 24.5.1.3   Reaching Static Equilibrium

Where all sensors are stationary, the state of equilibrium is defined as the condition where the movement based on the potential field will cease. Harward et al. [30] consider the network as a whole, and is able to reach a static equilibrium under the assumption that environment itself is static or is changing periodically or inter-mittently. Environment itself is static means that (1) there is no additional energy introduced into the system and (2) the network space is not changed such that previously unreachable place become reachable or vice-versa. For example, the en-vironment is changed if energy is added to or subtracted from the system, meaning that objects (i.e., sensors or obstacles) are moved by some agencies other than the network itself. Furthermore, a new place becomes reachable by mobile sensors, e.g., one of previously closed doors might be opened in an office environment.

## 24.5.2   Movement-Assisted Deployment

In the Movement-Assisted Deployment approach, the WSN network domain is di-vided into a set of Voronoi polygons. Each sensor is enclosed by its polygon as shown in Fig. 24.8. The Voronoi polygon has a unique characteristic that any point

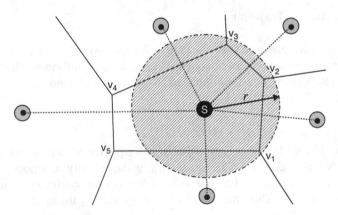

**Fig. 24.8** A Voronoi polygon enclosing mobile sensor node $S$ in the network. *Shaded circle* rep-resents the sensing range of node $S$

in the polygon is closer to the sensor enclosed by it than any other sensors. This allows each sensor to examine the coverage hole locally and monitor a specific area around it. To achieve a global maximum coverage, the movement-assisted deployment employs a protocol as following:

1. *Detecting coverage hole.* Each sensor constructs Voronoi polygon and examines the coverage hole locally.
2. *Improving local coverage.* If there is a coverage hole, it moves to a better location within its Voronoi polygon so as to improve its local coverage.
3. Steps (1) and (2) are repeated until there is no coverage hole or no further improvement with respect to network coverage in the previous iteration.

### 24.5.2.1  Detecting Coverage Hole

The Voronoi Polygon is constructed by a procedure of exchanging location information among neighboring sensors. A Voronoi polygon can be defined by a set of vertices and edges as shown in Fig. 24.8. For example, the Voronoi polygon for sensor $S$ can be modeled as $G_p(S) = (V_p(S), E_p(S))$, where $V_p(S) = \{v_1, v_2, v_3, v_4, v_5\}$ as shown in Fig. 24.8, representing the set of Voronoi vertices, and $E_p(S)$ is the set of Voronoi edges and each edge connects a pair of Voronoi vertices.

Once the location of each vertex $v_i \in V_p(S)$ is identified by sensor $S$, it compares the physical distance $d(S, v_i)$ and the sensor sensing range $r$. In other words, it checks $d(S, v_i) < r$, for all $v_i \in V_p(S)$, which enables sensor $S$ to determine where exists coverage hole, e.g., the place near $v_4$ and $v_5$ as shown in Fig. 24.8.

### 24.5.2.2  Improving Local Coverage

The coverage hole is fixed by moving close to the farthest Voronoi vertex. At the same time, the movement should be controlled in a manner to avoid a situation that the vertex, which was originally covered, becomes the farthest vertex. For this purpose, Wang et al. [30] have developed a *Minimax algorithm* that chooses the targeted moving location as the point inside the Voronoi polygon whose distance from the sensor to the farthest Voronoi vertex is minimized. The circumcircle is a triangle's circumscribed circle, i.e., the unique circle that passes through each of the triangle's three vertices. For example, two circumcircles in Fig. 24.9a are determined by $\{v_1, v_2, v_3\}$ and $\{v_1, v_2, v_5\}$, respectively. Let us define $C(v_a, v_b, v_c)$ is the circle passing through vertex set $\{v_a, v_b, v_c\}$. At first, *Minimax* finds all circumcircles of any three Voronoi vertices like $C(v_1, v_2, v_5)$ and $C(v_1, v_2, v_3)$ as shown in Fig. 24.9a. Then, it selects the one with the smallest radius covering all vertices $v_i \in V_p(S)$, which is referred as the *Minimax* circle. The center of that circle is selected as the targeted location, marked as point $O$ for $G_p(S)$. After moving to the new location as shown in Fig. 24.9b, the local coverage is improved compared to previous location in Fig. 24.8.

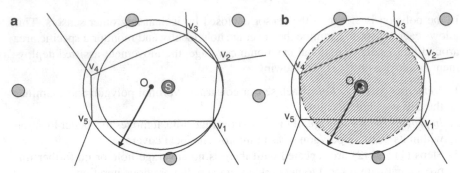

**Fig. 24.9** (a) Circumcircle $C(v_1, v_2, v_5)$ covers all Voronoi vertexes while all another circumcircles like $C(v_1, v_2, v_3)$ only cover some vertexes. (b) The coverage is improved after moving to the targeted location, i.e., $O$

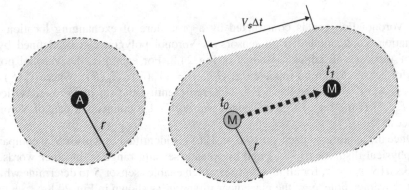

**Fig. 24.10** Area coverage over time for sensor node $A$ and $M$. Node $A$ is a static sensor. Node $M$ is a mobile sensor

### 24.5.3 Random Mobility Strategy

In the Random Mobility Strategy, each mobile sensor moves completely in an independent and random fashion. However, the random mobility can also improve the network coverage over time. The reason behind this is that the covered area of a mobile sensor increases with time. For example, as shown in Fig. 24.10, a static sensor $A$ constantly covers approximately a given area of $\pi r^2$ unit. Considering a time period $\Delta t = t_1 - t_0$, in contrast, a mobile sensor $M$ is able to cover $\pi r^2 + 2rV_s\Delta t$ unit area as shown in Fig. 24.10. It can be seen that $2rV_s\Delta t$ is the extra covered area compared to the static sensor $A$. Lemmas 1 and 2 model the coverage improvement in the random mobility strategy.

**Lemma 1.** *If the initial locations of sensors are uniformly and independently distributed in the network domain, the network coverage remains the same as initial network coverage in the random mobility strategy.*

*Proof.* At $t = 0$, $f_a(t = 0)$ is equivalent to the area coverage of stationary sensor network. Let $X(x)$ be the zero/one variable,

$$X(x) = \begin{cases} 1, & x \text{ is not covered} \\ 0, & \text{otherwise} \end{cases}. \tag{24.10}$$

The uncovered region $V$, representing the area that is not covered by any sensor, is

$$V = \int_R X(x) \, dx. \tag{24.11}$$

By Fubini's theorem [31], the expected uncovered area (i.e., $E(V)$) is,

$$E(V) = \int_R E(X(x)) \, dx. \tag{24.12}$$

Consider an arbitrary point $x$ in the region $R$ (i.e., the entire network domain) and denote the number of sensors which cover the point as $N$. Therefore, point $x$ is covered if any sensor is located within distance $r$ of $x$. The sensor deployment follows the Poisson point process that has a Poisson distribution with parameter $\lambda \pi r^2$, where $\lambda$ is the node density and $\pi r^2$ is the coverage of a sensor. Therefore, it has

$$E(X(x)) = P(x \text{ is not covered}) = P(N = 0) = e^{-\lambda \pi r^2} \tag{24.13}$$

and

$$E(V) = \|R\| e^{-\lambda \pi r^2}, \tag{24.14}$$

where $\|R\|$ is the area of the network domain $R$. Furthermore, the fraction of the covered area at the initial deployment is

$$f_a(t = 0) = 1 - \frac{E(V)}{\|R\|} = 1 - e^{-\lambda \pi r^2}. \tag{24.15}$$

At any time instant $t$, the locations of sensors still admit a two-dimensional Poisson point process with the same density $\lambda$. Therefore, the fraction of the area covered at time $t$ remains the same as in the initial configuration, i.e., $f_a(t) = 1 - e^{-\lambda \pi r^2}$.

**Lemma 2.** *In the random mobility strategy, the covered area increases over time interval* $[0, t]$.

*Proof.* During the time interval $[0, t]$, each sensor covers a shape of a racetrack whose expected area is $\pi r^2 + 2r E[V_s]t$, where $E[V_s]$ represents the expected sensor moving speed of the mobile sensor. The covered area depends on the distribution of the random shapes only through its expected area. Therefore, the covered area over a time interval is

$$f_i(t) = 1 - e^{-\lambda(\pi r^2 + 2r E[V_s]t)}.$$

It further has

$$1 - e^{-\lambda(\pi r^2 + 2rE[V_s]t)} > 1 - e^{-\lambda \pi r^2},$$

which indicates the correctness of the lemma.

Based on Lemmas 1 and 2, the random mobility strategy could be used to improve the network coverage over time. Once mobile sensors are preprogrammed to move in a random way, there is no extra communication overhead required between the sensors for collaborating their mobility and no high computation overhead to decide their next movement. Unlike other sensor mobility protocols, it is very scaleable since each sensor's movement is completely independent of others.

## 24.6 Network Connectivity

The network connectivity is hard to be guaranteed for every sensor in the random deployed WSN. It is even harder if the deployed WSN network is an unleveled terrain. Even if the initial network is fully connected due to structured placement or densely deployed static sensors, the network topology may be disconnected due to functional failure of sensors. The group of disconnected sensors is called isolated cluster (or island). This could be due to hardware failures of some sensors or unreliable wireless communication medium. As a result, the data packet of these sensors could not successfully deliver to the BS. If all sensors are mobile, the network topology can be maintained as a connected graph by a collaborative scheme. One the other hand, the disconnected islands can be also reconnected by only using some mobile sensors or relay sensor. In this part, we illustrate such a scheme called Maintain Connectivity by Dynamic Programming [32].

If the partitioned network is deployed with many mobile sensors as the relay nodes, the disconnected islands can be easily reconnected with the BS by placing them as the bridge between every island. However, due to the cost of the controllable mobile sensor, it is necessary to find a way to provide full connectivity by using minimal number of sensors. For this purpose, dynamic programming [32] is used to find the optimal set of islands to be connected with limited number of mobile relay nodes. For example, the existence of partition is first determined by the flooding from BS B as shown as Fig. 24.11. Then, it groups the sensors into a set of islands by validating if they have received the flooded message from the BS. Each sensor detects its neighbors and all directly and indirectly connected sensors are labeled the member of an island. Hereafter, it calculates $D_{B,C}$ and $D_{B,A}$, the minimum number of mobile sensors required to connect island B and A, or island B and A, respectively. There is two alternatives in connecting islands. We can connect island B and A, or B and C as shown as Fig. 24.11a, b, respectively. Since we have only two mobile relay nodes, we cannot connect all three islands. What would be the best choice? Obviously, it is Fig. 24.11b in the given example since island C is larger than A which achieves a higher overall coverage.

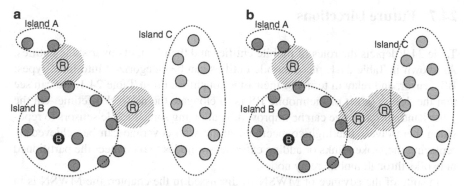

**Fig. 24.11** Connecting partitioned network using two mobile relay nodes $R$. (**a**) Connecting island A and B. (**b**) Connecting island B and C

The selection of the optimal set of island to be connected by the limited number of relay nodes is performed by using dynamic programming [32] as follows. Let $I$ be a set of islands that is not connected to BS but reachable by relay nodes. Denote $b$ is an island with BS. $W(G, m)$ denotes the maximum network coverage obtained by connecting optimal set of island $\{a_1, a_2, \ldots, a_k\} \subset I$ by $m$ relay nodes, where

$$G = \bigcup_{i=1}^{k} \{a_i\} \bigcup \{b\}.$$

The solution of finding optimal set of island can be expressed as a sequence of decisions, where $i$th decision is to select an island $a_i$ as the $(i + 1)$th island to be connected with BS, $i = 1, 2, \ldots, k$. The derived problem, resulting from having made the first decision (to select island $a_1$) by $D_{G,a_1}$ relay nodes, is to find subset of island from $I - \{a_i\}$ which give maximum network coverage. It is easily verified that if coverage of set $\{b, a_1\}$ is optimal then the union of coverage by set $\{a_2, \ldots, a_k\} \subset I - \{a_1\}$ is maximum coverage. Hence, the Specialized Principle of Optimality holds. Thus, we obtain following recurrence relation,

$$W(G, m) = \max_{\forall A \in I} \{C_{G \cup A} + W(G \cup A, m - D_{G,A})\}, \qquad (24.16)$$

where $C_{G \cup A}$ is the coverage obtained by connecting islands $G$ and $A$, and $m$ is the number of mobile sensors available. Recurrence relation (24.16) yields an algorithm for selecting optimal set of islands by $m$ relay nodes, in order to give maximum network coverage. Note that we are not concerned with the energy efficiency of the mobile relay node. Therefore, using the shortest path to the BS or efficient routing is not necessarily required.

## 24.7  Future Directions

Table 24.1 depicts the roles of mobile entities and their benefits over static entities. As shown in Table 24.1, these mobile entities can be categorized into three types: BS, sensor, and relay in terms of their functionalities. From Table 24.1, we can see that the design target of the mobile BS is to elongate the network lifetime. On the other hand, the coverage can be improved by allowing some mobile sensors to reposition themselves. The high connectivity along with coverage can be achieved by replacing the broken links or adding extra sensor nodes to reconnect the partitioned networks through mobile relay nodes.

In spite of the advance of MWSN as discussed in the chapter, the MWSN is in its infancy and many issues are open for further research. Compared to the SWSN, a MWSN has additional complexity of controlling mobile entities and its communication between mobile units or between mobile units and static units. Therefore, it is imperative to develop mobile sensor devices with capabilities of flexible mobility control, appropriate computational unit and memory, and easily rechargeable energy resources. In some specific applications, sensors are expected to be in dust size in near future, and thus the design of small size mobile unit, which is comparable to size of sensor, must evolve together.

Due to diversified applications of WSNs, a general MWSN design cannot satisfy the application requirements such as delay. For example, a real-time WSN application requires that the mobile sensor (i.e., mobile BS, mobile relay) could collect and deliver the sensed data to the administrative center without significant delay. In this case, it is an open issue to design the mobility pattern of mobile units and the collaborative scheme to maintain the connectivity toward the BS. In a large-scale sensor application, the network scalability has to be addressed from many aspects. The number of mobile relay nodes is expected to grow in the similar order of static sensors. In addition, the MWSN would be structured in a hierarchical manner as the network size grows. For a hybrid MWSN where some sensors are mobile while others are static, how to effectively integrate mobile components with static components is a challenge. For instance, when the BS is mobile and controllable, the routing path must be jointly designed such that the transmission through any long distance routes to the moving BS can be maximally avoided. In a homogeneous MWSN where every sensor is a mobile unit, how to collaborate each sensor to achieve the performance improvement is another challenge if the mobil-

**Table 24.1**  Role and advantages of mobile wireless sensor network

| Mobile entities | Main roles | Advantages over stationary sensor network |
| --- | --- | --- |
| BSs | Collect information | Elongate network lifetime |
| Sensors | Sense environment | Increase coverage |
| | Forward information to BS | Improve detection time |
| | | Adapt to changing environment |
| Relays | Inherit roles of other sensors | Inherit the roles of other sensors |
| | | Improve connectivity |

ity of each sensor is controllable. For example, we need to design the algorithm to spread the sensor fast and effectively such that it reduces coverage holes without high computational complexity. In addition, the security of mobile unit is also a challenging issue when the mobile node is deployed in a hostile environment as sensors. The adversary can capture and compromise the mobile node as well as the sensor. Therefore, authentication and other security strategies need to be taken into account in the MWSN to protect the communication between mobile unit as well as the static sensors.

## 24.8   Thoughts of Practitioners

A MWSN has its advantages in many applications due to its strength in conserving the energy and improving the network coverage and connectivity. For instance, the Joint Routing and Mobility Strategy could be used for environmental monitoring that needs to operate for a long period of time. A WSN for structural health monitoring is an example where sensors are mixed into the concrete right from the beginning of the construction. Such a network targets to function for a long period of time because the structural defect could incur many years later. Therefore, using mobile BSs to collect the data may be the only practical option to enhance the lifetime of the network since it is impractical to recharge the battery or replace sensors which are imbedded inside of the building.

Random Mobility Strategy is very useful for emergency situations or when mobile units are uncontrollable and their movement is unpredictable. In an emergency situation, there is no prior knowledge about the environment and the network must be setup immediately. In the Random Mobility Strategy, each mobile sensor node is moving in an independent and random way. Their movements improve the coverage over time and facilitate the search of the target in the network. In the monitoring of the natural disaster, the network topology could be changed frequently due to unexpected events. For example, in the forest fire monitoring application, the initial connected network may be disconnected due to damage of sensors by the fire. In such application, the usage of the relay node may be applicable to reconnect the partitioned network.

## 24.9   Conclusion

In this chapter, we categorize mobile units into three distinctive sets depending on its mobility characteristics and described and proposed several mobility based algorithm for life time elongation, extend coverage, and increase connectivity.

Mobility in WSN elongates life time since it can spread out the hot spot over time by continuously relocating the mobile BS or mobile relay nodes can be used to share the traffic load between static sensors with low life time. It extends coverage by moving mobile nodes to uncovered region while spreading out the network as much as it can while connectivity is maintained. In case mobile units are uncontrollable, coverage can be still extended over time since one of the mobile sensors covering the undetected target is strictly greater than zero. Whereas, in SWSN, uncovered region always stay uncovered. Also, mobile relay units are especially beneficial in reconnecting broken links in the network in case of network partition.

We show that having mobile units are always an advantage whether mobile units are controllable, predictable, or uncontrollable. Therefore, the mobility of sensors and base station in the MWSN are valuable capability which give an extra dimension in designing effective algorithms to improve the WSN network performance from many aspects, which may be constrained in the traditional WSN (i.e., SWSN).

## Questions

1. Assume that 64-sensors are arranged in a two-dimensional mesh. The distance between two neighboring sensors is 10 units. What is the minimum sensing range required so that any point in the area can be covered by at least one sensor?
2. Please analyze the coverage if the sensing range in Q1 is doubled?
3. What is the minimum communication range of sensors required in Q1 so as to maintain the full connectivity of all sensors in ideal 0–1 communication environment (i.e., two sensors assumed to be connected if their Euclidean distance is less than their communication radius)?
4. Assume the sensing and wireless connectivity is increased to reach diagonally placed sensors in two-dimensional gid with $n$ nodes ($n \gg 4$). If 10% of the sensors run out of energy, what is the probability of the area to be fully covered and sensors fully is connected if the sensors out of energy are distributed uniformly?
5. The sensor configuration of Q1, is augmented by adding 49 more sensors at the center of each rectangle. What is the impact on the sensing and communication capabilities? Explain clearly.
6. Assume that a BS is located in one corner of the two-dimensional mesh of Q1. How many messages in terms of number of hops are needed to be sent to BS if each sensor has to send exactly one packet to the BS?
7. What is the impact on Q6, if four BSs are located, one at each corner of the network? Explain clearly.
8. The approach taken in Q5 does provide fault tolerance to the sensor network. But, there is no reduction in the number of packets to be transmitted by the sensors to a BS, located at one corner as shown in Q6. Instead, the BS is allowed to move around sensors. Can you devise an efficient way of making BS visit different sensors in order to (a) reduce the total number of packets about three time less (b) enhance the lifetime? Assume traffic is flowing continuously.
9. What is the net impact on Q8(a), if a mobile BS is moving randomly around?
10. In the two-dimensional mesh of Q1, an index is calculated as an addition of its $x$–$y$ co-ordinates and a sensor is placed if the index is prime number. Using such a distribution, prepare Voronoi polygons for such a sensor distribution. Are there obvious "holes" in the network? Explain clearly.

# References

1. W. Ye, J. Heidemann, and D. Estrin, An energy-efficient MAC protocol for wireless sensor networks, in Proceedings of IEEE INFOCOM, 2002. New York, NY: IEEE.
2. G. Lu, B. Krishnamachari, and C.S. Raghavendra, An adaptive energy-efficient and low-latency MAC for data gathering in sensor networks, in International Workshop on Algorithms for Wireless, Mobile, Ad Hoc and Sensor Network (WMAN), 2004, Santa Fe, NM.
3. A.E. Cerpa and D. Estrin, ASCENT: Adaptive self-configuring sEnsor networks topologies, IEEE Transactions on Mobile Computing, 2004, 3(3): 272–285.
4. B. Chen, et al., Span: an energy-efficient coordination algorithm for topology maintenance in ad hoc wireless networks, ACM Wireless Network Journal, 2002, 8(5): 481–494.
5. S.S. Dhillon and K. Chakrabarty, Sensor placement for effective coverage and surveillance in distributed sensor networks, in Proceeding of IEEE Wireless Communications and Networking Conference, 2003, pp. 1609–1614.
6. X. Wang, et al., Integrated coverage and connectivity configuration in wireless sensor networks, in Proceeding of the First International Conference on Embedded Networked Sensor Systems, 2003. Los Angeles, CA: ACM, pp. 28–39.
7. J. Bachrach and C. Taylor, Localization in Sensor Networks, in Handbook of Sensor Networks Algorithms and Architectures (I. Stojmenovic, Ed.), 2005. Hoboken, NJ: Wiley, pp. 277–310.
8. K. Romer, P. Blum, and L. Meier, Time synchronization and calibration in wireless sensor networks, in Handbook of Sensor Networks Algorithms and Architectures (I. Stojmenovic, Ed.), 2005. Hoboken, NJ: Wiley, pp. 199–238.
9. J. Butler, Robotics and Microelectronics: Mobile Robots as Gateways into Wireless Sensor Networks, in Technology@Intel Magazine, 2003.
10. A. LaMarca, et al., Making sensor networks practical with robots, in Pervasive '02: Proceedings of the First International Conference on Pervasive Computing, 2002. London: Springer, pp. 152–166.
11. Intel Sensor Nets/RFID. [cited July 2007]; Available from: http://www.intel.com/research/exploratory/wireless_sensors.htm.
12. V. Bychkovsky, et al., Data management in the CarTel mobile sensor computing system, in Proceedings of the 2006 ACM SIGMOD International Conference on Management of Data, 2006. Chicago, IL: ACM, pp. 730–732.
13. S. Shukla, N. Bulusu, and S. Jha, Cane-toad monitoring in kakadu national park using wireless sensor networks, in Proceedings of APAN, 2004, Cairns, Australia.
14. I. Stojmenovic, Handbook of sensor networks algorithms and architectures.
15. J. Chang and L. Tassiulas. Routing for maximum system lifetime in wireless ad hoc networks, in Proceedings of 37th Annual Allerton Conference on Communication, Control and Computing, 1999, Monticello, IL.
16. M. Bhardwaj, T. Garnett, and A.P. Chandrakasan, Upper bounds on the lifetime of sensor networks, IEEE International Conference on Communications, 2001, 26: 785–790.
17. D.M. Blough and P. Santi, Investigating upper bounds on network lifetime extension for cell-based energy conservation techniques in stationary ad hoc networks, in Proceedings of the Eighth ACM Mobicom, 2002.
18. J. Luo and J.-P. Hubaux. Joint mobility and routing for lifetime elongation in wireless sensor networks, in INFOCOM 2005, 24th Annual Joint Conference of the IEEE Computer and Communications Societies, Proceedings IEEE, 2005, Miami, USA.
19. B. Bolob', Modern Graph Theory, 1998. New York, NY: Springer.
20. C. Bettstetter, 2002. On the minimum node degree and connectivity of a wireless multihop network, in Proceedings of the Third ACM International Symposium on Mobile Ad Hoc Networking and Computing (Lausanne, Switzerland, June 09–11, 2002), MobiHoc '02. New York, NY: ACM pp. 80–91.
21. S.R. Gandham, et al., Energy efficient schemes for wireless sensor networks with multiple mobile BSs, in IEEE GLOBECOM, 2003.

22. Z.M. Wang, et al., Exploiting sink mobility for maximizing sensor networks lifetime, in Proceedings of the 38th Annual Hawaii International Conference on System Sciences (Hicss'05), 2005. Washington, DC: IEEE Computer Society.

23. A. Chandrakasan, A. Sabharwal, and B. Aazhang. Using predictable observer mobility for power efficient design of sensor networks, in The Second International Workshop on Information Processing in Sensor Networks (IPSN), 2003.

24. R.C. Shah, et al., Data MULEs: modeling a three-tier architecture for sparse sensor networks, in Proceedings of the IEEE Workshop on Sensor Network Protocols and Applications, 2003.

25. W. Wang, V. Srinivasan, and K.-C. Chua, Using mobile relays to prolong the lifetime of wireless sensor networks, in MobiCom '05: Proceedings of the 11th Annual International Conference on Mobile Computing and Networking, 2005. Cologne, Germany: ACM.

26. T. Benerjee, B. Xie, J.H. Jun, and D.P. Agrawal, LIMOC: Enhancing the LIfetime of a Sensor Network with MObile Clusterheads, IEEE Vehicular Technology Conference (VTC), 2007.

27. G.T. Sibley, M.H. Rahimi, and G.S. Sukhatme, Robomotes: a tiny mobile robot platform for large-scale ad-hoc sensor networks, Robotics and Automation, 2002; Proceeding ICRA '02; IEEE International Conference, 2002. 2(2): pp. 1143–1148.

28. B. Liu, et al., Mobility improves coverage of sensor networks, in Proceedings of the Sixth ACM International Symposium on Mobile Ad Hoc Networking and Computing, MobiHoc '05, 2005. Urbana-Champaign, IL: ACM.

29. A. Howard, M.J. Mataric, and G.S. Sukhatme. Mobile sensor network deployment using potential fields: A distributed, scalable solution to the area coverage problem, in DARS 02, 2002, Fukuoka, Japan.

30. G. Wang, G. Cao, and T.F.L. Porta, Movement-assisted sensor deployment, IEEE Transactions on Mobile Computing, 2006, 5(6): 640–652.

31. R.M. Aarts and E.W. Weisstein, Fubini Theorem. From MathWorld–A Wolfram Resource. http://mathworld.wolfram.com/FubiniTheorem.html.

32. S. Zhou, M.-Y. Wu, and W. Shu, Finding optimal placements for mobile sensors: wireless sensor network topology adjustment, Emerging Technologies: Frontiers of Mobile and Wireless Communication, 2004; Proceedings of the IEEE 6th Circuits and Systems Symposium on, 2004. 2: pp. 529–532.

# Chapter 25
# Analysis Methods for Sensor Networks

Peter J. Hawrylak, J.T. Cain, and Marlin H. Mickle

**Abstract** Sensor networks are complex systems incorporating a variety of different devices. As with any system, simulation of the system, or key components, reduces design time. With simulation, a designer can investigate performance and system correctness without having to build a device and test-bed. As a result, simulation is well suited to sensor networks, saving time and money, because to construct a test-bed hundred or even thousands of devices may be required to be produced and deployed. However, simulating sensor networks involves conflicting tradeoffs in the development and running of a simulation. This chapter explores the two different methods, discrete event simulation and analytical modeling, available to simulate sensor networks and pros and cons of each method. The chapter concludes with a comparison of the two methods.

## 25.1 Introduction

A collection of hundreds, thousands, or even millions of small inexpensive devices with sensing capabilities scattered or placed throughout an area is commonly referred to as a sensor network. The devices that compose the sensor networks typically communicate via wireless communication equipment. The ability of the sensor network to monitor an area for extended periods of time is a critical application requirement. As a result of technological advances in the past 10 years, creation and deployment of sensor networks are not only technologically, but also economically feasible. Some specific advances, which enable the feasible deployment, are the availability of powerful, low-power, and inexpensive processors and accompanying

P.J. Hawrylak (✉)
Department of Electrical and Computer Engineering, University of Pittsburgh, 3700 O'Hara Street, 348 Benedum Hall, Pittsburgh, PA 15261, USA
e-mail: phawrylak@engr.pitt.edu

S. Misra et al. (eds.), *Guide to Wireless Sensor Networks*, Computer Communications and Networks, DOI: 10.1007/978-1-84882-218-4_25,
© Springer-Verlag London Limited 2009

hardware. The reduction of energy consumption both of individual nodes (or tags) and of the network as a whole is a key to the wide-scale use of sensor networks.

The concept of a sensor network is not new as wired sensor networks have been in place for many years. The availability of low cost electronics for sensors has made the cost of the wired connections between the sensors and the central controller a significant part of the overall system cost [1]. The need for a wired connection to exist between the sensor and the central controller places significant limitations on where sensors can be placed and deployment in general [1]. For example, sensors cannot be placed in numerous applications having moving parts because the wire connection would either break as a result of the movement or the wire would get in the moving parts and cause the machine that the sensor is to monitor to fail [1]. The limited ability to resist interruptions in communication links is another drawback of wired sensor networks [1]. In the event of a failure or death of some sensors in a wireless sensor network, it is still possible for the remaining sensors to reroute their messages through those sensors still in operation [1]. In a wired sensor network once a communication link is broken, all communication that traveled through the broken link is lost, thus multiple wired links are needed to provide a comparable level of resiliency [1].

The military has performed extensive research into and deployed a number of sensor networks. One of the most notable of these networks is the SOund SUrveillance System (SOSUS), deployed in the Cold War to monitor movement of submarines and today this system is used to monitor ocean conditions [2]. Sensor networks have a wide range of possible military, commercial, and public safety uses. Widespread use of sensor networks can provide solutions to problems in the areas of environmental monitoring, security, traffic control, and monitoring conditions within an area or structure. Monitoring the environment on a large scale is an ideal use of sensor networks as ecologists require frequent readings but cannot access a site due to remoteness or to prevent disturbing a habitat [3]. Sensor networks provide a means to monitor the environment with the required granularity while causing little disturbance of the organisms under study [4]. Providing the ability to monitor large areas, sensor networks allow possible intruders to be quickly discovered and apprehended [2]. The use of sensor networks to monitor traffic conditions in order to reduce congestion is another promising area of development [5]. Indoor climate control systems continue to advance, incorporating an increasingly complex amount of technology in the system. With respect to climate control the more temperature readings available to the system, the better the system can modulate the amount of heat or cooling generated and the distribution of that heat or cooling, providing a more efficient and conformable environment.

The field of sensor networks is concerned with designing, deploying, and investigating sensor networks. Of critical importance to sensor and Radio Frequency IDentification (RFID) networks is the power consumption of individual nodes and of the network as a whole. The nodes can derive the power needed for operation from an onboard power supply, harvest energy from the environment, or from an external power source. Nodes deriving power from an external source place severe limitations on the applications available to networks consisting of such nodes. Energy harvesting, while promising, still requires significant development to enable enough energy to be harvested to power a node. Therefore, the node must be powered from an onboard power supply, specifically a battery. The node remains operational as long as the battery can provide the needed operational power.

Minimizing the power consumption of an individual node will increase the lifetime of that node. Depending on what other factors are altered in minimizing the power consumed of a single node, the network lifetime could increase, decrease, or remain unchanged. Therefore, it is important to investigate power consumption on an individual node and over the entire network. Minimizing power consumption over the network as a whole will result in increased lifetime of the network. Simulation of proposed sensor network designs plays an important role in analyzing these and other design alternatives.

## 25.1.1   Background

Current work in sensor networks focuses on several different areas. As current technology allows most data gathering and communication requirements to be met, lifetime of the sensor network remains the critical problem. The main problem with increasing the lifetime of a given sensor network involves several tradeoffs between areas of operation. Tradeoffs create a large design space that must be evaluated to determine the best choice for a given problem.

In addition to optimizing lifetime of a network, initialization, communication link construction, routing protocols, mobility, and security are important issues in sensor networks and have received the much attention. Optimizing the operation of the individual nodes making up the sensor network is a key area of research focused on extending lifetime. Network architectures for selective activation of nodes or for operations performed on data to reduce the amount of data transmitted through the network are a growing area of interest. Energy harvesting is one method to augment the energy supply of the battery but with current technology cannot provide a dependable substitute for the battery. Mobility of nodes within the sensor networks presents significant problems to maintaining communication paths within the network. However, networks with mobile nodes are being investigated as mobile nodes can provide additional benefits and open new areas for use of sensor networks. Of particular interest is combining the use of mobility with energy harvesting. For example, a node with a small solar panel could move to a nearby location where more sunlight is available in order to harvest more solar energy.

Routing and initialization are critical to the operation of sensor networks. While some sensor networks are deployed by hand and use predefined routing tables. In state of the art networks nodes will be randomly deployed over an area and must build and connect the sensor network themselves without outside assistance. Initialization is the building and connection of the individual sensors to form the sensor network. Initialization consists of achieving two goals. The first is to construct the network by finding and connecting nodes together to form a network. The second objective is to setup the initial communication paths, routes, by which information will be disseminated. Routing focuses on the paths that messages take between entities in the network. Routing schemes for sensor networks must be flexible to handle changes (additions and deletions) of nodes in the network.

The first step in initializing a sensor network is for the individual sensor nodes to identify their neighbors. Several methods are possible for discovery, but a node must either listen for messages from other nodes announcing their presence, or broadcast a message announcing its presence. In one algorithm for network construction, new nodes first listen for an invitation to join the network [6]. Receiving an invitation within a certain time causes the new node to broadcast a request to join the network [6]. Nodes form local networks with the other nodes they discover. This process continues; the local networks continue to grow and the network gradually becomes connected [6].

The location of each sensor node is an extremely useful piece of information to determine the routing paths within the network. Knowledge of location allows individual nodes to identify nodes that are closer to the destination node of a particular message. With that information, the message can be addressed only to those nodes that are closer to the destination. Another use of location is to base the routing paths on minimizing the distance between all the nodes since received power is inversely proportional to the distance squared.

However, obtaining the location of each sensor node presents significant problems. Clearly if each node is placed intentionally, the location of each node would be known and could easily be programmed into the node. Placing each node when a network may contain thousands of nodes is not feasible, and this solution is limited only to very small sensor networks. Location finding devices such as GPS are costly in terms of energy use, and take up valuable real estate on the node. Further, the error range of a few meters, of the GPS system, significantly reduces the usefulness of the location provided, because the distance between nodes in dense sensor networks will usually be only a meter or two at the most. Relative distances between two nodes can be estimated through measurements of the signal strength of messages from the other nodes. While this relative distance does not give an exact position (i.e., latitude, longitude, and altitude) knowing just the relative distance allows for significant improvement and optimization for during topology formation and routing. Using three special nodes the location of a fourth node can be obtained through triangulation based on the received signal strength of a message from a fourth node [7]. Mondinelli and Kovacs-Vajna present another location finding method using just one special node, but that method requires complicated computations and other regular nodes to participate in the process as well where the method using triangulation does not require the participation of other regular nodes. These techniques provide location information but have an added overhead of transmitting and receiving multiple messages, all consuming additional energy.

The topologies used in sensor networks can be generalized into two categories; cluster topologies and flat topologies. Cluster topologies partition the network into a set of smaller local networks, or clusters. Each cluster has a cluster head that is responsible for communication among clusters. Further, establishment of a local network within each cluster links each node to the cluster head. Routing within a cluster is simplified as the number of nodes and possible paths are reduced to only those in the cluster. Clustering also allows for the application of higher-level optimizations based on clusters rather than individual nodes. One such optimization is to partition all the sensor nodes into mutually exclusive sets of nodes that cover the entire area (coverage area is defined by the application and node) being monitored [8]. Only one of the sets must be active at any given time; thus for a network of homogeneous nodes, the lifetime of the network is directly related to the number of sets, increasing as the number of sets increases [8]. In a flat topology, the nodes are not grouped together.

Once the topology is defined communication links and routing paths must be setup to allow information to flow through the network. The primary objective of any routing algorithm is to discover communication paths that enable all nodes in the network to communicate their data to the network. The secondary objective of the routing algorithm is to minimize the energy consumption of the entire network keeping the data delay within acceptable limits, and ensuring that the network remains connected.

Networks that employ a cluster topology can perform higher-level optimizations to setup the data routes. For example, the LEACH protocol randomly changes the designated cluster head distributing the additional energy consumed being a cluster head throughout the cluster and the PEGASIS protocol, which extends LEACH, can be used if enough global information is known about the network [9].

Other protocols support data aggregation in a network, or data centric networks. Data aggregation attempts to reduce the amount of data sent throughout the network by combining the data received with the sender's data, for example through averaging or compression. Data aggregation trades the additional processing required on the sensor node for a reduction in the number or length of messages transmitted. The majority of research focusing on data centric sensor networks sees the sensor network as a large database. Data centric networks view the request similar to an SQL query in a database. The request will contain information specifying what data are requested and only those nodes with data matching the requested data type will respond.

Sensor networks must continue to function as nodes die. From a routing viewpoint, the simplest method to increase the fault-tolerance of a network is to send data along multiple paths. As long as one path is intact, the data will arrive at the destination. One routing algorithm forms multiple routes between a source and a destination by providing intermediate nodes several choices of where to forward the packet [10]. With multiple choices, most, or ideally all intermediary nodes, the network can tolerate significant numbers of node failures before the network becomes completely disconnected, and the network can better balance the communication load through selective routing [10]. These benefits can increase the lifetime of the network, but sending a single message along multiple paths obviously requires more energy than sending the message along a single path. The drawbacks to this solution are that as the number of nodes in the network increases so does the potential size of the routing tables stored in the nodes and the overhead associated with finding and using multiple routes.

Optimizations applied to individual nodes are another area of research interest. Most optimization strategies focus on reducing the power consumption of the processor. Optimization of the other hardware contained on the sensor node, such as the sensor, analog to digital converter (ADC), etc., also reduces power, but short of custom designed components selecting more efficient off the shelf parts is the only solution.

Reducing power consumption at the node level can be achieved through more efficient processing and routing of messages. Sensor networks generate huge numbers of messages and a message arriving at a node falls into one of the three categories; the message is for that node, the node must relay the message, or the message can be ignored. The wireless interface normally handles the physical reception and collection of the message, and the processor decodes and processes the message. In this scheme, the processor determines which of the three categories the message falls into and takes appropriate action. In the case that the message is to be relayed to another node, custom logic has been developed and integrated onto the wireless interface board to relay the message to the next node if the message is to be relayed [11]. This reduces the node power usage as the message is not transferred to the processor board, decoded, and then back to the wireless interface board for retransmission [11].

Another node level optimization is to ignore all packets not destined for the node. Particularly useful in networks made up of active RFID tags, the ignoring of messages addressed to other tags saves power. Waking the processor up only when a message arrives addressed to the tag will allow the processor to remain in a dormant state for the maximum amount of time. This results in significantly less energy consumption per unit time over a tag, that requires the processor to decode every message to check the address. This optimization works best when the network is frequently interrogated.

Management of the processor state is critical to the energy consumption of a node. The ideal node would keep the processor in the lowest level sleep mode unless work must be done. However, entering and leaving each level of sleep mode requires a time delay to power down or up, as required information must be stored or retrieved by the processor to transition between modes [12]. During this time, no useful work can be performed, so the energy consumed during this time is wasted; thus the processor must remain in that state long enough to save at least as much power as was wasted. Sinha and Chandrakasan have developed an algorithm that attempts to predict the future workload, and by extension predict how long the processor can remain in a given state and use this to determine when and what level of sleep mode to enter. If the workload is predicted to be high in the immediate future, the algorithm will either stay in the active processing mode or enter a sleep mode saving minimum power, but requiring minimum delay and energy to enter and leave [12].

Sensor networks have tasks that must be performed throughout the network. Some or all of the tasks may be able to be processed on several different nodes, while still providing the required level of detail and accuracy. In such a case, tasks can be assigned to different nodes to prevent some nodes from quickly draining their energy reserve and failing. One algorithm has been developed to spread the workload over a group of nodes. Assigning a reward and an energy cost to each task as a part of a network policy, each node is then able to determine if it should perform a given task based on the reward for performing it, the network policy, and the remaining energy in the node [13]. Nodes that are low on energy will perform less tasks or tasks with higher rewards than nodes with significant energy remaining. One drawback of this solution is that if all the nodes that are issued a request to perform a certain task are low on power, there is a possibility that none of the nodes performs the task. If the tasks were critical, the network would fail to meet the operational requirements, which would be a failure of the same magnitude as a significant number of nodes fail due to lack of energy.

Sensor networks containing a large number of nodes also contain and ultimately report huge amounts of information. Rarely are all the available data required, and usually only more specific data are required, such as the temperature, or the temperature in a certain area of the network. Sending back readings from all the nodes wastes the energy of those nodes whose readings are not needed, and communication is greatly increased, further wasting energy. One solution is to reduce the amount of information reported. This can be accomplished by activation of only a selected number of nodes, or the combination or compression of results as they are making their way through the network.

Selectively requesting and reporting data reduces energy consumption of the sensor network and reduces the amount of processing required on the entity that must process the readings and make decisions based on those readings. One way to reduce the amount of data is to query the sensor network for only the desired information. Several papers have compared a sensor network to a large database, distributed over a large number of nodes [14, 15]. These papers envision the querying of the sensor network with SQL-like commands where only those nodes with the requested data would reply. This is similar to the use of the SELECT SQL command used in a database to return those records matching the query. This solution would require additional processing to be performed by the nodes to determine if they have the requested information, but it reduces the amount of data sent back. The energy consumption of the additional processing must be weighed against the energy savings of reducing the amount of data sent through the network.

Sensor networks that track targets require only those nodes in the area of the target to be actively monitoring. Research has been conducted by Zhao et. al. for the development of a network that determines which sensors should be activated and how to handoff the monitoring requirements as conditions change [16]. Target tracking is an ideal application for this type of network, because only those sensors within the range of the target need to be actively monitoring it, the other sensors can remain dormant until the target enters their area. They propose methods to estimate the useful information added with the addition of information from another sensor. Those sensors that maximize the estimates will be activated. Keeping sensors without much useful information dormant increases the lifetime of those sensors and possibly that of the network.

Selective querying and activation reduces the number of replies, but in larger networks, the number of replies can still be large. Data fusion and compression can reduce the amount of data transmitted while keeping a high number of replies. One solution is to average the reported data over a small geographic area and then send only that average on to the destination. This reduces the amount of data that must be processed and the number of messages sent to the destination. This solution is easily implemented in a network that can be divided into clusters by having each cluster head average the data and sending only that average. Another solution is compression of the data in order, to send the same information but with fewer bits. One such solution is to divide up the data space of the response into indexed bins, each containing a portion of the response space, and then only send the index of the bin back [17]. As long as the length of the index is less than the length of the data, sending the index reduces the amount of total number of bits transmitted. Energy is saved by transmitting fewer bits. The sizing of the bins has impacts on accuracy and precision as well as the amount of energy saved.

Security in sensor networks has received less attention than research focusing on implementation and deployment of the network. However, security will be an issue as sensor networks are deployed. Chan and Perrig list eavesdropping, data privacy, and attacks on networks themselves as some of the security concerns with sensor networks [18]. Monitoring the messages transmitted through the network can be as simple as listening on the same frequency that the nodes use, or by inserting nodes into the network to collect data. Private data can now be accessed and read by third parties. Encryption is one possible solution, but it requires significant strength, robust key distribution, and more energy [18]. Another solution suggested is to fragment the data and send each fragment through a different route [18]. This would require the attacker to successfully monitor all or a majority of the possible routes. This could reduce the required strength of the encryption scheme. Attacks on sensor networks can take the form of jamming communications or depleting the remaining energy of the nodes [18]. Jamming prevents communication as the channel becomes unusable, but using communication schemes specifically targeted at overcoming noisy channels can counter jamming, and jamming can be quickly discovered [18]. Draining the remaining energy of enough of the nodes will result in the network becoming too disconnected to function. Injecting large numbers of commands into the network can quickly drain the batteries of the nodes and may not be immediately detected [18]. Requiring authentication of commands will counter this attack [18]. However, these counter measures must be carefully designed as they will consume more energy than they can save.

### 25.1.2 Energy Harvesting

A few researchers have looked at using energy harvesting to power wireless networks, or recharge batteries in the nodes [19–22]. This work can be applied to sensor networks as well. Energy harvesting methods viable for use in sensor networks include RF energy harvesting, thermoelectric generation, solar power, and harvesting energy from vibrations in the environment.

The addition of RF energy harvesting circuitry to the node requires minimal additional hardware as the node is already equipped with an antenna for communication. The large amount of available RF in the environment both from the sensor network and outside sources (i.e., radio stations) makes RF energy harvesting one of the leading choices for energy harvesting in a sensor network. However, the drawback of RF energy harvesting is that it is limited to the amount of power that can be received by the antenna. The received power in free space is determined by the following equation.

$$P_r = \frac{P_t \lambda^2}{(4\pi)^2 d^\alpha}, \tag{25.1}$$

where $\lambda$ is the wavelength of the carrier frequency; $d$ is the distance between the transmitter and the receiver; $\alpha$ is the power that the distance, $d$, is raised to; $P_t$ is the transmit power; and $P_r$ is the received power. The free space model uses 2 for $\alpha$, while in a real-world environment $\alpha$ will be larger due to interference. Thus, the received power decreases quickly as distance increases making it nearly impossible to power sensor networks with current technology using RF energy harvesting.

Thermoelectric generation is another method of energy harvesting applicable to sensor networks. Thermoelectric devices generate energy in response to a difference of temperature across the generator, the larger the difference the more the energy generated. However, for small temperature differences, little energy can be generated. An individual node in a sensor network will normally not experience large temperature differences as the internal temperature of the node will be close to the temperature of the surrounding environment. The resulting small temperature difference is not enough to provide the energy required by the node for operation.

Solar power is another alternative for powering sensor networks. Solar cells harvest energy from light in the environment. Currently solar cells can harvest sufficient energy for small devices such as a handheld calculator to operate. The drawback of solar cells is their need to be in a brightly illuminated environment. In a low light environment, the solar cell generates little energy. Another drawback is that the light level is rarely constant, for example, node outdoors should have enough light in the day but not during the night. Uncertainty with respect to the maximum light level and the consistency of the light level available to each node after deployment of the network makes solar power ineffective.

Using vibration to generate energy has long been used in the manufacture of self-winding watches [20]. Piezoelectric materials are commonly used to harvest energy from vibration. This type of energy harvesting suffers from similar drawbacks as solar cells as they are dependent on the environmental conditions. Most sensor networks are targeted at deployment over a stationary terrain such as a forest or desert that experiences very little vibration. Further, the vibration would have to exist for a long period to recharge the batteries that power the node. For these reasons, harvesting energy from vibrations provides limited support for sensor networks.

Several networks have been proposed and studied in literature utilizing energy harvesting [21, 23]. The work done by Kansal and Srivastava proposes a network that attempts to divide tasks among the nodes based on the amount of energy each node can harvest from the environment. The goal is to distribute the workload such that those nodes that can harvest more energy will perform more work. This extends previous work investigating distributing work based only on remaining battery energy. Dividing the workload in such a manner requires that nodes be able to determine or estimate the amount of energy they can harvest from the environment and communicate this value to other nodes. Determination of the available energy and communication of that value with other nodes must be done in an energy efficient manner. Scheduling tasks based on the energy available to each node and remaining battery power theoretically increases the lifetime of the network simulated. Rahimi et. al. describe a network architecture consisting of two types of node, the first having the ability to harvest energy, move around and replace depleted batteries with batteries with a full charge. The second are regular nodes that cannot recharge their batteries or move. The number of energy harvesting nodes required to cover an area is determined by the total area, energy for the energy harvesting node to move, available energy, energy consumption of regular nodes, and the number of regular nodes [21]. Implementation of such a network requires that enough available energy can be harvested, and that the terrain over which the network is deployed allows movement of the energy harvesting nodes to replace batteries. Even with high efficiency energy harvesting technology and an abundant source of ambient energy, the inability to move that energy over rough terrain prevents replacement of depleted batteries.

## 25.2  Thoughts for Practitioners

The previous sections illustrate the numerous design tradeoffs that are present in sensor networks. With so many tradeoffs it is difficult to determine which design alternative will provide the sensor network that meets the requirements while providing for the longest lifetime. The sensor network with the longest lifetime is preferable because that sensor network will be better suited to handle unexpected occurrences such as receiving more messages or interference than anticipated during design.

Clearly a means is required to evaluate the performance of a number of design alternatives to identify the best alternative and to evaluate the performance of a given sensor network before deployment. A simplistic method is to construct a small prototype network of each design alternative and collect experimental performance results. However, this method is costly, both in terms of money and time, and for large networks may not capture the larger dynamics at work in the larger networks. Simulation and modeling is an alternative to building small prototype networks. Both, enable the designer to evaluate sensor networks with large number of entities (an entity is a device in the sensor network) and can be used to explore the design space. Two techniques commonly used are discrete event simulations and analytic modeling. Both techniques have benefits and drawbacks.

## 25.2.1   Discrete Event Simulation

Discrete event simulations provide a model platform for a system by determining and updating a set of state variables describing the system at each time increment. Discrete simulations can be expanded into a number of different types. Two of the most common are event-driven and time-stepped simulations [24]. In time-stepped simulations, time is advanced an equal amount at the time step at which point the simulation states are updated [24]. In a time-stepped simulation, all events occurring within one-time step are assumed to happen at the same time, thus the choice of the length of the time step in a time-stepped simulation is critical to the accuracy and precision of the results [24].

Event-driven simulations update the state only when something of interest, called an *event*, happens [24]. Each event contains information indicating the time that the event is to occur in the simulation [24]. The simulation state is only updated in response to the occurrence of an event requiring that the order in which events are processed be maintained in a chronological ordering (i.e., earliest event is processed first) [24]. Ensuring this chronological ordering of events results in the primary over-head in the simulation and thus causes the simulation to slow down considerably [4]. Even with optimizations, the number of events generated per second for a small- or medium-sized sensor network is considerable. Time-stepped simulations are usually faster but provide less accurate results than the event-driven simulations described above [4].

Discrete event simulations use events to pass messages between two simulation entities. Events represent such things as one entity sending a data message to another entity. Events in discrete event simulations contain a time stamp indicating the time that an event is to occur. Events must be processed in order based on their timestamps [24]. This is termed, "the synchronization problem" [24]. In a parallel simulation, the events must be processed in such a manner as to generate the same result as given by a sequential system processing events one-by-one in time stamp order [24]. Processing events out of order can lead to situations in which some entities have advanced to a time ahead of still unprocessed events leading to "causality errors" [24].

Simulations of wireless networks typically use the event-driven discrete model for their improved accuracy. Simulation entities typically represent nodes or sinks in the sensor network, and events typically represent the messages sent between nodes or sinks. For each message sent, two events may be required, one event for the transmission of the message, and a second event for the receipt of the message [25]. While not a significant problem for simulations of networks containing a small number of nodes, the event ordering overhead becomes an issue as the size of the network increases. One study reports that for a network of 3,200 nodes more than 5.3 million events per second were generated, using their proposed optimizations the number of events per second was reduced to slightly more than 210,000 [26]. Adding more processors to the simulation environment only elevates the problem to a point, where the extra messages passed between the processors will begin to reduce or negate gains achieved.

The synchronization algorithms used by discrete event simulators to determine when and which events can be processed generally fall into one of the two categories; conservative synchronization and optimistic synchronization. In conservative synchronization, an event is processed if and only if the simulation determines that there is no event with a timestamp before the event in question, and that no event with a timestamp for a time before the timestamp of the event in question will be received in the future [24]. The possibility exists where all simulation entities are waiting on possible events from other entities and all entities block, resulting in deadlock [24]. There are a number of methods to recover from a deadlocked situation. All deadlock recovery methods require additional overhead in the form of first detecting a deadlock situation, then determining what messages to send to break the deadlock, and finally the actual sending of messages to break the deadlock. These additions increase the amount of communication, the amount of processing, and ultimately the running time of a parallel discrete event simulation.

Parallel simulations must prevent messages from being delivered to a simulation entity with a time stamp in that entity's past. Messages that are still propagating through the network and have not been delivered to the receiving entity yet, called transient messages, must be taken into account when a simulation entity wishes to advance the simulation time [24]. The problem surfaces when a transient message exists for an entity having time stamp, $T1$ and the entity, not aware of the transient message, finds that it is safe to advance to time $T2$, and $T2 > T1$. Causality of the simulation is violated when the transient messages arrives at the entity with time stamp $T1$, since the entity is now at time $T2$, and $T2 > T1$. Detection of the presence of transient messages requires entities to keep track of the number of messages they have sent and the number of messages they have received [24]. This requirement adds to the simulation overhead as two counters must be maintained in the system, and additional processing must be performed by the simulation controller when an entity wishes to advance its simulation time. Further, some entities will experience additional delays because they must wait until no transient messages are present to advance their simulation time.

The amount of time that the simulation time can be advanced must be determined for each time advance. Determination of the amount of time the simulation can be advanced requires information from other simulation entities. In the worst case, all entities require information from all other entities. In a simulation consisting of $N$ entities, in the worst case $N^2$ messages containing the required information are sent through the simulation environment [24]. While methods to improve the required number of messages are possible, determination of the amount of time the simulation can advance still requires significant overhead [24]. Thus, maximizing the amount simulation time advances at each advance is critical to the performance of the simulation [24]. Simulations in which the maximum amount of time that the simulation can advance at each step are limited to a small value will not scale well as entities are added to the simulation [24].

Optimistic synchronization algorithms form the second major class of synchronization algorithms for discrete event parallel simulations. Optimistic synchronization algorithms allow for the processing of messages without first determining if they are safe allowing for a possible violation of causality in the simulation [24]. The simulation is rolled back to a previous simulation time in response to a causality violation [24]. Pipelining microprocessor architectures are one example of an optimistic synchronization algorithm, and provisions are made to remove those instructions that were issued when a hazard is detected [24]. Optimistic algorithms do not require that messages be sent or received in order based on their time stamps, nor do communication links between simulation entities need to be explicitly determined and defined [24].

Supporting simulation roll back requires the storage of information describing the state of the system in the past [24]. Two popular methods to store state information are copy state saving, where all state variables are stored before processing each event, and incremental state saving, where for each event the prior value of every state variable modified by that event is recorded in a log [24]. Better performance is achieved using copy state saving in simulations where a high percentage of the state variables are modified by each event, and incremental state saving results in better performance when only a small percentage of the state variables are modified with each event [24]. Both methods require additional overhead for storing the redundant state information needed to support roll back. The additional memory needed grows as the number of simulation entities increases and as the number of events increases because events cause changes to the state variables. This memory requirement can severely limit the ability to simulate large networks.

In addition to state saving, support of simulation, roll back requires a method to cancel messages that were sent incorrectly [24]. The messages requiring deletion were sent by simulation entities that did not process a message with a time stamp earlier than their current time [24]. The data contained within these messages may be altered by the earlier message (message causing the roll back) or the message may not have been sent at all. Processing of these messages can initiate a causality violation in the simulation, and possibly lead to incorrect results. In the Time Warp system, antimessages are used to destroy messages that must be unsent [24]. When an antimessage and a message appear together in the input queue they cancel each other out and thus the message is deleted [24]. An antimessage must be sent for each message that must be deleted due to a roll back [24]. The use of antimessages to destroy previously sent messages increases the number of messages sent and increases the strain on the messaging scheduling task. In cases were large numbers of messages must be deleted per roll back, this can add significant overhead to the simulation.

To be valid, a simulation must produce identical results over multiple runs of the same simulation [24]. For a parallel processor, this requires that all events must be processed in the same order for each execution of the simulation [24]. Further, differences in the hardware used to execute the simulation can affect the results. For example, a difference in floating point calculations computed on two different processors can result in different results if the messages are not processed on the same processor for each execution of the simulation [24]. The difficulty in producing repeatable simulations significantly increases the difficulty in obtaining accurate comparisons of different systems using results of a simulation.

Messages transmitted in the network are not point to point as with a wired network, but are transmitted in all directions from the transmitting node. The only requirement is that the designated receiving node(s) be within range of the signal to "hear" it. One implication of this communication strategy is that many nodes will overhear messages not designated for them. The nodes overhearing the message may still be required to process the message adding overhead to those nodes, and the simulator must decide the nodes within range and will hear the message sent by the transmitting node. This causes another performance bottleneck in discrete event simulations. In the worst-case in a simulation containing $N$ nodes, for each node in the simulation the simulator must determine if a signal sent by any of the other $N - 1$ nodes is received at the $N$th node [27]. This leads to $O(N^2)$ checks per node [27]. This problem magnifies itself both as the number of nodes and the density of the nodes increase. Even though the number of checks per node may be able to be reduced, the increasing of this overhead with the increasing size of the network under simulation does not provide a platform that scales well. Although optimizations have been proposed [27], these come at a cost of reduced accuracy in the simulation results [26].

### 25.2.2 Discrete Event Simulation of Sensor Networks

There currently exists a vast number of simulators for networks. A survey conducted by Akhtar lists a total of 42 different network simulators [28]. The three most commonly used discrete event simulators used for analyzing sensor networks are GloMoSim, ns-2, or PARSEC. However, custom simulators are also popular.

GloMoSim is a library providing the capability for discrete-event simulation of wireless networks using PARSEC [29, 30]. The GloMoSim simulator was developed to provide an environment capable of simulating wireless networks containing thousands of nodes [29]. The GloMoSim simulator supports both sequential and parallel simulations [30]. The GloMoSim environment consists of a number of layers with each layer having interfaces to the layers immediately above and below the layer in question making it compatible with the seven-layer OSI model [29]. The layered structure of GloMoSim allows several different models to be evaluated at one level without requiring reimplementation of the other layers. The primary focus of GloMoSim is the modeling and evaluating the performance of protocols at the different layers of the OSI stack.

GloMoSim supports parallel simulation. As mentioned above, a critical factor influencing the efficiency of a parallel simulation is the number of messages sent between simulation entities. The simple solution of representing each node in a wireless network with an individual entity would quickly lead to simulations containing thousands of entities. This will result in several thousand or possibly more messages being sent between entities. Such a large number of messages will quickly saturate the communication infrastructure of the parallel system and performance will degrade significantly as the message delay increases. In order to reduce the number of messages passed between entities and to allow for scalability, GloMoSim divides the area where the wireless network exists into smaller areas [29]. Each smaller area is represented by a single simulation entity and contains all nodes placed within the smaller area [29]. Messages sent between nodes within a smaller area are routed locally by the simulation entity representing that area [29]. Only those messages from a node within one smaller area to a node within another smaller area are transmitted across the system communication structure [29].

Configuration of the GloMoSim simulation is achieved by altering an input file read by the simulator [29]. The area covered by the network is approximated by a rectangle with the length in both the $x$ and $y$ direction being customizable by the user [29]. The number of smaller areas is specified by specifying the number of divisions in the $x$ direction (columns) and the number of divisions in the $y$ direction (rows) [29]. An example of dividing the total area into nine smaller areas with three divisions in the $x$ direction (rows) and three divisions in the $y$ direction is shown in Fig. 25.1. For example, for nodes $N1$ and $N3$ (or $N2$) to communicate, the message is routed internally by the simulation entity. However, if nodes $N7$ and $N9$ wished to communicate, a message must be sent from the simulation entity containing node $N7$ into the simulation entity that contains node $N9$ over the communication structure of the system.

In addition to the number of smaller areas, the user can alter the maximum transmitter range of the wireless interface. Because the goal of dividing the entire area into smaller areas is to allow messages to be handled locally within the entity having a large enough smaller area that the majority of nodes cannot transmit outside of the smaller area best utilizes this optimization. Thus, the maximum transmitter range is critical to deciding the size of each smaller area, and by extension, the

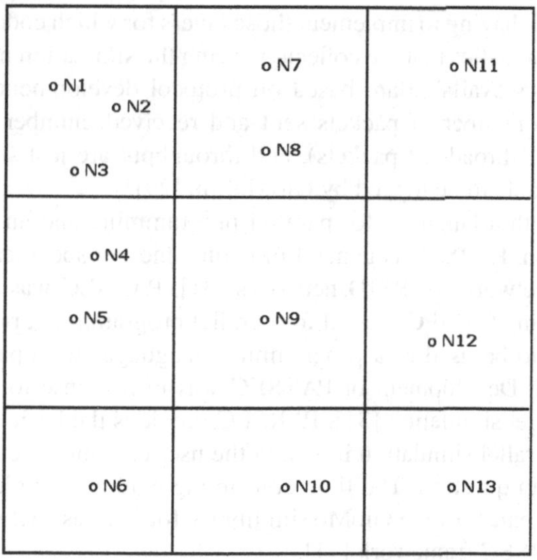

**Fig. 25.1** A network of 13 nodes, divided into nine smaller areas

number of smaller areas. During the simulation, there are many times when one of a number of events may occur, and one must be chosen (i.e., is a message corrupted by environmental noise). In most cases, a random number is obtained and used to determine which of several events occurs. GloMoSim utilizes a random number generator that requires a seed to initialize the random number generator. The seed affects the numbers generated and is configurable by the user. The user must also specify the maximum time that the simulation must execute.

Further, details about the nodes that make up the wireless network must be specified. First, the user must specify the total number of nodes in the wireless network. The nodes can be placed within the area randomly, uniformly spaced within the area, in a 2D grid (must be a full 2D grid), and manually placed where the location ($x$ and $y$ coordinate) of each node is specified by the user [29]. Second, the model for signal propagation selected can be the free space model where received power is inversely proportional to the square of the distance between sender and receiver, the Rayleigh fading distribution, or the Ricean fading distribution [29]. Third, the data rate of the network must be specified.

GloMoSim provides support for mobile nodes. Two different types of mobility can be simulated within GloMoSim. The first type is random mobility in which a node moves one unit in either the $x$ or the $y$ direction. The second type of mobility simulates the node moving to a randomly selected waypoint. The speed at which the node moves, and the time that the node stays at the waypoint, once reached, are configurable by the user [29]. The simulation randomly selects the waypoint [29]. Finally, the simulation allows the user to set the position precision which determines how often the location of the mobile nodes must be updated [29].

As GloMoSim focuses mainly on the development and evaluation of protocols a number of built in protocols and algorithms are provided with the library [29]. These components can be added to simulation, allowing the developer to investigate other areas without having to implement those layers for which code is already available. Finally, the statistics that are collected during the simulation must be specified. Again the statistics available are based on protocol development and evaluation. Statistics such as number of packets sent and received, number of each type of packet (UDP, TCP, broadcast packets), and throughput are just some of the many statistics that are able to be logged by GloMoSim [29].

PARSEC is both a language for parallel programming and an environment for parallel simulation. PARSEC is general in nature and not specifically designed for wireless sensor networks or RFID networks [31]. PARSEC was designed around the message-passing C (MPC) kernel for parallel programming, providing the ability for PARSEC to be used as a programming language and a parallel simulation environment [31]. Development of PARSEC was in response to the lack of tools focusing on parallel simulation [31]. PARSEC provides the basic environment and framework for parallel simulation in which the user can implement models to simulate the system in question. The time-consuming nature of implementing models lead to the development of the GloMoSim library for wireless networks that is built on top of the PARSEC framework [31].

### 25.2.2.1   Drawbacks of Discrete Event Simulation

The large number of simulators available and a difficulty in guaranteeing repeatability increases the difficulty for comparison of results obtained from different simulators. Ideally, different simulators should present similar results, or trends for a given system. A study by Cavin et al. investigating the variation of results simulating a network on ns-2, GloMoSim, and OPNET showed that the results were significantly different [32]. The results obtained showed significant differences not only in values, but also in behavior of the network in general [32]. These differences lead Calvin et al. to conclude that simulations provide little improvement to the design process [32]. Given the variation in results of the simulators investigated, making accurate comparisons requires the developer to implement not only their work, but also the network(s) they wish to compare their solution to, in the simulator used to evaluate their solution. While possible, the amount of extra work and time needed to implement other solutions becomes prohibitive to thorough testing and evaluation of a particular solution.

With many design alternatives to consider designers require a discrete event simulation with a short run-time to enable a thorough exploration of the design space. Most simulators currently in use have a discrete event simulation at the backend that scales very poorly with the size of the system in question. For example, a leading discrete event simulator ns-2 scales poorly, thus limiting simulations to a networks with at most a few thousand nodes [32]. The use of a discrete event simulator at the core is problematic if a large number of messages are sent because the discrete event simulator must schedule each event and ensure delivery of events in the proper order. In sensor or RFID networks, the number of messages sent (and thus needing scheduling) often becomes very large. Such a model should execute significantly faster and scale well as the network size and the number of messages increases.

## 25.2.3  Analytic Modeling

Analytic modeling is an alternative to discrete event simulator to analyze a sensor network. Analytic modeling attempts to find probabilistic or closed form expressions to describe various aspects of a system. Markov processes or Markov chains are a commonly used analytic modeling tool. Queuing networks, popular in analyzing network protocols, are another type of analytic modeling.

Markov processes are an efficient way to analyze systems. Markov processes can be used to model a system when the next state of the system depends on the current state of the system, and not on the past states [33]. A Markov process consists of a set of states, a set of transition probabilities, and a set of reward values. The states represent different conditions in a system. The transition probabilities provide the likelihood, or probability, that the system will transition from state i to state k. The reward values denote the gain or profit, or the loss or penalty from moving from state i to state k.

### 25.2.3.1  Drawbacks of Analytical Modeling

Markov processes allow analysis to be performed at either the steady state, or with a transient component that changes with time. The steady-state analysis requires that the system "stabilizes" after some time. By stabilize it is meant that the state probabilities remain constant after some period of time. The steady-state solution can be achieved using simple linear algebra techniques. If the system does not "stabilize" after some time then the transient component must be taken into account. This requires solving a series of differential equations and Z-transforms can be employed to solve the system [34]. In the steady-state analysis, the transient component goes to zero after some finite amount of time. Clearly, it is more efficient to analysis a system described by a steady-state Markov process than a system that is described by a Markov process with a transient component that does not go to zero after some finite time.

Markov processes with or without a transient component reduce to solving a system of equations and efficient linear algebra and numerical computation techniques are required. While many efficient numerical computation algorithms exist for solving systems of equations the problem of dimensionality is still a major issue. Dimensionality refers to the number of state variables, and is directly proportional to the size of the probability and reward matrices used in Markov processes. As the dimensionality increases, the time required to solve the system of equations also increases.

The dimensionality of a Markov process often grows as the complexity of the system being analyzed grows. Thus, Markov processes used to model very complex systems often suffer for a large dimensionality and require long run times. Therefore, when using Markov processes to analyze a large and complex system such as a sensor network keeping dimensionality to a minimum is critical.

Often it is difficult to obtain the closed form or probabilistic equations to describe the behavior of a system. Sometimes simplifications or assumptions must be made to obtain these equations. The use of these simplifications and assumptions can adversely effect the value of the results obtained because cases where those simplifications or assumptions do not hold may occur and have not been analyzed.

For example, Chiasserini and Garetto describe an analytical model describing the energy consumption in sensor networks in which nodes can be put to sleep to conserve power [35]. This work limits the model of the node to that of the processor and the communication hardware, while the network is modeled as a queuing network in which all messages are received without error [35]. Such simplifications neglect critical facts such as how nodes receive messages when a node is asleep. An analytical model allowing modeling of individual components within nodes and the network connecting them using assumptions based on real-world behavior and performance is needed. While this model is useful for looking at a single entity, it is also important to investigate how this alternative will function in the larger network.

## 25.3 Directions for Future Research

Developments in both the discrete event simulation and analytical modeling areas are needed to improve the simulation and analysis of sensor networks during the design stage. The run-time of the analysis whether discrete event simulation or an analytical model is important and must be within a reasonable time to allow thorough exploration of the design space. Discrete event simulations do not scale well with the size of the network, so research is required to improve the scalability of discrete event simulations. Markov processes describing complex systems suffer from an explosion of dimensionality. Because large sensor networks are very complex systems research is required to limit the dimensionality of the Markov process while providing useful and adequate information.

Discrete event simulators, in various forms, currently exist that can simulate a sensor network. These simulators are sometimes difficult to learn, require substantial time to develop the simulation environment, and can require a significant amount of time to produce results. Because simulators require so much time to produce results, the number of different possible solutions that can be investigated is limited.

Both OPNET and ns-2, two simulators commonly used to simulate sensor networks, originated as simulators for wired networks that have been extended to simulate wireless networks [32]. While sensor networks have some similarities to wired networks, sensor networks are focused on minimizing energy consumption in order to extend lifetime. While OPNET, ns-2, and Parsec are good tools for analyzing a protocol they are not inherently designed to analyze energy consumption. GloMoSim provides added support for energy consumption, but still focuses on the protocol or communication aspect of the sensor network. Future research is needed to add support for analyzing the network outside of the protocol or communication framework. This will provide a better picture of what is going on in a sensor network.

While some analytic models exist, they are usually targeted at a specific protocol, algorithm, or some other area of interest in the field of sensor networks. Analytic models to model entire sensor networks can be extremely limited in scope, often to the single network protocol in question, and are not well suited to allow customization by users or to evaluate alternatives wishing to investigate variations of the original idea. Research is needed to develop a "toolbox" of customizable analytical models for sensor networks that allow the designer control over all parameters of interest. With such a "toolbox," a designer can quickly construct an analytical model with the capability to alter the parameters to investigate design alternatives.

The ability to compare two or more different networks accurately, i.e., compare apples to apples, is needed to decide between different design alternatives. With a modeling approach, it is easier to verify if the implementation of the network follows the specifications of the network being investigated. Further, an analytical model is more transparent than the elaborate code written for a simulator, allowing other researchers to quickly see and understand how the model was implemented and to determine if the implementation closely follows the specifications for the particular network evaluated. This increased transparency will increase the ability to compare results obtained for two or more different networks.

Analytical models should be faster to evaluate and scale better than simulated models, especially when looking at large networks (10,000+ nodes) or longer time frames (i.e., months or years). Once solved the results from analytical models can be reused to very quickly and efficiently investigate performance of several design alternatives, such as using different sets of components to construct entities. Discrete event simulator-based methods typically require that the entire simulation be rerun to get this performance information.

## 25.4  Conclusions

Sensor network design requires evaluating several design tradeoffs to identify the set of alternatives that will provide a sensor network that meets the requirements having the longest lifetime. Maximizing lifetime of a sensor network adds robustness because that sensor network is better equipped to handle unforeseen circumstances. Discrete event simulation and analytical modeling, specifically Markov processes, are two methods enabling designers to explore the design space.

Work is required to construct and expand libraries for discrete event simulations for sensor networks. These extensions should focus on adding more advanced and involved capabilities to investigate noncommunication or protocol-based behavior of the entities making up a sensor network. A library of model frameworks for different types of sensor networks or situations would be helpful because one of the major tasks of developing a Markov process is to determine the state space and then the system of equations to describe the system. A library of generic or specific Markov process structures would greatly simplify this step.

Taking into account the criteria discussed in this chapter, Markov processes are a powerful and efficient tool for analysis of the sensor networks. The two main problems of identification of the state space and restraining of the dimensionality can be solved by the use of topologies as described in [36]. The use of topological entities yields a Markov processes that scale well with both the size of the network and amount of time the network is analyzed for. Thus, using the method of topological entities, designers can in a reasonable amount of time (run-time) analyze a very large sensor network for a very long time.

Topological entities automatically generate a library of Markov processes describing different types of smaller sensor networks. With this library of topological entities, designers can quickly generate a Markov process describing large new types of sensor networks. The topological entity approach can handle mobility by looking at how the topologies change as entities move around within the network [36]. Further, the topological entity framework enables exploration of the numerous design alternatives with only a handful of scalar arithmetic operations once the model is solved once [36].

Markov processes can be described by state machines which are simple to code into a discrete event simulator. Therefore, the Markov processes can be easily transferred into a discrete event simulator and the sensor network can be analyzed using the discrete event simulator [36]. This will limit the number of entities in the discrete event simulation, keeping the run-time within a reasonable time. Topological entities are a powerful tool for analyzing sensor networks.

## Terminologies

*RFID.* Radio Frequency IDentification, is a wireless device that associates a unique identification number to an object. RFID systems are composed of tags, which are attached to objects or assets, and readers which communicate with tags and transmit the tag identification numbers to the larger back-end system.

*Discrete event simulation.* A method of simulation where a system is broken up into a set of objects. All objects in the simulation are updated with every time advance. The time advance is defined in advance.

*Time-stepped discrete event simulation.* A discrete event simulation in which every time advance is the same amount of time, i.e. 1 min.

*Event-driven discrete event simulation.* A discrete event simulation in which time is advanced to the time of the next event in the system. In an event-driven discrete event simulation, time does not have to advance in fixed intervals.

*Deadlock.* Deadlock is when at least two simulation entities are waiting on events from the other entity before they can issue new events. In this case, neither entity will be able to advance their time. Deadlock occurs only in an event-driven discrete event simulation and can be recovered from.

*Markov process.* A process where the next state depends only on the current state of the system and the current input to the system.

*PARSEC.* PARSEC is a programming language specifically designed for both single processor, sequential, or and multiple processor, or parallel, discrete event simulation.

*GloMoSim.* GloMoSim is a library of routines for simulating sensor networks built on the PARSEC platform.

*Node.* A node is the most numerous of the devices in a sensor network. A node typically contains some processing capability, a wireless communication interface, and one or more sensors.

*SOSUS.* SOund SUrveillance System is a sensor network that is used to monitor the world's oceans.

## Questions

1. Calculate the received power, $P_r$, received 23 km away from a radio transmitter (FM band in the USA) transmitting a 10 W signal at 92.1 MHz. Assume that the transmission is into free space making $\alpha = 2$.

2. Assuming that the radio station 2 km away is transmitting in free space, meaning $\alpha = 2$, what is the transmitted power if the received power, $P_r$, is 1 W. The radio station transmits at 92.1 MHz.

3. If a radio receiver needs a received signal of at least 1 W to receive the transmission (play music) what is the maximum range of a 10-W transmitter when using this receiver. The radio station transmits at 92.1 MHz and is in free space so $\alpha = 2$.

4. Transportation of energy in a sensor network could be used to recharge depleted nodes, but requires an infrastructure that is not feasible for a sensor network. Based on the results from the previous three questions would it be economical for one node to transmit a signal to a depleted node to recharge the depleted node's battery?

5. What is the primary difference between time-stepped and event-driven discrete event simulations?

6. What is the primary bottleneck in an event-driven discrete event simulation?

7. Describe the difference between evaluating sensor networks using discrete event simulation and analytic modeling using Markov processes.

8. What is the primary drawback of analytic modeling using Markov process?

9. When using a Markov process why does the number of states, or the dimensionality, need to be kept to a minimum?

10. Why must several design alternatives be investigated during development of a sensor network?

# References

1. E. H. Callaway, Jr., Wireless Sensor Networks Architectures and Protocols. Auerbach, Boca Raton, FL, 2004.
2. C. Chee-Yee and S. P. Kumar, Sensor networks: evolution, opportunities, and challenges, Proceedings of the IEEE, vol. 91, pp. 1247–1256, 2003.
3. R. Szewczyk, E. Osterweil, J. Polastre, M. Hamilton, A. Mainwaring, and D. Estrin, Habitat monitoring with sensor networks, Communications of the ACM, vol. 47, pp. 34–40, 2004.
4. F. Zhao and L. Guibas, Wireless Sensor Networks An Information Processing Approach. Morgan Kaufmann, San Fransisco, CA, 2004.
5. H. Tim Tau, Using sensor networks for highway and traffic applications, Potentials IEEE, vol. 23, pp. 13–16, 2004.
6. K. Sohrabi and G. J. Pottie, Performance of a novel self-organization protocol for wireless ad-hoc sensor networks, Presented at Vehicular Technology Conference, 1999.
7. F. Mondinelli and Z. M. Kovacs-Vajna, Self-localizing sensor network architectures, IEEE Transactions on Instrumentation and Measurement, vol. 53, pp. 277–283, 2004.
8. S. Slijepcevic and M. Potkonjak, Power efficient organization of wireless sensor networks, Presented at IEEE International Conference on Communications (ICC 2001), 2001.
9. J. Qiangfeng and D. Manivannan, Routing protocols for sensor networks, Presented at First IEEE Consumer Communications and Networking Conference, 2004.
10. S. De, Q. Chunming, and W. Hongyi, Meshed multipath routing: an efficient strategy in sensor networks, Presented at IEEE Wireless Communications and Networking (WCNC 2003), 2003.
11. V. Tsiatsis, S. A. Zimbeck, and M. B. Srivastava, Architecture strategies for energy-efficient packet forwarding in wireless sensor networks, Presented at International Symposium on Low Power Electronics and Design, 2001.

12. A. Sinha and A. Chandrakasan, Dynamic power management in wireless sensor networks, Design and Test of Computers, IEEE, vol. 18, pp. 62–74, 2001.
13. A. Boulis and M. B. Srivastava, Node-level energy management for sensor networks in the presence of multiple applications, Presented at Proceedings. of the First IEEE International Conference on Pervasive Computing and Communications (PerCom 2003), 2003.
14. J. Agre and L. Clare, An integrated architecture for cooperative sensing networks, Computer, vol. 33, pp. 106–108, 2000.
15. C.-C. Shen, C. Srisathapornphat, and C. Jaikaeo, Sensor information networking architecture and applications, Personal Communications, IEEE [see also IEEE Wireless Communications], vol. 8, pp. 52–59, 2001.
16. Z. Feng, S. Jaewon, and J. Reich, Information-driven dynamic sensor collaboration, Signal Processing Magazine, IEEE, vol. 19, pp. 61–72, 2002.
17. S. S. Pradhan, J. Kusuma, and K. Ramchandran, Distributed compression in a dense microsensor network, Signal Processing Magazine, IEEE, vol. 19, pp. 51–60, 2002.
18. C. Haowen and A. Perrig, Security and privacy in sensor networks, Computer, vol. 36, pp. 103–105, 2003.
19. A. D. Joseph, Energy harvesting projects, Pervasive Computing, IEEE, vol. 4, pp. 69–71, 2005.
20. R. Want, K. I. Farkas, and C. Narayanaswami, Guest editors' introduction: Energy harvesting and conservation, Pervasive Computing, IEEE, vol. 4, pp. 14–17, 2005.
21. M. Rahimi, H. Shah, G. S. Sukhatme, J. Heideman, and D. Estrin, Studying the feasibility of energy harvesting in a mobile sensor network, Presented at IEEE International Conference on Robotics and Automation (ICRA '03), 2003.
22. J. A. Paradiso and T. Starner, Energy scavenging for mobile and wireless electronics, Pervasive Computing, IEEE, vol. 4, pp. 18–27, 2005.
23. A. Kansal and M. B. Srivastava, An environmental energy harvesting framework for sensor networks, International Symposium on Low Power Electronic and Design, 2003.
24. R. M. Fujimoto, Parallel and Distributed Simulation Systems. Wiley, New York, NY, 2000.
25. G. F. Riley, Large-scale network simulations with GTNetS, Presented at Proceedings of the 2003 Winter Simulation Conference, 2003.
26. Z. Ji, J. Zhou, M. Takai, and R. Bagrodia, Scalable simulation of large-scale wireless networks with bounded inaccuracies, in Proceedings of the Seventh ACM International Symposium on Modeling, Analysis and Simulation of Wireless and Mobile Systems, Venice, Italy. ACM, New York, NY, 2004.
27. V. Naoumov and T. Gross, Simulation of large ad hoc networks, in Proceedings of the Sixth ACM International Workshop on Modeling Analysis and Simulation of Wireless and Mobile System, San Diego, CA, USA. ACM, New York, NY, 2003.
28. H. Akhtar, An overview of some network modeling, simulation and performance analysis tools, Presented at Proceedings of Second IEEE Symposium on Computers and Communications, 1997.
29. GloMoSim Manual Version 1.2, UCLA Parallel Computing Laboratory, http://pcl.cs.ucla.edu/projects/glomosim/GloMoSimManual.html.
30. X. Zeng, R. Bagrodia, and M. Gerla, GloMoSim: a library for parallel simulation of large-scale wireless networks, Presented at Proceedings of Twelfth Workshop on Parallel and Distributed Simulation (PADS 98), 1998.
31. R. Bagrodia, R. Meyer, M. Takai, Y.-A. Chen, X. Zeng, M. Jay, and H. Y. Song, Parsec: a parallel simulation environment for complex systems, Computer, vol. 31, pp. 77–85, 1998.
32. D. Cavin, Y. Sasson, and A. Schiper, On the accuracy of MANET simulators, in Proceedings of the Second ACM International Workshop on Principles of Mobile Computing, Toulouse, France. ACM, New York, NY, 2002.
33. D. L. Isaacson and R. W. Madsen, Markov Chains, Theory and Applications. Wiley, New York, NY, 1976.
34. M. H. Mickle and T. W. Sze, Optimization in Systems Engineering. Intext Educational Publishers, Scranton, PA, 1972.

35. C. F. Chiasserini and M. Garetto, Modeling the performance of wireless sensor networks, Presented at 23rd Annual Joint Conference of the IEEE Computer and Communications Societies (INFOCOM 2004), 2004.
36. P. J. Hawrylak, Analysis and Development Of A Mathematical Structure To Describe Energy Consumption Of Sensor Networks, Ph.D. dissertation, University of Pittsburgh, Pittsburgh, PA, 2006.

# Chapter 26
# Bio-inspired Communications in Wireless Sensor Networks

**Barış Atakan, Özgür B. Akan, and Tuna Tuğcu**

**Abstract** Wireless-sensor networks (WSN) are expected to enable connection between physical world and the Internet to provide access to vast amount of information from anywhere and anytime through any kind of communication devices and services. However, this vision poses significant challenges for WSN. Due to the pervasion in its nature, centralized control of WSN is not a practical solution. Instead, WSN and its communication protocols must have the capabilities of scalability, self-organization, self-adaptation, and survivability. In nature, the biological systems intrinsically have these capabilities such that billions of blood cells, which constitute the immune system, can protect the organism from the pathogens without any central control of the brain. Similarly, in the insect colonies insects can collaboratively allocate certain tasks according to the sensed information from the environment without any central controller. Therefore, the natural biological systems may give great inspiration to develop the communication network models and techniques for WSN. In this chapter, we introduce potential solution avenues from the biological systems toward addressing the challenges of WSN such as scalability, self-organization, self-adaptation, and survivability. After the existing biological models are first investigated, biologically inspired communication approaches are introduced for WSN. The objective of these communication approaches is to serve as a roadmap for the development of efficient scalable, adaptive, and self-organizing bio-inspired communication techniques for WSN.

## 26.1 Introduction

Recent developments in electronics have enabled low cost and low powered sensor nodes with limited wireless communications capability [1]. These multifunctional sensor nodes have enabled wireless sensor networks (WSN) consisting of large

B. Atakan (✉)
Department of Electrical and Electronics Engineering, Middle East Technical University, Ankara 06531, Turkey
e-mail: atakan@eee.metu.edu.tr

S. Misra et al. (eds.), *Guide to Wireless Sensor Networks*, Computer Communications and Networks, DOI: 10.1007/978-1-84882-218-4_26,
© Springer-Verlag London Limited 2009

number of densely deployed sensor nodes for any monitoring task. Due to spatial deployment of sensor nodes in an environment, WSN provide great monitoring capabilities beyond traditional monitoring mechanisms in this environment. Together with the developments in WSN domain, Internet is getting to include more information obtained from physical world via small sensor nodes. Therefore, WSN are expected to enable connection between physical world and the Internet to provide access to vast amount of information from anywhere and anytime through any kind of communication devices and services. However, this vision poses significant challenges for WSN. Due to the pervasion in its nature, centralized control of WSN is not a practical solution. Instead, WSN and its communication protocols must have the capabilities of scalability, self-organization, self-adaptation, and survivability.

WSN are event-based systems, which enable the sensor nodes to communicate the properties of observed event to the sink. This communication from event to sink has to be reliable and timely to enable the sink node to reliably estimate the event properties and to timely perform the appropriate action if required. However, in WSN, it cannot be easily ensured that observed event information can be timely and reliably communicated to the sink node because most of the information may be lost on the forward path from the event to the sink node. Therefore, centralized solutions are not practical for WSN since coordination between the sink node and sensor nodes cannot be ensured. Furthermore, due to the pervasion of sensor nodes over large geographic observation areas, it cannot be possible to engineer the centralized controllers to reliably and timely communicate the observed event information. Thus, self-organized protocols that do not need any central controller are imperative for WSN.

The notion of the self-organizations provides some important capabilities such as self-adaptation, survivability, and scalability.

- *Self-Adaptation*. Self-organization allows WSN to adapt any change in the environment or network by regulating communication parameters of the sensor nodes.
- *Survivability*. Self-organization allows WSN to survive in any state of the network. For example, when some node failures are experienced, self-organization provides the capability of survivability such that it enables sensor nodes to survive for reliable and timely communications of event information.
- *Scalability*. Self-organization allows each sensor node to locally interact with its neighbor nodes to communicate the event information. Therefore, as size of the network increases, the self-organization allows WSN to pursue their normal operations, that is, growing network size does not negatively affect the normal operations of the network.

As needed for WSN, self-organization is imperative for almost all of infrastructureless network architectures. Especially, ad hoc network architectures heavily need self-organization for several network-wide tasks such as topology formation and routing. WSN are event-based systems, which trigger the event-to-sink data communications as soon as an event is detected and this communication is heavily affected by properties of the event. For example, size of the event area in which source nodes[1] are selected is critical to determine how many sensors are selected as source nodes to cover the event area and which sensor nodes should be selected as source nodes. This event-based communication paradigm in WSN imposes unique challenges that necessitate great self-organization capability to WSN beyond the traditional self-organization mechanisms in ad hoc network architectures. This self-organization capability can enable the network to organize according to the properties of the event, dynamically changing environment, and network conditions.

In nature, biological systems are intrinsically self-organized systems and most of the biological systems can organize to react to biological events without any need to a central controller. For example, in the immune system, when a pathogen enters the body, this event is detected by T-cells in immune system, and T-cells trigger B-cells. Then, these triggered B-cells organize to determine which B-cells are most appropriate to react to the pathogen. This way, the pathogens are reliably eliminated within a specific time delay before the pathogen results in an infection in the organism.

In the homeostatic system, nervous, endocrine, and immune systems organize to bring the organism from an unstable state to a stable state. Gland cells in endocrine system, neurons in nervous system, and blood cells in immune system collaboratively react to an unstable state of the organism to bring the organism back to the stable state. During this operation, without any need of a central controller, homeostatic system enables the organism to reliably bring into a stable state within a specific time delay before some functions of the organism are corrupted due to the unstable state.

In this chapter, we introduce potential solution avenues from the biological systems toward addressing the challenges of WSN such as scalability, self-organization, self-adaptation, and survivability. After the existing biological models are first investigated, biologically inspired communication approaches are introduced for WSN. The objective of these communication approaches is to serve as a roadmap for the development of efficient scalable, adaptive, and self-organizing bio-inspired communication techniques for WSN.

The remainder of this chapter is organized as follow. In Sect. 26.2, we briefly introduce the concept of WSN and we discuss how biological systems can enable a solution avenue for the challenges in WSN domain and then, we give previous bio-inspired solutions in the literature for WSN. In Sects. 26.3 and 26.4, we introduce immune-system-based WSN establishing some analogy and mapping between these concepts such that this can enable an efficient bio-inspired communication protocol for WSN. In Sects. 26.5 and 26.6, we introduce a homeostatic-system-based Wireless Multimedia Sensor Networks (WMSN) based on the established analogies between a homeostatic system and WMSN to provide an energy-efficient, reliable, and distributed communications algorithms for WMSN. In Sect. 26.7, we discuss the potential analogies between insect colonies and Wireless Sensor and

---

[1] Here, we consider a source node as a sensor node which senses and samples the event signal and forwards to the sink node.

Actor Networks (WSAN) and we adopt biological task allocation phenomenon in an insect colony to enable an energy-efficient, delay-aware, and reliable communications algorithm for WSAN. In Sect. 26.8, we give thoughts for practitioners and directions of future research on bio-inspired communication techniques for WSN. In Sect. 26.9, we conclude the envisioned bio-inspired communication methods for WSN.

## 26.2 Wireless Sensor Networks and Biological Systems

A WSN consists of large number of sensor nodes that are densely deployed either inside the phenomenon or around the phenomenon [1]. Sensor nodes in the network are randomly positioned. This necessitates sensor network protocols and algorithms having the capability of self-organization among the sensor nodes. Due to unique challenges in WSN domain, self-organization is one of the most critical network capabilities in WSN. However, every self-organized algorithm may not be appropriate to enable an efficient event-to-sink data communication. For example, a self-organized algorithm that excessively depends on coordination among sensor nodes is not an appropriate algorithm because the coordination among sensor nodes imposes excessive energy cost to sensor nodes. Moreover, the coordination also imposes a delay burden and prevents timely event-to-sink communication of sensor nodes. Thus, it is essential to design a self-organized algorithm, which can enable reliable and delay-aware event-to-sink data communication with minimum energy consumption.

As a result of natural evolution, biological systems acquire great self-organization capabilities that can be adopted and modeled to overcome the challenges in WSN domain. In essence, since almost every biological system is composed of small entities, self-organization is the most important capability for biological systems to organize the small entities for an ultimate aim. For example, human immune system is composed of white blood cells named as B-cell and T-cell. The task of defending the body toward pathogens can be achieved by means of the self-organization among B-cells and T-cells. As in biological systems, in WSN, self-organization among sensor nodes is critical to enable energy-efficient, reliable, and delay-aware communications. In fact, energy-efficiency, delay-awareness, and reliability are common for biological systems and WSN. In biological systems, these must be achieved to survive by exploiting self-organization phenomenon. For example, in insect colonies, billions of insects can organize to effectively react to an event, which occurs in the colony so that minimum and uniformly distributed energy consumption can be provided while they reliably react to the event within a specific time delay. For example, if the task is larval feeding, larvae are fed by the insects that can sense the larval demand and have sufficient energy such that these insects can feed the larvae within a specific time delay before the larvae starve.

Due to the parallelism between biological systems and WSN, bio-inspired algorithms for the challenges in WSN have attracted the researchers in the area of computer networks. There exist several research efforts on bio-inspired algorithms and protocols for WSN. The bio-inspired protocols in the literature are mainly based on several biological phenomena such as ant colony, fireflies, quorum sensing, symbiotic, cellular and molecular process, genetic system, and self-healing.

### 26.1.1 Background

Biological colonies of ants and bees typically consist of tens of thousands of dynamic elements [2]. In the ant colonies, each ant has relatively little intelligence, while the collaborative behavior of the colony provides a great deal of global intelligence capable of optimizing certain tasks [2]. This global intelligence capability can certainly be modified to apply to virtually any kind of challenge in WSN domain. Especially, ant colonies have great routing capabilities to reach available food sources. In the literature, this great capability of ant colonies can be exploited to enable efficient routing algorithms, which are self-organized, fault-tolerant, and scalable. In [3], an energy efficient routing algorithm is proposed that emulates the direction finding capability of ants. In [4], an energy efficient and delay-aware routing algorithm is proposed emulating ant-colony-based algorithms. In [5], AntHocNet routing algorithm is shown to outperform most routing algorithms in the literature in terms of packet delivery ratio, scalability, average end-to-end delay, and average jitter.

For WSN, synchronization of the sensor nodes is essential for almost all of WSN applications. Furthermore, distributed and scalable synchronization algorithms are imperative to enable the sensor nodes to timely and distributively perform a given task such as instantaneous monitoring of a phenomena, velocity measurement of moving objects, etc. Biological synchronization phenomena have great potential to enable distributed and scalable synchronization algorithms for WSN. In [6], a bio-inspired scalable network synchronization protocol for large scale sensor networks is proposed, which is inspired by the simple synchronization strategies in biological phenomena such as flashing fireflies and spiking of neurons. In [7], a biologically inspired distributed synchronization algorithm is introduced based on a mathematical model explaining how neurons and fireflies spontaneously synchronize.

For a biological organism, genes are the most critical actors of all biological operations related with proliferation and protein synthesis. Genes manage these operations without any need of central control of the brain. Therefore, this management process is performed by means of self-organization. In [8], the principles of genetics and evolution are adopted to enable service-oriented, autonomous, and self-adaptive communication systems for pervasive environments such as WSN and mobile ad hoc networks.

In an organism, all vital biological operations are performed by means of self-organization among components of the organism from a single cell up to complex organs. In [9], the principles of cell and molecular biology are adopted to enable efficient and scalable communication architectures. After the mapping between computer networks and cell and molecular biology is drawn, efficient, scalable, and self-organized autonomous communication network models are introduced based on the similarity between cell and molecular biology and computer networks. In [10], efficient bio-inspired communication paradigm for WSN is proposed based on the feedback loop mechanism developed by inspiration from the principles of cell biology. In [11], a bio-inspired congestion control mechanism for WSN is proposed with inspiration from signaling pathways in cell and molecular biology such that the bio-inspired algorithm does not need any topology and address knowledge.

Clustering is a useful technique in sensor networks that prevents the sensor nodes from using a lot of energy to transmit their data over large distances to the base station and reduces the dependency on individual nodes that are prone to failure [12]. Clustering phenomenon in WSN is an emergent behavior of the network. Clusters are established according to characteristic of the event such as size of the event area, location of the event, etc. Therefore, clustering should be distributively performed by sensor nodes according to the characteristic of the event. In biology, clustering is an indispensable phenomenon, which enables biological entities to establish emergent clusters to perform any kind of biological operations. Quorum sensing is a biological process used by bacterial cells to monitor when the cell density in their vicinity exceeds a certain threshold that leads to a change in their behavior [12]. Based on biological quorum sensing mechanism, the authors propose a clustering algorithm to enable the sensor nodes to form clusters according to spatial characteristics of the observed event signal.

Next, we introduce three different bio-inspired models that enable efficient self-organized, scalable, and self-adaptive algorithms for WSN. For each model, we first give biological model and then, we draw an analogy, which can provide a mapping between the biological system and WSN. According to this mapping, we introduce possible roadmap that allows efficient bio-inspired network protocols for WSN. Then, we discuss advantages of the envisioned network protocols in terms of their capabilities such as self-organization, scalability, and self-adaptation.

## 26.3  Immune System and Wireless Sensor Networks

In this section, we first briefly introduce the biological immune system. Then, we draw an analogy from the immune system to WSN [13].

### 26.3.1  Biological Immune System

The biological immune system is a natural defense mechanism to recognize foreign substances (pathogens) and to respond to them producing antibodies [14].

The operation of immune system to eliminate the pathogen consists of two main tasks performed by white blood cells named as B-cells and T-cells. These two main operations are known as B-cell stimulation and antibody secretion.

### 26.3.1.1   B-cell Stimulation

The immune system consists of white blood cells named as B- and T-cells. Each of B-cells has distinct molecular structure and produces antibodies from its surfaces. B-cells have the capability of antibody secretion such that the secreted antibodies recognize antigen produced by the pathogen and eliminate it. When an antibody of a B-cell binds to an antigen, the B-cell becomes stimulated. The level of B-cell stimulation depends on the success of the match to the antigen and other B-cells in the immune network [14]. The level of B-cell stimulation is determined by three factors [14]:

- First factor is the affinity between the B-cell and pathogen. This factor is defined as follows

$$ps = (1 - d), \tag{26.1}$$

where ps denotes the stimulation effect of the pathogen, $d$ is the distance between B-cell and pathogen, and $d$ is normalized between 0 and 1 ($0 \leq d \leq 1$).
- Second factor is the affinity between B-cell and its neighbors that stimulate it. This factor is expressed as

$$ns = \sum_{i=1}^{n} (1 - d_i), \tag{26.2}$$

where ns denotes the effect of stimulation from the neighbors, which stimulate the B-cell, $d_i$ is the normalized distance ($0 \leq d_i \leq 1$) between the B-cell and its $i$th neighbor, which stimulates it, $n$ is the number of neighbors, which stimulate the B-cell.
- A B-cell is also suppressed by some loosely connected neighbors. Therefore, third factor is the affinity between the B-cell and its neighbors, which suppress it. This factor can be expressed as

$$nn = -\sum_{i=1}^{m} d_i, \tag{26.3}$$

where nn denotes the effect of suppression from the neighbors, which suppress it; $d_i$ is the normalized distance ($0 \leq d_i \leq 1$) between the B-cell and its $i$th neighbor, which suppress it; and $m$ is the number of suppressing neighbors. Hence, the total stimulation of a B-cell, i.e, sl, can be expressed as follows [14]

$$sl = ps + ns + nn. \tag{26.4}$$

Consequently, when total stimulation level of a B-cell (sl) exceeds a certain threshold, this B-cell is thought to be stimulated by the pathogen and starts secreting its antibodies to eliminate the antigen produced by the pathogen [14].

As will be explained in Sect. 26.4, we adopt the natural B-cell selection mechanism briefly introduced above to develop the efficient source node selection model that enables sensor nodes to distributively select source nodes, which provide efficient event signal reconstruction performance.

### 26.3.1.2 Antibody Secretion

After B-cell stimulation, the stimulated B-cells secrete free antibodies to eliminate the antigen produced by the pathogen. Based on the stimulation and suppression level of B-cells and the natural extinction of antibodies, the antigen concentration is collaboratively kept at a desired level by regulating the antibody concentrations [15]. This model is analytically given as [16]

$$\frac{dS_i(t+1)}{dt} = \left( \alpha \sum_{j=1}^{N} m_{ij}s_j(t) - \alpha \sum_{k=1}^{N} m_{ki}s_k(t) + \beta g_i - k_i \right) s_i(t), \qquad (26.5)$$

where $s_i$ is the concentration of antibody $i$, $m_{ij}$ is mutual coefficient of antibody $i$ and $j$, and $N$ is the number of B-cell type. $\sum_j m_{ij}s_j(t)$ denotes the effect of stimulated neighbors of B-cell $i$ and $\sum_k m_{ki}s_k(t)$ denotes the effect of suppressing neighbors of B-cell $i$, $g_i$ is the affinity between antibody $i$ and antigen, $k_i$ is the natural extinction of the antibody $i$, and $\alpha$ and $\beta$ are constants. $S_i$ denotes the total stimulation of antibody $i$ secreted by B-cell $i$. Based on $S_i$, the concentration of antibody $i$ ($s_i$) is given as

$$s_i(t+1) = \frac{1}{1 + e^{(0.5 - S_i(t+1))}}. \qquad (26.6)$$

The basic operation of the model given in (26.5) and (26.6) can be outlined as follows:

- When a pathogen enters the body, affinity between antigen and B-cell $i$ ($g_i$) increases since antigen concentration increases. This results in increase in the antibody secretion of B-cell $i$ and $s_i$ increases.
- If B-cell $i$ starts to be mostly suppressed by its neighbors, B-cell $i$ decreases antibody secretion and $s_i$ decreases.
- If B-cell $i$ starts to be mostly stimulated by its neighbors and antigens, B-cell $i$ increases the antibody secretion and $s_i$ increases.
- If the natural extinction of antibody $i$ secreted by B-cell $i$ ($k_i$) increases, B-cell $i$ decreases the antibody secretion and $s_i$ decreases.

In Sect. 26.4, we propose an effective frequency rate selection model for sensor nodes based on the antibody secretion mechanism given in (26.5) and (26.6). In Sect. 26.3.2, we introduce some analogies between Immune System and WSN.

### 26.3.2  Immune-System-Based Sensor Networks

The immune system and WSN are in different concepts. However, many analogies between them may be established in terms of their functions. When a pathogen enters a body, immune system is triggered by the pathogen such that it stimulates some B-cells and allows these stimulated B-cells to secrete antibodies in different densities to eliminate the antigens produced by the pathogen. Similarly, when an event occurs in WSN environment, some sensor nodes referred to as source node sense the event and send the event information to sink with a reporting frequency rate ($f$) to achieve a certain event signal reconstruction distortion at the sink.

In WSN, it is essential to select source nodes to reconstruct the event signal at the sink node within a certain distortion level as well as to conserve energy over the network. For a fixed number of source nodes, the minimum distortion can be achieved by choosing these nodes such that (1) they are located as close to the event source as possible and (2) they are located as farther apart from each other as possible [17]. Similarly, as given in [14], it is most possible for a B-cell to be excited if it is located as close to the pathogen and its stimulated B-cells and located as farther apart from its suppressed neighbors. After the stimulation, the stimulated B-cells secrete the antibodies in different densities to keep the antigen densities at a desired level. Similarly, after the selection of source nodes, the selected source nodes send the event information to the sink with a certain reporting frequency to achieve the distortion constraint at the sink node.

## 26.4  Immune-System-Based Distributed Node and Rate Selection

In this section, based on the analogy between natural immune system and WSN, we discuss a possible roadmap to develop an efficient protocol, which enables the sensor nodes to distributively select source nodes and to regulate their frequency rate. The aim of this protocol is to enable the sensor nodes to distributively achieve an application-specific event signal reconstruction distortion at the sink node with minimum energy consumption. We first discuss source node selection based on the principles of B-cell stimulation and then, we discuss frequency rate selection to successfully reconstruct the event signal at the sink node using the principles of antibody secretion.

### 26.4.1 Distributed Source Node Selection

In WSN, the minimum distortion can be achieved by choosing the source nodes such that (1) they are located as close to the event source as possible and (2) they are located as farther apart from each other as possible [17]. (1) implies that source nodes should be highly correlated with event source. Therefore, a source node should be selected as a sensor node that has maximum correlation with event source. To express the correlations between event source and the sensor nodes, we use the correlation coefficients, $\rho_{s,i}$. $\rho_{s,i}$ indicates the correlation between sensor node $i$ and event source. (2) implies that source nodes should be uncorrelated with each other as possible. Therefore, a source node should be selected as a sensor node that has minimum correlation with its neighbors. To express the correlation between a sensor node and its neighbors we use the correlation coefficients $\rho_{i,j}$. $\rho_{i,j}$ indicates the correlation between sensor node $i$ and sensor node $j$. We use the power exponential form to model the correlation coefficients $\rho_{s,i}$ and $\rho_{i,j}$ as [18]

$$\rho_{s,i} = K_\vartheta(d_{s,i}) = e^{(-d_{s,i}/\theta_1)^{\theta_2}} ; \quad \theta_1 > 0, \theta_2 \in (0,2], \tag{26.7}$$

$$\rho_{i,j} = K_\vartheta(d_{i,j}) = e^{(-d_{i,j}/\theta_1)^{\theta_2}} ; \quad \theta_1 > 0, \theta_2 \in (0,2], \tag{26.8}$$

where $d_{s,i}$ and $d_{i,j}$ are the distances between event source and sensor node $i$ and between sensor nodes $i$ and $j$, respectively. The correlation coefficients are assumed to be nonnegative and decrease with the distance, with limiting values of 1 at $d = 0$ and of 0 at $d = \infty$.

Based on these selection criteria, the selection of source nodes is influenced from the three factors similar to B-cell stimulation principles given in Sect. 26.3.1.1:

- First factor is the affinity between sensor node and event source and can be modeled as $\rho_{s,i}$.
- Second factor is the affinity between the sensor node $i$ and its uncorrelated neighbors. Here, we define an application-specific correlation radius $r$ and we assume that for a sensor node, the neighbor nodes in its correlation radius $r$ are the correlated neighbors for this sensor node, while the neighbor nodes, which are not in correlation radius $r$ are uncorrelated neighbors for this sensor node. The second factor can be modeled as

$$\sum_j (1 - \rho_{i,j}). \tag{26.9}$$

Here, sensor node $j$ is selected as the neighbor node, which is not in the correlation radius of sensor node $i$. Therefore, $\sum_j (1 - \rho_{i,j})$ is large for a sensor node that has more uncorrelated neighbors (out of $r$). Hence, it is more possible to become a source node for such sensor nodes.

- Third factor, which is the affinity between the sensor node and its correlated neighbor sensor nodes (in $r$) $k, \forall k$ such that sensor node $k$ is in the correlation radius of sensor node $i$ and can be given as

$$\sum_k (-\rho_{i,k}). \qquad (26.10)$$

Here, $\sum_k (-\rho_{i,k})$ is small for a sensor node that has more correlated neighbors. Hence, it is the least possible for such sensor nodes to become a source node.

As the combination of these three factors, the source node selection weight of sensor node $i$ $(T_i)$ can be given as

$$T_i = \rho_{s,i} + \sum_j (1 - \rho_{i,j}) + \sum_k (-\rho_{i,k}). \qquad (26.11)$$

Since each sensor node knows locations of its and its neighbors, each sensor node can compute its source node selection weight $(T_i)$ according to the correlation coefficients. Here, we assume that sensor node $i$ becomes the source, when $T_i$ exceeds a certain threshold. While the threshold increases, number of source node decreases because the number of nodes whose weight $(T_i)$ exceeds the threshold decreases. Conversely, while the threshold decreases, number of source nodes increases. Therefore, every selected threshold imposes a number of source nodes to the network. In fact, number of source nodes should be carefully selected to effectively reconstruct the event signal at the sink node to control the generated traffic load over the network in terms of energy consumption and reliable reconstruction of event signal at the sink node. Furthermore, regulation of reporting frequency rate of selected source nodes is essential for energy efficient and reliable communication of event signal. In Sect. 26.4.2, based on the relations between immune system and WSN, we discuss a possible roadmap of developing an efficient self-organized rate selection algorithm for WSN.

### 26.4.2 Distributed Frequency Rate Selection of Source Nodes

In WSN, reporting frequency of a source node is defined as the number of packets transmitted per unit time by this node. It is critical for WSN to regulate the reporting frequency rate of the source nodes in terms of energy conservation and event signal reconstruction distortion at the sink. To achieve a certain distortion level at the sink, certain number of packets must be delivered to the sink per unit time as well as minimum amount of data traffic that imposes minimum energy consumption have to be generated. Therefore, reporting frequency rate of source nodes should be regulated based on packet-loss rate of source nodes and event signal reconstruction distortion at the sink node such that it should be ensured that a certain number of packets can be delivered to achieve certain distortion level at the sink. Similarly, in the natural immune system, stimulated B-cells can collaboratively regulate their antibody secretion to keep antigen concentration in a certain level based on antigen concentration and natural extinction of their antibodies according to model given in

(26.5) and (26.6). Using this natural mechanism, we introduce an efficient reporting frequency rate selection mechanism for source nodes in the following way [13]:

- Each source node is considered as a stimulated B-cell.
- The source node packets are considered as the antibodies secreted by B-cells, and we model the reporting frequency of the source nodes ($f_i$) as the antibody concentration given by $s_i$. $f_i$ denotes the reporting frequency identifier of source node $i$ and it is determined by normalizing the actual reporting frequency of source node $i$ to a number between 0 and 1.
- To regulate the reporting frequency rate of source nodes according to the reconstruction distortion ($D$) and the packet-loss rate $\lambda_i$, we consider $S_i$ as the rate control parameter $F_i(t_k)$ and calculate it as

$$F_i(t_{k+1}) = F_i(t_k) + \left( \frac{1}{K} \sum_{j=1}^{K} f_j(t_k) + aD - b\lambda_i \right) f_i(t_k), \qquad (26.12)$$

where $a$ and $b$ are constants, $K$ is the number of source nodes, $t_k$ is the $k$th time interval, $f_i(t_k)$ is the reporting frequency identifier of source node $i$ at time interval $t_k$, and $F_i(t_{k+1})$ is the rate control parameter at $t_{k+1}$.

- Based on the rate control parameter $F_i(t_k)$ given in (26.12), the reporting frequency identifier $f_i(t_{k+1})$ is given as

$$f_i(t_{k+1}) = \frac{1}{1 + e^{(0.5 - F_i(t_k))}}. \qquad (26.13)$$

This kind of frequency rate regulation scheme provides a new congestion control mechanism for sensor networks. While at the data paths in which the packet losses arise, $f$ is decreased; at the noncongested paths $f$ is increased. Furthermore, this kind of rate control mechanism enables sensor nodes to organize without any need for a central control such that the event signal can be satisfactorily reconstructed at the sink node while providing great energy conservation by means of the collaborative effort of sensor nodes. This collaborative effort of sensor nodes also enables scalable and fault-tolerant communication algorithms for WSN.

## 26.5 Biological Homeostasis and Wireless Multimedia Sensor Networks

### 26.5.1 Biological Homeostatic System

A vital functionality of many biological organisms is the ability to maintain a stable internal state although the external environmental conditions may change rapidly [19]. This functionality is called homeostasis, and it is the leading feature of an organism to sustain its autonomy. The scientific approach, which is the most direct representation of the autonomy in a homeostatic mechanism, is the dynamical systems approach. In this approach, the state of an organism is represented by some state space, and homeostasis is usually assumed to be located on a cyclic path around some attractor point that represents the normal condition for the organism.

By means of homeostatic mechanisms, the organism self-regulates its growth and development, and maintains itself in a stable condition. To maintain homeostatic stability within an organism, the nervous system, the endocrine system, and the immune system behave as one large, unified, and complex system. The interaction and communication among all the three systems are provided by the specific receptors on the cells [19].

A biological organism is open to various external stimuli. The nervous system of the organism takes the stimuli, e.g., taste, smell, vision, etc., via the sensory parts, and triggers an output reaction at the effectors, e.g., tissues and muscles. Two types of cells that take part in this reactive process are neurons and neuroglia [19]. Neurons generate electrical impulses in response to an input stimulus, and neuroglia support neurons in terms of nutrition, development, and maintenance. After the nervous system detects any change in the internal state of the organism, the endocrine system produces and releases hormones through gland cells. Thus, the interaction between the nervous system and the endocrine system is the homeostatic response behavior of the organism to maintain its stable internal state. Any malfunction that adversely affects the operation of the organism is detected by the immune system of the organism.

In conclusion, each of the three systems is constantly interacting with each other and the collaboration of the neural, immune, and endocrine systems provides a proper model for the construction and development of self-organizing, highly functional, and adaptable intelligent systems.

## 26.5.2  Homeostasis-Based Wireless Multimedia Sensor Networks

In nature, it is critical for the organisms to keep themselves in stable states, by means of homeostatic mechanisms. Similar to an organism, WMSN must keep itself within a stable state. This state provides WMSN with minimum packet-loss rate and minimum energy consumption under varying spectral characteristics of the multimedia event signal.

In biological homeostatic system, the aim of the neural system is to perceive the external environment and manage the endocrine and immune systems to maintain a biologically stable state based on the interaction with the endocrine and immune systems. Likewise, in WMSN, some sensor nodes must sense the spectral characteristic of the multimedia event signal and manage source nodes and intermediate nodes to keep WMSN within a stable state providing minimum packet loss and minimum energy consumption.

In WMSN, the spectral characteristics of event signal impose a sampling frequency rate on source nodes to accurately reconstruct the multimedia event signal according to Nyquist Sampling Theory [20]. The spectral characteristics of the multimedia event signal determine the total number of samples transmitted per unit time over the network and hence, the traffic loads on the forward paths. When the traffic load over the network is excessively high, this increases possible congestion, collision, and channel errors on the forward paths. Thus, it is imperative to estimate the spectral bandwidth of the multimedia signal for the efficient multimedia transport providing energy efficient and reliable communication for sensor nodes. Based on this similarity, we consider some sensor nodes as neurons in neural system and call them as N-Sensors. Like a neuron in neural system, in WMSN the aim of each N-Sensor is to estimate the spectrum of the sensed multimedia event signal and manage the source and intermediate nodes to effectively reconstruct the multimedia event signal at the sink.

In the endocrine system, gland cells secrete hormones to keep the organism in the biologically stable state based on the interaction with the neural and the immune system. Similarly, in WMSN, source nodes sample and transmit the multimedia event signal to sink node to fulfill the application objective, i.e., successful reconstruction of event signal at the sink with minimum energy consumption. According to this similarity, we consider the source nodes as gland cells and call them G-Sensors.

In immune system, T-cells sense any malfunction in the organism and trigger the neural and endocrine systems to keep the organism within the biologically stable state. Similarly, in WMSN, some sensor nodes must detect any malfunction such as congestion, collision, and channel error to effectively reconstruct the multimedia event signal. With this regard, we consider the intermediate nodes from sources to the sink as T-cells and call them as T-Sensors.

## 26.6 Homeostasis-Based Multimedia Communication in Wireless Multimedia Sensor Networks

In this section, based on the mapping given above between biological homeostatic system and WMSN, we detail the bio-inspired operations performed by N-Sensors, G-Sensors, and T-Sensors such that these operations can enable a transition from biological domain to WMSN domain emulating the principles of a homeostatic system for efficient communication in WMSN.

- *N-Sensor selection.* Sensor nodes, which detect the event signal, select an N-Sensor. Using event source location [21], sensor nodes collaboratively select the most appropriate sensor node that is nearest to the event source as N-Sensor. Apart from proximity of N-Sensor, it is important for N-Sensor that it can capture maximum signal power from the multimedia event signal with respect to its neighbors.

- *Spectrum estimation.* N-Sensor estimates the spectrum bandwidth of the multimedia event signal [22] to determine how many samples must be delivered by sensor nodes to satisfactorily reconstruct the event signal at the sink.
- *G-Sensors selection.* G-Sensors are the source nodes that sample and communicate the multimedia event signal to the sink node. N-Sensor first determines the number of G-Sensors and their reporting frequency rate such that the multimedia event signal can be satisfactorily reconstructed. Then, N-Sensor selects the most appropriate sensor nodes which are nearest to the event source as G-Sensors.
- *Path determination.* Every sensor node that intends to transmit selects one of its neighbors as its next hop such that this neighbor is the closest node to the sink. This provides minimum-hop packet delivery, which is crucial for real-time multimedia communication with minimum delay from sources to the sink. Furthermore, in the selection of next hop nodes, each sensor node selects its next hop as a sensor node having small number of packets in its queue such that this selection should not cause any congestion. Thus, this kind of routing provides minimum-hop packet delivery and smaller congestion rate on the forward paths.
- *Loss detection.* T-Sensors detect any congestion, collision, and channel error on the forward path, which cause possible packet losses. Then, T-Sensors inform G-Sensors about their packet losses.
- *Reporting frequency update.* According to the packet loss information from T-Sensors, each G-Sensor regulates its reporting frequency such that its reporting frequency is decreased to avoid excessive traffic load over the network if it has higher packet-loss rate.
- *New G-Sensors assignment.* G-Sensors inform the N-Sensors about the decrease in their reporting frequencies. Since the decrease in the reporting frequency results in decrease in the number of samples delivered to the sink, N-Sensor decides how many new G-Sensors should be assigned and which sensors should be selected as the new G-Sensors to ensure that the multimedia event signal can be satisfactorily reconstructed.

In biological homeostatic system, a duty cycle is performed by three biological systems (i.e., nervous, immune, and endocrine systems) to keep an organism in a stable state such that each biological system performs its specific operations in a self-organized manner. Similarly, in WMSN, N-sensors, T-sensors, and G-sensors perform the above operations to keep the network in a stable state with minimum packet-loss rate, delay, and energy consumption to satisfactorily reconstruct the multimedia event signal at the sink node. Therefore, these operations need a duty cycle to be performed in a self-organized manner. This allows a homeostatic-system-based communication algorithm for WMSN. In the following, we give a possible duty cycle to enable these operations to be performed in a self-organized manner.

1. *N-Sensor selection* is performed to determine a sensor as an N-Sensor.
2. *Spectrum estimation* is performed to enable the N-Sensor to estimate the spectral bandwidth of the multimedia event signal.

3. *G-Sensors selection* is performed to enable the N-Sensor to select some sensor nodes as G-Sensors and to determine their reporting frequencies according to the spectral bandwidth of the multimedia event signal.
4. *Path determination* is performed to allow the G-Sensors to determine their paths toward the sink.
5. *Loss detection* is performed to enable the T-Sensors to detect packet losses of each G-Sensor and to inform the G-Sensors about how many packets they have lost.
6. *Reporting frequency update* is performed to enable the G-Sensors to update their reporting frequencies.
7. *New G-Sensor assignment* is performed to enable the N-Sensor to decide whether new G-Sensor(s) is needed to satisfactorily reconstruct the multimedia event signal. For the newly assigned G-Sensors, above steps from 4 to 6 are repeated until the multimedia event signal is satisfactorily reconstructed.

The homeostatic-system-based communication model introduced above has several unified features that enable sensor nodes to communicate the multimedia event signal to the sink based on the principles of biological homeostasis. Based on the spectral bandwidth of the multimedia event signal, it first determines number of samples that must be delivered by the sensor nodes for accurate reconstruction at the sink. Furthermore, without need for any coordination between sensor nodes and the sink, it provides the sensor nodes with self-organization capability, which enables sensor nodes to distributively control the congestion on the forward paths to ensure that the sensor nodes to reliably communicate the required number of samples to the sink.

As the biological systems, biological colonies give great inspiration to develop efficient self-organized communication algorithms for the challenge of WSN. Next, based on the task allocation phenomenon in insect colonies, we introduce a possible roadmap for developing an efficient bio-inspired coordination and communication algorithm for WSAN.

## 26.7 Biologically Inspired Coordination Models for Wireless Sensor and Actor Networks

In this section, we first introduce a task allocation model in an insect colony. Then, based on the task allocation model, we derive the sensor–actor and actor–actor coordination models for WSAN.

### 26.7.1 Task Allocation Model of Insect Colony

Biological colonies of ants and bees typically consist of tens of thousands of dynamic elements [2]. In ant colonies, each ant has relatively little intelligence while the collaborative behavior of the colony provides a great deal of global intelligence capable of optimizing certain tasks. In social insect colonies, different activities are often performed simultaneously by individuals, which are better equipped for the task. This phenomenon is called as division of labor [2].

In the task allocation problem in insect colony, every individual has a response threshold for every task. Response threshold refers to the likelihood of reacting to a task-associated stimulus. Task-associated stimulus, $s$, is defined as the intensity of an activator associated with a particular task, and it can be a number of encounters such as a chemical concentration or any cue sensed by individuals [2]. For example, if the task is larval feeding, the task-associated stimulus, $s$, may be larval demand expressed through the emission of the pheromone, i.e., a chemical substance deposited by real-life ants [2]. Individuals perform the task when the level of $s$ exceeds their threshold. Therefore, a response threshold, $\theta$, expressed in units of stimulus intensity, $s$, is an internal variable that determines the tendency of an individual to respond to stimulus $s$ and perform the associated task [2]. Thus, based on the definitions above, for any individual, the probability of performing task as a function of $s$ and $\theta$, is given by

$$T_\theta(s) = \frac{s^n}{s^n + \theta^n},\qquad(26.14)$$

where $n > 1$ determines the steepness of the threshold [2]. As observed in (26.14), for $s \ll \theta$, the probability of performing the task is close to zero, and for $s \gg \theta$, this probability is close to 1.

Let $x_i$ be the binary variable representing the state of individual $i$ such that $x_i = 0$ corresponds to inactivity and $x_i = 1$ corresponds to performing the task. Also, let $\theta_i$ be the response threshold of individual $i$. An inactive individual starts to perform the task with a probability $P$

$$P(x_i = 0 \to x_i = 1) = T_{\theta_i}(s) = \frac{s_i^n}{s_i^n + \theta_i^n}.\qquad(26.15)$$

The task allocation model of insect colony briefly introduced here can certainly be modified to apply to virtually any kind of task allocation [2]. In the following sections, we apply the task allocation model of insect colony to establish the sensor–actor and actor–actor coordination models in WSAN.

## 26.7.2 Biologically Inspired Sensor–Actor Coordination Model

WSAN refer to a group of sensors and actors linked by wireless medium to perform distributed sensing and acting tasks [23]. Sensors gather information about the physical world while actors take decisions and then perform appropriate actions upon the environment. Sensors are low-cost, low-power devices with limited sensing, computation, and wireless communication capabilities [23]. Actors are generally assumed to be relatively resource-rich nodes equipped with better processing capabilities [23].

Sensor–actor and actor–actor coordination, and the coordinated efficient communication among them are the main challenges for the realization of WSAN [23]. In particular, due to the limited battery capacity of the sensor nodes, sensor–actor communication must provide minimum and uniformly distributed energy consumption for sensor nodes. Furthermore, sensor–actor communication must meet an application-specific delay bound to enable actor nodes to timely act upon the environment. Sensor–actor communication must also provide an application-specific packet-loss rate to enable actor nodes to reliably estimate the event properties. Therefore, in WSAN, it is essential to provide reliable and delay-aware sensor–actor communication to enable actor nodes to timely and reliably act upon the environment. Moreover, after receiving event information, actors need to coordinate with each other in order to make decisions on the most appropriate way to successfully accomplish the given task. Hence, the coordinated communication protocols are imperative among sensor–actor and actor–actor connections, which provide energy-efficient, delay-aware, and reliable communication and acting.

In WSAN, the coordinated behavior among the nodes is essential to provide effective communication between sensor and actor nodes. In insect colonies, individuals have great coordination capability to optimize certain tasks. This natural coordination capability among individuals enables each individual to perform a task that is the most appropriate task in terms of the sensed task-associated stimulus and its response threshold. Similar to the individuals in an insect colony, in WSAN, sensor nodes coordinate with each other to effectively communicate the sensed event information to actor nodes. In this regard, to establish the sensor–actor coordination model, we consider a sensor node as an individual in an insect colony. For this coordination, we introduce a node-to-node transmission probability ($P_{ij}$) by adopting the task performing probability given in (26.14). Here, we define the node-to-node transmission probability, $P_{ij}$, as the probability that sensor node $j$ is the next hop from sensor node $i$ to reach actor node $k$.

The aim of the node-to-node transmission probability ($P_{ij}$) is to enable sensor nodes to effectively communicate sensed information to actor nodes. Using $P_{ij}$, each sensor can select its most appropriate next hop node that provides minimum and uniformly distributed energy consumption while allowing reliable and delay-aware sensor–actor communication. In order to drive the node-to-node transmission probability ($P_{ij}$), we map the concept in the task performing probability given in (26.15) to $P_{ij}$.

In insect colonies, each insect is stimulated by stimulus $s_i$. While $s_i$ increases, the likelihood of reacting to task, $T_\theta(s_i)$, increases as in (26.15). Similarly, in WSAN, each sensor node is stimulated by its residual energy level. If it has higher residual energy, the likelihood of reacting to an event as a source node or an intermediate node to carry the event information increases. In this regard, we consider the stimulus intensity, $s_i$, given in (26.15) as the residual energy of sensor node $i$, $E_i$.

**Table 26.1** Relationship between insect colony and WSAN

| Insect colony | WSAN |
|---|---|
| Insects | Sensor nodes |
| Stimulus | Residual energy sensor node |
| Response threshold | Energy cost |

In insect colony, the likelihood of reacting to a task also depends on the response threshold of the individuals, $\theta_i$. As in (26.15), while $\theta_i$ increases, the likelihood of reacting to the task $T_\theta(s_i)$ decreases. Similarly, in WSAN, the likelihood of transmitting to a possible next hop node depends on the cost of required energy for transmission to this possible next hop node. While this energy cost increases, the likelihood of transmitting to this next node decreases. Therefore, we consider the response threshold, $\theta_i$, given in (26.15) as the energy cost, $\theta_{ij}$ needed for transmission from sensor node $i$ to sensor node $j$ in order to reach the actor node $k$. Here, we assume that sensor node $j$ is a possible next hop for sensor node $i$ to reach actor node $k$. The relationship between insect colony and WSAN, introduced above, is also shown in Table 26.1.

In addition to minimum and uniformly distributed energy consumption, reliable sensor–actor communication is essential to enable actor nodes to reliably estimate event properties. Therefore, sensor nodes should achieve delivering sufficient information to actor nodes for reliable sensor–actor communication. To this end, each sensor node must select its next hop node, which provides an appropriate packet-loss rate to its packets. Moreover, delay-aware sensor–actor communication is also crucial to enable the actor nodes to timely act upon the environment. Therefore, each sensor node must select its next hop node, which imposes minimum delay to its packets such that delay-aware sensor–actor communication can be achieved. Thus, to enable reliable and delay-aware sensor–actor communication, we also incorporate a learning component into the node-to-node transmission probability, $P_{ij}$. The aim of the learning component is to allow each sensor node to learn and select its next hop node, which can impose appropriate packet-loss rate and delay to achieve reliable and delay-aware sensor–actor communication.

Next, in Sect. 26.7.3, we first derive and incorporate energy cost, $\theta_{ij}$, into the task performing probability given in (26.15) to derive $P_{ij}$ providing minimum and uniformly distributed energy consumption. Then, in Sect. 26.7.4, we incorporate the learning component into $P_{ij}$ to provide reliable and delay-aware sensor–actor communication. The aim of the final node-to-node transmission probability is to provide minimum and uniformly distributed energy consumption while achieving reliable and delay-aware sensor–actor communication.

## 26.7.3  Minimum and Uniformly Distributed Energy Consumption

As introduced in Sect. 26.7.2, $\theta_{ij}$, is the energy required for transmission from sensor node $i$ to sensor node $j$. For node-to-node transmission, the energy consumption heavily depends on the distance between two nodes [24]. Therefore, we give the energy cost $\theta_{ij}$, as the distance between sensor node $i$ and $j$, $d_{ij}^m$, $(2 \leq m \leq 5)$. Hence, following the relationship summarized in Table 26.1, we give the node-to-node transmission probability as

$$P_{ij} = \frac{E_i^n}{E_i^n + \theta_{ij}^n} = \frac{E_i^n}{E_i^n + \alpha(d_{ij}^m)^n}, \qquad (26.16)$$

where $n > 1$ determines the steepness of the transmission probability and $\alpha$ is a positive constant that regulates the respective influence between $E_i$ and $d_{ij}$. Each sensor node evaluates the node-to-node transmission probability given in (26.16) in the following way:

- While $d_{ij}$ decreases, $P_{ij}$ increases. Therefore, for all hops, it is most probable that each sensor node transmits to closest sensor node that provides minimum energy consumption for this sensor node. This provides minimum energy consumption for sensor nodes.
- For smaller $E_i$ and $d_{ij}$, $P_{ij}$ is higher than the case for smaller $E_i$ and higher $d_{ij}$. Therefore, while $E_i$ decreases, it is most probable that sensor node $i$ transmits to the closest sensor node such that sensor node $i$ expends lower energy to deliver its packets. However, when sensor node $i$ has higher $E_i$, $P_{ij}$ is still higher for higher $d_{ij}$. Therefore, in this case, it may be possible that sensor node $i$ can transmit to distant nodes, which impose higher energy consumption to sensor node $i$. Thus, while sensor nodes having higher residual energy can transmit to near or far nodes, sensor nodes having smaller residual energy can transmit only to near nodes. This enables the sensor nodes to uniformly distribute energy consumption throughout the network.
- Using the node-to-node transmission probability given in (26.16), each sensor node can distributively regulate its hop distance. When it has higher residual energy, it transmits to far nodes and consumes higher energy. On the other hand, when it has lower residual energy, it transmits to the nearest nodes and consumes lower energy. However, as the hop distances decrease over the network, number of hops from sensor to actor nodes increases. Furthermore, while the number of hops increases, network traffic load on the forward path increases and possible bottlenecks, packet losses, and delay over the network increases. This prevents reliable and delay-aware sensor–actor communication. Thus, calculation of the hop distance over the network depends on whether the network can achieve reliable and delay-aware sensor–actor communication. If reliable and delay-aware communication cannot be provided, hop distances over the network should be further regulated such that some hops that impose higher packet losses and delay should be deferred. For this purpose, in the following, we incorporate the learning component into the node-to-node transmission probability given in (26.16).

## 26.7.4   Reliable and Delay Aware Communication

In addition to the minimum and uniformly distributed energy consumption, reliable and delay-aware communication in WSAN is essential to enable actor nodes to reliably and timely act upon the environment. More specifically, a sensor node should deliver its packets to its closest actor node within a delay bound. Furthermore, to enable actor nodes to reliably estimate event properties and act upon the environment, sensor nodes should deliver an application-specific number of packets within a specific time interval. Here, we assume that for reliable and delay-aware sensor–actor communication, each sensor node must deliver its packets to its next hop with an application-specific reliable and delay-aware packet delivery ratio (PD). PD is the number of packets that each sensor node should deliver to its next hop within an application-specific delay bound in order to provide reliable and delay-aware sensor–actor communication.

To allow each sensor node to achieve PD, we incorporate a learning factor into the node-to-node transmission probability given in (26.16). The aim of the learning factor is to allow sensor nodes to coordinate each other such that each sensor node learns and selects its next hop to achieve PD. Node-to-node transmission probability with the learning factor is calculated as

$$P_{ij} = \frac{E_i^n}{E_i^n + \alpha(d_{ij}^m)^n + L_{ij}^n},$$   (26.17)

where $L_{ij}$ is the learning factor that allows sensor node $i$ to learn or forget sensor node $j$ as its next hop in the following way:

- Initially, sensor node $i$ sets all of its learning factors to the same value for all of its neighbors.
- Each sensor node computes its node-to-node transmission probabilities ($P_{ij}$) for all of its neighbors toward its closest actor node. Then, each sensor node selects and transmits to its next hop node according to the computed node-to-node transmission probabilities. Let $j$ be the next hop node of sensor node $i$.
- If sensor node $i$ achieves PD, it learns sensor node $j$ and updates $L_{ij}$ as $L_{ij} = L_{ij} - \xi_0$ as long as $j$ has not been learnt by any sensor node up to that time. This update decreases $L_{ij}$ and increases $P_{ij}$, where $\xi_0$ is the positive learning coefficients. Therefore, it is more probable that sensor node $i$ again transmits to sensor node $j$.
- Once sensor node $i$ learns sensor node $j$, it cannot forget sensor node $j$ anymore. This enables sensor nodes to permanently allocate its next hop node providing reliable and delay-aware sensor–actor communication.
- If sensor node $i$ cannot achieve PD and sensor node $j$ has not been learnt by sensor node $i$, sensor node $i$ forgets $j$ and updates $L_{ij}$ as $L_{ij} = L_{ij} - \xi_1$, where $\xi_1$ is the positive learning coefficients. This increases $L_{ij}$ and decreases $P_{ij}$ and it is less likely that sensor node $i$ selects sensor node $j$.

- Hence, sensor nodes, which cannot achieve PD, forget their next hops. This forgetting strategy allows these sensor nodes to defer their transmissions to other sensor nodes that may allow the sensor nodes to achieve PD. Thus, the bottleneck over the network, which can impose higher packet-loss rate and delay to sensor–actor communication, can be avoided to enable reliable and delay-aware sensor–actor communication.
- While hop distances between sensor nodes ($d_{ij}$) decrease, node-to-node transmission probabilities ($P_{ij}$) increase. Therefore, initially it is most probable that each sensor node transmits to the sensor nodes that are closer since node-to-node transmission probabilities are maximum for these nodes. These next hops increase number of hops from source nodes to actor nodes while imposing smaller energy consumption. However, while the number of hops increases, network traffic load also increases. Increasing network traffic load imposes higher packet-loss rate and delay over the network and source nodes cannot achieve PD. Thus, sensor nodes frequently forget their nearest next hop nodes and they start to select the next hop nodes that are farther. These next hop nodes decrease number of hops from source to actor nodes. Therefore, the network traffic load decreases with smaller packet-loss rate and delay over the network, and source nodes can achieve PD. However, these next hop nodes necessitate higher energy consumption. Hence, this learning strategy enables sensor nodes to trade off energy consumption for reliable and delay-aware sensor–actor communication. For a smaller PD, sensor nodes can achieve PD with smaller hop distance and higher number of hops. Therefore, this provides smaller energy consumption. However, as PD increases, source nodes can achieve PD with higher hop distance and smaller number of hops, leading to higher energy consumption.

### 26.7.5 Biologically Inspired Actor–Actor Coordination Model

In WSAN, the coordination among actors is needed to enable the actor nodes to perform the most appropriate action upon the environment. We introduce an actor–actor coordination model based on the biologically inspired task allocation scheme overviewed in Sect. 26.7.1. We define the task performing probability $A_i$ for actor node $i$ such that $A_i$ denotes the likelihood of performing the task for actor node $i$. To map the concept introduced in Sect. 26.7.1 to actor–actor coordination in WSAN, we consider an actor as an individual in an insect colony.

In insect colonies, based on the sensed task-associated stimuli, each insect performs the task if sufficient stimuli belonging to a task can be sensed by the insect. Similar to insect colonies, in WSAN based on collected information belonging to an event, actor nodes perform the task associated with the event. Therefore, we consider the stimuli intensity $s$ as the number of packets collected by actor node $i$ ($S_i$).

The response threshold $\theta$ given in (26.14) is associated with the likelihood of reacting to a task. For actor node $i$, the likelihood of reacting to the event depends on whether actor node $i$ can reliably estimate the event properties or not. Here, we assume that to reliably estimate the event properties, actor node $i$ collects a number of packets, denoted by rp, within each time interval $\tau$. Thus, if actor node $i$ can collect rp packets within $\tau$, it can reliably estimate the event properties and perform the task associated with the event. Therefore, we consider the response threshold $\theta$ as the number of packets (rp) that must be collected to reliably estimate the event properties. Thus, following (26.15), we give the task performing probability $A_i$ as

$$A_i = \frac{S_i^n}{S_i^n + \text{rp}^n + N_i^n},\qquad(26.18)$$

where $n > 1$ determines the steepness of the task performing probability and $N_i$ is the learning factor that enables actor node $i$ to learn or forget the event. The task performing probability given in (26.18) is evaluated by the actor nodes in the following way:

- When actor node $i$ can collect higher number of data packets ($S_i$) belonging to the event within $\tau$ such that $S_i > \text{rp}$, $A_i$ is higher and it is more probable that actor node $i$ performs the task associated with the event. Thus, actor nodes that can reliably estimate the event properties perform the task.
- When actor node $i$ can collect smaller number of data packets ($S_i$) belonging to the event within $\tau$ such that $S_i < \text{rp}$, $A_i$ is close to zero and it is almost impossible that actor node $i$ performs the task. Thus, actor nodes that cannot reliably estimate the event properties can be prevented from performing the task.
- Learning factor $N_i$ enables actor node $i$ to learn or forget the event. If actor node $i$ can collect sufficient information from the sensor nodes to reliably estimate the event properties ($S_i > \text{rp}$), it learns the event and updates $N_i$ as $N_i = N_i - \xi_2$ where $\xi_2$ is the positive learning coefficients. This update increases $A_i$ and the probability that actor node $i$ performs the event increases. If actor node $i$ cannot collect sufficient information to reliably estimate the event properties ($S_i < \text{rp}$), it forgets the event and updates $N_i$ as $N_i = N_i + \xi_3$, where $\xi_3$ is the positive forgetting coefficients. This update decreases $A_i$ and the probability that actor node $i$ performs the task decreases. Thus, the task is performed by the actor nodes that can collect sufficient information about the event.
- While sensor nodes transmit higher number of packets to actor nodes, the collected number of packets ($S_i$) increases as well as task-performing probabilities ($A_i$) for all actor nodes. This enables higher number of actors to perform the task associated with the event. Thus, number of actor nodes performing the task can be regulated according to the data transmitted by sensor nodes. For example, for an event that diffuses to a large area and necessitates higher number of source nodes to cover its area, collected number of data packets ($S_i$) is higher in all actor nodes. This increases task performing probabilities ($A_i$) for all actor nodes and higher number of actor nodes performs the task. Conversely, in the event for which smaller number of source nodes sense the event and transmits their data packets, collected number of data packets $S_i$ is smaller in actor nodes and task-performing probabilities ($A_i$) is smaller for all actor nodes. This enables smaller

number of actor nodes to perform the task. Thus, using the task-performing probabilities $(A_i)$, number of actor nodes performing the task can be regulated according to the properties of the event area.

## 26.8 Thoughts for Practitioners

Using the bio-inspired communication models introduced above, efficient communication techniques and algorithms, which are self-organized, survivable, scalable, and self-adaptable can be developed. To develop an immune-system-based node and rate selection algorithm, it is critical to select and normalize some parameters to efficiently exploit the bio-inspired mathematical equations (26.11)–(26.13). For example, it is essential to select appropriate source node selection threshold and correlation radius $r$ for an efficient node selection algorithm. The source node selection threshold and correlation radius determine number of source nodes and event signal distortion at the sink node. Therefore, the aim of source node selection threshold and correlation radius is to determine minimum number of source nodes achieving desired event signal distortion at the sink node such that the minimum number of source nodes imposes minimum energy consumption to the network. However, some analytical efforts are required to determine appropriate source node selection threshold and correlation radius based on network density and spectral properties of event signal.

To develop an efficient coordination protocol using the bio-inspired mathematical equation (26.17), it is also critical to select and regulate some parameters. For example, optimal value of $\alpha$ can be found to extend lifetime of the network by means of some analytical efforts. Furthermore, transmission range of the sensor nodes should be considered to provide an efficient trade off between minimum and uniformly distributed energy consumption.

Despite determination of some optimal parameters, regulation of these parameters can enable the sensor nodes to conduct themselves according to any state of the network. For example, regulation of $\alpha$ can enable each sensor node to regulate its hop distance according to its residual energy level. While its residual energy decreases, it can decrease its average hop distance by increasing $\alpha$. Hence, each sensor node can extend its lifetime. However, some analytical efforts are required to develop a regulation rule for this regulation.

### 26.8.1 Directions for Future Research

Like immune system homeostasis, and insect colony, plenty of biological mechanisms can lead to develop efficient algorithm for the problems imposed by WSN.

- Embryonics can lead to develop efficient security mechanisms for WSN.
- Gene-regulatory networks can be useful to develop efficient medium access control mechanism for WSN.
- Biological-switching-based routing can be an efficient routing scheme for WSN.

## 26.9   Conclusion

In this chapter, we introduce the potential analogies between biological systems and colonies and WSN, then, based on these analogies, we introduce biologically inspired communication approaches for WSN using some principles of biological systems and colonies. The objective of these communication approaches is to serve as a roadmap for the development of efficient scalable, adaptive, and self-organizing bio-inspired communication techniques for WSN. The proposed approaches are clearly promising to address the challenges in WSN domain in terms of energy consumption, fault-tolerance, delay-awareness, and reliability. From the point of view that the chapter introduces, further analogies between biological systems and WSN can be established to develop further efficient self-organized algorithm for any of challenges in WSN domain.

## Terminologies

*Inspiration.* Something that gives you ideas for doing something.

*Immune system.* The various cells and tissues in the body which make it able to protect itself against infection.

*Antibody.* A protein produced in the blood which fights diseases by attacking and killing harmful bacteria.

*Affinity.* An attraction or sympathy for someone or something, especially because of shared characteristics.

*Stimulation.* When something causes something to become more active or enthusiastic, or to develop or function.

*Distort.* To change something from its usual, original, natural, or intended meaning, condition, or shape.

*Event.* Anything that happens, especially something important or unusual.

*Correlation.* A connection between two or more things, often one in which one of them causes or influences the other.

*Endocrine system.* A biological system consisting of any of the organs of the body, which produce and release hormones into the blood to be carried around the body.

*Nervous system.* An animal's nervous system consists of its brain and all the nerves in its body which together makes movement and feeling possible by sending messages around the body.

*Multimedia.* Using a combination of moving and still pictures, sound, music, and words, especially in computers or entertainment.

*Spectrum.* A range of waves, such as light waves or radio waves.

*Action.* The process of doing something, especially when dealing with a problem or difficulty.

# Questions

1. Why does the next generation vision of WSN impose significant challenges to enable a connection between physical world and Internet?
2. How can one overcome these challenges imposed by the next generation vision of WSN?
3. How can the biological systems give great inspiration to develop efficient communication algorithms for WSN?
4. Explain the analogy between regulation of antibody density in biological immune system and regulation of reporting frequency rate in WSN.
5. Comment on how the number of source nodes changes when the correlation radius $r$ increases according to the distributed source node selection model given in Sect. 26.4.1?
6. Which state in WMSN is compliance with the stable state in biological homeostatic system according to the analogy drawn in Sect. 26.5?
7. Propose an efficient loss detection strategy to enable T-Sensors to detect any packet loss in the forward paths for the homeostasis-based multimedia communication model given in Sect. 26.6.
8. Comment on how the hop distance of sensor node $i$ ($d_{ij}$) changes while its residual energy ($E_i$) decreases according to the node-to-node transmission probability given in (26.16). Based on your comments, explain how the biologically inspired coordination model given in Sect. 26.7 can enable a sensor–actor communication scheme with minimum and uniformly distributed energy consumption in WSAN.
9. Explain how the learning factor $L_{ij}$ can enable sensor node $i$ to select its next hop node providing minimum packet-loss rate.
10. Explain how the task performing probability given in (26.18) can regulate number of actor nodes performing the task according to amount of data transmitted by sensor nodes.

# References

1. Akyildiz I F, Su W, Sankarasubramaniam Y, Cayirci E (2002) A survey on sensor networks. IEEE Communications Magazine 40:102–114.
2. Bonabeau E, Dorigo M, Theraulaz G (1999) Swarm Intelligence, From Natural to Artificial System. Oxford University Press, Oxford.
3. Muraleedharan R, Osadciw L A (2003) Sensor communication network using swarm intelligence. IEEE Upstate New York Workshop, Syracuse, NY, USA.
4. Muraleedharan R, Osadciw L A (2003) Balancing the performance of a sensor network using an ant system. Annual Conference on Information Sciences and Systems, Baltimore, MD, USA.
5. Caro G D, Ducatelle F, Gambardella L M (2005) AntHocNet: an adaptive nature-inspired algorithm for routing in mobile ad hoc networks. European Transactions on Telecommunications 16:443–455.

6. Hong Y W, Scaglione A (2005) A scalable synchronization protocol for large scale sensor networks and its applications. IEEE Journal on Selected Areas in Communications 23:1085–1099.
7. Werner-Allen G, Tewari G, Patel A, Welsh M, Nagpal R (2005) FireflyInspired Sensor Network Synchronicity with Realistic Radio Effects. SenSys'05.
8. Carreras I, Chlamtac I, Woesner H, Kiraly C (2005) BIONETS: Bio-inspired next generation networks. Lecture Notes in Computer Science 3457:245–252.
9. Dressler F (2005) Efficient and Scalable Communication in autonomous networking using bio-inspired mechanisms – An overview. Informatica 29:183–188.
10. Dressler F, Krüger B, Fuchs G, German R (2005) Self-Organization in Sensor Networks Using Bio-Inspired Mechanism. ARCS'05.
11. Dressler F (2005) Locality Driven Congestion Control in Self-Organizing Wireless Sensor Networks. SASO-STEPS'05.
12. Wokoma T, Shum L L, Sacks L, Marshall I (2005) A biologically inspired clustering algorithm dependent on spatial data in sensor networks. Second European Workshop on Wireless Sensor Networks.
13. Atakan B, Akan O B (2007) Immune system based energy efficient and reliable communication in wireless sensor networks. In: Dressler F and Carreras I (eds.) Advances in Biologically Inspired Information Systems, Springer, New York, NY.
14. Timmis J, Neal M, Hunt J (2000) An artificial immune system for data analysis. Biosystems 55:143–150.
15. Jerne N K (1984) Idiotypic network and other preconceived ideas. Immunological Review 79:5–24.
16. Farmer J D, Packard N H, Perelson A S (1986) The immune system, adaptation, and machine learning. Physica 22:187–204.
17. Vuran M C, Akan O B, Akyildiz I F (2004) Spatio-temporal correlation: theory and applications for wireless sensor networks. Computer Networks Journal (Elsevier) 45:245–261.
18. Berger J O, Oliviera V, Sanso B (2001) Objective bayesian analysis of spatially correlated data. Journal of the American Statistical Association 96:1361–1374.
19. Neal M, Timmis J (2005) Once more unto the breach towards artificial homeostasis. Recent Advances in Biologically Inspired Computing, Idea Group, pp. 340–365.
20. Oppenheim A V, Schafer R W, Buck J R (1999) Discrete-Time Signal Processing, Prentice Hall, Upper Saddle River, NJ.
21. Hightower J, Borriello G (2001) Location systems for ubiquitous computing. IEEE Computer 34:57–66.
22. Welch P D (1967) The use of fast Fourier transform for the estimation of power spectra: A method based on time averaging over short modified periodogram. IEEE Transaction on Audio and Electroacoustics 15:70–73.
23. Akyildiz I F, Kasimoglu I H (2004) Wireless sensor and actor networks: research challenges. Ad Hoc Networks 2:351–367.
24. Heinzelman W, Chandrakasan A, Balakrishnan H (2002) An application-specific protocol architecture for wireless microsensor networks. IEEE Transaction on Wireless Communications 1:660–667.

# Chapter 27
# Mobile Ad Hoc and Sensor Systems for Global and Homeland Security Applications

Raffaele Bruno, Marco Conti, and Antonio Pinizzotto

**Abstract** Communications infrastructures are a critical asset in today's information society. However, legacy telecommunication systems easily collapse in case of disruptions that may occur due to security incidents or crises. In this chapter, we first elaborate on the major shortcomings of the current communications networks for security applications to identify the key missing requirements for such networks. Then, we show that the ad hoc networking technologies, coupled with disruptive-tolerant techniques, are the best suited paradigm to build the next generation of dependable, secure, and rapidly deployable communications infrastructures. In particular, we focus on mesh, opportunistic, vehicular, and sensor networks giving an overview of the most recent advances and summarizing the challenges facing the design and the deployment of these networks. Finally, we conclude this chapter presenting the open research issues to realize the vision of a dependable communications infrastructure, with special attention to aspects such as interoperability among multiple heterogeneous networks, autonomic network management, and QoS protection.

## 27.1 Introduction

Today's modern society is considered an information society because the creation, circulation, and manipulation of information are activities that pervade many aspects of our cultural, economical, and social life. Consequently, governments, economy and society in general, are becoming increasingly dependent on *Information and Communication Technologies* (ICT), which are the means of providing information. For these reasons, the communications infrastructures used to transport information

R. Bruno (✉)
Institute for Informatics and Telematics (IIT), Italian National Research Council (CNR),
Via G. Moruzzi 1, 56124 Pisa, Italy
e-mail: r.bruno@iit.cnr.it

S. Misra et al. (eds.), *Guide to Wireless Sensor Networks*, Computer Communications and Networks, DOI: 10.1007/978-1-84882-218-4_27,
© Springer-Verlag London Limited 2009

are considered a critical asset of our society, such as the transportation and power supply infrastructures, and they should be protected and secured. The need to ensure resiliency, security, and dependability of our communications systems is made more compelling by the tight interdependence between the information infrastructure and other critical infrastructures. For instance, security problems, breakdowns, and failures in the information systems may create widespread damage in transportation or energy infrastructures. In addition, the nature and the extent of the threats jeopardizing our communications infrastructures are considerable higher today than in earlier times. As well explained by the European Security Research Advisory Board in its 2006 report "modern crises are progressively changing their character from 'predictable' emergencies... to unpredictable catastrophic events" [1], and current communications networks are not designed to withstand unplanned and unexpected disruptive events such as natural or manmade disasters. In fact, in assessing the communication breakdowns that have taken place in the aftermath of events of the magnitude of 9/11, Katrina hurricane or London bombings, when many mission-critical networks were down and unavailable, it has been observed that "telecommunications was the greatest single area of concern" [2, 3]. It is also important to highlight that, during a crisis or an emergency situation, the availability of a reliable and dependable communications system is also fundamental to allow first responders, rescue teams, and public safety agencies operating in the disaster area to carry out disaster relief operations. In fact, all the disaster and crisis management activities rely on the exchange of information between government entities, operators of critical infrastructures, and rescue teams, as well as on the interaction of first responders with citizens and victims. In the following discussion, we will primarily concentrate our attention on this communication scenario, i.e., the provision of resilient and flexible communication services in a disaster zone for Public Protection Disaster Relief (PPDR) missions.

The experiences gathered after the most recent large disasters (e.g., Indian Ocean tsunami in 2004) or massive terrorist attacks (e.g. 9/11 airplane crashes in 2001 or Madrid train bombings in 2004) have permitted the clear identification of the missing capabilities of existing communications systems to provide the necessary support for PPDR applications. Among the most important shortcomings that have been identified by various forums and committees [2–6], it is useful to note: the lack of sufficient robustness and resiliency to disruptive events, the limitations in the interoperability between private networks operated by public safety agencies, the difficulties for integrating private networks with the core communications infrastructures, the lack of flexibility and versatility in the communication services, and the limited support of priority communications in public networks. To effectively address the above issues, we advocate the use of self-organizing architectures exploiting the ad hoc networking paradigm to realize a resilient and versatile communications system meeting the requirements of a disaster response system. Traditionally, mobile multihop ad hoc networks (also MANETs) are conceived as groups of devices that self-organize into peer-to-peer networks by establishing multihop wireless connections [7, 8]. Therefore, it is intuitive that first responders may use the ad hoc networking technologies to quickly set up on-demand communication services between their handheld devices, enabling a reliable dissemination of vital information, as well as an effective collaboration in time-critical relief operations. However, in the recent years, the MANET research has achieved important results in successfully exploiting the multihop ad hoc networking to build various types of specialized networks, such as mesh networks, vehicular networks, sensor networks, and opportunistic networks, which have been designed to support well-defined application requirements [9]. For instance, mesh networks provide rapidly deployable wireless extension to legacy communications infrastructures; vehicular networks apply the MANET technology to the intervehicles communications; sensor networks are designed to support monitoring applications in general; and opportunistic networks are an extension of MANET technology to cope with intermittently connected networks. We expect that these emerging technologies will provide most of the missing communications capabilities needed to develop a dependable, secure, and rapidly deployable communications system for mission-critical scenarios and emergency response.

In this chapter, we present the main characteristics and properties of these emerging technologies with special emphasis on mesh, vehicular, sensor, and opportunistic networks. The focus of our discussion is to explain how these networking solutions will facilitate the development of flexible and easily deployable communications systems that would be resilient to disruptive and unplanned events. While the maturity of these technologies is sufficient to predict the readily deployment in all the typical situations characterizing PPDR scenarios, there are still several open research and technical challenges that have to be addressed to realize an information sharing system for disaster response fully integrated with the existing communications infrastructures. In particular, in our discussion we will give special attention to aspects such as interoperability among multiple heterogeneous networks, autonomic network management, and QoS protection.

The remaining of this chapter is organized as follows. Section 27.1 illustrates the reference disaster scenarios that exemplify the communications challenges that characterize first responders' emergency response operations. In Sect. 27.2, we analyze the missing technological capabilities necessary to develop the next-generation of resilient, rapidly deployable, and secure communications systems for PPDR applications. In Sect. 27.3, we outline the most consolidated international initiatives aiming at promoting the security research in the PPDR area. Section 27.4 reviews the most recent advances in the deployment of mesh, opportunistic, vehicular, and sensor networks. In Sect. 27.5, we discusses some of the most important research challenges. Finally, Sect. 27.6 draws concluding remarks.

## 27.2  Background

To identify the communications challenges that emerge after a security incident, and to highlight the communications capabilities needed during disaster relief operations, we consider a reference scenario, where a natural or man-made disaster devastates the communications infrastructures and first responders are involved in the emergency response.

First of all, we observe that the today public telecommunications networks are characterized by the considerable heterogeneity of the technologies and architectures adopted to provide communication services, either at the local or geographical scale. At one extreme, these networks are based on wired and wireless narrowband technologies (e.g., leased telephone lines, cellular and satellite systems, etc.), and they are mainly used to provide voice communications and a limited support of data transmissions. On the other extreme, these networks employ broadband wired and wireless technologies (e.g., WiFi, Wi-MAX, optical networks, etc.) to support more complex multimedia communications. However, these systems have common characteristics such as the dependence on dedicated infrastructures, the adoption of a centralized management for the communications resources, and the use of point-to-point links to interconnect the devices to other devices or control units. In case of an incident that causes partial damages to the network infrastructures (either turning some point-to-point links down or making some devices nonfunctioning), large portions of these communications systems may stop working properly. To reduce the risk of suffering interruptions of communication services during a disruptive event, the most critical components of large-scale telecommunications networks are usually replicated. However, the experiences gathered from the most recent security incidents and disasters (e.g., 9/11 attacks or Katrina hurricane) have highlighted that this approach is not effective to ensure communications system resiliency because these backup systems are generally unable to handle the huge traffic volumes generated in the wake of a crisis situation. The solution we envisage for dealing with the damages that an incident may cause to the legacy communications systems is to reuse what remains available of the infrastructure by establishing additional wireless backup links if possible (e.g., satellite links), and substituting point-to-point links with multihop wireless connections to form a more reliable wireless mesh backbone. This scenario is illustrated in Fig. 27.1, which exemplifies an urban

**Fig. 27.1** Communications infrastructure partially damaged: backup wireless links are established to activate mesh-mode communications

environment where an incident has interrupted wired links in the picture a link inter-
ruption is represented by a cross, and communicating devices establish alternative
wireless links using satellites or terrestrial antennas.

In addition to re-establish the public communications systems in a disaster area,
it is fundamental to rapidly deploy a communications platform that may guarantee an acceptable level of communication to first responders, rescue workers, and
any other Public Safety user operating in the disaster area. This temporary on-
demand communications network may be created by establishing multihop ad hoc
communications between the handheld devices carried by first responders and/or
communicating devices (i.e., wireless routers) transported by rescue land vehicles
or helicopters deployed on the disaster area. These specialized networks may be
operated in parallel to the legacy networks or tightly integrated with them as an extension or replacement of a too seriously damaged communications infrastructure
(see Fig. 27.2). Note that, for first responders, it is necessary to have also access to
the legacy wireless infrastructure networks to stay in contact with remote command
and control centers.

In addition to deploying powerful wireless communications devices, the emergency response personnel may spread out across the disaster area tiny sensing
devices. These sensing devices will form a sensor network that may provide a useful
tool to remotely monitor a location or situation in real time, assisting first responders in the decision process and coordination activities during emergency response
and security operations, as well as to detect and predict threats (e.g., the presence
of toxic substances after a chemical plant explosion, or the imminent collapse of a
building after an earthquake).

**Fig. 27.2** Communications infrastructure heavily compromised: heterogeneous and interoperable
self-organizing wireless networks are deployed

In extreme cases, a disruptive event may produce so extensive damages to bring down almost all the existing network infrastructures. Moreover, because of the prohibitive environmental conditions, it might be impractical to spread around a sufficient number of rescue vehicles so as to create well-connected ad hoc networks. In this context, it is more likely to envisage the case of *"clouds"* of connected hand-held devices (e.g., palmtops carried by first responders) that will be just sporadically connected to each other, and, possibly, to the surviving part of the infrastructure. These communication clouds will be extremely dynamic, as the rescue teams will move, and wireless links will appear and disappear. In the extreme case, a single, disconnected, user can form a communication cloud. Traditional networking approaches will fail to preserve the communication services in such scenario because they require a continuous end-to-end path between communicating endpoints, computed by a routing protocol, while such continuous paths will seldom be available in a security incident area. On the contrary, opportunistic networking techniques enable end-to-end paths even when communication endpoints are not connected at the same time to the same network by exploiting the *store-carry-and-forward* approach. It is evident that devices should have highly versatile communications capabilities to efficiently operate in a network that would be extremely dynamic, heterogeneous, and mainly disconnected, formed by possibly isolated devices. In addition, in this disaster scenario, where communications will be extremely challenged as a consequence of infrastructure disruptions, communications opportunities will be a scarce resource to be sparingly managed. It is then critical to ensure that critical data are made available to the right set of users, by avoiding congestion and data unavailability.

## 27.3 Thoughts for Practitioners

From the analysis of the previous reference scenario, as well as the analysis of other global and homeland security scenarios, we can identify the user requirements associated to typical public safety, emergency, and disaster applications. These user requirements will be the basis to derive the technical requirements for the design of resilient, rapidly deployable, and secure communications systems for PPDR applications [10]. The most important technical requirements we have identified are the following:

- *Ubiquitous access.* Public safety mobile radio networks must function in all areas served by first responders and involving disaster victims. This should include underground places, rural areas, remote or under-served areas, and challenged environments that were subjected to devastations. In addition, the seamless support of user mobility should be an integral part of the system design.
- *Resiliency.* Natural and man-made disasters may cause partial, or even extensive, disruptions of the terrestrial communications infrastructures. However, a resilient communications system must be designed to survive to damages and failures, and to ensure the continuity of communication services, at least for

critical applications. To this end, centralized architectures should be avoided because more prone to failures and clearly less reconfigurable.

- *Fast deployment.* To effectively deal with emergency situations, a communications system for PPDR applications should be easily and rapidly deployable, and the communication services should be operational very quickly.
- *Self-organization.* It is crucial that public safety networks implement advanced self-management capabilities in order to limit as much as possible human operations and maintenance, guaranteeing that the network properly operates despite unplanned and unexpected events. Self-organization is also a prerequisite to provide fast and dynamic deployment of temporary, on demand, communications network in disaster areas.
- *Interoperability.* Emergency operations require the involvement of several groups of first responders operating for different agencies and authorities. Seamless communications between different units do not require only common procedures, but also interoperable equipments and communication protocols. In addition, private networks owned by public safety agencies should be easily integrable with the public networks used by citizens to favor the information collection and distribution.
- *QoS.* Emergency response management and disaster relief operations very often rely on the timely exchange of critical information (e.g., via voice or images/video) between first responders, and on providing correct and updated information to people. Therefore, the communications system used by first responders should provide QoS support to meet the stringent requirements of real-time flows. In addition, priority schemes should be integrated in the public communications networks to ensure that vital communications for first responders are not hindered by legacy data transmissions during emergency situations.
- *Security.* Standard security properties should be assured also in a disruptive environment. However, in addition to protecting the privacy of the communications, in emergency scenarios it is also important to provide a reliable establishment of trust relationships among users in order to guarantee the secure identification of devices and users.

Although the technical requirements for reliable communications infrastructures to be used in PPDR operations are well defined, the recent disaster experiences have revealed that the existing solutions are unable to provide an adequate support for these situations. Traditionally, public safety agencies have relied on dedicated wireless systems to support communications between teams of first responders. In particular, it was generally believed that the reliability and security of the public Internet is inadequate for mission-critical functions. On the contrary, the allocation of dedicated spectrum for public safety applications, as well as the adoption of more stringent reliability and security requirements than the ones considered in commercial networks, should make dedicated systems sufficiently robust to operate also during emergency situations. For these reasons, industry standards for implementing narrow-band private mobile radio systems, e.g., ETSI standard TETRA in Europe or APCO25 in the USA, have been developed in the last decade, facilitating the deployment of these networks. However, a central lesson underscored by recent disruptive events (e.g., Katrina hurricane or London bombings) is that private mobile radio systems maintained by public safety agencies were outdated and incompatible [2–6]. Specifically, these aging technologies were too limited to meet the growing demands of emergency communication services, because they were designed primarily for voice communications and lack other important capabilities such as high-speed data communications. Moreover, teams of first responders from different agencies were not able to communicate due to lack of interoperability between their private networks. This severely hindered the capability of first responders to acquire, process, and disseminate vital information. In addition, the wireless communications systems used by the first responders and law enforcement communities were unable to support seamless and interoperable communications with the legacy telecommunications networks used by citizens. This made impossible to distribute early warnings and updated information to people at disaster areas.

The inefficiencies in the design or deployment of their private networks led first responders and emergency managers to switch to public mobile networks to provide emergency services during large-scale disasters. However, terrestrial communications infrastructures (also called Land Mobile Radio systems, or LMR), such as traditional 3G cellular systems or emerging metro-scale broadband wireless access technologies, are generally based on centralized architectures where central units have full control over each cell. Thus, fundamental system functionalities, such as access control, connection establishment, support of mobility, etc., relay on the existence and the availability of the network infrastructure itself. Consequently, centralized architectures suffer the main drawback of collapsing when the centralized infrastructure is out of order, and when unplanned or unexpected disruptive events occur. For example, disasters such as New Orleans flooding destroyed all available network infrastructures. Nowadays, the only practical solution to deal with partial or total unavailability of LMR systems is to use satellite communications. However, satellite systems are seen as a fallback technology, suitable only for outdoor communications and subject to the availability of a satellite to act as a relay station between earth terminals. The lack of radio communications ability within buildings represented a notable failing of public safety LMRs and one that has led to tragic results during emergency situations such as 9/11 [3,5].

Even if available, commercial telecommunications systems often were severely overloaded during emergencies. All the reports from governments and experts investigating the causes of the communication failure during recent natural or manmade disasters highlighted that commercial systems are often the most unreliable during critical incidents when public demand overwhelms the system [6]. Unfortunately, prioritization schemes to reserve dedicated resources to emergency calls or to limit resource usage by low priority users are rarely implemented in commercial systems, or have not appropriate objectives. In fact, if congestion occurs in normal conditions, network operators assign greater importance to flows that have greater revenue-generating capability. On the contrary, during exceptional conditions like emergencies or disasters, network operators should consider more valuable the traffic generated by users involved in disaster relief operations.

The above analysis of the shortcomings of the existing, either public or private, communications systems for PPDR applications points out that the development of new networking technologies capable of providing the needed degree of reliability and dependability is fundamental. This need, as well as the growing threat perception, has boosted both private and public investments in researching novel security solutions. As explained in the following, these research initiatives have rapidly converged to an increasing consensus about the fact that the most mature and best-suited networking paradigm fulfilling the requirements of PPDR applications is the ad hoc networking paradigm [1]. In fact, being peer-to-peer networks formed by mobile devices with self-organizing capabilities, multihop ad hoc networks represent a key technological driver to deploy more resilient communications systems. To support this claim, in the following sections, we first outline the most important national and international research programs that have been established in the sector of national and civil security, with special attention to the communications concerns. Then, we discuss how the recent advances in ad hoc networking may be successfully applied to realize a practical communications system for PPDR applications.

## 27.4 International Initiatives

A series of national and international initiatives have been established to bring together national governments, international organizations, industrial stakeholders, academia, and emergency response communities and to set up the agenda of long term PPDR research. All these initiatives have identified the development of novel IT solutions to deploy dependable, versatile, and secure communications infrastructures, as a key investment area.

One of the first examples of this new approach to address global security challenges is represented by the establishment in the USA of the Department of Homeland Security (DHS), whose primary aim is to define a high-level strategic plan to coordinate all the organizations and institutions involved in the security missions and emergency response. To accomplish this ambitious goal, the DHS has created, among the others, the Directorate for Science and Technology (S&T Directorate) that aims at driving the development of technologies and capabilities in support of the homeland security. To this end, a variety of agencies and programs have been established to promote the research on the security challenges identified by the DHS strategic plan. In particular, the SAFECOM program has been activated to improve interoperable communications nationwide through the definition of nonproprietary standards, open architectures, common operational procedures, and communications systems ensuring interoperable voice and data capabilities for emergency response. In addition, the Homeland Security Advanced Research Projects Agency (HSARPA) is launching new solicitations and funding programs on a broad range of topics to promote the research and development efforts of innovative security solutions. In particular, HSARPA is now promoting the development of novel communications and information systems supporting more effective and coordinated decision-making processes and crisis management through reliable information acquisition and assessment. In this context, the development of more robust and flexible sensor networks is considered of paramount importance, because much of the security mission involves the monitoring of various environments, and the prediction and detection of threats to these environments.

Collaborative programs have been also established between Europe and USA for the coordination and development of joint specification of standards for Public Safety and Emergency (PS&E) scenarios. The most important example of these joint initiatives is the Project MESA (Mobility for Emergency and Safety Applications). Specifically, MESA is a standardization Partnership Project established between the European Telecommunications Standards Institute (ETSI) and the Telecommunications Industry Association (TIA) in the USA, whose original purpose was to elaborate a joint specification of next-generation mobile broadband technology to be deployed for the PS&E. Since 2002, the vision has evolved toward the definition of a set of interconnection standards between heterogeneous systems, i.e., following the so-called "systems of systems" approach.

Another relevant European initiative jointly funded by the European Commission and the European Space Agency is Global Monitoring for Environment and Security (GMES). Since 2001, the GMES group is working on the implementation of European-level policies and information services dealing with environmental monitoring and security needs. The GMES approach is based on the observation and the understanding of the phenomena of the terrestrial environment through satellite and ground systems. This information is then provided to all the organizations involved in environmental management and security enforcement.

Although these programs have obtained important results, the European states felt the need to develop a longer-term perspective in the field of security research. For these reasons, in April 2005, the European Security Research Advisory Board (ESRAB) was created to draw the strategic lines for European security research and to recommend the most adequate instruments to implement it. The key findings of ESARB [1], and the experience formed with the Preparatory Action for Security Research (PASR, 2004–2006), have been taken into account in the definition of the Security theme in the Sevent Framework Programme (FP7). Specifically, four priority missions have been identified: protection against terrorism and organized crime, border security, critical infrastructure protection, and restoring security in case of crisis [11]. Then, from the analysis of the requirements of these security missions, the technology capabilities needed to meet these requirements have been identified, such as robust communications capabilities, improved situation awareness, and interoperable command and control capabilities. For these reasons, ad hoc networking technologies, by providing decentralization, flexibility, reliability, and adaptability as intrinsic features, should be key components of future communications systems for PPDR applications.

## 27.5  MASS Solutions for Public Safety Applications

The ad hoc networking concept is not new, having been around in various forms for over 30 years. The initial development of ad hoc wireless communications for military and tactical purposes can be dated back to 1972, when the DARPA agency initiated the Packet Radio Network (PRN) program. The initial concept was then expanded in follow-up programs such as the Survivable Radio Network (SURAN) initiative in 1983 and the Global Mobile (GloMo) Information program in 1994. However, a real boost in the ad hoc networking research was given by the creation in 1997 of an Internet Engineering Task Force (IETF) working group, called MANET WG. The mission of this working group was to "standardize IP routing protocol functionality suitable for wireless routing application" in multihop dynamic network topologies. A decade of intensive research in this field has generated a considerable number of different routing algorithms, although only a few of them have been successfully deployed in real ad hoc networks. In parallel, several research projects in the area of mobile ad hoc networks had been lunched by academia. The extensive research activities conducted in the ad hoc networking field have developed both the theoretical and technical background for the deployment of multihop ad hoc networks [8, 12]. However, despite the massive research efforts that have been dedicated to this field in the last two decades, it is quite recent the successful application of the ad hoc networking paradigm in real-world applications that are appearing on the mass market. The explanation of this apparent contradiction is that, initially the research on MANETs adopted quite unrealistic assumptions: large-scale and totally decentralized networks capable of supporting any type of legacy TCP/IP applications. On the contrary, as discussed in [9], the recent success of the ad hoc networking technologies is due to the adoption of a more pragmatic approach and the exploitation of the ad hoc networking paradigm to extend the Internet and to support well-defined application requirements. Among the various classes of ad hoc networks that are under deployment, we believe that mesh, vehicular, sensor, and opportunistic networks are of particular interest and importance for PPDR scenarios, because they can be considered fundamental building blocks of the next-generation of dependable, and rapidly deployable communications systems for mission-critical scenarios. In the following, we present an overview of the most recent advances in the design and the deployment of these emerging networks, and we discuss their relevance to the PPDR scenario.

### 27.5.1  Mesh Networks

Mesh networks are hybrid MANETs, where dedicated nodes, namely mesh routers, communicating wirelessly through multihop paths construct a wireless backbone. The wireless backbone may have a (limited) number of connections with the existing wired infrastructure to provide a flexible and "low cost" extension of the Internet [13]. Mobile/nomadic users obtain a multihop connectivity through the wireless backbone to communicate directly to each other, or to access the Internet via the closest mesh router. The use of multiple independent paths increases the availability and dependability of the wireless backbone through resilience to operational anomalies or security attacks. Therefore, the mesh technology can be used to rapidly deploy a high-capacity backbone in an area where the terrestrial infrastructures are partially collapsed, as shown in Fig. 27.1.

The growing interest in mesh applications has boosted the industrial efforts to offer diverse wireless mesh solutions. Some vendors have focused on standard wireless technologies, such as IEEE 802.11 (aka WiFi) and IEEE 802.16 (aka WiMax) [14]. However, on top of the standard 802-based wireless connectivity, they adopted proprietary networking software solutions that cannot interoperate. For these reasons, various IEEE standardization groups are also actively working on including wireless mesh networking techniques in the specifications of wireless technologies. The most mature example of these standardization activities is the IEEE 802.11s working group that is working to introduce advanced meshing capabilities in the WiFi technology [15]. Another limitation of the existing solutions for building mesh network, as we will extensively discuss in Sect. 27.6.1, is the lack of reliable self-configuration procedures that can dynamically adapt to varying network conditions. Nevertheless the ability to use traditional wireless technologies, e.g., 802.11, for mesh networking, makes their development easier and less expensive. The RoofNet project at MIT [16] demonstrated that it is possible to provide a city such as Boston, with broadband access with an 802.11b-based wireless network backbone infrastructure. Specifically, RoofNet consists of a limited number of nodes, positioned on rooftops operated on a volunteer basis, which dynamically create the backbone and support mesh networking. Another example of a real mesh application, which is relevant for our reference scenario, is the Quail Ridge Reserve Wireless Mesh Network project [17], an effort to provide a wireless communications infrastructure to a wildlife reserve. Aim of the project is to benefit on-site ecological research and to provide continuous and real-time monitoring of the environment. Finally, CalMesh, which is deployed on the UCSD campus and the San Diego County, is a specific example of an experimental mesh network for emergency and crisis scenarios, which provides first responders with a local network to communicate to each other and, in case, to the Internet [18].

## 27.5.2 Vehicular Ad Hoc Networks

Vehicular Ad hoc NETworks (VANETs) are emerging as one of the most successful specializations of (pure) MANETs, which is expected to rapidly penetrate the market. Traditional VANETs use ad hoc communications for performing efficient driver assistance and car safety. In this sense, VANETs can be viewed as fundamental components of any Intelligent Transportation System (ITS) [19, 20]. However, a vehicular network may be also used to perform efficient data distribution between vehicles and users during emergency situations as shown in Fig. 27.2. Note that

VANETs have a relevant advantage compared to traditional MANETs, as they rarely have constraints related to the devices' capabilities (in terms of space, computation, and power). Moreover VANETs research is pushed by both industrial and government organizations. Thus, VANET systems are one of the fields where MANET research can achieve its full potential. Examples of this effort can be found in projects such as the European FleetNet. In FleetNet, vehicles exchange short messages with local information. These messages inform the drivers about obstacles or traffic jams ahead, beyond the view of the driver's vision or the vehicle sensors. Additional projects, such as the European Project CarTALK 2000 exploited the development of cooperative driver assistance systems and the development of self-organizing ad hoc radio network as a communication basis with the aim of preparing a future standard. CarTALK uses both direct and multihop communications for the data transfer, empowered with position and spatial awareness. Similarly, in the US, several projects are involved in this area, in some cases integrating VANET into a broader view, including mesh or grid networking as in VMesh/VGrid or the PORTAL project. There is also a large involvement from the military and, since 2004, DARPA sponsors the Urban Challenge, where fully autonomous ground vehicles must conduct simulated military supply missions in an urban area. It is evident that all the knowledge developed in these projects will be useful also for the development of VANETs in emergency and crisis scenarios, when the equipments forming the vehicular network are transported by rescue land (e.g., trucks) or flying (e.g., helicopters) vehicles. For instance, in [21] an intervehicular communication system is described, which is able to quickly discover and transmit real-time multimedia information from around a crisis area to approaching first responders' vehicles.

## 27.5.3   Sensor Networks

Among ad hoc networks, wireless sensor networks have a special role. The aim of a sensor network is to collect information about events occurring in the sensor field. To this end, sensor nodes, which are tiny, low-power, and low-resources communicating devices with sensing capabilities, are deployed in the monitoring area, and the information collected by sensor nodes is generally delivered to collecting centers, also called sinks, by exploiting a wireless multihop ad hoc network. In some applications, the retrieval of sensors' readings can be implemented in a more efficient way by introducing mobile nodes inside the network (e.g., robots) that move inside the sensors field collecting the information from sensor nodes via ad hoc wireless communications and then move close to the collecting center for delivering the sensed data. Alternatively, the sink node can move in the sensor field (e.g., unmanned helicopters flying over the sensor) collecting data from each sensor node. In addition, the robots (actuators) can be used not only to collect data but also to perform actions on the sensor field depending on the detected events. For example, a robot can be used to remove explosives. Therefore, sensors and actuators networks can be successfully applied in several security scenarios. In military and tactical contexts, one of the major applications of sensor network is considered the target localization and target tracking. To this end, a variety of different physical measurements have been developed to detect the target presence and its position [22]. In parallel, many sensor networks have been developed for civilian applications, mainly for habitat and environmental monitoring. A very famous example of this type of applications is the Great Duck Island Habitat Monitoring project, a collaborative project between Intel and the University of California at Berkeley to deploy a sensor network on Great Duck Island, Maine, for monitoring migratory seabirds and the microclimates in and around nesting burrows. Another more recent example is the CitySense project, which is deploying an urban scale sensor network for monitoring weather conditions and air pollutants in the city of Cambridge, MA, USA. Note that the technologies and protocols developed to deploy these real-world sensor networks for environmental monitoring represent also the basis for sensor networks targeting mission-critical application scenarios, such as surveillance; intruders' reconnaissance and tracking; tracking of goods and vehicles; detection of nuclear, biological, and chemical attack; underwater surveillance for harbor control; etc. [23].

### 27.5.4 Opportunistic Networks

Opportunistic networks constitute a medium-term application of general-purpose MANETs for providing connectivity opportunities to pervasive devices when no direct access to the Internet is available. One of the main limitations of legacy MANETs is the fact that partitioning causes the failure of ongoing communications, and/or nodes that are temporarily disconnected from the network cannot communicate. In opportunistic networks, the information delivery is still multihop, but intermediate nodes store the messages when no forwarding opportunity toward the final destination(s) exists and exploit any contact opportunity with other mobile devices to forward information. In other words, this evolution of MANETs opportunistically exploits mobility, which resulted "hostile" for legacy ad hoc networks, and local forwarding in order to take advantage of the temporary wireless links when distributing information. Therefore, this networking paradigm has a huge potential for significantly improving the capability of first responders to re-establish effective communications in a crisis area, as shown in Fig. 27.2 and discussed in [24]. Note that, the opportunistic networking has several application scenarios beyond the PPDR scenarios, especially for pervasive computing and autonomic environments [25]. For instance, the IRTF Delay Tolerant Networking (DTN) Research Group is working to standardize architecture and protocols for enabling Internet services in networks with intermittent connectivity where continuous end-to-end connectivity cannot be assumed. The DTN architecture is suitable to interconnect systems of different scales, ranging from small-size networks formed by single mobile devices sparsely deployed in the environment, to interplanetary networks bringing together Internet-like network trunks sporadically connected through satellite links. DakNet or Saami network connectivity (SNC) are good examples of the potential applications of opportunistic and delay tolerant networks. DakNet aims at providing low-cost connectivity to rural villages in India, by exploiting mobile relays (i.e. access points mounted on buses, motorcycles, and even bicycles) passing by the village kiosks and exchanging data with them wirelessly. SNC uses DTN architecture to provide network connectivity to the nomadic Saami population. The KiosNet Project is another example of opportunistic network application in developing countries to provide a variety of services such as birth, marriage, and death certificates; land records; and consulting on medical and agricultural problems.

## 27.6 Directions for Future Research

In the last two decades, the research on MANET technologies has laid the foundations to understand the intrinsic limitations and constraints introduced by multihop wireless communications and the absence of an authority managing and controlling the network. As discussed in Sect. 27.4, these extensive research activities not only have generated a considerable amount of technical papers, but also have contributed to the development of several classes of real ad hoc networks, namely mesh networks, VANETs, WSN, and delay tolerant networks, which will have a key role in the deployment of disaster-response communications systems. However, the specific requirements of safety applications pose new technical challenges that have not been adequately addressed so far. In the following, we elaborate on the research issues that still need to be solved to realize practical and efficient systems.

### 27.6.1 Autonomic Network Management

The development of self-organizing capabilities is a fundamental prerequisite of any resilient communications system, because the communications devices should be able to react to the variations in the operating conditions without human intervention. In a sense, wireless multihop networks, being infrastructure-less peer-to-peer networks, represent an excellent example of self-organized networks, because computing devices must coordinate with each other to perform all the networking functions. However, most of the research efforts in the MANET community have been dedicated to the development of routing protocols for mobile multihop ad hoc networks, producing an incredible number of algorithms. On the contrary, the self-organization property is a multifaceted concept that incorporates a variety of capabilities. Specifically, self-organization includes self-healing, which refers to the ability of the network to detect, localize, and repair failures automatically;

self-configuration, which is the capacity of automatically generating the set of appropriate configurations parameters to operate in the current environment; and self-optimization, which is the capability to adapt the network in order to achieve relevant objectives (e.g., desired QoS levels). Consequently, the deployment of a truly self-organized network requires the adoption of a holistic approach that takes into account the interplay between all the various self-capabilities.

The ultimate objective of an autonomic network-management module should be to design an autonomic network management architecture, where the network itself helps to detect, diagnose, and repair failures, as well as to adapt its configuration and optimize its performance. However, the management of wireless networks in general is by far more complex than the management of wired networks, because wireless communications are affected by the irregularity and instability of the channel conditions that cause nonuniform and variable radio coverage areas. In addition, radio interference may lead to unpredictable behaviors and dramatical performance degradations. Moreover, in a disaster scenario, additional complexities arise because the parts of the communications network are deployed on demand in an unplanned manner. Thus, nodes may malfunction, be incorrectly configured or isolated. Individual link and node failures can easily cause network partitions. Network monitoring is a key tool to build the knowledge of the current status of the network and to discover the operating environment characteristics. Each device should not only collect local information, but also cooperate with other devices to build a representation of the entire network status. The collected information is the fundamental basis to detect anomalies and to trigger alerts to neighboring nodes or control units. The diagnostic tool responsible for the interpretation of the network state may adopt various policies such as a rule-based (i.e., the normal network state is codified though a set of admissible behaviors) or traffic-based (i.e., a set of normal traffic signatures characterizes the proper behavior of the network) analysis engine. After an alert, additional diagnostic tests should be executed to verify the root cause of the problem and to automatically trigger the most appropriate countermeasure, such as to isolate trouble links and nodes, to reallocate channels, to find alternative multihop paths, or to balance network loads.

Since the research on the self-management of ad hoc networks is in a very preliminary phase, a few solutions can be identified, which are usually tailored for mesh networks. One example is the Distributed Ad hoc Monitoring (DAMON) [26] system, which uses agents to monitor network behaviors and send collected measurements to central data repositories. However, the use of centralized analysis does not make this system suitable for challenged environments. A more recent proposal is described in [27], which describes a diagnostic system that employs trace-driven simulations to detect faults and perform root cause analysis in mesh networks. While a simulation-based approach may be useful to model the complex interaction between the several factors that affect the network behavior, the time required to simulate a large-scale network impedes the utilization of this solution for real-time network management.

## 27.6.2   Network Interoperability

Ensuring interoperable wireless communications among the devices belonging to first responders is a key requirement to effectively respond to man-made and natural disasters. The harmonization of the various standards employed by public safety agencies, as well as the shift towards open architectures and nonproprietary standards will both be crucial factors in favor of the device interoperability. However, due to the different national and international regulations on spectrum allocation, it is extremely difficult to predict a global harmonization of radio systems in the short/medium term. For instance, US and other developed countries are planning to allocate parts of the frequency bands now used for analog TV for public safety purposes [28], while those bands will continue to be used in many developing countries for broadcasting analog TV signals. A promising technological approach to overcome these constraints is to promote the use of cognitive radios and software-defined radios (i.e., software reconfigurable radios, or SDR) in the devices used by first responders. Specifically, cognitive radios are special SDRs that can adjust their transmission and reception parameters and algorithms according to multiple factors, such as radio spectrum occupancy or current state of the environment. This concept opens the way to more efficient radio resource management, but it also represents a potential solution for frequency coordination issues, limitations of available spectrum, and problems of incompatible equipments. For these reasons, the design of cognitive radios for public safety applications is emerging as a very active research area [29, 30], and two major research directions can be identified. On the one hand, there are still technological obstacles to build cheap and highly flexible SDR equipments supporting different modulation schemes and operating on large spectrum. On the other hand, the development of efficient spectrum sensing capabilities and the design of conflict resolution algorithms are still open issues where insufficient results have been obtained. For instance, in [31] a cooperative spectrum sensing framework is proposed, where cognitive radios can exchange local sensing results to obtain an accurate estimate of unused frequency bands, and even the locations of the other radios, as well as to reduce detection times. In [32], a game theoretic framework is developed to model the efficiency of adaptive and distributed channel allocation for cognitive radios. However, it is not clear that the tradeoff between the overheads needed to coordinate the frequency allocations and the network performance improvement.

## 27.6.3   QoS Protection

Until recently, the design of mechanisms and policies to support QoS levels and the design of a resilient communications infrastructure appeared as two separated and uncorrelated research domain areas. However, after the analysis of communication breakdowns during recent disasters it is clearly emerged that the survivability of the communications infrastructure and end-to-end connectivity is not sufficient to guarantee the survivability of the communication services. For instance, in the final report of the 9/11 Commission it was pointed out that, although the cellular telecommunications networks were not destroyed by the terrorist attacks, the first responders where unable to use them because severely congested by the huge number of simultaneous connection attempts. In other words, in crisis response the network workloads can overwhelm the available network capacity such that the minimum application requirements of real-time traffic (e.g., voice communications) cannot be met. On the contrary, in emergency situations it is fundamental to ensure that critical data are made available to the right set of users, avoiding congestion and data unavailability [33]. For these reasons, novel mechanisms are needed to support QoS in ad hoc networks to guarantee different QoS levels, which are appropriate to the information criticality and the network mission. It is evident that a system-wide QoS notion requires that the QoS support be implemented in each MANET protocol. However, it is also true that a QoS-aware routing protocol is the basis of any QoS solution for MANETs, because the ad hoc routing protocol is responsible for finding the relaying nodes that can meet the applications' requirements. For these reasons, especially in the last years, the MANET research focus has shifted from routing protocols maintaining best-effort end-to-end connectivity between mobile devices to the provision of diverse and more complex QoS attributes. These research activities have produced a considerable number of solutions, and the major contributions are outlined in [34]. However, most of these potential solutions have neglected the importance of QoS robustness, namely the capacity of maintaining with high probability the QoS guarantees regardless of network variations such as individual link or node failures. Thus, the design of policies and mechanisms to obtain reliable and adaptive QoS support is still an open issue. An interesting direction for future research in the area of reliable QoS is the use of preemptive strategies. For instance, in [35] the authors proposed to use preemptive selection of routes according to predictive stability measures. Admission control strategies and segregation of dedicated network resources are also promising areas of investigation. As an example, Beard and Frost [33] described an architecture composed of geographically distributed ticket servers to identify the priority that should be given to a flow in stressed networks, and to limit resource usage by low priority users.

## 27.7 Conclusions

In this chapter, we have advocated the adoption of ad hoc networking technologies to address the fragility of our communications infrastructure, which has been dramatically exposed in the aftermath of recent natural and man-made disasters. In fact, in the recent years the significant advances in ad hoc networking technologies have led to the development of various types of specialized networks, such as mesh networks, vehicular networks, sensor networks, and opportunistic networks, which are of particular interest and importance for PPDR scenarios. In addition, the ad hoc networking paradigm intrinsically provides flexibility, self-configurability, and fully decentralized operations, which are necessary requirements to deploy the future generation of dependable, versatile, and secure communications systems for PPDR applications. However, there are several open technical challenges that have to be addressed to realize this vision of a survivable communications system in disaster scenarios. For instance, it is unacceptable to have a communications network that partially stops working correctly during a crisis. Therefore, the focus is on providing continuous communication services, even with degraded performance. In other words, for modern disaster scenarios the focus should move from traditional QoS provision to QoS protection, with a native support of prioritization of emergency-related traffic. Second, interoperability between devices, communication paradigms, and network architectures is a prerequisite for an effective implementation of PPDR operations. However, the design of very specialized MANET-based networks has largely neglected the interoperability concerns. Finally, in disaster scenarios the human intervention for the bootstrap, configuration, maintenance, and adaptation of the communications infrastructures is impossible. Therefore, self-management capabilities should be native functionalities and an integral part of the network design, so that the network itself may help to detect, diagnose, and repair failures, as well as to adapt its configuration and optimize its performance.

# Terminologies

*Actuator.* Robot able to perform actions on the sensor field as a reaction to the events detected by the sensor nodes.

*Ad Hoc Network.* Wireless network where mobile devices communicate directly without relying on any predeployed infrastructure. This type of network is also referred to as infrastructure-less network.

*Delay tolerant network (DTN).* Architecture and protocols standardized by IRTF *Delay Tolerant Networking* (DTN) Research Group for enabling communications in networks with intermittent connectivity where continuous end-to-end connectivity cannot be assumed such as interplanetary networks, military/tactical networks, disaster/emergency networks, and some forms of ad hoc sensor/actuator networks.

*Infrastructure-based network/systems.* Wireless network/system with a preinstalled fixed infrastructure. Mobile device access the fixed infrastructure through Access Points.

*Mesh network.* A multihop ad hoc network that uses dedicated nodes (called *mesh routers*) communicating wirelessly to construct a *wireless backbone* that has a (limited) number of connections with the wired Internet. Mobile users obtain a multihop connectivity to the Internet by connecting to the closest mesh router.

*Mobile ad hoc network (MANET).* Ad hoc network where the source and destination nodes are not within the transmission range of each other. Communication occurs through intermediate nodes. Nodes in a MANET thus act as nodes and router at the same time.

*Opportunistic network* Heterogeneous multihop ad hoc networks that exploit any contact opportunity to forward data. Forwarding is performed by opportunistically exploiting the network interfaces (by wired and wireless) a node has. When no forwarding opportunity exists (e.g., no other nodes in the transmission range, or neighboring nodes are considered not useful for that communication) a node locally stores the messages. This type of network is suitable for sparse and frequently disconnected networks.

*Public protection and disaster relief (PPDR).* Public protection is the act of protecting people from dangers. It involves activities such as risk identification, prevention, and response to critical situations. Disaster relief is the process of returning the community to the normal state (recovery).

*Self-organization.* The capability of a system or network to automatically and dynamically generating the set of appropriate and/or optimized configurations parameters to operate in the current environment.

*Sensor network.* A network of sensor nodes, densely and randomly deployed inside the area in which a phenomenon is being monitored. Each sensor node delivers the collected data to one (or more) neighbor node, one hop away. By following a multihop communication paradigm the data are routed to a special node (sink) and, through this, to the user.

*Sensor node.* Tiny device with computing, wireless communication and sensing capabilities that can be used for various purposes. Typical sensing tasks could be temperature, light, sound, etc.

*Vehicular networks.* Mobile ad hoc networks used for car-to-car communications. Cars located between the source and destination cars operate as traffic relays.

## Questions

1. Elaborate on the major differences between infrastructure-based networks and infrastructure-less networks.
2. Explain the main features of a self-organizing network.
3. Describe the main characteristics of a mesh network.
4. Describe the main characteristics of a wireless sensor network.
5. Describe the main characteristics of a vehicular ad hoc network.
6. Describe the main characteristics of an opportunistic network.
7. Elaborate on the technical challenges associated to QoS protection in PPDR scenarios.
8. Elaborate on the technical challenges associated to autonomic network management in ad hoc networks.
9. Motivate the importance of ad hoc networking technologies in PPDR scenarios.
10. Discuss the technical limitations of private and public mobile radio systems in a typical emergency scenario.

**Acknowledgment**  This work was supported in part by the European Commission through project EU-MESH (Enhanced, Ubiquitous, and Dependable Broadband Access using MESH Networks), FP7 ICT-215320.

# References

1. ESARB, "Meeting the Challenge: the European Security Research Agenda – A report from the European Security Research Advisory Board", September 2006.
2. London Regional Resilience Forum, "Looking Back, Moving Forward – The Multi-Agency Debrief", September 2006.
3. UK Government (J. Reid, T. Jowell), "Addressing Lessons from the Emergency Response to the 7 July 2005 London Bombings", September 2006.
4. US Homeland Security, "Hurricane Katrina: A Nation Still Unprepared", May 2006.
5. US National Task Force on Interoperability, "Why Can't We Talk?", February 2005.
6. D. Hatfield, P. Weiser, "Toward A Next Generation Strategy – Learning from Katrina and Taking Advantage of New Technologies", 2005.
7. N. Ahmed, K. Jamshaid, O.Z. Khan, SAFIRE: A self-organizing architecture for information exchange between first responders, in Proceedings of IEEE Workshop on Networking Technologies for SDR Networks, San Diego, CA, USA, 8 June 2007.
8. I. Chlamtac, M. Conti, J. Liu, Mobile ad hoc networking: imperatives and challenges, Elsevier Ad Hoc Networks Journal, 1(1), 2003, 13–64.
9. M. Conti, S. Giordano, Multihop ad hoc networking: The reality, IEEE Communications Magazine, 45(4), 2007, 88–95.
10. B.S. Manoj, A. Hubenko-Baker, Communication challenges in emergency response, Communications of the ACM, 50(3), 2007, 51–53.
11. European Commission, "FP7 Cooperation Work Programme – Theme 10: Security (Call 1)", 22 December 2006.
12. M. Conti, S. Giordano, Multihop ad hoc networking: The theory, IEEE Communications Magazine, 45(4), 2007, 78–86.
13. R. Bruno, M. Conti, E. Gregori, Mesh networks: Commodity multi-hop ad hoc networks, IEEE Communications Magazine, 43, 2005, 123–131.
14. C. Eklund, R.B. Marks, K.L. Stanwood, S. Wang, IEEE standard 802.16: A technical overview of the WirelessMAN air interface for broadband wireless access, IEEE Communications Magazine, June 2002, 98–107.
15. IEEE TGs, "Joint SEE-Mesh/Wi-Mesh Proposal to 802.11 TGs," IEEE 802.11s-06/0328r0, March 2006. [Online]. Available: http://grouper.ieee.org/groups/802/11/
16. J. Bicket, S. Biswas, D. Aguayo, R. Morris, Architecture and evaluation of an unplanned 802.11b mesh network, in Proceedings of ACM MobiCom 2005, Cologne, Germany, 28 August 28–2 September 2005, pp. 31–42.
17. D. Wu, D. Gupta, P. Mohapatra, Quail ridge reserve wireless mesh network: Experiences, challenges and findings, Proceedings of the TRIDENTCOM 2007, Florida, USA, May 2007.
18. R.B. Dilmaghani, R.R. Rao, Future wireless communication infrastructure with application to emergency scenarios, in Proceedings of IEEE WoWMoM 2007, Helsinki, Finland, 18–21 June 2007.
19. S. Yousefi, M.S. Mousavi, M. Fathy, Vehicular ad hoc networks (VANETs): Challenges and perspectives, in Proceedings of sixth International Conference on ITS Telecommunications, Chengdu, China, 21–23 June 2006, pp. 761–766.
20. M. Torrent-Moreno, M. Killat, H. Hartenstein, The challenges of robust inter-vehicle communications, in Proceedings of IEEE VTC-Fall 2005, Dallas, TX, USA, 28–25 September 2005, pp. 319–323.

21. M. Roccetti, M. Gerla, C.E. Palazzi, S. Ferretti, G. Pau, First Responders' crystal ball: How to scry the emergency from a remote vehicle in Proceedings of IEEE IPCCC 2007, New Orleans, LA, USA, 11–13 April 2007, pp. 556–561.
22. S. Liang, D. Hatzinakos, A cross-layer architecture of wireless sensor networks for target tracking, IEEE/ACM Transactions on Networking, 15(1), 2007, 145–158.
23. M. Lopez-Ramos, J. Leguay, V. Conan, Designing a novel SOA architecture for security and surveillance WSNs with COTS, in Proceedings of IEEE MASS-GHS'07, Pisa, Italy, 8 October 2007.
24. L. lilien, A. Gupta, Z. Yang, Opportunistic networks for emergency applications and their standard implementation framework, in Proceedings of IEEE IPCCC 2007, New Orleans, LA, USA, 11–13 April 2007, pp. 588–593.
25. L. Pelusi, A. Passarella, M. Conti, Opportunistic networking: Data forwarding in disconnected mobile ad hoc networks, IEEE Communications Magazine, 44(11), 2006, 134–141.
26. K.N. Ramachandran, E.M. Belding-Royer, K.C. Almeroth, DAMON: A distributed architecture for monitoring multi-hop mobile net-works, in Proceeding of IEEE SECON 2004, Santa Clara, CA, USA, 4–7 October 2004, pp. 601–609.
27. L. Qiu, P. Bahl, A. Rao, L. Zhou, Troubleshooting wireless mesh networks, Computer Communications Review, 36(5), 2006, 19–28.
28. J.M. Peha, The digital TV transition: A chance to enhance public safety and improve spectrum auctions, IEEE Communications Magazine, 44(6), 2006, 22–23.
29. T.W. Rondeau, C.W. Bostian, D. Maldonado, A. Ferguson, S. Ball, S.F. Midkiff, B. Le, Cognitive radios in public safety and spectrum management, in Proceeding of 33rd Research Conference on Communication, Information and Internet Policy, Arlington, VA, USA, 23–25 September 2005.
30. P. Pawelczak, R.V. Prasad, X. Liang, I.G. Niemegeers, Cognitive radio emergency networks – requirements and design, in Proceeding of DySPAN 2005, Baltimora, MD, USA, 8–11 November 2005, pp. 601–606.
31. G. Ganesan, Y. Li, Cooperative spectrum sensing in cognitive radio networks, in Proceeding of DySPAN 2005, Baltimora, MD, USA, 8–11 November 2005, pp. 137–143.
32. N. Nie , C. Comaniciu, Adaptive channel allocation spectrum etiquette for cognitive radio networks, Mobile Networks and Applications, 11(6), 2006, 779–797.
33. C.C. Beard, V.S. Frost, Prioritization of emergency network traffic using ticket servers: A performance analysis, Simulation: Transactions of the Society for Modeling and Simulation, 80(6), 2004, 289–299.
34. L. Hanzo II, R. Tafazolli, A survey of QoS routing solutions for mobile ad hoc networks, IEEE Communications Surveys and Tutorials, 9(2), 2007, 50–70.
35. M. Ayyash, K. Alzoubi, Y. Alsbou, Preemptive quality of service infrastructure for wireless mobile ad hoc networks, in Proceeding of IWCMC'06, Vancouver, British Columbia, Canada, 3–6 July 2006, pp. 707–712.

# Biography

**Dr. Sudip Misra** is an assistant professor in the School of Information Technology at the Indian Institute of Technology, Kharagpur, India. Prior to this, he worked in Cornell University (USA), Yale University (USA), Nortel Networks (Canada), and the Government of Ontario (Canada). He received his Ph.D. degree in Computer Science from Carleton University, in Ottawa, Canada, and the masters and bachelors degrees, respectively, from the University of New Brunswick, Fredericton, Canada, and the Indian Institute of Technology, Kharagpur, India. Dr. Misra has several years of experience working in the academia, government, and the private sectors in research, teaching, consulting, project management, architecture, software design, and product engineering roles.

His current research interests include wireless ad hoc, sensor, and mesh networks. Dr. Misra is the author/editor of over 90 scholarly research papers and books. He has won *five research paper awards* in different conferences. He was also the recipient of several academic awards and fellowships such as the *(Canadian) Governor General's Academic Gold Medal* at Carleton University, the *University Outstanding Graduate Student Award* in the Doctoral level at Carleton University, and the Canadian Government's prestigious *NSERC Post Doctoral Fellowship*. His biography was also selected for inclusion in the 2006–2007 edition of Marquis *Who's Who in Science and Engineering*, and the 25th edition of the Marquis *Who's Who in the World*, California, USA. A mention about him and his work has also appeared in a Canadian newspaper.

Dr. Misra is the *editor-in-chief* of two journals – the *International Journal of Communication Networks and Distributed Systems* (IJCNDS) and the *International Journal of Information and Coding Theory* (IJICoT), UK. He is an *associate editor* of the *Telecommunication Systems Journal* (Springer SBM), *Security and Communication Networks Journal* (Wiley), *International Journal of Communication Systems* (Wiley), and the *EURASIP Journal of Wireless Communications and Networking*. He is also an *editor/editorial board member/editorial review board member* of the *IET Communications Journal, Computers and Electrical Engineering Journal* (Elsevier), *International Journal of Internet Protocol Technology*, the *International Journal of Theoretical and Applied Computer Science*, the *International Journal of Ad Hoc and Ubiquitous Computing, Journal of Internet Technology*, and the *Applied Intelligence Journal* (Springer). He was invited to chair several international conference/workshop programs and sessions. He has been serving in the program committees of over a dozen international conferences. Dr. Misra was also invited to deliver *keynote lectures* in around a dozen international conferences in the USA, Canada, Europe, Asia, and Africa.

**Dr. Isaac Woungang** received his M.A.Sc. and Ph.D. degrees, all in Applied Mathematics from the Université du Sud, Toulon-Var, France, in 1990 and 1994, respectively. In 1999, he received an M.A.Sc from INRS-Materials and Telecommunications, University of Quebec, Canada. From 1999 to 2002, he worked as a software engineer at Nortel Networks, Ottawa, Canada. Since 2002, he has been with Ryerson University, where he is now an assistant professor of computer science. In 2004, he founded the DABNEL (the Distributed Applications and Broadband NEtworks Laboratory) R&D group at Ryerson University, Canada. His research interests are telecommunications network design, network security, and computational intelligence applications in telecommunications.

**Dr. Subhas C. Misra** is currently a visiting faculty in the Indian Institute of Technology, Kanpur, India. He was earlier a visiting scientist at Harvard University, Boston, USA. He received his Ph.D. degree from Carleton University, in Ottawa, Canada, and M.S. and M.Tech. degrees, respectively, from the University of New Brunswick, in Fredericton, Canada, and the Indian Institute of Technology (IIT), at Kharagpur, India. Dr. Misra has several years of experience working in the academia, and the public and private sectors in research, teaching, consulting, project management, architecture, software design, and product engineering roles. His current research interests include telecommunications systems management, software management, and information systems security. Dr. Misra has authored over 50 scholarly research papers and published (yet to appear) five books.

He has won *Best Research Paper Award* in an international conference held in the United States. He was also the recipient of more than 15 academic awards and fellowships such as the *Achievement Award* at the 2007 World Congress held in Las Vegas in the United States for outstanding research contributions to his field, and the Canadian Government's *NSERC Post Doctoral Fellowship*. A mention about him and his work has also appeared in the June 8, 2007 issue of the *Carleton Now* newspaper. His biography has also been selected to appear in the *Cambridge Blue Book*, Cambridge, UK, 2008.

Dr. Misra is the managing editor/associate editor of more than ten international journals. He was invited to serve as program chair, organizing chair, and session chair in different international conferences. He has been serving in the program committees of over a dozen international conferences.

Dr. Misra was also invited to offer keynote lectures in more than a dozen of international conferences in different countries of North America, Europe, and Asia. Recently, he was invited to deliver a keynote lecture in the World Congress on Engineering and Computer Science held in California, USA.

# Index

**A**

Access-based energy efficient (ABEE), 38
Acknowledgement (ACK), 471, 476, 480, 482, 484
Acknowledgment spoofing, 495
Acoustic sensor network, 354–355, 357–363, 365, 368, 370, 372
Adaptive distributed resource allocation (ADRA), 349, 351–353, 355, 357, 362–375
Adaptive modulation and coding (AMC), 462
Adaptive rate control (ARC), 233–234
Ad hoc networks, 183, 184
ADRA. *See* Adaptive distributed resource allocation
ADRA-dc, 348, 354, 358–362, 365–370, 372, 373
ADV, 456, 457
AEWMA. *See* Attenuation method with exponentially weighted moving-average
Aggregation, 192–198, 492, 495, 496, 498, 502, 503, 506–509
Aggregation function, 534
Agilla, 598
Agimone, 598
Air interface, 450
AMC. *See* Adaptive modulation and coding
Analytic modeling, 651
Anchor-based localization, 115
Anchor-free localization, 115
Angular relaying, 96
Application-specific approaches, 222
ARC. *See* Adaptive rate control
ARQ. *See* Automatic repeat request
Artificial immune system (AIS), 299

Attenuation, 472, 475, 476, 478–480, 482, 484–486
Attenuation method with exponentially weighted moving-average (AEWMA), 482–484
Attestation path, 550
Authentication, 492, 494, 499–500, 503, 506, 507, 509
Automatic repeat request (ARQ), 446, 465
Autonomic network management, 701–702
Available, 274

**B**

Backpressure routing, 310, 311
Base station, 608, 632
Basic probability assignment (BPA), 18
Beaconless forwarder planarization, 96
Beaconless routing, 94–97
Beacon vector routing, 128
Bio-inspired, 659, 661–664, 672, 674, 682, 683
Birthday attacks, 515
Birthday threshold, 516
Bit error rate (BER), 444, 446, 447
B-MAC, 470
Bottlenecked sensors, 609, 614, 621
Bottleneck nodes, 621
Bounded buffer producer-consumer problem, 332, 344
Breakdown point, 539
Broadcast, 240, 257, 475, 485, 499

**C**

Carrier sense multiple access (CSMA), 421
Carrier sense multiple access with collision avoidance (CSMA/CA), 449

Carrier sensing, 452–454
CCA. *See* Clear channel assessment
CC-MAC. *See* Correlation-based collaborative
          medium access control
CCP. *See* Coverage and configuration protocol
CDMA. *See* Code division multiple access
Central node (CN), 296
Channel fading, 440, 445, 447, 462, 464
Channel quality, 446, 447, 462, 463
Channel state information (CSI), 447, 461, 462
Channel variations, 446, 461, 462
C language, 342
Classification
    fusion, majority voting, 16
    maximum likelihood classifier, 7
    supervised, 6
    unsupervised, 6
Clear channel assessment (CCA), 475, 479
Clear-To-Send frame (CTS), 421
Clock frequency, 442
Clock-tick, 335
Closed-loop flow control, 209
Cluster-based energy conservation (CEC), 34
Cluster head(s), 29, 458–460, 465, 608,
          614, 615
Clustering, 28
Clustering algorithms, 36–42
Clustering protocol (CP), 40
Cluster radius decision point (CRDP), 36
Cluster topologies, 638
CODA. *See* Congestion detection and
          avoidance
Code division multiple access (CDMA), 380
Coding, 473, 486, 487
Cognitive radio, 464
Coherence time, 447
Coherent signal processing, 296
Collision(s), 423, 435, 473
Combinatorial Delaunay graph, 125
Combining strategies, 164–167
    equal ratio combining (ERC), 165
    fixed ratio combining (FRC), 165
    maximum ratio combining (MRC),
          166–167
    selection combining (SC), 165–166
    threshold combining (TC), 166
Commitment forest, 547
Commitment tree, 545
Communication latency, 444, 445, 451,
          454, 461
Communication load, 442, 444, 445, 455,
          462, 465
Compact routing, 130
Concave node, 84

Concurrency model, 327, 331
Confidentiality, 498, 506, 509
Congestion
    avoidance, 209
    control, 208–209
    control goals, 214
    detection, 208, 214, 216, 218, 220, 222,
          224, 226, 231
    management, 206–208
    mitigation, 209, 217
Congestion control for multi-class traffic
          (COMUT), 224–225
Congestion detection and avoidance (CODA),
          216–217, 241
Connected dominating set (CDS), 271
Connectivity, 628
Consensus set, 542
Conservative synchronization, 646
Constraint satisfaction problem (CSP), 349,
          350, 374
Contamination, 538
Contention-based MACs, 420, 436
Contention window (CW), 450
Context switches, 328, 333, 334, 338, 339
Contik, 335–337, 343
Convergecast, 238, 255–257
Cooperative diagnosis, 264
Cooperative relaying, 159–181
CoProVe, 517
Copy state saving, 647
Correlation-based collaborative medium
          access control (CC-MAC), 308
Correlation radius, 308, 668, 682
Correlation regions, 308
Corruption, 445, 446, 453, 464
Coverage, 608–614, 616, 621–623, 625–632
Coverage and configuration protocol (CCP),
          50
Coverage hole(s), 613, 622, 625, 631
Covering with disk, 264
Cramer-Rao lower bound (CRLB), 4
Critical head, 274
Crossbow MICA2 motes, 348, 370
Cross layer design, 460, 462, 464, 466
Cross-layer QoS solutions, 309
Cryptography, 492, 498
CSI. *See* Channel state information
CSMA/CA. *See* Carrier sense multiple access
          with collision avoidance
CSP. *See* Constraint satisfaction problem
CTS, 447, 449, 450, 452, 453
Cut and paste attack, 572

**D**

Data aggregation, 402, 458, 465, 534, 639
Database, 187–188
Data-centricity, 183–201
Data-centric routing, 97, 114, 188–192
Data-centric storage, 198–200
Data fusion, 534
Data gathering and aggregation, 280
Data link layer (DLL), 440, 445–447, 465
Data redundancy, 213, 458
Data transport control, 239, 240, 251
DCSP. *See* Distributed constraint satisfaction
    problem
Decision fusion, 275
Delivery rate, 85, 99, 100, 106
Delivery speed, 310
Deluge, 256
Dempster-Shafer theory of evidence, 17
Denial of service (DoS), 571
Dense deployment, 294
Density control, 48
Department of Homeland Security (DHS), 695
Design tradeoffs, 644
Detached cooperative diversity (DCD)
    protocol, 171
DFuse, 195
DHS. *See* Department of Homeland Security
Differentiated contention control, 249
Differentiated services (DiffServ), 305
Digital right management (DRM), 570
DIM. *See* Distributed index for
    multidimensional data
Directed diffusion, 98, 136, 190–192, 299
Discrete center hierarchy, 143
Discrete event simulation, 645–648
Distance-based clustering algorithm, 37
Distributed coding, 564, 576, 577
Distributed constraint satisfaction problem
    (DCSP), 349, 350, 374
Distributed coordination function (DCF), 449
Distributed detection, 279
Distributed index for multidimensional data
    (DIM), 199
Distributed resource allocation, 347–374
Distributed system(s), 325, 343
Diversity
    frequency, 161–162, 180
    spatial, 162, 180
    temporal, 161, 180
DMS. *See* Dynamic modulation scaling
Double rulings, 137
Drawbacks of analytical modeling, 652
Drawbacks of discrete event simulation, 651
DVM. *See* Dynamic velocity monotonic

DVS. *See* Dynamic voltage scaling
Dynamic modulation scaling (DMS), 440,
    441, 444, 445, 464
Dynamic reconfiguration, 331, 334
Dynamic velocity monotonic (DVM), 309
Dynamic voltage scaling (DVS), 440–442,
    444, 464

**E**

Eavesdropping, 503, 506, 509
Efficient channel utilization, 425, 435
ElGamal on elliptic curves, 552–553
E-MAC. *See* Event medium access control
Embedded wireless interconnect (EWI), 312
Endocrine system, 661, 671, 672
End-to-end QoS, 309
Energy, 469–472, 474, 483–486
    awareness, 465
    consumption, 294
    efficiency, 324, 328, 338, 339, 440, 442,
        444, 445, 448, 451, 454, 458,
        464, 465
    harvesting, 642–644
    saving, 440, 442, 443, 447, 463, 466
Energy efficient clustering scheme (EECS), 39
Error protection, 446–448
Error recovery, 447
Error tolerance and budget, 313, 314
ESRT. *See* Event-to-sink reliable transport
Estimation
    best linear unbiased estimator (BLUE), 6
    maximum likelihood estimator (MLE), 5
    minimum variance unbiased (MVU), 4
    unbiased, 4
European Security Research Advisory Board
    (ESRAB), 696
Event-driven model, 325, 327, 331, 339
Event driven simulations, 645
Event medium access control (E-MAC), 308
Event signal, 664, 666, 667, 669, 674, 682
Event-to-sink reliable transport (ESRT),
    222–224, 241
EWI. *See* Embedded wireless interconnect
EWMA. *See* Exponentially weighted
    moving-average
Exclusion basis system (EBS), 523
Exponentially weighted moving-average
    (EWMA), 482

**F**

Face routing, 89, 118
Face tracing, 123

FACTS, 597
Fairness and/or QoS, 235
False data injection, 495
Fast deployment, 693
Fault tolerance, 262
FDMA. *See* Frequency division multiple
    access
FEC. *See* Forward error correction
FIFO scheduler, 328, 344
Flat topology, 639
Flooding, 188–189, 455, 456
Flooding rate, 99
Flow control, 205–236
Flush, 230–232
Formal verification, based on constraint
    solving, 515
Forward error correction (FEC), 440, 447–449,
    465
Frequency division multiple access (FDMA),
    380
Funnelling, 194
Fusion, 217–218


G
Gabriel graph, 90, 118, 119
Gain,
GANGS, 431
GEIDR. *See* Geographical distance routing
Geographic Adaptive Fidelity (GAF), 34
Geographical distance routing (GEDIR), 85
Geographic(al) hash table (GHT), 136, 199
Geographic(al) routing, 82, 96, 97, 99, 102,
    104, 106, 117
Geographic forwarding, 309
Geographic routing with cooperative relaying
    and leapfrogging, 175
GFG routing, 92
GHT. *See* Geographic(al) hash table
Global Monitoring for Environment and
    Security (GMES), 696
Global rigidity, 116
GloMoSim, 648
GMES. *See* Global Monitoring for
    Environment and Security
GOAFR. *See* Greedy other adaptive face
    routing
Gossiping, 188–189, 456
GPSR. *See* Greedy perimeter stateless routing
Gradient, 191, 299
Gradient landmark-based routing, 124
Graph labeling, 382
Greedy embedding, 121
Greedy forwarding, 99–100

Greedy other adaptive face routing (GOAFR),
    94
Greedy perimeter stateless routing (GPSR),
    118, 133, 136
Grid location service (GLS), 145
Group TDMA, 432, 436


H
HELLO flood, 495, 508
HELLO flood attack, 573
Heterogeneous sensor network, 562
Heterogeneous WSN, 36
Hidden collision problem, 452
Hierarchical in-network aggregation, 546–548
Hierarchical routing, 114
Holistic view of QoS, 315
Homeland Security Advanced Research
    Projects Agency (HSARPA), 695
Homeostasis, 670–674, 682
Homogeneous sensor network, 562
Homogeneous WSN, 36
Homomorphic public-key cryptosystem, 551
Hood, 592
HSARPA. *See* Homeland Security Advanced
    Research Projects Agency
Hybrid approach/method, 225–227, 480–482,
    484
Hybrid ARQ-based cooperative relaying
    (HACR) protocol, 174


I
Idle listening, 395, 423, 435, 450, 451,
    470, 487
Idle time, 338
IEEE 802.11, 449, 450, 452, 453
IFRC. *See* Interference-aware fair rate control
Immune system, 659, 661, 662, 664–672, 682
Impala, 599
Implementation(s), 325, 331, 339–341
Implosion, 455, 456
Incremental state saving, 647
Information and Communication Technologies
    (ICT), 687
Information assurance, 513
Information brokerage, 136
Information consumer, 136
Information producer, 136
Infrastructure network, 183
Infuse, 256, 257
Initiator, 294, 299
In-network processing, 534
Insect, 659, 661, 662, 674–680, 682

Integrated Services (IntServ), 305, 306
Interactive proof algorithm, 544
Interest, 299
Interface(s), 324, 332, 334, 337, 340, 341, 343
Interference, 380
Interference-aware fair rate control (IFRC),
    219–220
Interoperability, 693, 703
Interrupts, 333–335
Intrusion detection, 500
IntServ. *See* Integrated Services
Iterative method, 476–478, 480, 481

J
Jamming, 492, 497, 507, 508
Jist-in-time-scheduling (JiTS), 311
Jump table, 330, 331

K
Kairos/Pleiades, 595–596
k-coverage, 58
Kernel, 326, 328, 330–334, 336, 337
Key, 491, 492, 498–500, 504, 509
    cluster, 519
    global, 519
    pairwise, 520
Key agreement protocol, 517
Key distribution, 492, 498, 499, 503, 507
Key establishment protocol, 517
Key pre-distribution protocol, 517
Key refreshment protocol, 523–524
Key revocation protocol, 524–526
Key ring, 520
Key transport protocol, 517
*K* vertex-disjoint, 270

L
Landmark routing, 124–131
Land Mobile Radio (LMR) systems, 694
Latency, 425, 435, 473, 488
LEACH, 639. *See* Low-energy adaptive
    clustering hierarchy
Leaf node, 338
Leakage energy, 443
LEAP+, 523
LESOP. *See* Low energy self-organizing
    protocols
Lifetime, 288
Lightweight clustering, 298
Linear programming (LP), 288
Listening state, 451

Load balancing, 309
Localization, 115, 492, 493, 496, 501, 502,
    507, 508, 509
Localization ambiguity, 116
Location, 496, 501, 502, 507
    server, 145
    service, 114
Logical Neighborhoods, 592–593
Low-energy adaptive clustering hierarchy
    (LEACH), 37, 431, 458–460, 465,
    565, 566
Low-energy localized clustering (LLC), 36
Low energy self-organizing protocols
    (LESOP), 312
Low power listening, 426

M
MAC. *See* Medium access control; Message
    authentication code
MANET. *See* Mobile ad hoc networks
Mantis OS (MOS), 332, 333, 338, 339,
    342, 343
Market-based macro-programming (MBM),
    350, 374
Markov processes, 651
Mate, 596–597
MBM. *See* Market-based macro-programming
Medial axis, 131
Medial-axis based routing, 131
Medium access control (MAC), 419, 422, 435,
    449–454, 459, 461–463
Medium contention, 472
MEMS. *See* Microelectronic-mechanical
    systems
MESA. *See* Mobility for emergency and safety
    applications
Mesh networks, 689, 690, 697–698
Message authentication code (MAC), 499,
    500, 504, 509
Message passing, 451, 453–454, 457
Meta-data, 456, 457
MFR. *See* Most forward within radius
Microelectronic-mechanical systems (MEMS),
    439, 441, 465
Min-disk-cover scheme, 266
Minimal virtual dominating set (MVDS), 32
Minimum connected dominating set (MCDS),
    254
Minimum mean square error (MMSE), 194
Minimum placement of relay nodes, 265
Minimum square error (MSE), 194
Min-weight *k*-OutConnectivity, 271
MMAC, 432

MMSPEED. *See* Multi-path multi-SPEED
MNP. *See* Multihop network reprogramming
Mobile ad hoc networks (MANET), 306,
        688, 697
Mobile wireless sensor networks, 608
Mobile WSN. *See* Mobile wireless sensor
        networks
Mobility, 477, 485
Mobility for emergency and safety
        applications, 696
Modularity, 328
Modulation, 473, 485–487
Modulation level, 441, 442, 444, 445, 464
MOS. *See* Mantis OS
Most forward within radius (MFR), 84
Moving average, 482
Moving Picture Experts Group (MPEG), 313
Multicast, 485
Multihop network reprogramming (MNP), 256
Multimedia, 661, 670–674
Multimedia based sensing, 316
Multi-path routing, 306, 309, 311
Multi-path multi-SPEED (MMSPEED), 310


N
Naps, 35
NAV. *See* Network allocation vector
N-centering, 50
Neighbor-based clustering algorithm, 37
Nervous system, 661, 671
nesC, 339, 340, 343
Network allocation vector (NAV), 450,
        452, 453
Network-centric approaches, 215–222
Network layer, 440, 454, 461
Network medium access control (N-MAC),
        308
Network monitoring, 702
Neyman-Pearson test, 71
N-MAC. *See* Network medium access control
Node capture, 515
Node capture attack, 573
Node compromise, 493
Non-available, 274
Noncoherent signal processing, 296
Non-malleability, 551
ns-2, 648


O
OGDC. *See* Optimal geographical density
        control
Okamoto-Uchiyama cryptosystem, 553–554

One-incremental, 62
One-to-many protocols, 232–233
One-to-one protocols, 230–232
Open-loop flow control, 209
Open systems interconnect (OSI), 460
Operating system(s), 323–331, 335
Opportunistic networks, 689, 692, 700–701
Optimal geographical density control
        (OGDC), 50
Optimistic synchronization, 646
OSI. *See* Open systems interconnect
Outlier, 538
Overhearing, 424, 436
Overlay, 455


P
Packet collision, 450, 452, 453
Packet fragmentation, 440, 445–447, 453,
        461, 465
Packet overhearing, 450
PARSEC, 650
Passive participation, 520
PEGASIS, 639
Perimeter routing, 118
Pervasive sensor network, 297
p-hop critical node, 273
p-hop sub-graph, 273
Physical layer, 440–445, 453, 461–464
Physical SINR model, 406
Piconet, 297
Piezoelectric, 643
Planar subgraph construction, 90–91
Point-to-point routing, 114
Positive and negative, 287
Positive and retreat, 287
Positive, negative and retreat, 287
Positive-only, 287
POSIX threads, 333
Power-efficient gathering in sensor
        information systems (PEGASIS), 42
Power management, 441, 448
Power routing, 88
Power save mode (PS), 296
Preamble sampling, 426
Prediction-based monitoring (PREMON), 313
Prediction model, 313
Preemption, 325, 334, 337, 343
PREMON. *See* Prediction-based monitoring
Prioritized MAC, 310
Privacy homomorphism, 551–554
Private mobile radio systems, 694
Probabilistic key sharing, 520
Program loader, 336

Proof of concept, 167–170, 179
Protocol model, 406
Protocol overhead, 424, 435
PSFQ, 229–230
Public protection and disaster relief (PPDR), 688
Publish/subscribe, 186–187
Pull diffusion, 192
Push diffusion, 192

**Q**
Q-digest, 540
Q-MAC. *See* QoS aware medium access control
QoD. *See* Quality of data
QoS. *See* Quality of service
QoS aware medium access control (Q-MAC), 308
QoS centric operating system, 316
QoS robustness, 704
Quality of data (QoD), 313
Quality of service (QoS), 305–318, 693, 703
Quality-of-service specific information retrieval (QUIRE), 307
Quantiles aggregation, 539–541
Quasi unit disk graph
Query language, 187
QUIRE. *See* Quality-of-service specific information retrieval

**R**
Radio circuits, 448
Radio frequency identification (RFID), 636
Radio jamming, 493
RANBAR, 541–542
Randomized, 278
Random key pre-distribution, 520
Random sample consensus (RANSAC), 542
RANSAC. *See* Random sample consensus
RAP. *See* Realtime communication architecture
Rate-controlled reliable transport (RCRT), 220–222
Rate control mechanisms, 214
Rate monitoring, 496, 505, 509
RBC. *See* Reliable bursty convergecast
RCRT. *See* Rate-controlled reliable transport
Realtime communication architecture (RAP), 309
Real-time systems, 325
Received signal strength indicator (RSSI), 474, 475, 479, 480

Receiver initiated channel adaptive routing (RICA), 461
Rectilinear double rulings, 137
Reduction energy, 471
Regiment, 594–595
Re-keying protocol, 524–526
Relative neighborhood graph (RNG), 90, 118
Relaying strategies
    amplify and forward, 163
    decode and forward, 163
    decode and re-encode, 163
Reliable bursty convergecast (RBC), 229, 241, 255
Replay attack, 572
Reporting frequency, 667, 669, 670
Reprogramming, 251, 256, 257
REQ, 456–457
Request-To-Send frame (RTS), 421
Resiliency, 692
Resilient aggregation, 537–542
Response threshold, 22, 24, 25, 29
Restricted directional flooding, 86
Result verification, 544–550
RETOS, 334–335
Revised sensing area, 286
RF energy harvesting, 643
RICA. *See* Receiver initiated channel adaptive routing
Right-hand rule, 91
Rigidity, 116
RMBTS, 228
RMST, 228
Routing, 471, 472, 484–486, 492, 494–498, 500, 501, 504–509
    protocol, 440, 454, 456, 458, 461, 462, 465
    tree, 338, 345
*r*-sampling, 128
RSSI. *See* Received signal strength indicator
RTS, 447, 449, 450, 452, 453
Rubberband representation, 120
Rumor routing, 141
Runes middleware, 598–599

**S**
SAFECOM, 695
Scalability, 295
Schedule-based MACs, 420, 436
Scheduling, 380
SCIPUFF. *See* Second-order closure integrated Gaussian puff
SDAP. *See* Secure hop-by-hop data aggregation protocol
Sdlib, 600

Second-order closure integrated Gaussian puff
        (SCIPUFF), 64
Secure data aggregation, 534
Secure hop-by-hop data aggregation protocol
        (SDAP), 548–550
Security, 693
Selective forwarding, 494, 572
Self-configurable, 294
Self-configuration, 702
Self-diagnosis, 264
Self-healing, 701
Self-optimization, 702
Self-organization, 663, 664, 674, 693, 701
Semantic security
    under adaptive chosen-ciphertext attacks
        (IND-CCA2), 551
    under chosen-plaintext attacks (IND-CPA),
        551
    under non-adaptive chosen-ciphertext
        attacks (IND-CCA1), 551
Sensor field, 294
Sensor monitoring and surveillance, 285
Sensor networks, 689, 691, 699–700
Sensor node(s), 324, 333
Sensor placement, 48
Sensor protocol(s) for information via
        negotiation (SPIN), 189–190,
        456–458, 565, 566
Sensor topology retrieval at multiple
        resolutions (STREAM), 32
Service differentiation, 310
SetSpeed, 310, 311
Shared-key discovery, 521
Signal-based clustering algorithm, 37
Signal strength, 472, 474, 475, 478, 479, 487
Signal-to-noise ratio (SNR), 447, 462–464,
        479, 487
Simple cooperative diversity (SCD) protocol,
        172
Sink, 294
Sinkhole attack(s), 494, 509, 571, 572
Slack time, 312
Sleep-and-awake protocol, 298
Sleep cycle management, 28
Sleep state, 451, 453, 454
Sleep-wake cycle, 28
S-MAC, 429, 451–454
SNR. See Signal-to-noise ratio
Solar, 643
SOS, 330–331
Sound surveillance system (SOSUS), 636
Source coding, 564–565, 576, 577

Source node, 661, 665–672, 676, 680–682
    selection threshold, 682
    selection weight, 669
Span, 35
Sparse topology and energy management
        (STEM), 33
Spatial correlation, 308
Spatial reuse, 29
SPEED, 310, 566, 567, 574
Spherical double rulings, 138
SPIN. See Sensor protocols for information via
        negotiation
Sprinkler, 256, 257
Stack(s), 328, 333, 334, 336, 337, 342
Stansfield algorithm, 355–357, 373
Static velocity monotonic (SVM), 309
Stationary wireless sensor networks (SWSN),
        607
STCP, 230
Steinerization, 269
Steinerized graph, 269
Steiner tree problem with minimum number of
        Steiner points (STP-MSP), 266
Stimulation, 665–668
Stimuli, 671, 680
Store-carry-and-forward, 692
STP-MSP. See Steiner tree problem with
        minimum number of Steiner points
Strand space model, 516
Supply voltage, 442, 443, 464
SVM. See Static velocity monotonic
Switching energy, 443
Sybil, 494
Sybil attack, 572, 573
SYNC, 451, 452
Synchronization algorithms, 646
Synchronization problem, 645
Synopsis diffusion, 280


T
Target and event detection, 275
Target application model, 215
TeenyLime, 593
Temporal coherency, 313
Temporal coherency-aware in-network
        aggregation (TiNA), 313
$\mu$TESLA, 518
Thermoelectric generation, 643
Thread-driven model(s), 325–327, 331,
        335, 339
Threat profile, 49
Threshold, 278

Throughput, 425, 435, 471–473
Time correlation, 496, 505
Time-division multiple access (TDMA), 253, 420
Timeout-MAC (T-MAC), 430
Time stepped simulations, 645
Time-varying, 442, 446, 460, 461
TiNA. *See* Temporal coherency-aware in-network aggregation
TinyDB, 136, 594
TinyOS, 327, 331, 338, 340, 343, 495
Topologies, 638
  control, 271, 470, 484–486
  discovery, 28
  management, 27, 28
  request, 28
Topology and energy control algorithm (TECA), 41
Topology discovery algorithm (TopDisc), 28
Traffic analysis, 496, 505–507, 509
Traffic patterns, 212
Transceiver, 295
Transducer, 295
Transmission power, 442, 444, 462
Transmission power control, 469–486
Trilateration, 115
Trust, 501, 504, 507, 509
Two-tiered architecture, 265

U
Ubiquitous access, 692
Unit disk graph, 90
Unit disk graph embedding, 117
UNIX, 333

V
Value fusion, 275
Vehicular networks, 689, 698, 699
Virtual cluster, 298, 452
Virtual polar coordinate routing (VPCR), 122
Virtual ring routing (VRR), 133
Voting, 543–544

W
Weighted random waypoint (WRW), 68
Windowless acknowledgment, 246–249
Wireless sensor network(s) (WSNs), 1, 183–201, 261, 323–344
Wiring, 328–330
WiseMAC, 426
Witness nodes, 543
Wormhole(s), 495, 508, 573

Z
Zebra-MAC (Z-MAC), 429, 436
ZigBee, 564, 568, 569